Spanish Relativity Meeting (ERE 2014): Almost 100 Years After Einstein's Revolution

Journal of Physics: Conference Series Volume 600

Valencia, Spain
1 – 5 September 2014

ISBN: 978-1-5108-0357-2
ISSN: 1742-6588

Printed from e-media with permission by:

Curran Associates, Inc.
57 Morehouse Lane
Red Hook, NY 12571

Some format issues inherent in the e-media version may also appear in this print version.

Printed by Curran Associates, Inc. (2015)

Institute of Physics
Dirac House, Temple Back
Bristol BS1 6BE UK

Phone: 44 1 17 929 7481
Fax: 44 1 17 920 0979

techtracking@iop.org

Additional copies of this publication are available from:

Curran Associates, Inc.
57 Morehouse Lane
Red Hook, NY 12571 USA
Phone: 845-758-0400
Fax: 845-758-2634
Email: curran@proceedings.com
Web: www.proceedings.com

TABLE OF CONTENTS

PLENARY TALKS

CONTRIBUTED TALKS

POSTERS

Proceedings of the Spanish Relativity Meeting - ERE2014

Pablo Cerdá-Durán[1], José A. Font[1,2], José M. Ibáñez[1,2], Maria A. Lledo[3], J. Navarro-Salas[3], Gonzalo J. Olmo[3]

[1] Departamento de Astronomía y Astrofísica, Universitat de Valéncia, Dr. Moliner 50, 46100, Burjassot (Valéncia), Spain

[2] Observatori Astronómic, Universitat de Valéncia, C/ Catedrático José Beltrán 2, 46980, Paterna (Valéncia), Spain

[3] Departamento de Física Teórica and IFIC, Centro Mixto Universitat de Valéncia - CSIC, Dr. Moliner 50, 46100, Burjassot (Valéncia), Spain

This volume presents the proceedings of the international scientific conference "Spanish Relativity Meeting (ERE 2014): almost 100 years after Einstein's revolution". The conference was devoted to discussing the current state-of-the-art of a wide variety of topics of research in the fields of Gravitation and General Relativity in the "pre-centennial" year of General Relativity. The name of the conference was chosen to highlight the importance of the upcoming one hundredth anniversary of Einstein's theory of General Relativity, officially established by the Internal Society on General Relativity and Gravitation in November 25th, 2015. In particular, the conference was organized along three main lines of present-day research and applications of the theory, namely, Relativistic Astrophysics, Mathematical Relativity, and the interface between Gravitation and Quantum Field Theory.

The conference attracted a lot of attention and interest, as shown by the large number of delegates, numbeirng over one hundred. The ambitious original scientific program put forward by the Local Organizing Committee (LOC) under the supervision of the International Scientific Committee$*$ was fulfilled in its entirety, without any unexpected modification. As a result, the quality of all plenary talks was excellent. The LOC comprised staff from the Departamento de Astronomía y Astrofísica, the Departamento de Física Teórica, and the Instituto de Física Corpuscular, affiliated with the Universitat de Valéncia (Spain) and the Spanish Research Council (Consejo Superior de Investigaciones Científicas). The conference was held at ADEIT's Fundació Universitat Empresa \Universitat de Valéncia, and spanned five days, from September 1st to 5th, 2014. Each of these five days was devoted to discuss matters of general interest in the context of the aforementioned three main themes of the meeting. The morning sessions were dedicated to one-hour long invited plenary talks, three a day, leading to a total of 15 plenary talks. Owing to the high quality of the invited speakers, the plenary talks were recorded on video and are publicly available at the conference web site: <http://www.uv.es/ere2014/>.

As these proceedings certify, the results presented in the conference offer a precise general overview of the current state-of-the-art in research at the forefront of Gravitation and General Relativity worldwide. In

$*$ Miguel Alcubierre (ICN, Universidad Nacional Autónoma de México, Mexico), Abhay Ashtekar (Penn State University, USA), Robert Beig (Institute of Theoretival Physics, University of Vienna, Austria), Thibault Damour (Institut Hautes Études Scientifiques, Paris, France), Roy Maartens (Institute of Cosmology and Gravitation, Portsmouth, UK), Fernando Quevedo Rodríguez (DAMTP, University of Cambridge, UK), and Matt Visser (Victoria University of Wellington, New Zealand).

particular, Prof. Ashtekar discussed fundamental open issues in gravitational physics when considering the effects of a positive cosmological constant in Einstein's field equations. Prof. Shibata presented a comprehensive overview of the current status of the theoretical investigation (using sophisticated techniques from Numerical Relativity) in the field of binary neutron star mergers, a topic which was further discussed during the plenary talk of Prof. Stergioulas. Both plenary talks highlighted the importance of the potential wealth of information to be extracted on the physics of neutron stars from the study of the gravitational radiation produced during binary neutron star mergers. In the same astrophysical context, Prof. Gorosabel discussed observational implications of short gamma-ray bursts. The use of pulsar timing arrays to detect gravitational radiation in scenarios involving supermassive black holes was reviewed in the plenary talk of Prof. Sesana. Such technique provides a complementary (and entirely different) approach to that of laser interferometric detectors to directly detect gravitational radiation. Prof. Parker presented an overall view of the quantum phenomenon of particle creation from a vacuum by the expansion of the Universe. He explained how his discovery came about and how it relates to the current observations of temperature anisotropies in the CMB radiation and the potential detection of B-mode polarization patterns originated from primordial gravitons.

The physics of black holes was one of the key topics of the conference. Prof. Fabbri discussed the analogy to Hawking's radiation (black hole evaporation) in Bose-Einstein condensates, and Prof. Cardoso discussed various issues related with turbulence, superradiance, and the possibility of the existence of black hole bombs. Geometrical aspects of black holes were analyzed in detail in Prof. Zanelli's plenary talk, where the effects that SU(2) hair might have in axially symmetric black holes were considered. Correspondingly, Prof. Charmousis examined scalar-tensor black holes in modified gravity theories, which are receiving increasing attention in recent years. In a similar context, Prof. Lobo presented a very interesting review talk of current theories that go beyond Einstein's General Relativity. Ultimately, such theories attempt to fill the gaps and concerns in our cosmological understanding of the Universe, chiefly exemplified in the so-called dark matter and dark energy "issues". Those issues were analyzed from a rather unconventional point of view during the plenary talk of Prof. Portilla, whose model challenges the very concepts of dark matter and dark energy.

Prof. Senovilla's plenary talk dealt with the discussion of the so-called "gravitational double layers", a type of distribution of the energy-momentum tensor that, contrary to what happens in electromagnetism, was deemed nonexistent (actually forbidden) in gravitation until his recent study and discovery. Likewise, the plenary talk of Prof. Leffloch was also devoted to consider mathematical details of Einstein's field equations, in particular for weakly regular spacetimes with T2 symmetries. The type of mathematical computations the General Relativity practitioner has to face in his/her daily work is notoriously complicated. In this context, the use of mathematical software to alleviate the burden of technical work is certainly highly welcome. The plenary talk of Prof. Gourgoulhon was precisely devoted to discussing the recent implementation in Sage of a software package named SageManifolds that allows to carry out tensor calculus in an accurate and efficient way. The advantages of using such mathematical software became apparent during the conference.

Finally, Prof. Aloy was the speaker responsible for the Spanish Society of Gravitation and Relativity (SEGRE)'s public lecture, entitled "Relatividad y erupciones de rayos gamma: un matrimonio muy bien avenido". This public lecture was held at the historical headquarters (Aula Magna) of the Universitat de Valéncia, and the attendance of non-specialists was significant.

We would like to thank the participants who submitted their papers to the Proceedings of the "Spanish Relativity Meeting - ERE2014". We also are grateful to the personnel of ADEIT - Fundació Universitat Empresa \Universitat de Valéncia, in particular Cristina Mascarell, María José García, and Leonor de Arizón, for the excellent management of the conference.

The conference was possible thanks to the support of the Universitat de Valéncia, Consejo Superior

Journal of Physics: Conference Series **600** (2015) 011001 doi:10.1088/1742-6596/600/1/011001

de Investigaciones Científicas, Spanish Society of Gravitation and Relativity (SEGRE), Consellería d'Educació, Cultura i Esport of the Generalitat Valenciana (grant AORG/2014/100), ADEIT, La Nau, and the grants of the Spanish Ministerio de Economía y Competividad: CPAN- Ingenio2010 and Red Temática de Relatividad y Gravitación (FIS2011-15363-E).

The Local Organizing Committee:

Pablo Cerdá-Durán, José A. Font, José M. Ibáñez, María A. Lledó, José Navarro-Salas, and Gonzalo J. Olmo.

Peer review statement

All papers published in this volume of *Journal of Physics: Conference Series* have been peer reviewed through processes administered by the proceedings Editors. Reviews were conducted by expert referees to the professional and scientific standards expected of a proceedings journal published by IOP Publishing.

Journal of Physics: Conference Series **600** (2015) 012001 doi:10.1088/1742-6596/600/1/012001

Creation of quantized particles, gravitons, and scalar perturbations by the expanding universe[1]

Leonard Parker

Center for Gravitation, Cosmology and Astrophysics,
Physics Department, University of Wisconsin-Milwaukee, USA

E-mail: `leonard@uwm.edu`

Abstract. Quantum creation processes during the very rapid early expansion of the universe are believed to give rise to temperature anisotropies and polarization patterns in the CMB radiation. These have been observed by satellites such as COBE, WMAP, and PLANCK, and by bolometric instruments placed near the South Pole by the BICEP collaborations. The expected temperature anisotropies are well-confirmed. The B-mode polarization patterns in the CMB are currently under measurement jointly by the PLANCK and BICEP groups to determine the extent to which the B-modes can be attributed to gravitational waves from the creation of gravitons in the earliest universe.

As the original discoverer of the quantum phenomenon of particle creation from vacuum by the expansion of the universe, I will explain how the discovery came about and how it relates to the current observations. The first system that I considered when I started my Ph.D. thesis in 1962 was the quantized minimally-coupled scalar field in an expanding FLRW (Friedmann, Lemaitré, Robertson, Walker) universe having a general continuous scale factor $a(t)$ with continuous time derivatives. I also considered quantized fermion fields of spin-$1/2$ and the spin-1 massless photon field, as well as the quantized conformally-invariant field equations of arbitrary integer and half-integer spins that had been written down in the classical context for general gravitational metrics by Penrose.

It was during 1962 that I proved that quanta of the minimally-coupled scalar field were created by the general expanding FLRW universe. This was relevant also to the creation of quantized perturbations of the gravitational field, since these perturbations satisfied linear field equations that could be quantized in the same way as the minimally-coupled scalar field equation. In fact, in 1946, E.M. Lifshitz had considered the classical Einstein gravitational field in FLRW expanding universes and had shown that the classical linearized Einstein field equations reduced, in what is now known as the Lifshitz gauge, to two separate classical minimally-coupled massless scalar field equations. These field equations of Lifshitz, when quantized, correspond to the field equations for massless gravitons, one equation for each of the two independent polarization components of the spin-2 massless graviton. I will discuss this further in this article.

1. Introduction

In 1962, at Harvard when I began my Ph.D. thesis, I wanted to work at the interface of general relativity and quantum field theory. I had already studied general relativity quite carefully as an undergraduate before coming to Harvard. I had the good fortune to learn quantum field theory

[1] This article is based on the Plenary Lecture of the author given on September 2, 2014 at the ERE2014 Conference in Valencia, Spain.

and particle physics at Harvard from Wendell Furry, Roy Glauber (Nobel Prize 2005), Sidney Coleman, Sheldon Glashow (Nobel Prize 1979), and Julian Schwinger (Nobel Prize 1965).

I wanted to find new consequences of the quantum field theory of elementary particles in the context of Einstein's theory of general relativity. At the time, I felt that *quantizing* the nonlinear gravitational field *itself* was so difficult that I would not be able to make significant progress in trying to go beyond the deep work that had already been done in that area. Nevertheless, I felt that it would be valuable to study quantized elementary particle fields in the curved space-times that were solutions of the classical nonlinear Einstein gravitational field equations. Luckily, Sidney Coleman agreed to be my thesis advisor on such a project, which was outside the main stream of the time.

I started by looking for new consequences of quantum field theory in the isotropically expanding cosmological space-times that were solutions of the nonlinear classical equations of general relativity. To investigate the consequences of quantum field theory in the expanding universes of general relativity, it was first necessary to extend the known quantum field theory of elementary particles from the well-established flat Minkowski space-time of special relativity to the context of a classical general relativistic isotropically expanding universe with a general expansion scale-factor, $a(t)$. Some examples that I studied were the dust-filled universe, the radiation-dominated universe and the exponentially-expanding universe of deSitter, but my main result was for general $a(t)$.

The quanta (particles, scalar perturbations, and tensor gravitational wave perturbations) that reach a steady-state density during the inflationary stage of the expansion are created or maintained by the quantum field theory process of particle creation that I discovered and will discuss in this lecture. The quantum field theory vacuum state in the exponentially expanding universe is known as the Bunch-Davies[1] vacuum. That vacuum state is symmetric under the deSitter invariance group of the exponentially expanding FLRW universe. The Bunch-Davies vacuum is the natural vacuum state having the 10 deSitter group symmetries[2, 3], just as the natural vacuum state in flat Minkowski spacetime has the 10 Poincaré group symmetries.

The constant, non-zero, average number density of particles, scalar perturbations and/or tensor gravitational wave perturbations is consistent with the result of spontaneous particle creation from vacuum of an FLRW universe that undergoes the deSitter exponential expansion. I will discuss this further in Section 3, and will show that the creation rate of these perturbations and/or particles follows immediately from the result that I obtained in my Ph.D. thesis for general continuous expansion factors $a(t)$ having a sufficient number of continuous time derivatives. This follows very simply from my *general* equation for the average number of particles created, and from the deSitter invariance of the exponentially expanding universe.

I chose to first consider the spatially-flat FLRW universes because I felt that one could *dynamically propagate* the experimentally verified theory of quantized fields from flat spacetime into the curved spacetime of the spatially-flat isotropically expanding universes. It was not necessary in this framework to go beyond the already tested assumptions of classical general relativity and of quantum field theory in *flat* spacetime. The dynamical equations of the quantized fields were taken to be the simplest generally-covariant ones obtained from the flat-spacetime quantized field equations. It was then possible to use the field equations in the curved spacetime to propagate the fields and their commutation relations into curved spacetime. Any fundamentally new results coming from this minimal set of assumptions would have to be taken very seriously. As it turned out, this procedure also gave an independent new proof of the spin-statistics theorem; for bosons only the commutation relations were propagated consistently from the initial to the final flat spacetime, and for fermions only the anticommutation relations were propagated consistently.

First, I considered *the quantized generally-covariant minimally-coupled free scalar-field equation* in an isotropically-expanding spatially-flat universe with a continuous and smooth scale factor $a(t)$. Because there was no ambiguity as to the definition of the measurable particle number operator of the quantized field in flat spacetime, it was natural to consider what happened to the particle number when the scale factor $a(t)$ went smoothly from an initial constant value a_1 at early times to a final constant value a_2 at late times. During the periods of time when a_1 and a_2 were constant, simple constant rescalings of the spatial coordinates at early and late times showed that spacetimes before and after the expansion of the universe were Minkowskian, each exhibiting the 10-dimensional Poincaré group of symmetries. The number density of particles of the quantized field was taken to be 0 at early times. Evaluating the expectation value of the number density at late times, after the universe had undergone a smooth expansion, I found that the number density of created particles was finite and nonzero! This demonstrated that real particles were created from the initial vacuum state by the expansion of the universe.

To briefly summarize the calculation, I used the Heisenberg picture to evolve the quantized field during the expansion of the universe. The state vector of the quantized field was chosen to be the well-defined vacuum state in the initial Minkowski spacetime. As a result of the time-dependent unitary evolution of the quantized field, the creation and annihilation operators of the quantized field at *early* times had evolved into linear combinations of the creation and annihilation operators of the quantized field at *late* times. Consequently, the state vector having no particles of the quantized field at early times does have a non-zero density of particles of the quantized field at late times. This implies that particles are created by the expansion of the universe.

My results first appeared in my Ph.D. Thesis[4] ("The Creation of Particles by the Expansion of the Universe", Harvard University, 1966) and also in [5, 6, 7]. My thesis also included results I had obtained for the spin-1/2 fermion field, including particle creation for fermions of non-zero mass, as well as the absence of particle creation for the *massless* Pauli neutrino field. Those results for fermion fields appeared in [7].

I also found that particles of 0-*mass* satisfying the minimally-coupled scalar field equation *would be created* from the initial vacuum state as a result of the expansion from an initial Minkowski spacetime to a final Minkowski spacetime. This result implied that quantum gravitons would be created by the quantized linearized Einstein gravitational field equations. E.M. Lifshitz[8] in 1946 had published an important paper on the *classical* linearized Einstein gravitational field equations, in which he proved that there is a gauge (the Lifshitz gauge) in which the linearized Einstein field equations reduce to a set of two separate minimally-coupled scalar field equations. One could then use the Lifshitz gauge to directly apply my results to the Einstein gravitational field to show that quantized gravitons (tensor perturbations) were created by the expanding universe. A paper on this was published by Leonid P. Grischuk[9], in which he used the Lifshitz gauge to apply my earlier quantization of a minimally-coupled scalar-field, including my expression for particle creation in a general expanding universe. He used the Lifshitz gauge to obtain an expression for the creation of gravitons in a universe with $a(t)$ proportional to a power of t.

A more detailed quantization of the linearized Einstein gravitational field was published in 1977 by Lawrence H. Ford and me in [10]. One of the things we showed in our paper was that there were spurious infrared (IR) divergences present in Grishchuk's result that were not present when we imposed more appropriate late-time boundary conditions. In our paper, we also reviewed the derivation of the Lifshitz gauge and performed an explicit *quantization*, defining creation and annihilation operators for massless gravitons in each of their two polarization states. (We carried this out for more than one way of specifying the pair of gravitational wave polarization states.)

The minimally-coupled scalar field equations that appear in the Lifshitz gauge are the same ones I studied [4, 5, 6] in the 1960's for general $a(t)$. In the exponentially expanding inflationary universe, there are well-known analytic solutions of these equations that can be applied to the creation of gravitons and gravitational wave perturbations present in the inflationary universe. These form the basis for obtaining the inflationary universe predictions of linearized gravitational waves and gravitons created in the early universe, as well for scalar gravitational perturbations that are related to the temperature variations of the CMB.

In Section 3, I will show how one obtains directly from my published results of the 1960's, the average number of quantized particles created as a function of time in an expanding physical volume of the exponentially expanding universe. This result follows directly from deSitter invariance of the exponentially expanding universe. In addition, if there are similar particles already present at the time that the exponential expansion begins, then the initial presence of these particles can significantly increase the rate of particle creation. This stimulated creation of identical bosonic particles occurs also for general $a(t)$. (For fermions the initial presence of identical fermions would decrease the rate of particle creation.)

Scalar perturbations of the gravitational field were also created by the exponential expansion of the universe, before reheating occurred. Reheating resulted in the formation of a plasma of charged particles in a high- temperature, radiation-dominated expanding universe. The acoustic waves that these scalar gravitational perturbations produced in the plasma were present for about 400,000 years, until the universe cooled enough for electrons and protons to form neutral atoms, after which the photons of the CMB were able to propagate freely (for the most part). However, temperature differences in the plasma acoustic waves present at the time of last scattering resulted in temperature differences in the free-streaming CMB radiation. These temperature differences in the CMB radiation at the time of last scattering were responsible for the temperature anisotropies of about 1 part in 10^5 that were ultimately measured by the COBE satellite.

If the recent possible gravitational wave observations are confirmed and if the intensities of the B-mode polarization patterns they produce are consistent with the inflationary universe model, including the observed temperature anisotropies in the CMB, then this consistency would be further evidence for an early inflationary expansion of the universe.

In the next Section, I will briefly explain how I proved in my Ph.D. thesis[4] (see also [5, 6, 7]) that minimally-coupled scalar particles are created by expanding FLRW universes.

But first let me quote the Conclusion of my 1968 paper[5]: "The particle creation in the expanding universe at the present time is quite negligible. However, for the early stages of a Friedmann expansion it may well be of great cosmological significance, especially since it seems inescapable if one accepts quantum field theory and general relativity. In considering the large amount of particle creation taking place in the early stages of an expansion, it is necessary to take into account the reaction of the matter created back on the gravitational field. Furthermore, it may be necessary to consider the effects of the quantization of the gravitational field. Therefore, no conclusive quantitative result can yet be reported here concerning the primeval creation."

This conclusion of my paper emphasized the strong effects that particle creation would have on the early expansion of the universe, so the reaction-back would have to be taken into account. The first such calculation was done by me and Steven A. Fulling [11], in which we showed by numerical evolution of the semi-classical Einstein equations that a bounce could occur as a result of the particle creation process, thus avoiding the cosmological singularity. Our result also demonstrated that the classical energy-conditions can be avoided through the creation of particles by quantum field theory in an FLRW universe.

In my thesis [4], I also used quantum measurement theory to develop the method of adiabatic regularization (or renormalization) as applied to the particle number operator during the expansion of the universe. In 1974, Fulling and I extended the adiabatic regularization method to renormalize the expectation value of the energy-momentum tensor [12, 13, 3].

An interesting sidelight, is that I was hired as an Instructor to teach at the University of

North Carolina (UNC) for 1966 and 1967. My thesis advisor was away in Europe, so I had to wait until he returned to Harvard in 1966 to fly to Boston for my thesis defense. The thesis defense committee consisted of Sidney Coleman, Sheldon Glashow, and Walter Gilbert (Nobel Prize 1980).

While teaching at UNC, I attended a Colloquium given at UNC in 1966 or 1967 by Fred Hoyle. After his talk, I explained to him my as yet unpublished work. He was an advocate of the steady state theory (of Bondi, Gold, and Hoyle), which assumed a permanent exponentially expanding universe having a constant density of particles, maintained by a process that they had postulated, in which particles or atoms were constantly being created, including at the present time, at a sufficient rate to maintain a steady-state universe that did not change significantly on cosmological time scales. I explained to him that my quantum field theory mechanism of particle creation would not support such a steady-state universe at the present time because the particle creation rate would be much too small. However, I explained to him that my process of quantum field theory particle creation could support an exponentially expanding steady-state universe at very early times with a much higher particle density. He listened, but was not enthusiastic about the idea because at that time he believed in a universe that was eternal and unchanging.

In Section 3 of this article, I will show how the equations that I first derived in my Ph.D.thesis[4, 5, 6, 7] for the general case of particle creation by an FLRW universe directly yields the rate per unit physical volume at which particles or quanta are created in an exponentially expanding, spatially-flat FLRW universe that is in (or has evolved to) a deSitter invariant state. For example, the inflationary universe is in such a state prior to reheating. This very high rate of particle creation is what I was referring to in connection with the exponentially expanding steady-state universe in the discussion I had with Fred Hoyle after his Colloquium at UNC.

2. Minimally-coupled scalar field in FLRW universe

Let us now consider the case of a free quantized scalar field ϕ in a smoothly changing spatially flat FLRW universe with line element

$$ds^2 = dt^2 - a^2(t)(dx^2 + dy^2 + dz^2), \tag{1}$$

and equation of motion

$$(\Box + m^2)\phi = 0, \tag{2}$$

where $\Box = g^{\mu\nu}\nabla_\mu\nabla_\nu$, with ∇_λ denoting the covariant derivative.

For now we are considering the so-called minimally-coupled scalar field of mass m (which may be 0). More generally, one could include a coupling to the Ricci scalar curvature, $R = g^{\mu\nu}R_{\mu\nu}$, namely, $(\Box + m^2 + \xi R)\phi = 0$, where ξ is a dimensionless constant. We work in units with $\hbar$ and c each equal to 1. (For brevity, I will not go into spin-1/2 and spin-1 fields, and spatially-curved FLRW universes, which I also considered in my Ph.D. thesis[4, 5, 6, 7].

Writing (2) with the metric of (1), one finds that

$$a^{-3}\partial_t(a^3\partial_t\phi) - a^{-2}\sum_i \partial_i^2\phi + m^2\phi = 0. \tag{3}$$

It is convenient to impose periodic boundary conditions in a cube having sides of coordinate length L and coordinate volume $V = L^3$. The physical volume of this cube is expanding, with the physical volume proportional to $a(t)^3$. As in Minkowski spacetime, this is a mathematical device, with L taken to infinity after physical quantities such as the density of created particles have been calculated. Then we can expand the field operator ϕ in the form

$$\phi = \sum_{\vec{k}} \left\{ A_{\vec{k}} f_{\vec{k}}(\vec{x}, t) + A^\dagger_{\vec{k}} f^*_{\vec{k}}(\vec{x}, t) \right\}, \tag{4}$$

where

$$f_{\vec{k}} = (2Va(t)^3)^{-1/2} e^{i\vec{k}\cdot\vec{x}} h_k(t). \tag{5}$$

Here $k^i = 2\pi n^i/L$ with n^i an integer, $k = |\vec{k}|$, and $h_k(t)$ satisfies

$$\frac{d^2}{dt^2} h_k(t) + \frac{k^2}{a^2} h_k(t) + m^2 h_k(t) - \frac{3}{4}\left(\frac{\dot{a}}{a}\right)^2 h_k(t) - \frac{3}{2}\frac{\ddot{a}}{a} h_k(t) = 0. \tag{6}$$

Examples in which this equation has analytic solutions are power law expansions and the exponential expansion. For the latter case, $a(t) \propto \exp(Ht)$ with H being the Hubble constant during the exponential inflationary stage of the expansion, the solutions are Hankel functions. The metric has the 10 symmetries of deSitter spacetime, and a solution having these symmetries is

$$f_{\vec{k}} = (2Va(t)^3)^{-1/2} e^{i\vec{k}\cdot\vec{x}} H^1_\nu(v). \tag{7}$$

with

$$v = kH^{-1}\exp(-Ht) \tag{8}$$

$$\nu = (9/4 - m^2/H^2)^{1/2} \tag{9}$$

Note that v decreases with increasing t. Alternatively, one could replace $H^1_\nu(v)$ in (7) by $H^2_\nu(u)$ with $u = -v$, where the variable u increases with t. (See [3], Sec.2.10, for a discussion of deSitter spacetime and the Bunch-Davies vacuum state.)

As already mentioned in Section 1, the linearized Einstein equations for a gravitational wave field in the Lifshitz gauge reduce to a pair of massless minimally-coupled scalar field equations, one equation for each polarization of the gravitational wave. It is now clear that quanta of the gravitational wave field, i.e., gravitons, would be created by the expansion of the universe, in the same way that massless, minimally-coupled scalar particles would be created. For detailed quantization of linearized gravitational waves, see [10].

Now let us continue with our discussion of the minimally-coupled scalar field to summarize the method that I used to arrive at the number density of particles created from the initial Minkowski vacuum state (i.e., spontaneous particle creation) and from a non-vacuum state (i.e., stimulated particle creation). Since the reaction back of the created particles would be very significant in the early expansion of the universe, I chose to deal with an arbitrary expansion of the universe that began in a Minkowski spacetime and ended in a Minkowski spacetime (to within a constant rescaling of the coordinates) and that had continuous functions $a(t)$, $\dot{a}(t)$, and $\ddot{a}(t)$.

From the theory of ordinary differential equations there are two linearly independent solutions of (6). One can prove that the particle number at late times is an adiabatic invariant and that the solutions that approach positive and negative frequency solutions in the late time Minkowski spacetime approach the asymptotic forms of the Liouville adiabatic solutions of the harmonic oscillator equation with slowly changing frequency. (This is also the form of the JWKB solutions for a slowly changing quantum mechanical scattering potential in one spatial dimension, with the particle energy above that of the scattering potential - but in the present case, the spatial coordinate x is replaced by the time coordinate t.)

The adiabatic approximation of Liouville for the harmonic oscillator with a slowly changing frequency gives the following approximation to the solution of (6):

$$h_k(t) \sim (\omega_k(t))^{-1/2} \exp(\pm i \int^t \omega_k(t')\, dt'), \tag{10}$$

where

$$\omega_k(t) = \sqrt{(k/a(t))^2 + m^2} \tag{11}$$

This approximation is good in the limit that all time-derivatives of $a(t)$ smoothly approach 0. The two solutions in (10) are linearly independent. Then the general solution of the second-order ordinary differential equation (6) can be written in the adiabatic approximation as a linear combination

$$h_k(t) \sim \alpha_k\,(\omega_k(t))^{-1/2}\exp(-i\int^t \omega_k(t')\,dt') + \beta_k\,(\omega_k(t))^{-1/2}\exp(+i\int^t \omega_k(t')\,dt'), \tag{12}$$

where α_k and β_k are complex constants that must satisfy

$$|\alpha_k|^2 - |\beta_k|^2 = 1 \tag{13}$$

because of the conserved Wronskian or scalar product. We will assume that $a(t)$ and all its time-derivatives are smooth and well-defined.

In order to determine the expectation value of the number of particles that are created as a result of a smooth expansion of the universe, we take $a(t)$ such that $\lim_{t\to-\infty} a(t) = a_1$, and $\lim_{t\to\infty} a(t) = a_2$, where $a(t)$ and all its time-derivatives exist and are well-defined and continuous, and $0 < a_1 < a_2$.

We can take the solution of (6) in the early-time Minkowski spacetime such that $|\alpha_k|^2 = 1$ and $|\beta_k|^2 = 0$. Then the expression for the quantized field in (4) is the standard expression for the quantized field in the early time Minkowski spacetime, for which $a(t) = a_1$.

In the late-time Minkowski spacetime, when one uses the late-time asymptotic form in (12), and separates the positive frequency and negative frequency terms in the late-time expression for quantized field ϕ in (4), one finds that the field ϕ at late times can be rewritten in the form:

$$\phi = \sum_{\vec{k}} \left\{ B_{\vec{k}} g_{\vec{k}}(\vec{x},t) + B^{\dagger}{}_{\vec{k}} g^{*}{}_{\vec{k}}(\vec{x},t) \right\}, \tag{14}$$

where

$$g_{\vec{k}} = (2Va_2{}^3)^{-1/2} e^{i\vec{k}\cdot\vec{x}} \exp(-i\int^t \omega_k(t')\,dt'). \tag{15}$$

Here $k^i = 2\pi n^i/L$ with n^i an integer, and $k = |\vec{k}|$, and

$$B_{\vec{k}} = \alpha_k\,A_{\vec{k}} + \beta_k\,A^{\dagger}{}_{-\vec{k}}.$$

This transformation corresponds to the creation of particle-antiparticle pairs of total momentum 0 from the vacuum, as required by conservation of momentum. This linear transformation is an example of a Bogoliubov transformation, as I explained in [6] on page 1061, in connection with particle creation by expanding universes. Later, analogous Bogoliubov transformations involving sets of creation and annihilation operators were used by Fulling [15] to describe particles in the spacetime (Rindler spacetime) corresponding to a set of accelerating observers and by Hawking [16] to describe the distribution of particles created by a black hole.

In the expanding universe under consideration here, the expectation value of the number of particles created by the expanding universe from the early-time vacuum state $|0\rangle$, present at late times is

$$\langle 0|B^{\dagger}{}_{\vec{k}} B_{\vec{k}}|0\rangle = |\beta_k|^2 \tag{16}$$

(Recall that we are using the Heisenberg picture to describe the dynamics, so the state vector does not change with time.) Hence, if β_k is non-zero, there is particle creation as a result of the expansion of the universe between the initial and final Minkowski spacetimes.

Except in very special cases, quanta of the minimally-coupled scalar field are created by the expansion of the universe. From our earlier discussion of the quantized linearized-Einstein gravitational field in the Lifshitz gauge, the same should be true for the creation of gravitons by the expansion of the universe. For the isotropically expanding FLRW universes, the particles are created in pairs to conserve the 3-momentum. I also calculated the detailed probability distribution of the created particles as a function of momentum. I used similar methods in [14] to obtain the probability distribution of particles created by a Schwarzschild black hole.

I also considered the creation of particles if the initial state is not the Minkowski vacuum state, but instead has a non-zero density of minimally-coupled scalar particles present at early times. More generally, I considered a mixed state described by a statistical density matrix operator ρ_{stat} in the initial Minkowski spacetime. I showed that there would be stimulated creation of these bosonic particles by the expansion of the universe. The details of the calculation are described in [4], and in [6] on pp. 1063–1064.

Taking an isotropic distribution of particles initially, and an FLRW expansion, the average number of particles present in mode $\vec{k}$ at late times in the final Minkowski spacetime is

$$\langle N_{\vec{k}} \rangle_{late} = Tr[\rho_{stat} B^{\dagger}{}_{\vec{k}} B_{\vec{k}}] = \langle N_{\vec{k}} \rangle_{early} + |\beta_k|^2 (1 + 2\langle N_{\vec{k}} \rangle_{early}) \tag{17}$$

The stimulated creation of bosons is evident in the factor involving $|\beta_k|^2(1 + 2\langle N_{\vec{k}}\rangle_{early})$. The average number of these spin-0 bosons present at late times is increased by stimulated creation resulting from the presence of a non-zero average number of these particles at early times. I also showed in [4], and in [7] that the opposite is true in the case of fermionic particles created by the expanding universe.

But as we shall see in Section (3), when we discuss the rapid rate of increase of $|\beta_k|^2$ during the exponential stage of the expanding universe, the initial number of particles present in the coordinate volume L^3 can greatly influence the creation rate and the magnitude of the number of particles present at late times.

A further treatment of the stimulated creation of scalar gravitational perturbations during inflation is given by Ivan Agullo and me in [19, 20] where we discussed its possible effect on the CMB bispectrum. In [21] with J. Navarro-Salas, we considered the CMB trispectrum.

3. Exponentially Expanding FLRW Universe

Next we show how the above equations, (16) and (17), for the expectation value of the number of minimally-coupled scalar particles created in each mode $\vec{k}$ determine the time-dependence of $|\beta_k|^2$ in an exponentially-expanding, spatially-flat, FLRW universe.

Let us start with a quantized minimally-coupled scalar field in an initial flat spacetime invariant under the 10-dimensional Poincaré group. Suppose that the quantized scalar field is in its vacuum state having the 10 Poincaré symmetries. Let us assume that our universe undergoes a spontaneous fluctuation (with $a(t)$, $\dot{a}(t)$, and $\ddot{a}(t)$ continuous) that eventually takes it to an exponentially-expanding, spatially-flat universe with $a(t) \propto \exp(Ht)$, with H constant. After a time, suppose the quantized scalar field in this exponentially expanding deSitter universe has settled into a state having the 10 deSitter symmetries. Because we are working in the Heisenberg picture, the state vector does not change with time, so it is the state $|0\rangle$ that is annihilated by the operator $A_{\vec{k}}$, the annihilation operator for the scalar quanta in the original early-time Minkowski spacetime. However, the state $|0\rangle$ is not annihilated by the operators $B_{\vec{k}}$ that annihilate particles in the late-time Minkowski spacetime. Recall that In the Heisenberg picture, the quantized field operator ϕ evolves with time, and the $B_{\vec{k}}$ in the late-time Minkowski spacetime are obtained from the field operator at late-times. The particles that are present in the *late-time* Minkowski spacetime are the *real* (as opposed to virtual) particles that were created during the exponential de-Sitter portion of the expansion and thus survived the gradual smooth slow-down of the expansion to approach the late-time Minkowski spacetime.

Journal of Physics: Conference Series **600** (2015) 012001 doi:10.1088/1742-6596/600/1/012001

We will show from this and the equation for particle creation that $|\beta_{\vec{k}}(t)|^2$ must be proportional to $\exp(Ht)/\exp(Ht_1)$, where t_1 is approximately the time that deSitter symmetry was reached by the particles being created, and $t > t_1$. We have from (16), in the Heisenberg picture, for particle creation by the expanding universe from the Minkowski vacuum state in the initial Minkowski space that

$$\langle 0|B^{\dagger}{}_{\vec{k}}B_{\vec{k}}|0\rangle = |\beta_k|^2 \tag{18}$$

Suppose that the expansion $a(t)$ approaches deSitter, with $a(t) \propto \exp(Ht)$ smoothly, with at least $a(t)$, $\dot{a}(t)$, and $\ddot{a}(t)$ continuous. (This is so that the minimally-coupled scalar field equation is well-defined at all times.) Here $k^i = 2\pi n^i/L$ with n^i an integer, and $k = |\vec{k}|$, and

$$B_{\vec{k}}(t_1) = \alpha_k(t_1)\, A_{\vec{k}} + \beta_k(t_1)\, A^{\dagger}{}_{-\vec{k}}.$$

This transformation corresponds to the creation of pairs of total momentum 0 from the vacuum, as required by conservation of momentum. We also have as a consequence of the field equation that, as in (13),

$$|\alpha_k(t_1)|^2 - |\beta_k(t_1)|^2 = 1,$$

which is the condition for the commutation relation of boson creation and annihilation operators to obey Bose-Einstein statistics, i.e., to satisfy the correct commutation relations.

After deSitter invariance has been reached, we expect that the number of particles present per unit *physical* volume $(La(t)^3)$ is constant for $t > t_1$. Hence, if $t > t_1$, it follows that

$$|\beta_k(t)|^2/a(t)^3$$

is constant. Therefore, during the period for which inflationary equilibrium has been reached, we must have

$$|\beta_k(t_2)|^2/a(t_2)^3 = |\beta_k(t_1)|^2/a(t_1)^3,$$

which implies that

$$|\beta_k(t)|^2 \propto (\exp(Ht))^3, \tag{19}$$

after inflationary equilibrium has been reached.

This result in (19) was confirmed numerically by Glenz and Parker [18] by means of an explicit numerical integration, in which we took an expansion scale factor $a(t)$ that started expanding from flat spacetime, eventually becoming an exponentially expanding inflationary universe that expanded for more than 50 e-foldings and then gradually approached a flat spacetime at late times. We joined the three segments of the expansion such that $a(t)$, $\dot{a}(t)$, and $\ddot{a}(t)$ were continuous at the two joining points. We used the known analytic solutions of the quantum field differential equations in each of the three segments of the expansion of the universe. As a check, we confirmed that (13) was satisfied to at least 500 significant figures at all times throughout the expansion of the universe.

As an interesting aside, let me mention that the form of the parametrized scale factors $a(t)$ that we used in [18] to smoothly join the exponential inflationary expansion of the universe to Minkowski spaces at early and at late times are instances of the same four-parameter scale factor that I had used in a 1976 paper in Nature[17] to demonstrate that an expanding universe can create free particles from vacuum, with the created particles having a thermal distribution in the absence of any thermalizing interactions.

The time-dependence of $|\beta_k(t)|^2$ in (19) also follows in the case when an initial distribution of particles is present, as in (17). This is clear because the average number of created particles that is proportional to the average number, $\langle N_{\vec{k}}\rangle_{early}$, of particles already present at early times in (17) is $(1 + 2|\beta_k|^2)\langle N_{\vec{k}}\rangle_{early} \approx 2|\beta_k|^2\langle N_{\vec{k}}\rangle_{early}$, so the argument given above leads in this case also to the same time-dependence, (19), of $|\beta_k|^2$. However, we see that the average number of particles created during the exponential expansion can be greatly increased by the stimulated creation of particles, if there were a significant average number $\langle N_{\vec{k}} \rangle_{early}$ of these bosonic particles already present in the physical expanding volume when the exponential stage of the expansion first began.

References

[1] Bunch T S and Davies P C W, Proc. R. Soc. London A **360**, 117 (1979); see also Birrell D N and Davies P C W, *Quantum Fields in Curved Space* (Cambridge University Press, 1982)
[2] Higuchi A, Nucl. Phys. B **282**, 397 (1987).
[3] Parker L and Toms D J, *Quantum Field Theory in Curved Spacetime: Quantized Fields and Gravity* (Cambridge University Press, 2009).
[4] Parker L, *The Creation of Particles in an Expanding Universe,* Harvard University, Cambridge, Massachusetts 1966.
[5] Parker L, *Phys. Rev. Lett.* **21**, 562 (1968).
[6] L. Parker, *Phys. Rev.* **183**, 1057 (1969).
[7] Parker L, *Phys. Rev.* **D3**, 346 (1971).
[8] Lifshitz E M, *J. Phys. USSR* **10**, 116 (1946).
[9] Grishchuk L P, *Sov. Phys. JETP* **40**, 409 (1975).
[10] Ford L H and Parker L, *Phys. Rev.* **D16**, 1601 (1977).
[11] Parker L and Fulling S A, *Phys. Rev.* **D7**, 2357 (1973).
[12] Parker L and Fulling S A, *Phys. Rev.* **D9**, 341 (1974).
[13] Anderson P R and Parker L, *Phys. Rev.* **D36**, 2963 (1987).
[14] Parker L, *Phys. Rev.* **D12**, 1519 (1975).
[15] Fulling S A, *Phys. Rev.* **D7**, 2850 (1973).
[16] Hawking S W, *Nature* **248**, 248 (1974).
[17] Parker L, *Nature* **261**, 20 (1976).
[18] Glenz M M and Parker L, *Phys. Rev.* **D 80**, 063534 (2009).
[19] Agullo I and Parker L, *Phys. Rev.* **D83**, 063526 (2011).
[20] Agullo I and Parker L, *International Journal of Modern Physics* **D 20**, no. 14: 2861 (2011).
[21] Agullo I, Navarro-Salas J and Parker L, *Journal of Cosmology and Astroparticle Physics* 2012, no. 05: 019 (2012).

Tensor calculus with open-source software: the SageManifolds project

Eric Gourgoulhon[1], **Michał Bejger**[2], **Marco Mancini**[1]

[1] Laboratoire Univers et Théories, UMR 8102 du CNRS, Observatoire de Paris, Université Paris Diderot, 92190 Meudon, France

[2] Centrum Astronomiczne im. M. Kopernika, ul. Bartycka 18, 00-716 Warsaw, Poland

E-mail: `eric.gourgoulhon@obspm.fr`, `bejger@camk.edu.pl`, `marco.mancini@obspm.fr`

Abstract. The SageManifolds project aims at extending the mathematics software system Sage towards differential geometry and tensor calculus. Like Sage, SageManifolds is free, open-source and is based on the Python programming language. We discuss here some details of the implementation, which relies on Sage's parent/element framework, and present a concrete example of use.

1. Introduction

Computer algebra for general relativity (GR) has a long history, which started almost as soon as computer algebra itself in the 1960s. The first GR program was GEOM, written by J.G. Fletcher in 1965 [1]. Its main capability was to compute the Riemann tensor of a given metric. In 1969, R.A. d'Inverno developed ALAM (for *Atlas Lisp Algebraic Manipulator*) and used it to compute the Riemann and Ricci tensors of the Bondi metric. According to [2], the original calculations took Bondi and collaborators 6 months to finish, while the computation with ALAM took 4 minutes and yielded the discovery of 6 errors in the original paper. Since then, numerous packages have been developed: the reader is referred to [3] for a review of computer algebra systems for GR prior to 2002, and to [4] for a more recent review focused on tensor calculus. It is also worth to point out the extensive list of tensor calculus packages maintained by J. M. Martin-Garcia at [5].

2. Software for differential geometry

Software packages for differential geometry and tensor calculus can be classified in two categories:

(i) Applications atop some general purpose computer algebra system. Notable examples are the xAct suite [6] and Ricci [7], both running atop Mathematica, DifferentialGeometry [8] integrated into Maple, GRTensorII [9] atop Maple and Atlas 2 [10] for Mathematica and Maple.

(ii) Standalone applications. Recent examples are Cadabra (field theory) [11], SnapPy (topology and geometry of 3-manifolds) [12] and Redberry (tensors) [13].

All applications listed in the second category are free software. In the first category, xAct and Ricci are also free software, but they require a proprietary product, the source code of which is closed (Mathematica).

Journal of Physics: Conference Series **600** (2015) 012002 doi:10.1088/1742-6596/600/1/012002

As far as tensor calculus is concerned, the above packages can be distinguished by the type of computation that they perform: abstract calculus (xAct/xTensor, Ricci, Cadabra, Redberry), or component calculus (xAct/xCoba, DifferentialGeometry, GRTensorII, Atlas 2). In the first category, tensor operations such as contraction or covariant differentiation are performed by manipulating the indices themselves rather than the components to which they correspond. In the second category, vector frames are explicitly introduced on the manifold and tensor operations are carried out on the components in a given frame.

3. An overview of Sage

Sage [14] is a free, open-source mathematics software system, which is based on the Python programming language. It makes use of over 90 open-source packages, among which are Maxima and Pynac (symbolic calculations), GAP (group theory), PARI/GP (number theory), Singular (polynomial computations), and matplotlib (high quality 2D figures). Sage provides a uniform Python interface to all these packages; however, Sage is much more than a mere interface: it contains a large and increasing part of original code (more than 750,000 lines of Python and Cython, involving 5344 classes). Sage was created in 2005 by W. Stein [15] and since then its development has been sustained by more than a hundred researchers (mostly mathematicians). Very good introductory textbooks about Sage are [16, 17, 18].

Apart from the syntax, which is based on a popular programming language and not a custom script language, a difference between Sage and, e.g., Maple or Mathematica is the usage of the *parent/element pattern*. This framework more closely reflects actual mathematics. For instance, in Mathematica, all objects are trees of symbols and the program is essentially a set of sophisticated rules to manipulate symbols. On the contrary, in Sage each object has a given type (i.e. is an instance of a given Python class[1]), and one distinguishes *parent* types, which model mathematical sets with some structure (e.g. algebraic structure), from *element* types, which model set elements. Moreover, each parent belongs to some dynamically generated class that encodes informations about its *category*, in the mathematical sense of the word (see [19] for a discussion of Sage's category framework). Automatic conversion rules, called *coercions*, prior to a binary operation, e.g. $x + y$ with x and y having different parents, are implemented.

4. The SageManifolds project

4.1. Aim and scope

Sage is well developed in many areas of mathematics but very little exists for differential geometry and tensor calculus. One may mention differential forms defined on a fixed coordinate patch, implemented by J. Vankerschaver [20], and the 2-dimensional parametrized surfaces of the 3-dimensional Euclidean space recently added by M. Malakhaltsev and J. Vankerschaver [21].

The aim of SageManifolds [22] is to introduce smooth manifolds and tensor fields in Sage, with the following requirements: (i) one should be able to introduce various coordinate charts on a manifold, with the relevant transition maps; (ii) tensor fields must be manipulated as such and not through their components with respect to a specific (possibly coordinate) vector frame.

Concretely, the project amounts to creating new Python classes, such as `Manifold`, `Chart`, `TensorField` or `Metric`, to implement them within the parent/element pattern and to code mathematical operations as class methods. For instance the class `Manifold`, devoted to real smooth manifolds, is a parent class, i.e. it inherits from Sage's class `Parent`. On the other hand, the class devoted to manifold points, `ManifoldPoint`, is an element class and therefore inherits from Sage's class `Element`. This is illustrated by the inheritance diagram of Fig. 1. In this

[1] Let us recall that within an object-oriented programming language (as Python), a *class* is a structure to declare and store the properties common to a set of objects. These properties are data (called *attributes* or *state variables*) and functions acting on the data (called *methods*). A specific realization of an object within a given class is called an *instance* of that class.

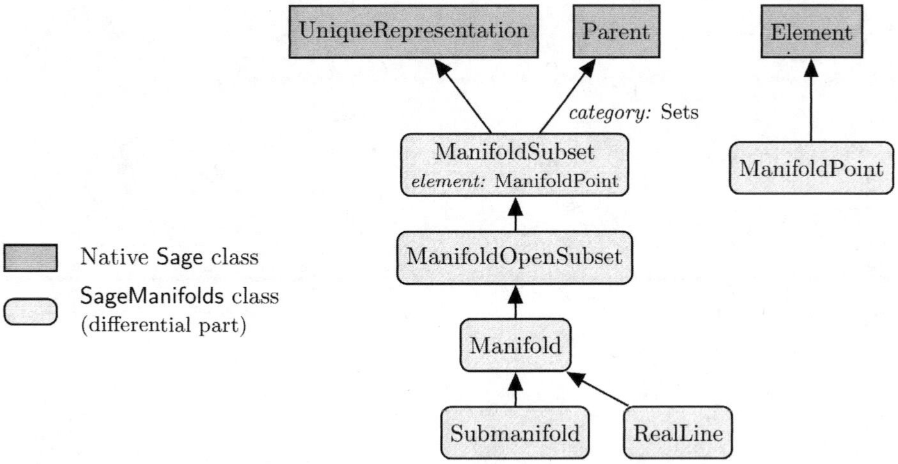

Figure 1. Python classes for smooth manifolds (`Manifold`), generic subsets of them (`ManifoldSubset`), open subsets of them (`ManifoldOpenSubset`) and points on them (`ManifoldPoint`).

diagram, each class at the base of some arrow is a subclass (also called *derived class*) of the class at the arrowhead. Note however that the actual type of a parent is a dynamically generated class taking into account the mathematical category to which it belongs. For instance, the actual type of a smooth manifold is not `Manifold`, but a subclass of it named `Manifold_with_category`, reflecting the fact that `Manifold` is declared in the category of `Sets`[2]. Note also that the class `Manifold` inherits from Sage's class `UniqueRepresentation`, which ensures that there is a unique manifold instance for a given dimension and given name.

4.2. Implementation of charts

Given a smooth manifold $\mathcal{M}$ of dimension n, a coordinate chart on some open subset $U \subset \mathcal{M}$ is implemented in SageManifolds via the class `Chart`, whose main data is a n-tuple of Sage symbolic variables $(x_1, \ldots, x_n)$, each of them representing a coordinate. In general, more than one (regular) chart is required to cover the entire manifold. For instance, at least 2 charts are necessary for the n-dimensional sphere $\mathbb{S}^n$ $(n \geq 1)$ and the torus $\mathbb{T}^2$ and 3 charts for the real projective plane $\mathbb{RP}^2$ (see Fig. 6 below). Accordingly, SageManifolds allows for an arbitrary number of charts. To fully specify the manifold, one shall also provide the transition maps (changes of coordinates) on overlapping chart domains (SageManifolds class `CoordChange`).

4.3. Implementation of scalar fields

A *scalar field* on manifold $\mathcal{M}$ is a smooth mapping

$$\begin{array}{rccl} f: & U \subset \mathcal{M} & \longrightarrow & \mathbb{R} \\ & p & \longmapsto & f(p), \end{array} \tag{1}$$

[2] A tighter category would be topological spaces, but such a category has been not implemented in Sage yet.

Journal of Physics: Conference Series **600** (2015) 012002 doi:10.1088/1742-6596/600/1/012002

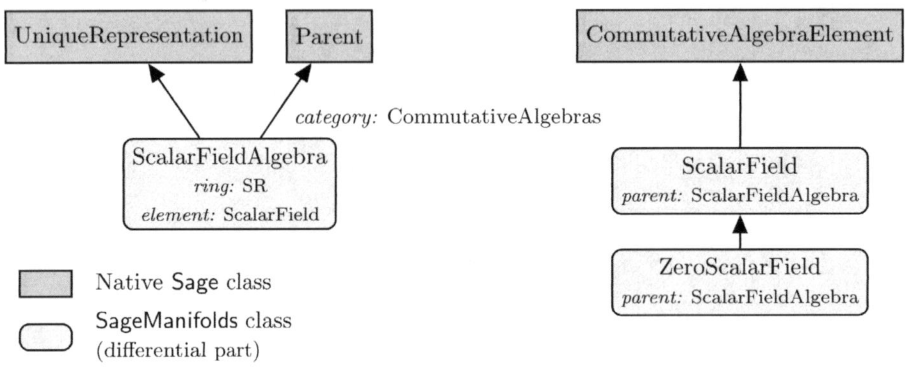

Figure 2. Python classes for scalar fields on a manifold.

where U is some open subset of $\mathcal{M}$. A scalar field has different coordinate representations F, $\hat{F}$, etc. in different charts X, $\hat{X}$, etc. defined on U:

$$f(p) = F(\underbrace{x^1, \dots, x^n}_{\substack{\text{coord. of } p \\ \text{in chart } X}}) = \hat{F}(\underbrace{\hat{x}^1, \dots, \hat{x}^n}_{\substack{\text{coord. of } p \\ \text{in chart } \hat{X}}}) = \dots \quad (2)$$

These representations are stored in some attribute of the class `ScalarField`, namely a Python dictionary[3] whose keys are the various charts defined on U:

$$f.\texttt{\_express} = \left\{X : F,\ \hat{X} : \hat{F}, \dots\right\}. \quad (3)$$

Each representation F is an instance of the class `FunctionChart`, which resembles Sage native symbolic functions, but involves automatic simplifications in all arithmetic operations.

Given an open subset $U \subset \mathcal{M}$, the set $C^\infty(U)$ of scalar fields defined on U has naturally the structure of a commutative algebra over $\mathbb{R}$: it is clearly a vector space over $\mathbb{R}$ and it is endowed with a commutative ring structure by pointwise multiplication:

$$\forall f, g \in C^\infty(U), \quad \forall p \in U, \quad (f.g)(p) := f(p)g(p). \quad (4)$$

The algebra $C^\infty(U)$ is implemented in SageManifolds via the parent class `ScalarFieldAlgebra`, in the category `CommutativeAlgebras`. The corresponding element class is of course `ScalarField` (cf. Fig. 2).

4.4. Modules and free modules

Given an open subset $U \subset \mathcal{M}$, the set $\mathscr{X}(U)$ of all smooth vector fields defined on U has naturally the structure of a *module over the algebra* $C^\infty(U)$. Let us recall that a *module* is similar to a *vector space*, except that it is based on a *ring* (here $C^\infty(U)$) instead of a *field* (usually $\mathbb{R}$ or $\mathbb{C}$ in physical applications). Of course, every vector space is a module, since every

[3] A *dictionary*, also known as *associative array*, is a data structure that generalizes the concept of array in the sense that the key to access to some element is not restricted to an integer or a tuple of integers.

field is a ring. There is an important difference though: every vector space has a basis (as a consequence of the axiom of choice), while a module does not necessarily have any. When it possesses one, it is called a *free module*. Moreover, if the module's base ring is commutative, it can be shown that all bases have the same cardinality, which is called the *rank* of the module (for vector spaces, which are free modules, the word *dimension* is used instead of *rank*).

If $\mathscr{X}(U)$ is a free module (examples are provided in Sec. 4.5 below), a basis of it is nothing but a *vector frame* $(\boldsymbol{e}_a)_{1\le a\le n}$ on U (often called a *tetrad* in the context of 4-dimensional GR):

$$\forall \boldsymbol{v} \in \mathscr{X}(U), \quad \boldsymbol{v} = v^a \boldsymbol{e}_a, \quad \text{with } v^a \in C^\infty(U). \tag{5}$$

The rank of $\mathscr{X}(U)$ is thus n, i.e. the manifold's dimension[4]. At any point $p \in U$, Eq. (5) gives birth to an identity in the tangent vector space $T_p\mathcal{M}$:

$$\boldsymbol{v}(p) = v^a(p)\, \boldsymbol{e}_a(p), \quad \text{with } v^a(p) \in \mathbb{R}, \tag{6}$$

which means that the set $(\boldsymbol{e}_a(p))_{1\le a\le n}$ is a basis of $T_p\mathcal{M}$. Note that if U is covered by a chart $(x^a)_{1\le a\le n}$, then $(\partial/\partial x^a)_{1\le a\le n}$ is a vector frame on U, usually called *coordinate frame* or *natural basis*. Note also that, being a vector space over $\mathbb{R}$, the tangent space $T_p\mathcal{M}$ represents another kind of free module which occurs naturally in the current context.

It turns out that so far only free modules *with a distinguished basis* were implemented in Sage. This means that, given a free module M of rank n, all calculations refer to a single basis of M. This amounts to identifying M with R^n, where R is the ring over which M is based. This is unfortunately not sufficient for dealing with smooth manifolds in a coordinate-independent way. For instance, there is no canonical isomorphism between $T_p\mathcal{M}$ and $\mathbb{R}^n$ when no coordinate system is privileged in the neighborhood of p. Therefore we have started a pure algebraic part of SageManifolds to implement generic free modules, with an arbitrary number of bases, none of them being distinguished. This resulted in (i) the parent class `FiniteRankFreeModule`, within Sage's category `Modules`, and (ii) the element class `FiniteRankFreeModuleElement`. Then both classes `VectorFieldFreeModule` (for $\mathscr{X}(U)$, when it is a free module) and `TangentSpace` (for $T_p\mathcal{M}$) inherit from `FiniteRankFreeModule` (see Fig. 3).

4.5. Implementation of vector fields

Ultimately, in SageManifolds, vector fields are described by their components with respect to various vector frames, according to Eq. (5), but without any vector frame being privileged, leaving the freedom to select one to the user, as well as to change coordinates. A key point is that not every manifold admits a global vector frame. A manifold $\mathcal{M}$, or more generally an open subset $U \subset \mathcal{M}$, that admits a global vector frame is called *parallelizable*. Equivalently, $\mathcal{M}$ is parallelizable if, and only if, $\mathscr{X}(\mathcal{M})$ is a free module. In terms of tangent bundles, parallelizable manifolds are those for which the tangent bundle is trivial: $T\mathcal{M} \simeq \mathcal{M} \times \mathbb{R}^n$. Examples of parallelizable manifolds are [23]

- the Cartesian space $\mathbb{R}^n$ for $n = 1, 2, \ldots$,
- the circle $\mathbb{S}^1$,
- the torus $\mathbb{T}^2 = \mathbb{S}^1 \times \mathbb{S}^1$,
- the sphere $\mathbb{S}^3 \simeq \mathrm{SU}(2)$, as any Lie group,
- the sphere $\mathbb{S}^7$,
- any orientable 3-manifold (Steenrod theorem [24]).

[4] Note that the dimensionality of $\mathscr{X}(U)$ depends of the adopted structure: as a vector space over $\mathbb{R}$, the dimension of $\mathscr{X}(U)$ is infinite, while as a free module over $C^\infty(U)$, $\mathscr{X}(U)$ has a finite rank. Note also that if $\mathscr{X}(U)$ is not free (i.e. no global vector frame exists on U), the notion of rank is meaningless.

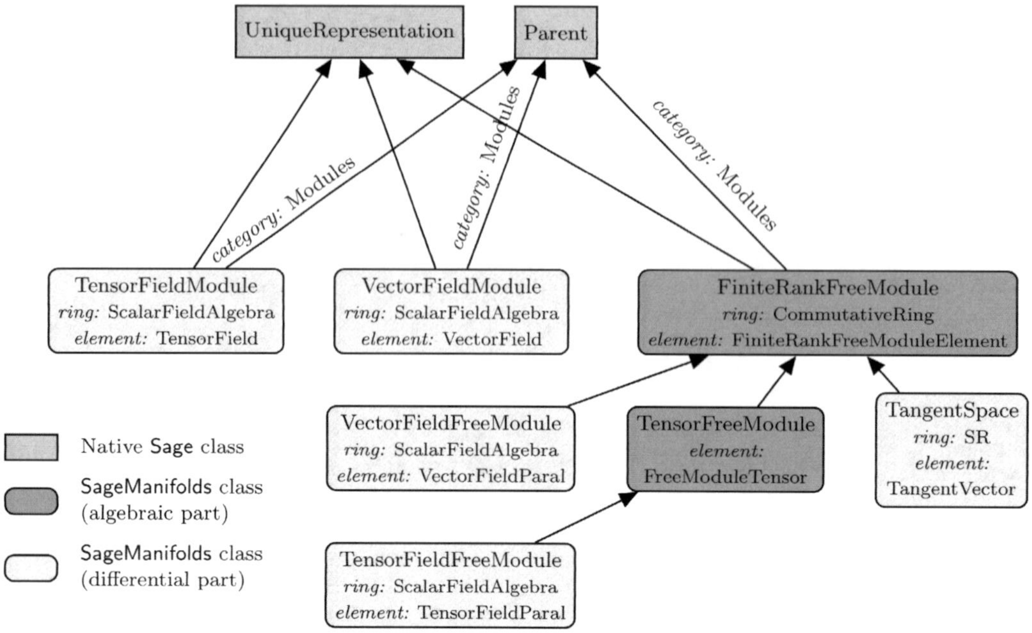

Figure 3. Python classes for modules. For each of them, the class of the base ring is indicated, as well as the class for the elements.

On the other hand, examples of non-parallelizable manifolds are

- the sphere $\mathbb{S}^2$ (as a consequence of the hairy ball theorem), as well as any sphere $\mathbb{S}^n$ with $n \notin \{1, 3, 7\}$,
- the real projective plane $\mathbb{RP}^2$.

Actually, "most" manifolds are non-parallelizable. As noticed above, if a manifold is covered by a single chart, it is parallelizable (the prototype being $\mathbb{R}^n$). But the reverse is not true: $\mathbb{S}^1$ and $\mathbb{T}^2$ are parallelizable and require at least two charts to cover them.

If the manifold $\mathcal{M}$ is not parallelizable, we assume that it can be covered by a finite number N of parallelizable open subsets U_i ($1 \leq i \leq N$). In particular, this holds if $\mathcal{M}$ is compact, for any compact manifold admits a finite atlas. We then consider the restrictions of vector fields to the U_i's. For each i, $\mathscr{X}(U_i)$ is a free module of rank $n = \dim \mathcal{M}$ and is implemented in SageManifolds as an instance of `VectorFieldFreeModule` (cf. Sec. 4.4 and Figs. 3 and 4). Each vector field $\boldsymbol{v} \in \mathscr{X}(U_i)$ has different sets of components $(v^a)_{1\leq a\leq n}$ in different vector frames $(\boldsymbol{e}_a)_{1\leq a\leq n}$ introduced on U_i [cf. Eq. (5)]. They are stored as a Python dictionary whose keys are the vector frames:

$$\boldsymbol{v}.\texttt{\_components} = \{(\boldsymbol{e}) : (v^a),\ (\hat{\boldsymbol{e}}) : (\hat{v}^a), \ldots\}\,. \qquad (7)$$

4.6. Implementation of tensor fields

The implementation of tensor fields in SageManifolds follows the strategy adopted for vector fields. Consider for instance a tensor field $\boldsymbol{T}$ of type (1,1) on the manifold $\mathcal{M}$. It can

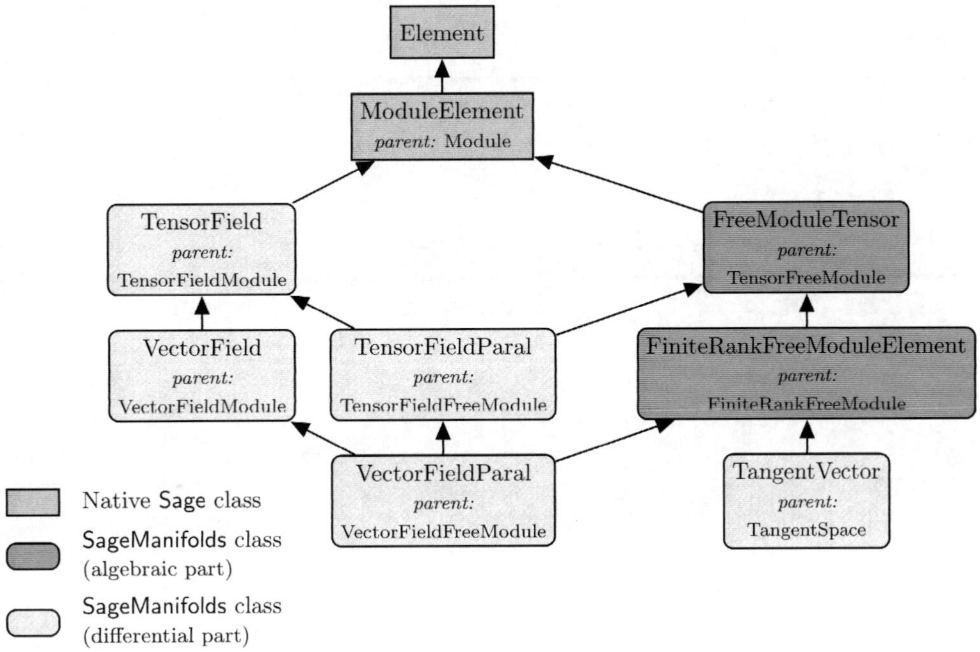

Figure 4. Python classes implementing tensors and tensor fields.

be represented by components $T^a{}_b$ only on a parallelizable open subset $U \subset \mathcal{M}$, since the decomposition

$$\boldsymbol{T}|_U = T^a{}_b \, \boldsymbol{e}_a \otimes \boldsymbol{e}^b, \tag{8}$$

which defines $T^a{}_b$, is meaningful only when a vector frame $(\boldsymbol{e}_a)$ exists[5]. Therefore, one first decomposes the tensor field $\boldsymbol{T}$ into its restrictions $\boldsymbol{T}|_{U_i}$ on parallelizable open subsets of $\mathcal{M}$, U_i $(1 \leq i \leq N)$ and then considers the components on various vector frames on each subset U_i. For each vector frame $(\boldsymbol{e}_a)$, the set of components $(T^a{}_b)$ is stored in a devoted class (named `Components`), which takes into account all the tensor monoterm symmetries: only non-redundant components are stored, the other ones being deduced by (anti)symmetry. This is illustrated in Fig. 5, which depicts the internal storage of tensor fields in SageManifolds. Note that each component $T^a{}_b$ is a scalar field on U_i, according to the formula

$$T^a{}_b = \boldsymbol{T}(\boldsymbol{e}^a, \boldsymbol{e}_b). \tag{9}$$

Accordingly, the penultimate level of Fig. 5 corresponds to the scalar field storage, as described by (3). The last level is constituted by Sage's symbolic expressions (class `Expression`).

5. Current status of SageManifolds

5.1. Functionalities

At present (version 0.6), the functionalities included in SageManifolds are as follows:

[5] Using standard notation, in Eq. (8), $(\boldsymbol{e}^b)$ stands for the coframe dual to $(\boldsymbol{e}_a)$

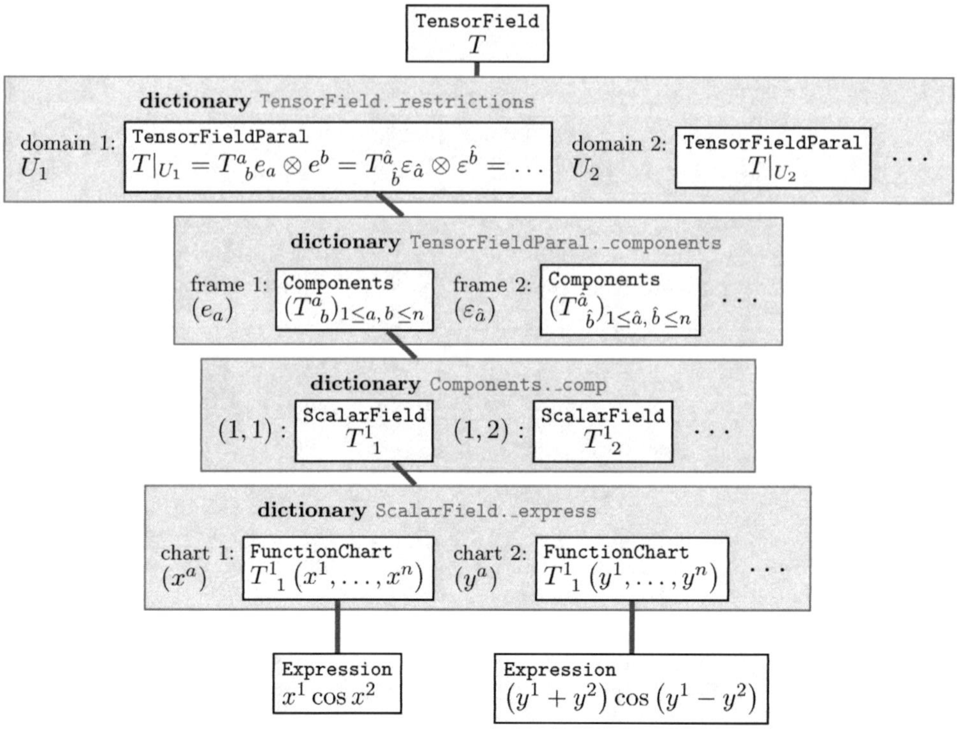

Figure 5. Storage of tensor fields in SageManifolds. Each red box represents a Python dictionary; the dictionary values are depicted by yellow boxes, the keys being indicated at the left of each box.

- maps between manifolds and pullback operator,
- submanifolds and pushforward operator,
- standard tensor calculus (tensor product, contraction, symmetrization, etc.), even on non-parallelizable manifolds,
- arbitrary monoterm tensor symmetries,
- exterior calculus (wedge product and exterior derivative, Hodge duality),
- Lie derivatives along a vector field,
- affine connections (curvature, torsion),
- pseudo-Riemannian metrics (Levi-Civita connection, Weyl tensor),
- graphical display of charts.

5.2. Parallelization

To improve the reactivity of SageManifolds and take advantage of multicore processors, some tensorial operations are performed by parallel processes. The parallelization is implemented by means of the Python library `multiprocessing`, via the built-in Sage decorator `@parallel`. Using it permits to define a function that is run on different sub-processes. If n processes are used, given a function and a list of arguments for it, any process will call the function with an element of the list, one at time, spanning all the list.

Currently[6], the parallelized operations are tensor algebra, tensor contractions, computation of the connection coefficient and computation of Riemann tensor.

The parallelization of an operation is achieved by first creating a function which computes the required operation on a subset of the components of a tensor; second, by creating a list of $2n$ (twice the number of used processes) arguments for this function. Then applying this function to the input list, the calculation is performed in parallel. At the end of the computation a fourth phase is needed to retrieve the results. The choice to divide the work in $2n$ is a compromise between the load balancing and the cost of creating multiple processes. The number of processors to be used in the parallelization can be controlled by the user.

6. SageManifolds at work: Kerr spacetime and Simon-Mars tensor

We give hereafter a short illustration of SageManifolds focused on tensor calculus in 4-dimensional GR. Another example, to be found at [25], is based on the manifold $\mathbb{S}^2$ and focuses more on the use of multiple charts and on the treatment of non-parallelizable manifolds. Yet another example illustrates some graphical capabilities of SageManifolds: Figure 6 shows the famous immersion of the real projective plane $\mathbb{RP}^2$ into the Euclidean space $\mathbb{R}^3$ known as the *Boy surface*. This figure has been obtained by means of the method `plot()` applied to three coordinate charts covering $\mathbb{RP}^2$, the definition of which is related to the interpretation of $\mathbb{RP}^2$ as the set of straight lines Δ through the origin of $\mathbb{R}^3$: (i) in red, the chart X_1 covering the open subset of $\mathbb{RP}^2$ defined by all lines Δ that are not parallel to the plane $z = 0$, the coordinates of X_1 being the coordinates (x, y) of the intersection of the considered line Δ with the plane $z = 1$; (ii) in green, the chart X_2 covering the open subset defined by all lines Δ that are not parallel to the plane $x = 0$, the coordinates of X_2 being the coordinates (y, z) of intersection with the plane $x = 1$; (iii) in blue, the chart X_3 covering the open subset defined by all lines Δ that are not parallel to the plane $y = 0$, the coordinates of X_2 being the coordinates (z, x) of intersection with the plane $y = 1$. Figure 6 actually shows the coordinate grids of these three charts through the Apéry map [26], which realizes an immersion of $\mathbb{RP}^2$ into $\mathbb{R}^3$. This example, as many others, can be found at [25].

Let us consider a 4-dimensional spacetime, i.e. a smooth 4-manifold $\mathcal{M}$ endowed with a Lorentzian metric $\boldsymbol{g}$. We assume that $(\mathcal{M}, \boldsymbol{g})$ is stationary and denote by $\boldsymbol{\xi}$ the corresponding Killing vector field. The *Simon-Mars tensor w.r.t.* $\boldsymbol{\xi}$ is then the type-(0,3) tensor field $\boldsymbol{S}$ defined by [27]

$$S_{\alpha\beta\gamma} := 4\mathcal{C}_{\mu\alpha\nu[\beta}\, \xi^{\mu}\xi^{\nu}\, \sigma_{\gamma]} + \gamma_{\alpha[\beta}\, \mathcal{C}_{\gamma]\rho\mu\nu}\, \xi^{\rho}\, \mathcal{F}^{\mu\nu}, \tag{10}$$

where

- $\gamma_{\alpha\beta} := \lambda\, g_{\alpha\beta} + \xi_\alpha \xi_\beta$, with $\lambda := -\xi_\mu \xi^\mu$;
- $\mathcal{C}_{\alpha\beta\mu\nu} := C_{\alpha\beta\mu\nu} + \frac{i}{2}\epsilon^{\rho\sigma}{}_{\mu\nu}\, C_{\alpha\beta\rho\sigma}$, with $C^{\alpha}{}_{\beta\mu\nu}$ being the Weyl curvature tensor and $\epsilon_{\alpha\beta\mu\nu}$ the Levi-Civita volume 4-form;
- $\mathcal{F}_{\alpha\beta} := F_{\alpha\beta} + i\, {}^{*}F_{\alpha\beta}$, with $F_{\alpha\beta} := \nabla_\alpha \xi_\beta$ (Killing 2-form) and ${}^{*}F_{\alpha\beta} := \frac{1}{2}\epsilon^{\mu\nu}{}_{\alpha\beta} F_{\mu\nu}$;
- $\sigma_\alpha := 2\mathcal{F}_{\mu\alpha}\xi^{\mu}$ (Ernst 1-form).

The Simon-Mars tensor provides a nice characterization of Kerr spacetime, according the following theorem proved by Mars [27]: if $\boldsymbol{g}$ satisfies the vacuum Einstein equation and $(\mathcal{M}, \boldsymbol{g})$ contains a stationary asymptotically flat end $\mathcal{M}^\infty$ such that $\boldsymbol{\xi}$ tends to a time translation at infinity in $\mathcal{M}^\infty$ and the Komar mass of $\boldsymbol{\xi}$ in $\mathcal{M}^\infty$ is non-zero, then $\boldsymbol{S} = 0$ if, and only if, $(\mathcal{M}, \boldsymbol{g})$ is locally isometric to a Kerr spacetime.

[6] in the development version of SageManifolds; this will become available in version 0.7 of the stable release.

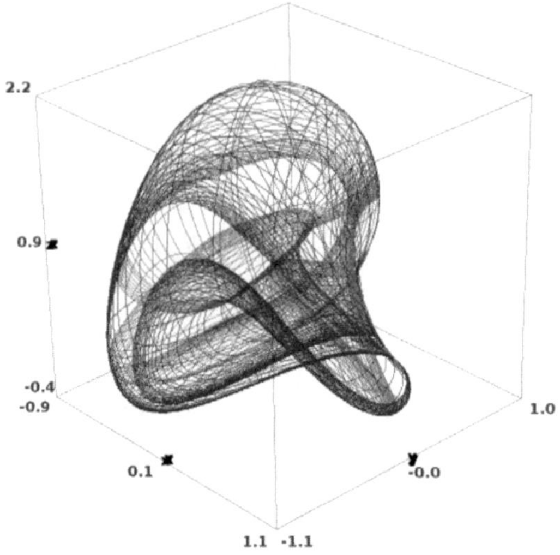

Figure 6. Boy surface depicted via the grids of 3 coordinate charts covering $\mathbb{RP}^2$ (see the text for the color code).

In what follows, we use SageManifolds to compute the Simon-Mars tensor according to formula (10) for the Kerr metric and check that we get zero (the "if" part of the above theorem). The corresponding worksheet can be downloaded from
http://sagemanifolds.obspm.fr/examples/html/SM_Simon-Mars_Kerr.html.
For the sake of clarity, let us recall that, as an object-oriented language, Python (and hence Sage) makes use of the following postfix notation:

result = object.function(arguments)

In a functional language, this would correspond to result = function(object,arguments). For instance, the Riemann tensor of a metric g is obtained as riem = g.riemann() (in this case, there is no extra argument, hence the empty parentheses). With this in mind, let us proceed with the computation by means of SageManifolds. In the text below, the blue color denotes the outputs as they appear in the Sage notebook (note that all outputs are automatically LaTeX-formatted by Sage).

The first step is to declare the Kerr spacetime (or more precisely the part of the Kerr spacetime covered by Boyer-Lindquist coordinates) as a 4-dimensional manifold:

```
M = Manifold(4, 'M', latex_name=r'\mathcal{M}')
print M
```

4-dimensional manifold 'M'

The standard Boyer-Lindquist coordinates (t, r, θ, ϕ) are introduced by declaring a chart X on $\mathcal{M}$, via the method chart(), the argument of which is a string expressing the coordinates names, their ranges (the default is $(-\infty, +\infty)$) and their LaTeX symbols:

```
X.<t,r,th,ph> = M.chart('t r:(0,+oo) th:(0,pi):\\theta ph:(0,2*pi):\\phi')
print X ; X
```

chart (M, (t, r, th, ph))
$(\mathcal{M}, (t, r, \theta, \phi))$

We define next the Kerr metric $\boldsymbol{g}$ by setting its components in the coordinate frame associated with Boyer-Lindquist coordinates. Since the latter is the current manifold's default frame (being the only one defined at this stage), we do not need to specify it when referring to the components by their indices:

```
g = M.lorentz_metric('g')
m = var('m') ; a = var('a')
rho2 = r^2 + (a*cos(th))^2
Delta = r^2 -2*m*r + a^2
g[0,0] = -(1-2*m*r/rho2)
g[0,3] = -2*a*m*r*sin(th)^2/rho2
g[1,1], g[2,2] = rho2/Delta, rho2
g[3,3] = (r^2+a^2+2*m*r*(a*sin(th))^2/rho2)*sin(th)^2
g.view()
```

$$g = \left(-\frac{a^2\cos(\theta)^2 - 2\,mr + r^2}{a^2\cos(\theta)^2 + r^2}\right) \mathrm{d}t \otimes \mathrm{d}t + \left(-\frac{2\,amr\sin(\theta)^2}{a^2\cos(\theta)^2 + r^2}\right) \mathrm{d}t \otimes \mathrm{d}\phi + \left(\frac{a^2\cos(\theta)^2 + r^2}{a^2 - 2\,mr + r^2}\right) \mathrm{d}r \otimes \mathrm{d}r + \left(a^2\cos(\theta)^2 + r^2\right) \mathrm{d}\theta \otimes \mathrm{d}\theta + \left(-\frac{2\,amr\sin(\theta)^2}{a^2\cos(\theta)^2 + r^2}\right) \mathrm{d}\phi \otimes \mathrm{d}t + \left(\frac{2\,a^2mr\sin(\theta)^4 + \left(a^2r^2 + r^4 + \left(a^4 + a^2r^2\right)\cos(\theta)^2\right)\sin(\theta)^2}{a^2\cos(\theta)^2 + r^2}\right) \mathrm{d}\phi \otimes \mathrm{d}\phi$$

The Levi-Civita connection $\boldsymbol{\nabla}$ associated with $\boldsymbol{g}$ is obtained by the method `connection()`:

```
nab = g.connection() ; print nab
```

Levi-Civita connection 'nabla_g' associated with the Lorentzian metric 'g' on the 4-dimensional manifold 'M'

As a check, we verify that the covariant derivative of $\boldsymbol{g}$ with respect to $\boldsymbol{\nabla}$ vanishes identically:

```
nab(g).view()
```

$\nabla_g g = 0$

As mentionned above, the default vector frame on the spacetime manifold is the coordinate basis associated with Boyer-Lindquist coordinates:

```
M.default_frame() is X.frame()
```

True

```
X.frame()
```

$\left(\mathcal{M}, \left(\frac{\partial}{\partial t}, \frac{\partial}{\partial r}, \frac{\partial}{\partial \theta}, \frac{\partial}{\partial \phi}\right)\right)$

Let us consider the first vector field of this frame:

```
xi = X.frame()[0] ; xi
```

$\frac{\partial}{\partial t}$

```
print xi
```

vector field 'd/dt' on the 4-dimensional manifold 'M'

The 1-form associated to it by metric duality is

```
xi_form = xi.down(g)
xi_form.set_name('xi_form', r'\underline{\xi}')
print xi_form ; xi_form.view()
```

1-form 'xi_form' on the 4-dimensional manifold 'M'

$$\underline{\xi} = \left(-\frac{a^2\cos(\theta)^2 - 2\,mr + r^2}{a^2\cos(\theta)^2 + r^2}\right)\mathrm{d}t + \left(-\frac{2\,amr\sin(\theta)^2}{a^2\cos(\theta)^2 + r^2}\right)\mathrm{d}\phi$$

Its covariant derivative is

```
nab_xi = nab(xi_form)
print nab_xi ; nab_xi.view()
```

tensor field 'nabla_g xi_form' of type (0,2) on the 4-dimensional manifold 'M'

$$\nabla_g\underline{\xi} = \left(\frac{a^2m\cos(\theta)^2 - mr^2}{a^4\cos(\theta)^4 + 2\,a^2r^2\cos(\theta)^2 + r^4}\right)\mathrm{d}t\otimes\mathrm{d}r + \left(\frac{2\,a^2mr\cos(\theta)\sin(\theta)}{a^4\cos(\theta)^4 + 2\,a^2r^2\cos(\theta)^2 + r^4}\right)\mathrm{d}t\otimes\mathrm{d}\theta + \left(-\frac{a^2m\cos(\theta)^2 - mr^2}{a^4\cos(\theta)^4 + 2\,a^2r^2\cos(\theta)^2 + r^4}\right)\mathrm{d}r\otimes\mathrm{d}t + \left(\frac{\left(a^3m\cos(\theta)^2 - amr^2\right)\sin(\theta)^2}{a^4\cos(\theta)^4 + 2\,a^2r^2\cos(\theta)^2 + r^4}\right)\mathrm{d}r\otimes\mathrm{d}\phi + \left(-\frac{2\,a^2mr\cos(\theta)\sin(\theta)}{a^4\cos(\theta)^4 + 2\,a^2r^2\cos(\theta)^2 + r^4}\right)\mathrm{d}\theta\otimes\mathrm{d}t + \left(\frac{2\left(a^3mr + amr^3\right)\cos(\theta)\sin(\theta)}{a^4\cos(\theta)^4 + 2\,a^2r^2\cos(\theta)^2 + r^4}\right)\mathrm{d}\theta\otimes\mathrm{d}\phi + \left(-\frac{\left(a^3m\cos(\theta)^2 - amr^2\right)\sin(\theta)^2}{a^4\cos(\theta)^4 + 2\,a^2r^2\cos(\theta)^2 + r^4}\right)\mathrm{d}\phi\otimes\mathrm{d}r + \left(-\frac{2\left(a^3mr + amr^3\right)\cos(\theta)\sin(\theta)}{a^4\cos(\theta)^4 + 2\,a^2r^2\cos(\theta)^2 + r^4}\right)\mathrm{d}\phi\otimes\mathrm{d}\theta$$

Let us check that the Killing equation is satisfied:

```
nab_xi.symmetrize().view()
```

0

Equivalently, we check that the Lie derivative of the metric along $\boldsymbol{\xi}$ vanishes:

```
g.lie_der(xi).view()
```

0

Thanks to Killing equation, $\boldsymbol{\nabla}\underline{\boldsymbol{\xi}}$ is antisymmetric. We may therefore define a 2-form by $\boldsymbol{F} := -\boldsymbol{\nabla}\underline{\boldsymbol{\xi}}$. Here we enforce the antisymmetry by calling the method `antisymmetrize()` on `nab_xi`:

```
F = - nab_xi.antisymmetrize()
F.set_name('F')
print F ; F.view()
```

2-form 'F' on the 4-dimensional manifold 'M'

$$F = \left(-\frac{a^2m\cos(\theta)^2 - mr^2}{a^4\cos(\theta)^4 + 2\,a^2r^2\cos(\theta)^2 + r^4}\right)\mathrm{d}t\wedge\mathrm{d}r + \left(-\frac{2\,a^2mr\cos(\theta)\sin(\theta)}{a^4\cos(\theta)^4 + 2\,a^2r^2\cos(\theta)^2 + r^4}\right)\mathrm{d}t\wedge\mathrm{d}\theta + \left(-\frac{\left(a^3m\cos(\theta)^2 - amr^2\right)\sin(\theta)^2}{a^4\cos(\theta)^4 + 2\,a^2r^2\cos(\theta)^2 + r^4}\right)\mathrm{d}r\wedge\mathrm{d}\phi + \left(-\frac{2\left(a^3mr + amr^3\right)\cos(\theta)\sin(\theta)}{a^4\cos(\theta)^4 + 2\,a^2r^2\cos(\theta)^2 + r^4}\right)\mathrm{d}\theta\wedge\mathrm{d}\phi$$

The squared norm of the Killing vector is

```
lamb = - g(xi,xi)
lamb.set_name('lambda', r'\lambda')
print lamb ; lamb.view()
```

scalar field 'lambda' on the 4-dimensional manifold 'M'

$$\begin{array}{llll} \lambda: & \mathcal{M} & \longrightarrow & \mathbb{R} \\ & (t,r,\theta,\phi) & \longmapsto & \frac{a^2\cos(\theta)^2 - 2\,mr + r^2}{a^2\cos(\theta)^2 + r^2} \end{array}$$

Instead of invoking $\boldsymbol{g}(\boldsymbol{\xi},\boldsymbol{\xi})$, we could have evaluated λ by means of the 1-form $\underline{\boldsymbol{\xi}}$ acting on the vector field $\boldsymbol{\xi}$:

```
lamb == - xi_form(xi)
```

True

---

or, using index notation as $\lambda = -\xi_a\xi^a$:

```
lamb == - ( xi_form['_a']*xi['^a'] )
```

True

The Riemann curvature tensor associated with $\boldsymbol{g}$ is

```
Riem = g.riemann()
print Riem
```

tensor field 'Riem(g)' of type (1,3) on the 4-dimensional manifold 'M'

The component ${R^0}_{123} = {R^t}_{r\theta\phi}$ is

```
Riem[0,1,2,3]
```

$$-\frac{\left(a^7m - 2\,a^5m^2r + a^5mr^2\right)\cos(\theta)\sin(\theta)^5 + \left(a^7m + 2\,a^5m^2r + 6\,a^5mr^2 - 6\,a^3m^2r^3 + 5\,a^3mr^4\right)\cos(\theta)\sin(\theta)^3}{a^2r^6 - 2\,mr^7 + r^8 + (a^8 - 2\,a^6mr + a^6r^2)\cos(\theta)^6 + 3\,(a^6r^2 - 2\,a^4mr^3 + a^4r^4)\cos(\theta)^4 + 3\,(a^4r^4 - 2\,a^2mr^5 + a^2r^6)\cos(\theta)^2} + \frac{2\left(a^7m - a^5mr^2 - 5\,a^3mr^4 - 3\,amr^6\right)\cos(\theta)\sin(\theta)}{a^2r^6 - 2\,mr^7 + r^8 + (a^8 - 2\,a^6mr + a^6r^2)\cos(\theta)^6 + 3\,(a^6r^2 - 2\,a^4mr^3 + a^4r^4)\cos(\theta)^4 + 3\,(a^4r^4 - 2\,a^2mr^5 + a^2r^6)\cos(\theta)^2}$$

Let us check that the Kerr metric is a vacuum solution of Einstein equation, i.e. that the Ricci tensor vanishes identically:

```
g.ricci().view()
```

$\mathrm{Ric}(g) = 0$

The Weyl conformal curvature tensor is

```
C = g.weyl() ; print C
```

tensor field 'C(g)' of type (1,3) on the 4-dimensional manifold 'M'

Let us exhibit the component ${C^0}_{101} = {C^t}_{rtr}$:

```
C[0,1,0,1]
```

$$\frac{3\,a^4mr\cos(\theta)^4 + 3\,a^2mr^3 + 2\,mr^5 - \left(9\,a^4mr + 7\,a^2mr^3\right)\cos(\theta)^2}{a^2r^6 - 2\,mr^7 + r^8 + (a^8 - 2\,a^6mr + a^6r^2)\cos(\theta)^6 + 3\,(a^6r^2 - 2\,a^4mr^3 + a^4r^4)\cos(\theta)^4 + 3\,(a^4r^4 - 2\,a^2mr^5 + a^2r^6)\cos(\theta)^2}$$

To form the Simon-Mars tensor, we need the fully covariant form (type-(0,4) tensor) of the Weyl tensor (i.e. $C_{\alpha\beta\mu\nu} = g_{\alpha\sigma}{C^\sigma}_{\beta\mu\nu}$); we get it by lowering the first index with the metric:

```
Cd = C.down(g) ; print Cd
```

tensor field of type (0,4) on the 4-dimensional manifold 'M'

The (monoterm) symmetries of this tensor are those inherited from the Weyl tensor, i.e. the antisymmetry on the last two indices (position 2 and 3, the first index being at position 0):

Journal of Physics: Conference Series **600** (2015) 012002 doi:10.1088/1742-6596/600/1/012002

```
Cd.symmetries()
```

no symmetry; antisymmetry: (2, 3)

Actually, `Cd` is also antisymmetric with respect to the first two indices (positions 0 and 1), as we can check:

```
Cd == Cd.antisymmetrize(0,1)
```

True

To take this symmetry into account explicitly, we set

```
Cd = Cd.antisymmetrize(0,1)
Cd.symmetries()
```

no symmetry; antisymmetries: [(0, 1), (2, 3)]

The starting point in the evaluation of Simon-Mars tensor is the self-dual complex 2-form associated with the Killing 2-form $\boldsymbol{F}$, i.e. the object $\boldsymbol{\mathcal{F}} := \boldsymbol{F} + i\,{}^{*}\boldsymbol{F}$, where ${}^{*}\boldsymbol{F}$ is the Hodge dual of $\boldsymbol{F}$:

```
FF = F + I * F.hodge_star(g)
FF.set_name('FF', r'\mathcal{F}')
print FF ; FF.view()
```

2-form 'FF' on the 4-dimensional manifold 'M'

$$\mathcal{F} = \left(-\frac{a^{2} m \cos(\theta)^{2} + 2i\, a m r \cos(\theta) - m r^{2}}{a^{4}\cos(\theta)^{4} + 2\, a^{2} r^{2} \cos(\theta)^{2} + r^{4}}\right) \mathrm{d}t \wedge \mathrm{d}r + \left(\frac{\left(i\, a^{3} m \cos(\theta)^{2} - 2\, a^{2} m r \cos(\theta) - i\, a m r^{2}\right)\sin(\theta)}{a^{4}\cos(\theta)^{4} + 2\, a^{2} r^{2} \cos(\theta)^{2} + r^{4}}\right) \mathrm{d}t \wedge \mathrm{d}\theta$$
$$+ \left(\frac{-4i\, a^{4} m^{2} r^{2} \cos(\theta)\sin(\theta)^{4} + \left(a^{3} m r^{4} - 2\, a m^{2} r^{5} + a m r^{6} - \left(a^{7} m - 2\, a^{5} m^{2} r + a^{5} m r^{2}\right)\cos(\theta)^{4}\right)\sin(\theta)^{2}}{a^{2} r^{6} - 2\, m r^{7} + r^{8} + \left(a^{8} - 2\, a^{6} m r + a^{6} r^{2}\right)\cos(\theta)^{6} + 3\left(a^{6} r^{2} - 2\, a^{4} m r^{3} + a^{4} r^{4}\right)\cos(\theta)^{4} + 3\left(a^{4} r^{4} - 2\, a^{2} m r^{5} + a^{2} r^{6}\right)\cos(\theta)^{2}}\right.$$
$$\left. - \frac{\left(\left(2i\, a^{6} m r + 2i\, a^{4} m r^{3}\right)\cos(\theta)^{3} + \left(-4i\, a^{4} m^{2} r^{2} + 2i\, a^{4} m r^{3} - 4i\, a^{2} m^{2} r^{4} + 2i\, a^{2} m r^{5}\right)\cos(\theta)\right)\sin(\theta)^{2}}{a^{2} r^{6} - 2\, m r^{7} + r^{8} + \left(a^{8} - 2\, a^{6} m r + a^{6} r^{2}\right)\cos(\theta)^{6} + 3\left(a^{6} r^{2} - 2\, a^{4} m r^{3} + a^{4} r^{4}\right)\cos(\theta)^{4} + 3\left(a^{4} r^{4} - 2\, a^{2} m r^{5} + a^{2} r^{6}\right)\cos(\theta)^{2}}\right) \mathrm{d}r \wedge \mathrm{d}\phi$$
$$+ \left(-\frac{\left(i\, a^{4} m + i\, a^{2} m r^{2}\right)\sin(\theta)^{3} + \left(-i\, a^{4} m + i\, m r^{4} + 2\left(a^{3} m r + a m r^{3}\right)\cos(\theta)\right)\sin(\theta)}{a^{4}\cos(\theta)^{4} + 2\, a^{2} r^{2} \cos(\theta)^{2} + r^{4}}\right) \mathrm{d}\theta \wedge \mathrm{d}\phi$$

Let us check that $\boldsymbol{\mathcal{F}}$ is self-dual, i.e. that it obeys ${}^{*}\boldsymbol{\mathcal{F}} = -i\boldsymbol{\mathcal{F}}$:

```
FF.hodge_star(g) == - I * FF
```

True

To evaluate the right self-dual of the Weyl tensor, we need the tensor $\epsilon^{\alpha\beta}{}_{\gamma\delta}$:

```
eps = g.volume_form(2)  # 2 = the first 2 indices are contravariant
print eps ; eps.symmetries()
```

tensor field of type (2,2) on the 4-dimensional manifold 'M'
no symmetry; antisymmetries: [(0, 1), (2, 3)]

The right self-dual Weyl tensor is then:

```
CC = Cd + I/2*( eps['^rs_..']*Cd['_..rs'] )
CC.set_name('CC', r'\mathcal{C}') ;
print CC ; CC.symmetries()
```

tensor field 'CC' of type (0,4) on the 4-dimensional manifold 'M'
no symmetry; antisymmetries: [(0, 1), (2, 3)]

```
CC[0,1,2,3]
```

$$\frac{\left(a^5 m \cos(\theta)^5 + 3i\, a^4 m r \cos(\theta)^4 + 3i\, a^2 m r^3 + 2i\, m r^5 - \left(3\, a^5 m + 5\, a^3 m r^2\right)\cos(\theta)^3\right)\sin(\theta)}{a^6 \cos(\theta)^6 + 3\, a^4 r^2 \cos(\theta)^4 + 3\, a^2 r^4 \cos(\theta)^2 + r^6} + \frac{\left(\left(-9i\, a^4 m r - 7i\, a^2 m r^3\right)\cos(\theta)^2 + 3\left(3\, a^3 m r^2 + 2\, a m r^4\right)\cos(\theta)\right)\sin(\theta)}{a^6 \cos(\theta)^6 + 3\, a^4 r^2 \cos(\theta)^4 + 3\, a^2 r^4 \cos(\theta)^2 + r^6}$$

The Ernst 1-form $\sigma_\alpha = 2\mathcal{F}_{\mu\alpha}\,\xi^\mu$ (0 = contraction on the first index of $\boldsymbol{\mathcal{F}}$):

```
sigma = 2*FF.contract(0, xi)
```

Instead of invoking the method `contract()`, we could have used the index notation to denote the contraction:

```
sigma == 2*( FF['_ma']*xi['^m'] )
```

True

```
sigma.set_name('sigma', r'\sigma')
print sigma ; sigma.view()
```

1-form 'sigma' on the 4-dimensional manifold 'M'

$$\sigma = \left(-\frac{2\, a^2 m \cos(\theta)^2 + 4i\, a m r \cos(\theta) - 2\, m r^2}{a^4 \cos(\theta)^4 + 2\, a^2 r^2 \cos(\theta)^2 + r^4}\right) \mathrm{d}r + \left(\frac{\left(2i\, a^3 m \cos(\theta)^2 - 4\, a^2 m r \cos(\theta) - 2i\, a m r^2\right)\sin(\theta)}{a^4 \cos(\theta)^4 + 2\, a^2 r^2 \cos(\theta)^2 + r^4}\right) \mathrm{d}\theta$$

The symmetric bilinear form $\boldsymbol{\gamma} = \lambda\, \boldsymbol{g} + \underline{\boldsymbol{\xi}} \otimes \underline{\boldsymbol{\xi}}$:

```
gamma = lamb*g + xi_form * xi_form
gamma.set_name('gamma', r'\gamma')
print gamma ; gamma.view()
```

field of symmetric bilinear forms 'gamma' on the 4-dimensional manifold 'M'

$$\gamma = \left(\frac{a^2 \cos(\theta)^2 - 2\, m r + r^2}{a^2 - 2\, m r + r^2}\right) \mathrm{d}r \otimes \mathrm{d}r + \left(a^2 \cos(\theta)^2 - 2\, m r + r^2\right) \mathrm{d}\theta \otimes \mathrm{d}\theta + \left(\frac{2\, a^2 m r \sin(\theta)^4 - \left(2\, a^2 m r - a^2 r^2 + 2\, m r^3 - r^4 - \left(a^4 + a^2 r^2\right)\cos(\theta)^2\right)\sin(\theta)^2}{a^2 \cos(\theta)^2 + r^2}\right) \mathrm{d}\phi \otimes \mathrm{d}\phi$$

The first part of the Simon-Mars tensor is $S^{(1)}_{\alpha\beta\gamma} = 4\mathcal{C}_{\mu\alpha\nu\beta}\,\xi^\mu\,\xi^\nu\,\sigma_\gamma$:

```
S1 = 4*( CC.contract(0,xi).contract(1,xi) ) * sigma
print S1
```

tensor field of type (0,3) on the 4-dimensional manifold 'M'

The second part is $S^{(2)}_{\alpha\beta\gamma} = \gamma_{\alpha\beta}\,\mathcal{C}_{\rho\gamma\mu\nu}\,\xi^\rho\,\mathcal{F}^{\mu\nu}$, which we compute using the index notation to perform the contractions:

```
FFuu = FF.up(g)
xiCC = CC['_.r..']*xi['^r']
S2 = gamma * ( xiCC['_.mn']*FFuu['^mn'] )
print S2
```

tensor field of type (0,3) on the 4-dimensional manifold 'M'

To get the Simon-Mars tensor, we need to antisymmetrize $\boldsymbol{S}^{(1)}$ and $\boldsymbol{S}^{(2)}$ on their last two indices; we choose to use the standard index notation to perform this operation (an alternative would have been to call directly the method `antisymmetrize()`):

```
S1A = S1['_a[bc]']
S2A = S2['_a[bc]']
```

The Simon-Mars tensor is then

```
S = = S1A + S2A
S.set_name('S')
print S ; S.symmetries()
```

tensor field 'S' of type (0,3) on the 4-dimensional manifold 'M'
no symmetry; antisymmetry: (1, 2)

We check that it vanishes identically, as it should for Kerr spacetime:

```
S.view()
```

$S = 0$

---

7. Conclusion and future prospects

SageManifolds is a work in progress. It encompasses currently $\sim 35,000$ lines of Python code (including comments and doctests). The last stable version (0.6), the functionalities of which are listed in Sec. 5.1, is freely downloadable from the project page [22]. The development version (to become version 0.7 soon) is also available from that page. Among future developments are

- the extrinsic geometry of pseudo-Riemannian submanifolds,
- the computation of geodesics (numerical integration via Sage/GSL or Gyoto [28]),
- evaluating integrals on submanifolds,
- adding more graphical outputs,
- adding more functionalities: symplectic forms, fibre bundles, spinors, variational calculus, etc.
- the connection with numerical relativity.

The last point means using SageManifolds for interactive exploration of numerically generated spacetimes. Looking at the diagram in Fig. 5, one realizes that it suffices to replace only the lowest level, currently relying on Sage's symbolic expressions, by computations on numerical data.

Let us conclude by stating that, in the very spirit of free software, anybody interested in contributing to the project is very welcome!

Acknowledgments

This work has benefited from enlightening discussions with Volker Braun, Vincent Delecroix, Simon King, Sébastien Labbé, José M. Martín-García, Marc Mezzarobba, Thierry Monteil, Travis Scrimshaw, Nicolas M. Thiéry and Anne Vaugon. We also thank Stéphane Méné for his technical help and Tolga Birkandan for fruitful comments about the manuscript. EG acknowledges the warm hospitality of the Organizers of the *Encuentros Relativistas Españoles 2014*, where this work was presented.

References

[1] Fletcher J G, Clemens R, Matzner R, Thorne K S and Zimmerman B A 1967 *Astrophys. J.* **148**, L91
[2] Skea J E F 1994 Applications of SHEEP *Lecture notes available at* http://www.computeralgebra.nl/systemsoverview/special/tensoranalysis/sheep/
[3] MacCallum M A H 2002 *Int. J. Mod. Phys. A* **17**, 2707
[4] Korol'kova A V, Kulyabov D S and Sevast'yanov L A 2013 *Prog. Comput. Soft.* **39**, 135
[5] http://www.xact.es/links.html

[6] Martin-Garcia J M 2008 *Comput. Phys. Commun.* **179**, 597; http://www.xact.es
[7] http://www.math.washington.edu/~lee/Ricci/
[8] Anderson I M and Torre C G 2012 *J. Math. Phys.* **53**, 013511; http://digitalcommons.usu.edu/dg/
[9] http://grtensor.phy.queensu.ca/
[10] http://digi-area.com/Maple/atlas/
[11] Peeters K 2007 *Comput. Phys. Commun.* **15**, 550; http://cadabra.phi-sci.com/
[12] Culler M, Dunfield N M and Weeks J R, SnapPy, a computer program for studying the geometry and topology of 3-manifolds, http://snappy.computop.org
[13] Bolotin D A and Poslavsky S V 2013 Introduction to Redberry: the computer algebra system designed for tensor manipulation *Preprint* arXiv:1302.1219; http://redberry.cc/
[14] http://sagemath.org/
[15] Stein W and Joyner D 2005 *Commun. Comput. Algebra* **39**, 61
[16] Joyner D and Stein W 2014 *Sage Tutorial* (CreateSpace)
[17] Zimmermann P et al. 2013 *Calcul mathématique avec Sage* (CreateSpace); freely downloadable from http://sagebook.gforge.inria.fr/
[18] Bard G V 2015 *Sage for Undergraduates* (Americ. Math. Soc.) in press; preprint freely downloadable from http://www.gregorybard.com/
[19] http://www.sagemath.org/doc/reference/categories/sage/categories/primer.html
[20] http://sagemath.org/doc/reference/tensor/
[21] http://sagemath.org/doc/reference/riemannian_geometry/
[22] http://sagemanifolds.obspm.fr/
[23] Lee J M 2013 *Introduction to Smooth Manifolds* 2nd edition (New York: Springer)
[24] Steenrod N 1951 *The Topology of Fibre Bundles* (Princeton: Princeton Univ. Press)
[25] http://sagemanifolds.obspm.fr/examples.html
[26] Apéry F 1986 *Adv. Math.* **61**, 185
[27] Mars M 1999 *Class. Quantum Grav.* **16**, 2507
[28] http://gyoto.obspm.fr/

Self tuning scalar tensor black holes

Christos Charmousis[1] and Damianos Iosifidis[2]

[1]Laboratoire de Physique Théorique (LPT), Univ. Paris-Sud, CNRS UMR 8627, F-91405 Orsay, France.

[2]Aristotle university of Thessaloniki 541 24, Thessaloniki, Greece

E-mail: christos.charmousis@th.u-psud.fr, diosifid@auth.gr

Abstract.
Studying a certain sub class of higher order Horndeski (scalar-tensor) theories we discuss a method discovered recently permitting analytic black hole solutions with a non trivial and regular scalar field. One of the solutions found has de Sitter asymptotics and self tunes the bulk cosmological constant. Using the aforementioned method we find and analyse new black hole solutions which can also have de Sitter asymptotics. By looking at small deviations of the integration constant responsible for self tuning we discuss the robustness of the self tuning mechanism. We find that neighboring solutions to the one previously found present also self tuning properties with unaltered effective cosmological constant.

1. Introduction-Motivation

General Relativity (GR) is a theoretically and experimentally consistent theory. It is also mathematically unique. Indeed theoretical consistency and mathematical uniqueness emanate as a corollary to Lovelock's theorem[1] [1] in 4 dimensions: consider a smooth Lagrangian depending up to second order derivatives of the metric tensor $\mathcal{L} = \mathcal{L}(\mathcal{M}, g, \nabla g, \nabla\nabla g)$ and endowed with a Levi Civita connection. Then in $D = 4$ dimensions GR with cosmological constant is the unique metric theory emerging from the action,

$$S_{(4)} = \int_{\mathcal{M}} d^4x\sqrt{-g^{(4)}}\,[R - 2\Lambda] \tag{1}$$

which upon metric variation yields,

- Equations of motion which are 2nd-order PDE's,
- given by a symmetric two-tensor, $G_{\mu\nu} + \Lambda g_{\mu\nu}$
- and admitting Bianchi identities.

Under these hypotheses GR with a cosmological constant is the unique massless, spin 2, 4 dimensional theory of gravity! In other words the Einstein-Hilbert action is the unique metric action which evades Ostrogradski no-go theorem of 1850 [4]. This no-go theorem states that any non-degenerate, higher than second order derivative theories inevitably lead to ghost degrees of freedom! In fact the only safe metric term we can add to the above action is the Gauss-Bonnet

[1] The four dimensional version of this theorem was already known early on [2] and mathematically stems from the mathematical works of Chern [3]

scalar, $R^2 - 4Ricci^2 + Riemann^2$, whose action is a 4 dimensional topological invariant the generalized Euler characteristic. Although this geometric term is not trivial its metric variation is not modifying thus GR equations of motion.

GR is a physical theory which has remarkable agreement with weak and strong gravity experiments from scales smaller than the millimeter and up to 30 times Earth to Sun distance... At astrophysical scales however, GR equations of motion, require dark matter interacting only gravitationally. The presence of dark matter although well founded theoretically still evades astroparticle and accelerator experimental detection. Furthermore, at cosmological scales the Universe is now observed to be in an accelerating phase requiring some source of yet unknown dark energy in order for Einstein's equations to agree with a multitude of differing cosmological and astrophysical observations [5]. In resumé, if we assume only ordinary (detected) sources of matter there is a flagrant disagreement between local, astrophysical and cosmological data. In fact roughly about 4% is matter of known source in the Universe. Even if we accept theoretically the presence of dark matter, dark energy of about 70% still has to be accounted for. For dark energy, which we will focus from here onwards, there is an almost trivial way out. Assume a cosmological constant of energy density $\rho_\Lambda = \frac{\Lambda_{obs}}{8\pi G} = (10^{-3}eV)^4$, ie. modify Einstein's equation by,

$$G_{\mu\nu} + \Lambda_{obs} g_{\mu\nu} = 8\pi G T_{\mu\nu} \tag{2}$$

Now this is a rather economic and consistent solution since a cosmological constant is the only other metric term allowed given the above uniqueness theorem (1). However this cosmological scale is tiny, the mass scale corresponds to neutrino mass scales. This is because a cosmological constant introduces a scale in the metric, $r_0 \sim \sqrt{\Lambda}$ which is as tiny, as the inverse size of the Universe today, $r_0 = H_0^{-1}$. Note that the quotient made out of Solar system scales divided by Cosmological Scales leads to an enormous number $\frac{10A.U.}{H_0^{-1}} = 10^{-15}$ at which we are probing the physical theory in question, GR, compared to the local experimentally robust scales. In fact suppose we were to use local experiments in order to bound the cosmological constant value (without taking into account cosmological data). That would involve constraining the cosmological constant from the (very) well known trajectories of planets. In essence, replacing Schwarzschild geometry by de Sitter Schwarzschild and taking the PPN limit. Local experiments would allow a cosmological constant roughly 10^{10} bigger than its current cosmological bound! Physical theories describe the physical world from good to remarkable precision but up to (or at) certain scales. Are we therefore beyond the scale of application of GR? Although deep in the UV, i.e. close to the Planck scale, GR is ill defined, could it be that also at vast and extremely dilute cosmological scales, of almost zero curvature, some other theory completes the theory of General Relativity? Furthermore, the cosmological constant is accompanied by the cosmological constant problem: its huge theoretical value, as expected from QFT considerations is at complete contrast with a finely tuned and tiny cosmological constant [6] (for a complete discussion and interesting developments see also [9]).

Summarizing: what if the need for exotic matter or a tiny cosmological constant is the sign for novel gravitational physics at very low energy scales or large distances. It is clearly a valid avenue of theoretical research in order to understand the robustness of GR at 10^{15} greater than local distances. In fact historically there is a very similar paradigm encountered at the advent of GR. More than 100 years ago, there were also observational short comings of Newton's theory of gravity. Indeed, Mercury the closest orbiting planet around the sun was deviating from Keplerian motion by a tiny yet definite advance in its perihelion. At the time, differing possibilities were pursued including most interestingly, the existence of a novel planet, Vulcan, which was postulated to orbit even closer to the sun, slightly deviating the perihelion of Mercury much like a dark source to Newtonnian equations. In fact a simple next order correction with one additional parameter (much like a cosmological constant) was enough to save Newton's laws (at the experimental precision of the time). Furthermore, although the

Journal of Physics: Conference Series **600** (2015) 012003 doi:10.1088/1742-6596/600/1/012003

presence of Vulcan may seem now days an ad hoc hypothesis it actually was not. The outermost planet of the solar system, Neptune, was in fact first theoretically predicted prior to being observed by the (same) astronomer Le Verrier. Indeed the theoretical prediction was triggered due to the reported inconsistencies in the Newtonian planetary trajectory of Uranus-which were explained by the presence of Neptune (see the exciting account in [7])! For the advance of the perihelion of Mercury, however, it was not the "simpler" solution that was the correct one, indeed GR, a far more involved than Newtonian theory had to be employed in order to explain strong gravity at local scales as was accepted by the scientific community only later by Eddington s famous eclipse experiment. Just as the eventual success of GR was not explaining the advance of Mercury's perihelion, modification of gravity cannot only be "an explanation" of the effective cosmological constant. Successes of GR underlies in all the theoretical and subsequently observational advances such as, the Big Bang Universe, inflation theory, black holes, binary pulsars, quasars etc.

So which avenues should one choose in an effort to modify such a well established theory, both theoretically and experimentally? For a start, we can deduce a set of basic facts/guidelines that can serve as a rough guide to possible research directions:

- Since GR is the unique 4 dimensional metric theory we will need to introduce new and genuine gravitational degrees of freedom!
- Novel and non degenerate gravitational degrees of freedom must not lead to higher derivative equations of motion. For then additional degrees of freedom are ghosts and vacuum is unstable due to Ostrogradski theorem 1850 as explained in [11].
- Matter must not directly couple to novel gravity degrees of freedom. Matter sees only the metric and evolves in metric geodesics. As such EEP is preserved and space-time can be put locally in an inertial frame.
- Novel degrees of freedom need to be screened from local gravity experiments. We need a well defined GR local limit such as the Chameleon, see for example[12], or the Vainshtein mechanism [13]. The equivalent of this in GR would be the weak gravity Newtonian limit.
- Exact gravitational solutions are essential in modified gravity theories in order to understand strong gravity regimes and novel characteristics. We will need to deal with the no hair paradigm and the absence of GR theorems such as Birkhoff's theorem in modified gravity theories.
- A modified gravity theory should tell us something about the short-comings of GR such as the cosmological constant problem and in particular how to screen an a priori enormous cosmological constant.

The above list of points are not exhaustive and one must treat them not dogmatically but rather as rough guidelines. For example f(R) theory has higher order derivative equations of motion but does not necessarily acquire ghost degrees of freedom. The theory can be easily transformed in a scalar-tensor theory. The spin 2 degrees of freedom are in fact not present for this (degenerate) case but only an extra scalar degree of freedom. Having this in mind we can sketch, again with precaution, differing directions to go beyond GR (for a full account see for example [10]):

- We can assume metric theories but in the presence of extra dimensions : Extension of GR to Lovelock theory with modified yet second order field equations [14]. Here belong braneworld models such as the DGP model [15], RS models [16] or the generic case of Kaluza-Klein compactification.
- One can assume that the graviton is not massless but massive. In fact in this case generically graviton potentials suffer from a Boulware Deser ghost mode [17]. The only possible theories avoiding this specific mode are non trivial and are parametrized by 3 independent couplings

(see for example [18]). Hence massive gravity and bigravity theories are greatly constrained already theoretically and the most general interaction (potential) term is known.

- One can assume a 4-dimensional scalar modification of GR. This is the case of Scalar-tensor theories [19], [20], $f(R)$ [24], Galileon/Hornedski theories [21], [22],[23] (for a review and relation to Lovelock theory see [25]).
- We can consider breaking symmetries present in GR such as Lorentz invariance symmetry. This has interesting implications in the UV as it is a wayof obtaining higher order theories which are power counting renormalizable and have a priori better ultra-violet properties: this is the case of Horava gravity [26], [27] and in its infra red limit, Einstein Aether theories [28], which are healthy and unique vector tensor theories with a consistent GR limit.
- Theories modifying geometric properties of spacetime such as with the inclusion of torsion or a different choice of geometric connection (see for example [29]).

In this paper we will consider the simpler of modifications with a single extra degree of freedom, scalar-tensor theory. Scalar tensor theories are in a certain sense a prototype of gravity modification and have some nice properties. In particular,

- They admit a uniqueness theorem due to Horndeski, 1973, which much like Lovelock's theorem determines the most general form of their action [21], [23].
- They contain or are limits of other modified gravity theories. $f(R)$ theory for example is a scalar tensor theory in disguise whereas massive gravity or bigravity are scalar tensor theories in their decoupling limit. This is also true of certain higher dimensional braneworlds such as DGP and even for Kaluza-Klein compactifications of Loveock theory [30], [31], [25]
- They can have insightful screening mechanisms as in the Chameleon [12] and Vainshtein [13] mechanisms.
- They include specific Galileon terms, dubbed Fab 4, that can screen classically a big cosmological constant [34].
- They specifically break GR black hole uniqueness theorems such as Birkhoff's theorem and suffer from no-hair black hole theorems.

It is this latter characteristic we will study and try to overcome for we will seek black hole solutions of such theories where the scalar field will not be trivial but well behaved. Before doing so let us point out the most general theory at hand. Horndeski showed several years ago with techniques similar to those of Lovelock that the most general scalar-tensor action with second order field equations is given by

$$S_H = \int d^4x\sqrt{-g}\,(L_2+L_3+L_4+L_5)$$

$$\begin{aligned} L_2 &= K(\phi,X),\\ L_3 &= -G_3(\phi,X)\nabla^2\phi,\\ L_4 &= G_4(\phi,X)R+G_{4X}\left[(\nabla^2\phi)^2-(\nabla_\mu\nabla_\nu\phi)^2\right],\\ L_5 &= G_5(\phi,X)G_{\mu\nu}\nabla^\mu\nabla^\nu\phi-\frac{G_{5X}}{6}\left[(\nabla^2\phi)^3-3\nabla^2\phi(\nabla_\mu\nabla_\nu\phi)^2+2(\nabla_\mu\nabla_\nu\phi)^3\right] \end{aligned}$$

where the G_i are unspecified functions of ϕ and $X\equiv -\frac{1}{2}\nabla^\mu\phi\nabla_\mu\phi$ and $G_{iX}\equiv\partial G_i/\partial X$. In fact this is the action of covariant Galileons found independently in [32] in a different and elegant manner. Due to the higher order kinetic terms in X the theory generically screens the scalar mode locally following the Vainshtein mechanism [13].

In the next section we will discuss scalar tensor black holes starting with a brief discussion of the no hair paradigm, and then certain solutions some of which are new and appear in the last subsection. We will then briefly conclude.

2. Scalar tensor black holes

Modified gravity theories have to deal with the absence of black hole uniqueness theorems (present in GR) as well as no hair theorems. Indeed, "Black holes have no hair", is the celebrated phrase conjectured long ago by Wheeler. The no hair paradigm takes concrete form as a set of mathematical theorems governing asymptotically flat (or de Sitter) and stationary black holes under certain hypotheses. The conjecture itself states, that apart from charges measured at infinity by a far away observer, no additional degrees of freedom can describe the black hole geometry (for a review see [35]). The physical picture is roughly the following: as black holes are formed they expel or eat up surrounding matter ending up, in their stationary phase, as rather "simple" bald objects characterized by a finite number of measurable charges at infinity: mass, angular momentum, electric and magnetic charge. A black hole, is a rather blunt gravitational object having specific charges and not allowing additional parameters primary hair which are not associated to a conserved charge, or, secondary hair of no additional parameters but non-trivial fields interacting with the black hole spacetime. An ideal observer measuring these charges out at infinity "knows" specifically which black hole she or he is dealing with. They do not have to venture close up in order to gather details about the black hole structure, its horizon properties etc. This is true for asymptotically flat and stationary black holes. As such, any scalar field, part of a scalar tensor theory is expected to be frozen to a constant in order to evade a naked singularity. In other words scalar-tensor theories in their prototype Brans Dicke form [19] basically have the same GR black holes with the scalar field fixed to a constant value. As such a spherically symmetric BD black hole is again a Schwarzschild black hole with a constant scalar field. Otherwise the scalar tensor spherically symmetric solutions are singular geometries describing rather stars which in BD theory are different solutions to those describing black holes. We emphasize here the difference with GR where the Schwarzschild solution is the exterior metric for a star and a black hole due to Birkhoff's uniqueness theorem. This is not true in scalar tensor theories and it is a source of problems of scalar tensor theories with local experiments. There is a non-trivial example of a black hole geometry that almost escapes this rule and that is the BBMB black hole which we will briefly turn to now.

2.1. An example: The BBMB black hole

Consider the following scalar tensor action,

$$S[g_{\mu\nu}, \phi, \psi] = \int_{\mathcal{M}} \sqrt{-g} \left(\frac{R}{16\pi G} - \frac{1}{2}\partial_\alpha\phi\partial^\alpha\phi - \frac{1}{12}R\phi^2 \right) \mathrm{d}^4x \tag{3}$$

The particular feature of this action is that the scalar ϕ couples conformally to the metric. Indeed there is an invariance of the ϕ equation of motion (EoM) under the conformal transformation

$$g_{\alpha\beta} \mapsto \tilde{g}_{\alpha\beta} = \Omega^2 g_{\alpha\beta} \tag{4}$$

$$\phi \mapsto \tilde{\phi} = \Omega^{-1}\phi \tag{5}$$

The whole action is not conformally invariant. In fact the Einstein-Hilbert term explicitly breaks conformal invariance. However, for this action for which there is a remnant of conformal symmetry there exists a vacuum black hole geometry with non-trivial scalar field and secondary black hole hair. This is the BBMB solution [36] a static and spherically symmetric solution,

$$\mathrm{d}s^2 = -\left(1 - \frac{m}{r}\right)^2 \mathrm{d}t^2 + \frac{\mathrm{d}r^2}{(1-\frac{m}{r})^2} + r^2\left(\mathrm{d}\theta^2 + \sin^2\theta \mathrm{d}\varphi^2\right)$$

with a non trivial scalar field

$$\phi = \sqrt{\frac{3}{4\pi G}}\frac{m}{r-m}$$

This particular geometry is also encountered in GR. It is that of an extremal Reisner-Nordtstrom solution with charge equal to mass. Our ideal observer sitting far out of this scalar tensor black hole would observe a static black hole of mass m other than Schwarzschild but with no electric charge. Indeed note that the scalar decays at infinity and does not carry a specific scalar charge. Hence we are not dealing with primary hair (by definition) rather secondary hair as the scalar couples non-trivially (conformally) to the metric modifying its geometry. Indeed the only parameter of the solution is the black hole mass m. Almost immediately however we encounter the shortcoming of the solution: the scalar field explodes at the horizon location at $r = m$. This can be remedied in the presence of a cosmological constant [37] and a ϕ^4 conformal potential. The black hole can be even in an accelerated phase or have a Taub NUT charge [39]. Here, for the asymptotically flat case, the singular scalar at the horizon is a problem and it is not clear whether the solution is a black hole i.e. an endpoint of matter collapse. Going to the Einstein frame we can actually see that the solution hides additional singularities there at $r = 2m$ where the conformal transformation in between the frames is singular. In fact at $r = 2m$ the ϕ dependent factor multiplying R in the action is exactly zero and for $r < 2m$ changes sign hinting on gravity acquiring a ghost kinetic term! Could self gravitating matter go beyond the $r = 2m$ surface? Or again, is the BBMB solution an endpoint of gravitational collapse? It would seem unlikely and this is an open problem which could be resolved numerically determining the exact nature of this solution. The BBMB metric is however a unique and simple 4 dimensional example with a non trivial scalar field. We will proceed now to remedy at least some of these problems with higher order Galileon fields.

3. Constructing Galileon black holes

We will now consider the action[2],

$$S = \int d^4x\sqrt{-g}\left[\zeta R - 2\Lambda - \eta\,(\partial\phi)^2 + \beta G^{\mu\nu}\partial_\mu\phi\partial_\nu\phi\right], \tag{6}$$

which is part of the Horndeski action given above (3) comprising of the Einstein-Hibert term with a cosmological constant as well as the canonical scalar kinetic term and a higher order scalar tensor interaction term. The latter term is a higher order Galileon, central in our forthcoming discussion, yielding second order field equations. This nice property boils down to the divergence freedom of the Einstein tensor. Finally ζ, η and β are coupling constants in order to keep track of the terms in the action. Indeed the metric field equations are,

$$\zeta G_{\mu\nu} \quad -\eta\left(\partial_\mu\phi\partial_\nu\phi - \tfrac{1}{2}g_{\mu\nu}(\partial\phi)^2\right) + g_{\mu\nu}\Lambda \tag{7}$$

$$+\tfrac{\beta}{2}\Big((\partial\phi)^2 G_{\mu\nu} + 2P_{\mu\alpha\nu\beta}\nabla^\alpha\phi\nabla^\beta\phi \tag{8}$$

$$+g_{\mu\alpha}\delta^{\alpha\rho\sigma}_{\nu\gamma\delta}\nabla^\gamma\nabla_\rho\phi\nabla^\delta\nabla_\sigma\phi\Big) = 0, \tag{9}$$

where $P_{\mu\alpha\nu\beta}$ is the double dual curvature tensor given by,

$$P_{\mu\rho\nu\sigma} = R_{\mu\rho\nu\sigma} + R_{\nu\rho}g_{\mu\sigma} + R_{\mu\sigma}g_{\nu\rho} - R_{\rho\sigma}g_{\mu\nu} - R_{\mu\nu}g_{\rho\sigma} + \frac{1}{2}R(g_{\mu\nu}g_{\rho\sigma} - g_{\mu\sigma}g_{\nu\rho}). \tag{10}$$

This tensor is divergence free has the same symmetry properties as the Riemann tensor and tracing two of its indices gives,

$$g^{\mu\nu}P_{\mu\rho\nu\sigma} = -G_{\rho\sigma}$$

the Einstein tensor.

[2] The details summarized in this section can be found in the original paper [8]

Journal of Physics: Conference Series **600** (2015) 012003 doi:10.1088/1742-6596/600/1/012003

An important property of the above action is that the scalar field has translational invariance: $\phi \to \phi$+const. This is important for then the scalar field equation can be written in terms of a conserved vector current,

$$\nabla_\mu J^\mu = 0, \quad J^\mu = \left(\eta g^{\mu\nu} - \beta G^{\mu\nu}\right) \partial_\nu \phi. \tag{11}$$

Note in the last expression of the current above, the explicit appearance of the metric field equations of the lower order Einstein-Hilbert and cosmological constant terms. Schematically we have that, $R \to G^{\mu\nu}\partial_\mu\phi\partial_\nu\phi$ whereas $\Lambda \to g^{\mu\nu}\partial_\mu\phi\partial_\nu\phi$ associating thus in this intriguing way the terms present in the action. Summarizing therefore, apart from the latter property, the above action crucially includes higher order Galileon terms and has translational invariance for the scalar field.

We will consider a static and spherically symmetric spacetime,

$$ds^2 = -h(r)dt^2 + \frac{dr^2}{f(r)} + r^2 d\Omega^2 \tag{12}$$

parametrized by the two radial functions f and g. It is then evident that if $\phi = \phi(r)$ then the scalar equation is integrable and reads,

$$(\eta g^{rr} - \beta G^{rr})\sqrt{g}\phi' = c \tag{13}$$

where c the integration constant stands for primary hair for a scalar-tensor black hole. A non zero value of c comes along with regularity problems since the current norm is singular $J^2 = J^\mu J^\nu g_{\mu\nu} = (J^r)^2 g_{rr}$ unless $J^r = 0$ at the horizon and hence $c = 0$. It is then easy to note that generically $\phi = constant$ everywhere as noted in [38] and we have the appearance of a no-hair argument i.e. the presence of a trivial scalar and a GR black hole solution. However, here there is a slight twist in this argument. Indeed, due to the higher order nature of the terms in the action we can instead of $\phi = constant$ require that,

$$\beta G^{rr} - \eta g^{rr} = 0. \tag{14}$$

and thus search for a different branch of solutions. Here note that the presence of the Einstein tensor (14) in the higher order term, is just the Hamiltonian constraint for a GR plus cosmological constant theory, and this is the reason for our Gallileon choice. A second point is that we can also include a linear time dependence in the scalar field $\phi(t, r) = q\,t + \psi(r)$ without affecting the scalar field equation, $-\partial_r[(\beta G^{rr} - \eta g^{rr})\partial_r\psi] - \partial_t[(\beta G^{tt} - \eta g^{tt})\partial_t(qt)] = 0$ which is still verified given (14). In fact, we can show that choosing $\phi(t, r) = q\,t + \psi(r)$ solves also the (tr)-equation of motion and this at the end tells us that our Anzatz is mathematically consistent. We must stress however that although the scalar is not necessarily constant we have no scalar charge i.e. $c = 0$. Furthermore we have a finite current. We have satisfied the no hair argument not in permitting a non trivial scalar charge in the face of the constant c, but still having a scalar field which is not trivial. This is due to the higher order nature of the Galileon term providing us with an alternative geometric resolution $\beta G^{rr} - \eta g^{rr} = 0$ to the scalar and tr-equation. Therefore the tr-equation as well as the scalar field equation boil down to the geometric constraint,

$$f = \frac{(\beta + \eta r^2)h}{\beta (rh)'},$$

fixing the spherically symmetric gauge and $\phi(t, r) = q\,t + \psi(r)$.

This solves therefore for the time dependence of ϕ and for the gauge function f. In order to find solutions we now need to solve for $\psi(r)$ and $h(r)$ with the last two independent ODE's, the (rr) and (tt) equations. From the (rr)-component we get ψ'

$$\psi' = \pm \frac{\sqrt{r}}{h(\beta + \eta r^2)} \left(q^2 \beta(\beta + \eta r^2) h' - \frac{\lambda}{2}(h^2 r^2)' \right)^{1/2}. \tag{15}$$

with

$$\lambda \equiv \zeta\eta + \beta\Lambda$$

. We see that for $\eta = \Lambda = 0$ time dependence of the scalar is essential in order to have a non trivial asymptotically flat solution. Finally the (tt)-component gives $h(r)$ via,

$$h(r) = -\frac{\mu}{r} + \frac{1}{r}\int \frac{k(r)}{\beta + \eta r^2} dr, \tag{16}$$

with

$$q^2\beta(\beta + \eta r^2)^2 - \left(2\zeta\beta + (2\zeta\eta - \lambda)\, r^2\right) k + C_0 k^{3/2} = 0, \tag{17}$$

Therefore any solution to the algebraic eq for $k = k(r)$ gives us a full solution to the system. For $\eta = \Lambda = 0$ we can easily find asymptotically flat solutions with a scalar field since then k is simply a constant which can be gauged away in a redefinition of time. The geometry is that of a Schwarzschild black hole [8] and the scalar field due to the linear time dependence is regular at the black hole future horizon. The scalar field linearly diverges at infinity but note that its derivatives appearing in the action are well behaved. The solution is therefore a valid and regular solution unlike the BBMB case where the scalar blows up at the horizon location.

One can then seek asymptotically de-Sitter solutions [8] setting $f = h$ whereupon

$$k(r) = \frac{(\beta + \eta r^2)^2}{\beta} \tag{18}$$

with

$$q^2 = (\zeta\eta + \beta\Lambda)/(\beta\eta), \qquad C_0 = (\zeta\eta - \beta\Lambda)\sqrt{\beta}/\eta. \tag{19}$$

The line element is that of de Sitter Schwarzschild $f = h = 1 - \frac{\mu}{r} + \frac{\eta}{3\beta} r^2$ with $\psi' = \pm\frac{q}{h}\sqrt{1-h}$ and $\phi(t,r) = qt + \psi(r)$. Going to Eddington-Filkestein coordinates we can check that the solution is regular at the horizon. Note again that the scalar itself diverges at the de Sitter horizon but not its derivatives.

The solution above has an interesting property : it is a GR like solution but crucially it does not directly depend on the vacuum cosmological constant Λ but rather on an effective cosmological constant $\Lambda_{eff} = -\eta/\beta$. This is true as long as the integration constant $q^2\eta = \Lambda - \Lambda_{eff} > 0$ which is just (19). Also note that the solution belongs to a different branch than the GR solution of de Sitter Schwarzschild, one cannot switch off the higher order Gallileon term. Hence any arbitrary bulk cosmological constant of the action $\Lambda > \Lambda_{eff}$, which can be as big as the Planck scale, fixes q our integration constant leaving us with an effective geometric acceleration Λ_{eff} which can be tuned once and for all in the theory to be small. In other words the scalar tensor solution self tunes the vacuum cosmological constant as long as $\Lambda > \Lambda_{eff}$. But what is the behavior of other solutions to the algebraic equation (17)? In particular is there a fine tuning in the integration constant q? If we deviate from the conditions (19) will we still have de Sitter Schwarzschild asymptotics? We will turn to this question now.

Journal of Physics: Conference Series **600** (2015) 012003 doi:10.1088/1742-6596/600/1/012003

4. General Solutions

In the previous section we summarized the method and certain key solutions of spherical symmetry found in [8]. We would now like to go further and study more generic solutions of the algebraic master equation (17) governing the general solution of the system. The first important thing to note is that the solution of the cubic $k = k(r)$, as given by (17) does not depend on the mass integration constant μ of the solution (16). The mass of the black hole solution does not interfere in the equation itself. The algebraic equation (17) specifies the asymptotic behavior of the solution via the integration constant C_0. As such, solution (18) corresponds to de Sitter space with a non trivial scalar field ϕ and can be promoted to a black hole for $\mu \neq 0$. We will refer to it as a de Sitter solution from now on (see in that aspect [40]). What other solutions $k = k(r)$ of de Sitter asymptotics exist other than (18)? It is easy to verify that asymptotically (anti) de Sitter solutions will be obtained by fixing the constant C_0 to its found de Sitter value given above for the de Sitter Schwarzschild geometry (18). The fake hair charge q remains a free parameter. We will look into these solutions in order to understand their behaviour beyond the GR stealth limit (18) and in particular to see locally around (18) how self tuning is influenced in this class of solutions.

But first of all there is also a second class of solutions we can obtain relatively easily. They are simply given by dividing the third order polynomial by the de Sitter root (18). The resulting quotient polynomial will yield a second degree equation for $k(r)$ which can have real roots and which can be easily solved. Indeed we know that the parameter choice $q^2 = \frac{\zeta\eta+\beta\Lambda}{\beta\eta}$ and $C_0 = (2\zeta\eta - \lambda)\sqrt{\beta}/\eta$ yields (18) as a solution of (17).

In what follows we will consider the cases $\beta < 0$ and $\beta > 0$ separately. For the second order equation it is $\beta > 0$ that presents more interest in terms of regular solutions. For the former de Sitter solutions they will be in the same family as (18) hence $\beta < 0$. Now that we have understood the influence of each term of the action in the field equations we will set natural values for the couplings in the action $\zeta = 1$, $\eta = 1/2$ and consider $\Lambda > 0$ while defining $R = \frac{r}{\sqrt{2|\beta|}}$ and $u = 2|\beta|\Lambda$ without loss of generality. Hence the u parameter is dimensionless and denotes the strength of the higher order coupling β for given cosmological constant.

For $\beta > 0$ we have $q^2|\beta| = (1+u)$, $C_0 = \sqrt{|\beta|}(1-u)$ and setting

$$x = \frac{\sqrt{\beta}}{k^{1/2}} \tag{20}$$

the cubic (17) takes the canonical form,

$$x^3 - \frac{\left[2 + (1-u)R^2\right]}{(1+u)(1+R^2)^2}x + \frac{(1-u)}{(1+u)(1+R^2)^2} = 0 \tag{21}$$

Solution (18) reads in the novel rescaled variables,

$$x = \frac{1}{1+R^2} \tag{22}$$

So, writing again the cubic as

$$\left(x - \frac{1}{1+R^2}\right)(x^2 + bx + c) = 0 \tag{23}$$

we conclude that

$$b = \frac{1}{1+R^2} > 0 \tag{24}$$

while

$$c = -v\frac{1}{1+R^2} = -vb \tag{25}$$

where we have set for simplicity

$$v = \frac{1-u}{1+u}. \tag{26}$$

It is then easy to show that the remaining two solutions of (21) are given by

$$x_{1,2}(R) = \frac{1}{2(1+R^2)}\left[-1 \pm \sqrt{1+4v(1+R^2)}\right] \tag{27}$$

For $v > 0$ both of the above roots are real which is true for $-1 < 2\Lambda\beta < 1$ and the solution avoids branch singularities (where the square root is zero). Computing the integral in (16) we get,

$$R\,h(r) = \sqrt{2\beta}\int \frac{k(r)}{\beta+\eta r^2}dr = 4\int \frac{(1+R^2)}{\left[-1 \pm \sqrt{1+4v(1+R^2)}\right]^2}dR \tag{28}$$

whereas

$$f(R) = h(R)\left[-1 \pm \sqrt{1+4v(1+R^2)}\right]^2 \tag{29}$$

Therefore f has at least the same zeros as the function h which will be event horizons. At the end the solution for h reads after direct integration,

$$h(R) = \frac{1}{v} - \frac{\mu}{R} + \frac{1}{2v^2R}\left(\arctan(R) \pm \arctan\left(\frac{R}{\sqrt{1+4v(R^2+1)}}\right)\right) \pm \frac{1}{v^{3/2}R}\sinh^{-1}\left(\sqrt{\frac{4vR^2}{1+4v}}\right) \tag{30}$$

Therefore, for large R, $h = h(R)$ is bounded and asymptotes $1/v$. For $\Lambda = 0$ we have $v = 1$ and this is the only case where the area of the sphere is not reduced. Otherwise we have a solid angle deficit which at the absence of an event horizon can lead to a naked singularity. For $\mu > 0$ there is always an event horizon covering the singularity at $R = 0$ and the above solution is a black hole with a well behaved scalar field. We have found therefore a regular scalar tensor black hole solution quite similar to the black hole embedded in an Einstein static universe [8].

A second case of interest to consider is to take $C_0 = (2\zeta\eta - \lambda)\sqrt{\beta}/\eta$ while allowing for arbitrary charge q. Our aim here is to seek solutions which asymptote de Sitter space and generalize the de-Sitter Schwarzschild solution found in [8] given by (18). So in accordance to this aim we take $\beta < 0$, that is $\beta = -|\beta|$. Now, the constant C_0 takes the value

$$C_0 = i\sqrt{|\beta|}(1+u) \tag{31}$$

so that

$$k \sim \frac{\eta^2}{\beta}r^4 = -\frac{\eta^2}{|\beta|}r^4 \tag{32}$$

as $r \to \infty$. Given the imaginary value of C_0, setting $ix = \frac{\sqrt{|\beta|}}{k^{1/2}}$ the cubic becomes

$$x^3 + \frac{\left[2-(1+u)R^2\right]}{q^2|\beta|(1-R^2)^2}x + \frac{(1+u)}{q^2|\beta|(1-R^2)^2} = 0 \tag{33}$$

Expressions are slightly more involved here and we make some redefinitions to render expressions more readable. Setting,

$$y = 1 - R^2, \qquad p^2 = \frac{q^2|\beta|}{u+1} \tag{34}$$

the cubic reads,

$$x^3 + \frac{y+v}{p^2y^2}x + \frac{1}{p^2y^2} = 0 \tag{35}$$

where v is given in (26) and we note that,

$$R\,h(R) = \int \frac{dR}{y(R)\,x_0^2(y)}, \qquad f(R) = h(R)y^2x_0^2(y) \tag{36}$$

where $x_0(y)$ is solution to the cubic (35). Customary to third order equations we define

$$Q = \frac{1}{p^2y^2}, \qquad P = (y+v)Q \tag{37}$$

The number of real solutions of (33) depends upon the sign of the discriminant

$$\Delta = Q^2 + 4\left(\frac{P}{3}\right)^3 \tag{38}$$

More specifically , for $\Delta > 0$ we have only one real root and a complex pair. Since we are looking for a static patch of De Sitter we take $0 < y < 1$. Indeed the de Sitter Schwarzschild solution is obtained for $p^2 = -v$ and $x_{dS} = \frac{1}{y}$. It is then clear that $\Delta > 0$ and the unique real solution is given by

$$x_0 = \frac{1}{2^{1/3}py}\left[-py + \sqrt{p^2y^2 + \frac{4}{3^3}(y+v)^3}\right]^{1/3} - \frac{2^{1/3}(y+v)}{3py\left[-py + \sqrt{p^2y^2 + \frac{4}{3^3}(y+v)^3}\right]^{1/3}} \tag{39}$$

The explicit solution can be obtained by a simple but tedious procedure. Here we are mostly interested in determining the asymptotic behavior of the solution close to the de Sitter like horizon in $y = 0$. Hence we consider the leading order behavior of the real solution(s) in small y for differing values of v respectively negative and positive. The stealth de-Sitter solution (18) belongs to the family with $v < 0$. In this case we obtain, $x_0 \sim \frac{\sqrt{-v}}{py}$ and this is again a de-Sitter Schwarzschild solution to leading order in y. Therefore, the de Sitter Schwarzschild solution is not a particularly special solution it is continuously related to a full branch of de Sitter like black hole solutions with similar characteristics. For $v > 0$ we get however $x_0 \sim \frac{1}{-v}$ which yields $Rh = v^2 \tanh^{-1} R - \mu$ which has a singular behavior asymptotically. Therefore we see that for $v < 0$ the solutions are indeed similar to (18).

In order to back up this conclusion without going to numerics we can also consider the following approximation. Suppose that we have a small deviation from the charge configuration valid around the de-Sitter solution (18), namely

$$q^2|\beta| = (u-1) + \epsilon \tag{40}$$

with $\epsilon << |u-1|$. We consider the linearized solution in ϵ to (33),

$$x = x_0 + \epsilon x_1 \tag{41}$$

where $x_0 = \frac{1}{R^2-1}$ and x_1 is the deviation from the de-Sitter solution (18) when we shift slightly the fake hair parameter. Now, upon substituting the above into (33) to linear order in ϵ we get,

$$x(R) = \frac{1}{(1-R^2)}\left[1 + \frac{\epsilon}{(3u-1)-(1+u)R^2}\right]. \tag{42}$$

Now, if $u \in (-1, 1/3)$ defining,

$$\frac{1+u}{1-3u} = \gamma^2 > 0 \tag{43}$$

we get

$$h(R) = \left(1 - \frac{2}{1+u}\epsilon\right) - \frac{\mu}{R} - \frac{R^2}{3} + \epsilon\frac{2(1+\gamma^2)}{(1+u)\gamma}\frac{1}{R}\arctan(\gamma R) \tag{44}$$

and

$$f(R) = \left(-\frac{\mu}{\sqrt{2|\beta|}}\frac{1}{R} + 1 - \frac{R^2}{3}\right)\left(1 - \frac{2\epsilon}{(3u-1)-(1+u)R^2}\right) - \frac{2\epsilon}{1+u} + \frac{2(1+\gamma^2)}{(1+u)\gamma}\frac{1}{R}\arctan(\gamma R) \tag{45}$$

As long as our approximation $\epsilon << |u-1|$ is true the linear perturbative solution is valid for large enough R and the solution is regular and very similar to a de Sitter Schwarzschild solution. Note that the correction in ϵ does not interfere with the effective cosmological constant which means that the self-tuning mechanism is stable under a small deviation of the coupling q. Alternatively for $u \in (-\infty, -1) \cup (1/3, +\infty)$ we can set,

$$\omega = i\gamma \equiv \sqrt{\frac{1+u}{3u-1}} \tag{46}$$

to get,

$$h(r) = \left(1 - \frac{2}{1+u}\epsilon\right) - \frac{\mu}{R} - \frac{R^2}{3} - \epsilon\left(\frac{\omega^2-1}{\omega}\right)\frac{1}{(1+u)R}\ln\left|\frac{1+\omega R}{1-\omega R}\right| \tag{47}$$

where we note now that the solution is again regular and the only difference is the slightly differing behavior of the logarithmic term.

5. Concluding remarks

In this paper we have considered a resolution method originally found in [8] giving regular black holes for scalar tensor theories. Following the analysis in [8] we showed how the complex system of field equations boiled down to solving a single third order algebraic equation (17) which basically fixed the asymptotic properties of the spacetime solution. We then presented some new solutions analyzing in some detail the algebraic equation at hand (17) which depends on two integration constants C_0 and q (but not the mass of the black hole). The two integration constants are the fake hair charge q and an integration constant C_0 determining the asymptotic properties of the solution at hand. Apart from novel regular black holes we also found that the particular de Sitter self tuning solution (18) found in [8] which is a self tuning black hole solution, is not some isolated point in the space of algebraic solutions, but is a generic and stable configuration with respect to small deviations in the fixed parameters of the solution. Indeed when we consider a small deviation in the self-tuning integration constant q the effective cosmological constant of the neighboring black hole solution remains the same not spoiling the self tuning mechanism.

The regularity and integrability of the method presented relies on three properties for the Galileon scalar field: first of all it has to have translational invariance and this symmetry yields a conserved charge and current with a special form for the scalar field equations. Secondly, the scalar tensor action considered has to include higher order Galileon terms so that the scalar field equation admits novel, geometric branches of solutions apart from the ones that lead to trivial or singular solutions. This enables us to satisfy the no hair arguments present for Galileons [38] while finding non trivial scalar tensor black holes (see also [41]). The regularity requirements required in [38] are still not sufficient to obtain regular solutions. We need a third ingredient property : the key to obtaining a regular scalar field at the horizon is to additionally allow a linear time dependence for the scalar field. Under these three important properties it is then relatively easy to find novel scalar tensor black hole solutions. It would be interesting to see how far can one generalize these principles to go to more general Horndeski theories (see [42] for progress in this direction) or again find hairy black holes with a genuine scalar charge. In the latter case hairy black holes with genuine scalar charge have been found in a bi scalar tensor theory [43] that are seen to elegantly generalize the BBMB black hole. Another interesting open question lies in the stability of such scalar-tensor black holes as those discussed here.

Acknowledgements
We are very happy to thank Eugeny Babichev and Tassos Petkou for discussions and support. CC thanks the theoretical Physics group of Thessaloniki for their hospitality during the course and completion of this work. CC finally thanks colleagues from the University of Valencia for the invitation to give a talk on this subject and for the organisation of a very nice and interesting ERE 2014 conference.

References
[1] D. Lovelock, J. Math. Phys. **12**, 498 (1971).
[2] E. Cartan Journal de Mathmatiques Pures et Appliques (1922): 141-204. H. Weyl, Space-Time-Matter, (Dover Publications, New York, 1922).
[3] S.-S. Chern, Ann. Math. 45 (1944) 747-752. S.-S. Chern, Ann. Math. 46 (1945) 674-684.
[4] M. Ostrogradski, Mem. Ac. St. Petersbourg 4, 385 (1850).
[5] M. Betoule *et al.* [SDSS Collaboration], [arXiv:1401.4064 [astro-ph.CO]]. G. Hinshaw *et al.* [WMAP Collaboration], Astrophys. J. Suppl. **208**, 19 (2013) [arXiv:1212.5226 [astro-ph.CO]].
[6] S. Weinberg, Rev. Mod. Phys. **61**, 1 (1989).
[7] Ferreira, P. G. (2014). The Perfect Theory: A Century of Geniuses and the Battle Over General Relativity. Houghton Mifflin Harcourt.
[8] E. Babichev and C. Charmousis, JHEP **1408**, 106 (2014) [arXiv:1312.3204 [gr-qc]].
[9] N. Kaloper and A. Padilla, Phys. Rev. Lett. **112**, 091304 (2014) [arXiv:1309.6562 [hep-th]].
[10] T. Clifton, P. G. Ferreira, A. Padilla and C. Skordis, Phys. Rept. **513**, 1 (2012) [arXiv:1106.2476 [astro-ph.CO]].
[11] R. P. Woodard, Lect. Notes Phys. **720**, 403 (2007) [astro-ph/0601672].
[12] J. Khoury, Class. Quant. Grav. **30**, 214004 (2013) [arXiv:1306.4326 [astro-ph.CO]].
[13] E. Babichev and C. Deffayet, Class. Quant. Grav. **30**, 184001 (2013) [arXiv:1304.7240 [gr-qc]].
[14] N. Deruelle and J. Madore, gr-qc/0305004. C. Charmousis, Lect. Notes Phys. **769**, 299 (2009) [arXiv:0805.0568 [gr-qc]].
[15] G. R. Dvali, G. Gabadadze and M. Porrati, Phys. Lett. B **485**, 208 (2000) [hep-th/0005016].
[16] L. Randall and R. Sundrum, Phys. Rev. Lett. **83**, 4690 (1999) [hep-th/9906064].
[17] D. G. Boulware and S. Deser, Phys. Rev. D **6**, 3368 (1972).
[18] C. de Rham, arXiv:1401.4173 [hep-th]. M. S. Volkov, arXiv:1405.1742 [hep-th].
[19] C. Brans and R. H. Dicke, Phys. Rev. **124**, 925 (1961).
[20] T. Damour and G. Esposito-Farese, Class. Quant. Grav. **9**, 2093 (1992). G. Esposito-Farese, AIP Conf. Proc. **736**, 35 (2004) [gr-qc/0409081]. T. P. Sotiriou, arXiv:1404.2955 [gr-qc].
[21] G. W. Horndeski, Int. J. Theor. Phys. **10**, 363 (1974).
[22] A. Nicolis, R. Rattazzi and E. Trincherini, Phys. Rev. D **79**, 064036 (2009) [arXiv:0811.2197 [hep-th]].
[23] C. Deffayet, S. Deser and G. Esposito-Farese, Phys. Rev. D **80**, 064015 (2009) [arXiv:0906.1967 [gr-qc]].
[24] T. P. Sotiriou and V. Faraoni, Rev. Mod. Phys. **82**, 451 (2010) [arXiv:0805.1726 [gr-qc]].
[25] C. Charmousis, Lect. Notes Phys. **892**, 25 (2015) [arXiv:1405.1612 [gr-qc]].
[26] P. Horava, JHEP **0903**, 020 (2009) [arXiv:0812.4287 [hep-th]]. P. Horava, Phys. Rev. Lett. **102** (2009) 161301 [arXiv:0902.3657 [hep-th]]. P. Horava, Phys. Rev. D **79** (2009) 084008 [arXiv:0901.3775 [hep-th]].
[27] D. Blas, O. Pujolas and S. Sibiryakov, Phys. Lett. B **688** (2010) 350 [arXiv:0912.0550 [hep-th]]. D. Blas, O. Pujolas and S. Sibiryakov, Phys. Rev. Lett. **104** (2010) 181302 [arXiv:0909.3525 [hep-th]].
[28] T. Jacobson and D. Mattingly, Phys. Rev. D **64** (2001) 024028 [gr-qc/0007031]. T. Jacobson, PoS QG **-PH** (2007) 020 [arXiv:0801.1547 [gr-qc]].

[29] G. J. Olmo, Chapter of the book 'Open Questions in Cosmology', edited by Gonzalo J. Olmo.InTech Publishing, Rijeka, Croatia, (2012) ISBN 978-953-51-0880-1. (Downloadable from http://www.intechopen.com/books/open-questions-in-cosmology) [arXiv:1212.6393].
[30] K. Van Acoleyen and J. Van Doorsselaere, Phys. Rev. D **83**, 084025 (2011) [arXiv:1102.0487 [gr-qc]].
[31] C. Charmousis, B. Gouteraux and E. Kiritsis, JHEP **1209**, 011 (2012) [arXiv:1206.1499 [hep-th]].
[32] C. Deffayet, G. Esposito-Farese and A. Vikman, Phys. Rev. D **79**, 084003 (2009) [arXiv:0901.1314 [hep-th]].
[33] C. Deffayet, X. Gao, D. A. Steer and G. Zahariade, Phys. Rev. D **84**, 064039 (2011) [arXiv:1103.3260 [hep-th]].
[34] C. Charmousis, E. J. Copeland, A. Padilla and P. M. Saffin, Phys. Rev. Lett. **108**, 051101 (2012) [arXiv:1106.2000 [hep-th]]. C. Charmousis, E. J. Copeland, A. Padilla and P. M. Saffin, Phys. Rev. D **85**, 104040 (2012) [arXiv:1112.4866 [hep-th]].
[35] J. D. Bekenstein, gr-qc/9808028.
[36] N. M. Bocharova, K. A. Bronnikov and V. N. Melnikov, "An exact solution of the system of einstein equations and mass-free scalar field," Vestn. Mosk. Univ. Fiz. Astro. **6** (1970) 706. J. D. Bekenstein, "Exact solutions of Einstein conformal scalar equations," Annals Phys. **82**, 535 (1974).
[37] C. Martinez, R. Troncoso and J. Zanelli, Phys. Rev. D **67**, 024008 (2003) [hep-th/0205319].
[38] L. Hui and A. Nicolis, Phys. Rev. Lett. **110**, no. 24, 241104 (2013) [arXiv:1202.1296 [hep-th]].
[39] Y. Bardoux, M. M. Caldarelli and C. Charmousis, JHEP **1405**, 039 (2014) [arXiv:1311.1192 [hep-th]]. A. Anabalon and H. Maeda, Phys. Rev. D **81**, 041501 (2010) [arXiv:0907.0219 [hep-th]]. Y. Bardoux, M. M. Caldarelli and C. Charmousis, JHEP **1405**, 039 (2014) [arXiv:1311.1192 [hep-th]].
[40] G. Gubitosi and E. V. Linder, Phys. Lett. B **703**, 113 (2011) [arXiv:1106.2815 [astro-ph.CO]].
[41] T. P. Sotiriou and S. -Y. Zhou, arXiv:1312.3622 [gr-qc].
[42] T. Kobayashi and N. Tanahashi, PTEP **2014**, no. 7, 073E02 (2014) [arXiv:1403.4364 [gr-qc]].
[43] C. Charmousis, T. Kolyvaris, E. Papantonopoulos and M. Tsoukalas, JHEP **1407**, 085 (2014) [arXiv:1404.1024 [gr-qc]].

Journal of Physics: Conference Series **600** (2015) 012004 doi:10.1088/1742-6596/600/1/012004

Double layers in gravity theories

José M M Senovilla

Física Teórica, Universidad del País Vasco, Apartado 644, 48080 Bilbao, Spain

E-mail: `josemm.senovilla@ehu.es`

Abstract. Gravitational double layers, unlike their classical electromagnetic counterparts, are thought to be forbidden in gravity theories. It has been recently shown, however, that they are feasible in, for instance, gravity theories with a Lagrangian quadratic in the curvature. This is surprising with many potential consequences and the possibility of new physical behaviours. While a clear interpretation seems elusive, several lines of research are open. I present the field equations for double layers, the new physical quantities arising, and several explicit examples.

1. Introduction

Double layers may be found in several disciplines:

(i) in biology separating two different forms of matter,

(ii) in chemistry as interfaces between different phases (e.g. liquid and solid),

(iii) in physics when two laminar parallel shells with opposite electric charges are found next to each other.

They are especially important in plasma and cellular physics, representing abrupt drops in the electric potential by which the cell, or plasma, "protect" itself from the environment. However, there were no known gravitational double layers —until recently [1, 2]. Their existence comes as a surprise because an analogy with electrostatics may seem to indicate a dipolar distribution of matter on double layers. At first sight, this sounds strange because only positive masses exist. Thus, the potential consequences are tremendous, and new physical behaviours can be described.

1.1. Reminder: layers and double layers in Classical Electrodynamics

In classical electromagnetism a surface charge distribution creates an electric field with a jump in its component normal to the surface S [3]. The electrostatic potential Φ for a volume (with charge density ρ_V) plus a surface (with charge density σ) charge distribution reads:

$$\Phi(\vec{r}) = \frac{1}{4\pi\epsilon}\int_V \frac{\rho_V(\vec{r}\,')}{|\vec{r}-\vec{r}\,'|}dV' + \frac{1}{4\pi\epsilon}\int_S \frac{\sigma(\vec{r}\,')}{|\vec{r}-\vec{r}\,'|}dA'$$

Recalling Poisson's equation for the electrostatic potential $\Delta\Phi = -\frac{\rho}{\epsilon}$ this implies that one can consider the following total charge density

$$\rho = \rho_V + \sigma\delta_S$$

Journal of Physics: Conference Series **600** (2015) 012004 doi:10.1088/1742-6596/600/1/012004

where δ_S is a Dirac delta with support on S:

$$\langle \delta_S, f \rangle = \int_S f dA$$

Observe that Φ is continuous, the electric field $\vec{E} = -$ grad Φ has a discontinuity (exclusively in its component normal to the surface) only if $\sigma \neq 0$.

Similarly, the electrostatic potential for a dipole layer charge distribution reads

$$\Phi(\vec{r}) = \frac{1}{4\pi\epsilon} \int_S D(\vec{r}\,')\, \vec{n} \cdot \vec{\nabla}' \frac{1}{|\vec{r} - \vec{r}\,'|} dA'$$

where $\vec{n}$ is the unit outward normal to the layer. Thus, Poisson's equation provides the following charge density

$$\rho = \Delta'_S$$

where Δ'_S is a 'Dirac delta prime' *with support on S and strength D*:

$$\langle \Delta'_S, f \rangle = -\int_S D\, \vec{n} \cdot \vec{\nabla} f \, dA$$

Observe that Φ is discontinuous now [3].

Thus, in electrodynamics, if one wishes to represent objects where the charge or current is very concentrated the sources are often idealized to charges and currents confined to points, curves or surfaces. A proper mathematical description for such idealized concentrated sources uses the theory of distributions. The sum of distributions and the derivative of a distribution are well defined. However, in general the product of two distributions is not well defined. This makes the use of distributions in a nonlinear theory, such as those governing gravity, problematic (Maxwell's equations are linear so that they do not pose any problem).

2. Spacetimes with distributional curvature

In gravity theories, one also wishes to have a description of concentrated sources, that is, of concentrated matter and energy. They represent cases with thin shells (or braneworlds, or domain walls) and impulsive waves. However, one cannot simply assume that the metric is a distribution, since then the curvature (and Einstein) tensor will, in general, not be well defined due to the products of distributions. The solution is to identify the class of metrics whose curvature is well defined as a distribution, and such that the field equations will make sense. For sources on thin shells, an appropriate class of metrics were identified in [4, 5, 6]. Essentially, these are the metrics which are smooth except on a hypersurface where the metric is only continuous. Although this hypersuface can have any (even changing) causal character [7], I will assume herein that it is timelike.

A wider class of metrics was defined in [8] and called "regular metrics". An important result is that if the curvature of a regular metric is concentrated on a submanifold in spacetime, then the submanifold must have co-dimension 1. In other words, regular metrics can represent thin shells but not thin strings, thin textures or thin lines in General Relativity. The question is, can there be any double layers? In principle they are fine, as they have the right co-dimension.

To answer this question, let Σ be a timelike hypersurface dividing the manifold V into two regions $V^{\pm}$, and let n^{μ} be the unit normal to Σ pointing from V^- to V^+. The metrics $g^{\pm}_{\mu\nu}$ are smooth on $V^{\pm}$ respectively. Observe that one can actually use two different coordinate systems on $V^{\pm}$ [7], but an indispensable requirement is that the inherited metrics on Σ from both sides agree. This implies that there exist local coordinate systems in which the metric is continuous across Σ [7]. Nevertheless, there will be two, in principle different, second fundamental forms of

Σ, one inherited from V^+ the other from V^-. Denote them by $K^{\pm}_{\mu\nu}$ and recall that $n^\mu K^{\pm}_{\mu\nu} = 0$. Define the discontinuity on Σ of the second fundamental form as usual

$$[K_{\mu\nu}] \equiv K^+_{\mu\nu} - K^-_{\mu\nu}, \qquad n^\mu [K_{\mu\nu}] = 0.$$

In general, $[P]$ will mean the discontinuity of any object P across Σ.

Consider the Σ-step function

$$\underline{\theta} = \begin{cases} 1 & V^+ \\ 1/2 \quad \text{on} & \Sigma \\ 0 & V^- \end{cases}$$

Its derivative as a distribution (distributions are distinguished by an underline) reads

$$\nabla_\mu \underline{\theta} = n_\mu \underline{\delta}^\Sigma$$

where $\underline{\delta}^\Sigma$ is a scalar distribution (a Dirac delta) with support on Σ acting on any test function Y as

$$\left\langle \underline{\delta}^\Sigma, Y \right\rangle = \int_\Sigma Y \, .$$

The continuity of the metric allows to compute the Riemann tensor in a distributional sense. The Riemann tensor acquires, in general, a "singular" part proportional to $\underline{\delta}^\Sigma$. The explicit expression for the Riemann tensor distribution is [7]

$$\underline{R}^\alpha{}_{\beta\mu\nu} = (1-\underline{\theta}) R^{-\alpha}{}_{\beta\mu\nu} + \underline{\theta} R^{+\alpha}{}_{\beta\mu\nu} + \underline{\delta}^\Sigma H^\alpha{}_{\beta\mu\nu}$$

where $R^{\pm\alpha}{}_{\beta\mu\nu}$ are the Riemann tensors of $V^\pm$ respectively. $H^\alpha{}_{\beta\mu\nu}$ is called the singular part of the Riemann tensor distribution and reads

$$H_{\alpha\beta\lambda\mu} = -n_\alpha [K_{\beta\mu}] n_\lambda + n_\alpha [K_{\beta\lambda}] n_\mu - n_\beta [K_{\alpha\lambda}] n_\mu + n_\beta [K_{\alpha\mu}] n_\lambda$$

which is obviously defined only on Σ. This also leads to the singular part of the Ricci tensor distribution

$$H^\rho{}_{\beta\rho\mu} \equiv H_{\beta\mu} = -[K_{\beta\mu}] - [K^\rho{}_\rho] n_\beta n_\mu,$$

to the singular part of the scalar curvature

$$H^\rho{}_\rho \equiv H = -2 [K^\mu{}_\mu] \tag{1}$$

and to the singular part of the Einstein tensor distribution

$$\mathcal{G}_{\beta\mu} = -[K_{\beta\mu}] + h_{\beta\mu} [K^\rho{}_\rho] \, , \qquad n^\mu \mathcal{G}_{\beta\mu} = 0 \, .$$

Here $h_{\beta\mu} = g_{\beta\mu} - n_\beta n_\mu$ is the projector to Σ, that is, the space-time version of the first fundamental form on Σ. The above formulae are purely geometric and therefore they are valid in any theory of gravity.

In the particular case of General Relativity (GR), the above provides —via the Einstein field equations— the singular part $\tau_{\mu\nu}$ of the energy-momentum tensor distribution

$$\kappa \tau_{\beta\mu} = -[K_{\beta\mu}] + h_{\beta\mu} [K^\rho{}_\rho] \, , \qquad n^\mu \tau_{\beta\mu} = 0 \quad \underline{\text{(only in GR)}}$$

where κ is the gravitational coupling constant. This is known as the Israel formula. Compare with electrodynamics: here $\tau_{\beta\mu}$ replaces σ and the discontinuity of the second fundamental form is the analogue of the discontinuity of the normal component of the electric field.

2.1. Thin shells in GR

From the 2nd Bianchi identity (which holds for curvature tensor distributions [7]) one deduces the formulas

$$(K^+_{\rho\sigma} + K^-_{\rho\sigma})\mathcal{G}^{\rho\sigma} = 2n^\beta n^\mu [G_{\beta\mu}] = 2n^\beta n^\mu [R_{\beta\mu}] - [R],$$
$$\overline{\nabla}^\beta \mathcal{G}_{\beta\mu} = -n^\rho h^\sigma{}_\mu [G_{\rho\sigma}] = -n^\rho h^\sigma{}_\mu [R_{\rho\sigma}]$$

where $\overline{\nabla}$ denotes the intrinsic covariant derivative within Σ associated to its first fundamental form. Again, these expressions are geometric and thus valid in general. In GR, *but not in general gravity theories*, these can be trivially rewritten in terms of the energy-momentum tensor $T_{\mu\nu}$ and its singular distributional part $\tau_{\mu\nu}$ via the Einstein field equations:

$$(K^+_{\rho\sigma} + K^-_{\rho\sigma})\tau^{\rho\sigma} = 2n^\beta n^\mu [T_{\beta\mu}]$$
$$\overline{\nabla}^\beta \tau_{\beta\mu} = -n^\rho h^\sigma{}_\mu [T_{\rho\sigma}] .$$

These are the Israel equations [4].

As one sees explicitly, there is nothing like a delta-prime distribution in GR. Thus, *gravitational double layers are absent in GR*. This seems very reasonable, because there are only positive masses, gravity is attractive.

3. Quadratic $F(R)$-gravity

Bur, what about other gravity theories? One can prove that double layers arise, at least, in any gravity theory with a Lagrangian quadratic on the curvature. In this communication, and in order to illustrate how thin shells behave in alternative gravity theories explicitly showing the feasibility of double layers, I will restrict myself to the simple case of a Lagrangian density given by a quadratic $F(R)$

$$F(R) = R - 2\Lambda + \alpha R^2$$

where Λ (the cosmological constant) and α are constants. GR is the linear case $\alpha = 0$. These theories include the important Starobinsky inflationary models [9] and they possess a positive mass theorem [10], a well-defined entropy formulation [11], and a well-posed Cauchy problem [12]. The field equations are [13, 14, 15]

$$(1 + 2\alpha R)R_{\mu\nu} - \frac{1}{2}(R - 2\Lambda + \alpha R^2)g_{\mu\nu} - 2\alpha\left(\nabla_\mu \nabla_\nu R - g_{\mu\nu}\nabla_\rho \nabla^\rho R\right) = \kappa T_{\mu\nu} \tag{2}$$

where κ is the gravitational coupling constant. Equations (2) make sense as distributions *only* if the quadratic terms R^2 do not contain singular parts, as one cannot multiply these singular parts. Therefore, these equations demand that $\underline{R}$ cannot have a singular distributional part: R can be, at most, a discontinuous function.

3.1. Thin shells in quadratic $F(R)$

Thus, the field equations make sense only if $H = 0$: the singular part of $\underline{R}$ must vanish. Taking into account formula (1) for the singular part H of the scalar curvature this entails inevitably

$$[K^\mu{}_\mu] = 0 \tag{3}$$

and implies radical differences with GR [1, 2]: it forbids the use of umbilical hypersurfaces Σ — characterized by $K_{\mu\nu} = f h_{\mu\nu}$— unless they are totally geodesic. It also restricts the possibility of having mirror symmetric branes to the case with zero mean curvature $K^\mu{}_\mu = 0$.

More importantly, *observe that in (2) there are no terms quadratic in* ∇R ! This allows for a *discontinuous* R and still the field equations make sense. When R is discontinuous, the terms $\nabla\nabla R$ provide $\underline{\delta}^\Sigma$-terms and derivatives of these: gravitational double layers!

Journal of Physics: Conference Series **600** (2015) 012004 doi:10.1088/1742-6596/600/1/012004

4. Gravitational double layers

The energy-momentum tensor distribution can be computed from the field equations (2) leading to

$$\underline{T}_{\mu\nu} = T^+_{\mu\nu}\underline{\theta} + T^-_{\mu\nu}(1-\underline{\theta}) + (\tau_{\mu\nu} + \tau_\mu n_\nu + \tau_\nu n_\mu + \tau n_\mu n_\nu)\,\underline{\delta}^\Sigma + \underline{t}_{\mu\nu} \tag{4}$$

Thus, the singular part of the energy-momentum distribution consists of two different parts, one of thin-shell type proportional to the $\underline{\delta}^\Sigma$, the other of 'delta-prime' type, or of dipole layer kind, represented by $\underline{t}_{\mu\nu}$. The former splits into three different objects tangent to Σ, unlike in GR where only the first term $\tau_{\mu\nu}$ arises.

The explicit expressions and names for the objects in (4) supported on Σ are:

(i) the *energy-momentum tensor* $\tau_{\mu\nu}$ on Σ, given by

$$\kappa\tau_{\mu\nu} = -\{1+\alpha(R^+ + R^-)\}[K_{\mu\nu}] + \alpha\left\{2bh_{\mu\nu} - [R](K^+_{\mu\nu} + K^-_{\mu\nu})\right\}, \; n^\mu\tau_{\mu\nu} = 0, \tag{5}$$

where b is a function on Σ measuring the jump on the normal derivative of R across Σ:

$$b = n^\mu\,[\nabla_\mu R]\,.$$

An alternative, useful, expression for (5) reads

$$\kappa\tau_{\mu\nu} = -[K_{\mu\nu}] + 2\alpha\,(bh_{\mu\nu} - [RK_{\mu\nu}])\,. \tag{6}$$

$\tau_{\mu\nu}$ is the only quantity usually defined in standard shells.

(ii) the *external flux momentum* τ_μ with

$$\kappa\tau_\mu = -2\alpha\overline{\nabla}_\mu[R], \qquad n^\mu\tau_\mu = 0. \tag{7}$$

This momentum vector measures the normal-tangent components of $\underline{T}_{\mu\nu}$ supported on Σ, so that its timelike component describes the normal flux of energy across Σ on Σ while its spatial components measure the normal-tangential stresses.

(iii) the *external pressure or tension* τ defined as

$$\kappa\tau = 2\alpha[R]K^\rho{}_\rho, \tag{8}$$

where $K^\rho{}_\rho$ is the trace of *either* $K^\pm_{\mu\nu}$ —they are equal due to (3). Taking the trace of (6) one obtains a relation between b, τ and the trace of $\tau_{\mu\nu}$ ($n = \dim(\Sigma)$):

$$\kappa\,(\tau^\rho{}_\rho + \tau) = 2\alpha nb$$

The scalar τ measures the total normal pressure/tension supported on Σ.

(iv) the *double-layer energy-momentum tensor distribution* $\underline{t}_{\mu\nu}$, whose components are defined by acting on any test function Y as follows

$$\boxed{\kappa\left\langle\underline{t}_{\mu\nu}, Y\right\rangle = -2\alpha\int_\Sigma [R]h_{\mu\nu}\,n^\rho\nabla_\rho Y\,.} \tag{9}$$

This is a symmetric ($\underline{t}_{\mu\nu} = \underline{t}_{\nu\mu}$) tensor distribution of Dirac "delta-prime" type. $\underline{t}_{\mu\nu}$ has support on Σ but its product with objects intrinsic to Σ is not defined unless their extensions off Σ are known. For instance, one cannot write $\underline{t}_{\mu\nu} = h_{\mu\nu}\underline{t}$ for some scalar distribution $\underline{t}$ unless n_μ is extended somehow outside Σ. This $\underline{t}_{\mu\nu}$ resembles the energy-momentum content of double layer surface charge distributions, or "dipole distributions", with strength

$$2\alpha[R]h_{\mu\nu}.$$

This is the analogue of the strength D in a electrostatic dipole layer.

The case $\alpha = 0$ is simply GR supplemented with the condition (3). If $\alpha = 0$, then the new terms τ_μ, τ and $\underline{t}_{\mu\nu}$ all vanish and the formulas collapse to those of GR with the extra condition of the continuity of the second fundamental form (3).

A term such as (9) is remarkable and very surprising. It seems to represent the idealization of an abrupt change on the scalar curvature R which, one has to bear in mind, acts as a source or dynamical variable in quadratic $F(R)$ gravity. Therefore, these gravitational double layers can be thought as an imbalanced strain on the border between two different cosmological constants.

For the readership who dislikes $F(R)$ theories, it is worth reminding that they are equivalent to a scalar-tensor theory. In particular, quadratic $F(R)$ is equivalent to Brans-Dicke with $\omega = 0$ [16, 12, 17, 18]. The scalar field is $\phi = 2\alpha R - 1$ with a potential

$$V(\phi) = 2\Lambda + \frac{1}{4\alpha}(\phi^2 - 2\phi - 3).$$

In this alternative point of view, double layers represent an abrupt discontinuity in ϕ. Thus, double layers can be thought of as the separation surface of two different phases of ϕ: a ϕ-phase transition.

An important remark is that *all the terms* in (4), including $\underline{t}_{\mu\nu}$, are indispensable to make the entire $\underline{T}_{\mu\nu}$ divergence free. The field equations derived from this divergence-free property read:

$$\overline{\nabla}^\nu \tau_{\mu\nu} + K^\rho{}_\rho \tau_\mu + \overline{\nabla}_\mu \tau = -n^\nu h^\rho{}_\mu [T_{\rho\nu}],$$

$$n^\mu n^\nu [T_{\mu\nu}] - \tau_{\mu\nu} K^{\mu\nu}_\Sigma + \overline{\nabla}^\mu \tau_\mu = 2\frac{\alpha}{\kappa}[R]\left(R^\Sigma_{\mu\nu} n^\mu n^\nu + K^{\mu\nu}_\Sigma K_{\Sigma\mu\nu}\right)$$

where for any function f, $f_\Sigma \equiv (f^+ + f^-)/2$. It may be observed that $\tau_{\mu\nu}$ is not divergence-free *even* in cases where there is no flux of energy from the bulk across Σ (i.e. even when $n^\nu h^\rho{}_\mu [T_{\rho\nu}] = 0$) due to the new terms τ_μ and τ. Notice that $\underline{t}_{\mu\nu}$ is decoupled from the rest of the objects in the field equations.

Two important observations are in order:

(i) *Pure double layers.*
Given that $\underline{t}_{\mu\nu}$ is does not arise in the field equations, there seems to be extreme exotic situations in which all $\tau_{\mu\nu}$, τ_μ and τ vanish while $\underline{t}_{\mu\nu}$ does not, describing thin shells with *double layer energy-momentum contribution exclusively.* This happens if

$$K^\rho{}_\rho = 0, \quad \overline{\nabla}_\mu [R] = 0, \quad n^\mu [\nabla_\mu R] = 0, \quad [K_{\mu\nu}] + 2\alpha [R K_{\mu\nu}] = 0.$$

The second and third imply that $R^\pm$ are constants on Σ, while the first one implies that Σ must have zero mean curvature. No examples of these pure double layers have been built hitherto.

(ii) *Matching hypersurfaces in GR are double layers.*
Any thin shell in GR with $[K^\mu{}_\mu] = 0$ is a double layer in quadratic $F(R)$. In fact, any properly matched space-time in GR contains a double layer at the matching hypersurface if the true theory is quadratic $F(R)$! To see this, recall that in GR a proper matching requires $[K_{\mu\nu}] = 0$, but $[R] \neq 0$ in general, and therefore on the matching hypersurface we would have

$$\kappa \tau_{\mu\nu} = 2\alpha \left(n^\rho [\nabla_\rho R] h_{\mu\nu} - [R] K_{\mu\nu}\right),$$
$$\kappa \tau_\mu = -2\alpha \overline{\nabla}_\mu [R], \qquad \kappa \tau = 2\alpha [R] K^\rho{}_\rho,$$
$$\kappa \left\langle \underline{t}_{\mu\nu}, Y \right\rangle = -2\alpha \int_\Sigma [R] h_{\mu\nu}\, n^\rho \nabla_\rho Y\, .$$

For instance, by matching perfect fluids to vacuum in GR we would have

$$[R] = \rho^{GR}|_\Sigma = (\text{GR-energy density of the fluid on } \Sigma)$$

This happens because the matching conditions in $F(R)$ theories are more demanding than in GR [19, 1].

5. Examples

Consider a hypersurface Σ separating two constant-curvature n-dimensional spacetimes with different cosmological constants $\Lambda^+ \neq \Lambda^-$. The line-element on each $\pm$ side reads ((anti)-de Sitter or flat)

$$ds_\pm^2 = -\left(k - \frac{R_\pm x_\pm^2}{n(n-1)}\right) dy_\pm^2 + \left(k - \frac{R_\pm x_\pm^2}{n(n-1)}\right)^{-1} dx_\pm^2 + x_\pm^2 d\Omega_k^{n-2}$$

where $k = \pm 1, 0$ and $d\Omega_k^{n-2}$ is the complete Riemannian $(n-2)$-dimensional metric of constant curvature k. The cosmological constants $\Lambda_\pm$ are related to the corresponding $\pm$ constant scalar curvatures $R_\pm$ by

$$\Lambda_\pm = R_\pm \frac{n-2+(n-4)\alpha R_\pm}{2n}.$$

In general, the ranges (and causal character) of the coordinates $\{x_\pm, y_\pm\}$ depend on the signs of the constants $R_\pm$ and k.

To build double layers, one needs to find corresponding hypersurfaces $\Sigma^\pm$ on both $\pm$ sides such that the first fundamental forms agree and the condition (3) is satisfied. Restricting to hypersurfaces respecting the spherical symmetry, Σ is given by means of the parametric expressions

$$x_\pm = x_\pm(t), \qquad y_\pm = y_\pm(t)$$

so that the equality of the inherited first fundamental forms on Σ requires

$$x_+(t) = x_-(t) \equiv a(t)$$

and

$$\frac{dy_\pm}{dt} = \frac{\varepsilon_\pm}{k - \frac{R_\pm}{n(n-1)} a^2} \sqrt{\dot{a}^2 + k - \frac{R_\pm}{n(n-1)} a^2}$$

where $\varepsilon_\pm^2 = 1$ and dots stand for derivatives with respect to t. Then, the first fundamental form on Σ is a Robertson-Walker metric

$$d\gamma^2 = -dt^2 + a^2(t) d\Omega_k^{n-2}.$$

The unit normals on each side of Σ read

$$n_\pm = -\epsilon_\pm(-\dot{a} dy_\pm + \dot{y}_\pm dx_\pm)$$

where the signs $\epsilon_\pm$ determine the part of the $\pm$-side of the bulk space-time to be matched. The second fundamental forms on each side read then

$$K_{\mu\nu}^\pm dx^\mu dx^\nu =$$

$$(\varepsilon\epsilon)_\pm \left(-\frac{\ddot{a} - \frac{R_\pm}{n(n-1)} a}{\sqrt{\dot{a}^2 + k - \frac{R_\pm}{n(n-1)} a^2}} dt^2 + a \sqrt{\dot{a}^2 + k - \frac{R_\pm}{n(n-1)} a^2}\, d\Omega_k^2 \right).$$

The condition $[K^\mu{}_\mu] = 0$ leads to two possible families of solutions for $a(t)$ depending on whether or not $\dot{a} = 0$.

(i) **Case (i):** a is a constant given by the relation

$$(n-1)^2(k - \frac{R_+}{n(n-1)}a^2)(k - \frac{R_-}{n(n-1)}a^2) = 1$$

with both factors $k - \frac{R_\pm}{n(n-1)}a^2 > 0$ and $k = -(\epsilon\varepsilon)_+/(\epsilon\varepsilon)_-$.

(ii) **Case (ii):** $\dot{a} \neq 0$, then $a(t)$ is the solution of the ODE

$$\dot{a}^2 + k = \frac{a^{2n}}{m^2}\left(\frac{[R]}{2n(n-1)}\right)^2 + a^2\frac{R^+ + R^-}{2n(n-1)} + \frac{m^2}{4a^{2(n-2)}} \tag{10}$$

where $m \neq 0$ is an arbitrary constant. This can also be written in any of the two alternative forms

$$\dot{a}^2 + k - \frac{R_\pm}{n(n-1)}a^2 - \left(\frac{m}{2a^{n-2}} \mp \frac{a^n}{2n(n-1)m}[R]\right)^2.$$

5.1. Minkowski/de Sitter double layer

As a specific example of case (i), consider the possibility that the double layer separates de Sitter from Minkowski space-time, and set the dimension $n = 4$. By choosing $R_- = 0$, then $k = 1$ and

$$a^2 = \frac{32}{3}\frac{1}{R_+} = \frac{8}{3}\frac{1}{\Lambda_+} > 0.$$

This fixes uniquely the constant radius of the double layer in Minkowski, and in de Sitter. The non-zero eigenvalues of $\tau_{\mu\nu}$ are easily computed to give the energy density and pressure of the double layer (setting $(\epsilon\varepsilon)_+ = -1$)

$$\kappa\varrho = \frac{8}{3a}(1 + 2\alpha R_+),$$
$$\kappa p = \frac{1}{3a}(4 + 2\alpha R_+)$$

which are constant. Concerning the other quantities on the double layer one has

$$\tau_\mu = 0,$$

and

$$\kappa\tau = 6\frac{\alpha}{a}R_+ = 24\frac{\alpha}{a}\Lambda_+ = 64\frac{\alpha}{a^3}.$$

Finally

$$\kappa\left\langle \underline{t}_{\mu\nu}, Y\right\rangle = -8\alpha\Lambda_+\int_\Sigma h_{\mu\nu}Y_{,x}$$

with a first fundamental form

$$h_{\mu\nu}dx^\mu dx^\nu = d\gamma^2 = -dt^2 + \frac{8}{3}\frac{1}{\Lambda_+}d\Omega^2.$$

This represents a 3-dimensional Einstein static spacetime.

Journal of Physics: Conference Series **600** (2015) 012004 doi:10.1088/1742-6596/600/1/012004

5.2. Analysis of the dynamical solution

In order to analyze the dynamical solution (case (ii)) for a double layer, one can look at the Eq.(10) in the standard "kinetic + potential energy" form

$$\dot{a}^2 + V(a) = 0.$$

In this way it is very simple to get the qualitative behaviour of the solutions $a(t)$. The "potential" $V(a)$ is an even function. In the physical region with $a > 0$, $V(a)$ has a unique maximum and its value is always negative for the cases with $k = 0, -1$. Therefore, in these cases the double layers describe a $(n-1)$-dimensional Universe starting from a big-bang followed by a decelerated expansion phase until $\dot{a}$ reaches a minimum value from where the Universe undergoes an accelerated expansion epoch leading to unbounded values of a and $\dot{a}$.

For the remaining, closed, case with $k = 1$, the maximum of $V(a)$ may be positive or negative depending on whether $|m|$ is small or large, respectively. In the latter case the solutions behave just as in the previous cases with $k = 0, -1$. If $|m|$ is small, on the contrary, the possible solutions for $a(t)$ have two branches. In one of them the double layer describes a Universe which, from a big-bang, reaches a re-collapsing time and then contracts to a big crunch. The other branch, however, is singularity-free, starting with a large Universe that contracts to a minimum volume and then re-expands with accelerated expansion. Of course, there is a critical value of $|m|$ between the two mentioned possibilities, where the Universe tends asymptotically to an Einstein static solution of the type given in case (i).

Given the properties of the double layer, let us consider that it is a "braneworld" in a 5-dimensional bulk, that is, set $n = 5$. The non-zero eigenvalues of $\tau_{\mu\nu}$ provide the energy density and pressure within the double layer:

$$\kappa\varrho = -(\epsilon\varepsilon)_+ \left\{ \frac{3m}{a^4}\left(1 + \alpha(R_+ + R_-)\right) + \alpha\frac{a^4}{4m}[R]^2 \right\},$$

$$\kappa p = -(\epsilon\varepsilon)_+ \left\{ \frac{m}{a^4}\left(1 + \alpha(R_+ + R_-)\right) - \alpha\frac{a^4}{20m}[R]^2 \right\}$$

Observe that these expressions have two terms, the one on the left dominates for small values of a, the other for large values of a. The former has an equation of state of radiation type $p = \varrho/3$, and is the only one surviving in the GR limit when $\alpha \to 0$. The latter has an equation of state $p = -\varrho/5$ and is proportional to the square of the difference between the constant curvatures at both sides of the double layer.

Observe that the constant k only is implicit entering only through $a(t)$. Notice also the signs of $R_\pm$ are totally free. Thus, any possible combination of de Sitter, anti-de Sitter and flat spacetimes are allowed. Concerning the new energy-momentum quantities we have

$$\tau_\mu = 0,$$

$$\kappa\tau = -(\epsilon\varepsilon)_+ \frac{2a^4}{5m}\alpha[R]^2$$

and

$$\kappa\left\langle \underline{t}_{\mu\nu}, Y \right\rangle = -2\alpha[R]\int_\Sigma h_{\mu\nu}\, n^\rho \nabla_\rho Y\, .$$

The last one is independent of the signs $\epsilon_\pm$ and $\varepsilon_\pm$. Here $h_{\mu\nu}$ is the standard Robertson-Walker metric with scale factor $a(t)$.

Observe that the energy density and pressure satisfy the following *generalized continuity equation*

$$\dot{\varrho} + 3\frac{\dot{a}}{a}(\varrho + p) = \dot{\tau}$$

which reflects the novel fact that the energy-momentum $\tau_{\mu\nu}$ of the double layer is not divergence free in general. Hence, the traditional behavior of

$$\varrho \sim a^{-3(1+w)}$$

for an equation of state $p = w\varrho$ does not hold for double-layer braneworlds in general. This opens the door for new physical behaviors.

As an amusement, one can wonder how advanced physicists living in such a double layer, but unaware of its braneworld/double-layer character, would interpret their measurements, in the same vein as in [20, 21]. In order to answer this, the energy density and pressure of the double layer must be compared with the GR values as they would be computed by those scientists on Σ. Imagine, to that end, that they manage to find the correct equation for the scale factor, that is (10) with $n = 5$:

$$\dot{a}^2 + k = \frac{a^{10}}{m^2}\left(\frac{[R]}{40}\right)^2 + a^2\frac{R_+ + R_-}{40} + \frac{m^2}{4a^6}\,.$$

This can be rewritten in terms of the Hubble parameter $H(t)$

$$H^2 + \frac{k}{a^2} = \frac{a^8}{m^2}\left(\frac{[R]}{40}\right)^2 + \frac{R_+ + R_-}{40} + \frac{m^2}{4a^8}$$

so this would just mean that they had been able to determine the behavior of the Hubble parameter with cosmic time t. They could then use the Einstein field equations to compute what they would believe are the matter contents of the Universe, and they would get:

$$8\pi G\varrho^{GR} = \frac{3}{4}\left(\frac{a^8}{m^2}\frac{[R]^2}{400} + \frac{m^2}{a^8}\right),$$

$$8\pi G p^{GR} = \frac{1}{4}\left(\frac{-11a^8}{m^2}\frac{[R]^2}{400} + \frac{5m^2}{a^8}\right),$$

$$\Lambda^{GR} = \frac{3}{40}\left(R_+ + R_-\right).$$

Now, one can speculate with these formulas...

Note, in any case, that the "fake" GR cosmological constant would be simply proportional to the sum of the constant scalar curvatures at both sides of the double layer. Hence, in order to have a positive Λ^{GR}, at least one side of the bulk should have a positive constant curvature. Notice also that the choice with one flat Minkowskian side is feasible. And the the value $\Lambda^{GR} = 0$ would require a too exact fine tuning between the two sides of the double layer.

6. Conclusions and prospects

I have shown that gravitational double layers are consistent in some quadratic theories. If, as usually claimed, the quadratic terms are really needed for a quantum theory of gravity, then double layers may represent an idealized good approximation to situations where the cosmological constant changes abruptly.

Furthermore, good genuine matching hypersurfaces in GR would actually become double layers in any regime where quadratic terms in the Lagrangian start to be non negligible. If, for instance, when approaching a quantum regime these quadratic terms are switched on, or become relevant, then any classical GR solution which can be approximately described by a matching procedure will become an approximate solution which contains, on the matching hypersurface, a gravitational double layer. The implications of this feature must be analyzed in depth.

Furthermore, the question of how these double layer solutions fit in a general gravitational theory must be addressed, to examine if they actually arise *generically* in regimes dominated by quadratic terms in the curvature, or not.

In order to give consistency and a solid background to the gravitational double layers described herein, one should try to find actual, regular, solutions that approach the idealized version described by distributions. Moreover, the stability of the solutions with double layers should also be analyzed.

Most importantly, a clear interpretation of the dipolar term is needed, and yet unclear.

Acknowledgements
The author is grateful to the organizers of the Spanish Relativity Meeting (ERE-2014) for their kind invitation. Supported by grants FIS2010-15492 (MICINN), GIU12/15 (Gobierno Vasco), P09-FQM-4496 (J. Andalucía—FEDER) and UFI 11/55 (UPV/EHU).

References
[1] Senovilla J M M 2013 *Phys. Rev.* D **88** 064015
[2] Senovilla J M M 2014 *Class. Quantum Grav.* **31** 072002
[3] Jackson J D 1962 *Classical Electrodynamics* (J. Wiley and sons, N. York).
[4] Israel W 1966 *Nuovo Cimento* **44**, 1; erratum 1967 **48**, 463
[5] Lichnerowicz A 1971 *C. R. Acad. Sci.* **273** 528
[6] Taub A H 1979 *J. Math. Phys.* **21** 1423
[7] Mars M and Senovilla J M M 1993 *Class. Quantum Grav.* **10** 1865
[8] Geroch R and Traschen J 1987 *Phys. Rev.* D **36** 1017
[9] Starobinsky A A 1980 *Phys.Lett.* B **91** 99-102
[10] Strominger A 1984 *Phys. Rev.* D **30** 2257
[11] Jacobson T, Kang G and Myers R C 1995 *Phys.Rev.* D **52** 3518
[12] Teyssandier P and Tourrenc Ph 1983 *J. Math. Phys.* **24** 2793
[13] Sotiriou T P and Faraoni V 2010 *Rev. Mod. Phys.* **82** 451
[14] Capozziello S and Faraoni V 2011*Beyond Einstein Gravity* (Springer, New York)
[15] Nojiri S and Odintsov S D 2011 *Phys. Rep.* **505** 59
[16] Higgs P W 1959 *Nuovo Cimento* **11** 816
[17] Chiba T 2003 *Phys. Lett.* B **575** 1
[18] de Felice A and Tsujikawa S 2010 *Living. Rev. Relativity* **13** 3
[19] Deruelle N, Sasaki M and Sendouda Y 2008 *Prog. Theor. Phys.* **119** 237
[20] Mars M, Senovilla J M M and Vera R 2007 *Phys. Rev.* D **76** 044029
[21] Mars M, Senovilla J M M and Vera R 2008 *Phys. Rev.* D **77** 027501

2+1 black hole with $SU(2)$ hair (and the theory where it grows)

Jorge Zanelli

Centro de Estudios Científicos, CECs - Valdivia, Chile and
Universidad Andres Bello, Av. Republica 440, Santiago, Chile

E-mail: z-at-cecs-dot-cl

Abstract. A black hole solution in three spacetime dimensions, endowed with an $SU(2)$ charge is presented. The construction is based on two main features of three dimensions: i) AdS_3 spacetime is locally Lorentz-flat, that is, it can be covered with a congruence of local inertial observers, just like flat Minkowski space; ii) The $SO(2,1)$ and $SU(2)$ groups are isomorphic, so that a flat connection of the first can be mapped to a flat connection of the second. The global nontrivial nature of the solution is a consequence of the topology produced by the identification in the covering space that gives rise to the 2+1 black hole.

It can be seen that this solution belongs to the vacuum (matter-free) sector of a supersymmetric theory based on the Chern–Simons action for the $su(1,2|2)$ superalgebra. The action for this system matches that of graphene in the long wavelength limit near the Dirac point. The $SU(2)$ gauge symmetry is interpreted as the freedom to choose locally the definition of spin quantization axis for the electrons.

1. The Lorentz flatness of AdS_3 spacetime

Three-dimensional gravity is a source of surprises. In the standard theory, all classical solutions have constant Riemannian curvature, with a radius determined by the cosmological constant in the action. Hence, given the action, the solutions have a uniquely fixed curvature and the geometry is completely determined by the global topology. The geometry of each solution is determined by two parameters that characterize the particular identification applied to the covering AdS space. If closed timelike curves and naked singularities are excluded a, family of black holes labeled by a mass M and angular momentum J are obtained [1, 2]. A few other solutions dual to the black holes can also be found [3] and, if moderate naked singularities like conical defects are allowed, also spinning point particles can be included in the spectrum [4, 5].

Due to its simple dynamical structure 2+1 gravity is a remarkable test ground to study gravitational phenomena, analogous to those in 3+1 dimensions, in a simplified setting.

This simplicity can be pushed one step further by the following little

Theorem: *A 2+1 locally AdS spacetime is Lorentz flat* [6].

Proof: Consider a three-dimensional manifold M endowed with a Lorentz connection $\omega^{ab} = -\omega^{ba}$, and an open simply connected region of M where the connection is flat,

$$R^{ab} = d\omega^{ab} + \omega^a{}_c\omega^{cb} = 0, \tag{1}$$

where R^{ab} is the Lorentz curvature two-form.[1] The metric structure of the space is determined by the one-form fields defining a local orthonormal frame e^a (also called the soldering form, vielbein or coframe field). The exterior covariant derivative of e^a defines the torsion two-form $T^a = De^a = de^a + \omega^a{}_c e^c$. From (25) it is straightforward to infer that the torsion is covariantly constant,

$$0 = R^a{}_b e^b = DDe^a = DT^a \,. \tag{2}$$

We therefore learn that the Lorentz flat geometry must have covariantly constant torsion $DT^a = 0$. In three dimensions, this equation is integrated as

$$T^a = \tau \varepsilon^a{}_{bc} e^b e^c \,, \tag{3}$$

where τ is an integration constant. It is always possible to split ω^{ab} as

$$\omega^{ab} = \tilde{\omega}^{ab} + \kappa^{ab}, \tag{4}$$

where $\tilde{\omega}^{ab}$ is the torsion-free (Riemannian) part,

$$de^a + \tilde{\omega}^a{}_c e^c = 0 \,, \tag{5}$$

and κ^{ab} is the contorsion. Therefore, $T^a = De^a = de^a + \omega^a{}_c e^c \equiv \kappa^a{}_c e^c$, and

$$R^{ab} = \tilde{R}^{ab} + \tilde{D}\kappa^{ab} + \kappa^a{}_c \kappa^{cb}, \tag{6}$$

where $\tilde{D}$ is the covariant exterior derivative in the connection $\tilde{\omega}^a{}_b$. Then, in view of (3) and (4), (6) becomes

$$R^{ab} = \tilde{R}^{ab} + \tau^2 e^a e^b \,, \tag{7}$$

which implies that anti-de Sitter spaces

$$\tilde{R}^{ab} = -\tau^2 e^a e^b \,, \tag{8}$$

are Lorentz-flat manifolds. (Q.E.D.)

This means that any simply connected open set of 2+1 AdS space can be covered with a family of local frames related with each other by Lorentz transformations. In particular, this implies that a by parallel transport of a vector with a Lorentz connection, a globally defined vector field can be constructed that completely covers any simply connected open region. This can be done with a basis of three vectors at one point, thus covering the entire region with three independent vector fields.

It is interesting to observe that the minus sign in the 0-0 component of the Lorentzian metric is responsible for the negative sign in the curvature. Had the metric been Euclidean, the resulting geometry would have been that of a three sphere instead of AdS_3. The little theorem presented here can then be seen as the Lorentzian version of Adams' theorem, which states that a Euclidean three sphere is parallelizable, namely, it can be globally covered by a family of three independent vector fields [7, 8]. Adams' theorem goes even further, stating that also the seven sphere is parallelizable, from which we can conjecture that a Minkowskian AdS_7 is also Lorentz flat.

[1] We implicitly assume exterior (wedge) products for forms except where confusion may arise.

Journal of Physics: Conference Series **600** (2015) 012005 doi:10.1088/1742-6596/600/1/012005

2. The 3D black hole as a Lorentz flat geometry

Since a 2+1 black hole is locally an AdS geometry with radius of curvature ℓ, it can be identified with a Lorentz flat spacetime with constant torsion given by (3) with $\tau = \eta\ell^{-1}$, where $\eta = \pm 1$. How can that be?

In order to illustrate the point, consider the geometry of a static BTZ black hole [9] (the rotating case is left as an exercise to the reader),

$$ds^2 = f^2 dt^2 + f^2 dr^2 + r^2 d\phi^2, \tag{9}$$

with $f^2 = M + r^2/\ell^2$, and M is the mass. The vielbein are given by

$$e^0 = f(r)dt, \quad e^1 = dr/f(r), \quad e^2 = rd\phi\,. \tag{10}$$

Then, the torsion-free (Riemannian) connection, satisfying (5), is given by

$$\tilde{\omega}^0{}_1 = \frac{r}{\ell^2}dt, \quad \tilde{\omega}^1{}_2 = -fd\phi, \quad \tilde{\omega}^2{}_0 = 0\,, \tag{11}$$

and can be directly confirmed that $\tilde{R}^{ab} = -\ell^{-2}e^a e^b$. It is also straightforward to see that the torsion given by (3) is

$$T^0 = -\eta\frac{r}{\ell f}dr \wedge d\phi, \quad T^1 = \eta\frac{rf}{\ell}d\phi \wedge dt, \quad T^2 = \frac{\eta}{\ell}dt \wedge dr\,. \tag{12}$$

The Lorentz connection, on the other hand, is given by (4), where

$$\kappa^0{}_1 = \eta\frac{r}{\ell}d\phi, \qquad \kappa^1{}_2 = -\eta\frac{f}{\ell}dt, \qquad \kappa^2{}_0 = -\eta\frac{dr}{\ell f}. \tag{13}$$

It is then simple algebra to show that

$$\omega^0{}_1 = \eta\frac{r}{\ell}\left[\eta\frac{dt}{\ell} + d\phi\right], \quad \omega^1{}_2 = -f\left[\eta\frac{dt}{\ell} + d\phi\right], \quad \omega^2{}_0 = -\eta\frac{dr}{\ell f}\,. \tag{14}$$

is Lorentz flat, $R^a{}_b = d\omega^a{}_b + \omega^a{}_c\omega^c{}_b \equiv 0$. Finally, direct computation shows that the torsion is indeed covariantly constant in the Lorentz connection, $dT^a + \omega^a{}_b T^b \equiv 0$.

It should be noted that the flatness of the Lorentz connection does not imply that parallel-transporting a vector round an arbitrary closed path will take it back to itself. That is only so for a path contained in a simply connected open patch of AdS_3 which, in the case of a black hole –or a point particle–, means a region that does not include the central singularity. In particular, a path that winds around the line $r = 0$ would yield a nontrivial holonomy because such a path is not contractible and the region where it lies is not simply connected.

Another important point is the fact that the flatness of the Lorentz connection (14) does not depend on the value of ℓ, so long as $\ell \neq 0$. In fact, ℓ can be set equal to 1 by the re-scaling of $r \to r' = r/\ell$, $t \to t' = t/\ell$.

The same discussion can be repeated for a rotating black hole, or for a point particle spinning or not (naked conical singularity). All of these are locally AdS_3 spacetimes, the only difference being the type of identification that is made to produce them and therefore they belong to different topological classes but with the same local geometry.

3. Adding SU(2) hair on a 2+1 BH

Black holes do no suffer hair gladly. The possibility of adding hair to a black hole is always a challenging problem as there are not many standard black hole solutions with minimally coupled nontrivial gauge fields. The previous discussion however, hints at a possibility of including an $SU(2)$ hair in 2+1 dimensions, based on the isomorphism of this gauge group with $SO(2,1)$. The point is that since these two groups are isomorphic, having a nontrivial locally flat configuration of the latter suggests the possibility of a nontrivial flat configuration for the former.

The isomorphism $SO(2,1) \cong SU(2)$ means that there exists a one-to-one correspondence between the elements of each group and their corresponding algebras,

$$[\mathbb{J}_a, \mathbb{J}_b] = \varepsilon_{ab}{}^c \mathbb{J}_c \qquad \leftrightarrow \qquad [\mathbb{G}_I, \mathbb{G}_J] = i\epsilon_{IJK}\mathbb{G}_K \,, \tag{15}$$

where $\varepsilon_{ab}{}^c = \eta^{cd}\varepsilon_{abd}$, and both ε_{abc} and ϵ_{IJK} are the three-dimensional Levi-Civita symbols, with $\varepsilon_{012} = \epsilon_{123} = 1$. An explicit representation for the SO(2,1) and SU(2) algebras is given by $\mathbb{J}_a = (\Gamma_a)^\alpha{}_\beta$, and $\mathbb{G}_I = (\sigma_I)^i{}_j$, where γ_a and σ_I are the 2×2 representations of the Dirac and Pauli matrices respectively.

In particular, starting from the connection (14) one can produce a flat $SU(2)$ connection given by

$$\mathbf{A}^3 = \eta' \frac{r}{\zeta}\left[\eta' \frac{dt}{\zeta} + d\phi\right] \sim \omega^0{}_1, \quad \mathbf{A}^2 = ih\left[\eta' \frac{dt}{\zeta} + d\phi\right] \sim \omega^1{}_2, \quad \mathbf{A}^1 = \eta' \frac{dr}{\zeta h} \sim \omega^2{}_0 \,, \tag{16}$$

where $h^2(r) = r^2/\zeta^2 - Q$ and $\eta' = \pm 1$. The global nontriviality is reflected by the fact that $d\phi$ is not an exact form in an open region containing $r = 0$.

It is straightforward to check that A^I defines a flat $SU(2)$ connection,

$$\mathbf{F}^I = d\mathbf{A}^I + \epsilon^{IJK}\mathbf{A}^J\mathbf{A}^K \equiv 0 \,, \tag{17}$$

that holds for all values of ζ and η'. Here ζ plays the role of the AdS radius, although there is no need to identify ζ with ℓ. Similarly, the charge Q is an integration constant of the solution, similar to the mass of the black hole. However, since there is no analogue for the metric in the $SU(2)$ sector, there is no restriction for the charge corresponding to the cosmic censorship that restricts M to non negative values in order to avoid naked singularities.

The analysis can be extended to the case of rotating solutions as reported in the forthcoming paper [9], where the BPS configurations are also discussed. The physical meaning of these solutions requires the identification of the physical observables of the system, the conserved charges and this cannot be discussed without knowing the action principle.

4. Action principle

The construction above would be a rather artificial exercise if the equations $R^{ab} = 0$ and $\mathbf{F}^I = 0$ did not originate in some action principle. The natural three-dimensional Lagrangians that give rise to these equations are the Chern-Simons (**CS**) forms for the connections ω^{ab} and $\mathbf{A}^I$, respectively, so one might tentatively consider

$$I_0[\omega, \mathbf{A}] = k_1 \int \left\langle \frac{1}{2}\mathbf{A}^I d\mathbf{A}^I + \frac{1}{3}\epsilon_{IJK}\mathbf{A}^I\mathbf{A}^J\mathbf{A}^K \right\rangle + k_2 \int \left[\frac{1}{2}\omega^a{}_b d\omega^b{}_a + \frac{1}{3}\omega^a{}_b\omega^b{}_c\omega^c{}_a\right] . \tag{18}$$

However this action is still somewhat uninteresting since there is no coupling between the fields and there seem to be no reason to put them together. In addition, the identification of the second term as something related to the spacetime geometry and gravity could only be justified if the metric properties of spacetime are to play a role, otherwise the earlier discussion about the AdS spacetime would be completely unjustified. The link between the spacetime symmetry represented by the Lorentz connection and the internal $SU(2)$ gauge symmetry is naturally provided by a spin-1/2 field. Fermions are intrinsically sensitive to the Lorentz symmetry because spinors are fundamental representations of the Lorentz group. In addition, fermions can be charged with respect to an $SU(N)$ gauge group as we know from the Standard Model of particle physics. So, the presence of spinning matter can provide a natural link between the two free fields in (18).

Moreover, introducing fermions turns out to be a very economical solution because, as Weyl noted long ago [10], spinors also naturally couple to the metric of the manifold on which they live. One way of seeing this is that in order to write down the Dirac equation in an arbitrary spacetime background it is necessary to know the metric structure of the spacetime geometry. The need can be seen in the fundamental structure for spin-1/2 representation, the Clifford algebra,

$$\{\gamma_a, \gamma_b\} = 2\eta_{ab}, \tag{19}$$

where η_{ab} is the Lorentz invariant metric of Minkowski space. Then, in order to define the Dirac operator $\partial\!\!\!/ = \gamma^\mu \partial_\mu$, it is necessary to transmute the tangent space index of the Dirac gamma matrix into a spacetime index. The instrument that performs this is the local Lorentz (co-)frame $e^a = e^a_\mu dx^\mu$ and its inverse, the local vector frame $E_a = E^\mu_a \partial_\mu$,

$$\partial\!\!\!/ = \gamma^a E_a \equiv \gamma^a E^\mu_a \partial_\mu \,. \tag{20}$$

This means that the manifold where spinors propagate must be endowed with a metric structure that projects the tangent space metric down to the spacetime manifold,

$$g_{\mu\nu} = \eta_{ab} e^a_\mu e^b_\nu \,. \tag{21}$$

Thus, a simple and natural way to link the $SU(2)$ and $SO(2,1)$ connections, bringing a metric structure at the same time is to add to (18) a minimally coupled spin-1/2 $SU(2)$-charged Dirac field

$$I[\omega, \mathbf{A}, e, \psi] = I_0[\omega, \mathbf{A}] + \int L_F \,, \tag{22}$$

where

$$L_F = -\overline{\psi}\left[\partial\!\!\!/ + i\mathbf{A}\!\!\!/ - \frac{1}{4}\gamma_a \omega\!\!\!/^{ab}\gamma_b\right]\psi |e| d^3x - \overline{\psi}\psi e^a T_a \,. \tag{23}$$

The first term in brackets on the right is just the usual Dirac Lagrangian for a complex spinor minimally coupled to the the $SU(2)$ gauge field and to the standard spin connection coupling required by covariance in a curved background. The last term results from the requirement of hermiticity,

$$1/2\left[\overline{\psi}\partial\!\!\!/\psi + h.c.\right] = \overline{\psi}\partial\!\!\!/\psi + \overline{\psi}\psi e^a T_a \,,$$

and the torsion is produced by the presence of the frame E^μ_a in $\partial\!\!\!/$, as noted by Weyl [10]. The field equations for this action are:

$$\delta\mathbf{A} \;:\; \mathbf{F}^I = \frac{i}{2}\epsilon^{IJK}\overline{\psi}^J \gamma_a\gamma_b \psi^K e^a e^b \tag{24}$$

$$\delta\omega \;:\; R^{ab} = -\frac{1}{2}\overline{\psi}^I \psi^I e^a e^b \tag{25}$$

$$\delta\overline{\psi} \;:\; \left[\partial\!\!\!/ + i\mathbf{A}\!\!\!/ - \frac{1}{4}\gamma_a \omega\!\!\!/^{ab}\gamma_b + e^a_\mu T^b_{\nu\lambda}\eta_{ab}\varepsilon^{\mu\nu\lambda}\right]\psi = 0 \tag{26}$$

$$\delta e^a \;:\; \overline{\psi}\varepsilon_{abc}\gamma^b\left[\overleftarrow{d}\, e^c - \overrightarrow{d}\, e^c + 2ie^c\mathbf{A}\right]\psi + 2\overline{\psi}\psi T^a = 0. \tag{27}$$

It is immediately obvious that in the limit of vanishing fermionic field, the $SU(2)$ and $SO(2,1)$ connections are flat, $\mathbf{F} = 0 = R^{ab}$. This means that the discussion of Sect. 1 refers to the matter-free sector of this system. It is interesting, however, that even if $\psi \neq 0$, the Lorentz curvature is such that $R^a{}_b e^b = 0$, so that the torsion is still covariantly constant. Therefore (3) would still be valid and (8) would be replaced by

$$\tilde{R}^{ab} = -\left[\tau^2 + \overline{\psi}^I \psi^I\right] e^a e^b \,. \tag{28}$$

This in turn means that the fermions can be included perturbatively as corrections to the curvature around the locally AdS spacetime background. The fact that even in the presence of fermions the torsion is covariantly constant and therefore of the form $T^a = \tau\varepsilon^{abc} e_b e_c$, implies that the last term in the Dirac equation is always a mass term,

$$e^a_\mu T^b_{\nu\lambda}\eta_{ab}\varepsilon^{\mu\nu\lambda} = -6\tau \equiv m \,. \tag{29}$$

Thus, in this model although there is no mass parameter and no cosmological constant in the action, the fermion mass and the cosmological constants result from the integration constant related to the torsion of the spacetime background.

5. Superconnection theory

The fact that fermionic fields couple naturally to the Lorentz connection ω^{ab} and to an internal gauge connection $\mathbb{A}$ suggests that there could be a way to combine the two symmetries $SO(2,1)$ and $SU(2)$ into a larger group G. However, some well-known no-go theorems make this possibility unlikely for a Lie group, except for the trivial case $G \sim SO(2,1) \times SU(2)$ [11].

As noted long ago by Haag, Lopuszanski and Sohnius [12] a way to circumvent this difficulty is to consider a *graded* Lie group, also referred to as a supergroup. The idea is that, together with the generators of the Lie groups $\mathbb{J}_a$ and $\mathbb{G}_I$ (bosonic sector), fermionic generators $\mathbb{Q}^\alpha_i$, $\overline{\mathbb{Q}}^i_\alpha$ that transform in a spin-1/2 representation of $SO(2,1)$ and a fundamental representation of $SU(2)$ are introduced. These supercharges obey

$$[\mathbb{J}_a, \mathbb{Q}] = -\frac{1}{2}\gamma_a\mathbb{Q}\,, \qquad [\mathbb{J}_a, \overline{\mathbb{Q}}] = \frac{1}{2}\overline{\mathbb{Q}}\gamma_a \tag{30}$$

$$[\mathbb{G}_I, \mathbb{Q}] = \frac{1}{2}\mathbb{Q}\sigma_I\,, \qquad [\mathbb{G}_I, \overline{\mathbb{Q}}] = -\frac{1}{2}\sigma_I\overline{\mathbb{Q}}\,, \tag{31}$$

where we have suppressed all but the relevant indices. The simplest 4×4 matrix representation for this is

$$\mathbb{J}_a = \begin{bmatrix} \frac{1}{2}(\gamma_a)^\alpha{}_\beta & 0 \\ 0 & 0 \end{bmatrix}, \mathbb{G}_I = \begin{bmatrix} 0 & 0 \\ 0 & \frac{1}{2}(\sigma_I)^i{}_j \end{bmatrix}; \mathbb{Q}^\alpha_i = \begin{bmatrix} 0 & 0 \\ \delta^\alpha_\beta\delta^j_i & 0 \end{bmatrix}, \overline{\mathbb{Q}}^i_\alpha = \begin{bmatrix} 0 & \delta^\beta_\alpha\delta^i_j \\ 0 & 0 \end{bmatrix}. \tag{32}$$

The superalgebra is completed by the anticommutator,

$$\{\mathbb{Q}^\alpha_i, \overline{\mathbb{Q}}^j_\alpha\} = i\delta^j_i\delta^\alpha_\beta\mathbb{Z} + \delta^j_i(\gamma^a)^\alpha_\beta\mathbb{J}_a + \delta^\alpha_\beta(\sigma^I)^j_i\mathbb{G}_I. \tag{33}$$

Here $\mathbb{Z}$ is an abelian central extension that can be identified with the $U(1)$ generator

$$\mathbb{Z} = \frac{-i}{2}\begin{bmatrix} \delta^\alpha_\beta & 0 \\ 0 & \delta^i_j \end{bmatrix}. \tag{34}$$

Note that all these generators have vanishing supertrace, and the superalgebra can be identified as $su(2,1|2) \oplus u(1)$.

At this point a legitimate question would be, what is the meaning of this enlarged symmetry? It turns out that the action (22) can be written in a very compact way, as a Chern-Simons form,

$$str\left[\frac{1}{2}\mathcal{A}d\mathcal{A}\frac{1}{3}\mathcal{A}\mathcal{A}\mathcal{A}\right] = I[\omega, A, e, \psi] + AdA\,, \tag{35}$$

where str stands for the graded trace (supertrace) in the superalgebra and $\mathcal{A}$ is the connection

$$\mathcal{A} = \omega^a \mathbb{J}_a + \mathbf{A}^I \mathbb{G}_I + \overline{\mathbb{Q}}^j \gamma_a e^a \psi_j - \overline{\psi}^i e^a \gamma_a \mathbb{Q}_i + A\mathbb{Z}, \tag{36}$$

(here $\omega^a \equiv \varepsilon^a{}_{bc}\omega^{bc}$). Since the $U(1)$ field A is decoupled from the rest, the theory can be consistently truncated dropping the $u(1)$ generator from the superalgebra. Alternatively, one could include a nontrivial coupling between the fermion and the $U(1)$ connection A.

An even simpler superalgebra can be constructed by accommodating the generators $\mathbb{J}_a$ of $SO(2,1)$ together with a complex two-component spinor supercharge $\mathbb{Q}$ and a $U(1)$ generator $\mathbb{K}$ dropping the $SU(2)$ part [13], so that instead of (36) one has

$$\mathcal{A} = \omega^a \mathbb{J}_a + A\mathbb{K} + \overline{\mathbb{Q}}\gamma_a e^a \psi - \overline{\psi} e^a \gamma_a \mathbb{Q}. \tag{37}$$

Here the generators are in a 3×3 matrix representation,

$$\mathbb{J}_a = -\frac{1}{2}\begin{bmatrix} \gamma_a & 0 \\ 0 & 0 \end{bmatrix}, \quad \mathbb{K} = i\begin{bmatrix} 1 & 0 & 0 \\ 0 & 1 & 0 \\ 0 & 0 & 2 \end{bmatrix}, \quad \mathbb{Q}^\alpha = \left\{\begin{bmatrix} 0 & 0 & 0 \\ 0 & 0 & 0 \\ 1 & 0 & 0 \end{bmatrix}, \begin{bmatrix} 0 & 0 & 0 \\ 0 & 0 & 0 \\ 0 & 1 & 0 \end{bmatrix}\right\}, \tag{38}$$

and $\overline{\mathbb{Q}}_\alpha = (\mathbb{Q}^\alpha)^\top$. The resulting superalgebra is $osp(2|2)$ and the corresponding CS action describes a complex Dirac field minimally coupled to a $U(1)$ gauge field in a curved geometry described by the Lagrangian

$$L(A,\omega,\psi) = AdA + \omega^a{}_b d\omega^b{}_a + \frac{2}{3}\omega^a{}_b\omega^b{}_c\omega^c{}_a - \overline{\psi}\left[\partial\!\!\!/ + iA\!\!\!/ - \frac{1}{4}\gamma_a \omega\!\!\!/^{ab}\gamma_b\right]\psi|e|d^3x - \overline{\psi}\psi e^a T_a\,. \tag{39}$$

Clearly, the recipe can be extended to dimensions $D = 2n+1$: take your favorite gauge group G, then define a supercharge $\mathbb{Q}^\alpha_K$, in a vector representation of G ($K = 1, 2, ..., s$) and a spin-1/2 representation of the Lorentz group ($\alpha = 1, 2, ..., 2^n$). The graded Lie algebra spanned by the supercharges together with Lorentz and G-generators can always be made to close by including additional bosonic generators. With this superalgebra, a connection one-form analogous to (36) can be constructed and with it the Chern-Simons form is uniquely defined. This CS Lagrangian is guaranteed to be quadratic and first order in fermions only for dimension three. In higher dimensions the CS form may include quartic or higher powers of the fermions.

The lack of CS forms in even dimensions, forces the next best option, the Yang-Mills Lagrangian for the superconnection. The resulting Lagrangian in four dimensions is the Dirac equation, minimally coupled to standard electromagnetism and Einstein gravity with positive cosmological constant, plus a four fermion Nambu–Jona-Lasinio term. The resulting theory has only a residual $SO(3,2) \times U(1)$ gauge symmetry and supersymmetry is explicitly broken [14].

6. Relation with graphene

Graphene is a remarkable system consisting of a one-atom thick layer of carbon atoms in a two-dimensional hexagonal array. This crystal is best described as a superposition of two dislocated triangular sublattices with an average of one free electron per atom that can hop between nearest neighbors [15].

Since nearest neighbor sites cannot be in the same triangular sublattice, hopping between nearest neighbors implies jumping between the two sublattices, which can be pictured as the destruction of an electron in one sublattice and simultaneous creation of one in the other. The electron field can be represented by a two-component spinor,

$$\psi = \begin{pmatrix} a \\ b \end{pmatrix}, \qquad \psi^\dagger = \begin{pmatrix} a^\dagger \\ b^\dagger \end{pmatrix}, \tag{40}$$

where a, b destroy electrons on each sublattice, and $a^\dagger$, $b^\dagger$ create them.

It can be shown that the dispersion relation for conduction electrons near the Fermi surface has the form of a massless relativistic particle, $E \simeq v_F|\mathbf{k}|$ [15, 16]. Hence, the action that describes the propagation of electrons in graphene is the Dirac Lagrangian for a massless spin-1/2 field, where the speed of light corresponds to the Fermi velocity, $v_F \sim c/300 = 1,000$ km/sec. (For comprehensive reviews on graphene see, e.g., [17, 18].)

Interestingly enough, a good description of graphene in the presence of electromagnetism is given by (39). If the geometry is fixed, one can ignore the dynamic features of the field ω. In that case, the Lorentz connection is a classical external field and therefore its variation in the action is irrelevant. Similarly, if the electromagnetic interaction comes from an external classical field, one can ignore the CS term AdA, assuming A to represent an external electromagnetic field in which the graphene sample is immersed. The fact that the CS term is intrinsically three-dimensional may seem incorrect from the point of view of the experimental setup, where the graphene sheet defines a 2+1 space embedded in a 3+1 laboratory and the electromagnetic field should therefore be described by the usual Maxwell action. However, in the presence of a weak magnetic field, the electrons in a graphene crystal behave as in the presence of a $U(1)$ field described by the CS Lagrangian like in (39) [19].

The fermionic character of the electrons in graphene is reflected in the Fermi-Dirac statistics of the creation and destruction operators for particles in each sublattice, $a, a^\dagger, b, b^\dagger$. The two components are not really related to the intrinsic angular momentum of the conduction electrons, but to the fact that the honeycomb array is not a single lattice but two interlocking sublattices.

The spin of the conduction electrons is an independent degree of freedom that requires a new label for the Dirac field. Its inclusion requires a field that can be denoted as ψ^α_a, where $\alpha = 1, 2$ correspond to the components displayed in (40), and $a = 1, 2$ is a new spin index that represents the two possible orientations along the z-axi. Clearly, the orientation of the "z axis" is conventional and nothing would change if one decides to modify its orientation arbitrarily at each lattice point. Thus, the inclusion of the spin degree of freedom requires introducing a local $SU(2)$ gauge freedom just as in the model described by (39)!

This idea of turning the freedom to choose locally the spin quantization axis and therefore inducing a coupling with an $SU(2)$ gauge field was examined in the context of the Hubbard model more than two decades ago [20].The $SU(2)$-mediated interaction between conduction electrons is likely to produce a strong correlation between electrons with antiparallel spins, analogous to the formation of Cooper pairs in superconductivity.

For certain values of the parameters of these black hole solution and the $SU(2)$ charge, the geometry admits globally defined Killing spinors. These configurations are therefore candidates for stable supersymmetric vacua of the theory. It would be interesting to see if there are experimental counterparts of these special states predicted by the field theoretic model.

7. Summary

We started out by observing that Lorentz flatness does not necessarily imply flat metric geometry. In particular, in three dimensions Lorentz flatness can correspond to anti-de Sitter spaces of arbitrary radii, including infinite radius, that is, Minkowski space. As has been known for some time now, locally AdS_3 spacetimes can take a number of different shapes, including black holes, static or rotating, so that Lorentz flat metrics may be quite nontrivial. This result can be seen as the Lorentzian counterpart of a classic theorem in geometry that states that a three sphere is parallelizable, i.e., it can be globally covered with a congruence of orthonormal frames. Another way of understanding a three-dimensional black hole then is as a geometry in which any simply connected patch can be covered by a congruence of "inertial observers", so that the Lorentz connection can be consistently chosen to vanish everywhere in that patch.

Since the three-dimensional Lorentz group is isomorphic to $SU(2)$, a black hole in 2+1 dimensions an also be endowed with a nontrivial $SU(2)$ hair. This solution with vanishing both Lorentz and $SU(2)$ curvatures, can be seen as a configuration in the matter-free sector of a system consisting of a Dirac field minimally coupled to a CS gauge field and CS gravity. This system can also be seen as a CS action constructed with a graded 1-form that combines bosonic and fermionic generators. The bosonic gauge fields are one-form connections, while the fermionic 1-forms are the product of a spin-1/2 zero-form and a dreibein.

Remarkably, the coupled CS $SU(2)$ and Dirac system is exactly what one would expect of a graphene action if the spin degree of freedom of the conduction electron were also taken into account. So, in the end, it may be that the vacuum sector of one of the simplest materials in nature, a two-dimensional array of identical atoms may have physical features that may resemble a 2+1 black hole with $SU(2)$ hair.

Acknowledgements
I have benefited from many enlightening discussions with my collaborators, Pedro Alvarez, Gastón Giribet, Luis Huerta, Cristián Martínez, Pablo Pais, Eduardo Rodríguez, Patricio Salgado and Mauricio Valenzuela. Also, very fruitful discussions with Nicolás Grandi, Marcelo Loewe, Guillermo Silva and David Valenzuela, have helped me enormously to understand what I report here. This work is partially supported by Fondecyt grant 1140155. The Centro de Estudios Científicos is funded by the Chilean Government through the Centers of Excellence Base Financing Program of Conicyt.

References
[1] Bañados M, Teitelboim C and Zanelli J 1992, The Black hole in three-dimensional space-time, *Phys. Rev. Lett.* **69** 1849. [hep-th/9204099]
[2] Bañados M, Henneaux M, Teitelboim C and Zanelli J 1993, Geometry of the (2+1) black hole, *Phys. Rev.* **D48** 1506. [gr-qc/9302012]
[3] Coussaert O and Henneaux M 1998, Selfdual solutions of (2+1) Einstein gravity with a negative cosmological constant in *The Black Hole, 25 Years Later*, edited by Teitelboim C and Zanelli J. World Scientific, Singapore. [hep-th/9407181]
[4] Deser S and Jackiw R 1984, Three-Dimensional Cosmological Gravity: Dynamics of Constant Curvature *Ann. Phys. (N.Y.)* **153** 405.
[5] Mišković O and Zanelli J 2009, Negative spectrum of the 2+1 Black Hole, *Phys. Rev.* **D79**105011. [arXiv:0904.0475. hep-th].
[6] Alvarez P D, Pais P, Rodríguez E, Salgado-Rebolledo P and Zanelli J 2014, The BTZ Black Hole as a Lorentz-flat Geometry, *Phys. Lett.* B **738**, 134.[arXiv:1405.6657. gr-qc]
[7] Adams J F 1958, On the Nonexistence of Elements of Hopf Invariant One, *Bull. Amer. Math. Soc.* **64**, 279
[8] Adams J F 1960,On the Nonexistence of Elements of Hopf Invariant One, *Ann. Math.* **72**, 20.
[9] Alvarez P D, Pais P, Rodríguez E, Salgado-Rebolledo P and Zanelli J 2015, (to appear).
[10] Weyl H, 1950, Remark on the coupling of gravitation and electron, *Phys. Rev.* **77**, 699.
[11] Coleman S and Mandula J 1967, All Possible Symmetries of the S Matrix, *Phys. Rev.* **159** 1251.
[12] Haag R, SohniusM and Lopuszański J T 1975, All possible generators of supersymmetries of the S-matrix, *Nucl. Phys.* **B 88**, 257.
[13] Alvarez P D, Valenzuela M and Zanelli J 2012, Supersymmetry of a different kind, JHEP **1204**, 058. [arXiv:1109.3944 (hep-th)]
[14] Alvarez P D, Pais P and Zanelli J 2014, Local supersymmetry without SUSY partners, *Phys. Lett.*, **B735**, 314. [arXiv:1306.1247 (hep-th)]

[15] Wallace P R 1947, Band Theory of Graphite, *Phys. Rev.* **71**, 622.
[16] Semenoff G W 1984, Condensed-Matter Simulation of a Three-Dimensional Anomaly, *Phys. Rev. Lett.* **53**, 2449.
[17] Guinea F, Katsnelson M I and Vozmediano M A H 2010, Gauge fields in graphene, *Phys. Rept.* **496**, 109.
[18] Iorio A, Lambiase G 2014, Quantum field theory in curved graphene spacetimes, Lobachevsky geometry, Weyl symmetry, Hawking effect, and all that, Phys. Rev. **D90**, 025006. [arXiv:1308.0265 (hep-th)]
[19] Valenzuela D, Hernández-Ortiz, Loewe M and Raya A 2014, Graphene transparency in weak magnetic fields, *J. Phys. A: Math. Gen.*, in press. [arXiv:1410.5501 (cond-mat.mes-hall)]
[20] Huerta L and Zanelli J 1993, Bose-Fermi Transformation in Three-Dimensional Space, *Phys. Rev. Lett.*, **71**, 3622. [cond-mat/9308006]

Journal of Physics: Conference Series **600** (2015) 012006 doi:10.1088/1742-6596/600/1/012006

Beyond Einstein's General Relativity

Francisco S. N. Lobo

Centro de Astronomia e Astrofísica da Universidade de Lisboa, Campo Grande, Edifício C8, 1749-016 Lisboa, Portugal
Instituto de Astrofísica e Ciências do Espaço, Universidade de Lisboa, OAL, Tapada da Ajuda, PT1349-018 Lisboa, Portugal.

E-mail: fslobo@fc.ul.pt

Abstract.
Modern astrophysical and cosmological models are plagued with two severe theoretical difficulties, namely, the dark energy and the dark matter problems. Relative to the former, high-precision observational data have confirmed with startling evidence that the Universe is undergoing a phase of accelerated expansion. This phase, one of the most important and challenging current problems in cosmology, represents a new imbalance in the governing gravitational equations. Several candidates, responsible for this expansion, have been proposed in the literature, in particular, dark energy models and modified gravity, amongst others. Outstanding questions are related to the nature of this so-called "dark energy" that is driving the acceleration of the universe, and whether it is due to the vacuum energy or a dynamical field. On the other hand, the late-time cosmic acceleration may be due to modifications of General Relativity, which introduce new degrees of freedom to the gravitational sector itself. We analyze some of the modified theories of gravity that address these intriguing and exciting problems facing modern physics, and explore the foundations of gravitation theory, essential for the construction of modified theories of gravity.

1. Introduction

1.1. Need for new gravitational physics?

A central theme in Cosmology is the perplexing fact that the Universe is currently undergoing an accelerating expansion [1]. In this context, during the last two decades Cosmology has evolved from being mainly a theoretical area of Physics to become a field supported by high precision observational data. Recent experiments call upon state of the art technology in Astronomy and in Astrophysics to provide detailed information on the contents and history of the Universe, which has led to the measuring of the parameters that describe our Universe with increasing precision. The standard model of cosmology is remarkably successful in accounting for the observed features of the Universe. However, there remain a number of fundamental open questions at the foundation of the standard model. In particular, we lack a fundamental understanding of the acceleration of the late universe. What is the so-called "dark energy" that is driving the acceleration of the universe? Is it a vacuum energy or a dynamical field? Or is the acceleration due to infra-red modifications of Einstein's theory of General Relativity (GR)? How is structure formation affected in these alternative scenarios? What will the outcome be of this acceleration for the future fate of the universe? Indeed, the fact that GR is facing so many challenges, namely, the difficulty in explaining particular observations, the incompatibility with other well established theories and the lack of uniqueness, may be indicative of a need for new gravitational physics.

The resolution of these fundamental questions is extremely important for theoretical cosmology, looking beyond the standard theory of gravity. The standard model of cosmology has favoured the dark energy models as fundamental candidates responsible for the cosmic accelerated expansion. However, it is clear that these questions involve not only gravity, but also particle physics. String theory provides a synthesis of these two branches of physics and is widely believed to be moving towards a viable quantum gravity theory. One of the key predictions of string theory is the existence of extra spatial dimensions. In the brane-world scenario, motivated by recent developments in string theory, the observed 3-dimensional universe is embedded in a higher-dimensional spacetime [2]. The new degrees of freedom belong to the gravitational sector, and can be responsible for the late-time cosmic acceleration [3, 4]. Of course, cosmology is also an ideal testing ground for GR (in particular, the problem of the late-time cosmic acceleration). Thus, a promising approach is to assume that at large scales GR breaks down, and a more general action describes the gravitational field.

1.2. Foundations of gravitation theory

GR is a classical theory, therefore no reference to an action is required [5]. Consider the Hilbert-Einstein action

$$S = \int d^4x \sqrt{-g} \left[\frac{R}{2\kappa^2} + L_m(g^{\mu\nu}, \psi) \right] , \tag{1}$$

where R is the curvature scalar and $L_m(g^{\mu\nu}, \psi)$ is the matter Lagrangian, in which matter is minimally coupled to the metric $g_{\mu\nu}$ and ψ collectively denotes the matter fields. However, the Lagrangian formulation is elegant, and has merits. First, at the quantum level, the action acquires a physical meaning, and a more fundamental theory of gravity will provide an effective gravitational action at a suitable limit. Secondly, it is easier to compare alternative gravitational theories through their actions rather than by their field equations. Thirdly, in many cases one has a better grasp of the physics as described through the action, by considering the couplings, kinetic and the dynamical terms.

In fact, Einstein, in deriving GR was not motivated to account for unexplained experimental results, but was driven by theoretical criteria of elegance and simplicity [6]. His primary goal was to produce a gravitation theory that incorporated the principle of equivalence and special relativity in a natural way. However, at the end of the day the theory had to be confronted with experiment, in particular, the "three classical tests": (i) Accounts for the perihelion advance of Mercury; (ii) Eddington's measurement of light deflection, in 1919; (iii) Pound and Rebka measure the gravitational redshift (in 1960).

It is interesting to note that in the late 1950s, Schiff and Dicke suggested that the gravitational redshift was not a true test of GR. It was purely a consequence of the equivalence principle, and did not test the field equations of gravitational theory. Note that by 1960, one may consider that the validity of GR rested upon the following empirical foundation [6] (i) One test of moderate precision (the perihelion shift of Mercury; precision: approx. 1%); (ii) One test of low precision (the deflection of light; precision: approx. 50%); (iii) One inconclusive test that was not a real test anyway (the gravitational redshift). In the meantime, alternative theories of gravitation were being constructed (Poincaré, Whitehead, Milne, Birkhoff, Belinfante, etc), that also laid claim to the viability of the above tests. This triggered the development of powerful tools for distinguishing and testing theories, such as the Parametrized Post-Newtonian (PPN) expansion, pioneered by Nordvedt and extended by Nordvedt and Will [6]. Indeed, the idea that experiments test principles and not specific theories, implies the need of exploring the conceptual basis of a gravitational theory.

Probably the most unbiased assumptions to start with, in developing a gravitation theory is the Dicke framework [6], which states that: (i) spacetime is a 4-dim manifold, with each point in the manifold corresponding to a physical event (note that a metric and affine connection is not necessary at this stage); (ii) the equations of gravity and the mathematical entities in them are to be expressed in a form that is independent of the coordinates used, i.e., in a covariant form. It is common to think of GR, or any other gravitation theory, as a set of field equations (or an action). However, a complete and coherent axiomatic formulation of GR, or any other gravitation theory, is still lacking. One needs to formulate "principles" and an important one is the covariance principal, present in the Dicke Framework [5].

Thus, the basic criteria for the viability of a Gravitation Theory, following C. Will [6], are the following. (i) It must be complete: The theory should be able to analyse from "first principles" the outcome of any experiment; (ii) It must be self-consistent: Predictions should be unique and independent of the calculation method; (iii) It must be relativistic: The theory should reduce to Special Relativity when gravity is "turned off"; (iv) It must have the correct Newtonian limit: In the limit of weak gravitational fields and slow motion it should reproduce Newton's laws.

Note also that the Einstein Equivalence Principle (EEP) is at the heart of gravitation theory [6]. Thus if the EEP is valid, then gravitation must be a curved spacetime phenomenon, i.e., it must obey the postulates of Metric Theories of Gravity: (i) Spacetime is endowed with a metric (second rank non-degenerate tensor); (ii) The world lines of test bodies are geodesics of that metric; (iii) In local freely falling frames, Lorentz frames, the non-gravitational laws of physics are those of Special Relativity.

1.3. Discriminating between dark energy and modified gravity models

On the other hand, generalizations of the Einstein-Hilbert Lagrangian, including quadratic Lagrangians, involving second order curvature invariants, such as R^2, $R_{\mu\nu}R^{\mu\nu}$, $R_{\alpha\beta\mu\nu}R^{\alpha\beta\mu\nu}$, $\varepsilon^{\alpha\beta\mu\nu}R_{\alpha\beta\gamma\delta}R^{\gamma\delta}_{\mu\nu}$, $C_{\alpha\beta\mu\nu}C^{\alpha\beta\mu\nu}$, have also been extensively explored [7, 8]. The main physical motivations for these modifications of gravity consist of a possibility of a more realistic representation of the gravitational fields near curvature singularities, and to create some first order approximation for the quantum theory of gravitational fields. While these modified theories of gravity, or dark gravity, offer an alternative explanation to the standard cosmological model for the expansion history of the universe [9], it offers a paradigm for nature fundamentally distinct from dark energy models of cosmic acceleration [10], even those that perfectly mimic the same expansion history. It is fundamental to understand how one may differentiate these modified theories of gravity from dark energy models.

An explicit distinction between models of "dark energy" and "modified gravity" is necessary. This is due to the fact that models consisting of modifications to the gravitational sector can be mapped into a specific subset of scalar-tensor theories, namely, to a class of interacting scalar field dark energy models universally coupled to the matter species. Thus, due to this ambiguity, a practical classification is needed. In fact, taking into account solely the expansion rate of the Universe, one cannot discriminate between the dark energy models and modified gravity. However, these two alternative models will affect structure formation differently. Therefore, information on the growth of structure, at different scales and redshifts, will break the degeneracy and will serve to discriminate between both models of dark energy and modified gravity. More specifically, using linear perturbation theory, these alternative models will affect differently the constraints imposed by the Einstein equations, and the second-order growth equation will be modified, consequently changing the growth factor [11]. Thus, generic modifications of the dynamics of scalar perturbations, with respect to the ΛCDM background, can be represented by the introduction of two new degrees of freedom in the Einstein constraint equations, by the functions $Q(a,k)$ and $\eta(a,k)$, where a is the scale factor and k the perturbation scale.

In the context of modified gravity, the function $Q(a,k)$ results from a mass-screening effect due to local modifications of gravity, and effectively modifies Newton's constant. In the context of dynamical dark energy models, the function $Q(a,k)$ incorporates additional clustering, or interaction with other fields, due to the perturbations. The function η, absent in ΛCDM, parameterizes the effective stresses due to the modification of gravity or specific dynamical dark energy models. Finally, the scale and time-dependence of both functions, Q and η, can be derived in the specific model considered and projected on a (Q,η) plane. Following [11], we denote the term "modified gravity" when there are additional contributions to the Poisson equation, which induces $Q \neq 1$, and where extra effective stresses arise, implying $\eta \neq 1$. Thus, "modified gravity" denotes models in which modifications are present in the gravitational sector and in which dark energy clusters or interacts with other fields. Thus, following this practical classification outlined in [11], in the context of first order perturbation theory, models with $Q = \eta = 1$ are denoted standard dark energy models, such as, a minimally-coupled scalar field with standard kinetic energy [11]. On the other hand, models for $Q \neq 1$ and $\eta \neq 1$ are denoted "modified gravity", such as scalar-tensor theories, $f(R)$ gravity [7, 8], the DGP model [3] and generalizations [4], massive gravity and generalized galileons [12], Horndeski interactions [13], bi-(multi-) gravity [14], etc. Thus, in the context of the EUCLID mission [11], the definitions of the functions Q and η are extremely convenient, for instance, EUCLID can distinguish between standard dynamical dark energy and modified gravity by forecasting the errors on Q and η, and several combinations of these functions, such as Q/η.

1.4. $f(R)$ gravity and extensions

In modified gravity, in particular, in $f(R)$ gravity, one may tackle the problem using the metric formalism, which consists in varying the action with respect to the metric, although other alternative approaches have been considered in the literature, namely, the Palatini formalism [15], where the metric and the connections are treated as separate variables; and the metric-affine formalism, where the matter part of the action now depends and is varied with respect to the connection [16]. Recently, a novel approach to modified theories of gravity that consists of adding to the Einstein-Hilbert Lagrangian an $f(R)$ term constructed à la Palatini, was proposed [17, 18]. It was shown that the theory can pass the Solar System observational constraints even if the scalar field is very light. These explicit models are consistent with local tests and lead to the late-time cosmic acceleration. This hybrid metric-Palatini theory will be briefly outlined and explored below.

The cosmological applications of generalized $f(R)$-type gravity models by assuming that the gravitational Lagrangian is given by an arbitrary function of the Ricci scalar R and of the matter Lagrangian L_m has also been extensively analysed [19, 20]. Specific models were explored in detail, and the gravitational field equations in the metric formalism, in the presence of a nonminimal coupling were obtained, as well as the equations of motion for test particles, which follow from the covariant divergence of the energy-momentum tensor [19, 20]. Generally, the motion is non-geodesic, and takes place in the presence of an extra force orthogonal to the four-velocity. The Newtonian limit of the equation of motion was also considered, and a procedure for obtaining the energy-momentum tensor of the matter was presented. On the other hand, the gravitational field equations are equivalent to the Einstein equations of the $f(R)$ model in empty spacetime, but differ from them, as well as from standard general relativity, in the presence of matter. Therefore the predictions of the model could lead to some major differences, as compared to the predictions of standard general relativity, or its extensions ignoring the role of matter, in several problems of current interest, such as cosmology, gravitational collapse or the generation of gravitational waves. The study of these phenomena may also provide some specific signatures and effects, which could distinguish and discriminate between the various theories of modified gravity.

1.5. Horndeski Lagrangian

In the context of scalar fields, these are popular building blocks used to construct models of present-day cosmological acceleration. They are appealing because such fields are ubiquitous in theories of high energy physics beyond the standard model, and, in particular, are present in theories which include extra spatial dimensions, such as those derived from string theories. The initial simplest model consisted of a minimally coupled single scalar field self-interacting through a scalar potential, and with canonical kinetic terms. This was named a quintessence field [21]. An interesting model described as coupled quintessence analysed the consequences of coupling the field responsible for acceleration to matter fields [22]. The model was rapidly extended to consider non-canonical kinetic terms, with higher powers of the field's velocity, denoted k-essence [23], and non-minimal couplings to gravity. Such models included scalar tensor theories and Gauss-Bonnet dark energy. A Lagrangian based on the Galileon symmetry, with second order derivative terms, has been shown to provide a working model with interesting cosmological implications [24, 25]. Given the large number of models, the question arises how we should study and compare them in a unified manner, and determine which if any is the origin of cosmic acceleration.

A particularly useful tool in this direction is the realisation that all these classes of models are special cases of the most general Lagrangian which leads to second order field equations. This Lagrangian was first written down by Horndeski in 1974 [13] and rediscovered by Deffayet *et al* in 2011 [26, 27]. It enables researchers to adopt a unifying framework, and to determine the regions within this general theory that have appealing theoretical properties. In combination with the need to fit observations such properties are helpful in preferring regions of this general theory, and hence particular models. One example of such an appealing theoretical property is the possibility that terms within the Horndeski Lagrangian can be used to partially explain the huge discrepancy between the value of the vacuum energy in particle physics, and the value of the cosmological constant as inferred from cosmological observations. In particular these terms have been investigated with the aim of finding a viable self-tuning mechanism, which screens the spacetime curvature from the net cosmological constant [28, 29, 30]. While not yet a full theory of cosmological evolution, such a mechanism is intriguing, and leads to the question of whether realistic cosmologies can be constructed which include this mechanism as well as the late time acceleration.

1.6. Massive gravity

Another gravitational modification that has recently attracted much interest is the massive gravity paradigm, where instead of introducing new scalar degrees of freedom such as in $f(R)$ gravity, it modifies the graviton itself. Massive gravity is a well-defined theoretical problem by its own and has important cosmological motivations, namely, if gravity is massive it will be weaker at large scales and thus one can obtain the late-time cosmic acceleration. Fierz and Pauli presented the first linear massive gravity. However, it was shown to suffer from the van Dam-Veltman-Zakharov (vDVZ) discontinuity [31], namely the massless limit of the results do not yield the massless theory, i.e., General Relativity. The incorporation of nonlinear terms cured the problem but introduced the Boulware-Deser (BD) ghost. This fundamental problem puzzled physicists until recently, where a specific nonlinear extension of massive gravity was proposed by de Rham, Gabadadze and Tolley (dRGT), in which the BD ghost is eliminated by a secondary Hamiltonian constraint [12]. This new nonlinear massive gravity has interesting cosmological implications, for instance, it can give rise to inflation, late-time acceleration, etc [12]. However, the basic versions of this theory exhibit instabilities at the perturbative level, and thus suitable extensions are necessary. These could be anisotropic versions, $f(R)$ extensions, bi-metric generalizations, partially-massive constructions, etc. An interesting possibility consists in a graviton mass varying function of a new scalar field [32], where the graviton mass would be effectively larger at early cosmological times, implying interesting phenomenology both at early and late times. The crucial issue is whether one can construct a massive gravity and cosmology that can be consistent as an alternative to dark energy or other models of modified gravity, and whether this theory is in agreement with high-accuracy cosmological data, such as the growth-index or the tensor-to-scalar ratio ones, remains to be explored in detail.

1.7. Modified teleparralel gravity and extensions

In modified gravitational theories one usually generalizes the Einstein-Hilbert action of General Relativity, namely, one starts from the curvature description of gravity. However, a different and interesting class of modified gravity arises when one generalizes the action of the equivalent formulation of GR based on torsion. As is well known, Einstein also constructed the "Teleparallel Equivalent of General Relativity" (TEGR) in which the gravitational field is described by the torsion tensor and not by the curvature. In TEGR the corresponding Lagrangian is given by the torsion scalar $\mathcal{T}$, and it results from contractions of the torsion tensor, in a similar way that the Einstein-Hilbert Lagrangian R results from contractions of the curvature (Riemann) tensor. Thus, in analogy with $f(R)$ gravity, one can start from TEGR and construct the $f(\mathcal{T})$ modified theories of gravity, by extending $\mathcal{T}$ to an arbitrary function in the Lagrangian [33]. The interesting feature is that although TEGR is completely equivalent to General Relativity at the level of the field equations, $f(\mathcal{T})$ differs radically from $f(R)$ gravity, in that they form different gravitational modifications. Hence, $f(\mathcal{T})$ gravity has novel and interesting cosmological implications, capable of describing inflation, the late-time acceleration [34], large scale structure, bouncing solutions, etc. There are many open issues in $f(\mathcal{T})$ gravity and cosmology. In particular, one may extend this class of theories by allowing for non-minimal couplings to matter, such as $f(L_m, \mathcal{T})$ [35] and $f(T, \mathcal{T})$ [36] theories, where here T is the trace of stress-energy tensor, and study their constraints from local tests and cosmology [37], focusing on the perturbations [38]. Additionally, one may further explore the theoretical issues of the extra degrees of freedom of the theory and the different vierbein choices.

1.8. Outline of the paper

In this work, we briefly consider the generalized curvature-matter couplings in Section 2, by analysing interesting aspects and the phenomenology of the theory. In Section 3, we present the recent and novel approach to modified theories of gravity, denoted hybrid metric-Palatini gravity, that consists of adding to the Einstein-Hilbert Lagrangian an $f(R)$ term constructed à la Palatini. In Section 4, we consider the generalized virial theorem in the scalar-tensor representation of the hybrid metric-Palatini gravity, which can be an efficient tool in observationally testing the viability of this class of generalized gravity models.

2. $f(R)$ gravity and extensions

2.1. $f(R)$ gravity

For instance, consider $f(R)$ gravity, for simplicity, which has an appealing feature, namely, it combines mathematical simplicity and a fair amount of generality [7]. The action is given by

$$S = \int d^4x\,\sqrt{-g}\left[\frac{f(R)}{2\kappa^2} + L_m(g^{\mu\nu}, \psi)\right] , \tag{2}$$

where $f(R)$ is an arbitrary function of the Ricci scalar R. Varying the action with respect to the metric $g_{\mu\nu}$ yields the field equations, given by

$$F(R)R_{\mu\nu} - \frac{1}{2}f(R)g_{\mu\nu} - \nabla_\mu\nabla_\nu\, F(R) + g_{\mu\nu}\nabla^\alpha\nabla_\alpha F(R) = \kappa^2 T^{(m)}_{\mu\nu} , \tag{3}$$

where we have denoted $F(R) = f'(R)$, and the prime represents the derivative with respect to the scalar curvature. The matter energy-momentum tensor is defined as

$$T_{\mu\nu}^{(m)} = -\frac{2}{\sqrt{-g}} \frac{\delta(\sqrt{-g}\, L_m)}{\delta(g^{\mu\nu})} \,. \tag{4}$$

Note that the Ricci scalar is a dynamical degree of freedom, which is transparent from the trace equation given by $FR - 2f + 3\nabla_\mu \nabla^\mu F = \kappa T$ (where $F = df/dR$). This introduces a new light scalar degree of freedom, which consequently produces a late-time cosmic acceleration. However, the light scalar strongly violates the Solar System constraints and the way out of this problem is through the 'chameleon' mechanism, i.e., the scalar field becomes massive in the Solar System, as mentioned in the Introduction. The approaches to $f(R)$ gravity consist in the metric, Palatini, and metric-affine formalisms (and one may mention a fourth approach, i.e., the hybrid metric-Palatini formalism which will be explored below, in some detail).

2.2. Geons in quadratic Palatini gravity

We mention an interesting application in the Palatini formalism, by considering the action [39]:

$$S[g,\Gamma,\psi_m] = \frac{1}{2\kappa^2} \int d^4x \sqrt{-g} \left[R + l_P^2 \left(aR^2 + R_{\mu\nu}R^{\mu\nu} \right) \right] - \frac{1}{16\pi} \int d^4x \sqrt{-g} F_{\mu\nu}F^{\mu\nu} \,, \tag{5}$$

where $\kappa^2 \equiv 8\pi G/c^4$, l_P^2 represents the Planck length squared, a is a free parameter, $F_{\mu\nu}$ is the electromagnetic field strength, $g_{\mu\nu}$ is the space-time metric, $R = g^{\mu\nu}R_{\mu\nu}$, $R_{\mu\nu} = R^\rho{}_{\mu\rho\nu} = R_{\nu\mu}$, and $R^\alpha{}_{\beta\mu\nu} = \partial_\mu \Gamma^\alpha_{\nu\beta} - \partial_\nu \Gamma^\alpha_{\mu\beta} + \Gamma^\alpha_{\mu\lambda}\Gamma^\lambda_{\nu\beta} - \Gamma^\alpha_{\nu\lambda}\Gamma^\lambda_{\mu\beta}$. The connection $\Gamma^\alpha_{\mu\nu}$ is *a priori* independent of the metric (Palatini formalism) and must be determined by the field equations.

More specifically, it was found that the end state of black hole evaporation can be represented by non-singular and without event horizon stable solitonic remnants with masses of the order the Planck scale [39]. Though these objects are locally indistinguishable from spherically symmetric, massive electric (or magnetic) charges, they turn out to be sourceless geons containing a wormhole generated by the electromagnetic field. The results were obtained by interpreting semiclassical corrections to Einstein's theory in the first-order (Palatini) formalism, which yields second-order equations and avoids the instabilities of the usual (metric) formulation of quadratic gravity.

This construction has interesting applications. For instance, it has recently been suggested that Einstein-Rosen (ER) bridges can be interpreted as maximally entangled states of two black holes that form a complex Einstein-Podolsky-Rosen (EPR) pair [40], in the context of the firewall debate [41]. This relationship has been dubbed as the $ER = EPR$ correlation. The latter conjecture was also considered in the context of the quadratic Palatini theory [42]. An important result, which stems from the underlying assumptions about the geometry on which the theory is constructed, is the fact that all the charged solutions of the quadratic Palatini theory possess a wormhole structure. The results show that spacetime may have a foam-like microstructure with wormholes generated by fluctuations of the quantum vacuum. This involves the spontaneous creation/annihilation of entangled particle-antiparticle pairs, existing in a maximally entangled state connected by a non-traversable wormhole. Since the particles are produced from the vacuum and therefore exist in a singlet state, they are necessarily entangled with one another, which provides further support to the $ER = EPR$ claim.

Furthermore, the collapse of a charged radiation fluid in a Planck-suppressed quadratic extension of GR was also considered [43]. Exact analytical solutions were obtained that extend the charged Vaidya-type solution of GR, which allows to explore in detail new physics at the Planck scale. Starting from Minkowski space, it was found that the collapsing fluid generates wormholes supported by the electric field. The relevance of the findings in relation to the quantum foam structure of space-time and the meaning of curvature divergences in this theory was also discussed. The above analysis was extended in [44], where the physical significance of curvature divergences in this theory and the topology change issue was discussed. This supports the view that space-time could have a foam-like microstructure pervaded by wormholes generated by quantum gravitational effects.

2.3. Nonminimal curvature-matter coupling

$f(R)$ modified theories of gravity can be generalized by introducing in the action a linear nonminimal coupling between matter and geometry, given by the following action

$$S = \int \left\{\frac{1}{2}f_1(R) + [1+\lambda f_2(R)]\, L_m\right\} \sqrt{-g}\, d^4x \,, \tag{6}$$

where $f_i(R)$ (with $i = 1, 2$) are arbitrary functions of the Ricci scalar R and L_m is the matter Lagrangian density [19] .

Varying the action with respect to the metric $g_{\mu\nu}$ yields the field equations, given by

$$\begin{aligned} F_1(R)R_{\mu\nu} - \frac{1}{2}f_1(R)g_{\mu\nu} - \nabla_\mu\nabla_\nu\, F_1(R) + g_{\mu\nu}\nabla_\alpha\nabla^\alpha F_1(R) = -2\lambda F_2(R)L_m R_{\mu\nu} \\ +2\lambda(\nabla_\mu\nabla_\nu - g_{\mu\nu}\nabla_\alpha\nabla^\alpha)L_m F_2(R) + [1+\lambda f_2(R)]T^{(m)}_{\mu\nu} \,, \end{aligned} \tag{7}$$

where we have denoted $F_i(R) = f_i'(R)$, and the prime represents the derivative with respect to the scalar curvature.

One verifies the non-conservation of the energy-momentum tensor

$$\nabla^\mu T^{(m)}_{\mu\nu} = \frac{\lambda F_2}{1+\lambda f_2}\left[g_{\mu\nu}L_m - T^{(m)}_{\mu\nu}\right]\nabla^\mu R \,. \tag{8}$$

Thus, the coupling between the matter and the higher derivative curvature terms describes an exchange of energy and momentum between both. Analogous couplings arise after a conformal transformation in scalar-tensor theories of gravity (and string theory). In the absence of the coupling, one verifies the conservation of the energy-momentum, which can also be verified from the diffeomorphism invariance of the matter part of the action.

In order to test the motion in the model, consider a perfect fluid, so that the equation of motion for a fluid element is given by

$$\frac{Du^\mu}{ds} \equiv \frac{du^\mu}{ds} + \Gamma^\mu_{\alpha\beta}u^\alpha u^\beta = f^\mu \,, \tag{9}$$

where f^μ is an extra-force given by

$$f^\mu \;=\; \frac{1}{\rho+p}\left[\frac{\lambda F_2}{1+\lambda f_2}\,(L_m - p)\,\nabla_\nu R + \nabla_\nu p\right] h^{\mu\nu} \,. \tag{10}$$

with $h_{\mu\lambda} = g_{\mu\lambda} + u_\mu u_\lambda$ the projection operator. The extra force f^μ is orthogonal to the four-velocity of the particle, $f^\mu u_\mu = 0$.

A particularly intriguing feature is that the extra force depends on the form of the Lagrangian density. Note that considering the Lagrangian density $L_m = p$, where p is the pressure, the extra-force vanishes [45]. However, it has been argued that this is not the unique choice for the matter Lagrangian density and that more natural forms for L_m, such as $L_m = -\rho$, do not imply the vanishing of the extra-force. Indeed, in the presence of the nonminimal coupling, they give rise to two distinct theories with different predictions [46]. We refer the reader to [47, 48] for a more detailed discussion.

2.4. Generalized curvature-matter couplings

2.4.1. $f(R, L_m)$ gravity: One may further generalize the linear curvature-matter coupling by assuming that the gravitational Lagrangian is given by an arbitrary function of the Ricci scalar R and of the matter Lagrangian L_m [20]. This essentially consists in a maximal extension of the Hilbert-Einstein action, which takes the following form

$$S = \int f(R, L_m) \sqrt{-g}\, d^4x \,. \tag{11}$$

Varying the action with respect to the metric, yields the following field equation

$$\begin{aligned} & f_R(R, L_m) R_{\mu\nu} + (g_{\mu\nu}\nabla_\alpha\nabla_\alpha - \nabla_\mu\nabla_\nu) f_R(R, L_m) \\ & -\frac{1}{2}\left[f(R, L_m) - f_{L_m}(R, L_m) L_m\right] g_{\mu\nu} = \frac{1}{2} f_{L_m}(R, L_m) T_{\mu\nu}. \end{aligned} \tag{12}$$

Considering $f(R, L_m) = R/2 + L_m$, the Hilbert-Einstein Lagrangian, we recover the standard Einstein field equation of general relativity, $R_{\mu\nu} - (1/2)g_{\mu\nu}R = T_{\mu\nu}$.

This theory possess extremely interesting properties, such as the covariant divergence of the energy-momentum tensor is non-zero

$$\nabla^\mu T_{\mu\nu} = 2\nabla^\mu \ln\left[f_{L_m}(R, L_m)\right] \frac{\partial L_m}{\partial g^{\mu\nu}} \,. \tag{13}$$

The requirement of $\nabla^\mu T_{\mu\nu} = 0$ yields an effective functional relation between the matter Lagrangian density and the function $f_{L_m}(R, L_m)$, given by

$$\nabla^\mu \ln\left[f_{L_m}(R, L_m)\right] \partial L_m/\partial g^{\mu\nu} = 0 \,. \tag{14}$$

Note that considering a specific matter Lagrangian density, by an appropriate choice of the function $f(R, L_m)$, one can construct, at least in principle, conservative models with arbitrary curvature-matter dependence. We refer the reader to [48] for more details.

2.4.2. $f(R,T)$ gravity: Consider another extension of GR, namely, $f(R,T)$ gravity, where the gravitational Lagrangian is given by an arbitrary function of the Ricci scalar R and of the trace of the energy-momentum tensor T [49]. The action takes the following form

$$S = \frac{1}{2\kappa^2}\int f(R,T)\sqrt{-g}\, d^4x + \int L_m\sqrt{-g}\, d^4x \,. \tag{15}$$

Note that the dependence from T may be induced by exotic imperfect fluids or quantum effects (conformal anomaly). This theory may be considered a relativistically covariant model of interacting dark energy. The $f(R,T)$ gravitational model mentioned above has been given a great amount of recent attention and we refer the reader to [48, 49] for more details. In particular, we mention a short-coming of the theory, in [50] using the standard energy-momentum conservation equation, the matter density perturbations yields different results from the ones derived in the frame of the concordance ΛCDM model, which severely constrains the viability of such theories.

2.4.3. $f(R, T, R_{\mu\nu}T^{\mu\nu})$ gravity: The $f(R,T)$ gravitational theory [49] for the specific case of a traceless energy-momentum tensor, $T = 0$, for instance, when the electromagnetic field is involved, reduces to $f(R)$ gravity and all non-minimal couplings of gravity to the matter field vanish. This fact motivated a further generalization of $f(R,T)$ gravity that consists in including an explicit first order coupling between the matter energy-momentum $T_{\mu\nu}$ and the Ricci tensor

[51, 52]. It is interesting to note that in contrast to $f(R,T)$ gravity, for $T = 0$, this extra coupling still has a non-minimal coupling to the electromagnetic field via the $R_{\mu\nu}T^{\mu\nu}$ coupling term in the action, which is non-zero in general.

The action, is given by

$$S = \frac{1}{2\kappa^2}\int d^4x\sqrt{-g}f\left(R,T,R_{\mu\nu}T^{\mu\nu}\right) + \int d^4x\sqrt{-g}L_m\,. \tag{16}$$

The only requirement imposed on the function $f\left(R,T,R_{\mu\nu}T^{\mu\nu}\right)$ is that it is an arbitrary analytical function in all arguments, although it is interesting to note that only linear contributions of the form $R_{\mu\nu}T^{\mu\nu}$ may escape the Ostrogradski instability [53]. We refer the reader to [51, 52] for more details on this theory.

3. Hybrid metric-Palatini gravity

Recently, a novel approach to modified theories of gravity that consists of adding to the Einstein-Hilbert Lagrangian an $f(R)$ term constructed à la Palatini, was proposed [17]. It was shown that the theory can satisfy the Solar System observational constraints even if the scalar field is very light. This implies the existence of a long-range scalar field, which is able to modify the cosmological and galactic dynamics, but leaves the Solar System unaffected. These explicit models are consistent with local tests and lead to the late-time cosmic acceleration, and also verify the absence of instabilities in perturbations. Criteria to obtain cosmic acceleration were discussed and the field equations were also analyzed as a dynamical system [18]. In this context, several classes of dynamical cosmological solutions, depending on the functional form of the effective scalar field potential, describing both accelerating and decelerating Universes were explicitly obtained. Furthermore, the cosmological perturbation equations were derived and applied to uncover the nature of the propagating scalar degree of freedom and the signatures these models predict in the large-scale structure [18]. Thus, this novel model proves to have extremely interesting properties, and deserves a further in depth study.

The action of the hybrid metric-Palatini theory is given by [17, 18]

$$S = \frac{1}{2\kappa^2}\int d^4x\sqrt{-g}\left[R + f(\mathcal{R})\right] + S_m\,, \tag{17}$$

where S_m is the matter action, $\kappa^2 \equiv 8\pi G$, R is the Einstein-Hilbert term, $\mathcal{R} \equiv g^{\mu\nu}\mathcal{R}_{\mu\nu}$ is the Palatini curvature, and $\mathcal{R}_{\mu\nu}$ is defined in terms of an independent connection $\hat{\Gamma}^\alpha_{\mu\nu}$ as $\mathcal{R}_{\mu\nu} \equiv \hat{\Gamma}^\alpha_{\mu\nu,\alpha} - \hat{\Gamma}^\alpha_{\mu\alpha,\nu} + \hat{\Gamma}^\alpha_{\alpha\lambda}\hat{\Gamma}^\lambda_{\mu\nu} - \hat{\Gamma}^\alpha_{\mu\lambda}\hat{\Gamma}^\lambda_{\alpha\nu}$.

The action (17) may be expressed as the following scalar-tensor theory [17]

$$S = \int \frac{d^4x\sqrt{-g}}{2\kappa^2}\left[(1+\phi)R + \frac{3}{2\phi}\partial_\mu\phi\partial^\mu\phi - V(\phi)\right] + S_m. \tag{18}$$

This action differs from the $w = -3/2$ Brans-Dicke theory in the coupling of the scalar to the curvature, which in the $w = -3/2$ theory is of the form ϕR. This simple modification will have important physical consequences.

3.1. Weak-field, slow-motion behaviour

Taking into account the effects of the scalar field ϕ on the solar system dynamics by studying the weak-field limit of the field equations, we consider an expansion of the metric and the scalar field about a cosmological solution. This sets the asymptotic boundary values, using a quasi-Minkowskian coordinate system, from which one deduces the effective Newton constant $G_{\rm eff}$ and the post-Newtonian parameter (PPN) γ, which are given by

$$G_{\rm eff} \equiv \frac{G}{1+\phi_0}\left[1 - (\phi_0/3)\,e^{-m_\varphi r}\right]\,, \qquad \gamma \equiv \frac{1 + (\phi_0/3)\,e^{-m_\varphi r}}{1 - (\phi_0/3)\,e^{-m_\varphi r}}\,, \tag{19}$$

respectively (we refer the reader to [17] for specific details). As is clear from the above expressions, the coupling of the scalar field to the local system depends on the amplitude of the background value ϕ_0. If ϕ_0 is small, then $G_{\rm eff} \approx G$ and $\gamma \approx 1$ regardless of the value of the effective mass m_φ^2.

This contrasts with the result obtained in the metric version of $f(R)$ theories:

$$G_{\rm eff} \equiv G\left(1 + e^{-m_f r}/3\right)/\phi_0\,, \qquad \gamma \equiv \left(1 - \frac{e^{-m_f r}}{3}\right) \Big/ \left(1 + \frac{e^{-m_f r}}{3}\right), \tag{20}$$

which requires a large mass $m_f^2 \equiv (\phi V_{\phi\phi} - V_\phi)/3$ to make the Yukawa-type corrections negligible in local experiments.

3.2. Late-time cosmic speedup

As a specific example of modified cosmological dynamics, consider the spatially flat Friedman-Robertson-Walker (FRW) metric, $ds^2 = -dt^2 + a^2(t)d\mathbf{x}^2$. The modified Friedmann equations are provided by

$$3H^2 = \frac{1}{1+\phi}\left[\kappa^2\rho + \frac{V}{2} - 3\dot{\phi}\left(H + \frac{\dot{\phi}}{4\phi}\right)\right], \tag{21}$$

$$2\dot{H} = \frac{1}{1+\phi}\left[-\kappa^2(\rho+P) + H\dot{\phi} + \frac{3}{2}\frac{\dot{\phi}^2}{\phi} - \ddot{\phi}\right], \tag{22}$$

respectively, and the scalar field equation is given by

$$\ddot{\phi} + 3H\dot{\phi} - \frac{\dot{\phi}^2}{2\phi} + \frac{\phi}{3}[2V - (1+\phi)V_\phi] = -\frac{\phi\kappa^2}{3}(\rho - 3P)\,. \tag{23}$$

The qualitative behavior of the scalar field can be read directly from the above by rewriting it as follows

$$\ddot{\phi} + 3H\dot{\phi} - \frac{\dot{\phi}^2}{2\phi} + M_\phi^2(T)\phi = 0\,, \tag{24}$$

where $T = -(\rho - 3P)$ and we have defined $M_\phi^2(T)$ as

$$M_\phi^2(T) \equiv m_\phi^2 - \frac{1}{3}\kappa^2 T = \frac{1}{3}[2V - (1+\phi)V_\phi - \kappa^2 T]\,, \tag{25}$$

We now propose a model that is consistent at Solar System and cosmological scales, and which is constructed on grounds of mathematical simplicity. Consider, for instance, the following potential

$$V(\phi) = V_0 + V_1\phi^2\,. \tag{26}$$

The trace of the field equation automatically implies $R = -\kappa^2 T + 2V_0$. As $T \to 0$ with the cosmic expansion, the solution naturally evolves into a de Sitter phase ($V_0 \sim \Lambda$) for consistency with observations. If V_1 is positive, the de Sitter regime represents the minimum of the potential. Note that the effective mass for local experiments, $m_\varphi^2 = 2(V_0 - 2V_1\phi)/3$, is positive if $\phi < V_0/V_1$. For $V_1 \gg V_0$, the amplitude is small enough to pass Solar System tests. Thus, the exact de Sitter solution is compatible with the dynamics of the scalar field in this model.

Journal of Physics: Conference Series **600** (2015) 012006 doi:10.1088/1742-6596/600/1/012006

4. Dark matter in modified gravity and the generalized virial theorem

In the context of dark matter, two observations, namely, the behaviour of the galactic rotation curves and the mass discrepancy in galactic clusters, suggest the existence of a (non or weakly interacting) form of dark matter at galactic and extra-galactic scales. The gravitational masses of galaxy clusters are estimated by assuming a hydrostatic equilibrium of both the hot intra-cluster gas and of the galaxies with the binding cluster potential. Therefore, the total mass of a cluster of galaxies can be estimated in two ways: First, by taking into account the dynamical motions of the member galaxies of the cluster, and with the application of the virial theorem, one obtains an estimate M_V for the mass of the cluster. Second, the total baryonic mass M_B can be determined by adding the mass of each individual galaxy member of the cluster. The mass discrepancy at the galactic cluster level arises as observations show that M_V is much greater than M_B, with typical values of $M_V/M_B \sim 20-30$.

Thus, generally, the dynamics of test particles around galaxies, as well as the corresponding mass deficit, is explained by postulating the existence of a hypothetical dark matter. In fact, the behavior of the rotation curves shows the existence of a constant velocity region, near the baryonic matter distribution, followed by a quick decay at large distances. In [54], the possibility was considered that the behavior of the rotational velocities of test particles gravitating around galaxies can be explained within the framework of hybrid metric-Palatini gravity.

In the intermediate galactic scale, it was shown explicitly that in the hybrid metric-Palatini model the tangential velocity can be explicitly obtained as a function of the scalar field of the equivalent scalar-tensor description. All the physical and geometrical quantities and the numerical parameters in the hybrid metric-Palatini model can be expressed in terms of observable/measurable parameters, such as the tangential velocity, the baryonic mass of the galaxy, the Doppler frequency shifts, and the stellar dispersion velocity, respectively. Therefore, the results obtained open the possibility of testing the hybrid metric-Palatini gravitational models at the galactic or extra-galactic scale by using direct astrophysical observations. We refer the reader to related work in other modified theories of gravity [55].

Here, we consider the generalized virial theorem in the scalar-tensor representation of the hybrid metric-Palatini gravity, which can be an efficient tool in observationally testing the viability of this class of generalized gravity models [56].

4.1. Generalized virial theorem in hybrid metric-Palatini gravity

Consider a self-gravitating system of identical, collisionless point particles in random motion. To obtain the basic field equations we will use the scalar-tensor representation of hybrid metric-Palatini gravity, given by action (18), which allows a clear physical interpretation of the model.

The metric of an isolated spherically symmetric cluster is given by

$$ds^2 = -e^{\nu(r)}dt^2 + e^{\lambda(r)}dr^2 + r^2\left(d\theta^2 + \sin^2\theta d\varphi^2\right). \tag{27}$$

The galaxies in the cluster are considered identical and collisionless point particles, and their space-time distribution is described by a distribution function f_B.

The distribution function f_B obeys the general relativistic Boltzmann equation, which is the transport equation for the distribution function for a system of particles in a curved arbitrary Riemannian space-time. The general relativistic Boltzmann equation is provided by

$$\left(p^\alpha \frac{\partial}{\partial x^\alpha} - p^\alpha p^\beta \Gamma^i_{\alpha\beta}\frac{\partial}{\partial p^i}\right) f_B = 0, \tag{28}$$

where p^α is the four-momentum of the particle, and $\Gamma^i_{\alpha\beta}$ are the Christoffel symbols associated to the metric. Note that the collissionless Boltzmann equation states that the local phase space density viewed by an observer co-moving with a star or galaxy is conserved.

Thus, the energy-momentum tensor of matter is given by

$$T_{\mu\nu} = \int f_B \, m \, u_\mu u_\nu \, du, \tag{29}$$

where m is the mass of the particle (galaxy), $u_\mu = (u_t, u_r, u_\theta, u_\varphi)$ is the four-velocity of the galaxy, with u_t denoting the temporal component, and $du = du_r du_\theta du_\varphi / u_t$ is the invariant volume element of the velocity space.

The energy-momentum tensor $T_{\mu\nu}$ can be represented in terms of an effective density $\rho_{\rm eff}$ and of two effective anisotropic pressures, the radial $p_{\rm eff}^{(r)}$ and the tangential $p_{\rm eff}^{(\perp)}$ pressures, respectively, given by

$$\rho_{\rm eff} = \rho \left\langle u_t^2 \right\rangle, p_{\rm eff}^{(r)} = \rho \left\langle u_r^2 \right\rangle, p_{\rm eff}^{(\perp)} = \rho \left\langle u_\theta^2 \right\rangle = \rho \left\langle u_\varphi^2 \right\rangle, \tag{30}$$

where ρ is the mass density of the ordinary baryonic matter, and $\langle u_i^2 \rangle$, $i = t, r, \theta, \varphi$ is the average value of u_i^2, $i = t, r, \theta, \varphi$.

By using this form of the energy-momentum tensor, and summing up the components of the gravitational field equations describing a cluster of galaxies in hybrid metric-Palatini gravity, provides the following relation

$$e^{-\lambda} \left(\frac{\nu''}{2} + \frac{\nu'^2}{4} + \frac{\nu'}{r} - \frac{\nu'\lambda'}{4} \right) \simeq 4\pi G \rho \left\langle u^2 \right\rangle + 4\pi G \rho_\phi^{(eff)}, \tag{31}$$

and the relativistic Boltzmann equation finally takes the form

$$\int_0^R 4\pi\rho \left[\left\langle u_1^2 \right\rangle + \left\langle u_2^2 \right\rangle + \left\langle u_3^2 \right\rangle \right] r^2 dr - \frac{1}{2} \int_0^R 4\pi r^3 \rho \left[\left\langle u_0^2 \right\rangle + \left\langle u_1^2 \right\rangle \right] \frac{\partial \nu}{\partial r} dr = 0. \tag{32}$$

We refer the reader to [56] for explicit details.

4.2. Geometric quantities

Since we are interested in astrophysical applications at the extra-galactic level, we may assume that the deviations from standard general relativity are small, i.e., $\phi \ll 1$. We assume that the approximations that apply to test particles in stable circular motion around galaxies, also apply to galactic clusters. First, assume ν and λ are slowly varying functions of r (i.e. ν' and λ' are small), so that all the quadratic terms are neglected. Secondly, assume that the motion of the galaxies is non-relativistic, so that they have velocities much smaller than the velocity of the light, i.e., $\langle u_1^2 \rangle \approx \langle u_2^2 \rangle \approx \langle u_3^2 \rangle \ll \langle u_0^2 \rangle \approx 1$.

Thus, the gravitational field equations reduce to

$$\frac{1}{2r^2} \frac{\partial}{\partial r} \left(r^2 \frac{\partial \nu}{\partial r} \right) = 4\pi G \rho + 4\pi G \rho_\phi^{(eff)}, \tag{33}$$

and the relativistic Boltzmann equation to

$$2K - \frac{1}{2} \int_0^R 4\pi r^3 \rho \frac{\partial \nu}{\partial r} dr = 0, \tag{34}$$

respectively, where

$$K = \int_0^R 2\pi\rho \left[\left\langle u_1^2 \right\rangle + \left\langle u_2^2 \right\rangle + \left\langle u_3^2 \right\rangle \right] r^2 dr, \tag{35}$$

is the total kinetic energy of the galaxies.

Consider that the gravitational potential energies of the cluster are defined as:

$$\Omega_B = -\int_0^R \frac{GM_B(r)}{r}\, dM_B(r), \qquad \Omega_\phi^{(eff)} = \int_0^R \frac{GM_\phi^{(eff)}(r)}{r}\, dM_B(r), \tag{36}$$

respectively, where R is the cluster radius. Since in hybrid metric-Palatini gravity, the quantity $M_\phi^{(eff)}$ has essentially a geometric origin, we tentatively denote it as the *geometric mass* of the cluster, defined as

$$M_\phi^{(eff)}(r) = 4\pi \int_0^r \rho_\phi^{(eff)}(r')r'^2 dr'. \tag{37}$$

Using the above equations, we arrive at the generalization of the virial theorem, which takes the familiar form

$$2K + \Omega = 0\,, \tag{38}$$

where Ω is total gravitational potential energy of the system defined by $\Omega = \Omega_B - \Omega_\phi^{(eff)}$, which contains a term consisting of a geometric origin, $\Omega_\phi^{(eff)}$.

The generalized virial theorem, can be represented in a more transparent physical form if we introduce the radii R_V and R_ϕ, defined by

$$R_V = M_B^2 \Big/ \int_0^R \frac{M_B(r)}{r}\, dM_B(r), \qquad R_\phi^{(eff)} = \left[M_\phi^{(eff)}\right]^2 \Big/ \int_0^R \frac{M_\phi^{(eff)}(r)}{r}\, dM_B(r), \tag{39}$$

respectively. In analogy to the geometric mass considered above, the quantity R_ϕ may be denoted as the *geometric radius* of the cluster of galaxies.

Thus, the baryonic potential energy Ω_B and the effective scalar field potential energy $\Omega_\phi^{(eff)}$ are finally given by

$$\Omega_B = -\frac{GM_B^2}{R_V}, \qquad \Omega_\phi^{(eff)} = \frac{G\left[M_\phi^{(eff)}\right]^2}{R_\phi^{(eff)}}, \tag{40}$$

respectively.

Now, we define the virial mass M_V of the cluster of galaxies as

$$2K = \frac{GM_B M_V}{R_V}. \tag{41}$$

After substitution into the virial theorem, we obtain the following relation between the virial and the baryonic mass of the galaxy cluster

$$\frac{M_V}{M_B} = 1 + \frac{\left[M_\phi^{(eff)}\right]^2 R_V}{M_B^2 R_\phi^{(eff)}}. \tag{42}$$

If $M_V/M_B > 3$, a condition which holds for most of the observed galactic clusters, then Eq. (42) provides the virial mass in hybrid metric-Palatini gravity, which can be approximated as

$$M_V \approx \frac{\left[M_\phi^{(eff)}\right]^2}{M_B} \frac{R_V}{R_\phi^{(eff)}}. \tag{43}$$

In the present model there is also a strict proportionality between the virial mass of the cluster and its baryonic mass, a relation which can also be tested observationally. From the point of view of the astrophysical observations the virial mass M_V is determined from the study of the velocity dispersion σ_r^2 of the stars and of the galaxies in the clusters. According to the virial theorem in hybrid metric-Palatini gravity, most of the mass in a cluster with mass M_{tot} is in the form of the geometric mass $M_\phi^{(eff)}$, so that $M_\phi^{(eff)} \approx M_{tot}$.

Thus, since galaxy clusters are "dark" matter dominated objects, the main contribution to their mass comes from the geometric mass M_ϕ, so that with a very good approximation we have $M_\phi \approx M_V \approx M_{tot}$. Therefore the virial theorem immediately provides the following mass scaling relation

$$M_V \approx M \frac{R_\phi}{R_V}. \tag{44}$$

This equation shows that the virial mass is proportional to the baryonic (normal) mass of the cluster, and that the ratio of the total mass and of baryonic mass is determined by a purely geometric quantity, the geometric radius R_ϕ. Hence the geometric radius of the cluster can be determined from observations, once the virial and baryonic masses and the virial radius, respectively, are known. We refer the reader to [56] for more details.

5. Conclusions

While dark gravity offers an alternative explanation to the standard cosmological model for the expansion history of the universe, it offers a paradigm for nature fundamentally distinct from dark energy models of cosmic acceleration, even those that perfectly mimic the same expansion history. It is fundamental to understand how one may differentiate these modified theories of gravity from dark energy models. All modified gravity models induce observational signatures at the post-Newtonian level, which are translated by the parameterized post-Newtonian (PPN) metric coefficients arising from these extensions of GR. Tests from the solar system, large scale structure and lensing, as well as laboratory and space-based Equivalence Principle experiments, essentially restrict the range of allowed modified gravity models.

Surveys such as the EUCLID space telescope, the Square Kilometre Array (SKA) radio telescope, the Dark Energy Survey (DES), and the Extended Baryon Oscillation Spectroscopic Survey (eBOSS) as part of the Sloan Digital Sky Survey III (SDSS) will provide new opportunities to test the different cosmological models. Indeed, with the wealth of unprecedented high precision observational data that will become available by these upcoming and planned surveys, we are dawning in a golden age of cosmology, which offers a window into understanding the perplexing nature of the cosmic acceleration, dark matter and of gravity itself.

Acknowledgements

I would like to deeply thank all my collaborators and co-athours related to the work presented in this paper. I am supported by a Fundação para a Ciência e Tecnologia Investigador FCT Research contract, with reference IF/00859/2012, funded by FCT/MCTES (Portugal), and acknowledge financial support of the Fundação para a Ciência e Tecnologia through the grant EXPL/FIS-AST/1608/2013.

References

[1] S. Perlmutter et al. [Supernova Cosmology Project Collaboration], Astrophys. J. 517, 565 (1999); A.G. Riess et al. [Supernova Search Team Collaboration], Astron. J. 116, 1009 (1998).
[2] R. Maartens, "Brane-world gravity," Living Rev. Rel. 7, 7 (2004).
[3] G.R. Dvali, G. Gabadadze and M. Porrati, Phys. Lett. B 485, 208 (2000).
[4] C. de Rham, S. Hofmann, J. Khoury and A.J. Tolley, JCAP 0802, 011 (2008).
[5] T. P. Sotiriou, "Modified Actions for Gravity: Theory and Phenomenology," arXiv:0710.4438 [gr-qc].

[6] C. M. Will, "Theory and experiment in gravitational physics," Cambridge, UK: Univ. Pr. (1993) 380 p
[7] T.P. Sotiriou and V. Faraoni, Rev. Mod. Phys. 82, 451 (2010); A. De Felice and S. Tsujikawa, Living Rev. Rel. 13, 3 (2010); S. Capozziello and M. De Laurentis, Phys. Rept. 509, 167 (2011); F. S. N. Lobo, Dark Energy-Current Advances and Ideas, **173-204**, Research Signpost, ISBN 978 (2009)
[8] S. Nojiri and S.D. Odintsov, Phys. Rept. 505, 59 (2011).
[9] S.M. Carroll, V. Duvvuri, M. Trodden and M.S. Turner, Phys. Rev. D 70, 043528 (2004).
[10] E.J. Copeland, M. Sami and S. Tsujikawa, Int. J. Mod. Phys. D 15, 1753 (2006).
[11] L. Amendola et al. [Euclid Theory Working Group Collaboration], Living Rev. Rel. 16, 6 (2013).
[12] C. de Rham, "Massive Gravity," to appear in Living Rev. Rel., arXiv:1401.4173 [hep-th].
[13] G.W. Horndeski, Int. J. Theor. Phys., 10, 363 (1974).
[14] M.S. Volkov, Class. Quant. Grav. 30, 184009 (2013).
[15] G.J. Olmo, Int. J. Mod. Phys. D 20, 413 (2011).
[16] T.P. Sotiriou and S. Liberati, Annals Phys. 322, 935 (2007).
[17] T. Harko, T. Koivisto, F. Lobo and G. Olmo, Phys. Rev. D 85, 084016 (2012).
[18] S. Capozziello, T. Harko, T. S. Koivisto, F.S.N. Lobo and G.J. Olmo. JCAP 1304, 011 (2013)
[19] O. Bertolami, C.G. Boehmer, T. Harko and F.S.N. Lobo, Phys. Rev. D 75, 104016 (2007).
[20] T. Harko and F. S. N. Lobo, Eur. Phys. J. C **70**, 373 (2010)
[21] I. Zlatev, Li-M. Wang, P.J. Steinhardt, Phys.Rev.Lett. 82 (1999) 896-899.
[22] L. Amendola, Phys.Rev. D62 (2000) 043511.
[23] C. Armendariz-Picon, Viatcheslav F. Mukhanov, Paul J. Steinhardt, Phys.Rev. D63 (2001) 103510
[24] A. Nicolis, R. Rattazzi, E. Trincherini, Phys.Rev. D79 (2009) 064036
[25] A. De Felice, S. Tsujikawa, Phys.Rev. D84 (2011) 124029
[26] C. Deffayet, X. Gao, D.A. Steer, G. Zahariade, Phys.Rev. D84 (2011) 064039.
[27] T. Kobayashi, M. Yamaguchi, J. Yokoyama, Prog.Theor.Phys. 126 (2011) 511-529.
[28] C. Charmousis, E.J. Copeland, A. Padilla, P.M. Saffin, Phys.Rev.Lett. 108 (2012) 051101.
[29] C. Charmousis, E.J. Copeland, A. Padilla, P.M. Saffin, Phys.Rev. D85 (2012) 104040.
[30] E.J. Copeland, A. Padilla, P.M. Saffin, JCAP 1212 (2012) 026.
[31] H. van Dam, M. Veltman, Nucl. Phys. B22, 397 411 (1970); V. Zakharov, JETP Lett. 12, 312 (1970).
[32] W. J. Percival, et al, Mon. Not. Roy. Astron. Soc. 381, 1053 (2007).
[33] R. Ferraro and F. Fiorini, Phys. Rev. D 75, 084031 (2007).
[34] E.V. Linder, Phys. Rev. D 81, 127301 (2010).
[35] T. Harko, F. S. N. Lobo, G. Otalora and E. N. Saridakis, Phys. Rev. D **89**, 124036 (2014).
[36] T. Harko, F. S. N. Lobo, G. Otalora and E. N. Saridakis, to appear in JCAP arXiv:1405.0519 [gr-qc].
[37] Y.-F. Cai, S.-H. Chen, J.B. Dent, S. Dutta and E.N. Saridakis, Class. Quant. Grav. 28, 215011 (2011).
[38] J.B. Dent et al , JCAP 1101, 009 (2011); S.-H. Chen et al, Phys. Rev. D 83, 023508 (2011).
[39] F. S. N. Lobo, G. J. Olmo and D. Rubiera-Garcia, JCAP **1307**, 011 (2013).
[40] J. Maldacena and L. Susskind, Fortsch. Phys. **61**, 781 (2013)
[41] S. L. Braunstein, S. Pirandola and K. yczkowski, Phys. Rev. Lett. **110**, no. 10, 101301 (2013); A. Almheiri, D. Marolf, J. Polchinski, and J. Sully, JHEP **1302**, 062 (2013); A. Almheiri, D. Marolf, J. Polchinski, D. Stanford, and J. Sully, JHEP **1309**, 018 (2013).
[42] F. S. N. Lobo, G. J. Olmo and D. Rubiera-Garcia, Eur. Phys. J. C **74**, no. 6, 2924 (2014)
[43] F. S. N. Lobo, J. Martinez-Asencio, G. J. Olmo and D. Rubiera-Garcia, Phys. Lett. B **731**, 163 (2014).
[44] F. S. N. Lobo, J. Martinez-Asencio, G. J. Olmo and D. Rubiera-Garcia, Phys. Rev. D **90**, 024033 (2014).
[45] O. Bertolami, F. S. N. Lobo and J. Paramos, Phys. Rev. D **78**, 064036 (2008).
[46] V. Faraoni, Phys. Rev. D **80**, 124040 (2009).
[47] O. Bertolami, J. Paramos, T. Harko and F. S. N. Lobo, arXiv:0811.2876 [gr-qc].
[48] T. Harko and F. S. N. Lobo, Galaxies 2 (2014) 3, 410-465
[49] T. Harko, F. S. N. Lobo, S. Nojiri and S. D. Odintsov, Phys. Rev. D **84**, 024020 (2011)
[50] F. G. Alvarenga, A. de la Cruz-Dombriz, M. J. S. Houndjo, M. E. Rodrigues and D. Sáez-Gómez, Phys. Rev. D **87**, no. 10, 103526 (2013).
[51] Z. Haghani, T. Harko, F. S. N. Lobo, H. R. Sepangi and S. Shahidi, Phys. Rev. D **88**, no. 4, 044023 (2013).
[52] S. D. Odintsov and D. Sáez-Gómez, Phys. Lett. B **725**, 437 (2013).
[53] I. Ayuso, J. B. Jimenez and A. de la Cruz Dombriz, arXiv:1411.1636 [hep-th].
[54] S. Capozziello, T. Harko, T. S. Koivisto, F. S. N. Lobo and G. J. Olmo, Astropart. Phys. **50-52**, 65 (2013).
[55] C. G. Boehmer, T. Harko and F. S. N. Lobo, Astropart. Phys. **29**, 386 (2008); C. G. Boehmer, T. Harko and F. S. N. Lobo, JCAP **0803**, 024 (2008); T. Harko, F. S. N. Lobo, M. K. Mak and S. V. Sushkov, Mod. Phys. Lett. A **29**, no. 9, 1450049 (2014); M. K. Mak and T. Harko, Phys. Rev. D **70**, 024010 (2004); T. Harko and K. S. Cheng, Phys. Rev. D **76**, 044013 (2007); T. Harko, Phys. Rev. D **81**, 084050 (2010).
[56] S. Capozziello, T. Harko, T. S. Koivisto, F. S. N. Lobo and G. J. Olmo, JCAP **1307**, 024 (2013).

The universe could be dark, ma non troppo

Miguel Portilla

Departament d'Astronomía i Astrofísica, Universitat de València, C/Dr. Moliner, Burjassot, València, Spain.

E-mail: `miguel.portilla@uv.es`

Abstract.
We will present new insights into the dark phenomena, i.e., the recent acceleration of the universe linked to a sort of dark energy, and to the unknown dark matter. The study is based only on Einsteins equations without cosmological constant, and on ordinary matter described as point masses. We will be limited, thereby, to the post -recombination epoch. We shall revise the outspread statement that a universe made of colission-less particles is well represented by dust, i.e; by an Einstein-de Sitter universe. Using well known results on the N-body problem expressed as a infinite series, and starting at zero order with the empty Milne universe, we shall get the aforementioned EdS universe at the first order, but at the next one we shall obtain a cosmological model whose energy density could explain the dark phenomena. No exotic dark components are necessary in principle, but we need to know the redshift of formation of the dominant particles in the present epoch, that we identify with the galaxies. Thus, assuming that redshift to be of the order of eleven, we shall get that the time evolution of the acceleration, and the supernovae luminous distance-redshift relation, are indistinguishable from the ones predicted by the ΛCDM model. However, if there was no realistic evolution model that could justify such an early galaxy formation epoch, then some quantity of dark energy would be necessary, ma non troppo.

1. Introduction

The necessity of a mysterious energy accounting for the recent acceleration of the universe comes from the outspread statement that a universe dominated by collision-less gravitating particles, should be described by a pressure-less perfect fluid, known as Einstein de Sitter universe (EdS); because, without a certain quantity of negative pressure, the universe will be by contrary decelerated. On the other hand, exotic matter is necessary for understanding the discrepancy between the content of mass traced by light and the traced by its gravitational effects. In this paper we accomplish a revision of the mentioned statement in two steps. We describe first properly the gravitational interaction of massive particles, arranged in a way so as satisfy the requirement of local isotropy everywhere, and then, we study the appropriate averaging in order to get a macroscopic description of the system of particles.

Our view is based on the fact that the energy and the metric tensors are untangled in the Einstein equations, and, their non linearity makes the averaging of the microscopic equations much more involved than in the linear case, as for example in the pure electromagnetic case. The average of both the energy and the metric tensors should be compatible with Einstein's equations, if we want they represent the system of auto gravitating particles. In section 3, we use a well known iterative method to obtain the metric and the energy tensor for a system of

Journal of Physics: Conference Series **600** (2015) 012007 doi:10.1088/1742-6596/600/1/012007

gravitating particles. In this way the average of the metric at each tieration is compatible with average of the energy tensor. We obtain the Einstein de Sitter universe in the first iteration , but the energy tensor coresponding to the second iteration is able to explain completely the observed dark energy phenomena (Subs. 5.1); and also, in a model dependent way, we can explain the dark matter content at cosmological scales (Subs. 5.2); and even, accepting an heuristic view, we can reproduce some aspects of the dark matter at galactic scales (Subs. 5.4).

This paper is on the wave of what generally is referred as backreaction. In the measure that we take care of the fact that, at each iteration, the averaged metric and the averaged energy tensor must satisfy the non linear Einstein equations, we are obtaining the backreaction of the metric on the macroscopic energy tensor. But, in the backreaction literature [2], the universe is considered as the real inhomogeneous universe, the averaging process is generally an spatial average, in order to reach an homogenous universe. One study the backreaction of the small scale structure onto the behavior of the universe at great scale. The difference with our view is that we consider an unreal homogeneous and isotropic universe, made of particles, and study the pass to the continuous description by means of an statistical average. In our case it is the backreaction of the metric on the macroscopic energy tensor what is relevant.

2. Gravitating point particles in a cosmological setting

Let us start recalling the main features of the Milne's cosmological model. It is a portion of a Minkowkian space, formed by a future light cone and a set of world lines inside it, whose intersection points with an hyperboloid of constant Minkowski proper time is a realization of an homogeneous Poisson process. Now we imagine that at some initial proper time τ_i the world lines represent the center of mass of galaxies, that will be considered as point particles. Havas-Goldberg [1] developed a method of solving the Einstein's equations for a finite number of point particles, taking as energy tensor a distribution with support over the parametrized world lines of the particles. They introduced the Minkowski metric as an auxiliary one and used as parameter the Minkowskian proper time. The energy tensor takes the form $\sqrt{-g}T^{\rho\sigma} = \sum_{j=1}^{N} \int M_j(\tau_j) v_j^\rho v_j^\sigma \delta^{(4)}(x^\rho - z_j^\rho(\tau_j)) d\tau_j$, with $v_j^\rho = \frac{dz_j^\rho}{d\tau_j}$. An iterative method allows determining the terms of the series for the metric $g_{\alpha\beta} = \eta_{\alpha\beta} + g_{\alpha\beta}^{(1)} + g_{\alpha\beta}^{(2)} \cdots$, and the mass function: $M_j(\tau_j) = M_j^{(0)}(\tau_j) + M_j^{(1)}(\tau_j) + M_j^{(2)}(\tau_j) \cdots$ Using de Donder coordinate conditions one put Einstein equations in the form

$$\sum_{m=1}^{n} (\Box\gamma_{\mu\nu}^{(m)} + 2\Lambda_{\mu\nu}^{(m)}) = -16\pi G \sum_{m=0}^{n-1} \overline{T}_{\mu\nu}^{(m)} \ , \quad \Lambda_{\mu\nu}^{(1)} = 0 \tag{1}$$

$$\sum_{m=1}^{n} \eta^{\nu\lambda} \partial_\lambda \gamma_{\mu\nu}^{(m)} = 0, \tag{2}$$

where the box represents the Dalambertian operator related to the auxiliary Minkowski metric, and appear two basic tensors:

$$\overline{T}_{\mu\nu}^{(m)} := \left(\sqrt{-g}T_\mu^\alpha\right)^{(m)} \eta_{\alpha\nu} \tag{3}$$

$$\gamma_{\mu\nu}^{(m)} := g_{\mu\nu}^{(m)} - \frac{1}{2}\eta_{\mu\nu}\eta^{\alpha\beta} g_{\alpha\beta}^{(m)} \tag{4}$$

The first order equations are

$$\Box\gamma_{\mu\nu}^{(1)} = -16\pi G \overline{T}_{\mu\nu}^{(0)} \tag{5}$$

$$\eta^{\nu\lambda} \partial_\lambda \gamma_{\mu\nu}^{(1)} = 0 \tag{6}$$

The field equation to first order gives the metric

$$g^{(1)}_{\mu\nu}(x) = -4Gm\sum_{j=1}^{N}\frac{(\eta_{\mu\alpha}\eta_{\nu\beta}v^{\alpha}_{j}v^{\beta}_{j} - \frac{1}{2}\eta_{\mu\nu})_r}{\eta(x - z_j, v_j)_r} \tag{7}$$

and the coordinate condition produces the motion and the mass to order zero:

$$\frac{dv^{\alpha}_{j}}{d\tau_j} = 0\ , \quad \dot{M}^{(0)}_{j} = 0 \quad \Rightarrow \quad M^{(0)}_{j} = m_j \tag{8}$$

The second order equations are

$$\Box(\gamma^{(1)}_{\mu\nu} + \gamma^{(2)}_{\mu\nu}) + 2\Lambda^{(2)}_{\mu\nu} = -16\pi G(\overline{T}^{(0)}_{\mu\nu} + \overline{T}^{(1)}_{\mu\nu}) \tag{9}$$

$$\eta^{\nu\lambda}\partial_{\lambda}(\gamma^{(1)}_{\mu\nu} + \gamma^{(2)}_{\mu\nu}) = 0 \tag{10}$$

the coordinate condition produces the motion and the mass to order one:

$$\frac{d}{d\tau_j}(mv^{\rho}_{j}(\eta_{\mu\rho} + g^{(1)}_{\mu\rho}) + M^{(1)}_{j}v^{\rho}_{j}\eta_{\rho\mu}) - \frac{1}{2}mv^{\rho}_{j}v^{\sigma}_{j}\partial_{x^{\mu}_{j}}g^{(1)}_{\rho\sigma} = 0 \tag{11}$$

$$\dot{M}^{(1)}_{j} = -\frac{1}{2}\frac{d}{d\tau_j}(mg^{(1)}_{\alpha\beta}v^{\alpha}_{j}v^{\beta}_{j}) \tag{12}$$

We shall assume the same mass for all the particles, hence from the last equation we get the mass function in the form

$$M^{(1)} = -\frac{1}{2}mg^{(1)}_{\alpha\beta}v^{\alpha}v^{\beta} + C^{(1)}\ , \quad C^{(1)} = m\alpha \tag{13}$$

where we have written the constant of integration in the form $m\alpha$. This is the basic result for solving the dark problems, and we proceed to obtain the sum appearing in the expression of the metric. A regularization is necessary because the metric is infinite on the world lines of the particles. This problem was solved with different methods and give for the metric on a particle k the expression

$$g^{(1)}_{\mu\nu}(z_k(\tau)) = -4Gm\sum_{j\neq k}^{n}\frac{(\eta_{\mu\alpha}\eta_{\nu\beta}v^{\alpha}_{j}v^{\beta}_{j} - \frac{1}{2}\eta_{\mu\nu})_r}{\eta(x - z_j, v_j)_r} + 4m(\dot{v}^{\alpha}_{k}v^{\beta}_{k} + v^{\alpha}_{k}\dot{v}^{\beta}_{k})\eta_{\mu\alpha}\eta_{\nu\beta}. \tag{14}$$

A major simplification comes from the fact that the world lines of the particles are straight lines, for $(z^0 = \tau, z^i = 0)$ is solution of the equations of motion (8,11), therefore only the sum in the right hand of (14) should be considered. Moreover, as all the world lines are equivalent due to the Lorentz invariance of the evolution equations, we only need to calculate the physical metric over a world line of reference L, and, as we want to obtain the mass function given in (13) we only need the component $g^{(1)}_{00}$ of the metric tensor, because using coordinates in which the particle of reference is at rest $v^a = 0$ for $a = 1, 2, 3$. Then the metric at any point p_o over the line of reference L after the birth of the galaxies will be the result of the sum

$$g^{(1)}_{00}(p_o) = -Gm\sum_{j}\frac{\gamma^2_j - 1/2}{r_j\gamma_j(1 + \bar{v}_j)} \tag{15}$$

At any point p_o over the reference line, the sum in equation (15), is always over a finite number of particles, even if the universe is made of infinite particles, for only intervene the world lines intersecting the past light cone of p_o limited by the initial hyperboloid. We assume that the intersection of the world lines of the particles with the hyperbolid $\tau = \tau_i$ define a uniform Poisson random process, the sum becomes an integral over the hyperboloid. This integral is in fact the statistical average. A change of variables (see appendix) allow us to integrate over the hyperplane $t = const.$.

$$\langle g_{00}^{(1)}(p_o)\rangle = -4GNm\int_0^{r_{mx}} \frac{t}{(t^2-r^2)^2}\frac{\gamma_j^2-1/2}{r_j\gamma_j(1+\bar{v}_j)}4\pi r^2 dr \tag{16}$$

The statistical average of the g_{00} metric component can be found in the Appendix of [2]

$$\langle g_{00}^{(1)}\rangle = -\frac{\Omega_{bai}}{4}H_0^2 t^2 g(t,t_i)\ , \quad \langle g_{0i}^{(1)}\rangle = 0 \tag{17}$$

$$g(t,t_i) = \left(1+\frac{t_i^2}{t^2}\right)^3 - 4\frac{t_i^2}{t^2}\left(\frac{3}{2}(1+\frac{t_i^2}{t^2}) - \frac{t_i}{t}\right) \tag{18}$$

$$\frac{dt}{ds} = 1-\frac{1}{2}\langle g_{00}^{(1)}\rangle = 1+\frac{\Omega_{bai}}{8}H_0^2 t^2 g(t,t_i) \tag{19}$$

The function $g(t,t_i)$ and its first derivative vanishes at the initial time t_i, and tends rapidly to unity when t increases. This is relevant to explain the acceleration, without use of a cosmological constant, as we will show bellow. Substituting into (13) we get the statistical average of the mass, up to the first order:

$$M(t) = m\frac{dt}{ds} + C^{(1)}\ , \quad C^{(1)} \equiv m\alpha \tag{20}$$

$$M(t) = m\left(1+\alpha+\frac{\Omega_{bai}}{8}H_0^2 t^2 g(t,t_i)\right) \tag{21}$$

This is all we need from the N-body gravitational problem. Now we pass to averaging the field equations in order to obtain a cosmological model.

3. The macroscopic energy tensor

In this section we will make the transition from a discrete system to a continuum and that implies to define a kind of averaging of the "microscopic" equations, in our case, the field equations (1). We have already given in Eq. (17) the statistical average of the first order metric. The average of non linear quantities as $\Lambda_{\mu\nu}^{(2)}$ reduces to the product of averages of first order quantities, for our statistical process is a uniform Poisson random process. It remains to define the average of distributional tensor densities of the kind $\sqrt{-g}P^{\mu\nu} = \sum\int p^{\mu\nu}(\lambda)\delta^{(4)}(x^\rho - z^\rho(\lambda))d\lambda$ with support on world lines of particles, where the parameter λ stand for the Minkowskian time τ or for the proper time s. Two comments will help to introduce a convenient procedure:

- The world lines of the particles are straight lines at each iteration.
- The surface $\tau = const.$ coincides with a surface $s = const.$; but with $\tau > s$, as can be derived from the expression of dt/ds given above in (19).

Let A_x be a convenient neighborhood of the point x, defined by two neighbor hypersurfaces: $\tau = t + \Delta t/2$, $\tau = t - \Delta t/2$ (that are also hypersurfaces of constant proper time: $s = s(t) + \Delta s/2, s = s(t) - \Delta s/2$) and a thin time like cone; and let S be the intersection of the hypersurface $\tau = t$ with A_x as shown in Figure 1.

Let φ_x be the characteristic function of the set A_x: $\varphi_x(u) = 1$ if $u \in A_x$, $\varphi_x(u) = 0$ otherwise. The average of a distributional tensor density $\sqrt{-g}P^{\mu\nu}$ is defined as follows

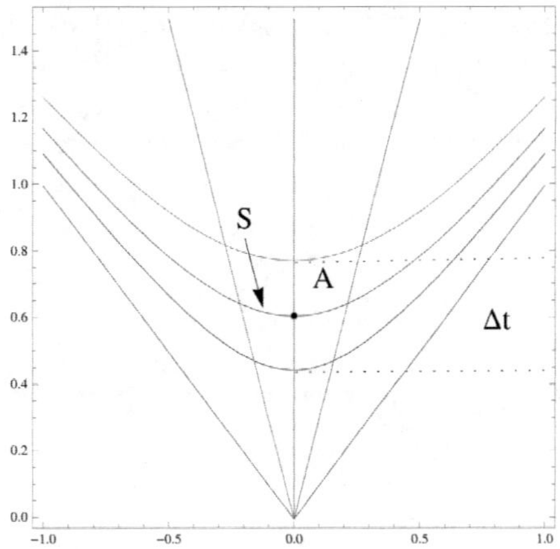

Figure 1: Element of volume A_x centered at a point over the reference line L, limited by two surfaces $\tau = const$. We show the 3-dimensional surface S necessary to get the density.

$$\langle\sqrt{-g}\; P^{\mu\nu}\rangle(x) = \lim_{\Delta\lambda\to 0}\frac{(\sqrt{-g}\; P^{\mu\nu},\varphi_x)}{Vol^{(3)}(S)\Delta\lambda}, \tag{22}$$

where the numerator is the action of the distribution on the characteristic function, and λ denotes τ or s. This average coincides with the statistical average of the quantities $p^{\mu\nu}$ in the case of a uniform random Poisson process over the surface $\lambda = const$.

Let us consider the distribution $\sqrt{-g}T^{\rho\sigma}_s = \sum_{j=1}^N \int m_j u^\rho_j u^\sigma_j \delta^{(4)}(x^\rho - z^\rho_j(s_j))ds_j$, $u^\rho_j = \frac{dz^\rho_j}{ds_j}$, where now the support are curves parametrized with the proper time. The value that this distribution takes on a test function is $(\sqrt{-g}T^{\rho\sigma}_s,\varphi) = \sum_{j=1}^N \int m_j u^\rho_j u^\sigma_j \varphi(s_j)s_j$, with $\varphi(s_j) = \varphi(z_j(s_j))$. The statistical average is $\langle\sqrt{-g}T^{\rho\sigma}_s\rangle(x) = \lim_{\Delta s\to 0}\frac{(\sqrt{-g}T^{\rho\sigma},\varphi_x)}{Vol^{(3)}(S)\Delta s} = nmu^\rho u^\sigma$. Therefore, taking this tensor as the macroscopic energy tensor for collision-less gravitating particles we get an Einstein de Sitter universe.

But, the macroscopic energy tensor should be compatible with the macroscopic metric tensor:

$$\sum_{m=1}^{n}(\Box\langle\gamma^{(m)}_{\mu\nu}\rangle + 2\langle\Lambda^{(m)}_{\mu\nu}\rangle) = -16\pi G\sum_{m=0}^{n-1}\langle\overline{T}^{(m)}_{(\tau)\mu\nu}\rangle \tag{23}$$

Therefore, we shall take the average of the tensor $\overline{T}^1_{\mu\nu} = \sum_{m=0}^1 \overline{T}^{(m)}_{\mu\nu}$, which according to definition (3) is $\overline{T}^1_{\mu\nu} := \sum_{m=0}^1(\sqrt{-g}T^\alpha_\mu)^{(m)}\eta_{\alpha\nu}$. The energy density for a universe made of collision-less particles is then: $\rho_{micro} = \overline{T}_{\alpha\beta}u^\alpha u^\beta = \overline{T}_{00}u^0u^0$. On the other hand $\overline{T}_{00} = \sqrt{-g}T^0_0 = g_{00}\sqrt{-g}T^{00}$, so, $\rho_{micro} = \sqrt{-g}T^{00}$. Averaging this quantity we get the macroscopic energy density $\rho = \langle\rho_{micro}\rangle$

$$\rho = \lim_{\Delta t\to 0}\frac{(\sqrt{-g}T^{oo},\varphi_x)}{Vol^{(3)}(S)\Delta t} = n(t)M(t) \tag{24}$$

with the mass given in Eq. (21).

$$\rho(t) = n(t)m\left(1 + \alpha + \frac{\Omega_{bai}}{8}H_0^2t^2g(t,t_i)\right) \tag{25}$$

Journal of Physics: Conference Series **600** (2015) 012007 doi:10.1088/1742-6596/600/1/012007

4. Construction of a sequence of FLRW cosmological models

Let us first recall some basic facts. Any space-time which admits a motion pattern, i.e., a set of world lines defining an observer, for which there exist local isotropy at each point, is a FLRW cosmological model [3]; and any FLRW model is completely characterized by a choice for the index of curvature k, and giving the energy density as a function of the expansion factor $\rho_F(a)$, that we shall write in the form

$$\rho_F(a) = \frac{3H_0^2}{8\pi G}\left(\frac{\Omega_m}{a^3} + f(a)\right), \tag{26}$$

The Friedmann equation

$$\frac{da}{ds} = H_0\sqrt{\frac{\Omega_m}{a} + \Omega_k + a^2 f(a)}\ , \quad \Omega_k = -k/H_0^2 R_0^2 \tag{27}$$

determines the expansion factor as function of the cosmological time s. Henceforth we shall take $k = 0$. Standard calculations produce the pressure and the acceleration of the model

$$p(a) = \frac{3H_0^2}{8\pi G}\left(-f(a) + \frac{1}{3}a\frac{df}{da}\right) \tag{28}$$

$$\frac{\ddot{a}}{a} = -H_0^2\left(\frac{\Omega_m}{2a^3} - f(a) + \frac{1}{2}a\frac{df}{da}\right) \tag{29}$$

In the two previous sections we have considered an algorithm $\eta \to \overline{T}^{(0)} \to g^{(1)} \to \overline{T}^{(1)} \to g^{(2)} \to \cdots$ which allows to obtain a sequence of approximate space-times :

$$\left(\mathsf{M}^+, \eta + \langle \mathsf{g}^{(1)}\rangle, \langle\overline{\mathsf{T}}^{(0)}\rangle\right), \left(\mathsf{M}^+, \eta + \sum_{\mathsf{n}=1}^{2}\langle \mathsf{g}^{(\mathsf{n})}\rangle, \sum_{\mathsf{k}=0}^{1}\langle\overline{\mathsf{T}}^{(\mathsf{k})}\rangle\right)\left(\mathsf{M}^+, \eta + \sum_{\mathsf{n}=1}^{\infty}\langle \mathsf{g}^{(\mathsf{n})}\rangle, \sum_{\mathsf{k}=0}^{\infty}\langle\overline{\mathsf{T}}^{(\mathsf{k})}\rangle\right) \tag{30}$$

In fact, we have obtained the first, and partially the second one, because we know the macroscopic energy tensor to the first oder but not the metric to second order $g^{(2)}$. If the series $\sum_{k=1}^{\infty} g^{(k)}$ converges to a metric g^{lim} it will be diffeomorfic to a FLRW metric g_F in the region $s \geq s_i$, because by construction it will be locally isotropic everywhere. We will not need to determine the whole change of coordinates $t = t(s, \bar{x}^m)$, $x^i = x^i(s, \bar{x}^m)$ that transforms de Donder coordinates (t, x^m) into the standard cosmological coordinates $(s, \bar{x}^m)$; the problem is quite simple if one takes into account that:

- We know that necessarily $n(t(a)) = n_o/a^3$.
- We only need the function $t = t(a)$ which relates the time like de Donder coordinate to the expansion factor of the universe; for, on the reference world line L we know $t = t(s)$, given by Eq. (19), and $s = s(a)$ given by Eq. (27).

We can get the function $t = t(a)$ as a series $\sum_{k=0}^{\infty} t^{(k)}(a)$, though as we will show here we shall only need the term $t^{(0)}$. Let us write (19) in the form

$$\frac{dt}{ds} = \sum_{k=0}^{\infty} F^{(k)}(t)\ , \quad F^{(0)}(t) = 1\ , \quad F^{(1)}(t) = -\frac{1}{2}\langle g_{00}^{(1)}\rangle \tag{31}$$

from Eq. (27), with $k = 0$, we get $\frac{ds}{da} = \frac{1}{H_o a}\left(\frac{\Omega_m}{a^3} + f(a)\right)^{-1/2}$, and from the relation $\frac{dt}{da} = \frac{dt}{ds}\frac{ds}{da}$ we get the equation that determines $t(a)$:

$$\frac{dt}{da} = \frac{1}{H_o a}\left(\frac{\Omega_m}{a^3} + \frac{\Omega_{ba}}{a^3}\sum_{k=1}^{\infty} F^{(k)}(t(a))\right)^{-1/2} \tag{32}$$

To order zero we obtain

$$H_o t^{(0)} = \frac{2a^{3/2}}{3\Omega_m^{1/2}} \tag{33}$$

But, each element of the sequence is also locally isotropic everywhere, hence we can associate to each term of the sequence $\left(\mathsf{M}^+, \eta + \sum_{\mathsf{n}=1}^{\mathsf{n}} \langle \mathsf{g}^{(\mathsf{n})} \rangle, \sum_{\mathsf{k}=0}^{\mathsf{n}-1} \langle \overline{\mathsf{T}}^{(\mathsf{k})} \rangle \right)$ the FLRW model that corresponds to the energy density $\sum_{k=0}^{n-1} \rho^{(k)}(t(a))$. We shall denote these space-times as:

$$(M, g_F^0)\ , (M, g_F^0 + g_F^1)\ , \cdots \tag{34}$$

Henceforth we shall refer all densities to the critical density. So, we shall write: $n_o m = \frac{3H_o^2}{8\pi G}\Omega_{ba}$, $\frac{n_o m}{a_i^3} = \frac{3H_o^2}{8\pi G}\Omega_{bai}$, $n_o(m+\alpha) = \frac{3H_o^2}{8\pi G}\Omega_m$, $\Omega_m = \Omega_{ba}(1+\alpha)$. We have obtained the two first terms of the sequence (34):

(i) $\rho_F^0(a) = \frac{3H_o^2}{8\pi G}\frac{\Omega_{ba}(1+\alpha)}{a^3}$, hence the first one is an Einstein de Sitter universe.

(ii) $\rho_F^{(0)}(a) + \rho_F^{(1)}(a)$ is obtained by replacing $n(t) \to n_o/a^3$, and substituting $t = t^{(0)}(a)$, into equation (25). Using henceforth the notation $\rho_F(a) = \rho_F^{(0)}(a) + \rho_F^{(1)}(a)$, we shall write the energy density (25) in the standard form of Eq. (26):

$$\rho_F(a) = \frac{3H_0^2}{8\pi G}\left(\frac{\Omega_m}{a^3} + f(a)\right)\ ,\quad f(a) = \frac{g(a,a_i)}{18a_i^3(1+\alpha)} \tag{35}$$

$$g(a,a_i) = \left(1 + \frac{a_i^3}{a^3}\right)^3 - 4\frac{a_i^3}{a^3}\left(\frac{3}{2}\left(1 + \frac{a_i^3}{a^3}\right) - \frac{a_i^{3/2}}{a^{3/2}}\right) \tag{36}$$

where the function $g(a,a_i)$ vanishes at $a = a_i$ and tends rapidly to a constant for increasing a. We assume that the series $\sum_{k=0}^{\infty} \rho_F^{(k)}(a)$ converge to the true metric $g_F^{lim}(a)$, though this has never been proved. But, the goal of this work is to show how well the two first terms of the series agree with the observations.

5. What is dark matter and what is dark energy

It is convenient to express Eqs. (35-36) in terms of the redshift. Substituting $a = \frac{1}{1+z}$ one gets:

$$\rho_F(z,z_i) = \frac{3H_o^2}{8\pi G}\left(\Omega_{ba}(1+\alpha)(1+z)^3 + f_z(z,z_i)\right)\ ,\quad f_z(z,z_i) = \frac{(1+z_i)^3}{18(1+\alpha)} g_z(z,z_i) \tag{37}$$

The component $f_z(z,z_i)$, is zero at the formation redshift z_i and tends to a constant when z diminishes, hence, it is likely that it could give predictions similar to the derived from the cosmological constant. But as it is not a constant, this component could be an explanation of what is been denoted by a dark energy. On the other hand, it is evident that the parameter α is related with the dark matter at cosmological scales: in terms of the critical density, the dark matter density is $\Omega_{dm} = \Omega_{ba}\alpha$. Let us explain separately the meaning of these components.

5.1. Dark energy

The energy density $\rho_F(z,z_i)$ depends on three parameters, namely, Ω_{ba}, α, z_i. We have obtained the moduli distance-redshift relation corresponding to this cosmological model. One can fit this expression to the supernovae data to get the parameter's values, to assess the goodness of the fitting, and compare with the ΛCDM model. But, here we shall proceed in a less systematic way to obtain a set of parameters that agree quite well with the supernova data. We shall take

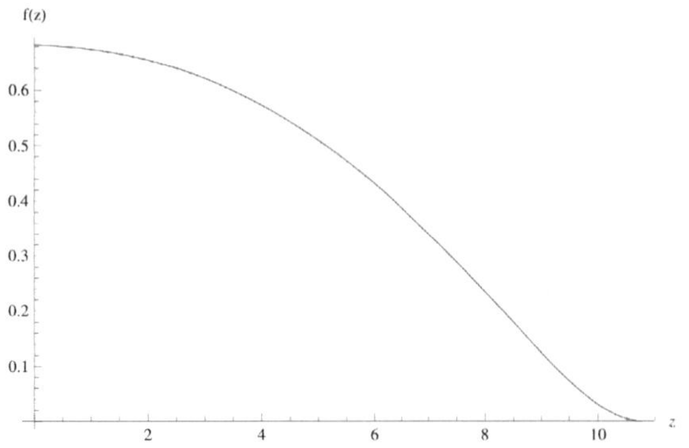

Figure 2: The energy density of our model versus redshift. It tends to a constant for decreasing redshift

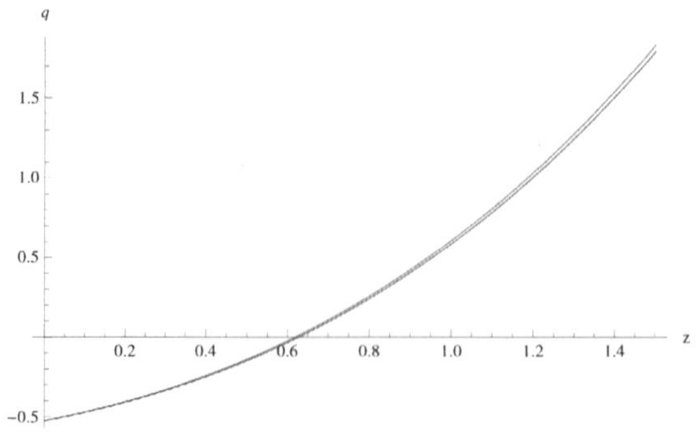

Figure 3: The deceleration factor becomes negative recently. Upper curb corresponds to our model, the lower one to the concordance model

$(\Omega_{ba}\ , \Omega_{dm}) = (0.049\ , 0.268)$ from the ΛCDM concordance model, and derive from them our parameter $\alpha = \Omega_{dm}/\Omega_{ba}$. Our third parameter z_i can be derived using the fact that our energy density component $f_z(z\ , z_i)$ tends rapidly to a constant for small redshifts, and, identifying the value of this function at zero redshift, with the energy density $\Omega_\Lambda = 0.683$ (related to the cosmological constant) obtained in the concordance model . The equation $f_z(0, z_i) = \Omega_\Lambda$ is a relation of the form $F_1(\Omega_{ba}, \Omega_{dm}, \Omega_\Lambda, z_i) = 0$ that **determines the redshift of galaxy formation** $z_i = 10.76$. With these values our component $f_z(z, z_i)$ is of the form shown in Fig. (2). Our cosmological model for the galactic phase is determined by the energy density given in Eq. (37), whose main consequences are the following:

(i) The moduli distance-redshift relation up to redshift 2, that we show in in Fig. (4). The agreement with the concordance model is quite impressive.

(ii) The deceleration factor as function of redshift, taking into account Eq. (29), that we show in Fig (3), where we see again the good agreement with the prediction of the concordance model. We have now a reason why the acceleration of the universe is a recent phenomena: It is due to the gravitational interaction between the galaxies, hence it must occur at redshift lesser than the galaxy formation redshift.

We still have a matter pending, that of justifying a non zero constant of integration α in our equations for the mass (13) and (21), that now will be convenient to express as function of the redshift. Using Eq. (33) and $a = 1/(1+z)$ we get:

$$M(z, z_i) = m + M^{(1)}(z, z_i)\ , \quad M^{(1)}(z_i, z_i) = m\alpha \tag{38}$$

$$M^{(1)}(z, z_i) = m\left(\alpha + \frac{(1+z_i)^3}{18(1+\alpha)(1+z)^3} g_z(z, z_i)\right) \tag{39}$$

The first summand is the baryonic mass, and, we shall call gravitational mass to the second one.

5.2. Dark matter at large scales

The only way to justify $\alpha \neq 0$ is to assume that before galaxy formation there was a phase dominated also by particles, "stars", formed at redshift z'_i, that collapsed at redshift z_i to form the galactic phase. We shall consider this phase described by a FLRW model too. This is certainly a rough model of evolution, used only to convey the main idea of this work, though qualitatively produces interesting results. In this phase we can write for the total mass of a

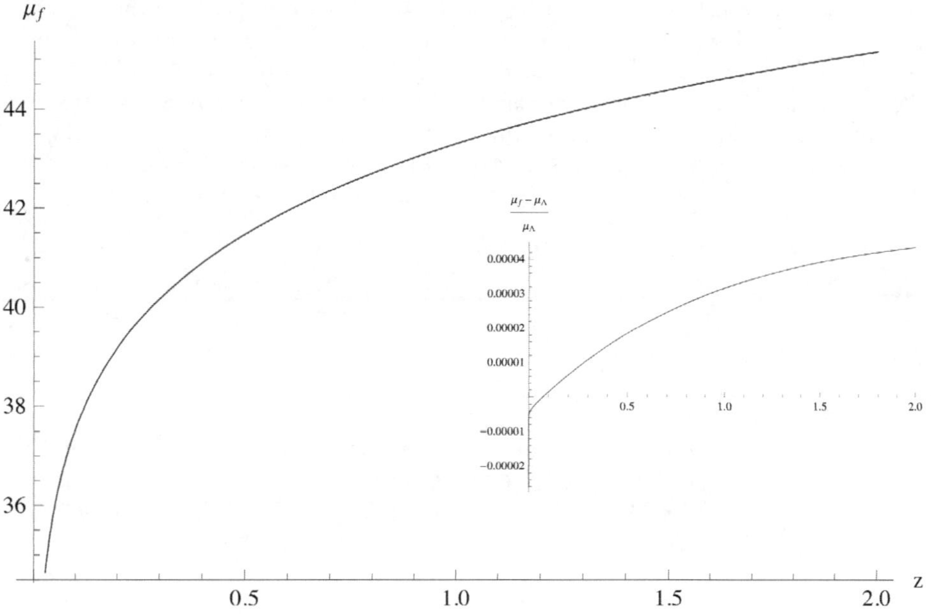

Figure 4: Our prediction for the moduli distance redshift relation is indistinguishable from the Λ CDM model.

star expressions like Eqs. (38, 39), but with magnitudes refered now to the star phase, that we denote by quotations marks:

$$M'(z, z'_i) = m' + M'^{(1)}(z, z'_i)\ , \quad m'\alpha' = M'^{(1)}(z'_i, z'_i) \tag{40}$$

$$M'^{(1)}(z, z'_i) = m'\left(\alpha' + \frac{(1+z'_i)^3}{18(1+\alpha')(1+z)^3} g_z(z, z'_i)\right) \tag{41}$$

We shall require at galaxy formation redshift z_i:

- Continuity baryonic mass: $m = N'm'$, where N' number of stars in a galaxy. Multiplying by the number density of galaxies at the formation redshift $n(z_i)$ we have $n(z_i)m = n(z_i)N'm'$, from which we get the number densities of stars $n'(z_i) = N'n(z_i)$.
- Continuity of gravitational mass:

$$m\alpha = M^{(1)}(z_i, z_i) = N'M'^{(1)}(z_i, z'_i) \tag{42}$$

This assumption, supports taken $\alpha \neq 0$, considering $m\alpha$ as the gravitational mass inherited from the previous phase: The stars that form a galaxy have being increasing its mass due to the gravitational interaction with all the stars crossing its past lift cone; in the same way that the mass of a galaxy augments in the galactic phase. As a consequence, the total mass is conserved $M(z_i, z_i) = N'M'(z_i, z'_i)$, and multiplying by $n(z_i)$ we obtain that **the energy density is conserved**: $n(z_i)M(z_i, z_i) = n(z_i)N'M'(z_i, z'_i) = n'(z_i)M'(z_i, z'_i)$. Taking into account Eqs. (38, 39), and (40, 41) we can write:

$$\alpha = \alpha' + \frac{(1+z'_i)^3}{18(1+\alpha')(1+z_i)^3} g_z(z_i, z'_i) \tag{43}$$

But Eq. (41) has introduced another constant of integration α' for the star phase. We can sustain again a non zero value for it by assuming that, before the star formation, there was a phase dominated also by free particles, that will be refered as " free atoms ", which appeared at redshift z_i'', and collapsed later at redshift $z_i' < z_i''$ to form the star phase. We can write equations, as we did in Eqs. (38) and (40), but using double quotation marks. Now we shall put $\alpha'' = 0$, because we shall consider that the phase before free atoms, dominated by electromagnetic interactions, can not be described as a system of auto gravitating particles:

$$M''(z, z_i'') = m'' + M''^{(1)}(z, z_i'') \ , \quad M''^{(1)}(z_i'', z_i'') = m''\alpha'' = 0 \tag{44}$$

Requiring continuity of the baryonic mass at star formation redshift z_i' and proceeding as in the previous case we get:

$$m' = N''m'' \quad n''(z_i) = N''n'(z_i') \tag{45}$$

$$M'^{(1)}(z_i', z_i') = m\alpha' = N''M''^{(1)}(z_i', z_i'') \tag{46}$$

where N'' number of atoms in a star, and $n''(z_i)$ its number density. The total mass is conserved $M'(z_i', z_i) = N''M''(z_i', z_i')$, and multiplying by $n'(z_i')$ we obtain again **the energy density is conserved**: $n'(z_i')M'(z_i', z_i') = n'(z_i')N''M''(z_i', z_i'') = n''(z_i')M''(z_i', z_i'')$. Taking into account Eqs. (38, 39), and (40, 41) we can write:

$$\alpha' = \frac{(1 + z_i'')^3}{18(1 + z_i')^3} g_z(z_i', z_i'') \tag{47}$$

We have needed to consider a rough model of evolution, distinguishing three phases dominated, by galaxies, stars and free atoms, to justify non null values for the constants α and α'; that are interpreted, after multiplying by Ω_{ba}, as the dark matter component at cosmological scales. That amounts to the introduction of six parameters, namely: $\Omega_{ba}, \alpha, \alpha', z_i, z_i', z_i''$. Each phase is described by a FLRW cosmological model. Matching conditions at the interphases are fulfilled because the energy density is continuous at the redshift of formation z_i, z_i' of the particles that define each phase. These conditions can be used to get a first estimation of the six parameters that defines our rough model.

5.3. A rough estimation of the parameters of our model

In subsection 5.1 we obtained the values of $(\Omega_{ba} \ , \Omega_{dm} \ , z_i) = (0.049 \ , 0.268 \ , 10.76)$, and being $\Omega_{dm} = \Omega_{ba}\alpha$, we obtained the first three parameters. Equations (43 , 47) put two conditions on the three remaining parameters. The equation (47) defines a function of the redshifts $\alpha'(z_i', z_i'')$, and substituting into Eq. (43)we get

$$\alpha'(z_i', z_i'')^2 + (1 - \alpha)\alpha'(z_i', z_i'') - \alpha + \frac{1}{18}\left(\frac{1 + z_i'}{1 + z_i}\right)^3 g(z_i, z_i') = 0) \tag{48}$$

The effective decoupling matter from radiation, has been discussed in [4], obtaining $z_i'' \approx 100$. Substituting $z_i = 10.76$, $\alpha = \Omega_{dm}/\Omega_{ba} = 0.268/0.049$, and $z_i'' = 100$ into Eq. (48) **we get an estimation for the redshift of formation of the star component:** $z_i' = 20.7$, when the universe was a hundred and sixty millions years old. It is funny that some models of star formation obtain similar redshift for the first generation of stars [5]. Using the Eq. (47) we get the dark matter content, at cosmological scales, of the star phase $\Omega_{ba}\alpha' = 0.267$. This is produced by the gravitational interaction between free atoms. This number is lesser than $\Omega_{ba}\alpha = 0.268$, but very proxim to it. The small difference is the contribution of the gravitational interaction between stars. So the major part of the "dark matter" in the galactic phase comes from the gravitational interaction between free atoms. Let us recall that we are explaining dark matter as the gravitational mass that have a group of particles at the moment that they form by collapse a new sort of particle.

5.4. Heuristic approach to dark matter at galactic scales

In order to explain dark matter at galactic scales it is necessary to consider inhomogeneous perturbations of our homogeneous model; but, we can get some insights from an "heuristic viewpoint", suggested by the vacuole solutions (Einstein-Strauss, Kottler). These are exact solutions in which, inside a vacuole, an inhomogenity is isolated from the cosmological expansion. So, we assume a gradual transition from a total screening from the cosmological effects at the center of the inhomogenity, to a total cosmological influence. Let us make the conjecture that the energy density at distance r from the center is well described by the expression:

$$\rho(r,z) = \mu_{ba}(r) + \chi(r)\mu_{ba}(r)\left(\alpha + \frac{(1+z_i)^3}{18(1+\alpha)(1+z)^3} g_z(z,z_i)\right), \tag{49}$$

where $\chi(r) = 1 - \frac{\Phi_{ba}(r)}{\Phi_{ba}(0)}$, μ_{ba} is the baryonic mass density, and $\Phi_{ba}(r)$ the galactic newtonian gravitational potential. With this we can calculate the total mass, the gravitational mass and the fraction of gravitational mass contained inside a sphere of radius r; and, the rotation velocity at radius r:

$$M(r,z) = \int_0^r \rho(r,z)(r))4\pi u^2 du\ , \quad M_{GR}(r,z) = \int_0^r (\rho(r,z) - \mu_{ba}(r))4\pi u^2 du \tag{50}$$

$$f_{GR}(r,z) = \frac{M_{GR}(r,z)}{M_{ba}(r) + M_{GR}(r,z)} \tag{51}$$

$$v^2(r) = \frac{GM(r,z)}{r} \tag{52}$$

Baryons and dark matter has been recently disentangled [6] in the spiral lens B1933+503, observed at redshift $z = 0.755$. We shall use a Plummer's model, with baryonic density $\mu_{ba} = \frac{3M}{4\pi R^3}(1 + \frac{r^2}{R^2})^{-5/2}$ and newtonian potential $\Phi_{ba} = -\frac{GM}{R}(1 + r^2/R^2)^{-1/2}$. We have taken $M = 2.6 \times 10^{11} M_\odot$ and scale factor $R = 12Kpc$ in order to get the intersection of dark matter and baryonic mass profiles just at $r = 10Kpc$, as it is manifest in Fig. (12) of [6]. We have represented in Fig. (5) the baryonic and the gravitational mass inside a radius r as calculated with Eqs.(50, 51). They agree qualitatively with the observations of baryonic and dark matter reported in [6].

The rotation curve calculated using Eq.(52) is shown in Fig. (6), and contrasted with some observations taken from [6]. We see that the baryonic component fails to explain the observed velocities, but adding the gravitational mass we get a good description.

The dependence on redshift of the gravitational mass fraction profile, given by Eq. (51), is illustrated in Fig. (7) for the redshifts $z = 0.755$ and $z = 0.138$. The fraction $f_{GR}(r,z)$ is greater for the older galaxies (lower z), for the content of gravitational mass increases with the age of the observed galaxy, as consequence of Eqs.(38 , 39). In Fig. (8) we give the graph of $f_{GR}(8.2, z)$, calculated with Eq. (51) to compare with the observed dark matter fraction in galaxies B 1933+503 and and SDSSJ2141-000, recently reported in [6] and [7] respectively. In both cases $8.2Kpc \approx 2.2R_d$ with R_d equal to the disk scale length. We conclude that our conjecture gives qualitative good results and could be a guide to start the study of the inhomogeneous case.

6. Conclusions

6.1. The more solid result

The equation (24) for the energy density, $\rho = n(t)M(t)$, comes from the second iteration of the Einstein equations (first order in the energy tensor, second order in the metric). It is

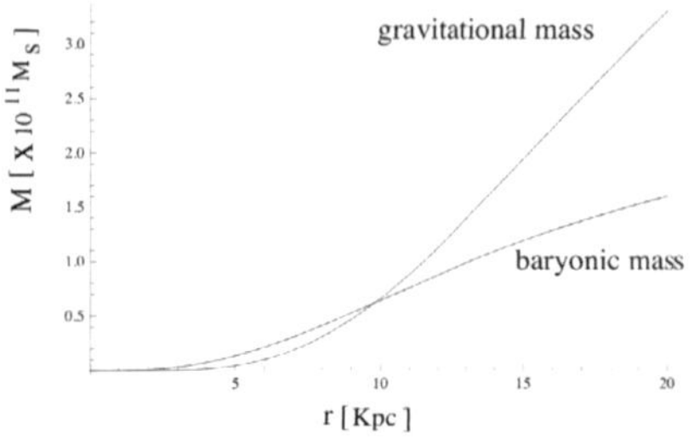

Figure 5: Gravitational and baryonic mass inside a sphere of radius r predicted by our model for the galaxy B 1933+503. It is similar to the DM observations showed in Fig 12 of [6].

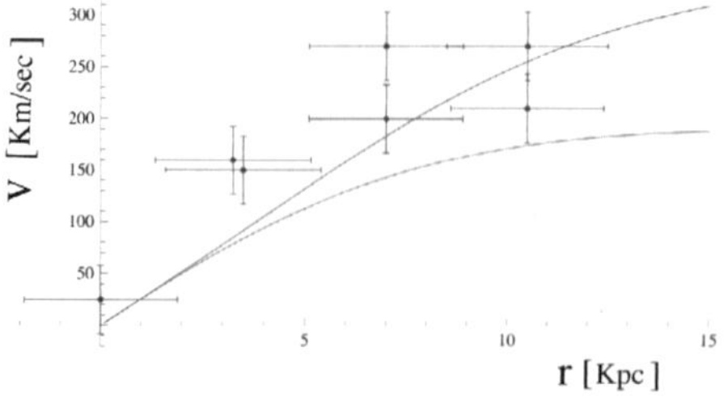

Figure 6: Predicted rotation curve considering baryonic plus gravitational mass (upper curve), and only baryonic mass (lower curve) in galaxy B 1933+503. Observed velocities taken from [6]. It is similar to Fig 8 in [6].

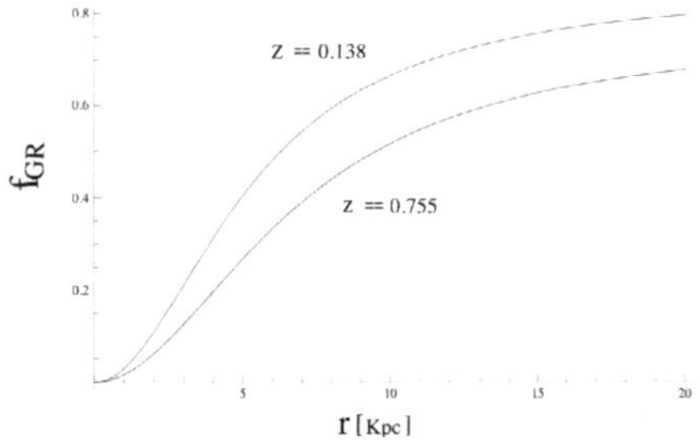

Figure 7: Gravitational mass fraction inside a sphere of radius r in a galaxy similar to B 1933+503 for two different redshifts.

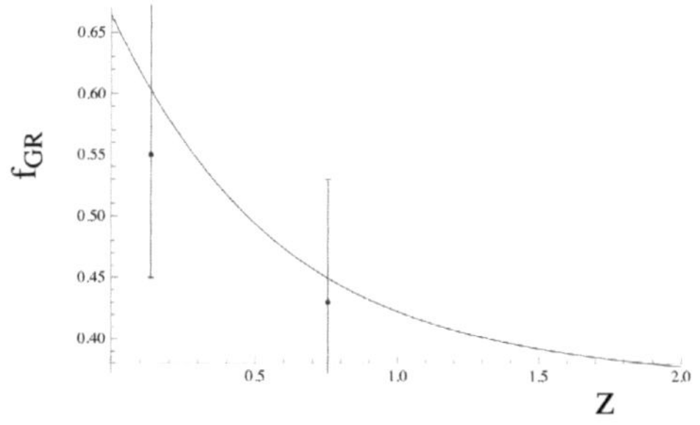

Figure 8: Gravitational mass fraction inside a sphere of 8.2 Kpc as function of redshift, and observed values in galaxies B 1933+503 (z=0.755), and SDSSJ2141-0001(z=0.138).

the basic result of this work, because it means that the mass of a particle, formed at instant t_i, is the monotonic increasing function $M(t)$ given in Eq. (21), or, expressed in terms of redshift, in Eqs.(38, 39).**These equations say that the mass of a particle depends on the interaction with all the particles over its past light cone**; and, as show Eqs.(19-20) it is composed of the baryonic mass plus the contribution of the metric. We have denoted this component as gravitational mass. This is clearly on the line of the Mach principle on the inertial mass, but here appears not as a principle, but as consequence of the Einstein equations. It has been an agreeable surprise to find a similar sentence to the outlined above in the Einstein's address to the Prussian Academy in 1921 [8]: **The general theory of relativity teaches that the inertia of a given body is greater if there are more ponderable masses in proximity to it; thus it seems very natural to reduce the total effect of inertia of a body to the action and reaction between it and the other bodies in the universe**. Dark matter and dark energy can be explained with this ignored gravitational mass. Recalling the Eqs.(38, 37)

$$M(z, z_i) = m + M^{(1)}(z, z_i)\ , \quad M^{(1)}(z_i, z_i) = m\alpha\ , \quad z < z_i \tag{53}$$

$$\rho(z, z_i) = n(z)M(z, z_i) = \frac{3H_o^2}{8\pi G}\left(\Omega_{ba}(1+\alpha)(1+z)^3 + \frac{(1+z_i)^3}{18(1+\alpha)(1+z)^3} g_z(z, z_i)\right) \tag{54}$$

will facilitate the following summary. The last addend in the right hand of Eq.(54), represented in Fig.(2), is responsible of the acceleration of the universe, due to its rapid tendency to a constant value. This fact justifies the cosmological constant, only as the limit value of the gravitational energy density. **There is no need of a mysterious dark energy, is just the metric, i.e., it is just gravitation**. Figs. 3, 4 show great agreement of our predictions with the obtained with the ΛCDM model, and therefore, the agreement with observations. From the supernova experiment we can infer the first three parameters of our model: $(\Omega_{ba}, \Omega_{ba}\alpha, z_i) = (0.049, 0.268, 10.76)$. **So we predict galaxy formation at redshift ≈ 11 when the universe was 400 million years old**. These results are based only on the existence of a phase of the universe, locally isotropic everywhere, dominated by auto-gravitating particles (galaxies).

6.2. Model dependent results
In order to explain the dark matter component at large scales we have needed two more phases dominated by particles. In subsection 5.2 the constant $m\alpha$ in Eq. (53) has been explained as the gravitational mass acquired by the particles forming the previous phases, namely, stars and atoms. Multiplying this constant by the number density of galaxies one explains the constant $\Omega_{ba}\alpha$ appearing in Eq. (54). **This can explain the dark matter component at cosmological scales, without necessity of exotic particles**. Our model depends on six parameters: the baryonic mass density Ω_{ba}, the gravitational mass densities in the galactic and the star phases $\Omega_{ba}\alpha$ and $\Omega_{ba}\alpha'$, the redshifts of formation of galaxies and stars: $z_i < z_i'$, and the redshift of the effective decoupling of atoms and radiation z_i''. Requiring continuity of total energy density through the surfaces of formation of the stars and galaxies, and assuming $z_i'' = 100$ as it is indicated in [4], we get the two remaining parameters $(\Omega_{ba}\alpha', z_i') = (0.267, 20.7)$. **Therefore, the major part of the gravitational mass was produced in the atomic phase ($\Omega_{ba}\alpha' = 0.267 < \Omega_{ba}\alpha = 0.268$) , and the redshift of formation of the star phase was $z_i'' \approx 21$ when the universe was a hundred millions years old.**

6.3. Results dependent on a conjecture
We have inferred funny results, making a conjecture, about the explanation of dark matter phenomena at galactic scales. Figures (5 - 8) show predictions for rotation curves, gravitational mass fraction profile, and its redshift dependence that seems to be in agreement with the observations.

Acknowledgments
The author is indebted to his friends for their patience and to the support of the Spanish Ministry of *Economía y Competitividad (sic)*, MICINN-FEDER project FIS2012-33582.

Appendix
Let us show how we have got the statistical average of the $g_{00}^{(1)}$ metric component, given in equations (17, 18). We must calculate the sum appearing in (9). We recall that the worldlines are parametrized with the Minkowskian time τ. Let p_o be a point on a world line of reference L, and choose coordinates $\{x^\alpha\}$ such that the tangent vector at p_o to this line be $v = \partial_t$, and the coordinates of $p_o = (t, 0, 0, 0)$. Let $\Sigma_t = \{x \in M / x^o = t = const.\}$,

$\Sigma_\tau = \{x \in M/t^2 - r^2 = \tau = const.\}$, and denote by $C_{p_o}^{past}$ the past light cone at p_o, limited by the hiperboloid Σ_{τ_i} which represents the birth of the galaxies at time τ_i. The summands in (9) are referred to the retarded points p_j where a particle world line L_j intersects the cone $C_{p_o}^{past}$. Let r_j be its radial coordinate and r the radial coordinate of the intersection point $L_j \cap \Sigma_t$. Thus, the tangent vector to the world line L_j at p_j will be $v_j = \gamma_j(\partial_t + \bar{v}_j\partial_r)$, with $\bar{v}_j = \frac{r}{t}$, and $\gamma_j = t(t^2 - r^2)^{-1/2}$. This is a unit vector with respect the Minkowski metric. With these notations we can write $g_{00}^{(1)}(p_o)$ as follows:

$$g_{00}^{(1)}(p_o) = -Gm \sum_j \frac{\gamma_j^2 - 1/2}{r_j\gamma_j(1 + \bar{v}_j)} \tag{A.1}$$

We have seen in section (3.1.1) that a Milne Universe has a uniform number density over hipersurfaces Σ_τ of constant Minkowski proper time, and in consequence a non uniform density $n^{(0)}(r,t) = Nt(t^2 - r^2)^{-2}$, on the hipersurfaces Σ_t. We shall make a statistical average by assuming a uniform random process over Σ_τ. This is equivalent to consider a point process on Σ_t with the density $n^{(0)}(r,t)$ given above. The statistical average is then given by the integral

$$\langle g_{00}^{(1)}(p_o)\rangle = -4GNm \int_0^{r_{mx}} \frac{t}{(t^2 - r^2)^2} \frac{\gamma_j^2 - 1/2}{r_j\gamma_j(1 + \bar{v}_j)} 4\pi r^2 dr \tag{A.2}$$

where r_{mx} is the radial coordinate of the point intersection of Σ_t with the world line L_{lim} as shown in Figure A1 . The set of worldliness like L_{lim} defines the boundary of the particles that have interacted with the reference particle L at time t.

To perform the integral we need the relation between the radial coordinates r_j and r. It is straightforward to derive it from the Figure A1: $r_j = \frac{rt}{r+t}$. The result of the integration is

$$\langle g_{00}^{(1)}(p_o)\rangle = -8\pi GNm \left(\frac{2}{3}t^2(t^2 - r_{mx}^2)^{-3/2} - (t^2 - r_{mx}^2)^{-1/2} + \frac{1}{3t}\right) \tag{A.3}$$

It remains to get the value of r_{mx}. We will shall show that it depends on both the present time t and the initial time τ_i. Let r^* be the radial coordinate of the point $p_{lim} = L_{lim} \cap \Sigma_t$. From the figure it is easy to get

$$r_{mx} = t\frac{r^*}{t - r^*} \tag{A.4}$$

Next we solve the system $\bar{t} + \bar{r} = t$,$\bar{t}^2 - \bar{r}^2 = \tau_i^2$ to get the radial coordinate $r^* = \frac{t^2 - \tau_i^2}{2t}$, and substituting into the previous equation we get

$$r_{max} = t\frac{t^2 - \tau_i^2}{t^2 + \tau_i^2} \tag{A.5}$$

But, the intersection of the reference world line L with the initial hipersurface Σ_{τ_i} is a point with coordinate time $t_i = \tau_i$, thus we can write finally $r_{max} = t\frac{t^2 - t_i^2}{t^2 + t_i^2}$. It is straightforward to get the relations:

$$t^2 - r_{mx}^2 = \frac{4t^4t_i^2}{(t^2 + t_i^2)^2}\ ,\ (t^2 - r_{mx}^2)^{3/2} = \frac{8t^6t_i^3}{(t^2 + t_i^2)^3}\ ,\ (t^2 - r_{mx}^2)^{1/2} = \frac{2t^2t_i}{t^2 + t_i^2} \tag{A.6}$$

Substituting into the expression (A.3), and writing $\mu_i = \frac{Nm}{t_i^3}$ for the initial mass density we get

$$\langle g_{00}^{(1)}(p_o)\rangle = -8\pi G\mu_i \left(\frac{1}{12}t^2\left(1 + \frac{t_i^2}{t^2}\right)^3 - t_i^2\left(\frac{1}{2}\left(1 + \frac{t_i^2}{t^2}\right) - \frac{1}{3}\frac{t_i}{t}\right)\right) \tag{A.7}$$

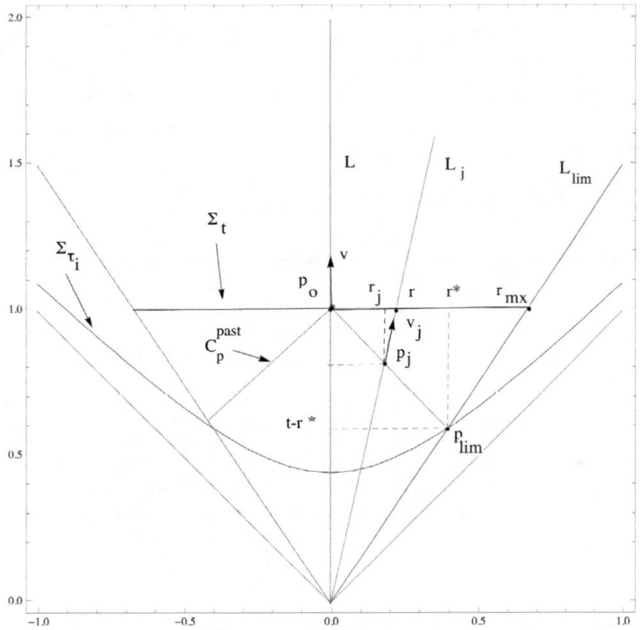

Figure A1: This figure shows the meaning of the radial coordinates used in the appendix:

(i) r_{mx} is the radial coordinate of the point $L_{lim} \cap \Sigma_t$

(ii) r^* is radial coordinate of the point $L_{lim} \cap \Sigma_{\tau_i}$

(iii) r_j is the radial coordinate of the point $L_j \cap C^{past}_{p_o}$: the retarded position of particle j.

(iv) r is the radial coordinate of the point $L_j \cap \Sigma_t$

References

[1] Havas P and Goldberg J 1962 *Phys. Rev.* **128** 398.
[2] Clarkson Ch, Ellis G, Larena J, Umeh O 2011: Does the growth of structure affect our dynamical models of the universe?. arXiv:1109.2314v1[astro-ph.CO].
[3] Portilla M 2014 *Universe made of baryonic gravitating particles behaves as a* ΛCDM *Universe* arXiv:1405.0399 [gr-qc].
[4] Rindler W 2006 *Relativity: Special, General and Cosmological* (Oxford)chapter 6 pp 170-180.
[5] Scott D and Moss A 2009 *Mon. Not. R. Astron. Soc.* **397** 445.
[6] Jeon et al. 2012 *Ap.J.***754** 34.
[7] Suyu S 2012 *Ap.J.***750** 10.
[8] Dutton S 2011 *Mon. Not. R. Astron. Soc.***1** 23.
[9] Einstein A *Sidelights on Relativity: Geometry and experience.* (Dover) pp 42.

Journal of Physics: Conference Series **600** (2015) 012008 doi:10.1088/1742-6596/600/1/012008

The analog of the Hawking effect in BECs

Alessandro Fabbri

Centro Studi e Ricerche Enrico Fermi, Piazza del Viminale 1, 00184 Roma, Italy
Dipartimento di Fisica dell'Università di Bologna, Via Irnerio 46, 40126 Bologna, Italy
Dep. de Física Teórica and IFIC, Universidad de Valencia-CSIC, C. Dr. Moliner 50, 46100 Burjassot, Spain

E-mail: afabbri@ific.uv.es

Abstract. The observation of the Hawking effect from black holes in the astrophysical context is unlikely. However, the analog of this effect is present in condensed matter systems. We focus on Bose-Einstein condensates, and on a proposal to detect it through correlation measurements.

1. Introduction

Gravity is by far the weakest interaction and its presence at the microscopic level is usually neglected. Quantum gravitational effects are extremely small, as typically one should go to the Planck scale to measure them. This makes the construction of a quantum theory of gravity a very difficult task and, mainly, a challenge for theorists.

Despite these facts, however, gravity is universal and always attractive. It is the dominant force in the macroscopic world and determines the large scale structure of the Universe. Moreover, when matter collapses inside its Schwarzschild radius $r_G = \frac{2GM}{c^2}$ it can form black holes, in which the gravitational field is so strong that light is trapped and cannot escape. They form in the gravitational collapse of massive stars with mass $M \geq 2 - 3M_{Sun}$, their outer boundary is the event horizon.

Black holes are important for many reasons, in particular because they allow unification of gravity with thermodynamics. In the early seventies Bardeen, Carter and Hawking [1] pointed out a surprising analogy between the laws of black hole mechanics

$$c^2\delta M = \frac{c^2}{8\pi G}\kappa\delta A_H + \Omega_H\delta J \ , \quad \delta A_H > 0 \ , .. \tag{1}$$

(where M, J are the black hole's mass and angular momentum and A_H, Ω_H, κ the horizon's area, angular velocity and surface gravity) and the laws of thermodynamics. This fact, in particular the area law [2] $\delta A_H > 0$ expressing that the surface of a black hole's event horizon never decreases, led Bekenstein [3] to propose that A_H gives a measure of the entropy of the black hole, i.e. $S_{BH} \propto A_H$. In order for this analogy to be taken seriously, however, a way was needed to assign a temperature to the black hole. At the classical level this is impossible since a black hole is a perfect absorber and has, therefore, $T = 0$.

This problem was solved by Hawking [4]. He used the formalism of quantum field theory in curved space, initially developed by Parker [5], to show that black holes quantum mechanically emit a thermal flux of particles at the characteristic temperature

$$T_H = \frac{\hbar\kappa}{2\pi c k_B} \ , \tag{2}$$

with κ the horizon's surface gravity. He also clarified the basic mechanism responsible for the emission. Virtual pairs are continuously created from the vacuum but annihilate almost instantaneously (after a time of the order of Planck time) leaving, in ordinary situations, no visible effect. Close to the event horizon, however, one member of the pair may get trapped inside the hole, leaving its partner free to propagate away where it is detected. Such particles, called the Hawking quanta, constitute the particles emitted by the black hole.

Eq. (2) is a remarkable prediction of a quantum effect in gravity. Unfortunately, in realistic situations in which black holes form from gravitational collapse

$$T_H \sim \frac{M_{Sun}}{M} K \ll T_{CMB} \sim 3\ K \ , \tag{3}$$

i.e. the Universe' cosmic microwave background temperature is far too high making the experimental detection of Hawking radiation from astrophysical black holes very unlikely. Alternative possibilities are based on the fact that mini black holes (being $T_H \propto \frac{1}{M}$ this effect becomes more important for small masses) may have been produced in the early Universe by quantum effects or at particle accelerators due to the existence of (large) extra dimensions.

2. Analog gravity

An important property of Hawking radiation, uncovered by Unruh [6], is that it is not specific to gravity. Indeed, this effect has a kinematical origin. Hawking did not use Einstein's field equations to derive it, his analysis only relies on the details of wave propagation around the black hole's horizon.

A stationary fluid undergoing supersonic motion traps sound waves in its supersonic region in the same way a gravitational black hole traps light. Such configurations, called acoustic black holes, possess an acoustic horizon, which is the surface where the fluid's velocity reaches the speed of sound. See Fig. 1.

There is a precise mathematical equivalence between light propagation around the event horizon of a gravitational black hole and sound propagation around the acoustic horizon of an acoustic black hole [7]. Unruh used it to predict that acoustic black holes will emit a thermal flux of phonons (analog Hawking radiation) from their acoustic horizon at the temperature

$$T_H = \frac{1}{4\pi k_B c_s}\frac{d(c_s^2 - \vec{v}_0^2)}{dn}|_{hor} \ , \tag{4}$$

where c_s is the sound speed, $\vec{v}_0$ the fluid's velocity and n is the normal to the horizon.

This application has various advantages. First of all, the above considerations are valid not only for fluids, but also for other condensed matter systems in the hydrodynamic (long wavelength) approximation.

From a theoretical viewpoint it is important because it allows to address one weak point in Hawking's derivation, the transplanckian problem [8]. Due to the huge redshift experienced by the field modes describing the asymptotic Hawking quanta in the propagation between the horizon and infinity, their initial (horizon's) frequency must have been enormous (i.e. transplanckian). In such a regime, we can no longer trust the formalism of QFT in curved space used by Hawking, we need a quantum gravity calculation. This difficulty makes Hawking's

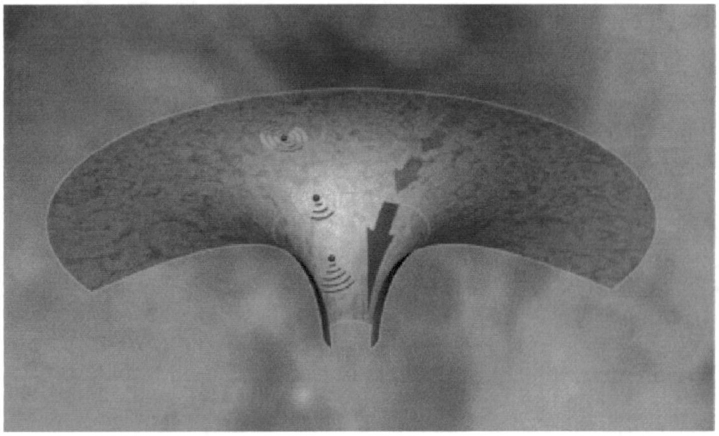

Figure 1. Pictorial representation of an acoustic black hole, with a subsonic and a supersonic region, respectively, above and under the yellow line (acoustic horizon). Sound waves are trapped in the supersonic region and cannot escape from it.

analysis dubious, and the results uncertain. In this context we have an analog problem (i.e. Hawking radiation relies on frequencies higher than the cutoff scale set by the validity of the hydrodynamical fluid description), however in many cases a microscopic description is available.

The other advantage of this application is to open up new possibilities to test experimentally the analog of the Hawking effect. Two classes of experiments have been realised, water tanks experiments [9] and quantum optics experiments using ultrashort laser pulse filaments [10]. The interpretation of the results of these interesting experiments is, however, controversial [11, 12].

3. Acoustic black holes in BECs

Bose-Einstein condensates (BECs) are dilute gas of bosons cooled down at extremely low temperatures (of the order of 100 nK) and characterized by having (almost) all its constituents in the same quantum state. In such a a state of matter macroscopic quantum phenomena become apparent. Predicted in 1924 by Bose and Einstein, they were discovered in 1995 by the groups of Cornell, Wieman and Ketterle using, respectively, atomic gases of rubidium and sodium.

BECs offer particularly favourable experimental conditions for the detection of the analog of the Hawking effect. Indeed on can reach situations in which the Hawking temperature of the BEC acoustic black hole is only one order of magnitude lower than that of the condensate

$$T_H \sim 10\ nK \leq T_C \sim 100\ nK\ . \tag{5}$$

This is a very significative improvement with respect to astrophysics (3), but it is not enough to attempt a direct detection of the thermal Hawking's emitted phonons.

Another important advantage of having an analog black hole in the lab is that unlike in the gravitational case, where measurements are necessarily restricted to the exterior of the black hole, no such restriction exists for the interior of an acoustic black hole (which is just a region of supersonic flow). Thus, measurements can be made in exterior as well as in the interior. This makes it possible to experimentally detect the analog of the Hawking effect via the pair creation phenomena that uniquely characterises it, i.e. Hawing quanta (in the exterior) - partner (in the interior).

Using QFT in curved space techniques, valid in the (quantum) hydrodynamic approximation of BEC theory, it was shown [13] that the correlator of the density fluctuations $\langle \delta\hat{n}(x)\delta\hat{n}(x')\rangle$ around an acoustic black hole in a BEC displays a characteristic peak for points, respectively, outside and inside the acoustic horizon and which clearly identifies the Hawking effect. This motivated an 'ab initio' numerical analysis within the microscopic BEC theory [14] whose results, displayed in Fig. 2, nicely confirmed the hydrodynamical prediction.

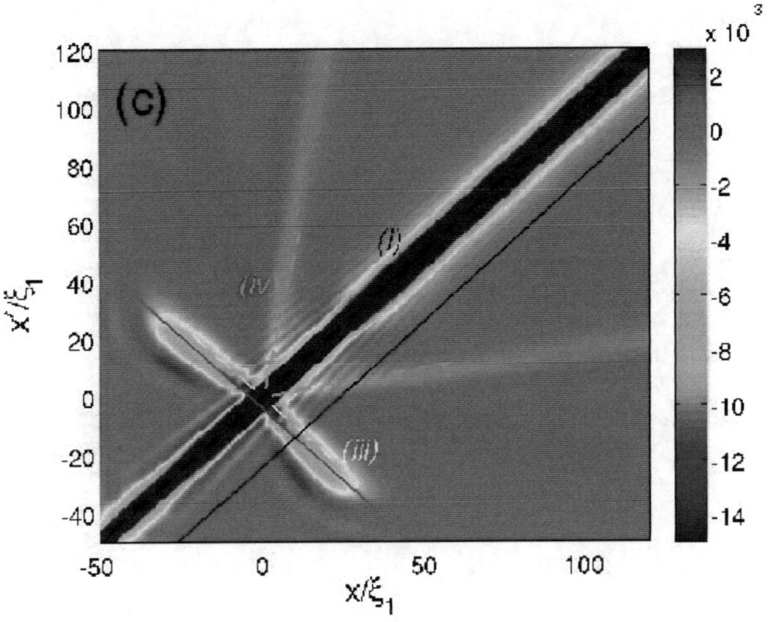

Figure 2. Density-density correlator in a 1D acoustic black hole. $x = 0$ is the acoustic horizon. Feature iii), (roughly) perpendicular to the main diagonal identifies the Hawking quanta - partner correlations for points on both sides of the acoustic horizon ($xx' < 0$).

The existence of Hawking radiation in BECs vs. the transplanckian problem is therefore established. Another interesting feature of the numerical results is that if we take the fluctuations to be initially in a thermal state (instead of the vacuum) at a temperature bigger than T_H the Hawking signal is not masked by thermal effects, a feature making the peak in the correlator the 'smoking gun' to experimentally identify the analog Hawking radiation.

In addition to black holes, in this context one can reproduce configurations which in gravity are problematic. It is the case of white holes, time reversal of black holes. Whereas black holes are perfect absorbers, white holes expel any matter inside them and do not allow anything to enter from the outside. The problem in gravity is that they require an initial singularity, and this makes physical predictions about them impossible.

The analog Hawking radiation requires the existence of negative energy states, and these exist in the supersonic region of both black holes and white holes [15]. The big difference between the two cases is that while in black holes the Hawking effect is a low energy effect (it does not depend on the details of the microscopic physics), in white holes it is a high-energy effect and, thus, its signal is system dependent.

In BECs, the Hawking effect in white holes is crucially determined by the short-scale modification of the linear (phononic) dispersion relation of the Bogoliubov fluctuations, which becomes supersonic at high momenta and develops, inside the white hole, a nontrivial (outgoing) zero mode [16]. This happens for $k \geq \frac{1}{\xi}$, where ξ is the healing length, the natural length scale of the condensate and the analog of the Planck length in gravity. We see in the density correlator of Fig. 3 that in the supersonic region $x, x' < 0$ a checkerboard pattern emerges, characterized by a series of oscillations with peaks parallel and antiparallel to the main diagonal. The overall amplitude of these oscillations grows with time, logarithmically if the initial state is the vacuum and linearly in the case of an initial thermal state. It describes the emission of (an infinite number of) zero-frequency waves, and signals a (weak) instability of the system.

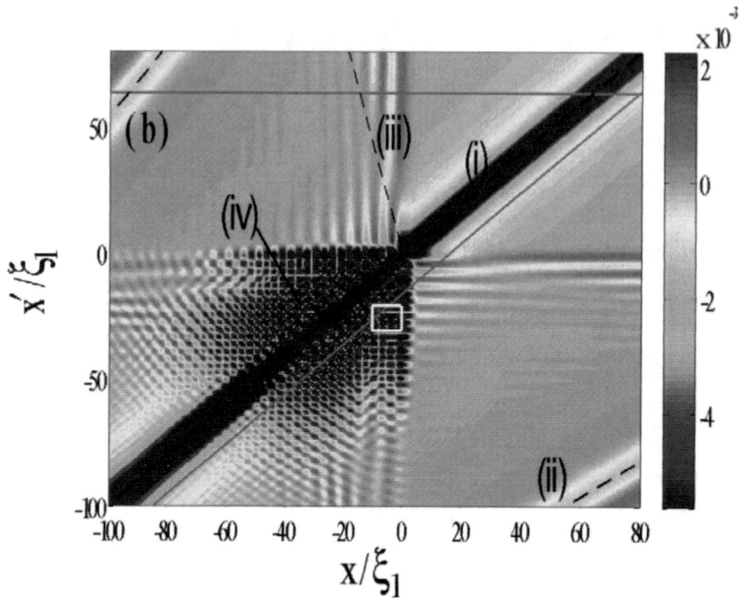

Figure 3. The checkerboard patter in the density correlator of an acoustic white hole in a BEC. The oscillations taking place in supersonic region (x, $x' < 0$) have an overall amplitude that grows with time, signalling a (weak) instability.

A dramatic manifestation of this instability takes place in a composite system, made of a black hole and a white hole horizon, that has been realised experimentally in [17]. In these systems, pairs are created at the black hole horizon, the Hawking quanta in the exterior and its partner in the interior. The partner propagates towards the white hole horizon where, due again to the supersonic character of Bogoliubov dispersion, it 'bounces' and comes back to the black hole horizon generating another Hawking quanta. This process repeats itself again and again and results in a laser-type instability [18]. This makes the system dynamically unstable and the emitted Hawking flux exponentially amplified. Evidence of this effect has been observed recently in [19].

4. Dynamical Casimir effect and momentum correlators

The numerical results for the density correlator in acoustic black holes in [14] displayed, at early times, a transient term, the feature ii) in Fig. 4. This additional peak does not depend on the inhomogeneities and the presence of the horizon. It is there because of time dependence, and appears because in the simulations the acoustic black hole was dynamically created. This feature is a manifestation a distinct pair creation taking place even in homogeneous systems, the dynamical Casimir effect [20].

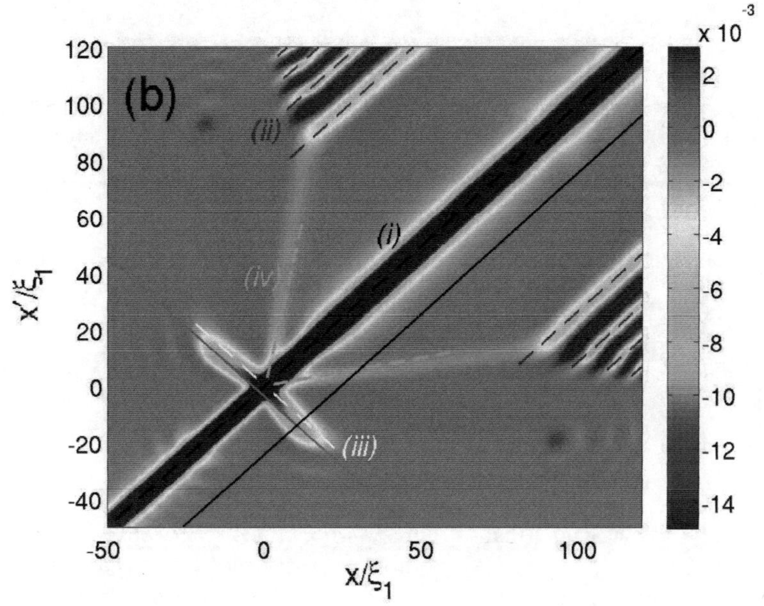

Figure 4. The early time simulations in acoustic black holes showed the existence of a transient term, the feature ii), which disappears at late time, Fig. 2, and is due to the dynamical Casimir effect.

An experiment was carried out to measure this effect in [21]. In it the confining potential of a homogeneous condensate was varied in time (both with a sudden compression and a periodic modulation). This created excitations around the condensate, which were measured with the time of flight technique. From the experimental data the correlation function with respect to the atoms' velocities was constructed. This displayed a peak along the anti-diagonal, $v = -v'$, which corresponds to the creation of correlated excitations at opposite momenta $k = -k'$ as predicted by the dynamical Casimir effect. It is worth mentioning that the same effect is responsible for particle creation in the early Universe. More specifically, it would be interesting to mimic it by reproducing in the lab the same de Sitter exponential expansion the Universe experienced during the (early) inflationary era.

The next step is to reproduce experimentally an acoustic black hole with only one horizon, using the 'waterfall' potential of [22] in which the the horizon and the supersonic region are created by an external step-like potential. At the theoretical level, we need to adapt the original proposal to study the features of the Hawking effect in the correlator of the fluctuations' occupation number operator $\langle \delta\hat{n}(k)\delta\hat{n}(k')\rangle$ in momentum space [23].

5. Infrared properties of the analog Hawking radiation

A black hole does not emit as a perfect black-body. Indeed, the modes of the Hawking quanta, from the the horizon, reach the far away detector with a probability $\Gamma(\omega)$ that depends on their frequency. Such quantity, called the gray-body factor, enters Hawking's formula for the particles emitted by the black hole and modulates the Planck distribution

$$N_\omega = \frac{\Gamma(\omega)}{e^{\frac{\hbar\omega}{k_B T_H}} - 1} . \tag{6}$$

Of particular interest is the low frequency emission. In gravity, for a black hole in asymptotically flat space Γ vanishes at small ω as $\Gamma \sim A_H \omega^2$, where A_H is the horizon's area [24]. This allows to eliminate the infrared divergence ($\frac{1}{\omega}$) in the Planck distribution and, moreover, indicates that the emission of low energy particles is suppressed.

Surpsingly, for a 1D acoustic black hole with just one horizon the graybody factor Γ goes to a nonzero constant at low frequency [25]. This behaviour, which is similar to that of a black hole immersed in an expanding de Sitter space, shows that the analog Hawking radiation is dominated by an infinite number ($\frac{1}{\omega}$) of soft phonons. One might worry that this infrared divergence will show up also in the density correlator, complicating its use to identify the Hawking effect. It is however not so, as the density-density correlator stays infrared finite.

6. Future directions

In gravity it is important to understand how the Hawking effect modifies the classical black hole geometry. The semiclassical Einstein equations, or backreaction equations, are useful to describe the details of the evaporation [26] at most until the black hole reaches the Planck mass, after which we enter in the realm of quantum gravity. By extrapolating his semiclassical results, Hawking predicted that black holes will disappear and, with them, the information of the star from which they were formed [27]. If so, it will have dramatic consequences, as it would imply that from a full quantum gravity perspective the whole process formation and evaporation of a black hole will be described by an S-matrix which is not unitary, in clear contradiction with the principles of quantum mechanics.

It is generally believed that unitarity will be preserved, however the difficult part is to understand how. Recently, it has been conjectured that under a set or reasonable assumptions unitarity of the emitted Hawking radiation will imply that at some (Page) time a high-energy barrier (a 'firewall' [28]) will form at the horizon destroying the Hawking quanta-partner correlations. These qualitative issues can be addressed concretely in BECs, where Hawking quanta-partner correlations can be measured and backreaction equations can be written down which are valid at all scales.

References

[1] J.M. Bardeen, B. Carter and S.W. Hawking, *Commun. Math. Phys.* **31** (1973), 161
[2] S.W. Hawking, *Commun. Math. Phys.* **25** (1972), 152
[3] J.D. Bekenstein, *Phys. Rev.* **D7** (1973), 2333
[4] S. W. Hawking, Nature **248**, 30 (1974); Commun. Math. Phys. **43**, 199 (1975)
[5] L. Parker, *Phys. Rev.* **183** (1969), 1057
[6] W.G. Unruh, *Phys. Rev. Lett.* **46** (1981), 1351
[7] C. Barceló, S. Liberati and M. Visser, *Living Rev. Relativity* **8** (2005), 12
[8] T. Jacobson, *Phys. Rev.* **D44** (1991), 1731
[9] S. Weinfurtner et al., *Phys. Rev. Lett.* **106** (2011), 021302
[10] F. Belgiorno et al., *Phys. Rev. Lett.* **105** (2010), 203901
[11] F. Michel and R. Parentani, *Phys. Rev.* **D90** (2014), 044033
[12] R. Schutzhold and W.G. Unruh, *Phys. Rev. Lett.* **107** (2011), 149401
[13] R. Balbinot et al., *Phys. Rev.* **A78** (2008), 021603
[14] I. Carusotto et al., *New J. Phys.* **10** (2008), 021603

[15] R. Balbinot et al., *Lect. Notes Phys.* **870** (2013), 181
[16] C. Mayoral et al., *New J. Phys.* **13** (2011), 025007
[17] O. Lahav et al., *Phys. Rev. Lett.* **105** (2010), 240401
[18] S. Finazzi and R. Parentani, *New J. Phys.* **12** (2010), 095015
[19] J. Steinhauer, *Nature Phys.* **10** (2014), 864
[20] I. Carusotto et al., *Eur. Phys. J.* **D56** (2010), 391
[21] J.C. Jaskula et al., *Phys. Rev. Lett.* **109** (2012), 220401
[22] P.E. Larré, A. Recati, I. Carusotto and N. Pavloff, *Phys.Rev.* **A85** (2012), 031621
[23] D. Boiron et al., arXiv:1406.5229 [cond-mat.quant-gas]
[24] D.N. Page, *Phys. Rev.* **D13** (1976), 198
[25] P.R. Anderson et al., *Phys. Rev.* **D90** (2014), 104044
[26] A. Fabbri and J. Navarro-Salas, *Modeling black hole evaporation*, Imperial College Press, London (2005)
[27] S.W. Hawking, *Phys. Rev.* **D14** (1976), 2460
[28] A. Almheiri, D. Marolf, J. Polchinski and J. Sully *JHEP* **1302** (2013), 062

Journal of Physics: Conference Series **600** (2015) 012009 doi:10.1088/1742-6596/600/1/012009

What can we learn from gravitational waves from nearby core-collapse supernovae?

Yudai Suwa[1,2], Takaaki Yokozawa[3], Mitsuhiro Asano[3], Tsubasa Kayano[4], Nobuyuki Kanda[3], Yusuke Koshio[4], and Mark R. Vagins[5,6]

[1]Yukawa Institute for Theoretical Physics, Kyoto University, Oiwake-cho, Kitashirakawa, Sakyo-ku, Kyoto, 606-8502, Japan
[2]Max-Planck-Institute für Astrophysik, Karl-Schwarzschild-Str. 1, 85748 Garching, Germany
[3]Graduate School of Science, Osaka City University, Sumiyoshi-ku, Osaka 558-8585, Japan
[4]Department of Physics, Okayama University, Okayama, Okayama, 700-8530, Japan
[5]Kavli Institute for the Physics and Mathematics of the Universe(WPI), Todai Institutes for Advanced Study, University of Tokyo, Kashiwa, Chiba 277-8583, Japan
[6]Department of Physics and Astronomy, University of California, Irvine, Irvine, CA 92697-4575, USA

E-mail: `suwa@yukawa.kyoto-u.ac.jp`

Abstract.
Core-collapse supernova is one of the expected sources of gravitational wave (GW). The GW detection can be a smoking gun to probe the still unknown explosion mechanism. In the coming era of "multi-messenger astronomy", we can use photons, neutrinos and GW simultaneously to investigate these objects. By performing multi-dimensional simulations of neutrino-radiation hydrodynamics systematically, we calculate the gravitational wave and neutrino signals from nearby (galactic) core-collapse supernova. Based on these signals we will discuss the extractable information about the very central part of core-collapse supernovae.

1. Introduction

Since there are various types of physics involved in core-collapse supernova explosions, detailed numerical simulations are indispensable. For instance, gravity describes how the matter collapses and how much energy is released during the collapse. In addition, a final outcome of supernovae are neutron stars (NSs), so that the nuclear force, which determines the structure of a NS, is critically important. Neutrino interaction rates, which give the cooling rates of the core as well as heating rates of post-shock material by neutrino absorption (this will be explained later), are treated in great detail to give a quantitatively correct answer. This is because the total amount of neutrino emission is $\sim 10^{53}$ erg, while the explosion energy itself is $\sim 10^{51}$ erg, therefore only a few percent of energy deposition by neutrinos can drive the explosion; indeed, this is the so-called standard scenario of core-collapse supernovae (see [1] for original idea and [2, 3, 4] for recent reviews). The electromagnetic interaction also plays an important role because the system consists of fully-ionized plasma, and so Coulomb scattering allows us to treat the stellar matter as if it were in the fluid state.

Since such complex phenomena are difficult to model by analytic calculations, we need to perform detailed numerical computations including all relevant physical interactions. Ever since the first pioneering numerical simulation was done by Colgate & White in 1966 [5], many scientists have put continuous efforts on producing the realistic explosion for almost 50 years, but yet nobody has succeeded, even by using supercomputers.

Not only the fact that all interactions involve the phenomena makes the system more complex, but it also makes them observable in various ways. This is because these interactions cause their own kind of emission (electromagnetic waves, neutrinos, gravitational waves, and cosmic rays) so that they can be seen with various kind of signals (particles and waves), which is called "multi-messenger astronomy". It should be noted that these signals are generated at different places due to different coupling constants with matter of the governing interactions. Hence, if an event occurs, we can get some information from different locations in the exploding stars using different signals (tomography). In addition, we can use this information from different epochs of the different times. This character makes it possible to follow up by different ways once we observe the early phase of explosion in any manner. Therefore, we can see these cosmic explosions using multi-messenger time domain astronomical observations.

In this study, we aim to develop a new strategy to give a constraint on the rotation of the core using GW and neutrino observations.

2. Hydrodynamical Simulations of Supernovae

The numerical methods used in this study are the same as [6, 7, 8, 9, 10]. In this code, we solve neutrino radiation hydrodynamic equations, which consist of hydrodynamic equations and neutrino radiative transfer equation (see [8] for more details). The nuclear equation of state [11] with the incompressibility parameter of K = 220 MeV. As for the initial condition, we employ $11.2M_\odot$ model from [12], which was used in several previous works.

The rotation is imposed with shellular rotation law as

$$\Omega(r) = \Omega_0 \frac{r_0^2}{r^2 + r_0^2}, \tag{1}$$

where $\Omega(r)$ is an angular velocity with r being the radius from the center, Ω_0 is the angular velocity at the center, and r_0 is a radius that determines the degree of differential rotation. Here, we employ $r_0 = 1000$ km. In order to investigate how the rotation affects emissions of gravitational waves and neutrinos, we perform simulations with different rotation strengths of $\Omega_0 = 0.0\ \pi$ and $1.0\ \pi$ radian s^{-1}.

In this study, we evaluate the gravitational wave emission from aspherical motion of fluids via the Newtonian quadrupole formulae of [13]. We do not discuss the gravitational wave emission from anisotropic neutrino emission [14, 15, 16] because this component contributes to GW at later time and does not affect GW around the bounce time.

The left panel of Figure 1 represents the time evolution of luminosities of electron-type neutrinos (ν_e). One can see the general trend does not depend on the initial rotation rate very much. In the first few tens of ms for ν_e, there is the so-called neutronization burst, which is generated by the rapid absorption of electrons by protons (electron capture; $\mathrm{e} + \mathrm{p} \rightarrow \mathrm{n} + \nu_e$) in the regime between the shock and protoneutron star, causing emission of large amounts of ν_e. After that the neutrino luminosity is gradually decreasing, but a large luminosity is still observed. It is found that the fast rotation model ($\Omega_0 = 1.0\pi$ rad s^{-1}) results in slightly smaller luminosity due to stronger centrifugal force and slower contraction of the PNS, while slower rotation models share similar, minimally-affected characteristics as compared to the non-rotating case. Therefore, we can argue that the currently employed rotation strength does not significantly change the generic picture about neutrinos.

The right panel of Figure 1 depicts the time evolution of the gravitational wave signal. Note that in the early phase the gravitational wave strength strongly depends on the initial

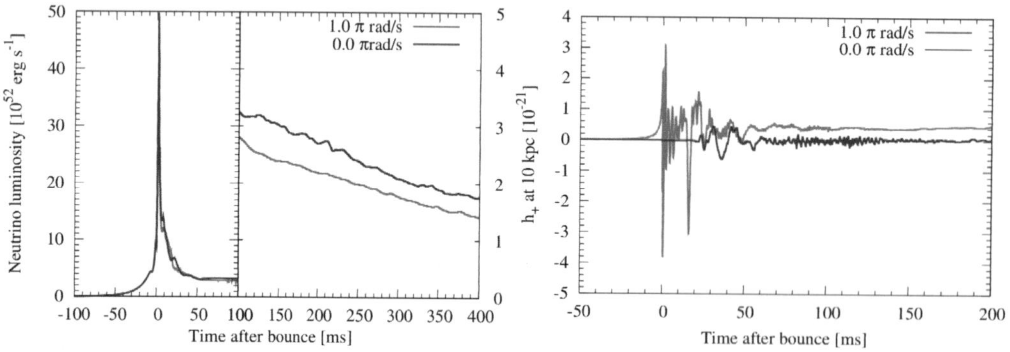

Figure 1. (*Left*): Time evolution of luminosities of electron-type neutrinos (ν_e). Each line represents models with different initial rotation rates, i.e., no rotation ($\Omega_0 = 0.0\pi$) (blue) and $\Omega_0 = 1.0\pi$ rad s^{-1} (red), respectively. (*Right*): GW amplitude as a function of time for a core-collapse supernova occurring 10 kpc from the observer. The fast rotation model exhibits a large amplitude GW at the bounce, while the slow rotation models have very small amplitudes. Since the later phase (i.e., a few tens of ms after the bounce) activity is dominated by convection motion, all models imply similar amplitude in the later phase.

rotation. As for models without rotation, the density structure is almost spherically symmetric so that there is no or only small GW emission. On the other hand, the strongly rotating model exhibits strong GW emission at the time of bounce because the centrifugal force makes the core asymmetric. Therefore, we can constrain the rotation strength by detecting GW at just the time of bounce. We can constrain the bounce time in turn using neutrino data as shown in the left panel, in which we show that the neutrino emission does not depend on the rotation strength so that neutrinos are a guaranteed signal from core-collapse supernova.

3. Detector Simulations and Detectabilities

In the following, we explain the detector simulations for KAGRA (GW detector) and EGADS (neutrino detector). The detail of the simulations and data analysis can be found in [17].

One of the important benefits of employing EGADS and KAGRA is the close proximity of these two detectors. This makes it possible to avoid any significant time-lag in arrival times of gravitational wave and neutrino signals due to distance between the detectors. Since both gravitational wave interferometers and Water Cherenkov neutrino detectors have much worse angular resolution than that of optical telescopes, the correction of arrival time based on direction for well-separated detectors will have a larger error than the timing accuracy which is required in this study, < 1 msec, for neutrino and gravitational wave signals only. Closely placed detectors do not suffer from this problem.

At first, we consider GW. To evaluate signal and noise power from obtained $s_{det}(t)$, the Excess Power Filter [18] and Short Time Fourier Transform (STFT) are used. The simulated signal is whitened via a FIR whitening filter to flaten the noise spectrum in the frequency domain. The whitened signal $\tilde{S}_w(f)$ is calculated by

$$\tilde{S}_w(t_s, f) = \frac{\tilde{S}(t_s, f)}{\langle \tilde{N}(f) \rangle}, \tag{2}$$

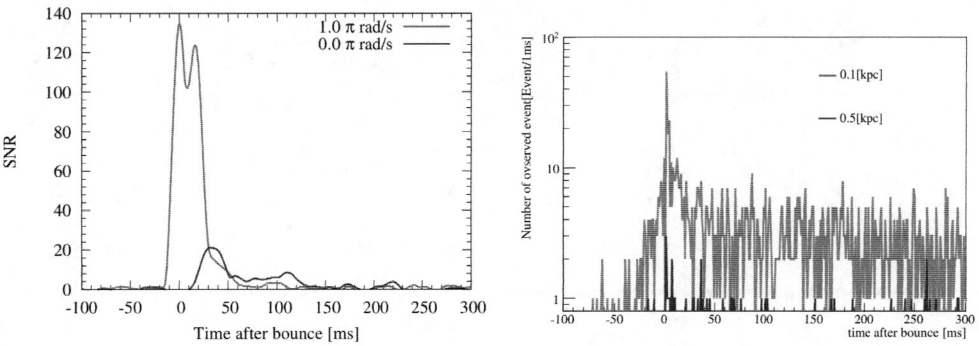

Figure 2. (*Left*): Time variation of obtained SNR. Each color shows one progenitor core rotation model, 0.0 π (blue) and 1.0 π (red) rad s^{-1}, respectively. The supernova distance is set to 1.0 kpc and on-direction for the KAGRA detector. (*Right*): Fluctuations of observed number of neutrinos for the distance 0.1 kpc (red) and 0.5 kpc (blue). The 1.0 πrad s^{-1} model is used for this figure.

and $\tilde{S}(t_s, f)$ is calculated by

$$\tilde{S}(t_s, f) = \int_{t_s}^{t_s+\Delta t} s_{det}(t')W(t' - t_s)\exp(-2\pi i f t')dt', \tag{3}$$

where $\langle \tilde{N}(f) \rangle$ is obtained from the running median of simulated noise data, and $W(t)$ is a Hann window function. The signal power P_s and Signal to Noise Ratio (SNR) can be defined as follows:

$$P_s(t_s) = \sqrt{\frac{\int \tilde{S}_w^*(t_s, f) \cdot \tilde{S}_w(t_s, f)df}{\int \langle \tilde{N}_w^*(f) \cdot \tilde{N}_w(f) \rangle\, df}}, \tag{4}$$

$$SNR(t_s) = \frac{P_s(t_s) - m}{\sigma}, \tag{5}$$

where m and σ are the normalized mean and deviation of $P_n = \sqrt{\int \tilde{N}_w^*(f) \cdot \tilde{N}_w(f)df}$ distribution, respectively. To obtain m and σ we first evaluate the noise behavior, without injecting a supernova signal, and evaluate P_n. The noise distribution shows m=1 and σ=0.06. The left panel of Figure 2 shows one example of the time variation of obtained SNR for each rotation model. The supernova distance is set to an on-direction 1.0 kpc for the KAGRA detector.

Next, we consider neutrinos. Using numerical results of neutrinos shown in the previous section and detector sensitivity, we evaluate the numbers of neutrons detectable by EGADS. The right panel of Figure 2 shows the expected number in EGADS for a supernova explosion near the center of our galaxy with $\Omega_0 = 1.0\ \pi$ rad s^{-1}.

Based on these two detector simulations, we investigate the method of how to determine progenitor core rotation from these detectors. The analysis path is as follows: (i) Run supernova detection simulation with KAGRA and EGADS detector, respectively. (ii) Extract the epoch time of GW, $\mathcal{T}_{GW}$, and neutronization burst, $\mathcal{T}_\nu$. (iii) Compare these times and determine rotation or not. (iv) Loop 10,000 times and evaluate P_r, which is the probability to of core rotation. The initial angular momentum of progenitor core rotation in the supernova models are given several different magnitudes: $\Omega_0 = 0.0\ \pi$ and $1.0\ \pi$ rad s^{-1}. The distance to the SN

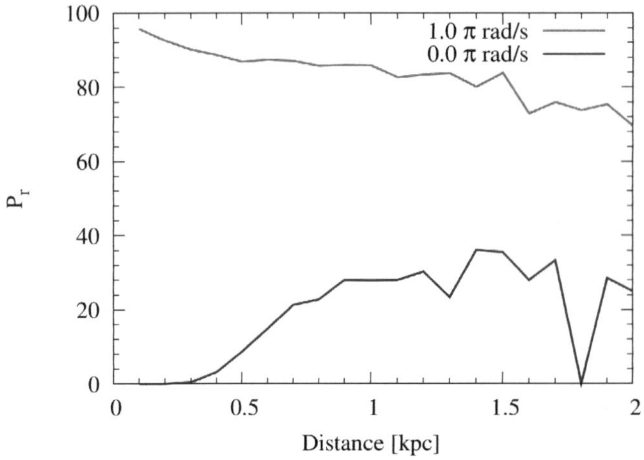

Figure 3. Model dependence of P_r distribution, which is the probability of core rotation. Horizontal axis shows distance. Each color shows one progenitor core rotation model, 0.0 π (blue) and 1.0 π (red) rad s^{-1}, respectively.

is chosen to be uniformly distributed between 0.1 to 2.0 kpc. The incident direction strongly affects GW detector response, but has little bearing on neutrino detection.

Figure 3 shows the distance dependence of P_r for each progenitor core rotation model, where P_r is defined as the probability of core rotation when both GW and neutronization burst signals have been observed. Once again, each color represents different progenitor core rotation models, 0.0 π (blue) and 1.0 π (red) rad s^{-1}, respectively. We have obtained these conclusions: (1) For weakly rotating models, the P_r value is expected to be close to 0%. In the near distance cases (0.1 and 0.2 kpc) the P_r value is almost 0 as expected. But in the far distance (>0.5 kpc) the P_r value becomes larger because of growing statistical uncertainty on the number of observed events, which leads in turn to reduced determination accuracy of the neutrinization burst time ($\mathcal{T}_\nu$). (2) For the strongly rotating model (1.0 π rad s^{-1}), the P_r value is as expected close to 100%. But, there are two peaks due to mis-identifying T_{GW}. This makes for a lower P_r for the far distance. Still, even for the 2.0 kpc case the P_r value exceeds 70%.

4. Summary

By using a consistent supernova explosion model emitting both GWs and neutrinos, we investigate the progenitor core rotation to compare the GW emission start time, $\mathcal{T}_{GW}$, obtained from KAGRA detector, and neutronization burst time, $\mathcal{T}_\nu$, which is obtained from EGADS detector. The results show if a nearby supernova is very close(<0.2 kpc), we can correctly determine no core rotation about 100% of the time if the progenitor core is indeed not rotating, and determine the presence of core rotation about 90% if the progenitor core is strongly rotating. Note that we investigated only nearby supernovae using only a single GW and neutrino detector for this analysis. For more detail, see [17].

Acknowledgments

This work was supported by the MEXT Grant-in-Aid for Scientific Research on Innovative Areas "New Developments in Astrophysics Through Multi-Messenger Observations of Gravitational Wave Sources" (Nos. 24103004, 24103005, 24103006, 25103511), JSPS postdoctoral fellowships for research abroad, MEXT SPIRE, and JICFuS. Numerical computations in this study were in part carried on XC30 at CfCA in NAOJ and SR16000 at YITP in Kyoto University.

[1] Bethe H A and Wilson J R 1985 ApJ **295** 14–23
[2] Kotake K, Takiwaki T, Suwa Y, Iwakami Nakano W, Kawagoe S, Masada Y and Fujimoto S i 2012 *Advances in Astronomy* **2012** 428757 (*Preprint* 1204.2330)
[3] Janka H T 2012 *Annual Review of Nuclear and Particle Science* **62** 407–451 (*Preprint* 1206.2503)
[4] Burrows A 2013 *Reviews of Modern Physics* **85** 245–261 (*Preprint* 1210.4921)
[5] Colgate S A and White R H 1966 ApJ **143** 626
[6] Suwa Y, Kotake K, Takiwaki T, Whitehouse S C, Liebendörfer M and Sato K 2010 PASJ **62** L49–L53
[7] Suwa Y, Kotake K, Takiwaki T, Liebendörfer M and Sato K 2011 ApJ **738** 165–+ (*Preprint* 1106.5487)
[8] Suwa Y, Takiwaki T, Kotake K, Fischer T, Liebendörfer M and Sato K 2013 ApJ **764** 99 (*Preprint* 1206.6101)
[9] Suwa Y 2014 PASJ **66** L1 (*Preprint* 1311.7249)
[10] Suwa Y, Yamada S, Takiwaki T and Kotake K 2014 *ArXiv e-prints* (*Preprint* 1406.6414)
[11] Lattimer J M and Swesty F D 1991 *Nuclear Physics A* **535** 331–376
[12] Woosley S E, Heger A and Weaver T A 2002 *Reviews of Modern Physics* **74** 1015–1071
[13] Mönchmeyer R, Schaefer G, Müller E and Kates R E 1991 A&A **246** 417–440
[14] Epstein R 1978 ApJ **223** 1037–1045
[15] Burrows A and Hayes J 1996 Phys. Rev. Lett. **76** 352–355 (*Preprint* astro-ph/9511106)
[16] Müller E and Janka H T 1997 A&A **317** 140–163
[17] Yokozawa T, Asano M, Kayano T, Suwa Y, Kanda N, Koshio Y and Vagins M R 2014 *ArXiv e-prints* (*Preprint* 1410.2050)
[18] Anderson W G, Brady P R, Creighton J D and Flanagan É É 2001 Phys. Rev. D **63** 042003 (*Preprint* gr-qc/0008066)

Journal of Physics: Conference Series **600** (2015) 012010 doi:10.1088/1742-6596/600/1/012010

General relativistic neutrino transport with spectral methods

Bruno Peres

Departament d'Astronomia i Astrofísica, Universitat de Valencia, Edifici d'Investigació Jeroni Munyoz, C/ Dr. Moliner, 50, 46100 Burjassot (Valencia)

E-mail: `bruno.peres@uv.es`

Abstract. Supernovae numerical simulations are highly demanding in computational power, yet they are also rapidly increasing in complexity. There is now a very high demand in new methods for neutrino treatment, due to the fact that neutrino is at the same time a crucial ingredient and the most expensive one in computational resources. A possibility to explore is the treatment by spectral methods, because they are known to require less resources. It is the purpose of a new code, named `Lorene's Ghost` (for Lorene's gravitational handling of spectral transport), developed to treat the problem of neutrino transport in supernovae. The code can handle full phase-space dependent problems (6D + time), retains the full energy dependence of the neutrinos, and can also handle general relativity in the conformally flat approximation. Several points of the formalism needed to obtain a formulation of the general relativistic Boltzmann equation adapted to our case are discussed, as well as some technical aspects of spectral methods.

1. Introduction

The explosion mechanism of supernovae remains a puzzle. Despite considerable effort and increasing complexity of the numerical models, no convincing consensus has emerged yet. For recent reviews, see e.g. [1–3] and references therein.

Neutrinos are one of the key ingredients in a supernova numerical simulation because they carry out more than 99% of the explosion energy, and because, in the neutrino driven explosion mechanism, they revive the shock by their deposition of energy in the gain layer. Because they decouple from the fluid before reaching this gain layer, a fluid description of the neutrinos would be wrong. One needs a description able to handle all the regimes, from fully opaque (where the neutrinos are coupled to the fluid, in the interior of the proto-neutron star) to fully transparent (after $\sim$ 1000km, where the neutrinos are completely decoupled). This mean solving the radiative transfert equation, or one simplified version of it. On the one hand, solving the radiative transfert equation, or Boltzmann equation, is computationnaly challenging, mostly because of its very high dimensionnality (seven phase space dimensions). On the other hand, it is the only way to have a quantitatively accurate treatment of the semi-transparent regime behind the shock, where the neutrinos can deposit energy to revive it.

One solution to overcome this issue may be the use of spectral methods. Spectral methods are known to have exponential convergence if the represented function is C^∞ (see section 3), meaning that they require less points, and so less computational power to represent the same

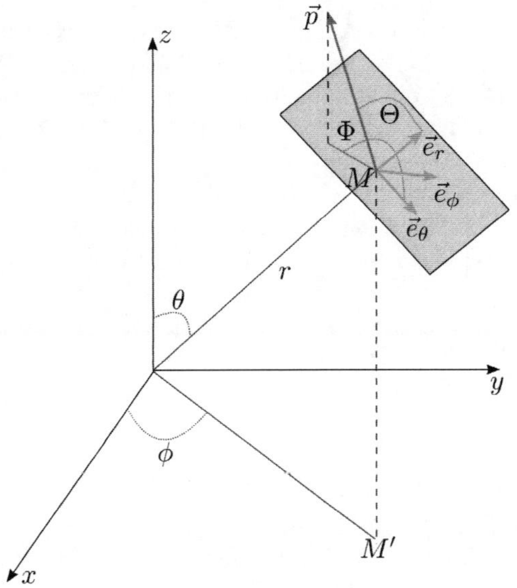

Figure 1. Representation of the 6D double spherical coordinate basis.

function. This idea lead to the writing of the `Lorene's Ghost` code [4]. `Lorene's Ghost` is an extension of the `Lorene` library [5], able to treat 7-dimensional scalar functions.

Greek indices run from 0 to 4, latin indices run from 1 to 4. I adopt the Einstein summation convention and a metric with a $(-,+,+,+)$ signature, and geometrical units in which $c = G = \hbar = 1$.

2. The Boltzmann equation

The general relativistic Boltzmann equation [4; 6; 7] can be written as

$$\epsilon\frac{\partial f}{\partial t} + p^i\frac{\partial f}{\partial x^i} - \Gamma^i{}_{\mu\nu}p^\mu p^\nu\frac{\partial f}{\partial p^i} = \mathcal{C}[f] \ , \qquad (1)$$

where f is the distribution function, x^i and t are the coordinate space and time, p^μ is the 4-momentum of the neutrinos, $\Gamma^i{}_{\mu\nu}$ are the Christoffel symbols and $\mathcal{C}[f]$ is the collision operator.

2.1. The distribution function

The distribution function is a scalar [6; 8], defined in principle on the phase space (x^μ, p^μ). Because of the mass shell condition $p^\mu p_\mu = 0$ (that holds with neutrinos, which we suppose have a negligibly small mass), the phase space reduces to seven independent dimensions, (t, x^i, p^i), as in eq.(1).

`Lorene's Ghost` uses the double spherical coordinate basis. Space is represented by a radius and two angles, as usual in a spherical coordinate basis, $x^i = (r, \theta, \phi)$, and momentum space is also represented by a radius and two angles, $p^i = (\epsilon, \Theta, \Phi)$. ϵ is the energy of the neutrinos. On figure 1, a representation of the double spherical coordinate basis is drawn, where we see clearly that the momentum space basis depends on the real space basis.

2.2. Mixing frames

In supernovae numerical simulations, most implementations of the Boltzmann equation or a simplified version of it [9–12] use the mixed frame formalism : while the spacetime variables belong to the coordinate basis (or, in non-general relativistic codes, the Eulerian basis, since both are the same if the metric is flat), the momentum space variables belong to the Lagrangian basis. This simplifies a lot the computation of the collision operator, the right hand side term of eq.(1). On the other hand, the connection coefficients are no longer the ones of the coordinate basis (the Christoffel symbols), as in eq.(1), but the connection coefficients of the Lagrangian frame, which are much harder to compute. A special treatment of these have been implemented in `Lorene's Ghost`, which avoids any time derivative (except for $\partial v/\partial t$), known to be hard to handle with spectral methods, and involves the computation of Ricci rotation coefficients (see [4]).

Physically, these connection coefficients of the Lagrangian basis include important terms for the neutrino treatment such as the Doppler shift, the gravitational redshift or the aberration due to the acceleration of the fluid.

3. Spectral methods

Spectral methods (see, *e.g.* [13; 14] and references therein) decompose a function using a global basis function (here, we use Fourier or Chebyshev bases). As a result, convergence is very rapid (exponential, unless the function is not C^∞ or close to it) and one can use fewer points than with finite differences to get to the same resolution. In the case of the neutrino treatment, this shall lead to a substantial decrease in the needed computational time.

3.1. `Lorene's Ghost`

`Lorene's Ghost` advances in time with an explicit Adams–Bashforth scheme of 3rd order. Another finite differencing in time has been tested, namely a Runge–Kutta 2nd order scheme. All six other dimensions are represented with spectral methods.

Two angles (ϕ and Φ) go from 0 to 2π and are fully periodic. These are well suited for a representation on a Fourier basis. For the θ angle, an analytic extension can be performed to recover a periodic dimension. This enables us to also use a Fourier representation. The remaining arguments (r, ϵ and Θ) are decomposed onto Chebyshev bases.

3.2. Gibbs phenomenon and filtering

When one tries to represent a discontinuous function with spectral methods, spurious oscillations appear that can potentially spoil the entire representation of the function. On figure 2 is an illustration of this phenomenon. The red analytical function $f(r)$ is discontinuous, and its representation using 33 points on a Chebyshev basis has several drawbacks : it is oscillating (in the whole domain), the exponential convergence is lost, and the discontinuity is considerably smoothed. Filtering helps getting rid of these problems. While the smoothing remains, filtering can get rid of the two other mentioned drawbacks.

Note also that a non linear operator often populates coefficients of higher order. It means that even when dealing with a smooth function well represented with a finite set of coefficients, non linearities can create oscillations, and so filtering is very often needed in spectral methods.

On figure 3, the same analytic function as on figure 2 is represented, but this time with 257 points in r and with an exponential filter applied to the result, meaning that the i-th coefficient is multiplied by

$$\exp\left(\alpha\left[\frac{i}{n_r}\right]^{2d}\right), \tag{2}$$

where n_r is the total number of coefficients in r, d is the order taken to be 1 here, and α is usually the machine precision (10^{-16} here).

On figure 3, one can see that no oscillation remains, and that the numerical representation is different from the analytical one only in the vicinity of the discontinuity.

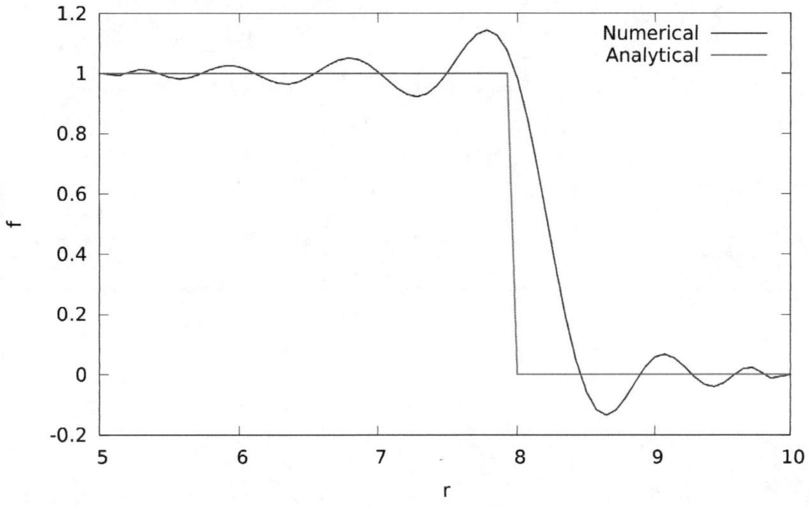

Figure 2. Representation of a discontinuity (red) by Chebyshev spectral methods (blue, 33 points). Gibbs phenomenon (oscillations) is clearly visible.

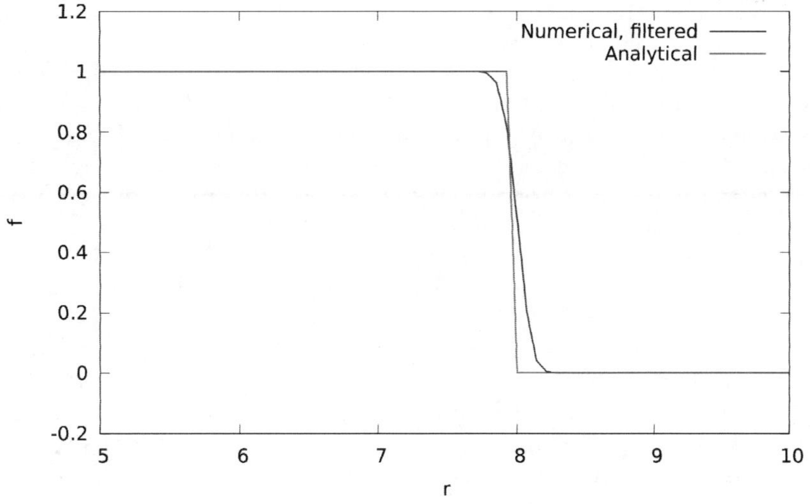

Figure 3. Representation of a discontinuity (red) by Chebyshev spectral methods (blue, 257 points) with a first order exponential filter.

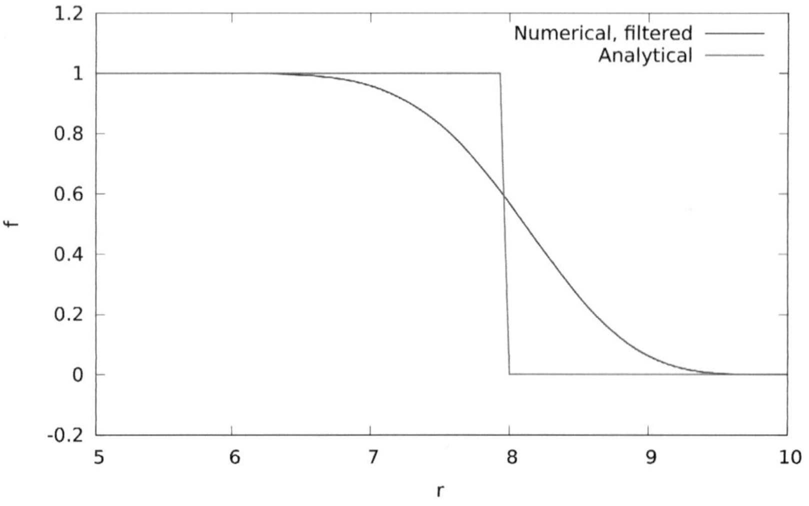

Figure 4. Representation of a discontinuity (red) by Chebyshev spectral methods (blue, 33 points) with a first order exponential filter.

On figure 4 is the same as on figure 3 but this time with 33 points in r. Note the much stronger smoothing that results in a smaller slope. This is because the first order exponential filter changes a lot the function. Other filters or higher order exponential filters would not change the function that much and can be sufficient in many cases, this test of a sharp discontinuity being an extreme case.

4. Conclusion

I showed that the treatment of the neutrino problem in supernovae is possible with spectral methods. They are viable methods with the capability of a strong speedup compared to finite difference methods. The code `Lorene's Ghost` is able to treat test cases, including 6D + time full angle and energy dependent configurations.

I addressed the main drawback of spectral methods, namely the appearance of Gibbs phenomenon due to either discontinuities, strong gradients in the represented function, or aliasing due to non linear terms in the equations. I showed that filtering is a good option to overcome this problem and is capable of getting rid of all oscillations.

`Lorene's Ghost` passed many tests, including the recovery of spectral convergence in all six dimensions, a particle number conservation with a collision term, a gravitational redshift test with a Schwarzschild metric and a so-called searchlight beam test (see [4]). All these tests, including the 7-dimensional ones, are able to run on a single core with a moderate spectral resolution.

A technical issue remains to be adressed, namely the fact that `Lorene's Ghost` cannot go down to $r = 0$ and has to stop at a finite $R_{\rm min}$. What is missing is the generalization (if possible) of the regularity conditions imposed in `Lorene` near the $r = 0$ point. If they do not exist, an alternative would be to couple `Lorene's Ghost` to an approximate solver that would be used only in the opaque regime. A diffusion equation, for instance, is sufficient in this regime. Another alternative is to treat the r dimension with finite differences.

Next steps include the coupling of `Lorene's Ghost` to an hydrodynamics solver to simulate supernovae, namely the `CoCoNuT` code [15]. Studies such as speed up tests and comparisons of approximated neutrino treatments with our Boltzmann solver will then be possible.

This work can also be extended to simplified treatments that would decrease even more the needed computational resources. It would be easy with `Lorene's Ghost` to reduce, for instance, to a Schwarzschild metric in the whole simulation, to reduce to one spatial dimension and use the so-called ray-by-ray approximation [9] to deal with multi-dimensional problems, or to reduce the velocity dependent terms in the connection coefficients (see section 2.2) to first order in the fluid velocity v.

Acknowledgments

The author acknowledges support from the European Research Council (grant CAMAP-259276), and from the Spanish Ministerio de Ciencia e Innovación (grant AYA2013- 40979-P Astrofísica Relativista Computacional) and from the Valencian Conselleria d'Educaciò (PROMETEO-2009-103).

References

[1] Kotake K, Sato K and Takahashi K 2006 *Rep. Prog. Phys.* **69** 971–1143

[2] Ott C D 2009 *Class. Quantum Grav.* **26** 204015

[3] Janka H T 2012 *Annual Review of Nuclear and Particle Science* **62** 407–451

[4] Peres B, Penner A J, Novak J and Bonazzola S 2014 *Classical and Quantum Gravity* **31** 045012

[5] Gourgoulhon E, Grandclément P, Marck J A and Novak J 1997–2012 LORENE: Langage Objet pour la RElativité NumériquE http://www.lorene.obspm.fr

[6] Stewart J M 1971 Non-equilibrium relativistic kinetic theory *Lecture Notes in Physics, Berlin Springer Verlag* vol 10 pp 1–113

[7] Cardall C Y, Lentz E J and Mezzacappa A 2005 *Phys. Rev. D* **72** 043007

[8] Ehlers J 1971 General relativity and kinetic theory. *General Relativity and Cosmology* ed Sachs R K pp 1–70

[9] Buras R, Rampp M, Janka H T and Kifonidis K 2006 *Astron. Astrophys.* **447** 1049–1092

[10] Müller B, Janka H T and Dimmelmeier H 2010 *Astrophys. J. Suppl.* **189** 104–133

[11] Bruenn S W, Lentz E J, Hix W R, Mezzacappa A, Harris J A, Bronson Messer O E, Endeve E, Blondin J M, Chertkow M A, Lingerfelt E J, Marronetti P and Yakunin K N 2014 *ArXiv e-prints* (*Preprint* `1409.5779`)

[12] Suwa Y, Kotake K, Takiwaki T, Whitehouse S C, Liebendörfer M and Sato K 2010 *Proc. Astron. Soc. Jap.* **62** L49

[13] Boyd J B 2001 *Chebyshev and Fourier Spectral Methods* 2nd ed (Mineola, New York: Dover Publications)

[14] Grandclément P and Novak J 2009 *Living Rev. Relat.* **12** 1

[15] Dimmelmeier H, Novak J, Font J A, Ibáñez J M and Müller E 2005 *Phys. Rev. D* **71** 064023

Journal of Physics: Conference Series **600** (2015) 012011 doi:10.1088/1742-6596/600/1/012011

Virial tests for post-Newtonian stationary black-hole–disk systems

Piotr Jaranowski

Wydział Fizyki, Uniwersytet w Białymstoku, Ciołkowskiego 1L, 15–245 Białystok, Poland

E-mail: p.jaranowski@uwb.edu.pl

Patryk Mach, Edward Malec, Michał Piróg

Instytut Fizyki im. Mariana Smoluchowskiego, Uniwersytet Jagielloński, Łojasiewicza 11, 30-348 Kraków, Poland

E-mail: patryk.mach@uj.edu.pl, malec@th.if.uj.edu.pl, michal.pirog@uj.edu.pl

Abstract. We investigated hydrodynamical post-Newtonian models of selfgravitating stationary black-hole–disk systems. The post-Newtonian scheme presented here and also in our recent paper is a continuation of previous, purely Newtonian studies of selfgravitating hydrodynamical disks rotating according to the Keplerian rotation law. The post-Newtonian relativistic corrections are significant even at the 1PN level. The 1PN correction to the angular velocity can be of the order of 10% of its Newtonian value. It can be expressed as a combination of geometric and hydrodynamical terms. Moreover, in contrast to the Newtonian Poincaré–Wavre theorem, it depends both on the distance from the rotation axis and the distance from the equatorial plane.

In the technical part of this note we derive virial relations valid up to 1PN order. We show that they are indeed satisfied by our numerical solutions.

1. Introduction

In a recent paper [1] we investigated post-Newtonian models of selfgravitating gaseous disks that rotate according to the Keplerian rotation law. The analysis presented there is a continuation of our previous studies, where such disk systems were investigated in Newtonian theory [2, 3].

In the Newtonian framework we posed the following problem: Suppose one observes a selfgravitating stationary gaseous disk around a central object (modeled by a point-mass) that rotates according to the Keplerian rotation law, that is with the angular velocity $v_0^\phi = \omega_0/r^{3/2}$, where r is the distance from the rotation axis, and ω_0 is a constant. For the disk consisting of test particles we have $\omega_0 = \sqrt{GM_\mathrm{c}}$, where M_c is the mass of the central object, and G is the gravitational constant. What does the observed value of ω_0 correspond to in the case where the mass of the disk is comparable with M_c? Is it still the central mass M_c, the sum of the two masses, or some nontrivial combination of them? It turns out that the selfgravity speeds up the rotation of the disk—it rotates faster than this would follow from the Keplerian formula involving the central mass M_c only. Moreover, the way in which ω_0 depends on the central mass and the mass of the disk is prescribed by the geometry of the disk. Thus, in principle, it is possible to measure the masses of Keplerian disks, whenever their geometry is known. This procedure was applied to the accretion disk in the AGN of NGC 4258, where the Keplerian rotation curve was measured in the maser emission [3].

In [1] we extended the Newtonian analysis to the first post-Newtonian approximation (1PN). Selfgravitating stationary gaseous disks were investigated before in full relativity (cf. [4, 5]). We decided on the post-Newtonian scheme, because of its conceptual simplicity. In particular, the notion of the Keplerian rotation has a clear meaning in the post-Newtonian scheme.

The main result obtained in the 1PN approximation is that the angular velocity profile is affected in two different ways—some parts of a disk can be speeded up and the others slowed down. Furthermore, the sum of the Newtonian and post-Newtonian components of the angular velocity is not anymore a function of the cylindrical radius only, but in general it depends on radial and vertical coordinates [1].

In this paper we sketch briefly the main equations that constitute the 1PN model and then derive virial-type relation that can be used to test the obtained numerical solutions. Suitable virial tests valid in the Newtonian case were presented in [6] and [2]. We discuss them here for clarity. The post-Newtonian virial identities given here are new and have not been discussed in [1]. In the last section of this paper we also show that they are satisfied by our numerical models with the accuracy similar to that of Newtonian solutions.

2. Description of the model

Our 1PN black-hole–disk models are constructed assuming the metric of the form

$$\begin{aligned} ds^2 &= g_{\mu\nu}dx^\mu dx^\nu = \left(-1 - 2\frac{U(x,y,z)}{c^2} - 2\frac{(U(x,y,z))^2}{c^4}\right)(dx^0)^2 - 2\frac{A_i(x,y,z)}{c^3}dx^i dx^0 \\ &+ \left(1 - 2\frac{U(x,y,z)}{c^2}\right)\left(dx^2 + dy^2 + dz^2\right), \end{aligned} \tag{1}$$

where we use Cartesian coordinates $x^0 = ct, x^1 = x, x^2 = y, x^3 = z$, and c is the speed of light.

We write the energy-momentum tensor as

$$T^{\alpha\beta} = \frac{M_c c^2}{\sqrt{g}} \frac{u^\alpha_{\rm BH} u^\beta_{\rm BH}}{u^0_{\rm BH}} \delta(\mathbf{x}) + \rho(c^2 + h)u^\alpha u^\beta + p g^{\alpha\beta}, \tag{2}$$

where the first component describes the point particle (it is proportional to the Dirac delta distribution and models the central black hole) at rest, located at the origin of the coordinate system; the second one is the energy-momentum tensor of the disk matter. Here M_c denotes the mass of the point particle; g is the determinant of the metric $g = -\det(g_{\mu\nu})$. The four-vectors $u^\alpha_{\rm BH}$ and u^α denote the four-velocities of the central point-mass and the fluid, respectively. The symbol ρ denotes the baryonic rest-mass density, h is the specific enthalpy, and p is the pressure. In the following sections we will also work with the three-velocity, defined as $v^i = cu^i/u^0$, $i = 1, 2, 3$.

In the remaining part of the article we use standard cylindrical coordinates (r, z, ϕ). We consider a stationary, selfgravitating, axially and equatorially symmetric polytropic disk, rotating around a central point mass M_c according to the Keplerian rotational law $v^\phi_0 = \omega_0 r^{-3/2}$. We assume that the disk is geometrically bounded by the inner and outer radius $r_{\rm in}$ and $r_{\rm out}$, respectively. We introduce the notation according to which any quantity ξ (if it is necessary) is separated into its Newtonian ξ_0 and post-Newtonian ξ_1 part according to the general pattern $\xi = \xi_0 + \xi_1/c^2$. Following [7] we derive basic equations which, separated into their Newtonian and post-Newtonian parts, read:

$$\Delta U_0 = 4\pi G\left(\rho_0 + M_c\delta(\mathbf{x})\right), \quad (3)$$

$$h_0 = -U_0 + \int dr (v_0^\phi)^2 r + C_0, \quad (4)$$

$$\Delta A_\phi = 2\frac{\partial_r A_\phi}{r} - 16\pi G r^2 \rho_0 v_0^\phi, \quad (5)$$

$$\Delta U_1 = 4\pi G\left(M_c U_0^{\mathrm{D}}(0)\delta(\mathbf{x}) + \rho_1 + 2p_0 + \rho_0\left(h_0 - 2U_0 + 2r^2(v_0^\phi)^2\right)\right), \quad (6)$$

$$h_1 = -U_1 - v_0^\phi A_\phi + 2h_0(v_0^\phi)^2 r^2 - \int dr (v_0^\phi)^4 r^3 - \frac{3}{2}h_0^2 - 4h_0U_0 - 2U_0^2 - C_1, \quad (7)$$

where Δ denotes the flat Laplacian with respect to coordinates (x^1, x^2, x^3), C_0 and C_1 are integration constants, and U_0^{D} is the gravitational potential due to the disk only, i.e., $U_0 = -GM_c/|\mathbf{x}| + U_0^{\mathrm{D}}$. They follow from the conservation law, $\nabla_\alpha T^{\alpha\beta} = 0$, the continuity equation $\nabla_\alpha(\rho u^\alpha) = 0$, and Einstein equations

$$R_{\mu\nu} - \frac{R}{2}g_{\mu\nu} = 8\pi\frac{G}{c^4}T_{\mu\nu}, \quad (8)$$

where $R_{\mu\nu}$ is the Ricci tensor and R denotes the Ricci scalar.

The above system of equations is closed by assuming an equation of state. Our numerical solutions are obtained for a polytropic equation of state of the form $p = K\rho^\gamma$, or equivalently $h_0 = K\gamma/(\gamma-1)\rho_0^{\gamma-1}$ (for the 0th order solution) and $h_1 = (\gamma-1)h_0\rho_1/\rho_0$ (for the 1PN part), where K and $\gamma > 1$ are constants.

Equations (3–7) should be solved with respect to the Newtonian gravitational potential $U_0(r,z)$, the post-Newtonian gravitational potential $U_1(r,z)$, the rotational potential $A_\phi(r,z)$ and the Newtonian and post-Newtonian enthalpy $h_0(r,z)$ and $h_1(r,z)$. Any other quantity (the density $\rho(r,z)$, the pressure $p(r,z)$, etc.) can be obtained from these five functions.

Numerical solutions are obtained as follows. We use the classic Self-Consistent Field (SCF) scheme (cf. [8]) to solve the set of Eqs. (3) and Eq. (4). Given the Newtonian potential Y_0 and the enthalpy h_0, we can solve Eq. (5) at once. Finally, we use again the SCF scheme to solve the set of Eqs. (6) and (7). For the Keplerian rotation law the above method converges for all values of the parameters that we have tested.

3. Results

The main result of our post-Newtonian scheme is the correction to the Newtonian angular velocity profile. The well-known theorem by Poincaré and Wavre states that Newtonian stationary barotropic disks (or stars) rotate with the angular velocity that can depend on the distance r from the rotation axis only [9]. It turns out that already the 1PN correction can be significant and in general it depends also on z. It can be expressed as

$$v_1^\phi(r,z) = -\frac{A_\phi}{2rv_0^\phi}\partial_r v_0^\phi + 2rh_0\partial_r v_0^\phi. \quad (9)$$

Note that the above formula involves both geometric and hydrodynamical factors.

Acceptable 1PN solutions should satisfy the following (quite stringent) conditions: *i)* $1 \gg |U_0|/c^2 \gg |U_1|/c^4$, *ii)* $2GM_c/c^2 \ll r_{\mathrm{in}}$, *iii)* $c \gg c_s$, where c_s is a speed of sound. Sample numerical models that do satisfy the above conditions are presented in [1]. It turns out that the 1PN correction to the velocity can be of the order of 10% of the Newtonian value. The reader interested in the details of these model may consult [1]. In the remainder of this paper we focus on the construction of virial tests that can be applied to our numerical scheme.

4. Virial identities

In this section we use Cartesian and cylindrical coordinates. It is implicitly assumed that Latin indices refer to Cartesian coordinates.

The virial identity that can be used to test the Newtonian solution (including the central point-mass) was obtained in [6]. It reads

$$E_{\rm pot} + 2E_{\rm kin} + 2E_{\rm therm} = 0, \tag{10}$$

where $E_{\rm pot} = \frac{1}{2}\int_{\mathbb{R}^3} d^3x\ \rho(U_0 - GM_{\rm c}/|\mathbf{x}|)$ is the total potential energy, $E_{\rm kin} = \frac{1}{2}\int_{\mathbb{R}^3} d^3x\ \rho v_{0i} v_0^i$ is the bulk kinetic energy and $E_{\rm therm} = \frac{3}{2}\int_{\mathbb{R}^3} d^3x\ p$ is the internal thermal energy. We assume, as a virial test parameter, the value

$$\epsilon = \left| \frac{E_{\rm pot} + 2E_{\rm kin} + 2E_{\rm therm}}{E_{\rm pot}} \right|.$$

In order to obtain the post-Newtonian relations we rewrite Eqs. (5) and (6) in a slightly different form. Equation Eqs. (5) can be written in Cartesian coordinates as

$$\Delta A^i = -16\pi G \rho_0 v_0^i.$$

Equation (6) is split in two parts:

$$\Delta U_1' = 4\pi G \left(\rho_1 + 2p_0 + \rho_0 \left(h_0 - 2U_0 + 2r^2 (v_0^\phi)^2 \right) \right), \tag{11}$$

$$\Delta U_1'' = 4\pi G M_{\rm c} U_0^{\rm D}(0) \delta(\mathbf{x}), \tag{12}$$

where $U_1 = U_1' + U_1''$. The solution for U_1'' is

$$U_1'' = -\frac{GM_{\rm c}U_0^{\rm D}(\mathbf{0})}{|\mathbf{x}|}.$$

Consider a vector

$$a^i = \left(x^l \partial_l A_k + \frac{1}{2} A_k \right) \partial^i A^k - \frac{1}{2} x^i \partial_l A^k \partial^l A_k.$$

Its divergence reads

$$\partial_i a^i = \left(x^l \partial_l A_k + \frac{1}{2} A_k \right) \Delta A^k = -16\pi G \left(x^l \partial_l A_k + \frac{1}{2} A_k \right) \rho_0 v_0^k. \tag{13}$$

For a finite disk (ρ_0 of compact support), A_k tends to zero sufficiently fast, and

$$|\mathbf{x}|^2 a^i \to 0, \quad \text{as} \quad |\mathbf{x}| = \sqrt{x^2 + y^2 + z^2} \to \infty.$$

Thus, by integrating Eq. (13) over $\mathbb{R}^3$, and making use of the Gauss theorem, we see that

$$0 = \int_{\mathbb{R}^3} d^3x \left(x^l \partial_l A_k + \frac{1}{2} A_k \right) \rho_0 v_0^k.$$

Integrating by parts, one can get rid of the term with $\partial_l A_k$. This yields

$$0 = \int_{\mathbb{R}^3} d^3x \left(-\frac{5}{2} \rho_0 A_k v_0^k - x^l \partial_l \left(\rho_0 v_0^k \right) A_k \right),$$

where only 0th order terms are differentiated.

The above relation can be also written in terms of the vector components in cylindrical coordinates. Because of symmetry assumptions, we have $A_k v_0^k = A_\phi v_0^\phi$, and

$$x^l \partial_l \left(\rho_0 v_0^k\right) A_k = x^l \nabla_l \left(\rho_0 v_0^k\right) A_k = \rho_0 A_\phi v_0^\phi + x^l \partial_l \left(\rho_0 v_0^\phi\right) A_\phi.$$

This yields

$$0 = \int_{\mathbb{R}^3} d^3x \left(-\frac{7}{2}\rho_0 A_\phi v_0^\phi - x^l \partial_l \left(\rho_0 v_0^\phi\right) A_\phi\right).$$

The virial relation following from Eq. (11) can be obtained in a similar way. It is enough to consider the divergence of the vector

$$b^i = \left(x^l \partial_l U_1' + \frac{1}{2} U_1'\right) \partial^i U_1' - \frac{1}{2} x^i \partial_l U_1' \partial^l U_1'.$$

It yields

$$\partial_i b^i = \left(x^l \partial_l U_1' + \frac{1}{2} U_1'\right) \Delta U_1',$$

and analogously

$$0 = \int_{\mathbb{R}^3} d^3x \left(x^l \partial_l U_1' + \frac{1}{2} U_1'\right) \left(\rho_1 + 2p_0 + \rho_0 \left(h_0 - 2U_0 + 2r^2 (v_0^\phi)^2\right)\right).$$

Many different forms of the above relation can be obtained by 'playing' with Eq. (7). A helpful relation that can be used here is

$$-x^l \partial_l U_1'' = U_1''.$$

In the analogy to the Newtonian case we choose as virial test parameters $\epsilon' = |(\epsilon_a' + \epsilon_b')/\epsilon_a'|$ and $\epsilon'' = |(\epsilon_a'' + \epsilon_b'')/\epsilon_a''|$, where

$$\epsilon_a' = \int_{\mathbb{R}^3} d^3x \, \frac{7}{2} \rho_0 A_\phi v_0^\phi, \tag{14}$$

$$\epsilon_b' = -\int_{\mathbb{R}^3} d^3x \, x^l \partial_l \left(\rho_0 v_0^\phi\right) A_\phi, \tag{15}$$

$$\epsilon_a'' = -\int_{\mathbb{R}^3} d^3x \, \frac{1}{2} U_1' \left(\rho_1 + 2p_0 + \rho_0 \left(h_0 - 2U_0 + 2r^2 (v_0^\phi)^2\right)\right), \tag{16}$$

$$\epsilon_b'' = \int_{\mathbb{R}^3} d^3x \, x^l \partial_l U_1' \left(\rho_1 + 2p_0 + \rho_0 \left(h_0 - 2U_0 + 2r^2 (v_0^\phi)^2\right)\right). \tag{17}$$

In Table 1 we report results of the convergence tests for a sample system. In our implementation, numerical precision is controlled by the resolution of the grid, the maximum number L of Legendre polynomials used in the angular expansion of the solutions of the scalar and vector Poisson equations, and a value of the maximal difference between density distributions obtained in the last two consecutive iterations ρ_{tol}. (In each iteration we compute the quantity $\rho_{\mathrm{err}} = \max_{i,j} |\rho_{i,j}^{(k+1)} - \rho_{i,j}^{(k)}|$. Here index k numbers subsequent iterations; indices i and j refer to different grid nodes. The iteration procedure is stopped, when $\rho_{\mathrm{err}} \le \rho_{\mathrm{tol}}$.)

The results of the virial test do depend on the grid resolution and almost do not depend on the numbers of Legendre polynomials and the precision in convergence procedure.

Table 1. Typical dependence of the results on the resolution of the numerical grid, the maximum number of the Legendre polynomials L, and the tolerance coefficient ρ_{tol}. These results are obtained for a polytropic fluid with polytropic index $\gamma = 5/3$ and the Keplerian rotation law $v_0^\phi = \omega_0/r^{3/2}$. The mass of the disk is $M_\text{d} = 0.327M_\text{c}$, the inner and outer radii are $r_\text{in} = 50R_\text{S}$ and $r_\text{out} = 500R_\text{S}$ respectively.

Resolution	L	ρ_{tol}	ϵ	ϵ'	ϵ''
100 × 100	100	10^{-6}	2.53×10^{-5}	2.29×10^{-4}	3.51×10^{-4}
200 × 200	100	10^{-6}	6.36×10^{-6}	5.87×10^{-5}	8.87×10^{-5}
400 × 400	100	10^{-6}	1.56×10^{-6}	1.93×10^{-5}	2.29×10^{-5}
800 × 800	100	10^{-6}	3.66×10^{-7}	3.65×10^{-6}	4.62×10^{-6}
1200 × 1200	100	10^{-6}	1.44×10^{-7}	1.56×10^{-6}	2.49×10^{-6}
1600 × 1600	100	10^{-6}	6.66×10^{-8}	5.43×10^{-7}	1.39×10^{-6}
400 × 400	50	10^{-6}	1.56×10^{-6}	1.93×10^{-5}	2.27×10^{-5}
400 × 400	75	10^{-6}	1.56×10^{-6}	1.93×10^{-5}	2.29×10^{-5}
400 × 400	100	10^{-6}	1.56×10^{-6}	1.93×10^{-5}	2.29×10^{-5}
400 × 400	125	10^{-6}	1.56×10^{-6}	1.93×10^{-5}	2.30×10^{-5}
400 × 400	150	10^{-6}	1.56×10^{-6}	1.93×10^{-5}	2.30×10^{-5}
400 × 400	100	10^{-5}	1.52×10^{-6}	1.94×10^{-5}	2.29×10^{-5}
400 × 400	100	10^{-6}	1.56×10^{-6}	1.93×10^{-5}	2.29×10^{-5}
400 × 400	100	10^{-7}	1.59×10^{-6}	1.93×10^{-5}	2.29×10^{-5}
400 × 400	100	10^{-8}	1.59×10^{-6}	1.93×10^{-5}	2.29×10^{-5}

Acknowledgments
This research was carried out with the supercomputer 'Deszno' purchased thanks to the financial support of the European Regional Development Fund in the framework of the Polish Innovation Economy Operational Program (contract no. POIG. 02.01.00-12-023/08). The work of PJ was partially supported by the Polish NCN grant *Networking and R&D for the Einstein Telescope.* PM and MP acknowledge the support of the Polish Ministry of Science and Higher Education grant IP2012 000172 (Iuventus Plus).

References
[1] P. Jaranowski, P. Mach, E. Malec, and M. Piróg, arXiv:1410.8527v1 [gr-qc] 30 Oct 2014
[2] P. Mach, E. Malec, and M. Piróg, Acta Phys. Pol. **B44**, 107 (2013)
[3] P. Mach, E. Malec, and M. Piróg, Acta Phys. Pol. **B43**, 2141 (2012)
[4] S. Nishida and Y. Eriguchi, ApJ. **427**, 429 (1994)
[5] S. Nishida, Y. Eriguchi, and A. Lanza, ApJ. **401**, 618 (1992)
[6] P. Mach, Mon. Not. R. Astron. Soc. **422**, 772 (2012)
[7] T. Damour, P. Jaranowski, and G. Schäfer, Phys. Lett. **B513**, 147 (2001)
[8] J.P. Ostriker and J.W-K. Mark, ApJ. **151**, 1075 (1968)
[9] J.L. Tassoul, *Theory of Rotating Stars*, Princeton Univ. Press, Princeton, NJ 1978

Journal of Physics: Conference Series **600** (2015) 012012 doi:10.1088/1742-6596/600/1/012012

Shocks in the low angular momentum accretion flow

Petra Suková and Agnieszka Janiuk

Center for Theoretical Physics, Polish Academy of Sciences, Warsaw, Poland

E-mail: psukova@cft.edu.pl

Abstract. We address the variability of low luminous galactic nuclei including the Sgr A* or other transient accreting systems, e.g. the black hole X-ray binaries, such as GX 339-4 or IGR J17091. These sources exhibit bright X-ray flares and quasi-periodical oscillations and are theoretically interpreted as the quasi-spherical accretion flows, formed instead of or around Keplerian accretion disks. In low angular momentum flows the existence of shocks for some range of leading parameters (energy, angular momentum and adiabatic constant of the gas) was studied semi-analytically. The possible hysteresis effect, caused by the fact that the evolution of the flow and the formation of the shock depends on its own history, was discovered. The presence of the shock in the accreted material is important for the observable properties of the out-coming radiation. In the shocked region the gas is dense and hot, thus much more luminous than in the other case. We study the appearance of standing shocks in low angular momentum gas accreting onto a black hole with numerical hydrodynamical simulations, using the ZEUS code with Paczynski-Wiita pseudo-Newtonian potential.

1. Introduction

The possible existence of shocks in low angular momentum flows has been studied from different points of view during the last thirty years. Quasi-spherical distribution of the gas endowed by constant specific angular momentum λ and the arisen bistability was studied already by [1]. More recently, the shock existence was found also for the disc-like structure in hydrostatic equilibrium with low angular momentum both in pseudo-Newtonian potential [2] and in full relativistic approach [3]. Two component advective model was proposed to describe the system with Keplerian disc accompanied by a low angular momentum component – the corona [4]. In this perspective the observed QPOs are also being linked with oscillations of the shock front [5].

In this contribution we follow up with our recent study [6], where more detailed discussion can be found. We proceed with the study of quasi-spherical configuration and perform "1.5-D" numerical simulations of the flow with changing specific angular momentum at the outer boundary in ZEUS code. Our main aim is to address the variability of the outgoing radiation.

2. Quasi-spherical slightly rotating flow

In our study we consider the non-viscous accretion flow with the polytropic equation of state for the gas pressure $p = K\rho^\gamma$, where γ is the adiabatic index, hence the local sound speed a is given by the relation $a^2 = \gamma p/\rho = \gamma K\rho^{\gamma-1}$, where ρ is the gas density.

The mass accretion rate $\dot{M}$, which is a constant linked to the chosen density units, and accordingly the entropy accretion rate $\dot{\mathcal{M}}$ is in the spherical geometry of the flow given by [4]

$$\dot{\mathcal{M}} = \dot{M}K^n\gamma^n = u\rho r^2 K^n\gamma^n = ur^2a^{2n}. \tag{1}$$

Energy conservation for the steady state is written in the form

$$\epsilon = \frac{1}{2}u^2 + \frac{a^2}{\gamma - 1} + \frac{\lambda^2}{2r^2} + \Phi(r), \tag{2}$$

where $\Phi(r)$ is the gravitational potential.

The radial gradient of the flow velocity is obtained from the equations after performing appropriate differentiation (for detailed derivation of all presented equations see [6])

$$\frac{\mathrm{d}u}{\mathrm{d}r} = \frac{\frac{\lambda^2}{r^3} - \frac{\mathrm{d}\Phi(r)}{\mathrm{d}r} + \frac{2a^2}{r}}{u - \frac{a^2}{u}} = \frac{\frac{\lambda^2}{r^3} - \frac{1}{2(r-1)^2} + \frac{2a^2}{r}}{u - \frac{a^2}{u}}. \tag{3}$$

The last equality holds for the Paczynski-Wiita potential, which in the form $\Phi(r) = \frac{1}{2(r-1)}$ mimics the most important features of strong gravitational field near the compact object (comparison of more pseudo-Newtonian potentials could be found in [2]).

The critical points position is given as the position r_c, where the numerator and denominator of equation (3) are both equal to zero, and is found as a solution of the equation

$$\epsilon - \frac{\lambda^2}{2r_c^2} + \frac{1}{2(r_c - 1)} - \frac{\gamma + 1}{2(\gamma - 1)}\left(\frac{r_c}{4(r_c - 1)^2} - \frac{\lambda^2}{2r_c^2}\right) = 0. \tag{4}$$

For a certain subset of the parameters $(\epsilon, \lambda, \gamma)$ equation (4) yields the existence of three different critical points, two of which r_{in} and r_{out} are of the saddle type, hence locally the accretion flow can pass through them (inner and outer branch of solution, respectively). However, because the inner critical point is homoclinic, the appropriate inner solution orbit returns to it after making a loop in the $r - \mathfrak{M}$ diagram. Hence, the accretion flow which is coming from infinity down onto the compact object cannot follow this branch only. Nevertheless, if Rankine-Hugoniot conditions, which are in our case expressed by the relation for the preshock $(-)$ and postshock $(+)$ Mach number $\mathfrak{M}_{\pm} = u_{\pm}/a_{\pm}$ as

$$\frac{\left(\frac{1}{\mathfrak{M}_+} + \gamma\mathfrak{M}_+\right)^2}{\mathfrak{M}_+^2(\gamma - 1) + 2} = \frac{\left(\frac{1}{\mathfrak{M}_-} + \gamma\mathfrak{M}_-\right)^2}{\mathfrak{M}_-^2(\gamma - 1) + 2}, \tag{5}$$

are fulfilled at certain radius r_s, then the shock can form at that position. This causes the flow jump from the outer branch with a lower entropy accretion rate $\dot{\mathcal{M}}_{\text{out}}$ to the inner branch with a higher entropy accretion rate $\dot{\mathcal{M}}_{\text{in}}$, while entropy is being produced at the shock front.

3. Numerical simulations

Using the above semi-analytical solution describing the accretion flow with shock as a initial condition we follow the dynamical evolution of the flow with Paczynski-Wiita pseudo-Newtonian potential included in the simulations with the hydrodynamical code ZEUS. The detailed description of the routine used for the initialization of the grid flow variables and the boundary conditions is given in [6].

In the multicritical region two different accretion solutions are possible, the outer branch without shock, and the solution with shock, connecting the outer and inner branch. These two types of accretion flows are depicted in Fig. 1, where the radial profiles of the most important variables $(\mathfrak{M}, u, \rho)$ are plotted for both. The shock location r_s is the point, where the supersonic flow $(\mathfrak{M} > 1)$ becomes suddenly subsonic $(\mathfrak{M} < 1)$, the velocity of the flow decreases and the density rises significantly. As can be seen in this Figure, the difference between the two possible

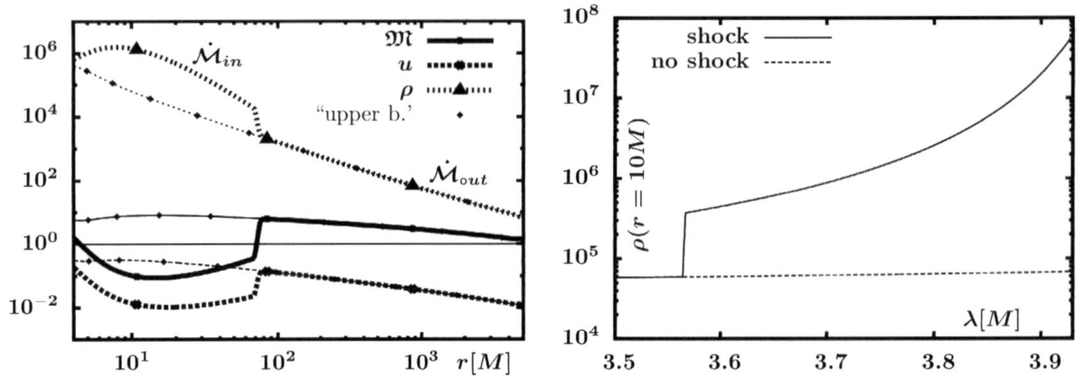

Figure 1. Radial profiles of $\mathfrak{M}$, ρ and u for flow with and without shock (denoted by diamonds), $\epsilon = 0.0001$, $\lambda = 3.75M$, $\gamma = 4/3$ (left). Comparison of density at radius $r = 10M$ for solution with shock (solid line) and without shock (dashed line) with the same accretion rate $\dot{M}$ (right).

densities is about one order of magnitude, hence the total mass of the gas enclosing the compact object up to a certain radius is very different.

The increase of density in the inner region of the flow with a constant mass accretion rate $\dot{M}$ depends on the specific angular momentum λ. In the right panel of Fig. 1 the density at a given radius ($r = 10M$) is plotted against λ for outer branch and the shock solution. For low value of λ no shock solution can be found and only the outer accretion flow exists. For λ lying in the shock existence interval the density of the shocked flow rises much faster with λ than the density of the outer branch. For a high λ their values differ by almost three orders of magnitude.

Hence for particular conditions at infinity given by the value of energy ϵ and mass accretion rate $\dot{M}$, which is related to the density at infinity ρ_∞ (see [6]), the accretion flow profile differs significantly with changing λ. Therefore we study the behaviour of the flow when the specific angular momentum of matter incoming through the outer boundary changes periodically with time. The boundary value of the angular velocity is given by the relation

$$v_3(t) = \lambda(t)/r, \qquad \lambda(t) = \lambda_0 + \Delta\lambda \sin\left(\frac{2\pi t}{T_v}\right), \tag{6}$$

where $\Delta\lambda$ is the amplitude of angular momentum variation and T_v is the period of the variation.

Due to the fact, that for parameters in the multicritical region two different steady solutions exist, the "hysteresis loop"-like behaviour can occur, as was suggested by [3]. In that case different initial conditions lead to different steady solution. Hence, in dynamical evolution the choice between these two states depends on the history of the flow. Namely when the flow is going through the outer type of critical point only (as is the only possibility for low λ), then the shock will not create even when the angular momentum rises into the multi-critical region. On the other hand, when the flow already had passed through the inner critical point (as the only solution for a very high λ does), then for decreasing λ the shock appears.

For simulating this effect we need the change of λ to be high enough to cross both lower and upper boundary of the multi-critical region. At the same time, we need to be aware of the numerical limitations given by the size of the grid and the chosen boundary conditions, which are not consistent with the inner type of solution. Therefore the angular momentum must not be too high. Otherwise the solution in the outer part would change and would interfere with the outer boundary, causing the production of some spurious reflecting waves.

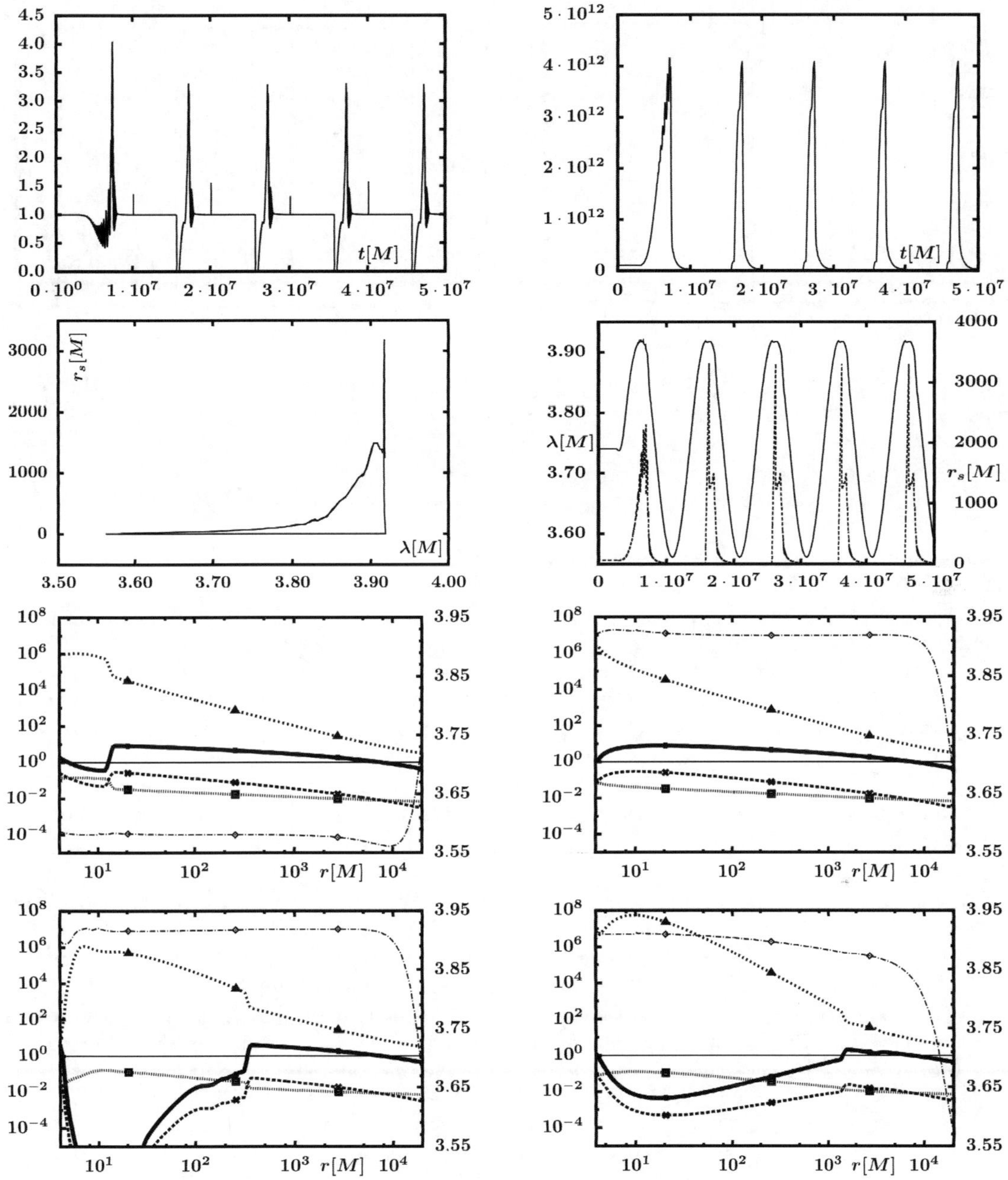

Figure 2. Accretion rate $\dot{M}_n(t)$ (1st row left) and total mass of gas enclosed inside the radius $r = 100M$ (1st row right). Position of shock r_s with respect to $\lambda(t, r = 100M)$ (2nd row left). $r_s(t)$ (dashed line) and $\lambda(r = 100M)$ (solid line, 2nd row right). For no shock we set $r_s = 0$. Snapshots of the accretion flow profiles at $t_1 = 2.01$ (3rd row left), $t_2 = 2.57$ (3rd row right), $t_3 = 2.575$ (4th row left), $t_4 = 2.695$ (4th row right) $\cdot 10^7 M$ – left y-axis stands for dimensionless $\mathfrak{M}$, $u[c]$ and $\rho[\rho_\infty]$, right y-axis stands for $\lambda[M]$. $\epsilon = 0.0001$, $\gamma = 4/3$, $\lambda_0 = 3.74M, \Delta\lambda = 0.18M$, $T_v = 10^7 M$.

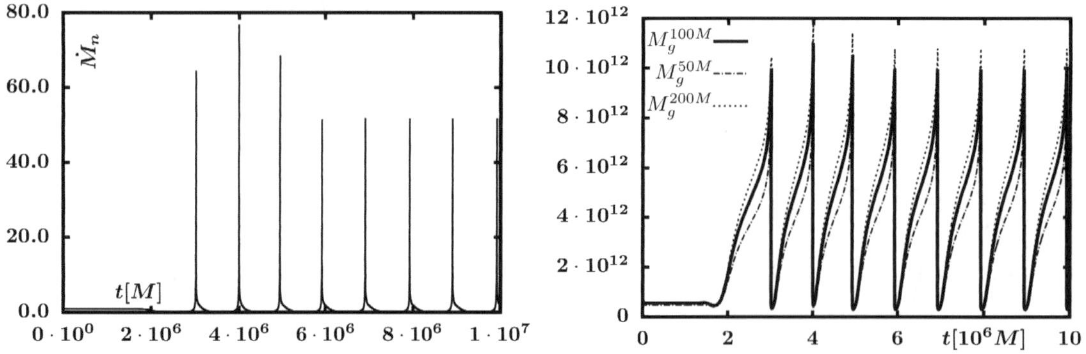

Figure 3. Normalized accretion rate through the inner boundary $\dot{M}_n(t)$ (left) and the total mass of gas $M_g^{r_i}(t)$ enclosed inside the radius $r_1 = 100M$, $r_2 = 50M$, $r_3 = 200M$ (right).

In Fig. 2 one example of such solution for $\lambda_0 = 3.74M, \Delta\lambda = 0.18M$ and $T_v = 10^7M$ is shown. In the top left panel the time dependence of the normalized accretion rate through the inner boundary of the grid, defined as $\dot{M}_n(t) = \dot{M}(t)/\dot{M}(t=0)$ is plotted. Broad and sharp peak could be seen, for more details see [6].

However, the changes of the accretion rate at the inner boundary are not the only important quantity for the out-coming radiation. Namely for higher λ the matter is gathered up in the innermost part of the accretion flow (within the inner several tens of Schwarzschild radii)), where the density increases significantly (see Fig. 1), even though the mass accretion rate remains constant. Thus for higher λ there is much more matter very close to the compact object, presumably being hot and emitting a lot of radiation. In the top right panel of Fig. 2 the total mass M_g^{100M} enclosed inside the radius $r = 100M$ is depicted. Its very sharp increase corresponds to the abrupt switch between the outer branch and shock solution, when the angular momentum in the inner region approaches its maximal value. (See the snapshots at time $t_2 = 2.57 \cdot 10^7M$ ($3^{\rm rd}$ row right) and $t_3 = 2.575 \cdot 10^7M$ (bottom left)). This is connected with a temporal switch off of the accretion rate, because the matter is stopped at several Schwarzschild radii. Then slower decline for decreasing angular momentum is seen accompanied by a peak in $\dot{M}_n$ owing to the gas overcoming the decreasing barrier and plunging into the black hole. Similar behaviour shows also the shock position r_s, which time dependence is plotted in the $2^{\rm nd}$ row right panel of Fig. 2. The second row left panel shows the loop behaviour of the shock, when the shock position decreases with angular momentum (measured at some chosen radius in the inner region, $r = 100M$), but is not present (then we set $r_s = 0$) when the angular momentum is increasing. The points $r_s(t)$ versus $\lambda(t, r = 100M)$ are plotted for $t > 1.025 \cdot 10^7M$, which covers four consecutive loops, during which repeating creation and disappearance of the shock can be seen.

In case when $\Delta\lambda$ is smaller so that the angular momentum does not leave the multicritical region, the evolution is a bit different, mainly the shock does not disappear or create, depending on the initial conditions. Such example is given in Figs. 3 and 4 with $\gamma = 4/3$, $\epsilon = 0.0001$, $\lambda_0 = 3.77M$, $\Delta\lambda = 0.2M$, $T_v = 10^6M$ (eventhough the amplitude is quite high, the period is so short, that the maximal value cannot pass through the last cells of the grid). The period was chosen with regard to the microquasar IGR J17091-3624 which in a "heartbeat" mode exhibits quasi-periodic flares with a period of a few tens of second. For the estimate of its mass $M \sim 6M_\odot$ [7], the period $T_v = 10^6M \sim 29.55s$. For such parameters we obtain very high and sharp peaks in $\dot{M}_n$ (Fig. 3 left). But the quantity $M_g^{r_i}$ shows periodical peaks with a shape very similar to the light curve of IGR J17091-3624, namely peaks with slow rise and sharp decline. The choice

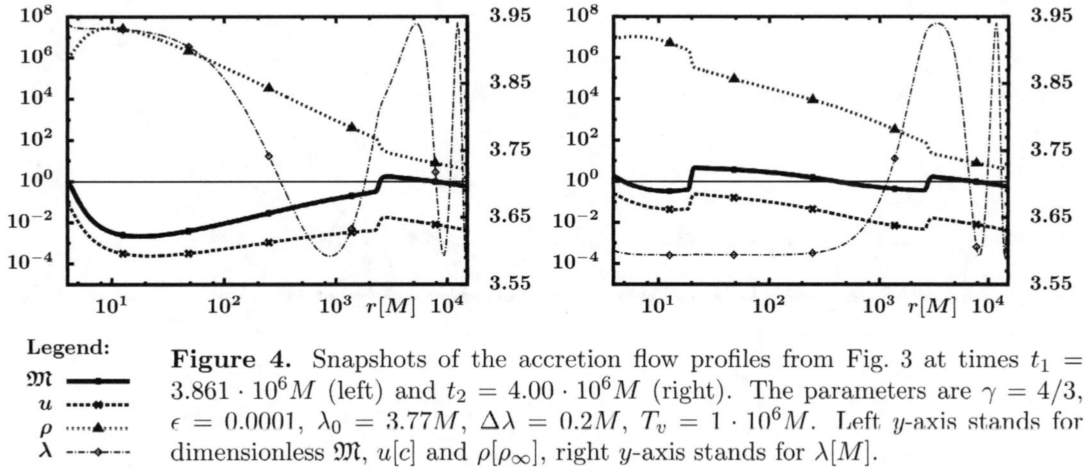

Figure 4. Snapshots of the accretion flow profiles from Fig. 3 at times $t_1 = 3.861 \cdot 10^6 M$ (left) and $t_2 = 4.00 \cdot 10^6 M$ (right). The parameters are $\gamma = 4/3$, $\epsilon = 0.0001$, $\lambda_0 = 3.77M$, $\Delta\lambda = 0.2M$, $T_v = 1 \cdot 10^6 M$. Left y-axis stands for dimensionless $\mathfrak{M}$, $u[c]$ and $\rho[\rho_\infty]$, right y-axis stands for $\lambda[M]$.

of the threshold radius r_i affects only slightly the height and the shape of the peak, and in Fig. 3 we show the comparison for $r_1 = 100M$, $r_2 = 50M$ and $r_3 = 200M$. Of course, to obtain a light curve predicted by our model more precisely, radiation characteristic at each radius and the possible contribution of another component of the accretion flow has to be taken into account.

We conclude that our simplified model offers a reasonable qualitative explanation of quasi-periodic X-ray flares originating in the low angular momentum flows. The periodic change of the angular momentum at some specified threshold radius may be responsible for the flaring activity behind this radius, with timescales and profiles' shape similar to the observed ones. In particular, for the microquasar IGR J17091-3624, the thermal instability of the Keplerian, geometrically slim accretion disk was proposed to account for the variability observed in soft X-rays [8]. The hard X-ray emitting corona, which forms above the disk in its innermost radial zone, may posses a sub-Keplerian quasi-spherical configuration. Since T_v at the radius where the coronal flow develops is of the same order as the flaring period, the hard X-ray flares will be connected with the conditions in the disk/corona boundary and the evolution of angular momentum in the coronal flow. The point that this phenomenon cannot be connected with e.g., the orbital motion of the companion star, which is by orders of magnitude longer, is further supported by the fact that in another microquasar showing this kind of activity, GRS 1915+105, the companion star and binary parameters are very different than in IGR J17091-3624 [9].

Acknowledgments
Authors wishing to thank Bozena Czerny, Tapas Das, Marat Gilfanov and Mikolaj Grzedzielski for helpful discussions. This work was supported in part by grant DEC-2012/05/E/ST9/03914 from the Polish National Science Center.

References
[1] Abramowicz M A and Zurek W H 1981 *ApJ* **246** 314–320
[2] Das T K 2002 *ApJ* **577** 880–892 (*Preprint* astro-ph/0212119)
[3] Das T K and Czerny B 2012 *NA* **17** 254–271 (*Preprint* 0906.4559)
[4] Chakrabarti S K 1989 *ApJ* **347** 365–372
[5] Das T K 2003 *ApJ Lett.* **588** L89–L92 (*Preprint* astro-ph/0302014)
[6] Suková P and Janiuk A 2014 *Preprint* 1411.7836 [astro–ph.HE]
[7] Rebusco P, Moskalik P, Kluźniak W and Abramowicz M A 2012 *A&A* **540** L4 (*Preprint* 1202.0483)
[8] Janiuk A, Grzedzielski M, Capitanio F and Bianchi S 2014 *Preprint* 1411.4434 [astro–ph.HE]
[9] Janiuk A and Czerny B 2005 *MNRAS* **356** 205–216 (*Preprint* astro-ph/0409671)

Hartle's model within the general theory of perturbative matchings: the change in mass

Borja Reina, Raül Vera

Dept. of Theoretical Physics and History of Science,
University of the Basque Country UPV/EHU,
644 PK, Bilbao 48080, Basque Country, Spain

E-mail: borja.reina@ehu.es, raul.vera@ehu.es

Abstract. Hartle's model provides the most widely used analytic framework to describe isolated compact bodies rotating slowly in equilibrium up to second order in perturbations in the context of General Relativity. Apart from some explicit assumptions, there are some implicit, like the "continuity" of the functions in the perturbed metric across the surface of the body. In this work we sketch the basics for the analysis of the second order problem using the modern theory of perturbed matchings. In particular, the result we present is that when the energy density of the fluid in the static configuration does not vanish at the boundary, one of the functions of the second order perturbation in the setting of the original work by Hartle is not continuous. This discrepancy affects the calculation of the change in mass of the rotating star with respect to the static configuration needed to keep the central energy density unchanged.

1. Introduction

One of the aims of our programme to study Hartle's model [1] within the general modern theory of perturbations is to analyse with rigour the implicit assumptions made in the original model regarding the continuity and/or differentiability of the functions in the metric of the perturbed configuration at the boundary of the star. In order to do that we have studied the problem of the perturbed matching by making use of modern spacetime and matching perturbation theory, only achieved in full generality and to second order in [2]. This work is a parallel continuation of the analysis started in [3] based on the theoretical analysis of a completely general perturbative approach to second order around static configurations of the exterior (asymptotically flat) vacuum problem of stationary and axisymmetric bodies with arbitrary matter content presented in [4].

In this short contribution to the ERE2014 we present the main result concerning the matching problem and continuity of functions in Hartle's model, which eventually implies a correction of the change in mass of the star needed by the perturbed configuration to keep the value of the central energy density unchanged with respect to the static configuration.

1.1. Hartle's model

Hartle's model [1] describes the equilibrium configuration of a slowly rotating isolated body in a perturbation scheme in the context of General Relativity. It is a global model, where the interior of the body is a perfect fluid with a barotropic equation of state, no convective motions and rigid rotation and the exterior is an asymptotically flat vacuum. These are matched across a timelike hypersurface that represents the common boundary between the star and the vacuum.

The whole model is stationary and axially symmetric, but also equatorial symmetry is assumed. The perturbations are computed up to second order around a static, spherically symmetric configuration and they are driven by a single parameter. In this section we present the basics of Hartle's model and the relevant results for the present work just as they are presented in the original paper [1], except that we make explicit the appearance of the perturbation parameter, denoted here as ε. In particular, Ω in [1] appears here as $\varepsilon\Omega$ for some constant Ω.

The "perturbed" metric used in [1] describes both the interior and exterior regions at once

$$\begin{aligned} g_\varepsilon &= -e^{\nu(r)}\left(1+2\varepsilon^2 h(r,\theta)\right)dt^2 + e^{\lambda(r)}\left(1+2\varepsilon^2\frac{e^{\lambda(r)}m(r,\theta)}{r}\right)dr^2 \\ &\quad + r^2(1+2\varepsilon^2 k(r,\theta))\left[d\theta^2+\sin^2\theta(d\varphi-\varepsilon\omega(r,\theta)dt)^2\right] + \mathcal{O}(\varepsilon^3), \end{aligned} \tag{1}$$

where r runs from 0 to ∞. It is *implicitly assumed* that in this set of coordinates this metric is, at least, continuous. The static and spherically symmetric background configuration is determined by the functions $\nu(r)$ and $\lambda(r)$. In this background, the common boundary of the interior and the exterior is located at $r = a$, so that the fluid region extends up to $r = a$ and the vacuum from there to ∞. The first order perturbation is described by the function $\omega(r,\theta)$ and the second order perturbation by $h(r,\theta)$, $m(r,\theta)$ and $k(r,\theta)$. As it is well known, the choice of the perturbation parameter ε is not relevant, since one can obtain other families of solutions by scaling. The physics of the model will restrict the scalability (see Eq. (1) in [1]). Therefore, the functions in (1) can be taken to correspond to the functions with the same name in [1].

The energy density E and pressure P of the background interior read from

$$\lambda' = \frac{1}{r}(1-e^\lambda) + re^\lambda 8\pi E, \quad \nu' = \frac{1}{r}(e^\lambda - 1) + re^\lambda 8\pi P \tag{2}$$

for the fluid with $\vec{u} = e^{-\nu/2}\partial_t$. These are combined with the pressure isotropy condition to give the usual TOV equation, which can be integrated given a value of the central energy density. It is useful to define $M(r) := r(1-e^{-\lambda})/2$ and $j(r) := e^{-(\lambda+\nu)/2}$ in order to cast the equations for the perturbations in a compact way. In vacuum, $j(r) = 1$ and $M(r) = M$, and thus

$$e^{-\lambda(r)} = e^{\nu(r)} = 1 - \frac{2M}{r}, \tag{3}$$

where M is a constant recognized as the mass of the nonrotating star by the assumption of continuity of $\lambda(r)$ at $r = a$. The first order perturbation for the fluid velocity reads $\vec{u}^{(1)} = \varepsilon\Omega\partial_\varphi$, for some constant Ω, and the field equations $(t\varphi)$ provide a PDE for $\omega(r,\theta)$. At this point, ω is *implicitly* assumed to be $\mathcal{C}^1$, so that regularity conditions at the origin together with asymptotic flatness are used to show that ω must be a function of r alone. Thence, $\omega(r)$ is integrated given its value at the origin. In vacuum $\omega(r) = 2Jr^{-3}$ for some constant J, which can be determined from the interior by continuity at $r = a$ [1].

At second order the fluid flow is proportional to ∂_t, and the dependence of ω propagates through the field equations leading to the finite expansions

$$h(r,\theta) = h_0(r)+h_2(r)P_2(\cos\theta),\ m(r,\theta) = m_0(r)+m_2(r)P_2(\cos\theta),\ k(r,\theta) = k_2(r)P_2(\cos\theta), \tag{4}$$

where r is chosen so that $k_0(r) = 0$. That corresponds to a choice of gauge at second order, which we call k-gauge. The problems for the $l = 0$ and $l = 2$ sectors separate and are studied independently. In this text we restrict ourselves to the $l = 0$ problem (for a complete analysis

Journal of Physics: Conference Series **600** (2015) 012013 doi:10.1088/1742-6596/600/1/012013

see [1, 5]). The perfect fluid equations for the set $\{h_0, m_0\}$ are presented in [1], instead, in terms of the set $\{p_0^*, m_0\}$, where p_0^* (called the pressure perturbation factor in [1]) is related to h_0 by

$$\gamma = h_0(r) + p_0^*(r) - \frac{1}{3}r^2 e^{-\nu(r)}(\omega(r) - \Omega)^2, \tag{5}$$

for some constant γ. The equations are written in [1] in terms of another radial coordinate R, which runs from 0 to a. However, at this point we can treat R as a dummy variable, which for convenience we recall here $r(\in (0, a])$. The system for $\{p_0^*(r), m_0(r)\}$ in $r \in (0, a]$ reads

$$p_0^{*\prime} = \tag{6}$$
$$\frac{r^4 j^2}{12(r - 2M(r))}\omega'^2 + \frac{1}{3}\left(\frac{r^3 j^2(\omega - \Omega)^2}{r - 2M(r)}\right)' - \frac{4\pi(E + P)r^2}{r - 2M(r)}p_0^* - \frac{m_0 r^2}{(r - 2M(r))^2}\left(8\pi P + \frac{1}{r^2}\right)$$
$$m_0' = 4\pi r^2 \frac{E'}{P'}(E + P)p_0^* + \frac{1}{12}j^2 r^4 \omega'^2 - \frac{2}{3}r^3 j j'(\omega - \Omega)^2, \tag{7}$$

where $'$ denotes a derivative with respect to the argument (r in this case). The system is solved imposing regularity conditions at the origin, plus $p_0^*(0) = 0$, thus forcing the central energy density to stay unchanged with respect to the static configuration [1]. Therefore, γ takes the value $h_0(0)$, which in principle is arbitrary (h_0 is determined up to an additive constant). In vacuum, the equations for m_0 and h_0 provide the solutions for $r \geq a$

$$m_0(r) = \delta M - \frac{J^2}{r^3}, \quad h_0(r) = -\frac{\delta M}{r - 2M} + \frac{J^2}{r^3(r - 2M)}, \tag{8}$$

where δM is a constant and the free additive constant in h_0 has been set so that h_0 vanishes at infinity. δM is recognized in the asymptotic region as the change in mass needed to keep the central energy density unchanged. The assumed continuity of m_0 at $r = a$ in [1] yields

$$\delta M = m_0(a) + \frac{J^2}{a^3}, \tag{9}$$

which determines δM in terms of the interior quantities, since $m_0(a)$ is the value of the function m_0 obtained by the integration of the system (6), (7) out from the origin. Continuity of $h_0(r)$ is used to fix the freedom left in h_0 in the interior and thus fix γ.

2. The perturbation method

The modern view of perturbation theory starts with a family of spacetimes $(\mathcal{V}_\varepsilon, \hat{g}_\varepsilon)$ with diffeomorphically related manifolds, from where a background spacetime say $(\mathcal{V}_0, g)$ is singled out, so that $\mathcal{V}_0 := \mathcal{V}_{\varepsilon=0}$ and $g := \hat{g}_{\varepsilon=0}$. The diffeomorphism $\psi_\varepsilon : \mathcal{V}_0 \to \mathcal{V}_\varepsilon$ used to identify points of the manifolds is employed to pull back $\hat{g}_\varepsilon$ onto the background spacetime, $g_\varepsilon := \psi_\varepsilon^*(\hat{g}_\varepsilon)$. The family of tensors g_ε defines the basic perturbative scheme on $(\mathcal{V}_0, g)$, where $g := \hat{g}_{\varepsilon=0} = g_{\varepsilon=0}$, in the particular gauge defined by ψ_ε. The metric perturbation tensors are now simply defined as the derivatives of g_ε with respect to ε at $\varepsilon = 0$ at each order of derivation. $K_1 := \partial_\varepsilon g_\varepsilon|_{\varepsilon=0}$ and $K_2 := \partial_\varepsilon \partial_\varepsilon g_\varepsilon|_{\varepsilon=0}$ will refer to the first and second metric perturbation tensors.

Given the above family (1), the spherically symmetric and static background metric reads

$$g = -e^{\nu(r)}dt^2 + e^{\lambda(r)}dr^2 + r^2(d\theta^2 + \sin^2\theta d\varphi^2), \tag{10}$$

while the perturbation tensors to second order take the form $K_1 = -2r^2\,\omega(r,\theta)\sin^2\theta dt d\varphi$ and

$$K_2 = \left(-4e^{\nu(r)}h(r,\theta) + 2r^2\omega^2(r,\theta)\sin^2\theta\right)dt^2 + 4\frac{e^{2\lambda(r)}}{r}m(r,\theta)dr^2$$
$$+4r^2 k(r,\theta)(d\theta^2 + \sin^2\theta d\varphi^2). \tag{11}$$

Matter fields are introduced as an ε-family of tensors T_ε on $(\mathcal{V}_0, g)$, with corresponding perturbations defined again by taking ε-derivatives at $\varepsilon = 0$. For an energy momentum tensor of the form $T_\varepsilon = (E_\varepsilon + P_\varepsilon) u_\varepsilon \otimes u_\varepsilon + P_\varepsilon g_\varepsilon$, those will be given in terms of the corresponding (unit) fluid flow u_ε energy density E_ε and pressure P_ε, expanded as $\vec{u}_\varepsilon = \vec{u} + \varepsilon \vec{u}^{(1)} + \frac{1}{2}\varepsilon^2 \vec{u}^{(2)} + \mathcal{O}(\varepsilon^3)$,

$$E_\varepsilon = E + \varepsilon E^{(1)} + \frac{1}{2}\varepsilon^2 E^{(2)} + \mathcal{O}(\varepsilon^3), \quad P_\varepsilon = P + \varepsilon P^{(1)} + \frac{1}{2}\varepsilon^2 P^{(2)} + \mathcal{O}(\varepsilon^3). \tag{12}$$

The Einstein's field equations for g_ε read $G(g_\varepsilon)_{\alpha\beta} = 8\pi T_{\varepsilon\alpha\beta}$, from where the first and second order field equations are obtained as $\partial_\varepsilon G(g_\varepsilon)_{\alpha\beta}|_{\varepsilon=0} = 8\pi \partial_\varepsilon T_{\varepsilon\alpha\beta}|_{\varepsilon=0}$ and $\partial_\varepsilon \partial_\varepsilon G(g_\varepsilon)_{\alpha\beta}|_{\varepsilon=0} = 8\pi \partial_\varepsilon \partial_\varepsilon T_{\varepsilon\alpha\beta}|_{\varepsilon=0}$, respectively.

In order to construct the global model we consider a perfect fluid interior (+) region to be matched to an asymptotically flat vacuum exterior (−) region. In the exact case, the matching of spacetimes with boundary, say $(\mathcal{V}^\pm, g^\pm, \Sigma^\pm)$ requires an identification of the boundaries, Σ^+ and Σ^-, that defines an abstract manifold Σ diffeomorphic to both. The matching conditions demand the existence of one such identification for which the first and second fundamental forms as computed from either side, $h^\pm$ and $\kappa^\pm$, agree. Let us assume that this construction holds for each pair $(\mathcal{V}^\pm_\varepsilon, \hat{g}^\pm_\varepsilon, \hat{\Sigma}^\pm_\varepsilon)$ (with ε fixed), so that there exists a family of diffeomorphically related hypersurfaces $\hat{\Sigma}_\varepsilon$ on which $\hat{h}^+_\varepsilon = \hat{h}^-_\varepsilon$, $\hat{\kappa}^+_\varepsilon = \hat{\kappa}^-_\varepsilon$ hold. By construction $\Sigma_0 = \hat{\Sigma}_{\varepsilon=0}$ is the matching hypersurface of the already matched background. These conditions are formulated as an ε-family of equations defined on Σ_0 as follows. The spacetime gauges $\psi^\pm_\varepsilon$ project (down or up [2]) the $\hat{\Sigma}^\pm_\varepsilon$ families to the background spacetime $(\mathcal{V}^\pm_0, g^\pm)$ giving rise to corresponding families of hypersurfaces $\Sigma^\pm_\varepsilon$ in $(\mathcal{V}^\pm_0, g^\pm)$. In general, $\Sigma^\pm_\varepsilon$ do not coincide with $\Sigma^\pm_0$. The prescription of how points in $\Sigma^\pm_0$ map to $\Sigma^\pm_\varepsilon$ is done through yet another diffeomorphism $\Phi_\varepsilon : \Sigma_0 \to \hat{\Sigma}_\varepsilon$ that identifies $\hat{\Sigma}_\varepsilon$ pointwise. This defines a second gauge freedom, the *hypersurface gauge*. The combination of the two gauges allows us (i) to construct the mapping of the families $\Sigma^\pm_\varepsilon$ to $\Sigma^\pm_0$ at each side, which describes the pointwise deformation of the hypersurfaces with respect to $\Sigma^\pm_0 (= \Sigma_0)$ in the gauges defined by $\psi^\pm_\varepsilon$ at either side, and (ii) to pull back $\hat{h}^\pm_\varepsilon$ and $\hat{\kappa}^\pm_\varepsilon$, down (or up) to (Σ_0, h) with $h = \hat{h}^+_0 = h^-_0$, and thus obtain the families of tensors $h^\pm_\varepsilon$ and $\kappa^\pm_\varepsilon$ on (Σ_0, h) [2]. The ε-derivatives evaluated on ε define $h^{(1)\pm}$, $h^{(2)\pm}$, $\kappa^{(1)\pm}$, and $\kappa^{(2)\pm}$. On the other hand, the first and second order of the deformation is encoded in the the first and second order deformation vectors [2]

$$\vec{Z}^\pm_1 = Q^\pm_1 \vec{n}^\pm + \vec{T}^\pm_1, \qquad \vec{Z}^\pm_2 = Q^\pm_2 \vec{n}^\pm + \vec{T}^\pm_2, \tag{13}$$

where $\vec{n}^\pm$ are the unit normals to $\Sigma^\pm_0$, $Q^\pm$ are normal components and $\vec{T}^\pm$ the tangent parts to Σ_0. Apart from the background configuration and the tensors $K^\pm_1$ and $K^\pm_2$, $\vec{Z}^\pm_1$ and $\vec{Z}^\pm_2$ are (unknown) ingredients needed to compute $h^{(1)\pm}$, $h^{(2)\pm}$, $\kappa^{(1)\pm}$, and $\kappa^{(2)\pm}$. The explicit expressions are found in Propositions 2 and 3 in [2] (see also [6, 7] for the first order case). The perturbed matching conditions to first and second order are

$$h^{(1)+} = h^{(1)-}, \quad \kappa^{(1)+} = \kappa^{(1)-}, \qquad h^{(2)+} = h^{(2)-}, \quad \kappa^{(2)+} = \kappa^{(2)-}. \tag{14}$$

The tensors $h^{(1)\pm}$, $h^{(2)\pm}$, $\kappa^{(1)\pm}$, and $\kappa^{(2)\pm}$ are, by construction, spacetime gauge invariant (they are defined on (Σ_0, h)), but not hypersurface gauge invariant [2]. However, *the equations* (14) are both spacetime and hypersurface gauge invariant [2]. Fulfilling the matching conditions at each order requires showing the existence of two vectors $\vec{Z}^\pm$ (at each order) such that the equations are satisfied.

The vectors $\vec{Z}$ depend fully on both spacetime and hypersurface gauges. Both (±) can be set to zero simultaneously using the appropriate spacetime gauges ψ_ε at each side. But one has to be careful, then. Let us remark the fact that generally in the literature on perturbed matching it has been common to use spacetime gauges for which $\vec{Z}^{\pm}_{1/2} = 0$, the so-called "co-moving" gauges. This is fine if no other condition has been used to restrict the form in which the families $g^{\pm}_{\varepsilon}$ are written, or equivalently, the form of $K^{\pm}_1$ and $K^{\pm}_2$. But this is precisely the case in using the k-gauge in (1). In order to compute the perturbed matching conditions in the k-gauge we need to consider the deformation vectors $\vec{Z}$, which, in fact, will determine how the surface gets perturbed form the point of view of the k-gauge.

3. Background configuration

The interior and exterior regions are described, respectively, by a couple of spacetimes $(\mathcal{V}^+_0, g^+)$ and $(\mathcal{V}^-_0, g^-)$ with corresponding boundaries Σ^+_0 and Σ^-_0 and coordinates $\{t_\pm, r_\pm, \theta_\pm, \varphi_\pm\}$, so that (10) holds with $+$ and $-$. Due to the symmetries the hypersurfaces Σ^+_0 and Σ^-_0 can be finally cast as (see e.g. [8]) $\Sigma^+_0 = \{t_+ = \tau, r_+ = a, \varphi_+ = \phi, \theta_+ = \vartheta\}$, $\Sigma^-_0 = \{t_- = \tau, r_- = a, \varphi_- = \phi, \theta_- = \vartheta\}$ without loss of generality. The coordinates $\{\tau, \phi, \vartheta\}$ parametrize $\Sigma_0 (\equiv \Sigma^+_0 = \Sigma^-_0)$. The unit normals to $\Sigma^\pm_0$ are taken to be $\vec{n}^\pm = -e^{-\frac{\lambda_\pm(a)}{2}} \partial_{r^\pm}|_{\Sigma^\pm_0}$, under the condition that $\vec{n}^+$ points $\mathcal{V}^+_0$ inwards and $\vec{n}^-$ points $\mathcal{V}^-_0$ outwards. This convention follows in order to call $\mathcal{V}^+_0$ the *interior* and $\mathcal{V}^-_0$ the *exterior*.

The matching conditions are found to be equivalent to the set of equations (see e.g. [8])

$$[\nu] = 0, \quad [\nu'] = 0, \quad [\lambda] = 0, \tag{15}$$

where $[f] = f^+|_{\Sigma^+_0} - f^-|_{\Sigma^-_0}$ for objects $f^\pm$ defined at either side, and a prime denotes differentiation with respect to the corresponding radial coordinate, i.e. r_+ or r_- accordingly. Given $f^\pm$ satisfying $[f] = 0$, we denote by $f|_{\Sigma_0}$ either of the equivalent $f^+|_{\Sigma^+_0}$ or $f^-|_{\Sigma^+_0}$.

The perfect fluid of the static background interior $(+)$ is already described above, with E and P given by (2), while in the vacuum region $(-)$ we have (3) for $\lambda_-(r_-)$ and $\nu_-(r_-)$. The matching conditions (15) imply, in particular, that $\nu_+(a) = -\lambda_+(a) = \log\left(1 - \frac{2M}{a}\right)$. The expressions for the differences of the derivative of the functions of the metric in terms of the differences of the fluid variables on Σ_0 are thus found to be

$$[\nu'] = a e^{\lambda(a)} 8\pi P(a) = 0, \quad [\lambda'] = a e^{\lambda(a)} 8\pi E(a), \quad [\nu''] = \left(1 + \frac{a\nu'(a)}{2}\right) e^{\lambda(a)} 8\pi E(a). \tag{16}$$

Let us remark that whereas the matching conditions imply that $P(r_+)$ must vanish on Σ_0, the energy density $E(a)$ stays free. Its value will be determined, if any, by the barotropic equation of state (EOS). In most cases the typical EOS's for neutron stars show a behaviour in which the pressure and energy density vanish together. This is also the case of the polytropic EOS, but one can consider many other cases where this behaviour no longer stands. An example is a fluid with constant density [9], and a more realistic case, strange quark matter EOS's [10].

4. Second order perturbations

The study of the first order perturbations is out of the scope of this paper, but as a summary, we find that one can set $[\omega] = 0$, whereas $[\omega'] = 0$ always holds. It can then be proven that the first order configuration described in [1] (see above) follows (see [3]). Also, $E^{(1)} = P^{(1)} = 0$.

At this point we assume that the functions $\{m^\pm, k^\pm, h^\pm\}$ in both regions satisfy (4). The second order field equations for the perfect fluid, whose second order perturbation vector is $\vec{u}^{(2)} = e^{-3\nu/2}\{\Omega^2 g_{\varphi\varphi} + 2\Omega K_{1t\varphi} + K_{2tt}/2\}\partial_t$ (let us drop the $+$ subindex for the interior quantities and r when not necessary) impose first the finite expansions for $E^{(2)}$ and $P^{(2)}$ in (12)

$$E^{(2)}(r,\theta) = E^{(2)}_0(r) + E^{(2)}_2(r) P_2(\cos\theta), \quad P^{(2)}(r,\theta) = P^{(2)}_0(r) + P^{(2)}_2(r) P_2(\cos\theta). \tag{17}$$

As known, the problems for $l = 0$ and $l = 2$ separate. Here we consider only the $l = 0$ problem. The field equations for the sector $l = 0$ provide the expressions for $E_0^{(2)}$ and $P_0^{(2)}$ plus an equation in the form $h_0'' = F(h_0', m_0', m_0)$. A convenient auxiliary redefinition of the second order pressure is given by $\mathcal{P}_0 := P_0^{(2)}/2(E+P)$. On the other hand, a barotropic EOS requires $E'P_0^{(2)} - P'E_0^{(2)} = 0$ (for $l = 0$), which, combined with the equation for $E_0^{(2)}$ yields the equation for m_0' given by (7) renaming $p_0^* \to \mathcal{P}_0$. Using this and differentiating the expression for $P_0^{(2)}$, the equation for h_0'' can be rewritten as an equation for $\mathcal{P}_0'$, which just reads (6) with $p_0^* \to \mathcal{P}_0$. In short, *the system for* $\{\mathcal{P}_0(r_+), m_0(r_+)\}$ *in* $r_+ \in (0, a]$ *corresponds indeed to the system for* $\{p_0^*(R), m_0(R)\}$ *in* $R \in (0, a]$ *given in [1], as expected.* Equation (5) is a first integral of the system and determines h_0. Clearly, the solutions for the exterior $m_0^-(r_-)$ and $h_0^-(r_-)$ are given by (8) with $r \to r_-$.

The boundary conditions for these functions in Σ_0 are obtained applying the perturbed matching theory reviewed in Section 2. The whole procedure is presented elsewhere [5]. In this contribution we only present one of the results. The second order matching conditions imply that $m_0^\pm$ do not agree in Σ_0 in general, since they must satisfy $[m_0] = -4\pi \frac{a^3}{M} E(a)\mathcal{P}_0(a)$. This fact contradicts the implicit assumption on the "continuity" of m_0, in particular, made in [1] (and other works, e.g. [9], [10]). As a result, using the exterior solution (8) for m_0^- the change in mass reads, in terms of the functions used in [1], as

$$\delta M = m_0(a) + \frac{J^2}{a^3} + 4\pi \frac{a^3}{M}(a - 2M)E(a)p_0^*(a), \tag{18}$$

where $m_0(a)(= m_0^+(a))$ and $p_0^*(a)$ are obtained from the interior problem. A jump in a "m_0" function had been found already in [11], but the discrepancy with [1] was not established at the moment. If the energy density of the static star E vanishes at the boundary expression (18) agrees with that found in [1], i.e. (9). Indeed, most of the models for neutron stars considered in the literature rely on EOS's for which the energy density vanishes whenever the pressure does. However, if E exhibits a non zero value at the boundary, one must take into account the third term, which may become important depending on the EOS considered. As an example, the aforementioned models for strange quark stars which exhibit a non zero value of E at the boundary [10], as well as those in [9] are under study at present.

Acknowledgments

We thank Marc Mars for his most valuable suggestions. We acknowledge financial support from projects IT592-13 (GIC12/66) of the Basque Government, FIS2010-15492 from the MICINN, and UPV/EHU under program UFI 11/55. BR acknowledges financial support from the Basque Government grant BFI-2011- 250.

References

[1] Hartle J B 1967 *The Astrophysical Journal* **150** 1005–1029
[2] Mars M 2005 *Classical and Quantum Gravity* **22** 3325–3348
[3] Reina B and Vera R 2014 *Progress in Mathematical Relativity, Gravitation and Cosmology* vol 60 ed García-Parrado A, Mena F C, Moura F and Vaz E (Springer) pp 377–381
[4] MacCallum M A H, Mars M and Vera R 2007 *Phys. Rev. D* **75** 024017
[5] Reina B and Vera R *In preparation*
[6] Battye R A and Carter B 1995 *Phys.Lett.* **B357** 29–35
[7] Mukohyama S 2000 *Physical Review D* **62** 84015
[8] Mars M, Mena F C and Vera R 2007 *Classical and Quantum Gravity* **24** 3673–3689
[9] Chandrasekhar S and Miller J C 1974 *MNRAS* **167** 63–80
[10] Colpi M and Miller J C 1992 *The Astrophysical Journal* **388** 513–520
[11] Bradley M, Eriksson D, Fodor G and Rácz I 2007 *Physical Review D* **75** 24013

Spanish Relativity Meeting (ERE 2014): almost 100 years after Einstein's revolution IOP Publishing
Journal of Physics: Conference Series **600** (2015) 012014 doi:10.1088/1742-6596/600/1/012014

Cracking of Anisotropic Spheres in General Relativity Revisited

Guillermo A. González[1], Anamaría Navarro[1] and Luis A. Núñez[1,2]

[1]Grupo de Investigación en Relatividad y Gravitación, Escuela de Física, Universidad Industrial de Santander, A. A. 678, Bucaramanga 680002, Colombia.
[2]Centro de Física Fundamental, Departamento de Física, Facultad de Ciencias, Universidad de Los Andes, Mérida 5101, Venezuela.

E-mail: guillego@uis.edu.co, ana.navarro1@correo.uis.edu.co and lnunez@uis.edu.co

Abstract. We have studied the stability in some anisotropic and isotropic matter configurations, by using the concept of cracking in General Relativity, which was conceived as an alternative and different approach to detect instabilities. It is based in the analysis of the radial forces that appear on the system after a perturbation capable of taking it out of its equilibrium state. We have studied the effects of local density perturbations on systems with "barotropic" equations of state. The considerations of the perturbations on the gradient of pressure lead to very different results than the previous works, and have found that not only anisotropic models may present cracking (or overturning), but it can also occur in isotropic matter configurations.

1. Introduction

The concepts of cracking and overturning were developed by Herrera and co-workers [1, 2, 3], with the aim of constituting an alternative way to determine instabilities on anisotropic matter configurations, and is complementary to the traditional formalism of Chandrasekhar [5, 6, 7]. The idea of studying instabilities around the appearance of cracking is very intuitive, it comes from the evaluation of the distribution of radial forces that appear on the system due to perturbations, and if there is a change of sign on this forces then it said that cracking occurs and we have an unstable configuration. The type of perturbation that are considered are capable of taking the system out of is equilibrium state so this distribution of forces may appear.

In the first contributions where the stability was studied with the cracking criterion [1, 2, 3], were analysed independent and simultaneous perturbations on the density and of the anisotropy of some spherical solutions, were it was assumed that the perturbations did not change the gradient of pressure. In a more recent contribution [8], were studied perturbations of the density on systems governed by barotropic equations of state, were the anistotropy was affected though the equations of state and not independently, however the type of perturbations were constant since they did not change the gradient of pressure.

In this paper, we will discuss how a different type of perturbations can be considered, following [8] we will also consider barotropic equations of state and will analyse perturbations of the density. The mean difference is that we want to consider local perturbations of the density, that are localized on a reduced region of the sphere. We will show the difference with the previous works when the effects of a local perturbation are considered. In section 2, we will describe the concepts of cracking and overturning, and will summarize the contributions around this topic. In section 3 will show the implications of local pertubations of the density in the stability of anisotropic and isotropic matter configurations using the concept of cracking.

2. Cracking of anisotropic compact objects

Let us consider an static spherically symmetric metric,

$$ds^2 = e^{2\nu(r)}\, dt^2 - e^{2\lambda(r)}\, dr^2 - r^2 \left(d\theta^2 + \sin^2\theta\, d\phi^2\right), \tag{1}$$

and an anisotropic fluid,

$$T_{\mu\nu} = (\rho + P_\perp)u_\mu u_\nu - P_\perp g_{\mu\nu} + (P - P_\perp)v_\mu v_\nu\,, \tag{2}$$

where

$$u_\mu = (e^\nu, 0, 0, 0)\,, \qquad v_\mu = (0, -e^\lambda, 0, 0)\,, \tag{3}$$

ρ represents the energy density, P the radial pressure, and $P_\perp$ the tangential pressure of the fluid. If the system is in equilibrium, the generalization of the Tolman-Oppenheimer-Volkoff equation for an anisotropic fluid is satisfied,

$$\mathcal{R} = \frac{dP}{dr} + (\rho + P)\frac{m + 4\pi r^3 P}{r(r - 2m)} - \frac{2(P_\perp - P)}{r}, \tag{4}$$

with $\mathcal{R} = 0$, since this term represents the net radial forces in the system and must vanish if it is in equilibrium. Herrera and co-workers [1, 2, 3] stated that if there is a perturbation in the system capable of taking it out of the equilibrium state, then (4) may not be satisfied and a distribution of radial forces $\delta\mathcal{R}$ might appear. This authors defined cracking or overturning as change of sign within the configuration of this distribution of forces that appear after a perturbation in one or more variables of the system. Cracking occurs when the force is directed inward in the inner part of the configuration and changes sing at some point and then it is directed outward. Overturning occurs in the opposite case.

In the series of papers [1, 2, 3], were analysed simultaneous and independent perturbations of the density and the anisotropy which occurred in all the configuration and leaved the gradient of pressure invariant. As a concluding remark, they found that the perturbations of the anisotropy and not in the density may lead to cracking. By a different approach, [8] studied perturbations of the density $\rho \to \rho + \delta\rho$ in systems with barotropic equations of stated; i.e, $P = P(\rho)$, $P_\perp = P_\perp(\rho)$ which perturbed the anisotropy of the system according to the equations and not independently. This perturbations were constant and as consequence did not affect the pressure gradient. As a result, they obtained the following stability criterion

$$-1 \le v_\perp^2 - v^2 \le 1 \;\Rightarrow \begin{cases} -1 \le v_\perp^2 - v^2 \le 0 & \text{Potencially stable} \\ 0 \le v_\perp^2 - v^2 \le 1 & \text{Potencially instable} \end{cases} \tag{5}$$

where v^2 and $v_\perp^2$ are defined by

$$v^2 = \frac{dP}{d\rho}, \qquad v_\perp^2 = \frac{dP_\perp}{d\rho}, \tag{6}$$

and denote the radial sound speed and the tangential sound speed, respectively.

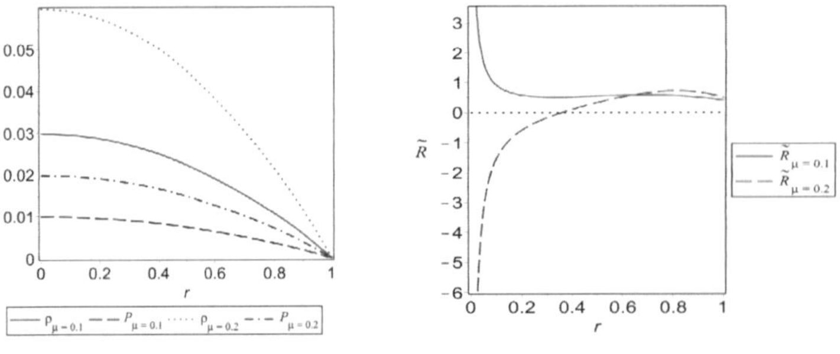

Figure 1. On the left density and pressure and on the right force distribution, $\widetilde{\mathcal{R}} = \delta\mathcal{R}/\delta\rho$, for the isotropic Mehra-Model [9] with $\mu = 0.1$ (solid line) and $\mu = 0.2$ (dashed line). The $\mu = 0.2$ curve presents a cracking point $r_c \approx 0.35$ which illustrates that cracking instability can be found for isotropic matter configurations.

3. Local Perturbations of the density

In this section, we will discuss the effects of local perturbations of the density in barotropic matter configurations and how the results differ from the results of the previous works. Following [8] we will consider systems with equations of state of the form $P = P(\rho)$ and $P_\perp = P_\perp(\rho)$, where a local perturbation of the density occurs; i.e, where the perturbation depends on the radial coordinate,

$$\rho \to \rho + \delta\rho(r), \tag{7}$$

and will be represented by a compact support function with the aims of representing a fluctuation that occurs in a finite and reduced part of the configuration, as any reasonable physical perturbation.

Now, since the pressures are related to the density through the equations of state, as well as the mass of the system,

$$m(\rho) = 4\pi \int_0^r \rho(\bar{r})\bar{r}^2 \mathrm{d}\bar{r}, \tag{8}$$

the effects of the perturbation on the radial pressure and its gradiente, the tangential pressure and the mass can be written as

$$P(\rho + \delta\rho) \approx P(\rho) + \delta P = P(\rho) + \frac{dP}{d\rho}\delta\rho = P(\rho) + v^2\delta\rho\,, \tag{9}$$

$$P_\perp(\rho + \delta\rho) \approx P_\perp(\rho) + \delta P_\perp = P_\perp(\rho) + \frac{dP_\perp}{d\rho}\delta\rho = P_\perp(\rho) + v_\perp^2\delta\rho\,, \tag{10}$$

$$P'(\rho + \delta\rho) \approx P'(\rho) + \delta P' = P'(\rho) + \frac{dP'}{d\rho}\delta\rho = P'(\rho) + \left[(v^2)' + v^2\frac{\rho''}{\rho'}\right]\delta\rho \tag{11}$$

and

$$m(\rho + \delta\rho) \approx m(\rho) + \delta m = m(\rho) + \frac{dm}{d\rho}\delta\rho = m(\rho) + \frac{4\pi r^2\rho}{\rho'}\delta\rho\,, \tag{12}$$

Now, we can find the distribution of radial forces that appear on the system due to the density perturbation by expanding equation (4) to first order,

$$\mathcal{R} \approx \mathcal{R}_0(\rho, P, P_\perp, P', m) + \delta\mathcal{R}, \tag{13}$$

where

$$\delta\mathcal{R} = \frac{\partial\mathcal{R}}{\partial\rho}\delta\rho + \frac{\partial\mathcal{R}}{\partial P}\delta P + \frac{\partial\mathcal{R}}{\partial P_\perp}\delta P_\perp + \frac{\partial\mathcal{R}}{\partial P'}\delta P' + \frac{\partial\mathcal{R}}{\partial m}\delta m, \tag{14}$$

$$\delta\mathcal{R} = \delta\rho\left\{\frac{\partial\mathcal{R}}{\partial\rho} + \frac{\partial\mathcal{R}}{\partial P}v^2 + \frac{\partial\mathcal{R}}{\partial P_\perp}v_\perp^2 + \frac{\partial\mathcal{R}}{\partial m}\left[\frac{4\pi r^2\rho}{\rho'}\right] + \frac{\partial\mathcal{R}}{\partial P'}\left[(v^2)' + v^2\frac{\rho''}{\rho'}\right]\right\}, \tag{15}$$

where the definition $\rho = \frac{m'}{4\pi r^2}$ has been considered. The derivatives of the force distribution, $\mathcal{R}$, are given by

$$\frac{\partial\mathcal{R}}{\partial\rho} = \frac{4\pi r^3 P + m}{r(r-2m)}, \qquad \frac{\partial\mathcal{R}}{\partial m} = \frac{(\rho+P)(1+8\pi r^2 P)}{(2m-r)^2}, \tag{16}$$

$$\frac{\partial\mathcal{R}}{\partial P} = \frac{m + 4\pi r^3(\rho+2P)}{r(r-2m)} + \frac{2}{r}, \qquad \frac{\partial\mathcal{R}}{\partial P_\perp} = -\frac{2}{r} \quad \text{and} \quad \frac{\partial\mathcal{R}}{\partial P'} = 1\,. \tag{17}$$

Now, it can be seen that the expression for the distribution of forces $\delta\mathcal{R}$ is independent of the form of the perturbation of the density $\delta\rho$, and will be different than zero only in the region where the function of compact support $\delta\rho$ is defined. On the other hand, it can be noticed that other terms appear on the expression for the distribution of forces than the ones that appear on [8], since we have considered local perturbations which affect the gradient of pressure and the function of mass. It can also be noticed that for the case of isotropic configurations a change of sign can occur since there are positive and negative terms.

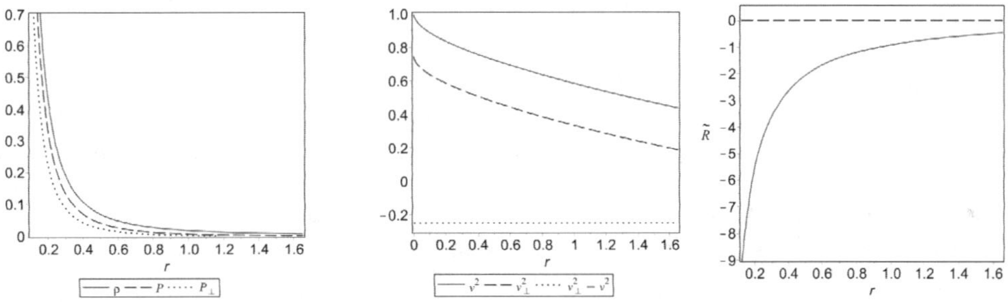

Figure 2. On the left density and pressures, on the middle speeds of sound, and on the right force distribution, $\tilde{\mathcal{R}} = \delta\mathcal{R}/\delta\rho$, for the anisotropic Herrera-Model [1]. $\tilde{\mathcal{R}}$ does not change its sign and the model could be considered as potential stable. This picture differs from the one presented in [1].

4. Detection of instabilities due to cracking

To study the effects of density perturbations on the stability of isotropic matter configurations, we examine a model proposed by Mehra [9] and considered physically viable by Delgaty and Lake[10]. Figure 1 displays the force distribution, $\tilde{\mathcal{R}} = \delta\mathcal{R}/\delta\rho$, for the Mehra-model [9] with two different values of the mass-radius μ. It can be noticed that the curve for $\mu = 0.2$ changes its sign around $r \approx 0.35$, which means that cracking instability can be found for this isotropic matter configurations.

Now, let us examine an anisotropic sphere that was considered unstable in [1] under simultaneous density and anisotropic perturbations. As it can be appreciated from Figure 2 the total distribution force $\tilde{\mathcal{R}}$ does not change its sign, thus the model could be considered as potential stable under the present criterion, the same way as it was stable under the criterion of [8] given in equation (5). Another anisotropic solution, derived by Gokhroo and Mehra [11], originally found by Florides[12] and later by Stewart [13], was analysed and found unstable under the criterion of [8]. It is clear from Figure 3 that with the present refinement of local density perturbation - and assuming the same set of parameters: $\mu = 0.42$, $K = 3/56\pi$ and $\gamma = K/4$- this solution does not present any cracking instability.

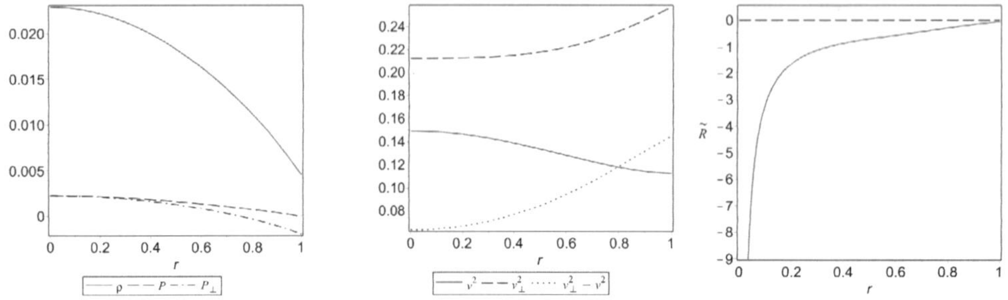

Figure 3. On the left density and pressures, on the middle speeds of sound and on the right force distribution, $\tilde{\mathcal{R}} = \delta\mathcal{R}/\delta\rho$, for the anisotropic Gokhroo/Mehra-Model [11] with: $\mu = 0.42$, $K = 3/56\pi$ and $\gamma = K/4$. Observe that it does not present any cracking instability reported in [8].

5. Remarks and conclusions

We have extended the cracking criterion for anisotropic and isotropic relativistic spheres under local perturbations of the density and have found that isotropic configurations may present cracking, in contrast to the results of the previous works. Also solutions that were unstable under the previous works are found to be stable according to our criterion, this demonstrates that local perturbations introduce changes in the distribution of forces, since they affect the gradient of pressure and the mass function.

It is worth to be mentioned that, the concept of cracking is complementary to the Bondi[4] and Chandrasekhar[5, 6, 7] stability criteria and it refers only to the tendency of the configuration to split (or to compress) at a particular point within the distribution but not to collapse or to expand. The cracking, overturning, expansion or collapse, has to be established from the integration of the full set of Einstein equations. Nevertheless, it should be clear that the occurrence of these phenomena could drastically alter the subsequent evolution of the system. If within a particular configuration no cracking (or overturning) is present, we could identify it as *potentially* stable, because other types of perturbations could lead to its expansions or collapse.

Acknowledgments

Two of us gratefully acknowledge the support from Vicerrectoría de Investigación y Extensión, Universidad Industrial de Santander. GAG was partially founded by Grants VIE1347 and 5705; and LAN by Grant VIE1342. AN wants to thank the financial support from the program of Jóvenes Investigadores of COLCIENCIAS, Colombia.

References

[1] L. Herrera. Cracking of self-gravitating compact objects. *Phys. Lett. A*, 165(206-210), 1992.

[2] A. Di Prisco, E. Fuenmayor, L. Herrera, and V. Varela. Tidal forces and fragmentation of self-gravitating compact objects. *Phys. Lett. A*, 195:23 – 26, 1994.

[3] A. Di Prisco, L. Herrera, and V. Varela. Cracking of homogeneous self-gravitating compact objects induced by fluctuations of local anisotropy. *Gen. Relativ. Gravitation*, 29(10):1239–1256, 1997.

[4] S. Chandrasekhar. Dynamical Instability of Gaseous Masses Approaching the Schwarzschild Limit in General Relativity. *Physical Review Letters*, 12(4):114–116, 1964.

[5] S. Chandrasekhar. The Dynamical Instability of Gaseous Masses Approaching the Schwarzschild Limit in General Relativity. *Astrophys. J.*, 140:417, August 1964.

[6] S Chandrasekhar. A general variational principle governing the radial and the non-radial oscillations of gaseous masses. *The Astrophysical Journal*, 139:664, 1964.

[7] H. Abreu, H. Hernández, and L. A. Núñez. Sound speeds, cracking and stability of self-gravitating anisotropic compact objects. *Class. Quant. Grav.*, 24:4631–4646, 2007.

[8] A. L. Mehra. Radially symmetric distribution of matter. *Journal of the Australian Mathematical Society*, 6:153–156, 5 1966.

[9] M.S.R. Delgaty and K. Lake. Physical acceptability of isolated, static, spherically symmetric, perfect fluid solutions of Einstein's equations. *Comput. Phys. Commun*, 115:395, 1998.

[10] M. K. Gokhroo and A. L. Mehra. Anisotropic spheres with variable energy density in general relativity. *Gen. Rel. Grav.*, 26(1):75 – 84, 1994.

[11] P. S. Florides. A new interior schwarzschild solution. *Proc. Roy. Soc. Lond*, A337:529 – 535, 1974.

[12] B. W. Stewart. Conformally flat, anisotropic spheres in general relativity. *J. Phys. A: Math. Gen.*, 15:2419–2427, 1982.

[13] H. Bondi. Massive Spheres in General Relativity. *Proceedings of the Royal Society of London. Series A, Mathematical and Physical Sciences*, 282(1390):303–317, 1964.

Journal of Physics: Conference Series **600** (2015) 012015 doi:10.1088/1742-6596/600/1/012015

Axially symmetric relativistic thin disks immersed in spheroidal matter haloes

Guillermo A González and Oscar M Pimentel

Escuela de Física, Universidad Industrial de Santander, A. A. 678, Bucaramanga 68002, Colombia

E-mail: guillego@uis.edu.co, OSCAR.PIMENTEL@correo.uis.edu.co

Abstract. Two infinite families of axially symmetric relativistic thin disks of dust immersed in spheroidal matter haloes are presented. The disks are obtained from solutions of the Einstein equations for an axially symmetric conformastatic spacetime in which the metric tensor is characterised only by one metric function. By introducing a finite discontinuity on the first derivatives of the metric tensor, solutions with a singularity of the delta function type are obtained, so describing thin disks. The nonzero components of the energy-momentum tensor, both for the disk and the halo, are obtained from the Einstein equations. In this way, the energy densities and pressures of the sources are determined. By imposing the fulfilment of all the energy conditions we obtain a constraint over the solutions, in such a way that the metric function can be properly expressed in terms of a solution of the Laplace equation. By using the solution to Laplace equation in cylindrical coordinates we find infinite disks and by using the solution to Laplace equation in oblate spheroidal coordinates we find finite disks. In both cases we obtain particular solutions with energy densities and pressures well behaved everywhere. We also show that the masses of de disks and the haloes are finite. Finally we solve the geodesic equation for circular orbits in the plane of the disk.

1. Introduction

Many isolated astrophysical objects in the universe have axial symmetry, some examples are the galaxies in thermodynamic equilibrium, stars, planets, accretion disks, black holes, etc. Since the study of these massive systems requires in some cases the use of the general relativity to give an appropriate description of the problem, it is very important to obtain as many exact solutions to the Einstein equations as possible, and to interpret them as astrophysical sources. For the case of a galactic system, the problem is not easy because those systems usually consists of many components like a thin disk, a central bulb, a black hole, a spheroidal matter halo, etc. Nevertheless, to simplify the problem, in this work we only consider the disk and the halo because they are the largest components. The problem of solving the Einstein equations with thin disks sources has been of great interest in theoretical physics due to their relevance in the description of galaxies and stars. Nevertheless, in literature there are only two references for the system consisting of a thin disk immersed in a fluid or matter halo. In the first one, Vogt and Letelier [1] apply the displacement, cut and reflect method to some known solutions to get disks with haloes. In the second work, Gutiérres-Piñeres, González and Quevedo [2] obtain a family of disks with halo by solving the Einstein-Maxwell equations for a conformastatic spacetime.

In this work we obtain new solutions to the electro-vacuum Einstein equations for a conformastatic metric, and interpret them as thin disks of infinite and finite extension immersed in spheroidal matter haloes. The paper is organised as follows. First we present, in Section 2, the solutions to the Einstein equations for the halo and for the disk, the expressions to compute the mass of the system, and the rotational curves. In Section 3 we choose a particular solution to obtain an family of infinite disks immersed in spheroidal matter haloes. Finally, In Section 4 we do the same as in Section 3 but this time we find the solutions in oblate spheroidal coordinates in order to define a finite radius for the disk.

2. Axially Symmetric Disklike Solutions With Halo

In order to find the energy-momentum tensor of a system that consists of a thin disk with a halo, the disk is modeled by a thin shell Σ with equation $\phi(x^\alpha) = z$ which separates the space in two regions: M^+ (above Σ) and M^- (below Σ). By using the distributional method [3], we can write the Einstein equations as,

$$T^{\pm}_{\alpha\beta} = R^{\pm}_{\alpha\beta} - \frac{1}{2} g_{\alpha\beta} R^{\pm}, \tag{1}$$

$$Q_{\alpha\beta} = H_{\alpha\beta} - \frac{1}{2} g_{\alpha\beta} H, \tag{2}$$

where $Q_{\alpha\beta}$ and $H_{\alpha\beta}$ are the energy-momentum tensor and the Ricci tensor in Σ. The equations (1) are the Einstein equations for the halo, and equations (2) are the Einstein equations of the disk.

Now, we introduce the conformastatic metric,

$$ds^2 = -e^{2\psi} dt^2 + e^{-2\psi}(dr^2 + r^2 d\varphi^2 + dz^2), \tag{3}$$

where we demand $\psi(r, -z) = \psi(r, z)$ in order to have axial symmetry and reflexion symmetry. With this metric and equation (1) we compute the energy density and the stress of the halo,

$$\rho = e^{2\psi}(2\nabla^2\psi - \nabla\psi \cdot \nabla\psi), \tag{4}$$

$$p = \frac{1}{3} e^{2\psi} \nabla\psi \cdot \nabla\psi, \tag{5}$$

and with equation (2) we compute the energy density of de disk

$$\sigma = 4e^{\psi} \psi_{,z}, \tag{6}$$

which is evaluated in $z = 0^+$. As the stress of the disk is zero, we have a fluid of dust.

In order to have physically well behaved solutions, the energy densities and the stress of the halo need to satisfy the energy conditions [4], that can be fulfilled if we consider ψ functions that are solutions to the equation

$$\nabla^2\psi = k\nabla\psi \cdot \nabla\psi, \tag{7}$$

with $k \geq 1$ and $\psi_{,z}\big|_{z=0^+} > 0$ (it can not be zero in order to have $\sigma \neq 0$). For the fluid in the halo,the above equation leads to

$$p = \frac{\rho}{3(2k-1)}, \tag{8}$$

the state equation of a gamma fluid. Equation (7) can be rewritten as $\nabla^2(e^{-k\psi}) = 0$, which is related to Laplace equation $\nabla^2 U = 0$ through the relation

$$e^{-k\psi} = 1 - U, \tag{9}$$

where U is the solution to Laplace equation.

Now, to compute the mass of the disk-halo system, we use the Komar formulae [5],

$$M = 2\int \left(T_{\alpha\beta} - \frac{1}{2}T g_{\alpha\beta}\right) m_\alpha \xi^\beta_{(t)} \sqrt{h} d^3y, \tag{10}$$

where m_α is the normal vector to the time-like hypersurface, $\xi^\beta_{(t)}$ is the time-like killing vector, and $\sqrt{h}d^3y$ is the volume element. Finally, to compute the circular velocity we solve the geodesic equation for a time-like particle that moves in the plane of the disk. By constraining the four-velocity of the particle to $u^\alpha = u^0(1, 0, \omega, 0)$, the resulting expression for the square of the circular velocity yield to

$$v_c^2 = \frac{rU_{,r}}{k(1-U) - rU_{,r}}. \tag{11}$$

3. Kuzmin-Toomre Disks With Halo (Infinite Disks)

To obtain solutions describing thin disks of infinite extension immersed in a spheroidal halo, we take U as the solution to Laplace equation in spherical coordinates (R, θ, φ),

$$U_n(r, \theta) = -\sum_{l=0}^{n} \frac{A_l}{R^{l+1}} P_l(\cos\theta), \tag{12}$$

where P_l are the Legendre Polynomials and A_l are constants. Here, in order to have $\psi|_{z=0^+} \neq 0$, we need to use the displacement, cut and reflexion method [6] which is equivalent to the transformation $z \to |z| + a$, so $R^2 = r^2 + (|z| + a)^2$, and $\cos\theta = (|z| + a)/R$.

With the solution to Laplace equation presented in (12) and the relation (9) we can compute the energy density of the disk (6). For the $n = 2$ model, σ takes the form

$$\tilde{\sigma}_2 = \frac{\left[(\tilde{A}_0 - \tilde{A}_1)m^2 + 3(\tilde{A}_1 - \tilde{A}_2)m + \frac{5}{2}\tilde{A}_2(2 - \tilde{r}^2)\right] m^{\frac{5-2k}{k}}}{\left[2m^{5/2} + 2\tilde{A}_0 m^2 + 2\tilde{A}_1 m + \tilde{A}_2(2 - \tilde{r}^2)\right]^{\frac{1+k}{k}}}, \tag{13}$$

where $m = 1 + \tilde{r}^2$, $\tilde{r} = r/a$, $\tilde{A}_0 = A_0/a$, $\tilde{A}_1 = A_1/a^2$, $\tilde{A}_2 = A_2/a^3$, and $\tilde{\sigma}_n = (ka/4)\sigma_n$. We can also compute the energy density of the halo (4) for $n = 2$,

$$\begin{aligned}\tilde{\rho}_2 \;=\;& \frac{\tilde{M}^{(5-2k)/k}\left[\tilde{A}_0 \tilde{r}\tilde{M}^2 + (3\tilde{A}_1\tilde{Z} + \tilde{A}_2)\tilde{r}\tilde{M} + \frac{5}{2}\tilde{A}_2(2\tilde{Z}^2 - \tilde{r}^2)\right]^2}{\left[\tilde{M}^{5/2} + \tilde{A}_0\tilde{M}^2 + \tilde{A}_1\tilde{M}\tilde{Z} + \frac{1}{2}\tilde{A}_2(\tilde{Z}^2 - \tilde{r}^2)\right]^{2(1+k)/k}} + \\ & \frac{\tilde{M}^{(5-2k)/k}\left[(\tilde{A}_0\tilde{Z} - \tilde{A}_1)\tilde{M}^2 + 3\tilde{Z}(\tilde{A}_1\tilde{Z} - \tilde{A}_2)\tilde{M} + \frac{5}{2}\tilde{A}_2\tilde{Z}(2\tilde{Z}^2 - \tilde{r}^2)\right]^2}{\left[\tilde{M}^{5/2} + \tilde{A}_0\tilde{M}^2 + \tilde{A}_1\tilde{M}\tilde{Z} + \frac{1}{2}\tilde{A}_2(\tilde{Z}^2 - \tilde{r}^2)\right]^{2(1+k)/k}},\end{aligned} \tag{14}$$

where $\tilde{M} = \tilde{r}^2 + \tilde{Z}^2$, $\tilde{Z} = |\tilde{z}| + 1$, $\tilde{z} = z/a$, $\tilde{r} = r/a$, $\tilde{A}_0 = A_0/a$, $\tilde{A}_1 = A_1/a^2$, $\tilde{A}_2 = A_2/a^3$, and $\tilde{\rho}_2 = [(ka)^2/(2k-1)]\rho_2$.

To analyse the behaviour of these quantities, we plot $\tilde{\sigma}_2$ in figure 1 and $\tilde{\rho}_2$ in figure 2. The constants values were chosen to show the different behaviours that can be obtained from the solutions. We can see in both figures that the energy densities have a maximum at the center of the system and go to zero at infinity. By varying the constants we can have different energy density profiles: we can manipulate its maximum at the center, its rate of decrease, and in some cases we can obtain more than one maximum.

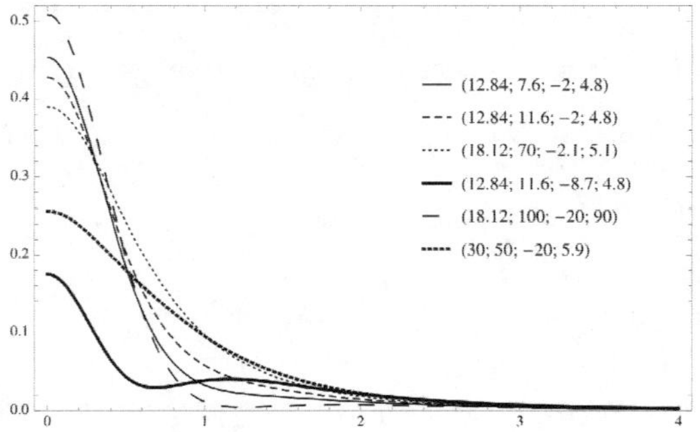

Figure 1. $\tilde{\sigma}_3$ as a function of $\tilde{r}$ for the $n = 3$ model. Each curve is labeled by the set of numbers $(k, \tilde{A}_0, \tilde{A}_1, \tilde{A}_2)$

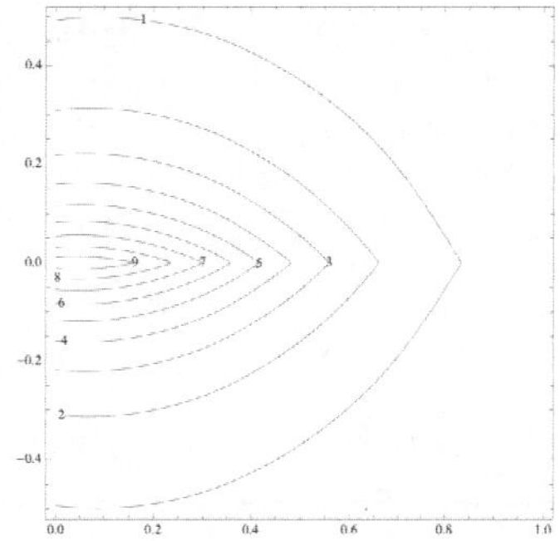

Figure 2. Contour plots of $\tilde{\rho}_2$ as a function of $\tilde{r}$ (horizontal axis) and $\tilde{z}$ (vertical axis) for the $n = 3$ model and for the constans $k = 18.12, \tilde{A}_0 = 100, \tilde{A}_1 = -20, \tilde{A}_2 = 90$.

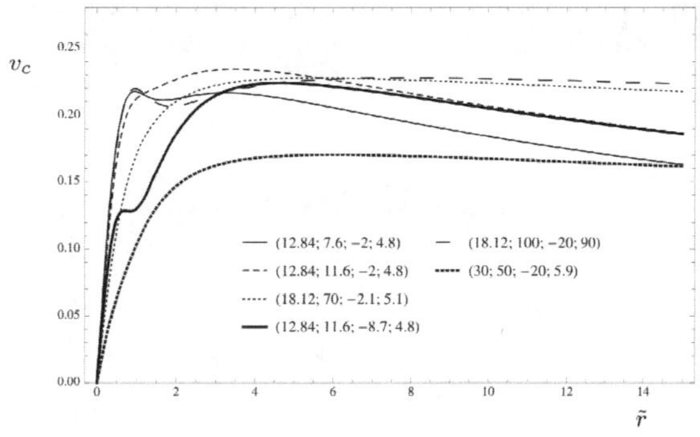

Figure 3. v_{c_2} as a function of $\tilde{r}$. Each curve is labeled by the set of numbers $(k, \tilde{A}_0, \tilde{A}_1, \tilde{A}_2)$

Now, we can show that the disk-halo system has finite mass, despite the fact that the matter distribution extends to infinity. To do this we use the limit comparison test to show that if the total mass of the system for the $n = 0$ model is finite, then all the family of solutions has finite masses. We compute the total mass of the system for $n = 0$ by solving the integral (10). This yields

$$M_{T_0} = M_{D_0} + M_{H_0} = \frac{8\pi A_0}{k}, \tag{15}$$

where M_{H_0} is the mass of the halo, and $M_{D_0} = \frac{8\pi a}{k} \ln(1 + A_0/k)$ is the mass of the disk. With this, it is possible to say that the masses of all the solutions are finite.

Finally, we compute v_c for the $n = 2$ model from equation (11) to obtain

$$v_{c_2}^2 = \frac{\tilde{r}^2[3\tilde{A}_2 + 3\tilde{A}_1 m^2 + \tilde{A}_0 m^4 + \frac{3}{2}\tilde{A}_2(3-m)m]}{k[m^2 + m\tilde{A}_1 + m^3\tilde{A}_0 + \frac{1}{2}\tilde{A}_2(3-m)]m^3 - \tilde{r}^2[3\tilde{A}_2 + 3\tilde{A}_1 m^2 + \tilde{A}_0 m^4 + \frac{3}{2}\tilde{A}_2(3-m)m]}. \tag{16}$$

In figure 3 we present the rotational curves asociated to v_{c_2}. We can see that the circular velocity increases very fast until a maximum value which remains approximately constant. As we can see, in some cases it is possible to obtain a rotational curve with more than one maximum. This kind of behavior is qualitatively similar to those observed for spiral galaxies.

Journal of Physics: Conference Series **600** (2015) 012015 doi:10.1088/1742-6596/600/1/012015

4. Kalnajs disks with halo (finite disks)

Another family of solutions can be obtained by taking U as the solution to Laplace equation in oblate spheroidal coordinates (ξ, η, ϕ),

$$U_m(\xi, \eta) = -\sum_{n=0}^{m} C_{2n} q_{2n}(\xi) P_{2n}(\eta), \tag{17}$$

where $q_{2n} = i^{2n+1} Q_{2n}(i\xi)$, with $Q_{2n}(i\xi)$ the Legendre functions of second kind of imaginary argument, and the constants C_{2n} are choosen properly to be [7]

$$C_{2n} = \frac{\tilde{M}/2}{(2n+1)q_{2n+1}(0)} \left[\frac{\pi^{1/2}(4n+1)(2m+1)!}{2^{2m}(m-n)!\Gamma(m+n+3/2)} \right] \tag{18}$$

if $n \leq m$, and zero if $n > m$ for $m \geq 1$. With U_m we can compute the energy density of the disk for the first three models of the family of solutions ($m = 1, 2, 3$) and the energy density of the halo for the $m = 3$ model. The plots of $\tilde{\sigma}_m = (ka/\tilde{M})\sigma_m$ for $m = 1, 2, 3$ are presented in figure 4, and the contours of $\tilde{\rho}_3 = [(k^2a^2)/((2k-1)\tilde{M}^2)]\rho_3$ are presented in figure 5. We can see in these figures that the energy densities have a maximum at the center of the system. In the disk, $\tilde{\sigma}_m$ goes to zero at the radious of the disk $\tilde{r} = 1$ (finite disk), while for the halo $\tilde{\rho}_3$ goes to zero at infinity.

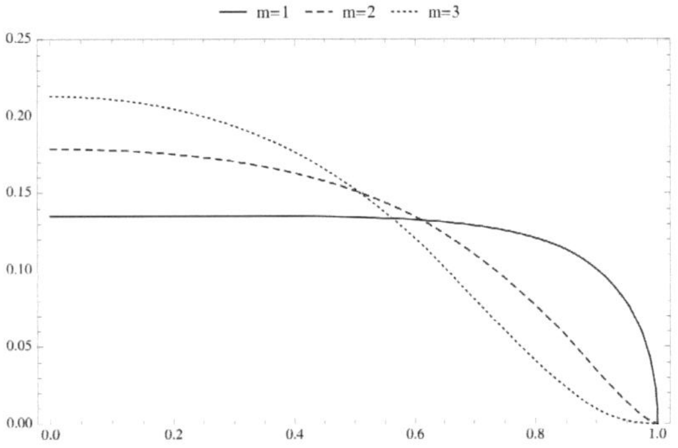

Figure 4. $\tilde{\sigma}_m$ as a function of $\tilde{r} = r/a$ for fixed values $k = 20$ and $\tilde{M} = 30$.

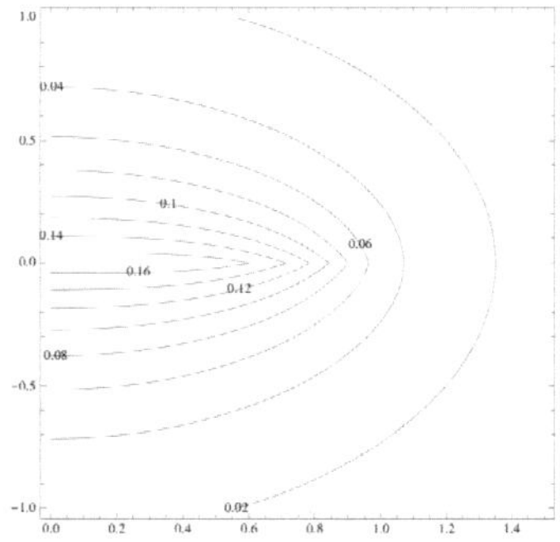

Figure 5. Contour plots of $\tilde{\rho}_3$ as a function of $\tilde{r} = r/a$ (horizontal axis) and $\tilde{z} = z/a$ (vertical axis) for the $m = 3$ model and for $k = 17.2$ and $\tilde{M} = 3.9$

Now, we can show, as in section 3, that the mass of the system is finite despite the fact that the halo matter distribution extends to infinity. To do that we use the limit comparison test and show that if the total mass is finite for $m = 1$, then all the family of kalnajs disks with halo are finite too. By using the integral (10) we find that $M_{T_m} = M_{D_m} + M_{H_m} = 8\pi a C_0/k$, where M_{H_m} is the mass of the halo and M_{D_m} is the mass of the disk. Here we can see that the total mass is the same for every m. For $m = 1$, we have

$$M_{D_1} = \frac{64a}{k} \left[1 - \sqrt{1 + \frac{8}{3\pi C_0}} arcot \sqrt{\frac{3\pi C_0}{3\pi C_0 + 8}} \right] \tag{19}$$

so the total mass of all the family of solutions are finite.

Finaly, in figure 6 we present the circular velocity profiles that we find by using the functions U_1, U_2 and U_3 in the expression (11). This curves correspond to the first three solutions for v_{c_m}.

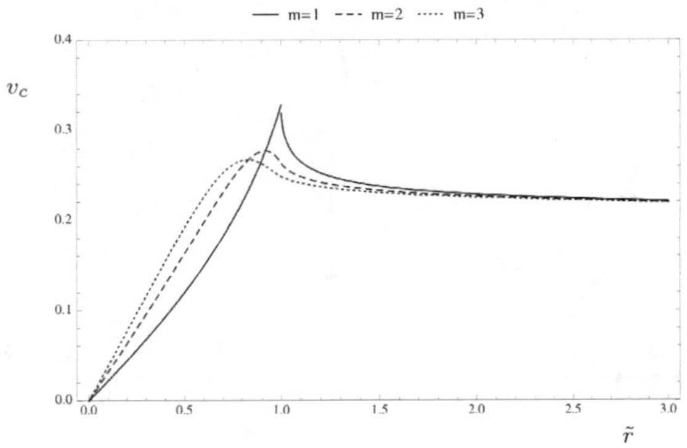

Figure 6. v_{c_m} as a function of $\tilde{r} = r/a$ for the $m = 1, 2, 3$ models, and for $k = 20$ and $\tilde{M} = 30$

In this figure we note that the rotational curve for $m = 1$ has a discontinuity in the derivative. This no longer happens when we take particular solutions with $m \geq 2$. It is clear that v_c increases until a maximum value that remains approximately constant.

5. Conclusions

We have obtained two infinite families of exact solutions to the Einstein equations that describe thin dust disks immersed in spheroidal matter haloes whose energy density and stress goes to zero at infinity. For the fliud in the halo we have obtained an equation of state of the form $p = \gamma\rho$, with $\gamma = [3(2k - 1)]^{-1}$. In the first family of solutions, the energy density of the disk-like source goes to zero at infinity, while for the second family of solutions the energy density of the disk goes to zero at the edge of the disk. All the solutions satisfy the energy conditions and yield to finite masses. Finaly, we have obtained the rotational curves by solving the geodesic equation of a test particle moving in circular orbits in the plane of the disk. This curves are qualitatively similar to the observed rotational curves for spiral galaxies.

References

[1] Vogt D and Letelier P S 2003 *Phys. Rev. D* **68** 084010
[2] Gutiérrez-Piñerez A C, García-Reyes G and González G A 2014 *Int.J.Mod.Phys.* **D23** 1450010
[3] Israel W 1966 *Nuovo Cim.* **B44S10** 1
[4] Hawking S and Ellis G 2004 *The Large scale structure of space-time* (Cambridge: Cambridge University Press) p 103
[5] Poisson E 2004 *A Relativist's Toolkit: The Mathematics of Black-Hole Mechanics* (Cambridge: Cambridge University Press) p 149
[6] Kuzmin G G 1956 *AZh* **33** 27
[7] Gonzalez G A and Reina J I 2006 *Mon.Not.Roy.Astron.Soc.* **371** 1873–1876

Quantum correlations across horizons

Ana Alonso-Serrano
Instituto de Física Fundamental (IFF-CSIC), Serrano 121, 28006 Madrid, Spain

Luis J. Garay
Departamento de Física Teórica II, Universidad Complutense de Madrid, 28040 Madrid, Spain
Instituto de Estructura de la Materia (IEM-CSIC), Serrano 121, 28006 Madrid, Spain

Guillermo A. Mena Marugán
Instituto de Estructura de la Materia (IEM-CSIC), Serrano 121, 28006 Madrid, Spain

Abstract. We analyze a simple minisuperspace model for the Schwarzschild-de Sitter spacetime and develop an exact quantization of this geometry. We determine the physiscal Hilbert space of the model and employ it to study the possible quantum influence of regions of the universe that are not classically accesible.

1. Introduction

In the classical study of spacetimes, the presence of horizons disconnect physically the regions that are separated by them, making any of these regions causally inaccessible from the outside. This is the case, for instance, for event and cosmological horizons, examples on which we will concentrate our attention in this work.

A new feature of our analysis is that, compared with many discussions carried out in the literature about quantum field theory on curved backgrounds (see e.g. [1]), we will construct a canonical quantization of the spacetime itself. So, our conclusions only depend on the quantum behavior of the geometry. For this purpose, we will consider a minisuperspace model for a spherically symmetric spacetime in the presence of a positive cosmological constant, situation described by a Kantowski-Sachs metric. The maximal analytic extension of the classical solutions, which describes Schwarzschild-de Sitter spacetimes, generically have black hole and cosmological horizons that isolate the spacetime region where we would pressumably live. We complete a canonical quantization of this model following an extension of the conventional Dirac quantization procedure (see e.g. [2]).

2. Classical solutions

We construct our model starting with a general spherically symmetric metric that depends on two variables, A and b, and on the lapse function N,

$$\sigma^{-2}\mathrm{d}s^2 = -\frac{N(r)^2}{A(r)}\mathrm{d}r^2 + A(r)\mathrm{d}T^2 + b(r)^2\mathrm{d}\Omega_2^2, \tag{1}$$

where all metric variables and coordinates are dimensionless, $\mathrm{d}\Omega_2^2$ is the line element on the unit two-sphere, and σ has units of length. Note that this is a Kantowski-Sachs metric with a suitably redefined lapse [3].

Generically, the horizons are located where $A = 0$. The range of the variables A and b are taken to be the whole real line. The change of sign in A is of much importance, because it corresponds to a change in the character of the radial coordinate from timelike to spacelike behavior, or vice versa.

It will prove convenient for our analysis to introduce a new variable $c = Ab$ [4], instead of A, which allows us to simplify the Hilbert-Einstein action, that can be written in terms of the new configuration variables and a positive cosmological constant, Λ, as

$$S = -\int \Big(\frac{\dot{b}\dot{c}}{N} + N\mathring{B}(b)\Big), \tag{2}$$

where the dot stands for the derivative with respect to r, we have defined $\lambda = \sigma^2\Lambda$, and

$$B(b) = \frac{\lambda}{3}b^3 - b, \qquad \mathring{B}(b) = \partial_b B(b) = \lambda b^2 - 1. \tag{3}$$

.

The variational principle for this action gives the classical equations of motion, that from the point of view of the metric (1) adopt as solutions the Schwarzschild-de Sitter metric,

$$b(r) = r, \qquad A(r) = -1 + \frac{2m}{r} + \frac{\lambda r^2}{3}, \tag{4}$$

where m is an integration constant. Another integration constant appears related with a rescaling of the coordinates r and t, but using the freedom in the definition of these coordinates it can be set equal to one.

The causal structure of these spacetimes depends on the value of m, and in all cases (except for $m = 0$) the solution presents a singularity at $r = 0$. The most interesting case is $0 < m < 1/\sqrt{9\lambda}$, where $A(r) = 0$ has two positive solutions that can be identified as a black hole horizon and a cosmological horizon.

A common feature of all solutions is that "our" spacetime region (between horizons) is characterized by $A < 0$, whereas $A > 0$ characterizes the regions outside.

In order to perform a canonical quantization, we adopt a Hamiltonian formulation of the system. The canonical action can be expressed as

$$S = \int dr\left(\dot{c}p_c + \dot{b}p_b - NC\right), \tag{5}$$

where the canonical conjugate momenta are $p_b = -\dot{c}/N$ and $p_c = -\dot{b}/N$. The variation with respect to the lapse function gives rise to the Hamiltonian constraint, $C = 0$, with

$$C = -p_b p_c + \mathring{B}(b). \tag{6}$$

3. Canonical quantization

3.1. Kinematical space and operator algebra

In order to quantize our system, we will start with the kinematical algebra of phase space functions constructed from the canonical variables b, c, p_b, and p_c (and the unit constant), which is closed under Poisson brackets. As kinematical space we choose the vector space spanned by simultaneous solutions to the equations

$$-i\partial_c \Psi_{hp} = p\Psi_{hp}, \qquad [\partial_c\partial_b + \mathring{B}(b)]\Psi_{hp} = h\Psi_{hp}, \tag{7}$$

with h and p being real. Thus, any kinematical state will be a linear combination of these solutions Ψ_{hp}. Explicitly, they take the form [5]

$$\Psi(b,c) = \int_{\mathbb{R}} \mathrm{d}h \int_{\mathbb{R}} \mathrm{d}p \tilde{\Psi}(h,p) e^{ipc+i[B(b)-bh]/p}, \tag{8}$$

where $\tilde{\Psi}(h,p)$ is a distribution. Actually we have introduced two kinematical representations that we can use: the metric (b,c)-representation and the (h,p)-representation. In the (h,p)-representation, basic operators act as

$$\hat{b} = -ip\partial_h, \quad \hat{c} = i\partial_p + \frac{B(-ip\partial_h)}{p^2} + \frac{i\partial_h h}{p},$$
$$\hat{p}_b = \frac{\mathring{B}(-ip\partial_h) - h}{p}, \quad \hat{p}_c = p, \tag{9}$$

as one can check by direct application of these operators on the kinematical states (8).

It is convenient to introduce an inner product in the kinematical space on which the operators $\hat{b}$, $\hat{c}$, $\hat{p}_c$, and $\hat{p}_b$ are self-adjoint, namely:

$$(\Psi_1, \Psi_2) = \int_{\mathbb{R}} \mathrm{d}h \int_{\mathbb{R}} \mathrm{d}p \tilde{\Psi}_1(h,p)^* \tilde{\Psi}_2(h,p), \tag{10}$$

where the symbol $*$ denotes complex conjugation. Then, the kinematical Hilbert space is $L^2(\mathbb{R}^2, \mathrm{d}h\mathrm{d}p)$.

Finally, the Hamiltonian constraint can be represented in this kinematical space by the operators

$$\hat{C} = \partial_b\partial_c + \mathring{B}(b), \qquad \hat{C} = h, \tag{11}$$

in the metric (b,c)-representation and in the (h,p)-representation, respectively.

3.2. Physical Hilbert space

The space of solutions can be obtained by imposing the constraint

$$\hat{C}\tilde{\Phi}(h,p) = 0. \tag{12}$$

Then, solutions turn out to be

$$\tilde{\Phi}(h,p) = \frac{1}{\sqrt{2\pi}}\delta(h)\phi(p), \tag{13}$$

where $\phi(p)$ is an arbitrary distribution. In this physical vector space, the operators $\hat{p}$ and $\hat{q}$ (which commute with the constraint $\hat{C}$) are represented as $\hat{p} = p$, and $\hat{q} = i\partial_p$.

We choose an inner product in this space of physical states so that the observables $\hat{p}$ and $\hat{q}$ be self-adjoint:

$$\langle\Phi_1, \Phi_2\rangle = \int_{\mathbb{R}} \mathrm{d}p\phi_1(p)^*\phi_2(p). \tag{14}$$

Hence the physical Hilbert space of quantum states for our system is $L^2(\mathbb{R}, \mathrm{d}p)$. We will refer to this representation as the p-representation.

These physical states can also be written in terms of the metric variables b and c:

$$\Phi(b,c) = \frac{1}{\sqrt{2\pi}}\int_{\mathbb{R}} \mathrm{d}p\phi(p)e^{i[pc+B(b)/p]}. \tag{15}$$

The inverse of this transformation is

$$\phi(p) = \frac{1}{\sqrt{2\pi}}e^{-iB(b)/p}\int_{\mathbb{R}} \mathrm{d}c\Phi(b,c)e^{-ipc}, \tag{16}$$

where we can see that the physical states $\Phi(b,c)$ depend on b only through $B(b)$.

This transformation allows us to write the inner product in terms of the metric variables on configuration sections of fixed value of b:

$$\langle\Phi_1, \Phi_2\rangle = \int_{\mathbb{R}} \mathrm{d}c\Phi_1(b,c)^*\Phi_2(b,c). \tag{17}$$

From the independece on b of the inner product, it immediately follows that we actually have not only one, but a whole family of p_b-representations labeled by b. This resembles a kind of transformation from the Heisenberg picture, $\phi(p)$, to the b-Schrödinger picture, $\phi(b,p)$, since

$$\phi(b,p) = \hat{U}(b)\phi(p) = e^{iB(b)/p}\phi(p), \tag{18}$$

where, for each value of b, we can define the observables $\hat{\pi}_b = \hat{p}$ and $\hat{c}_b = i\partial_p$.

We can adopt two alternative viewpoints analogous to the Schrödinger an Heisenberg pictures of standard quantum mechanics. From the first point of view, we can consider that the p_b-representations give the evolution of the p-representation, hence, the family of representations provide the corresponding Schrödinger "dynamics" in the parameter b.

From the second point of view, we can choose a fixed p_b-representation for a given value of b, and represent our family of observables in a kind of Heisenberg picture (see e.g. [6]). The family of observables corresponding to c at different values b_0 of b would be given by

$$\hat{c}_b^0 = \hat{U}^\dagger(b,b_0)\hat{c}_b\hat{U}(b,b_0) = i\partial_p + \frac{B(b_0)-B(b)}{p^2}, \tag{19}$$

where $\hat{U}(b,b_0) = e^{i[B(b)-B(b_0)]\hat{p}^{-1}}$. This observable gives, in the p_b-representation, the value of c when $b = b_0$.

It will be useful for our study to define a family of bases formed by eigenstates of the self-adjoint operator $\hat{c}_b^0$ in the p_b-representation with real eigenvalues c^0,

$$\phi_{c^0}(b,p) = \frac{1}{\sqrt{2\pi}} e^{-ipc^0 - i[B(b_0)-B(b)]/p}. \tag{20}$$

In this family of representations, the identity operator can be written as [5]

$$\mathbb{1}(b,p;b,p') = \int_{\mathbb{R}} \mathrm{d}c^0 \phi_{c^0}(b,p)\phi_{c^0}(b,p')^*. \tag{21}$$

Clearly, it is possible to decompose this identity operator in the sum of two orthogonal projectors, one for positive eigenvalues of c^0, $\hat{P}_+^0$, and the other for negative eigenvalues, $\hat{P}_-^0$,

$$\mathbb{1} = \hat{P}_+^0 + \hat{P}_-^0. \tag{22}$$

The superindex 0 indicates the dependence of the projection operators on the value of b_0 where c is evaluated.

4. Quantization and horizons

If we concentrate our attention on certain positive value b_0 of b, and there we observe only the region with negative values of c, we need to restrict our analysis just to states with null projection under $\hat{P}_+^0$. This restriction corresponds classically to considering only our region of the universe at the given "instant b_0 of the dynamical variable b".

The question about whether we can forget about the regions beyond the horizons quantum mechanically, as it occurs classically, is related now with the issue of whether this restriction to our region of the universe is robust dynamically. In our scheme, this is equivalent to ask whether observations of the values of c at different values of b are compatible, something which requires that the two considered observables $\hat{c}_b^0$ and $\hat{c}_b^1$ commute. A direct calculation shows that this is not the case, and therefore the family of considered observables are not mutually compatible. Then, the two projections (at the different values b_0 and b_1) would differ, and the restriction to our region of the universe between horizons would not be stable, in the sense that the projection at b_0 on negative values of c^0 would generally have a non-vanishing projection at b_1 on positive values of c^1, and vice versa. The restriction to the interior of the horizons would depend on the value of b, and would therefore be unstable under evolution in this variable [5].

We have seen that the dynamics in b mixes the projections, and thus it is not consistent to describe states as direct sum of positive and negative c^0-states. In consequence, general physical states rather belong to the tensor product of the projection subspaces for a particular instant of b:

$$\mathcal{H}^0 = \mathcal{H}_+^0 \otimes \mathcal{H}_-^0. \tag{23}$$

In this formalism, any observable $\hat{O}$ can then be decomposed in four operators between both projections:

$$\hat{O}_{\pm\pm}^0 : \mathcal{H}_\pm^0 \to \mathcal{H}_\pm^0, \qquad \hat{O}_{\pm\mp}^0 : \mathcal{H}_\pm^0 \to \mathcal{H}_\mp^0, \tag{24}$$

defined as $\hat{O}^0_{\pm\pm} = \hat{P}^0_\pm \hat{O} \hat{P}^0_\pm$ and $\hat{O}^0_{\pm\mp} = \hat{P}^0_\pm \hat{O} \hat{P}^0_\mp$.

The operators $\hat{O}^0_{\pm\mp}$ mix the two subspaces corresponding to the considered projections, $\mathcal{H}^0_\pm$, and lead to correlations between them. This is the case of $\hat{c}^1_b$, as we have seen. Finally, note that if $\hat{O}$ is a unitary observable, for instance $\exp(i\hat{c}^1_b)$, the existence of the two mixing components indicates that unitarity is not respected in each of the subspaces $\mathcal{H}^0_\pm$ separately.

5. Conclusion

We have analyzed a Kantowski-Sachs minisuperspace model of a spacetime with a positive cosmological constant, whose classical solutions are Schwarzschild-de Sitter universes. We have argued that quantum mechanics applied to the whole spacetime generically introduces quantum correlations between different classically disconnected regions (separated by horizons), so that we can not restrict ourselves to the observed classical region when we consider the spacetime quantum mechanically. Then, the physical structure is consistent only if we consider the whole spacetime [5]. This is explicitly shown by checking that unitarity is preserved only when the whole spacetime is taken into account. Another way to see that fact is because the two subspaces in which we have decomposed the physical Hilbert space of the system, corresponding to states with support either between or beyond the horizons, are not stable under the action of unitary operators that describe a natural concept of evolution on physical states. Therefore, these states are better conceived as belonging to the tensor product of both subspaces, which are not separable but entangled.

Acknowledgments: The authors acknowledge financial support from the research grant MICINN/MINECO FIS2011-30145-C03-02 from Spain.

References

[1] Birrell N D and Davies P C W 1982 *Quantum fields in curved space* (Cambridge University Press, Cambridge, UK)
[2] Ashtekar A and Tate R S 1994 *J.Math.Phys.* **35** 6434–6470
[3] Marugán G A M 1994 *Class.Quant.Grav.* **11** 589–608
[4] Campbell L and Garay L 1991 *Phys.Lett.* **B254** 49–54
[5] Alonso-Serrano A, Garay L J and Marugán G A M 2014 (*Preprint* 1410.2543)
[6] Ashtekar A, Pawlowski T and Singh P 2006 *Phys. Rev. D* **73**(12) 124038

Journal of Physics: Conference Series **600** (2015) 012017 doi:10.1088/1742-6596/600/1/012017

A formalism based on moments for classical and quantum cosmology

David Brizuela

Fisika Teorikoa eta Zientziaren Historia Saila, UPV/EHU, 644 P.K., 48080 Bilbao, Spain

E-mail: david.brizuela@ehu.es

Abstract. We compare the classical and quantum description of systems with one degree of freedom with an effective formalism based on statistical moments. Within this formalism the similarities and differences between the classical and quantum evolution of an initial probability distribution are made apparent. We revisit the application of this formalism to a simple cosmological model: the homogeneous and isotropic universe with a massless scalar field as matter content and with positive cosmological constant.

1. Introduction

The classical limit of a quantum system is a very subtle concept and much has been written about it. One usually has in mind that a semiclassical state, that is, a state that behaves almost classically, is given by a coherent state whose centroid follows a classical trajectory in phase space. In addition, this state is usually requested to be peaked, that is, to have small fluctuations both in position and momentum.

In this talk we will follow an idea pushed forward by Ballentine, Yang and Zibin [1] that the classical limit of a quantum system is not a single classical trajectory but an ensemble of classical trajectories described by a probability distribution on phase space. As it is well known, the evolution of such a distribution will be given by Liouville equation. We will see that indeed a system can behave almost classically even if the centroid of its wave function does not follow a classical orbit.

Another reason to consider classical probability distributions is that even if it is ideally possible to have exact initial conditions of the classical system under analysis (which would be given by a point on the phase space), in practice there are always errors in these initial data. These errors can then be described as a probability distribution. This might be of particular importance in quantum cosmology because, aside from multiverse theories, there is only one realization of the universe and if we were able to obtain, let us say the fluctuations of the volume of the universe, we would need to know whether these fluctuations are purely of quantum origin or they are due to experimental errors.

Therefore a question arises: how do we know if a dynamically evolving system (described by a probability distribution) is following its corresponding classical or quantum equations of motions?

In order to shed some light on this issue we will make use of a formalism based on statistical moments. Instead of using directly the probability distribution itself, which is not observable, we will decompose it into an infinite set of moments and study the evolution of these observable quantities directly. We will do so both for quantum and classical probability distributions and will point out their similarities and differences. In particular, special Hamiltonians regarding their classical and quantum behavior will be analyzed. Finally, applications of this formalism to a particular cosmological models will be considered.

2. General formalism

In the next subsection the general formalism for quantum moments as developed in [2, 3] will be briefly summarized. The classical counterpart of this formalism was presented in [4, 5] and will be reviewed in Subsection 2.2.

2.1. Quantum moments

Let us assume a quantum system with one degree of freedom described by the basic operators $(\hat{q}, \hat{p})$ and Hamiltonian operator $\hat{H}(\hat{q}, \hat{p})$. The observable information of this system is extracted through expectation values. Therefore, we define the following infinite set of moments

$$G^{a,b} := \langle (\hat{p} - p)^a \, (\hat{q} - q)^b \rangle_{\rm Weyl}, \tag{1}$$

with $p := \langle \hat{p} \rangle$ and $q := \langle \hat{q} \rangle$ for all nonnegative integers a and b. The subscript Weyl stands for completely symmetric ordering, which makes all moments to be real. The sum of the two indices $(a + b)$ of a given moment $G^{a,b}$ will be referred as its order.

The complete information of the state is contained in this infinite set of moments. The evolution is given by the following effective Hamiltonian that is defined as the expectation value of the Hamiltonian operator and then Taylor expanded around the position of the centroid (q, p):

$$H_{\rm Q}(q, p, G^{a,b}) := \langle \hat{H}(\hat{q}, \hat{p}) \rangle = H(q, p) + \sum_{a=0}^{\infty} \sum_{b=0}^{\infty} \frac{1}{a!b!} \frac{\partial^{a+b} H}{\partial p^a \partial q^b} G^{a,b}. \tag{2}$$

The dynamics generated by this Hamiltonian is completely equivalent to the Schrödinger flow of states. The equations of motion for expectation values (q, p) and moments are obtained by making use of the general relation $\{\langle \hat{X} \rangle, \langle \hat{Y} \rangle\} = -i\hbar \langle [\hat{X}, \hat{Y}] \rangle$ for any two operators $\hat{X}$ and $\hat{Y}$:

$$\frac{dq}{dt} = \frac{\partial H}{\partial p} + \sum_{a+b \geq 2} \frac{1}{a!b!} \frac{\partial^{a+b+1} H}{\partial p^{a+1} \partial q^b} G^{a,b}, \tag{3}$$

$$\frac{dp}{dt} = -\frac{\partial H}{\partial q} - \sum_{a+b \geq 2} \frac{1}{a!b!} \frac{\partial^{a+b+1} H}{\partial p^a \partial q^{b+1}} G^{a,b}, \tag{4}$$

$$\frac{dG^{a,b}}{dt} = \sum_{c+d \geq 2} \frac{1}{c!d!} \frac{\partial^{c+d} H}{\partial p^c \partial q^d} \{G^{a,b}, G^{c,d}\}. \tag{5}$$

As can ben seen, generically this infinite set of equations are highly coupled. In particular there are moments in the equations of motion for the position q and for the momentum p. [Note that without this moment terms Eqs. (3-4) are the usual Hamilton equations which describe the classical orbits on phase space.] This is the quantum back-reaction which prevents the centroid (q, p) from following a classical orbit on phase space.

Due to the infinite number of equations present in the system (3-5), usually a truncation is necessary for practical purposes. In fact this formalism is good as long as the state remains peaked so that the high-order moments are not too large and the commented truncation is valid.

2.2. Classical moments

Let us assume a classical system with one degree of freedom with Poisson brackets $\{\tilde{q},\tilde{p}\}_c = 1$ and Hamiltonian $H(\tilde{q},\tilde{p})$. A classical ensemble is then described by a probability distribution $\rho(\tilde{q},\tilde{p},t)$. The evolution of this distribution is given by the Liouville equation:

$$\frac{\partial\rho}{\partial t} = -\{\rho, H(\tilde{q},\tilde{p})\}_c. \tag{6}$$

With the probability distribution $\rho(\tilde{q},\tilde{p},t)$ at hand, it is possible to define the expectation value for any function f on phase space as follows:

$$\langle f(\tilde{q},\tilde{p})\rangle_c := \int d\tilde{q}d\tilde{p} f(\tilde{q},\tilde{p})\rho(\tilde{q},\tilde{p},t). \tag{7}$$

In this way, one can define the classical moments in complete analogy to their quantum counterparts:

$$C^{a,b} := \langle(\tilde{p}-p)^a\,(\tilde{q}-q)^b\rangle_c, \tag{8}$$

with $p := \langle\tilde{p}\rangle_c$ and $q := \langle\tilde{q}\rangle_c$.

As in the quantum case, the effective Hamiltonian is defined as the expectation value of the Hamiltonian and it is Taylor expanded around the position of the centroid (q,p):

$$H_C(q,p,C^{a,b}) := \langle H(\tilde{q},\tilde{p})\rangle_c = H(q,p) + \sum_{a=0}^{\infty}\sum_{b=0}^{\infty}\frac{1}{a!b!}\frac{\partial^{a+b}H}{\partial p^a\partial q^b}C^{a,b}. \tag{9}$$

The evolution equations for the moments can then be obtained by making use of the relation $\{\langle f\rangle_c\,,\langle g\rangle_c\} = -i\hbar\langle\{f,g\}_c\rangle_c$:

$$\frac{dq}{dt} = \frac{\partial H}{\partial p} + \sum_{a+b\geq 2}\frac{1}{a!b!}\frac{\partial^{a+b+1}H}{\partial p^{a+1}\partial q^b}C^{a,b}, \tag{10}$$

$$\frac{dp}{dt} = -\frac{\partial H}{\partial q} - \sum_{a+b\geq 2}\frac{1}{a!b!}\frac{\partial^{a+b+1}H}{\partial p^{a}\partial q^{b+1}}C^{a,b}, \tag{11}$$

$$\frac{dC^{a,b}}{dt} = \sum_{c+d\geq 2}\frac{1}{c!d!}\frac{\partial^{c+d}H}{\partial p^c\partial q^d}\{C^{a,b},C^{c,d}\}. \tag{12}$$

The only difference between the quantum (3-5) and the classical (10-12) set of equations is contained in the Poisson brackets between moments. The brackets between any two quantum moments is given by,

$$\{G^{a,b},G^{c,d}\} = a\,d\,G^{a-1,b}\,G^{c,d-1} - b\,c\,G^{a,b-1}\,G^{c-1,d} + \sum_{n\in 2\mathbb{N}+1}\left(\frac{i\hbar}{2}\right)^{n-1}K^n_{abcd}\,G^{a+c-n,b+d-n}, \tag{13}$$

with constants K^n_{abcd}, whereas for classical moments,

$$\{C^{a,b},C^{c,d}\} = a\,d\,C^{a-1,b}\,C^{c,d-1} - b\,c\,C^{a,b-1}\,C^{c-1,d} + (b\,c - a\,d)\,C^{a+c-1,b+d-1}. \tag{14}$$

In fact, this last equation is obtained from the previous one just by taking $\hbar = 0$.

Hence, it is possible to define two kind of quantum effects. On the one hand, there are *distributional* effects due to the fact that all moments can not be vanishing since this would violate the Heisenberg uncertainty relation. These effects are also present in the classical setting. On the other hand, there are *non-commutativity* or *purely-quantum* effects, which only appear in the quantum equations of motion as explicit $\hbar^{2m}$ factors. The origin of such terms lies in the non-commutativity of the basic operators. They first appear at third order and thus the evolution of the quantum and classical system is completely equivalent when truncating at second order.

3. Special Hamiltonians

In this section we will analyze two classes of Hamiltonians that have very peculiar properties regarding the classical and quantum dynamics they generate.

3.1. Harmonic Hamiltonians

We define a harmonic Hamiltonian as being at most quadratic in both q and p. For such kind of Hamiltonians, the equations of motion (3-5) read

$$\begin{aligned}\frac{dq}{dt} &= \frac{\partial H}{\partial p},\\ \frac{dp}{dt} &= -\frac{\partial H}{\partial q},\\ \frac{dG^{a,b}}{dt} &= b\frac{\partial^2 H}{\partial p^2}G^{a+1,b-1} + (b-a)\frac{\partial^2 H}{\partial p \partial q}G^{a,b} - a\frac{\partial^2 H}{\partial q^2}G^{a-1,b+1}.\end{aligned}$$

In this case all orders decouple. In particular there is no back reaction of the moments on the expectation values and therefore q and p follow a classical orbit on phase space. In addition, everything is independent of $\hbar$ and thus there is no purely-quantum effects. Therefore, the classical (distributional) evolution is exactly the same as the quantum one. This means that for harmonic Hamiltonians the classical and quantum evolutions are completely indistinguishable.

3.2. Linear Hamiltonians

Let us now analyze the case of a Hamiltonian that is linear in one of the basic variables, for instance on the position q,

$$H = q\,\varphi(p) + \xi(p)$$

with generic functions φ and ξ. In such a case there is indeed back reaction of the moments on the expectation values and the centroid does not follow a classical orbit not in the quantum neither in the classical setting. Nevertheless, the set $(q, p, G^{a,0}, G^{a,1})$, for all a, forms a closed (infinite) subsystem of equations with no $\hbar$ terms. Therefore, the departure of the centroid from the classical trajectory is uniquely due to distributional effects since the classical set $(q, p, C^{a,0}, C^{a,1})$ will follow exactly the same evolution as its quantum counterpart. Hence, in order to find purely-quantum effects in this system, the evolution of other moments must be checked.

4. Application to a cosmological model

In this section we will revisit the homogeneous and isotropic cosmological model with a positive cosmological constant Λ and filled with a scalar matter field ϕ studied in Ref. [6]. In that reference, the quantum evolution of this model was considered. Here, the classical distributional evolution will be studied and compared with the quantum one.

The Friedmann equation of the model under consideration is

$$\left(\frac{\dot{a}}{a}\right)^2 = \frac{4\pi G}{3}\frac{p_\phi^2}{a^6} + \Lambda. \tag{15}$$

The canonical gravitational variables are defined as $q := 2a^3/3$ and $p := \dot{a}/a$, which are essentially the volume of the universe and the Hubble parameter. The theory can then be deparameterized by using the scalar field ϕ as an internal time and solving the above equation for the conjugate momentum of the scalar field p_ϕ. Then this latter variable turns out to be the physical Hamiltonian for the system:

$$H := p_\phi = \frac{3}{2} q \sqrt{p^2 - \Lambda}. \tag{16}$$

This Hamiltonian is linear in the position variable q and thus all the properties commented in Subsection 3.2 apply. The limit of a vanishing cosmological constant ($\Lambda = 0$) leads to a harmonic Hamiltonian which is linear in both position and momentum variables. In this latter case, as commented in Subsection 3.1, the classical (distributional) and the quantum evolutions are completely equivalent.

Before further analyzing the distributional evolution of this system, let us briefly comment the behavior of the classical phase space orbits. Starting from finite values $q = q_0$ and $p = p_0$ at a time $\phi = \phi_0$, the position q increases reaching $q = \infty$, whereas p decreases up to $p = \Lambda^{1/2}$, at a finite value $\phi = \phi_{div}$.

4.1. Results: the harmonic ($\Lambda = 0$) case

As always happens when dealing with a harmonic Hamiltonian, in this case equations at different orders decouple. There is thus no quantum back-reaction on the classical orbits, and the solution at any order reads,

$$q(\phi) = q_0 \exp\left[\frac{3}{2}(\phi - \phi_0)\right], \tag{17}$$

$$p(\phi) = p_0 \exp\left[-\frac{3}{2}(\phi - \phi_0)\right], \tag{18}$$

$$G^{a,b}(\phi) = G^{a,b}(\phi_0) \exp\left[\frac{3}{2}(b-a)(\phi - \phi_0)\right]. \tag{19}$$

As can be seen in this solution, a given moment $G^{a,b}$ increases (decreases) exponentially if $b > a$ ($b < a$), whereas moments of the form $G^{a,a}$ are kept constant under evolution. In addition the combinations $\sqrt[b]{|G^{0,b}|}/q$ and $\sqrt[a]{|G^{a,0}|}/p$ are also constants of motion. Classical moments $C^{a,b}$ have exactly the same behavior as their quantum counterparts.

4.2. Results: the anharmonic ($\Lambda > 0$) case

In this ($\Lambda > 0$) case, the Hamiltonian is no longer harmonic but it is still linear in the position q. Therefore, as commented in Subsection 3.2, $q(\phi)$ and $p(\phi)$ have exactly the same evolution under the classical (distributional) or quantum dynamics. In addition to the expectation values, the following infinite set of moments have also the same classical and quantum behavior. That is, given the same initial distribution for both sectors,

$$G^{n,0}(\phi) = C^{n,0}(\phi) \quad \text{and} \quad G^{n,1}(\phi) = C^{n,1}(\phi), \tag{20}$$

for all integer n and all times ϕ.

In order to extract physical consequences for the rest of the moments, the numerical resolution for an initial unsqueezed Gaussian state has been performed. For such a state, the only non-vanishing moments are those with two even indices $G^{2n,2m}$. The numerical resolution has been done with different cut-offs, from zero up to tenth order, and the convergence of the method with the truncation order has been verified. Furthermore, several values of the cosmological constant have been considered.

The obtained results can be summarized as follows. Except moments of the form $G^{0,2n+1}$, the rest $G^{a,b}$ qualitatively have the same behavior as their classical counterparts $C^{a,b}$. Even so, odd fluctuations of the position $G^{0,2n+1}$ present also asymptotically the same evolution as their corresponding classical moments $C^{0,2n+1}$.

For a quantitative comparison, we define the ratio $r_{a,b} := G^{a,b}/C^{a,b}$. In this way, three different behaviors (independent of the value of Λ) have been observed:

- Moments with two even ($G^{2n,2m}$) and with two odd ($G^{odd,odd}$) indices have a very small ratio $r_{a,b} \approx 1 \pm 10^{-6}$. This kind of quantum moments show thus the smallest departure from their classical counterparts but this difference increases with time.
- Moments of the form $G^{2n+1,2m}$ are smaller than their classical counterparts and therefore $r_{a,b} < 1$. Interestingly this ratio is kept approximately constant under evolution.
- Finally, moments of the form $G^{2n,2m+1}$ are larger than their classical counterparts at the beginning of the evolution ($r_{a,b} > 1$), but this ratio decreases with time.

Note that the last two cases correspond to moments that are initially vanishing. Therefore, as soon as the evolution is started these moments are excited to certain value. But purely-quantum effects affect in an opposite way moments of the form $G^{2n+1,2m}$ and $G^{2n,2m+1}$. While for moments $G^{2n+1,2m}$ excitation happens to a smaller value than their classical counterparts, and thus purely-quantum effects reduce this excitation, the initial value for $G^{2n,2m+1}$ is enhanced by purely-quantum effects.

5. Conclusions

A formalism based on a decomposition of moments of the corresponding probability distributions has been presented for classical and quantum systems with one degree of freedom. This framework is very useful to compare both evolutions since it makes transparent the parallelism and differences between both. Harmonic and linear Hamiltonians have been analyzed and their special properties regarding the classical (distributional) and quantum evolution they generate have been discussed. Finally an application of this formalism to particular cosmological model has been considered.

Acknowledgments

This work is supported in part by Projects IT592-13 of the Basque Government and FIS2012-34379 of the Spanish Ministry of Economy and Competitiveness.

References

[1] L. E. Ballentine, Y. Yang, and J. P. Zibin, Phys. Rev. A **50**, 2854 (1994).
[2] M. Bojowald and A. Skirzewski, Rev. Math. Phys. **18**, 713 (2006).
[3] M. Bojowald, B. Sandhöfer, A. Skirzewski, and A. Tsobanjan, Rev. Math. Phys. **21**, 111 (2009).
[4] D. Brizuela, Phys. Rev. D **90**, 085027 (2014).
[5] D. Brizuela, arXiv:1411.1522 [quant-ph].
[6] M. Bojowald, D. Brizuela, H. H. Hernández, M. J. Koop, and H. A. Morales-Técotl, Phys. Rev. D **84** 043514 (2011).

Journal of Physics: Conference Series **600** (2015) 012018 doi:10.1088/1742-6596/600/1/012018

Calibrating Analytical Models for Semilocal Strings

A. Lopez-Eiguren[1]

Department of Theoretical Physics, University of the Basque Country UPV-EHU, 48040 Bilbao, Spain

E-mail: asier.lopez@ehu.es

Abstract. In this work we calibrate two different analytic models of semilocal strings by constraining the values of their free parameters. In order to do so, we use data obtained from the largest and most accurate field theory simulations of semilocal strings to date, and compare several key properties with the predictions of the models. As this is still work in progress, we present some preliminary results together with descriptions of the methodology we are using in the characterisation of semilocal string networks.

1. Introduction

Understanding the evolution of string networks is crucial for predicting their number densities, which in turn determine their potentially observable effects. However, the quantitatively accurate modelling of string network evolution is a difficult problem, requiring the combination of a range of techniques (both numerical and analytical), and interpolating between physics at very different energy scales.

Here, we present a calibration of analytical models for semilocal strings using field theory simulations. This work is part of an ongoing project where we tackle in turn different aspects of the calibration procedure by comparing the numerical simulations with predictions for the analytic models. In section 3, we study the large-scale properties of the simulated networks [1]. Then, in section 4, we outline the comparison between field theory simulations and the analytic models, anticipating work that will appear in [2]. Section 5 shows the last ingredient of our analysis, where we present the techniques we will use in a future work [3] to estimate the velocities of the semilocal strings. Prior to all this, in section 2 we introduce the semilocal model and the analytic models used in this work, as well as the numerical simulations performed.

2. The Model

2.1. Semilocal Model

Semilocal strings [5, 6, 7] were introduced as a minimal extension of the Abelian Higgs (AH) model with two complex scalar fields instead of just one, that form an $SU(2)$ doublet. This leads to $U(1)$ flux-tube solutions even though the vacuum manifold is simply connected. The strings of this extended model have some similarities with ordinary local $U(1)$ strings, but they are not purely topological and will therefore have different properties. For example, since they are not topological, they need not be closed or infinite and can have ends. These ends are effectively global monopoles which have long-range interactions[8].

[1] Work in collaboration with A.Achúcarro, A.Avgoustidis, A.M.M.Leite, C.J.A.P.Martins, A.S.Nunes, J.Urrestilla

The relevant action for the simplest semilocal string model, the one we will use in the numerical simulations, reads:

$$S = \int d^4x \Big([(\partial_\mu - iA_\mu)\Phi]^2 - \frac{1}{4}F^2 - \frac{\beta}{4}(\Phi^+\Phi)^2 \Big), \tag{1}$$

where $\Phi = (\phi, \varphi)$ (ϕ and φ are complex scalar fields), $F^2 = F_{\mu\nu}F^{\mu\nu}$ and $F_{\mu\nu} = \partial_\mu A_\nu - \partial_\nu A_\mu$. It can be easily seen that setting one of the two scalar fields to zero, we recover the AH model.

The stability of the strings is not trivial, and it will depend on the value of the parameter $\beta = m^2_{scalar}/m^2_{gauge}$: for $\beta < 1$ the string is stable, for $\beta > 1$ it is unstable, and for $\beta = 1$ it is neutrally stable [5, 6, 7]. Only low values of β will be of interest for the comparison, because otherwise the string network is either unstable or disappears very fast. It is also known that the lower the β, the more it looks like AH strings.

2.2. *Analytic Models*

Broadly speaking there are two ways to model the evolution of these defects [4] using effective models. The simplest one attempts to model the overall network, specifically focusing on the evolution of the monopoles (which are known to dominate the dynamics of the network). This is a 'classical' one-scale model description. On the other hand, given that the evolution of the semilocal segments is highly non-trivial, a more accurate description must necessary include the evolution of this segment population. In this subsection we will concentrate on the latter.

Numerical simulations indicate that while some segments shrink and disappear, there is also a significant probability that segments merge and form longer segments. This is due to the long range interactions of the global monopoles at the end of segments. To account for this behaviour we start from the evolution equations for segment size introduced in the velocity-dependent one-scale (VOS) model, where the equations come from the statistical averaging of the microscopic equations of motion [10, 11]:

$$\frac{dl_s}{dt} = Hl_s - v_s^2\frac{l_s}{l_d}, \quad \frac{dv_s}{dt} = (1 - v_s^2)\Big[\frac{k}{l_s} - \frac{v_s}{l_d}\Big], \tag{2}$$

where l_s is the length of the segment under consideration, v_s is the root mean square (RMS) velocity, k a free parameter describing string curvature(to be calibrated), H the Hubble parameter and l_d is the string damping length. We now modify those equations in two possible phenomenological ways [4]:

2.2.1. Scale-dependent Behaviour: A simple generalisation of the equations (2) would be

$$\frac{dl_s}{dt} = Hl_s - v_s^2\frac{l_s}{l_d} + \sigma\Big(1 - \frac{L}{l_s}\Big)v_m^2, \quad \frac{dv_s}{dt} = (1 - v_s^2)\Big[\frac{k}{l_s} - \frac{v_s}{l_d}\Big] \tag{3}$$

where L is the characteristic scale of the monopoles and σ is a free parameter controlling the importance of the newly introduced term. The new term was added on phenomenological reasoning that, to a first approximation, small segments should shrink and large ones should grow and merge [8]. This can be intuited as a competition between two characteristic timescales. Each segment will have an annihilation timescale, and each monopole will have a characteristic timescale in which to find its (anti)partner and annihilate, thereby producing a longer segment. The second process is expected to become relatively more likely as the segment size increases.

2.2.2. Balance Equation: The following modification for the evolution equations is considered

$$\frac{dl_s}{dt} = Hl_s - v_s^2\frac{l_s}{l_d} + dv_s\Big(\frac{l_s}{L} - 1\Big), \qquad \frac{dv_s}{dt} = (1 - v_s^2)\Big[\frac{k}{l_s} - \frac{v_s}{l_d}\Big] \tag{4}$$

In these we are assuming that the network of string segments has a Brownian distribution, something that can be tested in numerical simulations. The new term (including new free parameter, d) accounts for the probability that different segments intersect, which depends both on the length/number density and velocity of the segments.

2.3. Field Theory Simulations

We simulated numerically the semilocal model introduced in section 2.1 so as to provide us with data to be used for comparison with the analytic models. The parameter space we want to explore is rather large, so we carefully chose the cases to study, and tried to maximise the information we could obtain from simulations given the computer resources available to us. The cases for study chosen are $\beta = 0.01, 0.04, 0.09$; the cosmologies under consideration correspond to expanding universes, both in radiation and matter eras. We perform each one of these cases for two different lattice spacings $\delta x = 0.5, 1$.

We discretised the action given in equation (1) by standard techniques (using lattice-link variables and a staggered-leapfrog method) and evolved the discretised action in 1024^3 lattices with periodic boundary conditions, similar to [9].

Once the system reaches scaling, quantities of interest can be measured. Semilocal strings are non topological entities; therefore, we cannot use topology to detect them. This kind of strings can be thought of as concentrations of magnetic energy, and that is the strategy we follow: we first calculate the maximum of the magnetic field strength, and the radius, of a straight and infinite AH string for a given β. We use those values for the simulated semilocal string network: if the magnetic field strength of a simulated semilocal model measured at a point of the box exceeds the 25% of the maximum of the corresponding AH string, we consider that point to be part of a semilocal string segment. The output of our simulation is thus an array of points from the simulated box which have a considerable concentration of magnetic field strength.

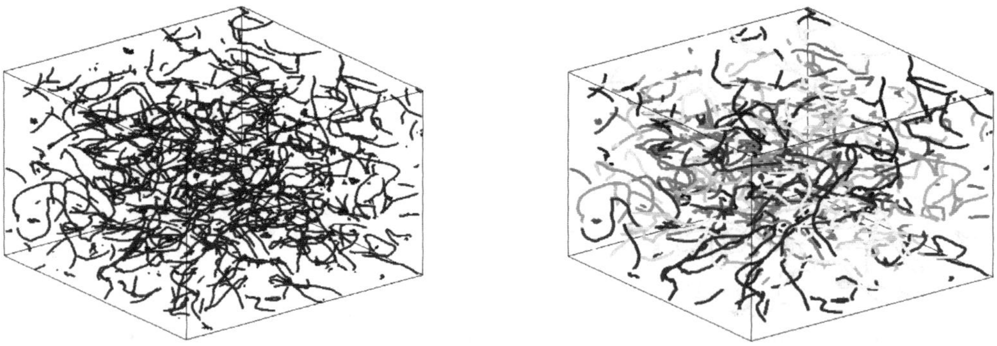

Figure 1. Semilocal string network, in matter domination with $\beta = 0.04$. The left figure shows two types of structures: on the one hand we have tube-like structures (proper strings) and on the other short blobs. These blobs we disregard in our analysis. The right figure shows the network without blobs, and also each segment has been identified and plotted with a different colour. As the number of segments is large, the colours are unfortunately used for more than one string segment. Note also that the blob removal procedure does sometimes fail to identify some sphere-like structures, since their volume is large.

We then group together the points that have been output by the simulations into segments. These segments are mostly tube-like, but some are sphere-like instead of tube-like, i.e., they are blobs of energy. These can be formed, for example, after a segment collapsed into itself. We do not wish to count these blobs as part of our network, and we introduce a lower cut-off: those segments that are not longer than a given factor times the typical radius of a string are considered to be blobs and are discarded. One typical simulation snapshot is shown in Fig. 1.

3. Large-scale Properties [1]

We have performed 12 simulations for each case of our set of parameters, and use the results to obtain basic statistics about the properties of the networks. All in all, for each one of those simulations, and for specific values of the simulation time, we obtain the total string length $\mathcal{L}(t)$ and monopole number $\mathcal{N}(t)$ in the box. Both of these provide simple diagnostics for the large-scale evolution of the network, and specifically for the presence of scaling.

Fig. 2 provides an example of the evolution of these quantities for the case $\beta = 0.09$, $\delta x = 1$ and matter era simulation. This is representative of all the sets of simulations we have performed. This analysis therefore shows that all the networks have reached the scaling solution.

The string lengths and number of monopoles obtained can easily be translated into VOS-type length scales using [10, 11],

$$\gamma_s \equiv \frac{L_s}{t} = \frac{1}{t}\sqrt{\frac{V}{\mathcal{L}}}, \tag{5}$$

$$\gamma_m \equiv \frac{L_m}{t} = \frac{1}{t}\left(\frac{V}{\mathcal{N}}\right)^{1/3}. \tag{6}$$

After analysing the VOS-type length scales, in all cases under study, we can see that in matter era γ_s is larger and γ_m smaller than in radiation era. This is because in radiation era the effective monopole velocity is higher and segments move faster to either grow and meet with other segments or collapse, giving a longer typical string length and smaller number of monopoles. Note that γ_s and γ_m are inversely proportional to $\mathcal{L}$ and $\mathcal{N}$, respectively.

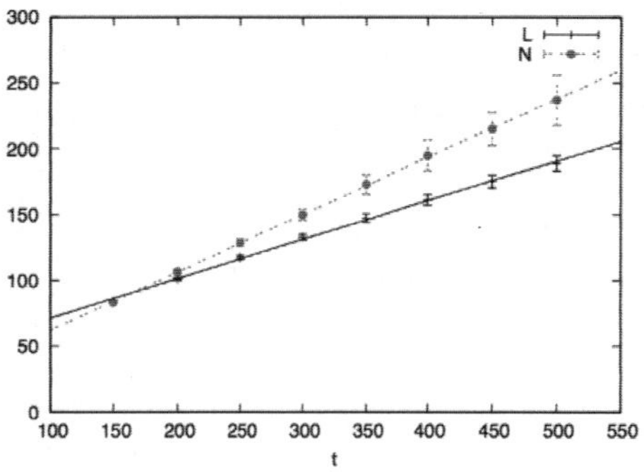

Figure 2. Scaling plots for L and N for $\beta = 0.09$, in the matter era, with $\delta x = 1$. The error bars show statistical errors over the 12 simulations.

However for a given cosmology, γ_s grows with β and γ_m gets smaller. This is also to be expected since for lower β we expect the system to behave more like an AH network, which has longer strings and fewer segments.

4. Segment Distribution [2]

Our simulations have sufficient resolution and dynamical range to allow us to perform quantitative diagnostics, and study in much more detail the structure and evolution of our string networks. While the previous section focused only on two important physical quantities, $\mathcal{L}$ and $\mathcal{N}$, in this section we will take this further and study the statistics of string segment length distributions. Starting from initial configurations of the semilocal string networks, we can group, at any given time, all string segments into suitable chosen length bins and study how the length distributions evolve in time.

Furthermore we can evolve these length distributions using our analytic models, so as to compare data from field theory simulations and analytic models. For example in Fig. 3 we can see length distributions for different time steps directly coming from the simulations, as well as, from analytic models. Now we can employ χ^2 minimisation to calibrate the free parameters of the analytic models.

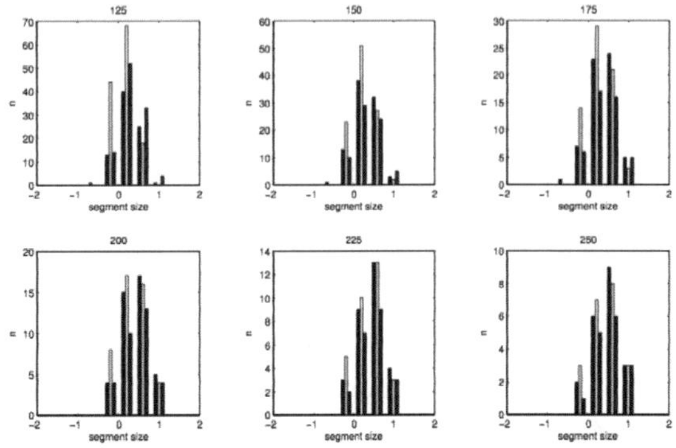

Figure 3. Number of strings in each bin for different time steps for $\beta = 0.01$, radiation era and $\delta x = 0.5$. Red columns show values coming from the simulations, while blue columns represents the 2.2.1 model and green columns the 2.2.2 model.

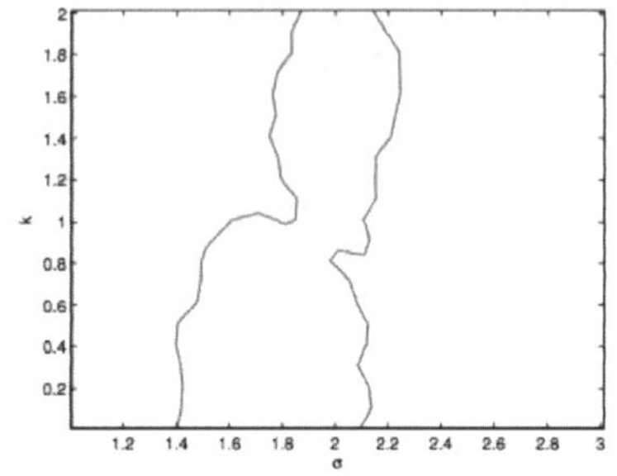

Figure 4. Contour plot showing the best values for the parameter space for $\beta = 0.01$, radiation era and $\delta x = 0.5$.

In Fig 4 we have a typical contour plot obtained after the χ^2 minimisation. The figure shown is a preliminary result for the scale-dependent behaviour case. There we can see that the parameter σ is constrained $[1.4, 2.2]$, but the parameter k is unconstrained.

5. Velocities [3]

In this last section we will reduce systematic uncertainties in our modelling to give a fully quantitative calibration. To do this, we will calculate transverse velocity of the string segments and also we will measure the monopole velocity by tracking monopoles during the evolution. The results obtained using these method will appear in [3].

5.1. String Transverse Velocity

In order to work out the transverse velocities we will use the method proposed by [12].

$$\gamma_F^2 \langle \dot{\mathbf{x}}^2 \rangle_F = \frac{\mathbf{E}_{\mathcal{L}}^2}{\mathbf{B}_{\mathcal{L}}^2} \qquad \gamma_G^2 \langle \dot{\mathbf{x}}^2 \rangle_G = \frac{\mathbf{\Pi}_{\mathcal{L}}^2}{(\mathbf{D}\Phi)_{\mathcal{L}}^2} \tag{7}$$

In the first case $E_i = F_{0i}$ and $B_k = \epsilon_{ijk} F_{ij}/2$ are the electric and magnetic fields respectively where ϵ_{ijk} is the Levi-Civita symbol. In the second case $\Pi = \dot{\Phi}$ is the canonical momentum and $\mathbf{D}\Phi = \partial\Phi/\partial\mathbf{x} + i\mathbf{A}\Phi$ represents the spatial gradients of the fields. γ^2 is calculated using $\langle \dot{\mathbf{x}}^2 \rangle$, and subscript $\mathcal{L}$ denotes a Lagrangian weighting of a field $\mathcal{X}$ according to

$$\mathcal{X}_{\mathcal{L}} = \frac{\int d^3x \; \mathcal{X} \; \mathcal{L}}{\int d^3x \; \mathcal{L}}. \tag{8}$$

5.2. Monopole Velocity

As said before semilocal strings can have ends. These ends are effectively global monopoles, and we investigate the velocity of these ends. In order to measure this velocity, we need to pinpoint where a monopole moves at every step, but it is difficult taking into account that there are lots of strings in the simulation box.

Firstly, we have to detect all the monopoles individually. As we know, monopoles are characterised by their topological charge. In order to detect them we compute the topological charge around a lattice point using the method described in [6].

Finally we track the path of all individual monopoles by taking into account that they cannot travel faster than the speed of light and that they conserve their topological charge. Thus, we can measure the monopole velocity directly using the path they have travelled during the evolution.

6. Summary

Here, we have presented a numerical study of semilocal string networks. Firstly, we have discussed the large-scale properties of simulated semilocal networks, covering couplings in the range $0.01 \leq \beta \leq 0.09$, and damping terms corresponding to expanding universes dominated by radiation and matter. We have demonstrated scaling behaviour for semilocal networks. Then we have described a comparison of our simulations with analytical models: starting from an initial configuration of a semilocal network we are gathering all string segments into length bins and we are comparing the evolution of the segments in each bin using our field theory simulation and our analytic models. An important source of uncertainty in this comparison is related to our lack of knowledge of the velocity in simulations of semilocal strings. Therefore, in the last part we have presented the methods that we will use to compute the transverse string velocity and the monopole velocity.

Acknowledgments

I am supported by the Basque Government grant BFI-2012-228. I acknowledge also financial support from the Basque Government (IT-559-10),the Spanish Ministry (FPA2012-34456), and the Spanish Consolider EPI CSD2010-00064. Additional support for the project was provided by grant PTDC/FIS/111725/2009 (FCT, Portugal).

References

[1] A. Achúcarro, A. Avgoustidis, A. M. M. Leite, A. Lopez-Eiguren, C. J. A. P. Martins, A. S. Nunes and J. Urrestilla, "Evolution of semilocal string networks: Large-scale properties," Phys. Rev. D **89** (2014) 6, 063503 [arXiv:1312.2123 [hep-ph]].

[2] A. Achúcarro, A. Avgoustidis, A. M. M. Leite, A. Lopez-Eiguren, C. J. A. P. Martins, A. S. Nunes and J. Urrestilla, "Evolution of semilocal string networks: II. Segment Distribution," In preparation

[3] A. Achúcarro, A. Avgoustidis, A. M. M. Leite, A. Lopez-Eiguren, C. J. A. P. Martins, A. S. Nunes and J. Urrestilla, "Evolution of semilocal string networks: III. Velocity Estimation," In preparation

[4] A. S. Nunes, A. Avgoustidis, C. J. A. P. Martins and J. Urrestilla, "Analytic Models for the Evolution of Semilocal String Networks," Phys. Rev. D **84** (2011) 063504 [arXiv:1107.2008 [hep-ph]].

[5] T. Vachaspati and A. Achúcarro, "Semilocal cosmic strings," Phys. Rev. D **44** (1991) 3067.

[6] A. Achúcarro and T. Vachaspati, "Semilocal and electroweak strings," Phys. Rept. **327** (2000) 347 [Phys. Rept. **327** (2000) 427] [hep-ph/9904229].

[7] M. Hindmarsh, "Existence and stability of semilocal strings," Phys. Rev. Lett. **68** (1992) 1263.

[8] M. Hindmarsh, "Semilocal topological defects," Nucl. Phys. B **392** (1993) 461 [hep-ph/9206229].

[9] A. Achúcarro, P. Salmi and J. Urrestilla, "Semilocal cosmic string networks," Phys. Rev. D **75** (2007) 121703 [astro-ph/0512487].

[10] C. J. A. P. Martins and E. P. S. Shellard, "Quantitative string evolution," Phys. Rev. D **54** (1996) 2535 [hep-ph/9602271].

[11] C. J. A. P. Martins and E. P. S. Shellard, "Extending the velocity dependent one scale string evolution model," Phys. Rev. D **65** (2002) 043514 [hep-ph/0003298].

[12] M. Hindmarsh, S. Stuckey and N. Bevis, "Abelian Higgs Cosmic Strings: Small Scale Structure and Loops," Phys. Rev. D **79** (2009) 123504 [arXiv:0812.1929 [hep-th]].

Spanish Relativity Meeting (ERE 2014): almost 100 years after Einstein's revolution IOP Publishing
Journal of Physics: Conference Series **600** (2015) 012019 doi:10.1088/1742-6596/600/1/012019

A new approach to the dynamics of AdS space-times

Daniel Santos-Oliván
Institut de Ciències de l'Espai (CSIC-IEEC), Campus UAB, 08193 Bellaterra, Spain
E-mail: santos@ice.cat

Carlos F. Sopuerta
Institut de Ciències de l'Espai (CSIC-IEEC), Campus UAB, 08193 Bellaterra, Spain
E-mail: sopuerta@ieec.uab.es

Abstract. In the last years, the stability of Anti-de Sitter (AdS) space-times has attracted a lot of attention. Not only because of its own importance but primarily due to the AdS/CFT correspondence that conjectures an equivalence between an string theory on an asymptotically AdS space-time and a conformally invariant quantum field theory (CFT). Recently, Bizon and Rostworowski (2011) showed that the boundary of AdS prevents energy from dispersing. As a consequence, AdS perturbations will collapse after bouncing back from the AdS boundary a sufficient number of times for their energy to be concentrated enough. The details of the mechanisms that triggers the instability are currently an active topic of discussion. Here, we present a new approach to the problem of the collapse of a massless scalar field that introduces a transition from a Cauchy-based evolution to a charateristic one that is able to track with greater accuracy the latest stages of the collapse.

1. Introduction

The interest in the study of space-times with negative cosmological constant is very recent. Although AdS is a maximally symmetric solution to Einstein's field equations known since a long time ago, physicists have ignored this space-time and even classical books of General Relativity rarely mention it (one notable exception is the Hawking and Ellis book [1]). The establishment of the gauge/gravity duality through the AdS/CFT correspondence shown by Maldacena [2] have changed the scenario completely, generating a lot of activity. As in the case of Minkowski and de Sitter space-times, AdS space-time has been shown to be stable under linear (infinitesimal) perturbations [3], but the problem of non-linear perturbations of AdS is still under debate. Different numerical simulations have shown, in contrast to the Minkowski and de Sitter cases, instabilities in the dynamics of real [4, 5] and complex [6] massless scalar fields in AdS space-times. However, some works indicate there may exist states free from instabilities [7, 8]. There have also been attempts to study the AdS stability with purely analytic means [9, 10, 11, 12]. In addition, there are some studies on the role of the AdS boundary that consider space-times without cosmological constant but with reflecting boundaries [13, 14, 15].

We address the problem of the collapse of a massless scalar field in an Asymptotically AdS (AAdS) space-times from a new point of view. We start with a Cauchy evolution scheme in order to follow the evolution through the different bounces at the AdS boundary, but then we introduce a transition to a characteristic evolution scheme to follow the last stages of the collapse with much more resolution (more details will appear in [16]).

2. Dynamics of AAdS space-times

The dynamics of a self-gravitating real massless scalar field ϕ in an AAdS space-time of dimension $d+1$ is described by the Einstein-Klein-Gordon field equations:

$$G_{\mu\nu} + \Lambda g_{\mu\nu} = (d-1)\left(\nabla_\mu\phi\,\nabla_\nu\phi - \frac{1}{2}g_{\mu\nu}\nabla_\alpha\phi\,\nabla^\alpha\phi\right)\,, \qquad g^{\mu\nu}\nabla_\mu\nabla_\nu\phi = 0\,, \tag{1}$$

where $g_{\mu\nu}$ $(\mu,\nu = 0,...,d)$ is the space-time metric (and ∇_μ the associated canonical covariant derivative), $\Lambda < 0$ is the cosmological constant, and $G_{\mu\nu}$ is the Einstein tensor.

We solve these equations by using a Cauchy-characteristic scheme. We start with Cauchy evolution using a system of coordinates where the AdS boundary (reachable in a finite time by the scalar field propagation) is at a finite coordinate distance. In this way we can follow the evolution of the scalar field, which typically will bounce at the AdS boundary a finite number of times before collapsing [4, 6]. When the collapse is approaching we need a high resolution to control the gradients in the field that signal the formation of an apparent horizon (AH). We have noticed that a multidomain pseudospectral method with adaptive refinement of the domains is still not enough for many purposes, although it provides a high accuracy for most parts of the evolution. To deal with this we have designed a numerical scheme that makes a transition from the Cauchy evolution to a characteristic one where each point follows an ingoing null geodesic. This allows us to follow much better the collapse.

2.1. Cauchy evolution

The metric of a $d+1$-dimensional spherically symmetric AAdS space-time can be written as [4, 5]

$$ds^2 = \frac{\ell^2}{\cos^2(x)}\left(-A(t,x)\mathrm{e}^{-2\delta(t,x)}\,dt^2 + \frac{dx^2}{A(t,x)} + \sin^2(x)\,d\Omega^2_{d-1}\right)\,, \tag{2}$$

where ℓ is the AdS length scale, defined as $\ell^2 = -d(d-1)/2\,\Lambda$, and $d\Omega^2_{d-1}$ is the line element of a $d-1$-sphere. The important point is that x is a radial compactified coordinate with range $[0,\pi/2]$, being $x=0$ the origin and $x=\pi/2$ the AdS boundary. We recover pure AdS by taking $A=1$ and $\delta=0$. Our evolution equations for (A,δ,ϕ) are a modification of the ones in [4, 5].

2.2. Characteristic evolution

We have adapted the scheme first introduced by Christodoulou [17] and later used by [18, 19] for AAdS space-times. The starting point is the spherically-symmetric metric

$$ds^2 = -g\bar{g}\,du^2 - 2g\,dudr + r^2\,d\Omega^2_{d-1}\,, \tag{3}$$

where u is an ingoing null coordinate and r is a radial one (the AdS boundary is at $r\to\infty$). The functions $g = g(u,r)$ and $\bar{g} = \bar{g}(u,r)$ are always greater that some normalization value at the origin that we choose to be 1. Pure AdS is recovered by taking $g=1$ and $\bar{g} = 1 + r^2/l^2$. The scalar field is described by the following two variables:

$$\bar{h} \equiv \phi\,, \qquad h \equiv \phi + \frac{2\,r}{d-1}\,\phi_{,r} \quad \Rightarrow \quad \bar{h}(u,r) = \frac{1}{r}\int_0^r dr'\,h(u,r')\,. \tag{4}$$

The metric functions can be recovered from the scalar field variables from the expressions:

$$g(u,r) = \exp\left\{\frac{(d-1)^2}{4}\int_0^r dr'\,\frac{\left(h(u,r')-\bar{h}(u,r')\right)^2}{r'}\right\}\,, \tag{5}$$

$$\bar{g}(u,r) = \frac{1}{r^{d-2}} \int_0^r dr' \left(d - 2 + d\frac{r'^2}{\ell^2} \right) r'^{d-3}\, g(u,r') \,. \tag{6}$$

The characteristic evolution consists in evolving the variables along ingoing geodesics from one $u = const.$ null slide to the next one. In this way the radial coordinate of a grid point obeys the following evolution equation [see Eq. (3)]:

$$\frac{dr}{du} = -\frac{1}{2}\bar{g} \,. \tag{7}$$

The idea is to evolve the scalar field variable h from one null slice $u = const.$ to the next one along the ingoing null geodesics of Eq. (7), while the rest of variables $(\bar{h}, \bar{g}, g)$ can be obtained from h by radial integration according to Eqs. (4), (5), and (6). The equation for h comes from the Klein-Gordon equation and using that along the ingoing null geodesics we have

$$\frac{dh(u,r(u))}{du} = \left(\frac{\partial h}{\partial u}\right)_{r=r(u)} + \left(\frac{\partial h}{\partial x}\right)_{r=r(u)} \frac{dr(u)}{du} \,, \tag{8}$$

where $r(u)$ is a solution of Eq. (7). Then

$$\frac{dh}{du} = \frac{d-2}{2\,r}\left(h - \bar{h}\right)\left[\left(1 + \frac{d}{d-2}\frac{r^2}{\ell^2}\right) g - \frac{d-1}{2(d-2)}\bar{g}\right] . \tag{9}$$

2.3. Transition from Cauchy to characteristic evolution

In order to use the characteristic evolution to follow the collapse, we need to construct initial data at an initial null slide $u = u_o = const.$ from the results of the Cauchy evolution. To that end we have to translate the information from a set of $t = const.$ slides coming from the Cauchy evolution to the values of the variable h, the one we evolve from one null slide to the next one, at the initial null slide $u = u_o$. A key ingredient to perform this transformation is the following relations between Cauchy and characteristics variables:

$$h = \phi + \sin(x)\cos(x)\frac{U_- + U_+}{2} \,, \qquad U_\pm = \partial_x\phi \pm \frac{e^\delta}{A}\partial_t\phi \,, \tag{10}$$

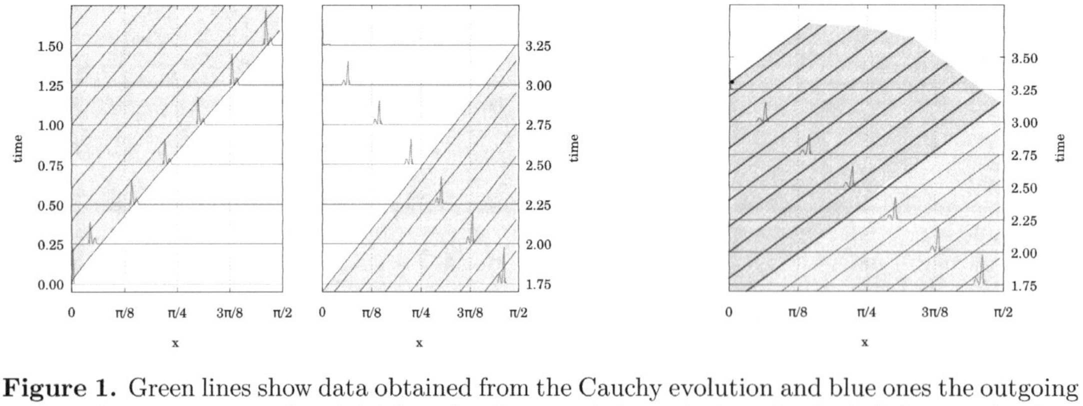

Figure 1. Green lines show data obtained from the Cauchy evolution and blue ones the outgoing null geodesics constructed to initialize the characteristic evolution. In the left figures the data generated by a complete Cauchy evolution is shown and, in the right one, the purple area indicates data obtained from the characteristic evolution.

where $U_\pm$ are characteristic fields of the Klein-Gordon equation (with propagation speeds $\pm Ae^{-\delta}$) so that the set of variables $(\phi, U_\pm)$ leads to a first-order hyperbolic system for the Cauchy evolution. Another key ingredient is the relation between the coordinate systems used for the Cauchy and characteristic evolutions. The relation between radial coordinates is quite simple: From Eqs. (2) and (3) it is just $r = \ell \tan(x)$. We do not have an expression for the null coordinate u in terms of (t, x), instead the $u = const.$ null slices have to be obtained from the equations for the outgoing null geodesics in (t, x) coordinates:

$$\frac{dx}{dt} = +Ae^{-\delta}, \tag{11}$$

which can be integrated from the origin using the Cauchy evolution data. From the coordinate change we can derive the relation

$$A = \frac{\bar{g}}{g\left(1 + \frac{r^2}{\ell^2}\right)}, \tag{12}$$

which can be used to track the formation of an AH in the characteristic evolution ($A \to 0$). The Cauchy-characteristic transition is illustrated in figure 1. In the left plot, the green lines (horizontal) represent the data generated by the Cauchy evolution. In each frame, we can start an outgoing null geodesic ($u = const.$) from $x = 0$ using Eq. (11) and compute where this geodesic will be in the next step and store the data for both x (or r) and the field h. We can construct in this way several null outgoing geodesics and then perform the Cauchy-characteristic transition using the one that has the optimal initial conditions.

3. Comparing the Cauchy and characteristic evolution schemes

In this section, for the case of $d = 3$, we compare a simulation done only with Cauchy evolution with another one that starts with Cauchy evolution but changes to characteristic evolution near the collapse. In the first case, as we approach the collapse, the scalar field develops a steep profile. Then, in order to maintain the accuracy and stability of the code we have to introduce more resolution, making the simulation much slower. In the other case, we can change before the profile becomes steep making the Cauchy evolution stage much faster. In the characteristic evolution stage we do not need to introduce more points because the null slicing adapts to the

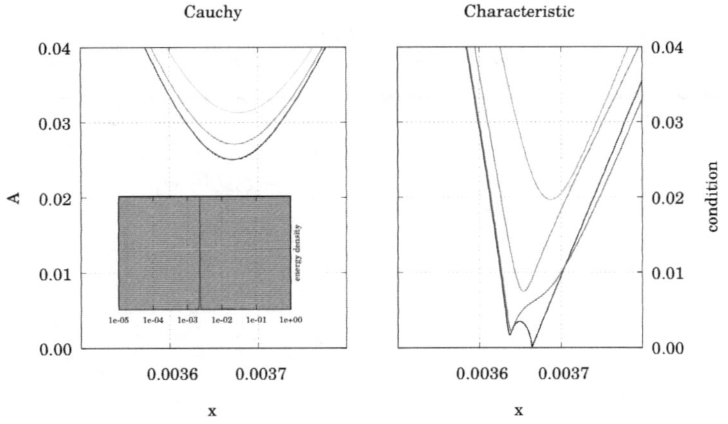

Figure 2. Comparison between the Cauchy and characteristic evolutions. AH formation occurs when $A \to 0$.

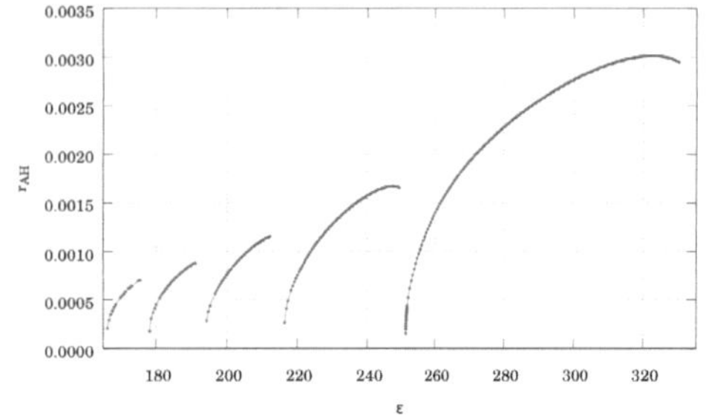

Figure 3. AH radius (r_{AH}) with respect to amplitude (ϵ). The number of bounces at the AdS boundary changes from branch to branch (see text for details).

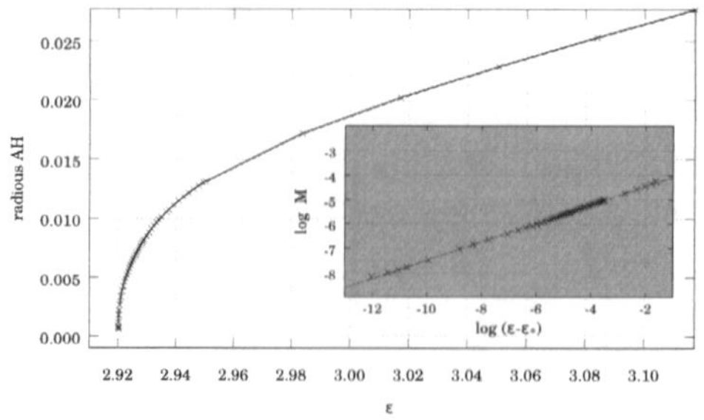

Figure 4. Critical exponent obtained using characteristic initial data and purely characteristic evolution. The value obtained is $\gamma \sim 0.378$.

horizon formation concentrating a large portion of the points in the region we are interested in. This is illustrated in figure 2, where the small plot in the left shows the steep profile generated in the energy density (notice the logarithmic scale in x). The plots shows A for both evolutions (in the characteristic case it is estimated using Eq. (12)). The computational time used for the two plots is similar. It is clear that the characteristic evolution adapts much better to our problem, allowing us to go much closer to the AH formation ($A \to 0$). The right plot shows how important is to get as close as possible to AH formation since at some point the profile of A generates structure at *small* scales.

4. Results and discussion

Using the method of the Cauchy-characteristic transition presented above, we can define our initial conditions as:

$$U_-(t=0,x) = +\epsilon \cos^{d-2}(x) e^{\frac{4\tan^2(x)}{\pi^2\omega^2}}, \qquad U_+(t=0,x) = -\epsilon \cos^{d-2} e^{\frac{4\tan^2(x)}{\pi^2\omega^2}}, \tag{13}$$

where ϵ and ω, the amplitude and width of the initial profile respectively, are free parameters. We have a adopted a fixed value for ω ($\omega = 0.05$) and have performed evolutions for a large number of values for the amplitude ϵ. For sufficient large amplitudes the field collapses in way very similar to what Choptuik described in the asymptotically-flat case [20]. This is problematic in our scheme because the time of collapse for this kind of simulations is smaller than the time that we typically need to construct an initial geodesic to initialize the characteristic evolution. For relatively small amplitudes, the field tries to collapse but fails and then it disperses. In the asymptotically-flat case the field will completely disperse and the end state of the evolution is Minkowski space-time. In the case of AAdS spacetimes the field eventually (in a finite time) bounces at the AdS boundary and comes back towards the origin where it will collapse provided certain conditions are satisfied, otherwise the field will disperse again and bounce at the AdS boundary. This will happen a number of times until the field will finally collapse. In figure 3 we can see some results from numerical simulation for initial configuration with amplitudes in the range $\epsilon \in [160, 330]$. The right branch correspond to initial conditions that collapse after just one bounce at the AdS boundary. Each branch to the left corresponds to configurations that go through one additional bounce (the left branch corresponds to configurations that went through five bounces). This behavior was reported in [4] and the evidence indicates that after each bounce the energy is transferred from lower to higher frequency modes [21, 22].

Up to now we have discussed simulations that use the Cauchy-characteristic transition. Another possibility is to use only the characteristic evolution, by prescribing initial configurations at a $u = u_o = const.$ null surface (an outgoing null geodesic) and by evolving it using our characteristic code. Since these simulations do not cover the whole space-time (see figure 1) we can only use them for configurations are always within the space-time region covered by the characteristic evolution. An example of this are configurations that collapse directly without bounces at the AdS boundary. For those we use initial conditions given by

$$\phi = \bar{h}(u=0,r) = \epsilon\, r^2\, e^{-\frac{(r-r_0)^2}{\omega^2}}\,. \tag{14}$$

Fixing $\omega = 0.05$ and $r_0 = 0.1$, we can vary the amplitude ϵ and only consider those configurations that form an AH. The others, which disperse, correspond to evolutions with bounces, but we cannot track them with the characteristic evolution. Around $\epsilon = \epsilon_* \sim 2.92$, we observe critical behavior: The black hole mass at the AH formation scales as: $M_{AH} \sim (\epsilon - \epsilon_*)^\gamma$, where $M_{AH} = r_{AH}\left(1 + (r_{AH}/\ell)^2\right)/2$. The analysis of this case can be seen in figure 4. By fitting our results in the region close to the critical amplitude, $\epsilon_* = 2.920151$, we obtain a critical exponent of $\gamma \sim 0.378$. This result is compatible with the one obtained first by Choptuik [20]. This is something that was anticipated by Bizon and Rostworowski [4] and that here we confirm with one more digit of precision.

To sum up, we have introduced a new scheme for the study of the dynamics of spherically symmetric AAdS space-times, in particular the fate of instabilities, which is based on a combined Cauchy-characteristic evolution that provides higher precision to follow collapse and the formation of an AH. We have also presented results from simulations that confirm recent findings in the literature obtained with other evolution schemes.

Acknowledgments

We acknowledge support from contracts 2009-SGR-935 (AGAUR), AYA-2010-15709 (MICINN), ESP2013-47637-P (MINECO) and the computational resources provided by CESGA (CESGA-ICTS-266). DSO acknowledges the FPI PhD contract BES-2012-057909.

References

[1] Hawking S W and Ellis G F R 1973 *The large scale structure of space-time* (Cambridge: Cambridge University Press)
[2] Maldacena J M 1999 *Int.J.Theor.Phys.* **38** 1113–1133 (*Preprint* `hep-th/9711200`)
[3] Ishibashi A and Wald R M 2004 *Class.Quant.Grav.* **21** 2981–3014 (*Preprint* `hep-th/0402184`)
[4] Bizon P and Rostworowski A 2011 *Phys.Rev.Lett.* **107** 031102 (*Preprint* `1104.3702`)
[5] Jalmuzna J, Rostworowski A and Bizon P 2011 *Phys.Rev.* **D84** 085021 (*Preprint* `1108.4539`)
[6] Buchel A, Lehner L and Liebling S L 2012 *Phys.Rev.* **D86** 123011 (*Preprint* `1210.0890`)
[7] Buchel A, Liebling S L and Lehner L 2013 *Phys.Rev.* **D87** 123006 (*Preprint* `1304.4166`)
[8] Maliborski M and Rostworowski A 2013 *arXiv* (*Preprint* `1307.2875`)
[9] Holzegel G and Smulevici J 2012 *Annales Henri Poincare* **13** 991–1038 (*Preprint* `1103.0712`)
[10] Ishibashi A and Maeda K 2012 *Phys.Rev.* **D86** 104012 (*Preprint* `1208.1563`)
[11] Holzegel G H and Warnick C M 2013 (*Preprint* `1312.5332`)
[12] Friedrich H 2014 *Class.Quant.Grav.* **31** 105001 (*Preprint* `1401.7172`)
[13] Maliborski M 2012 *Phys.Rev.Lett.* **109** 221101 (*Preprint* `1208.2934`)
[14] Okawa H, Cardoso V and Pani P 2014 *Phys.Rev.* **D89** 041502 (*Preprint* `1311.1235`)
[15] Okawa H, Cardoso V and Pani P 2014 *Phys.Rev.* **D90** 104032 (*Preprint* `1409.0533`)
[16] Santos-Oliván D and Sopuerta C F A new scheme for the numerical evolution of AAdS space-times In preparation
[17] Christodoulou D 1986 *Commun.Math.Phys.* **105** 337–361
[18] Goldwirth D S and Piran T 1987 *Phys.Rev.* **D36** 3575
[19] Garfinkle D 1995 *Phys.Rev.* **D51** 5558–5561 (*Preprint* `gr-qc/9412008`)
[20] Choptuik M W 1993 *Phys.Rev.Lett.* **70** 9–12
[21] Bizoń P and Jałmuna J 2013 *Phys.Rev.Lett.* **111** 041102 (*Preprint* `1306.0317`)
[22] Maliborski M and Rostworowski A 2014 *Phys.Rev.* **D89** 124006 (*Preprint* `1403.5434`)

Spanish Relativity Meeting (ERE 2014): almost 100 years after Einstein's revolution IOP Publishing
Journal of Physics: Conference Series **600** (2015) 012020 doi:10.1088/1742-6596/600/1/012020

On cosmic quantum tunneling from "nothing"

Ramon Lapiedra, Juan Antonio Morales-Lladosa
Departament d'Astronomia i Astrofísica, Universitat de València,
E-46100 Burjassot, València, Spain

E-mail: `ramon.lapiedra@uv.es; antonio.morales@uv.es`

Abstract. We extend to a general Λ-Friedmann-Lemaître-Robertson-Walker (ΛFLRW) a previous result by Vilenkin and others according to which a closed de Sitter universe could be created from "nothing". More specifically, our main result is that only the closed ΛFLRW universe (but not the open and flat ones) could be created from a corresponding *instanton*, that is, from the corresponding solution with signature +4 of the Einstein field equations. Before getting this result the suitable corresponding instantons are calculated. The result is in accordance with previous results by another authors obtained by different methods.

1. Introduction

In four seminal papers [1, 2, 3, 4] Vilenkin has estimated the non zero probability of quantically creating from "nothing" a closed de Sitter universe, here "nothing" meaning a state without time. More specifically, an Euclidean differentiable manifold (one with +4 signature) instead of a Lorentzian one (one with +2 signature). In the present paper an Euclidean manifold which is a solution of the Einstein field equations will be called an *instanton*.

An expression for the above creating probability, P, in the semiclassical approximation is [3, 4]:

$$P \propto \exp(-|S_E|),$$

with S_E the action of a suitable instanton associated to the Lorentzian metric whose creation probability we want to calculate.

In the present paper, we extend the Vilenkin results to the general Friedmann-Lemaître-Robertson-Walker universe with a cosmological constant, Λ, that is a ΛFLRW universe. To this end, we follow a general procedure that covers the treatment given by Vilenkin to his particular case.

Our result is that the closed ΛFLRW model is quantically creatable, but the open non flat and the flat ones are not. This is in agreement with the result in [5], whose authors do not follow the quantum creation from "nothing" procedure *à la* Vilenkin used for us, but one where the FLRW universe comes from some initial static space-time configuration.

2. The general procedure for the quantum creation of a universe: its two prescriptions

Let $T_{\alpha\beta}$ be the energy-momentum tensor in General Relativity, and $g_{\alpha\beta}$ a solution of the corresponding Einstein field equations ($\alpha, \beta, ... = 0, 1, 2, 3$, signature +2). In order to provide a general proposal for the creation of a given universe, we advance the following two prescriptions.

First prescription:
We define the instanton "energy-momentum" tensor $T^E_{\alpha\beta}$ as the one obtained putting in $T_{\alpha\beta}$ an instanton metric $g^E_{\alpha\beta}$ associated to the corresponding Lorentzian one. This associated instanton metric, $g^E_{\alpha\beta}$, is a suitable Euclidean (signature +4) solution of the Einstein field equations according to the following

Second prescription:
Given $g_{\alpha\beta}$, its quantum creation probability, P, can be estimated from $P \propto \exp(-|S_E|)$, by associating to $g_{\alpha\beta}$ some $g^E_{\alpha\beta}$ satisfying the Darmois continuity conditions. For a singular Lorentzian metric to be created these Darmois conditions are defined in the following way:

First, let it be a space-like 3-surface, Σ_3, where $g_{\alpha\beta}$ is singular, and let us consider the first, ${}^3g_{ij}$, and the second, K_{ij}, fundamental forms, whose meaning respectively is ${}^3g_{ij}$, the restriction of $g_{\alpha\beta}$ on Σ_3, and $K_{ij} = (\nabla_\mu n_\nu)\gamma^\mu_i \gamma^\nu_j$. Here, n_μ is the unit normal 4-vector to Σ_3 and γ^α_β the projector on Σ_3.

Then, for the associated instanton metric $g^E_{\alpha\beta}$, we similarly define its first and second fundamental instanton forms: ${}^3g^E_{ij}$, K^E_{ij}. Then, the imposed Darmois conditions are the vanishing of ${}^3g_{ij} - {}^3g^E_{ij}$, and $K_{ij} - K^E_{ij}$, through Σ_3, that we will write in a concise way:

$$[{}^3g_{ij}] = 0, \quad [K_{ij}] = 0.$$

3. A particular canonical Einstein equations for ΛFLRW metrics, including instantons

The Einstein equations for a ΛFLRW metric and its instantons

$$ds^2 = \epsilon\, du^2 + a^2\Big(\frac{d\rho^2}{1+\epsilon\, k\rho^2} + \rho^2 d\sigma^2\Big), \quad d\sigma^2 = d\theta^2 + \sin^2\theta d\phi^2$$

(a, the cosmic expansion factor, $\epsilon = -1$ for the Lorentzian solutions, and $\epsilon = +1$ for the instanton solutions) can be written

$$\frac{\dot a^2}{a^2} = \frac{\kappa}{3}\mu - \frac{k}{a^2} - \frac{\epsilon}{3}\Lambda, \quad \frac{\ddot a}{a} = -\frac{\kappa}{6}(\mu - 3\,\epsilon\, p) - \frac{\epsilon}{3}\Lambda, \quad (k = +1, 0, -1) \tag{1}$$

all in accordance with Ellis [6].

4. Matching ΛFRW universes with their instantons

The Darmois continuity conditions $[{}^3g_{ij}] = 0$, $[K_{ij}] = 0$, through the singular space-like 3-surface $u = t = 0$, now give

$$[a] = 0, \quad [\dot a] = 0, \quad [\epsilon k] = 0.$$

These conditions are satisfied in the particular case considered by Vilenkin [1, 2]: the quantum creation from "nothing" of a closed de Sitter universe. Thus, our procedure to create any ΛFLRW model extends this previous result.

In particular, the continuity condition $[\epsilon k] = 0$ associates the closed instanton:

$$ds^2_E = du^2 + a^2_E(u)\Big(\frac{d\rho^2}{1-\rho^2} + \rho^2 d\sigma^2\Big), \qquad E \text{ from "Euclidean"},$$

to the closed ΛFRW universe

$$ds^2 = -dt^2 + a^2(t)\Big(\frac{d\rho^2}{1-\rho^2} + \rho^2 d\sigma^2\Big),$$

where a_E and a satisfy Eqs. (1), for $k = -1$, $\epsilon = +1$ in the a_E case, and for $k = +1$, $\epsilon = -1$, in the a case, and with $a_E|_{u\to 0} = a|_{t\to 0}$, $\dot{a}_E|_{u\to 0} = \dot{a}|_{t\to 0}$ from the other two Darmois conditions $[a] = [\dot{a}] = 0$.

Since these two limiting conditions, near $u = t = 0$, because of the radiation dominance for the ΛFRW universe, we obtain for the approximate instanton behavior:

$$\mu_E(u) \simeq \mu(u), \qquad a_E \simeq a_{E_e}\sqrt{\frac{u}{u_e}},$$

a_{E_e} standing for the value of a_E in some "instant", u_e, of the "energy" instanton phase.

5. Quantum creation of a closed ΛFRW universe

We try a particular instanton model by imposing in a natural way that $\mu_E(u) = \mu(u)$, $\forall u$. On the other hand the instanton action S_E in evident notation is

$$S_E = \frac{1}{2\kappa}\int R_E\sqrt{\mathrm{g_E}}\, d^4x + \int L_E\sqrt{\mathrm{g_E}}\, d^4x.$$

Further, extending to instanton metrics a procedure by Hawking and Ellis [7], we obtain $L_E = \mu$, and then

$$\frac{1}{2\kappa}R_E + L_E = \frac{1}{2}(\mu - 3p_E) + 2\,\frac{\Lambda}{\kappa}.$$

For a closed FLRW model, the calculation of this integral over $\rho \in [0,1]$ gives a finite value. Then, in this model the creation probability becomes finite. Thus the closed ΛFRW universe becomes quantically creatable.

Differently, it can be seen that the open non flat and flat models are not quantically creatable, and so, as quoted above, we recover similar results by Atkatz and Pagels [5] obtained by a different method.

These results remain true for an arbitrary (out of the above unavoidable condition $\lim_{u\to 0}\mu_E(u) = \lim_{u\to 0}\mu(u)$) μ_E function, provided that for $a_E \to a_{E_m}$ we have

$$\frac{\dot{a}_E}{a_E} \propto (a_{E_m} - a_E)^n, \quad 0 \leq n < 1,$$

where a_{E_m} is the minimal positive value of a_E such that $\dot{a}_E = 0$.

6. Final considerations

The closed and flat ΛFLRW models have vanishing *intrinsic* 4-momenta [8] (in particular, have vanishing *intrinsic* energy), while the open non flat one has a minus infinite intrinsic energy. But, the flat ΛFLRW model, perturbed in the standard inflationary scenario, has an infinite *intrinsic* 4-momenta [9]. Thus the suggestion by Fomin [10] and Tryon [11], according to which a quantically creatable universe should have vanishing energy, becomes reinforced by our present result on the quantum creatability of the closed ΛFLRW model but not of the open and flat ones.

An enlarged and detailed treatment of the topic covered by the present article will be published elsewhere.

Acknowledgments

This work has been supported by the Spanish "Ministerio de Economía y Competitividad", MICINN-FEDER project FIS2012-33582. Useful discussions with J. Navarro-Salas are pleasingly recognized.

References

[1] Vilenkin A 1982, *Phys. Lett.* **117B**, 25
[2] Vilenkin A 1983, *Phys. Rev. D* **27**, 2848
[3] Vilenkin A 1984 *Phys. Rev. D* **30**, 509
[4] Vilenkin A 1985, *Phys. Rev. D* **32**, 2511
[5] Atkatz D and Pagels H 1982, *Phys. Rev. D* **25**, 2065
[6] Ellis G F R 1992, *Gen. Relativ. Gravit.* **24**, 1047
[7] Hawking S W and Ellis G F R 1973, *The large scale structure of the space-time* (Cambridge: Cambridge University Press), pages 69-70
[8] Lapiedra R and Morales–Lladosa J A 2012, *Gen. Relativ. Gravit.* **44**, 367
[9] Lapiedra R and Sáez D 2008, *Phys. Rev. D* **77**, 104011
[10] Fomin P I 1973, "Gravitational instability of vacuum and the cosmological problem" Preprint ITP-73-137P. English version printed in *Problems of Atomic Science and Technology* 2013, N3 (85). Series: Nuclear Physics Investigations (60), p.6-9. A published version of this preprint appeared in Dok. Akad. Nauk. SSSR **9A**, 831 (1975)
[11] Tryon E P 1973 *Nature* **246**, 396

Journal of Physics: Conference Series **600** (2015) 012021 doi:10.1088/1742-6596/600/1/012021

Extension of the adiabatic regularization method to spin-1/2 fields

Aitor Landete

Instituto de Física Teórica, UAM-CSIC. Universidad Autonoma de Madrid, Cantoblanco, 28049 Madrid, Spain

E-mail: aitor.landete@csic.es

Abstract. The adiabatic regularization method was designed by L. Parker [1] for scalar fields in order to to subtract the potentially UV divergences that appear in the particle number operator. After that the method was generalized [2] to remove, in a consistent way, the UV divergences that appear in the expectation value of the stress-energy tensor $\langle T_{\mu\nu}\rangle$ in homogeneous cosmological backgrounds. We are going to provide here the extension of the adiabatic regularization method to spin-1/2 fields first given in [3]. In order to achieve this extension we will show the generalization of the adiabatic expansion for fermionic fields which differs significantly from the WKB-type expansion that works for the scalar modes. We will also show the consistency of the extended method computing well-known results, computed by other renormalization methods for a Dirac field in a FLRW spacetime, like the conformal and axial anomalies. Finally we will compute the expectation value of the stress-energy tensor for a Dirac field in a de Sitter spacetime.

1. Introduction

It is well-know that in General Relativity matter exerts its gravitational influence by curving the spacetime. Thus, It is natural to study the propagation of quantum fields in curved spacetieme in order to search for new effects of gravitation in this approach, where the gravitational field is not quantized. So quantum field theory in curved spacetime could be seen as a first step to coalesce Einstein's theory of GR and QFT in Minkowski space within a self-consistent and successful framework [4, 5]. It plays a crucial role in the understanding of the low-energy regime in quantum cosmology [6]. and scrutinize the understanding of the power spectrum in inflationary cosmology [7]. Moreover it constitutes the driving mechanism to explain the quantum radiance of black holes [8]

Like in the well-known Minkowskian QFT, physical quantities of physical interest, such as the expectation value of the stress-energy tensor diverges. So it is necessary to employ regularization and renormalization methods to handle this divergences. But renormalization in Minkowski spacetime is different than the renormalization in curved spacetime because the inherent gravitatory interaction introduces additional divergences that cannot be renormalized in the same way as in Minkowski. As a toy example if we try to renormalize the expectation value of the component 00 of the stress energy tensor of a free scalar field, we see that normal ordering renormalizes the divergence that appear. Meanwhile we can see that in curved spacetime appear more divergent terms in the expectation value that cannot be renormalized by the usual techniques in Minkowski. The gravitationally created particles generate an energy density with new ultraviolet (UV) divergences, as compared with the UV divergences present in Minkowski space. This requires more sophisticated methods of renormalization, adapted to the time-dependent or curved background.

Adiabatic regularization was first introduced in Parker's pioneer work on particle creation in the expanding universe [1] as a way to overcome the rapid oscillation of the particle number operator and UV divergences during the expansion. The method was later systematized and generalized to consistently deal with the UV divergences of the stress-energy tensor of scalar fields. The adiabatic method identifies the UV subtraction terms by first considering a slowly varying expansion factor $a(t)$. This naturally leads to a Liouville or WKB-type asymptotic expansion for the modes characterized by the comoving momentum $\vec{k}$. The subtraction terms identified in this way are valid for arbitrary smooth expansions [9]. This method has the distinguishing features of renormalize the particle number operator and allow the use of numerical methods. It has been shown that when the method is applied to renormalize local expectation values, as the stress-energy tensor, it turns out to be equivalent to the DeWitt-Schwinger point-splitting method for scalar fields [10, 11].

2. Adiabatic expansion for spin-1/2 fields.

As we have seen, the adiabatic method is only applicable to scalar fields. Our main goal is to generalize the method in order to study the Dirac field in curved spacetime. First of all, consider the Dirac equation in a spatially flat FLRW spacetime $ds^2 = dt^2 - a(t)d\vec{x}^2$ (for simplicity we will develop all the computations in this background)

$$(i\gamma^0\partial_0 + \frac{3i}{2}\frac{\dot{a}}{a}\gamma^0 + \frac{i}{a}\vec{\gamma}\vec{\nabla} - m)\psi = 0 , \tag{1}$$

where γ^μ are the Dirac matrices in Minkowski spacetime. For our purposes it is convenient work with the standard Dirac-Pauli representation. After momentum expansion $\psi = \sum_{\vec{k}} \psi_{\vec{k}}(t)e^{i\vec{k}\vec{x}}$ it is convenient to write the Dirac field in terms of two two-components spinors

$$\psi_{\vec{k}}(t) = \begin{pmatrix} \frac{1}{\sqrt{L^3a^3}} h_k^I(t)\xi_\lambda(\vec{k}) \\ \frac{1}{\sqrt{L^3a^3}} h_k^{II}(t)\frac{\vec{\sigma}\vec{k}}{k}\xi_\lambda(\vec{k}) \end{pmatrix} \tag{2}$$

where $\vec{\sigma}$ are the usual Pauli matrices. $\xi_\lambda(\vec{k})$ is a constant normalized bispinor $\xi_\lambda^\dagger\xi_\lambda = 1$ such that $\frac{\vec{\sigma}\vec{k}}{2k}\xi_\lambda = \lambda\xi_\lambda$. $\lambda = \pm 1/2$ represents the eigenvalue for the helicity, or spin component along the $\vec{k}$ direction. Thus, with the expansion (2) we can express the Dirac equation (1) as first order coupled equation system of the scalar functions h_k^I and h_k^{II}

$$h_k^{II} = \frac{ia}{k}(\partial_t + im)h_k^I \tag{3}$$

$$h_k^I = \frac{ia}{k}(\partial_t - im)h_k^{II} , \tag{4}$$

we are able to uncouple the former system and obtain two Klein-Gordon-type equations: $(\partial_t^2 + \frac{\dot{a}}{a}\partial_t + im\frac{\dot{a}}{a} + m^2 + \frac{k^2}{a^2})h_k^I = 0$ and a similar one for h_k^{II}, obtained by changing $m \to -m$ in the equation for h_k^I. The normalization condition for the four-spinor is

$$|h_k^I(t)|^2 + |h_k^{II}(t)|^2 = 1 . \tag{5}$$

One could be tempted to use the above second order equations to generate a WKB-type expansion for h_k^I and h_k^{II}. However a WKB-type ansatz, like in the scalar case, is specifically designed to preserve the Klein-Gordon product, and hence the associated Wronskian condition, but not to preserve the Dirac product and the normalization condition (5). Therefore one should follow a different way.

The zeroth adiabatic order should generalize in a natural way the standard solution in Minkowski space. Thus, it has to be of the form

$$\begin{aligned} g_k^{I(0)}(t) &= \sqrt{\frac{\omega(t)+m}{2\omega(t)}}e^{-i\int^t \omega(t')dt'} \\ g_k^{II(0)}(t) &= \sqrt{\frac{\omega(t)-m}{2\omega(t)}}e^{-i\int^t \omega(t')dt'} . \end{aligned} \tag{6}$$

It is easy to see that the zeroth order obeys the normalization condition $|g_k^{I(0)}(t)|^2+|g_k^{II(0)}(t)|^2 = 1$. The form of the zeroth order and the field equations (3) and (4) suggests the following alternative ansatz for the adiabatic expansion (at adiabatic order n)

$$\begin{aligned} g_k^{I(n)}(t) &= \sqrt{\frac{\omega+m}{2\omega}}e^{-i\int^t (\omega(t')+\omega^{(1)}+...+\omega^{(n)})dt'} \\ &\times \ (1+F^{(1)}+...+F^{(n)}) \\ g_k^{II(n)}(t) &= \sqrt{\frac{\omega-m}{2\omega}}e^{-i\int^t (\omega(t')+\omega^{(1)}+...+\omega^{(n)})dt'} \\ &\times \ (1+G^{(1)}+...+G^{(n)}) , \end{aligned}$$

where $\omega^{(n)}$, $F^{(n)}$ and $G^{(n)}$ are local functions of adiabatic order n. Imposing equations (3) and (4) and keeping terms of fixed adiabatic order one gets a system of equations at each order that one has to solve. Moreover, the solution should also respect the normalization condition $|g_k^{I(n)}(t)|^2 + |g_k^{II(n)}(t)|^2 = 1$ (at the given adiabatic order n), which we impose as a new equation in our system. In order to show how the method works we are going to compute explicitly the first and the second adiabatic orders of the expansion.

For the first adiabatic order we obtain that $\omega^{(1)} = 0$. Moreover, the functions $F^{(1)}$, $G^{(1)}$ should have a vanishing real part and verify the relation $G^{(1)} = F^{(1)} + i\frac{m\dot{a}}{2\omega^2 a}$. This solution can be parametrized as $F^{(1)} = -Ai\frac{m\dot{a}}{\omega^2 a}$, $G^{(1)} = Bi\frac{m\dot{a}}{\omega^2 a}$, where A, B are arbitrary real constants obeying $A + B = 1/2$. We note that, although the solution at first order is not univocally determined, local observables are actually independent of the ambiguity in $A-B$. We find useful for simplifying expressions and for computational purposes to fix the parameters as $A = B$. This implies $F^{(1)}(-m) = G^{(1)}(m)$, we are able to apply this kind of simplification at every adiabatic order for simplicity.

If we go to second adiabatic order and we impose $F^{(2)}(-m) = G^{(2)}(m)$ for simplicity. The solutions are then

$$\omega^{(2)} = \frac{5m^4\dot{a}^2 - 3\omega^2 m^2\dot{a}^2 - 2\omega^2 m^2 \ddot{a}a}{8\omega^5 a^2} \tag{7}$$

$$F^{(2)} = \frac{m^2 R}{48\omega^4} - \frac{5m^4\dot{a}^2}{16\omega^6 a^2} - \frac{m^2\dot{a}^2}{32\omega^4 a^2} - \frac{mR}{48\omega^3} + \frac{5m^3\dot{a}^2}{16\omega^5 a^2} . \tag{8}$$

As in the scalar case, we obtain that $\omega^{(odd)} = 0$. The explicit solutions to third and fourth adiabatic orders are given in [12].

The nth adiabatic order for the fermionic modes defined by $g_k^{I(n)}$ and $g_k^{II(n)}$ allows us to define the subtraction terms to cancel the UV divergences. The covariance notion of adiabatic invariance ensures the underlying covariance of the subtraction procedure. In the next section we will show as a hint of the consistency of the extended method the agreement of the conformal and chiral anomalies computed with the adiabatic scheme for spin-1/2 fields with the well-known results computed by other regularization methods.

3. Consistency tests. Computation of the conformal al chiral anomalies

Renormalization methods are crucial in the computation of quantum anomalies in QFT in curved spacetime. The study of this methods is likely to obtain a deeper understanding of the known anomalies and, moreover, look for new ones [13]. It is well-known that the chiral and conformally anomalies have been computed by other renormalization methods. As a consistency test of the extension method shown before we are going to compute via our extended method this well-known results. Concerning local observables in a generic FLRW spacetime, the adiabatic regularization axioms tell us that the second adiabatic order is required to renormalize the two-point function and the fourth order is required when the observable is related to the stress-energy tensor.

Note that, due to the way that the adiabatic regularization method is constructed, the regularized expressions also allow for an efficient numerical estimation when the modes for the quantum state are difficult to manage analytically. [14]

3.1. Conformal anomaly

At the classical level we know that the trace of the stress-energy tensor for a Dirac field takes the simple expression

$$T^{\mu}_{\mu} = m\bar{\psi}\psi \ . \tag{9}$$

When the field is massless the trace classically vanishes, pointing the emergence of the conformal invariance. However, in the quantum theory the expectation value $\langle T^{\mu}_{\mu}\rangle = m\langle\bar{\psi}\psi\rangle$ takes a nonzero value even in the massless limit due to the conservation of the expectation value of the stress-energy tensor. Our purpose is to perform the calculation of this anomalous trace using the extension of the adiabatic regularization method for spin-1/2 introduced above. Since the expectation value $\langle\bar{\psi}\psi\rangle$ is now regarded as a piece of the average value of the stress-energy tensor $\langle T_{\mu\nu}\rangle$, the renormalization should be performed up to the fourth adiabatic order according to the adiabatic regularization axioms. So one have to evaluate the trace anomaly by taking the massless limit in the above expression

$$\begin{aligned}\langle T^{\mu}_{\mu}\rangle_r &= \lim_{m\to 0}\frac{-2m}{(2\pi)^3 a^3}\int d^3k(|h_k^I|^2 - |h_k^{II}|^2 \\ &- |g_k^{I(4)}|^2 + |g_k^{II(4)}|^2) \ .\end{aligned}$$

Using the adiabatic expansion of the fermionic field obtained in the former section and integrating we obtain

$$\begin{aligned}\left\langle T^{\mu}_{\mu}\right\rangle_r &= \frac{1}{2880\pi^2}\left[-11\left(R_{\alpha\beta}R^{\alpha\beta} - \frac{1}{3}R^2\right) + 6\Box R\right] \\ &= \frac{1}{2880\pi^2}\left[\frac{11}{2}G + 6\Box R\right] \ ,\end{aligned} \tag{10}$$

where in the second line we have rewritten the former expression in terms of the Gauss-Bonnet invariant G, which for a FLRW spacetime is given by $G = -2(R_{\mu\nu}R^{\mu\nu} - R^2/3)$. The conformal anomaly is generically given for a conformal field of spin 0, 1/2 or 1 in terms of three parameters

$$\langle T_\mu^{\ \mu}\rangle_r = \frac{1}{2880\pi^2}\left(AC_{\mu\nu\rho\sigma}C^{\mu\nu\rho\sigma} + BG + C\Box R\right) . \quad (11)$$

The result obtained for a Dirac spin-1/2 field by other renormalization procedures is $A = -9, B = 11/2, C = 6$ [5]. Our above result (10) agrees exactly with the results obtained from other methods. We note that in a FLRW spacetime the Weyl tensor $C_{\mu\nu\rho\sigma}$ vanishes identically. This can be regarded as a nontrivial test of the robustness of our proposal.

3.2. *Chiral anomaly*

In curved spacetime the axial vector current $J^\mu_A \equiv \bar{\psi}\gamma^\mu\gamma^5\psi$ obeys the covariant equation $\nabla_\mu J^\mu_A = 2im\bar{\psi}\gamma^5\psi$. For a massless Dirac field the classical axial current is conserved, due to the chiral symmetry. At the quantum level the expectation value $\langle\nabla_\mu J^\mu_A\rangle$ may acquire a non-zero value in the massless limit. We want to evaluate this quantity using the adiabatic regularization for fermions. The strategy is similar to the evaluation of the conformal or trace anomaly. Since the divergences of $\langle\nabla_\mu J^\mu_A\rangle$ are of fourth adiabatic order we have to compute $\langle 2im\bar{\psi}\gamma^5\psi\rangle$ also at fourth adiabatic order. In this case we have to compute

$$\langle\bar{\psi}\gamma^5\psi\rangle_r = \frac{-2}{(2\pi)^3a^3}\int d^3k(h_k^{I*}h_k^{II} - h_k^{II*}h_k^{I} - g_k^{I(4)*}g_k^{II(4)} + g_k^{II(4)*}g_k^{I(4)}) \quad (12)$$

Concerning the axial current anomaly, the subtraction terms of fourth adiabatic order cancel out while the third order terms are, after integration in momenta, still proportional to the mass. Therefore, in the massless limit $\langle\nabla_\mu J^\mu_A\rangle_r = 0$, in agreement with the fact that the axial current anomaly obtained from other renormalization prescriptions $\epsilon^{\mu\nu\alpha\beta}R_{\mu\nu}^{\ \ \lambda\xi}R_{\alpha\beta\lambda\xi}$ vanishes for a FLRW spacetime.

4. Expectation value of the stress-energy tensor of a free Dirac field in a de Sitter spacetime in a Bunch-Davies vacuum

For de Sitter spacetime with scale factor $a(t) = e^{Ht}$ and H constant, the coupled differential equations (3) and (4) take the form

$$h_k^{II} = \tfrac{ie^{Ht}}{k}(\partial_t + im)h_k^I \ , \ h_k^I = \tfrac{ie^{Ht}}{k}(\partial_t - im)h_k^{II}, \quad (13)$$

We define the following dimensionless variables

$$z \equiv kH^{-1}e^{-Ht} \qquad \mu \equiv \frac{m}{H} . \quad (14)$$

In terms of these variables, the exact modes are expressed in terms of the Hankel functions $H_\alpha^{(1)}$ in the following way

$$h_k^I = i\frac{\sqrt{\pi z}}{2}e^{\frac{\pi\mu}{2}}H^{(1)}_{\frac{1}{2}-i\mu}(z) \quad (15)$$

and

$$h_k^{II} = \frac{\sqrt{\pi z}}{2}e^{\frac{\pi\mu}{2}}H^{(1)}_{-\frac{1}{2}-i\mu}(z) . \quad (16)$$

Equations (15) and (16) determine a vacuum for spin-1/2 fields analogous to the Bunch-Davies vacuum for scalars, because it is the solution that reproduces the adiabatic modes for initial times. Note that $h_k^{II}(m) = h_k^I(-m)$. In the following computations we are going to simplify the ambiguity that appears in the adiabatic expansion assuming for simplicity the condition $F^{(n)}(m) = G^{(n)}(-m)$.

Due to the symmetries of de Sitter spacetime, the renormalized expectation value of the stress-energy tensor is related to the the renormalization of its quantum trace in the following way

$$\langle T^{\mu\nu}\rangle_r = \frac{1}{4} g^{\mu\nu} \langle T_\rho^\rho \rangle_r \ . \tag{17}$$

As analyzed in the Section 3.1, the unrenormalized formal expression for $\langle T_\rho^\rho\rangle$ contains UV divergences. So the unrenormalized stress-energy tensor is given by

$$\langle T^{\mu\nu}\rangle = \frac{1}{4} g^{\mu\nu} \frac{-mH^3}{\pi^2} \int_0^\infty dz \left[\left(\left|h_k^I\right|^2 - \left|h_k^{II}\right|^2 \right) z^2 \right] \tag{18}$$

If we substitute the exact modes in the former expression we see that contains quadratic and logarithmic UV divergences for $z \to \infty$ duc to

$$\left(\left|h_k^I\right|^2 - \left|h_k^{II}\right|^2 \right) z^2 \sim z - \frac{1}{z} + \mathcal{O}\left(\frac{1}{z^3}\right) \tag{19}$$

Therefore, in order to obtain the renormalized trace we have to substract the corresponding adiabatic expansion up to fourth order

$$\langle T_\rho^\rho\rangle_r = \frac{-mH^3}{\pi^2} \int_0^\infty dz z^2 \left(\left|h_k^I\right|^2 - \left|h_k^{II}\right|^2 - \left|g_k^{I\,(4)}\right|^2 + \left|g_k^{II\,(4)}\right|^2 \right) \ . \tag{20}$$

This integral is convergent and can be solved numerically. However, it can also be evaluated analytically by introducing an auxiliary regulator σ (20). The integration of (18) gives

$$\langle T_\rho^\rho\rangle = \lim_{\sigma\to 0} \left(\frac{H^2 m^2}{\pi^2\sigma^2} - \frac{m^2\left(H^2+m^2\right)}{4\pi^2} \cdot \frac{1 + 2\gamma + 2\log\left(\frac{\sigma}{2}\right) + 2\mathfrak{Re}\left[\psi\left(-2 - i\frac{m}{H}\right)\right]}{4\pi^2} \right) \tag{21}$$

which is divergent in the $\sigma \to 0$ limit as we already know. If we integrate the adiabatic counterterms that we have added in (20), we obtain

$$\langle T_\rho^\rho\rangle_{Ad} = \lim_{\sigma\to 0} \left(\frac{H^2 m^2}{\pi^2\sigma^2} - \frac{11H^4 + 190H^2m^2 + 60m^4}{240\pi^2} - \frac{m^2\left(H^2+m^2\right)\left(2\gamma + 2\log\left(\frac{\sigma}{2}\right) + 2\log\left(\frac{m}{H}\right)\right)}{4\pi^2} \right) . \tag{22}$$

Equation (22) is also divergent when $\sigma \to 0$. However, if (22) is subtracted from (21) the result is finite in the $\sigma \to 0$ limit and gives the quantum trace. From it we can immediately obtain an analytic expression for the renormalized stress-energy tensor

$$\langle T^{\mu\nu}\rangle_r = \frac{1}{960\pi^2} g^{\mu\nu} \left(11H^4 + 130H^2m^2 + 120m^2(H^2+m^2)\left(\log\left(\frac{m}{H}\right) - \mathfrak{Re}\left[\psi\left(-1 + i\frac{m}{H}\right)\right]\right)\right) . \tag{23}$$

where $\psi(z)$ is the digamma function.

Acknowledgments

A.L. would like to thank J. Navarro-Salas, I. Agullo and L. Parker for very useful discussions and also to the ERE2014 organizers for their hospitality during the congress. The work of A.L. is supported by the Severo Ochoa program of the Spanish ministry of economy and competitiveness.

References

[1] Parker L., *The creation of particles in an expanding universe*, Ph.D. thesis, Harvard University (1966).

[2] Parker L. and Fulling S.A., *Phys. Rev. D* **9**, 341 (1974). Fulling S. A. and Parker L., *Ann. Phys. (N. Y.)* **87**, 176 (1974). Fulling S. A., Parker L. and Hu B. L. *Phys. Rev. D* **10**, 3905 (1974).

[3] A. Landete, J. Navarro-Salas and F. Torrenti *Phys. Rev. D***88**, 061501 (2013).

[4] Parker L. and Toms D.J., *Quantum field theory in curved spacetime: quantized fields and gravity*, Cambridge University Press, (2009).

[5] Birrell N.D. and Davies P.C.W., *Quantum fields in curved space*, Cambridge University Press, (1982).

[6] Agulló I., Ashtekar A. and Nelson W., *Phys. Rev. Lett.* **109**, 251301 (2012).

[7] Parker L., Amplitude of perturbations from inflation, hep-th/0702216. Agulló I., Navarro-Salas J., Olmo G.J. and Parker L., *Phys. Rev. Lett.* **103**, 061301 (2009); *Phys. Rev. D* **81**, 043514, (2010).

[8] Hawking S. W. *Commun. Math. Phys.* **43**, 199 (1975).

[9] Parker L., *Phys.Rev.Lett.* **21** 562 (1968); *Phys. Rev.* **183**, 1057(1969); *Phys. Rev. D* **3**, 346 (1971)

[10] Birrell N.D. *Proc. R. Soc. Lond B* **361**, 513 (1978).

[11] Anderson P. R. and Parker L., *Phys. Rev. D* **36**, 2963 (1987).

[12] A. Landete, J. Navarro-Salas and F. Torrenti *Phys. Rev. D***89**, 044030 (2014).

[13] I. Agullo, A. Landete and J. Navarro-Salas *Preprint* gr-qc/1409.6406.

[14] Anderson. P.R. and Eaker W., *Phys. Rev. D* **61**, 024003 (1999); Habib S., Molina-Paris C. and Mottola E., *Phys. Rev. D* **61**, 024010 (1999); Anderson P.R., Molina-Paris C., Evanich D., Cook, G.B., *Phys. Rev. D* **78**, 083514 (2008); Bates, J.D. and Anderson, P.R., *Phys. Rev. D* **82**, 024018 (2010); and referenced cited in these papers.

Gravitational collapse with rotating thin shells and cosmic censorship

Térence Delsate[1], Jorge V Rocha[2] and Raphael Santarelli[3]

[1] Theoretical and Mathematical Physics Department, University of Mons, UMONS, 20, Place du Parc, B-7000 Mons, Belgium
[2] Centro Multidisciplinar de Astrofísica – CENTRA, Instituto Superior Técnico – IST, Universidade de Lisboa – UL, Avenida Rovisco Pais 1, 1049-001 Lisboa, Portugal
[3] Instituto de Física de São Carlos, Universidade de São Paulo, Caixa Postal 369, CEP 13560-970 São Carlos, São Paulo, Brazil

E-mail: terence.delsate@umons.ac.be, jorge.v.rocha@tecnico.ulisboa.pt, santarelli@ifsc.usp.br

Abstract. The study of gravitational collapse is a subject of great importance, both from an astrophysical and a holographic point of view. In this respect, exact solutions can be very helpful but known solutions are very scarce, especially when considering dynamical processes with rotation. We describe a setup in which gravitational collapse of rotating matter shells can be addressed with analytic tools, at the expense of going to higher dimensions and considering equal angular momenta spacetimes. The framework for an exact treatment of the dynamics, relying on a thin shell approximation, is developed. Our analysis allows the inclusion of a non-vanishing cosmological constant. Finally, we discuss applications of this machinery to the construction of stationary solutions describing matter around rotating black holes and to the cosmic censorship conjecture.

1. Introduction

The presence of rotation plays a major role in the formation and evolution of astrophysical systems and black holes are no exception. Half a century has passed since the discovery of the (arguably) most physically relevant black hole solution of the Einstein equations — the Kerr geometry [1]. We currently possess a good understanding of stationary, rotating black holes. However, we still have comparatively poor (analytic) control over highly dynamical scenarios involving rotation. In particular, very little is known about the associated non-spherical gravitational collapse [2]. Realistic collapses should include rotation and this issue becomes even more pressing since naked singularities are known to arise in fine-tuned non-rotating — and thus automatically non-generic — settings [3].

Clearly, the advantage of non-rotating setups is their large amount of symmetry. Imposing spherical symmetry reduces the study of the collapse to a problem in $1+1$ dimensions. Nevertheless, there exists a larger class of (rotating) black hole spacetimes that are stationary and whose metric depends on a single radial coordinate [4, 5]: these are the well-known Myers-Perry solutions [6] when restricted to equal angular momenta, which are blessed with the cohomogeneity-1 property. The price to pay for the convenience provided by cohomogeneity-1 spacetimes is the restriction to higher (odd) dimensions, namely $D=2N+3$ with $N=1,2,3,\ldots$.

Journal of Physics: Conference Series **600** (2015) 012022 doi:10.1088/1742-6596/600/1/012022

2. Background geometries

Myers-Perry(-AdS) black holes in $2N+3$ dimensions possess isometry group $\mathbb{R}\times U(1)^N$. When all spin parameters are set equal, $a_i = a$, this symmetry is enhanced to $\mathbb{R}\times U(N)$, and coordinates can be found that reflect this large amount of symmetry, such that the metric depends on just one (radial) coordinate. Constant time and radial sections are squashed $(2N+1)$-spheres, which can be written as an S^1 fibre over the N-dimensional complex projective space CP^N.

The line element describing these cohomogeneity-1 black holes can be written as follows [5],

$$ds^2 = -f(r)^2dt^2 + g(r)^2dr^2 + r^2\widehat{g}_{ab}dx^a dx^b + h(r)^2\left[d\psi + A_a dx^a - \Omega(r)dt\right]^2 , \tag{1}$$

where the various metric functions are given by

$$g(r)^2 = \left(1+\frac{r^2}{\ell^2} - \frac{2M}{r^{2N}} + \frac{2Ma^2}{r^{2N}\ell^2} + \frac{2Ma^2}{r^{2N+2}}\right)^{-1} , \tag{2}$$

$$h(r)^2 = r^2\left(1+\frac{2Ma^2}{r^{2N+2}}\right) , \qquad \Omega(r) = \frac{2Ma}{r^{2N}h(r)^2} , \qquad f(r) = \frac{r}{g(r)h(r)} . \tag{3}$$

The tensor $\widehat{g}_{ab}$ denotes the Fubini-Study metric on CP^N whereas $A = A_a dx^a$ is the associated Kähler potential, which for $N = 1$ are given respectively by

$$\widehat{g}_{ab}dx^a dx^b = \frac{1}{4}\left(d\theta^2 + \sin^2\theta\, d\phi^2\right) , \qquad A = \frac{1}{2}\cos\theta\, d\phi . \tag{4}$$

The angular coordinate ψ parametrizes the S^1 fibre and has 2π-periodicity.

These geometries accommodate a non-vanishing cosmological constant Λ. The parameter ℓ sets the (constant) curvature scale of the solutions and is related to the cosmological constant through $\Lambda = -(D-1)\ell^{-2}$ [5]. Thus, a (finite) real ℓ corresponds to asymptotically anti-de Sitter (AdS) black holes, and asymptotically flat spacetimes are recovered by taking the limit $\ell \to \infty$. In addition, the solutions (1) are fully determined by a mass parameter M and a spin parameter a, whose relation to the physical mass and angular momentum is explicitly given in Ref. [5].

From now on we shall restrict to the case $N = 1$ for concreteness, corresponding to five spacetime dimensions. However, the whole analysis can be generalised to higher odd dimensions.

3. Junction conditions and the stress-energy tensor

The cohomogeneity-1 property makes an exact (thin-shell) calculation possible, 'gluing' an interior to an exterior geometry [7]. We take advantage of the high degree of symmetry by considering shells that respect the full set of spatial isometries. Indeed, for test shells the dynamics on the $CP^1 \simeq S^2$ and on the S^1 separate [8] and the geometry of the shells only changes by a mere scaling of the 'radial' coordinate r. However, since their geometry is squashed — as opposed to a round sphere — one can anticipate that pressure anisotropies emerge during gravitational collapse of matter-filled rotating thin shells.

We take both the exterior (+) and interior (−) metrics to be of the form (1) and make use of the Darmois-Israel junction conditions [9] along a timelike hypersurface Σ, defined parametrically by $t = \mathcal{T}(\tau)$ and $r = \mathcal{R}(\tau)$, where τ is for the proper time for an observer comoving with the hypersurface.[1] The junction conditions determine the (dis)continuity of the first and second fundamental forms characterising the hypersurface. If $[C_{ij\ldots}] \equiv C^{(+)}_{ij\ldots} - C^{(-)}_{ij\ldots}$ denotes the jump in a quantity $C_{ij\ldots}$ across the shell, then the junction conditions read

$$[\mathfrak{g}_{ij}] = 0 \qquad \Longrightarrow \qquad \mathfrak{g}^{(+)}_{ij} = \mathfrak{g}^{(-)}_{ij} \equiv \mathfrak{g}_{ij} , \tag{5}$$

$$[k_{ij}] - \mathfrak{g}_{ij}[k] = -8\pi G S_{ij} , \tag{6}$$

[1] We take the coordinates $y^i = \{\tau, \psi, \theta, \phi\}$, with $i, j = 0, \ldots, 3$, to parametrize the surface Σ, but we shall use indices a, b to refer to the coordinates on CP^1, i.e., they run over $\{2, 3\}$.

where $\mathfrak{g}_{ij}^{(\pm)}$ is the induced metric on Σ, as inferred from either the exterior or the interior. Similarly, $k_{ij}^{(\pm)}$ are the extrinsic curvatures and $k^{(\pm)}$ their traces. $\mathcal{S}_{ij}$ stands for the surface stress-energy tensor. This formalism has been applied to rotating spacetimes in $(2+1)$ dimensions by Mann *et al.* [10]. This lower dimensional case is however very special, since general relativity possesses no propagating degrees of freedom in spacetime dimensions lower than four. In what follows we generalize to $D = 5$ following Ref. [11].

Compared to the $D = 3$ case, we get (for any $N \in \mathbb{N}$) one additional constraint from the first junction condition,[2]

$$h_+(\mathcal{R}) = h_-(\mathcal{R}) \equiv h(\mathcal{R}) \qquad \Longrightarrow \qquad M_+ a_+^2 = M_- a_-^2 \,. \tag{7}$$

This can be understood as the requirement that the areal radius of the shell, which is proportional to $r^{2N} h(r)$, must be the same when seen from the inside or outside. Note that Eq. (7) forbids the matching of an exterior black hole spacetime with a flat interior. However, working in the slow rotation limit (i.e., up to order $O(a^2)$) this constraint is avoided.

The second junction condition requires the (corotating) shell to possess the stress-energy tensor of an imperfect fluid,

$$\mathcal{S}_{ij} = (\rho + P) u_i u_j + P\, \mathfrak{g}_{ij} + 2\varphi\, u_{(i}\xi_{j)} + \Delta P\, \mathcal{R}^2 \widehat{g}_{ij} \,, \tag{8}$$

where $u = \partial_\tau$ is the fluid velocity, $\xi = h(\mathcal{R})^{-1}\partial_\psi$ is the normalised Killing angular vector field, and $\widehat{g}_{ij} dy^i dy^j = \widehat{g}_{ab} dx^a dx^b$. The quantities ΔP and φ are pressure anisotropy and intrinsic momentum terms, respectively, and we recover a perfect fluid when $\Delta P = \varphi = 0$.

The stress-energy tensor components are dictated by the metric components according to [11]

$$\rho = -\frac{(\beta_+ - \beta_-)(\mathcal{R}^2 h)'}{8\pi \mathcal{R}^3} \,, \qquad P = \frac{h}{8\pi \mathcal{R}^3} \left[\mathcal{R}^2 (\beta_+ - \beta_-)\right]' \,, \tag{9}$$

$$\varphi = -\frac{(\Omega_+' - \Omega_-') h^2}{16\pi \mathcal{R}} \,, \qquad \Delta P = \frac{(\beta_+ - \beta_-)}{8\pi} \left[\frac{h}{\mathcal{R}}\right]' \,, \tag{10}$$

where we have introduced the quantities

$$\beta_\pm \equiv f_\pm(\mathcal{R}) \sqrt{1 + g_\pm(\mathcal{R})^2 \dot{\mathcal{R}}^2} \,, \tag{11}$$

and where primes and overdots stand for $d/d\mathcal{R}$ and $d/d\tau$, respectively. Observe that the intrinsic momentum φ and the anisotropic pressure ΔP necessarily vanish in the non-rotating limit.

4. The shell's equation of motion

For simplicity we now assume a linear equation of state, $P = w\rho$, as in Ref [11]. Note, however, that other equations of state, e.g., polytropic, can be considered while still allowing the analytical integration of the equations of motion. Employing expressions (9) and (11), one obtains an effective equation governing the radial motion of the shell,

$$\dot{\mathcal{R}}^2 + V_{\text{eff}}(\mathcal{R}) = 0 \,, \tag{12}$$

[2] We could have taken different r coordinates for the interior and exterior spacetimes but the first junction condition would impose their equality.

where the effective potential $V_{\rm eff}$ is given by

$$V_{\rm eff}(\mathcal{R}) = 1+\frac{\mathcal{R}^2}{\ell^2}+\frac{2Ma^2}{\mathcal{R}^4}+\frac{2Ma^2}{\ell^2\mathcal{R}^2}-\frac{M_+ + M_-}{\mathcal{R}^2}-\frac{1}{4}\left(\frac{m_0}{\mathcal{R}^2}\right)^{2+3w}\left(1+\frac{2Ma^2}{\mathcal{R}^4}\right)^{1-w}$$
$$-\left(\frac{M_+ - M_-}{m_0}\right)^2\left(\frac{\mathcal{R}^2}{m_0}\right)^{3w}\left(1+\frac{2Ma^2}{\mathcal{R}^4}\right)^{w-1}. \qquad (13)$$

The parameter m_0 is a (positive) integration constant, which corresponds to the rest mass of the shell in the case of non-rotating dust. The second term in the potential is nothing but the usual infinite barrier (for positive ℓ^{-2}) responsible for the confining nature of AdS spacetime. The third term behaving like $\sim \mathcal{R}^{-4}$ represents the centrifugal repulsion due to rotation.

Solutions describing the radial trajectory of thin shells are found by simply integrating (12). If desired, the acceleration of the shell can be determined by

$$\ddot{\mathcal{R}} = -\frac{1}{2}V'_{\rm eff}(\mathcal{R}). \qquad (14)$$

5. Stationary solution in AdS

When considering a negative cosmological constant, the confining nature of the effective potential (when $-1 < w < 1/3$), in conjunction with the centrifugal barrier, suggests the possibility of stationary solutions consisting of a rotating shell hovering around an AdS black hole.

Indeed, these solutions exist and can be constructed as follows. First, the radial location of the shell must be constant, i.e., $\mathcal{R}(\tau) = \mathcal{R}_*$, and therefore Eqs. (12) and (14) imply

$$V_{\rm eff}(\mathcal{R}_*) = V'_{\rm eff}(\mathcal{R}_*) = 0. \qquad (15)$$

Second, for the shell to be stable against radial oscillations, $\mathcal{R}_*$ should correspond to a local minimum of the potential,

$$V''_{\rm eff}(\mathcal{R}_*) > 0. \qquad (16)$$

For the case of dust ($w = 0$) the conditions (15) reduce to high order polynomial equations in $\mathcal{R}_*^2$, but generically they are non-polynomial equations. However, the conditions are simple quadratic equations in M_+ and M_-, and so in practice it is more convenient to fix these mass parameters in terms of $\{w, m_0, \mathcal{R}_*, Ma^2\}$.

It is then just a matter of scanning the parameter space to obtain solutions such that:

(i) $\mathcal{R} = \mathcal{R}_*$ is a stable point, i.e., inequality (16) is satisfied;
(ii) The interior geometry has an event horizon located at $r = r_h^- < \mathcal{R}_*$;
(iii) The exterior geometry features no horizons, i.e., $r_h^+ < \mathcal{R}_*$;[3]
(iv) Energy conditions for the matter on the shell are obeyed.

In Ref. [11] the weak energy conditions were imposed and it was found that it is easy to satisfy simultaneously all the conditions above. An example is presented in Fig. 1.

While the confining potential is absent in the asymptotically flat case ($\ell \to \infty$), this does not exclude the existence of analogous stationary solutions describing thin shells around rotating black holes. In fact, we have been able to obtain such configurations, but in all cases found the shell sits at a local maximum of the potential and is therefore unstable against radial oscillations.

[3] The alternative would be to consider the exterior spacetime to be over-rotating, thus corresponding to a naked singularity. However, it can be shown using condition (7) that this would imply also a nakedly singular interior.

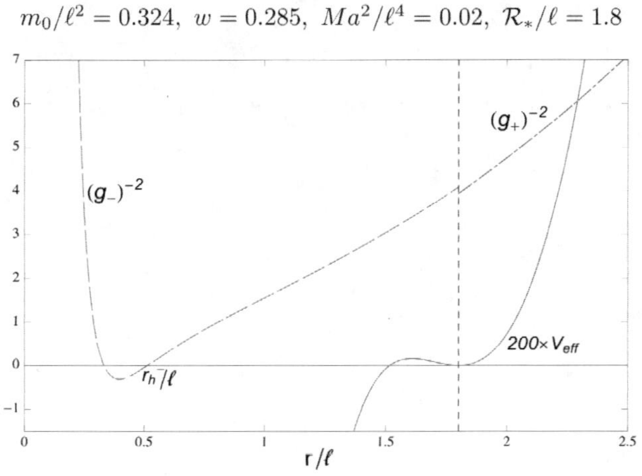

Figure 1. Stable configuration of a stationary rotating shell surrounding an asymptotically AdS black hole. The dashed vertical line identifies the location $\mathcal{R}_*$ of the shell. The metric function $g(r)^{-2}$ is shown in both the exterior domain, $r > \mathcal{R}_*$ (blue), and the interior domain, $r < \mathcal{R}_*$ (red). The largest real root of ${g_-}^{-2}$ marks the event horizon radius, r_h^-. The brown curve represents the effective radial potential (amplified to ease interpretation) and $r = \mathcal{R}_*$ corresponds to a local minimum of the potential.

6. Full collapse onto a black hole and cosmic censorship

Within the framework developed it is also possible to obtain collapse of a rotating thin shell onto a (previously existing) black hole. An interesting scenario is that of a shell collapsing from infinity, initially at rest, in an asymptotically flat spacetime ($\ell \to \infty$). The effective potential must then asymptote to zero as $\mathcal{R} \to \infty$. Using Eq. (13) it is easy to show that this can only occur for dust ($w = 0$) and only if the condition $m_0 = M_+ - M_-$ is satisfied. Having fixed these quantities it is not hard to find conditions on the remaining parameters, $\{Ma^2, M_+, M_-\}$, so that the effective potential is strictly negative — having no roots, and therefore no turning points. Simultaneously, the weak energy conditions can be obeyed (see Ref. [11]).

The solution thus obtained describes the collapse of a shell, initially located at $\mathcal{R} \to \infty$ with vanishing radial velocity. The spacetime in the interior of the shell consists of a rotating black hole with a horizon at $r = r_h^-$ and spin parameter $a_- = \sqrt{Ma^2/M_-}$. At larger radii, the spacetime is the same as that of a black hole with horizon radius $r_h^+ > r_h^-$ and spin parameter $a_+ = \sqrt{Ma^2/M_+} < a_-$. As the shell falls past the radius r_h^+, a new apparent horizon appears at that location. At this point nothing can prevent the shell from meeting its doom, as it will eventually fall onto the curvature singularity at $r = 0$. The end state is a black hole with larger mass and spin. Such an example (satisfying the weak energy conditions) is presented in Fig. 2. The centrifugal barrier at smaller radii is apparent, but not strong enough to prevent full collapse from occurring. However, the shell's radial trajectory turns into a bounce if one increases sufficiently the spin parameter a.

An analogous construction can be performed in AdS, i.e., with a finite ℓ, even though it is more laborious because it involves more parameters. The main difference is that it is no longer possible to have a shell at rest at infinity: either the AdS potential barrier dominates at large radii or the (negative) w-dependent terms in Eq. (13) make the potential diverge negatively as $\mathcal{R} \to \infty$. Fig. 3 shows a case in which a shell starting from rest at a finite radius fully collapses onto a black hole.

Finally, we point out that none of the collapsing solutions we have described can violate the cosmic censorship conjecture [12]. Indeed, it is easy to show, using identity (7) and $M_+ > M_-$, that if the interior solution is sub-extremal ($g_-(r)^{-2} = 0$ has a real and positive root) the global solution after full collapse cannot be over-extremal. Thus, the curvature singularity at $r = 0$ remains covered by a horizon at all times.

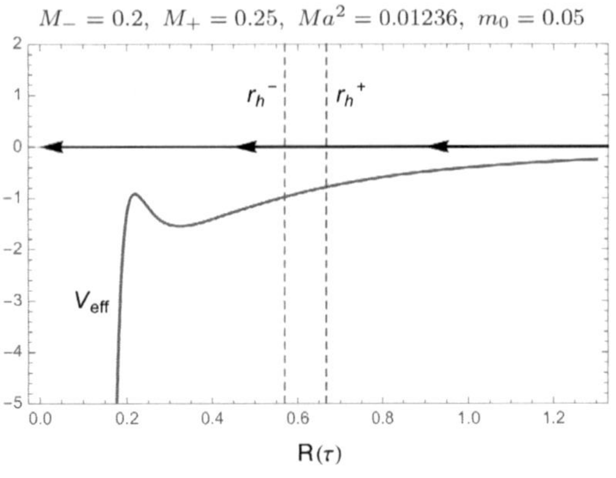

Figure 2. Full collapse of a rotating thin shell onto an asymptotically flat black hole. The vertical dashed lines mark the horizons of the interior and exterior geometries.

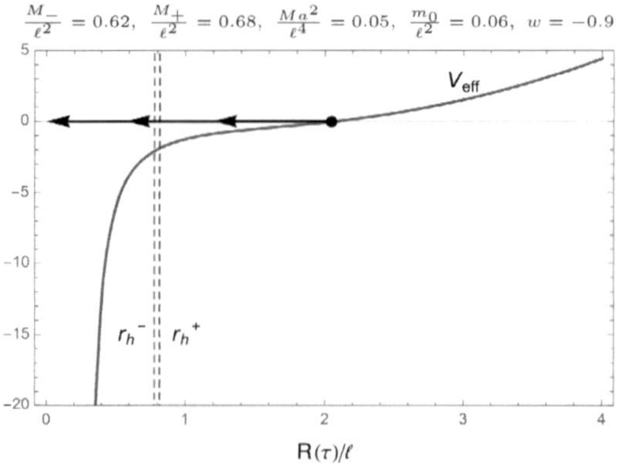

Figure 3. Full collapse of a rotating thin shell onto an asymptotically AdS black hole. The vertical dashed lines mark the horizons of the interior and exterior geometries.

7. Conclusion and outlook

We have described a simple framework to study the dynamics of thin matter shells around rotating black holes in five dimensions. In the presence of rotation the matter must take the form of an imperfect fluid. We constructed stationary solutions consisting of rotating shells surrounding spinning black holes in AdS and applied the formalism to address gravitational collapse with rotation, finding that cosmic censorship is not violated.

The extension of this work to higher (odd) dimensions and the exhaustive exploration of the parameter space is left for the future. Also, it would be interesting to consider the collapse of *null* shells in AdS. This would allow a confrontation with Ref. [13], which studied the (perturbative) black hole spin-up process with light-like test particles in AdS.

References

[1] Kerr R P 1963 Gravitational field of a spinning mass as an example of algebraically special metrics *Phys. Rev. Lett.* **11** 237

[2] Gundlach C and Martin-Garcia J M 2007 Critical phenomena in gravitational collapse *Living Rev. Rel.* **10** 5 (*Preprint* 0711.4620 [gr-qc])

[3] Joshi P S and Malafarina D 2011 Recent developments in gravitational collapse and spacetime singularities *Int. J. Mod. Phys.* D **20** 2641 (*Preprint* 1201.3660 [gr-qc])

[4] Frolov V P and Stojkovic D 2003 Quantum radiation from a five-dimensional rotating black hole *Phys. Rev.* D **67** 084004 (*Preprint* gr-qc/0211055)

[5] Kunduri H K, Lucietti J and Reall H S 2006 Gravitational perturbations of higher dimensional rotating black holes: tensor perturbations *Phys. Rev.* D **74** 084021 (*Preprint* hep-th/0606076)

[6] Myers R C and Perry M J 1986 Black Holes in Higher Dimensional Space-Times *Annals Phys.* **172** 304

[7] Boulware D G 1973 Naked singularities, thin shells, and the Reissner-Nordström metric *Phys. Rev.* D **8** 2363

[8] Rocha J V, Delsate T and Santarelli R 2014 Collapsing rotating shells in Myers-Perry-AdS$_5$ spacetime: a perturbative approach *Phys. Rev.* D **89** 104006 (*Preprint* 1402.4161 [gr-qc])

[9] Israel W 1966 Singular hypersurfaces and thin shells in general relativity *Nuovo Cim.* B **44** 1

[10] Mann R B, Oh J J and Park M I 2009 The role of angular momentum and cosmic censorship in the (2+1)-dimensional rotating shell collapse *Phys. Rev.* D **79** 064005 (*Preprint* 0812.2297 [hep-th])

[11] Delsate T, Rocha J V and Santarelli R 2014 Collapsing thin shells with rotation *Phys. Rev.* D **89** 121501(R) (*Preprint* 1405.1433 [gr-qc])

[12] Penrose R 1969 Gravitational collapse: The role of general relativity *Riv. Nuovo Cimento* **1** 252 [*Gen. Relativ. Gravit.* **34** 1141 (2002)]

[13] Rocha J V and Santarelli R 2014 Flowing along the edge: spinning up black holes in AdS spacetimes with test particles *Phys. Rev.* D **89** 064065 (*Preprint* 1402.4840 [gr-qc])

Spanish Relativity Meeting (ERE 2014): almost 100 years after Einstein's revolution IOP Publishing
Journal of Physics: Conference Series **600** (2015) 012023 doi:10.1088/1742-6596/600/1/012023

On the renormalization of ultraviolet divergences in the inflationary angular power spectrum

Adrian del Rio[1] **and Jose Navarro-Salas**[2]

Department of Theoretical Physics, IFIC. A mixed center University of Valencia - CSIC. Faculty of Physics, University of Valencia, Burjassot 46100, Valencia, Spain.

E-mail: [1]`adrian.rio@uv.es`, [2]`jnavarro@ific.uv.es`

Abstract.
We revise the role of ultraviolet divergences of cosmological observables and the corresponding renormalization from a space-time perspective. We employ the two-point function of primordial perturbations generated during inflation to derive an analytic expression for the multipole coefficients C_ℓ in the Sachs-Wolfe regime. We analyze the ultraviolet behavior and stress the fact that the standard result in the literature is equivalent to a renormalization of the two-point function at zeroth adiabatic order. We also argue that renormalization at second adiabatic order seems to be more appropriate from a physical point of view. This may change significantly the predictions for C_ℓ, while maintaining scale invariance.

1. Introduction and motivation

The theory of quantum fields interacting with gravity [1, 2], applied to the very early and rapidly expanding universe, explains considerably well the pattern of temperature anisotropies of the cosmic microwave background (CMB) as well as the large scale structure (LSS) of the universe [3]. The assumption of a slow-roll inflationary universe [4, 5] is particularly efficient to account for the huge and detailed cosmological data observed today [6]. Particle creation [7, 8] is the fundamental mechanism driving primordial perturbations that eventually seeded the tiny fluctuations in the temperature of the CMB. It can also be regarded as the source for the gravitational clumping of matter that gave rise to galaxies and structure formation.

In a curved space-time, new ultraviolet (UV) divergences arise in the computation of vacuum expectation values, and these infinities can not be removed by standard methods in Minkowski space-time. Specific methods to define regularization and renormalization in expanding universes have been constructed to account for the new UV divergences sourced by curved backgrounds [1, 2, 9] (for more recent works, see [10]).

Let φ represent a generic free field living in an homogeneous and isotropic *Fridmann-Lemaitre-Robertson-Walker* (FLRW) spacetime, with line element $ds^2 = dt^2 - a^2(t)d\vec{x}^2$. The field φ will later describe scalar (or tensor) perturbations during inflation. In the quantum theory the free field operator is most generally studied by its expansion in Fourier k-modes $\varphi_k(t)$,

$$\varphi(t,\vec{x}) = \int d^3k \left[A_{\vec{k}}\varphi_k(t) + A^\dagger_{-\vec{k}}\varphi^*_k(t)\right] e^{i\vec{k}\cdot\vec{x}} , \qquad (1)$$

where $A_{\vec{k}}$ and $A^\dagger_{\vec{k}}$ are creation and annihilation operators, such that $A_{\vec{k}}|0\rangle = 0$, and satisfy canonical commutation relations, $[A_{\vec{k}}, A_{\vec{k}'}] = 0$, $[A^\dagger_{\vec{k}}, A^\dagger_{\vec{k}'}] = 0$ and $[A_{\vec{k}}, A^\dagger_{\vec{k}'}] = \delta^3(\vec{k}-\vec{k}')$. A basic object in quantum field theory is the two-point function. For future purposes, we will consider it at equal times $t = t'$

$$\langle\varphi(t,\vec{x})\varphi(t,\vec{x}')\rangle = \int d^3k|\varphi_k(t)|^2 e^{i\vec{k}(\vec{x}-\vec{x}')} \ . \tag{2}$$

We can formally construct local physical observables from the above two-point function. For instance, the local quantum fluctuations of φ can be quantified by the mean square fluctuation in the vacuum state

$$\langle\varphi^2(t,\vec{x})\rangle = \int d^3k|\varphi_k(t)|^2 \ . \tag{3}$$

In either case, it is quite common in cosmology to refer the quantity $\Delta^2_\varphi(k,t) \equiv 4\pi k^3|\varphi_k(t)|^2$ as the primordial power spectrum.

A proper definition of the physical power spectrum in inflationary cosmology is not free of subtleties, as first pointed out in [11], and subsequently studied in [12, 13, 14]. In momentum-space, and for a single mode k, the power spectrum $\Delta^2_\varphi(k,t)$ is well defined. However, the formal variance $\langle\varphi^2(\vec{x},t)\rangle$, which is a sum in all modes, diverges in the ultraviolet. There is no doubt that the self-correlator needs renormalization when it is used to quantify the amplitude of quantum perturbations at a single space-time point, as in (3). However, the two-point function (2) may not need renormalization when used to quantify physical observables involving correlations. One could argue that the two-point function has a well-defined definition in the distributional sense and there is not mathematical need for any regularization [15]. However, as the spatial points get close together, the two-point function will grow without bound and produce divergences in physical observables [16, 1] (see expression (6) for this reasoning). In [11, 12, 13, 14, 16] it was argued that the physical power spectrum should be defined in terms of renormalized quantities. The challenging of this proposal for inflationary cosmology and quantum gravity has been recently stressed in [17]. The purpose of this work is to reanalyze these issues, specially from a spacetime viewpoint.

As we have said, (3) is UV divergent and needs to be renormalized according to standard rules

$$\langle\varphi^2(t,\vec{x})\rangle = \lim_{\vec{x}\to\vec{x}'}[\langle\varphi(t,\vec{x})\varphi(t,\vec{x}')\rangle - {}^{(N)}G_{Ad}((t,\vec{x}),(t,\vec{x}'))] \ , \tag{4}$$

where ${}^{(N)}G_{Ad}((t,\vec{x}),(t,\vec{x}'))$ is the Nth order adiabatic subtraction term. It is defined via the adiabatic regularization method, or, equivalently, using the DeWitt-Schwinger scheme (for more details see [18].) To properly cancel the UV divergences in our 4-dimensional spacetime the second adiabatic order $N = 2$ is the right choice for the mean square fluctuation $\langle\varphi^2(t,\vec{x})\rangle$. It must be noted that the proper adiabatic order of the subtraction term depends on the particular physical quantity to evaluate. For instance, the computation of the renormalized expectation value of the stress-energy tensor $\langle T_{\mu\nu}(t,\vec{x})\rangle$ needs subtraction up to the fourth adiabatic order, using $[\langle\varphi(t,\vec{x})\varphi(t,\vec{x}')\rangle - {}^{(4)}G_{Ad}((t,\vec{x}),(t,\vec{x}'))]$ instead of $\langle\varphi(t,\vec{x})\varphi(t,\vec{x}')\rangle$, and taking the coincident limit.

Apart from these fundamental objects, we can also be interested in integrated quantities from the two-point function, like

$$\langle\varphi_{\vec{p}}\varphi_{\vec{p}'}\rangle = \int d^3\vec{x}d^3\vec{x}' e^{i(\vec{p}\cdot\vec{x}+\vec{p}'\cdot\vec{x}')}\langle\varphi(t,\vec{x})\varphi(t,\vec{x}')\rangle = |\varphi_k(t)|^2\delta^3(\vec{p}+\vec{p}') \ , \tag{5}$$

or, assuming rotational invariance, we can also construct

$$
\begin{aligned}
C_\ell^{\varphi\varphi} &= \frac{1}{4\pi}\int d^2\hat{n}\, d^2\hat{n}' P_\ell(\hat{n}\cdot\hat{n}')\langle\varphi(t,\vec{x})\varphi(t,\vec{x}')\rangle \\
&= 2\pi\int_{-1}^{1} d\cos\theta P_\ell(\cos\theta)\langle\varphi(t,\vec{x})\varphi(t,\vec{x}')\rangle = 16\pi^2\int_0^\infty \frac{dk}{k}|\varphi_k(t)|^2 j_\ell^2(k|\vec{x}|) \ , \qquad (6)
\end{aligned}
$$

where P_ℓ are the Legendre Polynomials, and $\cos\theta = \vec{n}\cdot\vec{n}'$ is the angle formed by the two directions $\vec{n} = \vec{x}/|\vec{x}|$ and $\vec{n}' = \vec{x}'/|\vec{x}|$. For completeness, we have also added the equivalent and familiar expression in momentum space. It involves the spherical Bessel functions j_ℓ. Altough the former integral (5) is UV finite and no renormalization is needed, the latter (6) has an UV divergence. A detailed inspection reveals that the divergence is of $N = 0$ adiabatic order. This conclusion can be deduced from the short-distance (adiabatic) expansion ($\theta \to 0$)

$$
\langle\varphi(t,\vec{x})\varphi(t,\vec{x}')\rangle \sim \frac{1}{1-\cos\theta} - \frac{(\frac{1}{6}-\xi)R}{2}\log(1-\cos\theta) + ... \qquad (7)
$$

and $P_\ell(\cos\theta) \sim 1$. Therefore, while (5) should be kept unaltered, expression (6) must be modified according to the renormalization prescription. One should then replace $\langle\varphi(t,\vec{x})\varphi(t,\vec{x}')\rangle$ by $[\langle\varphi(t,\vec{x})\varphi(t,\vec{x}')\rangle - {}^{(0)}G_{Ad}((t,\vec{x}),(t,\vec{x}'))]$ in expression (6) to guarantee the UV finiteness of the integral.

However, we should remark that the appropriate choice of the adiabatic subtraction order N depend on the physically relevant object. In CMB cosmology the direct physical observables are temperature correlations, which are linked to the space-time two-point function $\langle\varphi(t,\vec{x})\varphi(t,\vec{x}')\rangle$. For instance, in the Sachs-Wolfe regime we have $\langle\Delta T(\vec{n})\Delta T(\vec{n}')\rangle_{SW} = \frac{T_0^2}{25}\langle\varphi(t,\vec{x})\varphi(t,\vec{x}')\rangle$, where $\varphi(t,\vec{x}) \equiv \mathcal{R}(t,\vec{x})$ is the comoving curvature perturbation. These temperature correlations are the ones upon which other observables like C_ℓ^{TT} are constructed

$$
C_\ell^{TT} = \int_{-1}^{1} d\cos\theta P_\ell(\cos\theta)\langle\Delta T(\vec{n})\Delta T(\vec{n}')\rangle \ . \qquad (8)
$$

Since physical correlations are direct observables that can be measured in an experiment, it seems natural to demand that they must always be finite, even at coincidence $\vec{x} = \vec{x}'$. To achieve that, one should then relate $\langle\Delta T(\vec{n})\Delta T(\vec{n}')\rangle$ with the quantity $[\langle\varphi(t,\vec{x})\varphi(t,\vec{x}')\rangle - {}^{(2)}G_{Ad}((t,\vec{x}),(t,\vec{x}'))]$ to ensure UV finiteness at coincidence. Therefore, the actual related multipole coefficients should be constructed with second order adiabatic subtractions

$$
C_{\ell,N=2}^{\varphi\varphi} = 2\pi\int_{-1}^{1} d\cos\theta P_\ell(\cos\theta)[\langle\varphi(t,\vec{x})\varphi(t,\vec{x}')\rangle - {}^{(2)}G_{Ad}((t,\vec{x}),(t,\vec{x}'))] \ . \qquad (9)
$$

2. Spacetime correlators in slow-roll inflation

Let us focus now on scalar perturbations during slow-roll inflation. For simplicity we will also consider the Sachs-Wolfe regime, for which the transfer functions are trivial. Scalar fluctuations are described through the comoving curvature perturbation field $\mathcal{R}(t,\vec{x})$. For single-field inflation, the modes $\mathcal{R}_k(t)$ defining the Bunch-Davies type vacuum, are [5]

$$
\mathcal{R}_k(t) = \sqrt{\frac{-\pi\eta}{4(2\pi)^3 z^2}} H_\nu^{(1)}(-\eta k) \ , \qquad (10)
$$

where $H \equiv \dot{a}/a$ is the Hubble rate, $H_\nu^{(1)}$ is the Hankel function with $\nu = \frac{3}{2} + \frac{2\epsilon+\delta}{1-\epsilon}$, $\epsilon \equiv -\dot{H}/H^2 \ll 1$, and $\delta \equiv \ddot{H}/2H\dot{H}$ is a second slow-roll parameter. In this expression we introduced the so-called proper time η, defined by $d\eta = \frac{dt}{a(t)}$. Moreover, $z \equiv a\dot{\phi}_0/H$, where $\phi_0(t)$ is the homogeneous part of the inflaton field, responsible for the period of inflation itself.

Using the modes (10) one can work out analytically the corresponding two-point function $\langle \mathcal{R}(t,\vec{x}), \mathcal{R}(t,\vec{x}')\rangle$. It is given by

$$\langle \mathcal{R}(t,\vec{x})\mathcal{R}(t,\vec{x}')\rangle = \frac{1}{16\pi^2 z^2\eta^2}\Gamma\left(\frac{3}{2}+\nu\right)\Gamma\left(\frac{3}{2}-\nu\right){}_2F_1\left(\frac{3}{2}+\nu, \frac{3}{2}-\nu; 2; 1-\frac{(\Delta x)^2}{4\eta^2}\right) , \quad (11)$$

where ${}_2F_1$ is the hypergeometric function, and Γ is the gamma function. The two-point separation is described by $\Delta x \equiv |\Delta\vec{x}| = 2^{\frac{1}{2}}|\vec{x}|(1-\cos\theta)^{1/2}$. For very short separations we get

$$\langle \mathcal{R}(t,\vec{x})\mathcal{R}(t,\vec{x}')\rangle = \frac{GH^2(1-\epsilon)^2}{4\pi\epsilon}\left\{\frac{4}{\Delta\bar{x}^2} + \left[\frac{1}{4}-\nu^2\right]\log\frac{\Delta\bar{x}^2}{4} + O(\Delta\bar{x}^0)\right\} \quad (12)$$

where we defined $\Delta\bar{x} = \Delta x\, a\, H(1-\epsilon)$, and G is the Newton's gravitational constant. It is clear then that, as expected, the scalar two-point function diverges when $\theta \to 0$. This produces an ultraviolet divergence in the (unrenormalized) expression for multipole coefficients in the Sachs-Wolfe regime, for which one makes the identification $\langle \Delta T(\vec{n})\Delta T(\vec{n}')\rangle_{SW} = \frac{T_0^2}{25}\langle \mathcal{R}(t,\vec{x})\mathcal{R}(t,\vec{x}')\rangle$,

$$C_\ell^{SW} = \frac{2\pi T_0^2}{25}\int_{-1}^{1} d\cos\theta P_\ell(\cos\theta)\langle \mathcal{R}(t,\vec{x})\mathcal{R}(t,\vec{x}')\rangle = \infty , \quad (13)$$

One could cure this divergence with the replacement $\langle \mathcal{R}(t,\vec{x})\mathcal{R}(t,\vec{x}')\rangle \to [\langle \mathcal{R}(t,\vec{x})\mathcal{R}(t,\vec{x}')\rangle - G_{Ad}^{(0)}((t,\vec{x}),(t,\vec{x}'))]$. One then gets

$$C_{\ell,N=0}^{SW} = \frac{2\pi T_0^2}{25}\int_{-1}^{1} d\cos\theta P_\ell(\cos\theta)[\langle \mathcal{R}(t,\vec{x})\mathcal{R}(t,\vec{x}')\rangle - G_{Ad}^{(0)}((t,\vec{x}),(t,\vec{x}'))] < \infty . \quad (14)$$

Evaluating this well-defined integral at late times yields the result [we define the new parameter $\bar{r}_L = r_L\, a\, H(1-\epsilon)$, with $|\vec{x}| = r_L$ the comoving radial coordinate of the last scattering surface]

$$C_{\ell,N=0}^{SW} = \frac{8\pi T_0^2}{25}\frac{4\pi G}{\epsilon}\frac{H^2(1-\epsilon)^2}{16\pi^2}\frac{\Gamma(3-n)\Gamma(\ell+\frac{n-1}{2})}{\Gamma(\ell+2-\frac{n-1}{2})}\bar{r}_L^{1-n} , \quad (15)$$

where we have used $\nu - \frac{3}{2} = \frac{1-n}{2}$, and n represents the scalar index of inflation $n = 1 - 4\epsilon - 2\delta + O(\epsilon,\delta)^2$. This is in exact agreement with the well-known standard result, obtained with the momentum-space power spectrum. It is worth to remark that the same result can be obtained if one consider the large-scale behavior [i.e. the late-time one during inflation $a|\vec{x}-\vec{x}'| \gg H^{-1}$] of the two-point function

$$\langle \mathcal{R}(t,\vec{x})\mathcal{R}(t,\vec{x}')\rangle \sim \frac{\Gamma(\frac{3}{2}-\nu)}{4\pi^2 z^2\eta^2}\frac{\Gamma(\nu)}{\sqrt{\pi}}\left(\frac{\Delta x}{-\eta}\right)^{2(\nu-3/2)} . \quad (16)$$

and use this expression in (13). This is indeed the standard approach, as reported in many textbooks [5]. Therefore, it is actually equivalent to the calculation of $C_{\ell,N=0}^{SW}$ with the renormalized two-point function at zero adiabatic order.

Even though this procedure provides a finite result, what we have really encountered in (12) is the typical quadratic and logarithmic short-distance behavior of a quantum field in a curved background. So, as argued before, it seems natural to remove both divergences, not only the leading one. Accordingly, it is natural to propose the following identification

$$\langle \Delta T(\vec{n})\Delta T(\vec{n}')\rangle_{SW} = \frac{T_0^2}{25}[\langle \mathcal{R}(t,\vec{x})\mathcal{R}(t,\vec{x}\,')\rangle - G_{Ad}^{(2)}((t,\vec{x}),(t,\vec{x}'))] \ , \tag{17}$$

instead of $\langle \Delta T(\vec{n})\Delta T(\vec{n}')\rangle_{SW} = \frac{T_0^2}{25}[\langle \mathcal{R}(t,\vec{x})\mathcal{R}(t,\vec{x}\,')\rangle - G_{Ad}^{(0)}((t,\vec{x}),(t,\vec{x}'))]$, as used in (14) to obtain (15).

The subtraction term at second adiabatic order is found to be

$$\begin{aligned} G_{Ad}^{(2)}((t,\vec{x}),(t,\vec{x}')) &= \frac{GH^2(1-\epsilon)^2}{4\pi\epsilon}\left\{\frac{4}{\Delta\vec{x}^2} + \left(\frac{1}{4}-\nu^2\right)\log\frac{\Delta\vec{x}^2}{4}\right. \\ &\left. +\frac{2-\epsilon}{3(1-\epsilon)^2} + \left(\frac{1}{4}-\nu^2\right)\left[2\gamma + \log\frac{\mu^2}{H^2(1-\epsilon)^2}\right]\right\} \ , \end{aligned} \tag{18}$$

where μ is a renormalization scale, and γ is the Euler constant. Now we should obtain a suitable expression for the two-point function. A good way to deal with this is to expand expression (11) as a power series of the "slow-roll" parameter ν around $\nu = 3/2$, and stay at first order (for details see [16]). The result reads:

$$\langle \mathcal{R}(t,\vec{x})\mathcal{R}(t,\vec{x}')\rangle \approx \frac{GH^2(1-\epsilon)^2}{4\pi\epsilon}\left\{\frac{4}{\Delta\vec{x}^2} - 1 + \frac{2}{(3/2-\nu)}\left(\frac{\Delta\vec{x}^2}{4}\right)^{\nu-3/2}\right\} \tag{19}$$

We can now proceed to do the subtraction. The renormalized two-point function then reads

$$\begin{aligned} \langle \mathcal{R}(t,\vec{x})\mathcal{R}(t,\vec{x}')\rangle - G_{Ad}^{(2)}((t,\vec{x}),(t,\vec{x}')) &\approx \frac{GH^2(1-\epsilon)^2}{4\pi\epsilon}\left\{\frac{2}{(3/2-\nu)}\left(\frac{\Delta\vec{x}^2}{4}\right)^{\nu-3/2} + 2\log\Delta\vec{x}^2\right. \\ &\left. -\frac{5}{3} + 4\gamma + 2\log\frac{\mu^2}{H^2}\right\} \ . \end{aligned} \tag{20}$$

We now proceed to compute the corresponding angular power spectrum from the renormalized two-point function and for the Sachs-Wolfe regime

$$C_{\ell,N=2}^{SW} = \frac{2\pi T_0^2}{25}\int_{-1}^{1} d\cos\theta P_\ell(\cos\theta)[\langle \mathcal{R}(t,\vec{x})\mathcal{R}(t,\vec{x}')\rangle - G_{Ad}^{(2)}((t,\vec{x}),(t,\vec{x}'))] \ . \tag{21}$$

To evaluate the logarithmic contributions of (20) to (21) we take into account that $\int_{-1}^{1} dy \log(1-y)P_\ell(y) = -2/\ell(\ell+1), \ell = 1,2,\ldots$. The final result is very well approximated by the following analytical expression:

$$C_{\ell,N=2}^{SW} \approx \frac{4\pi G}{\epsilon}\frac{8\pi T_0^2}{25}\frac{H^2(1-\epsilon)^2\bar{r}_L^{1-n}}{16\pi^2}\left\{\frac{\Gamma(\ell+\frac{n-1}{2})}{\Gamma(\ell+2-\frac{n-1}{2})} - \frac{\bar{r}_L^{n-1}}{\ell(\ell+1)}\right\} \ . \tag{22}$$

The above expression is valid for $\ell \geq 1$, as for $\ell = 0$ there would be present all the constant contributions from the renormalized two-point function (20), including the one depending on the renormalization scale. In fact, the renormalization scale may be fixed by imposing the natural condition $C_{0,N=2}^{SW} = 0$.

Notice that the first term in (22) comes form the renormalization procedure at zero adiabatic order, and reproduces the standard result (15). The second one comes from the purely second adiabatic order subtraction term, but it shows scale invariance as well. Therefore, Eq. (22) is also consistent with observations [6]. The first term is time-independent, while the second term depends slightly on time through the quantity $\bar{r}_L^{n-1} \equiv \alpha$. This parameter can be regarded as a phenomenological parameter, varying in the range $1 > \alpha > 0$. In the limiting case $\alpha \to 0$ one recovers the standard prediction, and this would correspond to the subtraction terms evaluated after the end of inflation. If the subtraction terms are evaluated a few e-foldings after the horizon exit of the scale r_L, the parameter α approaches 1 and the physical significance of the correction increases.

3. Conclusions

We have briefly reported a critical overview on the renormalization of cosmological observables. Firstly, we have analyzed two-point correlators and self-correlators of scalar primordial perturbations in quasi-de Sitter spacetime backgrounds. Then, the evaluation of the multipole coefficients C_ℓ has been specially scrutinised. We have stressed the fact that the standard result in the literature is equivalent to a renormalization of the two-point function at zero adiabatic order. We have argued that renormalization at second adiabatic order is more suitable from a physical point of view, it is necessary to ensure UV finiteness of direct physical observables such as temperature correlations. This may change significantly the predictions of inflation, provided the renormalization subtraction terms are evaluated a few e-foldings after the first horizon crossing of the scale r_L.

Acknowledgments

This work is supported by the Research Project of the Spanish MINECO FIS2011-29813-C02-02, and the Consolider Program No. CPANPHY-1205388. A. del Rio is supported by the Spanish Education Ministry Ph.D. fellowship FPU13/04948.

References

[1] Parker L and Toms D J, 2009 *Quantum Field Theory in Curved Spacetime: Quantized Fields and Gravity* (Cambridge University Press, Cambridge, UK).
[2] Birrell N D and Davies P C W, 1982 *Quantum Fields in Curved Space* (Cambridge University Press, Cambridge, UK).
[3] Mukhanov V F and Chibisov G V, 1981 *JETP Lett.* **33**, 532-53; Hawking S W, 1982 *Phys. Lett.* **115B**, 295-7; Guth A and Pi S-Y, 1982 *Phys. Rev. Lett.* **49**, 1110-3; Starobinsky A A, 1982 *Phys. Lett.* **117B**, 175-8; J.M. Bardeen, P.J. Steinhardt, and M.S. Turner, 1983 *Phys. Rev.*D**28**, 679-93.
[4] Kolb E and Turner M, 1994 *The Early Universe* (Addison-Wesley Publishing Company, USA).
[5] Weinberg S, 2008 *Cosmology* (Cambridge University Press, Cambridge, UK). Liddle A R and Lyth D H, 2006 *Cosmological Inflation and Large-Scale Structure* (Cambridge University Press, Cambridge, UK).
[6] Planck Collaboration 2013, *Planck 2013 Results. XV. CMB power spectra and likelihood*, arXiv:1303.5075.
[7] Parker L, 1966 *The creation of particles in an expanding universe* (Ph.D. thesis, Harvard University).
[8] Parker L, 1968 *Phys. Rev. Lett.* **21**, 562-4; 1969 *Phys. Rev.* **183**, 1057-68; 1971 *Phys. Rev.* D **3**, 346-56.
[9] Parker L and Fulling S A, 1974 *Phys. Rev.* D **9**, 341-54.
[10] Landete A, Navarro-Salas J and Torrenti F, 2013 *Phys. Rev.* D **88**, 061501; 2014 *Phys. Rev.* D **89**, 044030; del Rio A, Navarro-Salas J and Torrenti F, 2014 *Phys. Rev.* D **90**, 084017.
[11] Parker L, 2007 Amplitude of perturbations from inflation, *Preprint:* gr-qc/0702216
[12] Agullo I, Navarro-Salas J, Olmo G J and Parker L, 2009 *Gen. Rel. Grav.* **103**, 061301.
[13] Parker L and Glenz M M, 2009 *Phys. Rev. D* **80**, 063534
[14] Agullo I, Navarro-Salas J, Olmo G J and Parker L, 2009 *Phys. Rev. Lett.* **103**, 061301; 2010 *Phys. Rev. D***81**, 043514; 2011 *Phys. Rev. D***84**, 107304.
[15] Bastero-Gil M *et al.*, 2013 *Phys. Rev.* D **87**, 087302; Finelli F, Marozzi G, Vacca G P and Venturi G, 2007 *Phys. Rev.* D **76**, 103528.
[16] del Rio A and Navarro-Salas J, 2014 *Phys. Rev.* D **89**, 084037.
[17] Woodard R P, 2014 Perturbative quantum gravity comes of age, *Preprint:* gr-qc/1407.47481
[18] del Rio A and Navarro-Salas J, *On the equivalence between adiabatic and DeWitt-Schwinger renormalization schemes*, to appear.

Journal of Physics: Conference Series **600** (2015) 012024 doi:10.1088/1742-6596/600/1/012024

Viability of the Matter Bounce Scenario

Jaume de Haro and Jaume Amorós

Departament de Matemàtica Aplicada I, Universitat Politècnica de Catalunya, Diagonal 647, 08028 Barcelona, Spain

E-mail: jaime.haro@upc.edu,jaume.amoros@upc.edu

Abstract. It is shown that teleparallel $F(\mathcal{T})$ theories of gravity combined with Loop Quantum Cosmology support a Matter Bounce Scenario which is an alternative to the inflation scenario in the Big Bang paradigm. It is checked that these bouncing models provide theoretical data that fits well with the current observational data, allowing the viability of the Matter Bounce Scenario.

1. Introduction

It is well-known that inflation suffers from several problems (see [1] for a review about these problems), like the initial singularity which is usually not addressed, or the fine-tuning of the degree of flatness required for the potential in order to achieve successful inflation [2].

In order to avoid these problems, an alternative scenario to the inflationary paradigm, called *Matter Bounce Scenario* (MBS), has been developed in order to explain the evolution of our Universe (see [3]). Essentially, it depicts at very early times a matter dominated Universe in a contracting phase, that evolves towards the bounce and afterwards enters an expanding phase. This model, like inflation, solves the horizon problem that appears in General Relativity (GR) and improves the flatness problem in GR (where spatial flatness is an unstable fixed point and fine tuning of initial conditions is required), because the contribution of the spatial curvature decreases in the contracting phase at the same rate as it increases in the expanding one (see for instance [4]).

The aim of our work is to construct viable bouncing cosmologies where the matter part of the Lagrangian is composed of a single scalar field and, therefore, have to go beyond General Relativity, since GR forbids bounces when one deals with a single field. Hence, theories such as holonomy corrected Loop Quantum Cosmology (LQC) [5], where a big bounce appears owing to the discrete structure of space-time [6] or teleparalellism [7] must be taken into account. When dealing with these theories, in order to obtain a theoretical value of the spectral index and its running that may fit well with current experimental data, a quasi-matter dominated regime in the contracting phase termed by the condition $\left|w \equiv \frac{P}{\rho}\right| \ll 1$, where P and ρ are respectively the pressure and the energy density of the Universe, has to be introduced [8].

Since in Matter Bounce Scenario the number of e-folds before the end of the quasi-matter domination regime can be relatively small, the horizon problem does not exist in bouncing cosmologies and the flatness problem is neutralized [4]. This argues for the viability of such models, making it possible that for certain matter bounce scenarios the forecast values of the spectral index and of the running parameter agree well with the most accurate current observations.

In contrast, in slow roll inflation one must consider the running of the spectral index corresponding to N e-folds before the end of the inflation, which in general, is of the order of N^{-2}. This value turns out to be very small, when one substitutes for N the minimum number of e-folds which are needed to solve the horizon and flatness problem in inflationary cosmology (the usual accepted value is $N > 50$), as compared with its corresponding observational value -0.0134 ± 0.009 coming from the most recent Planck data [9]. This shows that these slow roll models are less favored by observations.

The units used in the paper are: $\hbar = c = 8\pi G = 1$.

2. $F(\mathcal{T})$ gravity in flat FLRW geometry

Teleparallel theories are based in the *Weitzenböck space-time*. This space is $\mathbb{R}^4$, with a Lorentz metric, in which a global, orthonormal basis of its tangent bundle given by four vector fields $\{e_i\}$ has been selected, that is, they satisfy $g(e_i, e_j) = \eta_{ij}$ with $\eta = \text{diag}\,(-1,1,1,1)$. The Weitzenböck connection ∇ is defined by imposing that the basis vectors e_i be absolutely parallel, i.e. that $\nabla e_i = 0$.

The Weitzenböck connection is compatible with the metric g, and it has zero curvature because of the global parallel transport defined by the basis $\{e_i\}$. The information of the Weitzenböck connection is carried by its torsion, and its basic invariant is the *scalar torsion* $\mathcal{T}$. The connection, and its torsion, depend on the choice of orthonormal basis $\{e_i\}$, but if one adopts the flat Friedmann-Lemaître-Robertson-Walker (FLRW) metric and selects as orthonormal basis $\{e_0 = \partial_0, e_1 = \frac{1}{a}\partial_1, e_2 = \frac{1}{a}\partial_2, e_3 = \frac{1}{a}\partial_3\}$, then the scalar torsion is

$$\mathcal{T} = -6H^2 \,, \tag{1}$$

where $H = \frac{\dot{a}}{a}$ is the Hubble parameter, and this identity is invariant with respect to local Lorentz transformations that only depend on the time, i.e. of the form $\tilde{e_i} = \Lambda_i^k(t) e_k$ (see [10, 11]).

With the above choice of orthonormal fields, the Lagrangian of the $F(\mathcal{T})$ theory of gravity is

$$\mathcal{L}_{\mathcal{T}} = \mathcal{V}(F(\mathcal{T}) + \mathcal{L}_M), \tag{2}$$

where $\mathcal{V} = a^3$ is the volume of the Universe, and $\mathcal{L}_M$ is the matter Lagrangian density.

The Hamiltonian of the system is

$$\mathcal{H}_{\mathcal{T}} = \left(2\mathcal{T}\frac{dF(\mathcal{T})}{d\mathcal{T}} - F(\mathcal{T}) + \rho\right)\mathcal{V}, \tag{3}$$

where ρ is the energy density. Imposing the Hamiltonian constrain $\mathcal{H}_{\mathcal{T}} = 0$ leads to the modified Friedmann equation

$$\rho = -2\frac{dF(\mathcal{T})}{d\mathcal{T}}\mathcal{T} + F(\mathcal{T}) \equiv G(\mathcal{T}) \tag{4}$$

which, as $\mathcal{T} = -6H^2$, defines a curve in the plane (H, ρ).

Equation (4) may be inverted, so a curve of the form $\rho = G(\mathcal{T})$ defines an $F(\mathcal{T})$ theory with

$$F(\mathcal{T}) = -\frac{\sqrt{-\mathcal{T}}}{2}\int \frac{G(\mathcal{T})}{\mathcal{T}\sqrt{-\mathcal{T}}} d\mathcal{T}. \tag{5}$$

To produce a cyclically evolving Universe, let us take the $F(\mathcal{T})$ theory arising from the ellipse that defines the holonomy corrected Friedmann equation in Loop Quantum Cosmology

$$H^2 = \frac{\rho}{3}\left(1 - \frac{\rho}{\rho_c}\right), \tag{6}$$

where ρ_c is the so-called *critical density.*

To obtain a parametrization of the form $\rho = G(\mathcal{T})$, the curve has to be split in two branches

$$\rho = G_\pm(\mathcal{T}) = \frac{\rho_c}{2}\left(1 \pm \sqrt{1 + \frac{2\mathcal{T}}{\rho_c}}\right), \tag{7}$$

where the branch $\rho = G_-(\mathcal{T})$ corresponds to $\dot{H} < 0$ and $\rho = G_+(\mathcal{T})$ is the branch with $\dot{H} > 0$. Applying Eq. (5) to these branches produces the model ([12, 13, 14])

$$F_\pm(\mathcal{T}) = \pm\sqrt{-\frac{\mathcal{T}\rho_c}{2}}\arcsin\left(\sqrt{-\frac{2\mathcal{T}}{\rho_c}}\right) + G_\pm(\mathcal{T}). \tag{8}$$

3. Matter Bounce Scenario

Matter Bounce Scenarios (see [3] for a recent review) are essentially characterized by the Universe being nearly matter dominated at very early times in the contracting phase (to obtain an approximately scale invariant power spectrum) and evolving towards a bounce where all the parts of the Universe become in causal contact [12], solving the horizon problem, to enter into a expanding regime, where it matches the behavior of the standard hot Friedmann Universe. They constitute an alternative to the inflationary paradigm.

According to the current observational data, in order to obtain a viable MBS model, the bouncing model has to satisfy some conditions that we have summarized as follows:

(i) The latest Planck data constrain the value of the spectral index for scalar perturbations and its running, namely n_s and α_s, to 0.9603 ± 0.0073 and -0.0134 ± 0.009 respectively [9]. The analysis of these parameters provided by Planck makes no slow roll approximation (in fact, the determination of cosmological parameters from the first year WMAP observations was done considering the ΛCDM model [15]), which means that the parameters n_s and α_s could be used to test bouncing models. On the other hand, it is well-known that the ways to obtain a nearly scale invariant power spectrum of perturbations with running are either a quasi de Sitter phase in the expanding phase or a nearly matter domination phase at early times, in the contracting phase [16]. Then, since for the MBS one has $n_s = 1$, if one wants to improve the model to match correctly with this observational data, one has to consider, at early times in the contracting phase, a quasi-matter domination period characterized by the condition $\left|w \equiv \frac{P}{\rho}\right| \ll 1$, being P and ρ the pressure and the energy density of the Universe.

(ii) The Universe has to reheat creating light particles that will thermalize matching with a hot Friedmann Universe. Reheating could be produced due to the gravitational particle creation in an expanding Universe [17]. In this case, an abrupt phase transition (a non adiabatic transition) is needed in order to obtain sufficient particle creation that thermalizes producing a reheating temperature that fits well with current observations. This method was used in the context of inflation in [18, 19], where a sudden phase transition from a quasi de Sitter phase to a radiation domination or a quintessence phase was assumed in the expanding regime. It is shown in [20] that gravitational particle production could be applied to the MBS, assuming a phase transition from the matter domination to an ekpyrotic phase in the contracting regime, and obtaining a reheating temperature compatible with current data.

(iii) Studies of distant type Ia supernovae ([21] and others) provide strong evidence that our Universe is expanding in an accelerating way. A viable model must take into account this current acceleration, which could be incorporated, in the simplest case, with a cosmological constant, or by quintessence models [22]. There are other ways to implement the current cosmic acceleration, for example using $F(\mathcal{R})$ gravity (see for instance [23]), but the current models that provide this behavior are very complicated, and the main objective in MBS is to present the simplest viable models.

(iv) The data of the seven-year survey WMAP ([24]) constrains the value of the power spectrum for scalar perturbations to be $\mathcal{P}_S(k) \cong 2\times 10^{-9}$. The numerical results (analytical ones will be impossible to obtain) calculated with bouncing models have to match with that experimental data.

(v) The constrain of the tensor/scalar ratio provided by WMAP and Planck projects ($r \leq 0.11$) is obtained indirectly assuming the *consistency* slow roll relation $r = 16\epsilon$ (where $\epsilon = -\frac{\dot{H}}{H^2} \cong \frac{1}{2}\left(\frac{V_\varphi}{V}\right)^2$ is the main slow roll parameter) [25], because gravitational waves are not longer detected by those projects. This means that, the slow roll inflationary models must satisfy this constrain, but not the bouncing ones, where there is not any consistency relation. This point is very important because some very complicated mechanisms are sometimes implemented to the MBS in order to enhance the power spectrum of scalar perturbation to achieve the observational bound provided by Planck [26]. In fact, in matter bounce scenario, to check if the models provide a viable value of the tensor/scalar ratio, first of all gravitational waves must be clearly detected in order to determine the observed value of this ratio. The authors hope that more accurate unified Planck-BICEP2 data (the B2P collaboration), which is going to be issued soon, may adress this point. In contrast, as we have pointed out in (i), the spectral index of scalar perturbations and its running could be calculated independently of the theory, which means that in order to check bouncing models, while in the absence of evidence of gravitational waves, one has to work in the space (n_s, α_s).

4. Perturbations in Matter Bounce Scenario

The Mukhanov-Sasaki equations (see [27] for a deduction of these equations in GR) for $F(\mathcal{T})$ gravity and LQC are given by [28, 29]

$$\zeta''_{S(T)} - c_s^2\nabla^2\zeta_{S(T)} + \frac{Z'_{S(T)}}{Z_{S(T)}}\zeta'_{S(T)} = 0, \tag{9}$$

where ζ_S and ζ_T denote the amplitude for scalar and tensor perturbations.

In $F(\mathcal{T})$ gravity one has

$$Z_S = \frac{a^2|\Omega|\dot{\varphi}^2}{c_s^2H^2}, \quad Z_T = \frac{a^2c_s^2}{|\Omega|}, \quad c_s^2 = |\Omega|\frac{\arcsin\left(2\sqrt{\frac{3}{\rho_c}}H\right)}{2\sqrt{\frac{3}{\rho_c}}H}, \quad \text{with} \quad \Omega = 1 - \frac{2\rho}{\rho_c}. \tag{10}$$

In contrast, for LQC,

$$Z_S = \frac{a^2\dot{\varphi}^2}{H^2}, \quad Z_T = \frac{a^2}{\Omega}, \quad c_s^2 = \Omega. \tag{11}$$

The power spectrum for scalar perturbations is given by [30]

$$\mathcal{P}_S(k) = \frac{3\rho_c^2}{\rho_{pl}}\left|\int_{-\infty}^{\infty} Z_S^{-1}(\eta)d\eta\right|^2, \tag{12}$$

where, in order to obtain this formula, the scale factor $a(t) \cong (\frac{4}{3}\rho_c t^2)^{1/3}$ at early times has been used. In the particular case of an exactly matter dominated universe during all the background evolution, i.e., when $a(t) = (\frac{4}{3}\rho_c t^2 + 1)^{1/3}$ for teleparalell $F(\mathcal{T})$ gravity one has $\mathcal{P}_S(k) = \frac{16}{9}\frac{\rho_c}{\rho_{pl}}\mathcal{C}^2$, [31] where $\mathcal{C} = 1 - \frac{1}{3^2} + \frac{1}{5^2} - ... = 0.915965...$ is the Catalan's constant, and for holonomy corrected LQC $\mathcal{P}_S(k) = \frac{\pi^2}{9}\frac{\rho_c}{\rho_{pl}}$ [32].

The ratio of tensor to scalar perturbations in MBS is given by

$$r = \frac{8}{3}\left(\frac{\int_{-\infty}^{\infty} Z_T^{-1}(\eta)d\eta}{\int_{-\infty}^{\infty} Z_S^{-1}(\eta)d\eta}\right)^2, \tag{13}$$

where the factor 8 appears due to the two polarizations of the gravitational waves and to the renormalization with respect to a canonical field [33].

The spectral index for scalar perturbations and its running are calculated in [8] given

$$n_s - 1 = 12w, \quad \alpha_s = -48\delta^2, \tag{14}$$

where the parameters w and δ^2, calculated in the quasi-matter domination, as a functions of the potential are

$$w \cong \frac{1}{3}\left(\frac{V_\varphi}{V}\right)^2 - 1, \quad \delta^2 \cong -\left(\frac{V_\varphi}{V}\right)_\varphi. \tag{15}$$

4.1. Comparison with observational data in the plane (n_s, α_s)

In slow-roll inflation, for the general models (monomial, natural, hilltop and plateau potentials), $1 - n_s$ is of the order N^{-1}, while the running parameter is of order N^{-2} and, consequently, one has $\alpha_s \sim (1 - n_s)^2$, which in most cases is incompatible with Planck and WMAP data, because the observed value of the running is not small enough [34, 35].

Thus, the observation of a large negative running implies that any inflationary phase requires multiple fields or the breakdown of slow roll. Following this second path, in [35] the authors consider the break of the slow-roll approximation for a short while, due to the inclusion of a quickly oscillating term in the potential. As a consequence, the theoretical value of the running parameter gets larger and could match well with observational data.

In contrast, in MBS the situation is completely different. For example, in [8] dealing with a perfect fluid whose Equation of State (EoS) is parametrized by the number of e-folds before the end of the quasi-matter domination period, namely N, the authors have shown that the theoretical values of the spectral index of scalar perturbations and its running fit well with their corresponding observational data. To be more precise, for the EoS $P = \frac{\beta}{(N+1)^\alpha}\rho$, $(\alpha > 0,\ \beta < 0)$ the following relation

$$\alpha_s = \frac{2\alpha}{N+1}(n_s - 1) \tag{16}$$

is obtained, which is perfectly compatible with the experimental data. In fact, for instance, if one takes $\alpha = 2$ and $N = 12$ (note that in bouncing cosmologies a large number of e-folds is *not* required, because the horizon problem does not exist, since at the bounce all parts of the Universe are already in causal contact, and also the flatness problem gets improved [4]), one obtains, for $n_s = 0.9603 \pm 0.0073$, the following value for the running parameter: $\alpha_s = -0.0122 \pm 0.0022$, which is compatible with the Planck data. Effectively, for these values of α and N one gets $n_s - 1 = \frac{12}{13^2}\beta \cong 0.071\beta$, which is indeed compatible with its observed value, by choosing $\beta \cong -\frac{1}{2}$.

Acknowledgements
The authors would like to thank Professor Sergei D. Odintsov for his valuable and useful comments. This investigation has been supported in part by MINECO (Spain), project MTM2011-27739-C04-01, MTM2012-38122-C03-01.

References
[1] R.H. Brandenberger, (2012) [arXiv:astro-ph/1206.4196].
R.H. Brandenberger, Int. J. Mod. Phys. Conf. Ser **01**, 67 (2008) [arXiv:0902.4731].
[2] F.C. Adams, K. Freese and A.H. Guth, Phys. Rev. **D43**, 965 (1991).
[3] Yifu Cai, SCIENCE CHINA: Phys. Mech. Astr. **57**, 1414 (2014) [arXiv: 1405.1369].
[4] R.H. Brandenberger, (2012) [arXiv:1204.6108].
[5] Yifu Cai and E. Wilson-Ewing, JCAP **03** 026, (2014) [arXiv:1402.3009].
M. Bojowald and G.M. Hossain, Phys. Rev. **D77**, 023508 (2008) [arXiv: 0709.2365].
T. Cailleteau, J. Mielzczarek, A. Barrau and J. Grain, Class. Quant. Grav. **29**, 095010 (2012) [arXiv:111.3535].
[6] A. Ashtekar and P. Singh, Class. Quant. Grav. **28** 23001, (2011) [arXiv:1108.0893].
P. Singh, Class. Quant. Grav. **26** 125005, (2009) [arXiv:0901.2750].
[7] K. Bamba, J. de Haro and S.D. Odintsov, JCAP 02 **008**, (2013) [arXiv:1211.2968].
[8] E. Elizalde, J. Haro and S.D. Odintsov (2014) [arXiv:1411.3475].
[9] P.A.R. Ade et al., Astronomy and Astrophysics **571**, A22 (2014) [arXiv:1303.5082].
[10] G. R. Bengochea and R. Ferraro, Phys. Rev. D **79**, 124019 (2009) [arXiv:0812.1205].
[11] J. de Haro and J. Amorós, Phys. Rev. Lett. **110**, 071104 (2013) [arXiv:1211.5336].
[12] J. Amorós, J. de Haro and S.D. Odintsov, Phys. Rev. **D87**, 104037 (2013) [arXiv:1305.2344].
[13] J. Amorós, J. de Haro and S.D. Odintsov, Phys. Rev. **D89**, 104010 (2014) [arXiv:1402.3071].
[14] A. Ashtekar and P. Singh, Class. Quantum Grav. **28**, 213001 (2011) [arXiv: 1108.0893].
[15] D.N. Spergel et al., Astrophys.J.Suppl. **148**, 175 (2003) [arXiv:0302209].
L. Verde et al., Astrophys.J.Suppl. **148**, 195 (2003) [arXiv:0302218].
[16] D. Wands, Phys Rev. **D 60**, 023507 (1999) [arXiv:9809062].
[17] A.A. Grib, S.G. Mamayev and V.M. Mostepanenko, Gen. Rel. Grav. **7**, 535 (1976).
A.A. Grib, S.G. Mamayev and V.M. Mostepanenko, *Vacuum Quantum Effects in Strong Fields*, St Petersburg: Friedmann Laboratory Publishing (1994).
L. Parker, Phys. Rev. **183**, 1057 (1969).
[18] L.H. Ford, Phys. Rev. **D35**, 2955 (1985).
[19] P.J.E. Peebles and A. Vilenkin, Phys. Rev. **D59**, 063505 (1999) [arXiv:9810509].
[20] J. Quintin, Yifu Cai and R. Brandenberger, Phys. Rev. **D90**, 063507 (2014)[arXiv:1406.6049]
[21] S. Perlmutter et al., Astrophys. J. **517**, 565 (1999) [arXiv:9812133].
A.G. Riess et al., Astron. J. **116**, 1009 (1999) [arXiv:9805201].
[22] R.R. Caldwell, R. Dave and P.J. Steinhardt, Phys. Rev. Lett. **80**, 1582 (1988).
I. Zlatev, L. Wang and P.J. Steinhardt, Phys. Rev. Lett.**82**, 896 (1999) [arXiv:9807002].
[23] S. Nojiri and S.D. Odintsov, Int.J.Geom.Meth.Mod.Phys. **4**, 115 (2007) [arXiv:0601213].
S. Nojiri and S.D. Odintsov, Phys.Rept.**505**, 59 (2011) [arXiv:1011.0544].
[24] D. Larson et al., Astrophys. J. Suppl. **192**, 16 (2011) [arXiv:1001.4635].
[25] H. Peiris and R. Easther, JCAP **0607**, 002 (2006) [arXiv:0603587].
S.M. Leach, A.R. Liddle, J. Martin and D.J. Schwarz, Phys.Rev. **D66**, 023515 (2002) [arXiv:0202094].
S.M. Leach and A.R. Liddle, Mon.Not.Roy.Astron.Soc. **341**, 1151 (2003) [arXiv:0207213].
[26] Yifu Cai, R. Brandenberger and X. Zhang, JCAP **1103**, 003 (2011) [arXiv:1101.0822].
Yifu Cai, D. A. Easson and R. Brandenberger, JCAP **1208**, 020 (2012) [arXiv:1206.2382].
[27] V.F. Mukhanov, JETP Lett. **41**, 493 (1985).
M. Sasaki, Prog. Theor. Phys. **76**, 1036 (1986).
[28] Yifu Cai, S-H. Chen, J.B. Dent, S. Dutta and E.N. Saridakis, Class.Quant.Grav. **28**, 215011 (2011) [arXiv:1104.4349].
[29] T. Cailleteau, A. Barrau, J. Grain and F. Vidotto, PRD **86**, 087301 (2012) [arXiv:1206.6736].
[30] J. de Haro and J. Amorós, JCAP **08**, 025(2014) [arXiv:1403.6396].
[31] J. Haro, JCAP **11**, 068 (2013) [arXiv:1309.0352].
[32] E. Wilson-Ewing, JCAP **03**, 026 (2013) [arXiv:1211.6269].
[33] D. Langlois, (2010) [arXiv:1001.5259].
[34] R. Easther and H. Peiris, JCAP **010**, 0609 (2006) [arXiv:0604214].
[35] Y. Wan, S. Li, M. Li, T. Qiu, Yifu Cai, and X. Zhang Phys. Rev. **D90**, 023537 (2014) [arXiv:1405.2784].

Journal of Physics: Conference Series **600** (2015) 012025 doi:10.1088/1742-6596/600/1/012025

Fitting BICEP2 with defects, primordial gravitational waves and dust

Joanes Lizarraga[1], Jon Urrestilla[1], David Daverio[2], Mark Hindmarsh[3,4], Martin Kunz[2,5], Andrew R. Liddle[6]

[1] Department of Theoretical Physics, University of the Basque Country UPV/EHU, 48080 Bilbao, Spain
[2] Département de Physique Théorique & Center for Astroparticle Physics, Université de Genève, Quai E. Ansermet 24, CH-1211 Genève 4, Switzerland
[3] Department of Physics & Astronomy, University of Sussex, Brighton, BN1 9QH, United Kingdom
[4] Department of Physics and Helsinki Institute of Physics, PL 64, FI-00014 University of Helsinki, Finland
[5] African Institute for Mathematical Sciences, 6 Melrose Road, Muizenberg, 7945, South Africa
[6] Institute for Astronomy, University of Edinburgh, Royal Observatory, Edinburgh EH9 3HJ, United Kingdom

E-mail: joanes.lizarraga@ehu.es

Abstract. In this work we discuss the possibility of cosmic defects being responsible for the B-mode signal measured by the BICEP2 collaboration. We also allow for the presence of other cosmological sources of B-modes such as inflationary gravitational waves and polarized dust foregrounds, which might contribute to or dominate the signal. On the one hand, we find that defects alone give a poor fit to the data points. On the other, we find that defects help to improve the fit at higher multipoles when they are considered alongside inflationary gravitational waves or polarized dust. Finally, we derive new defect constraints from models combining defects and dust. This proceeding is based on previous works [1, 2].

1. Introduction

The recent detection of B-mode polarization on large angular scales [3] has opened a new window to test and constrain models that predict primordial perturbations. The leading candidate, as claimed by the BICEP2 team, is primordial inflationary gravitational waves. For a tensor-to-scalar ratio r of around 0.2, these give a good match to the spectral shape in the region $\ell \approx [40\ 150]$.

An alternative mechanism of generating primordial B-modes is the presence of cosmic defects. Even though their relative contribution to the temperature power spectrum is expected to be sub-dominant, they can still contribute importantly to the B-mode polarization. We explore whether cosmic defects could explain or help other primary contributors fit the data better.

Recently some works have reported that the measurements made by BICEP2 can have a non-negligible astrophysical source: polarized dust foregrounds [4, 5, 6]. We extended our primary analysis including such a possible source.

2. Cosmic defects and defect zoo

Cosmic defects are extended objects that could be produced during cosmological phase transitions at the earliest stages of our universe, when spontaneous symmetry breaking occurred. Defects appear as a consequence of the development of a non trivial vacuum configuration.

Depending on the symmetry of the system, as well as on the vacuum manifold configuration, the breaking process leads to different defect networks. Among the wide range of defects analyzed in the literature, we have focused on the following types, which can be considered as the representative ones:

(i) Abelian Higgs strings (AH strings): consequence of a broken $U(1)_L$ local symmetry.
(ii) Textures (TX): products of the breaking of a global $O(4)_G$ symmetry.
(iii) Semi-Local strings (SL strings): produced after a $SU(2)_G \times U(1)_L$ symmetry is broken. In some sense, they are a mixture of local and global defects.

Cosmic defects are predicted by many inflationary scenarios and their cosmological implications have been widely analyzed. Cosmic defects not only generate CMB anisotropies, but they are also candidates for the generation of other phenomena such as gravitational waves or lensing. Their detection, therefore, would give invaluable information about the physics of the early universe.

2.1. CMB anisotropies

The possible contribution of defects to CMB anisotropies has been one of their most studied observable imprints, both in temperature and polarization channels. Their contribution has been highly constrained via the temperature anisotropies [7, 8, 9].

In contrast to ordinary inflationary perturbations, which are set by the initial perturbations, scaling of defects implies that they continuously induce perturbations to the background metric, *i.e.* they are active sources of perturbations.

One of the most important differences between primordial and defect induced anisotropies arises in the vector perturbations. Inflationary vector modes decay due to the cosmological expansion and do not create B-modes. On the contrary, defect vector perturbations are seeded continuously and consequently do not decay. The contribution of defects to B-modes, therefore, is twofold: through tensor *and* vector perturbations; and both ingredients contribute in a comparable amount. Scalar perturbations, in turn, do not induce directly B-modes, they only contribute through lensing of E-modes. This is the reason why it is plausible that the relative contribution of inflation and defects to B-modes can be similar in amplitude.

All defect networks produce CMB anisotropies in a very similar way, and even though the shape of the power spectra is very similar, there are some differences that must be pointed out. Beyond the shape and peak position, those differences come from the constraints imposed by CMB experiments. In opposition to the inflationary case, where scalars and tensors can vary almost independently (apart from the inflationary consistency relation), the amplitudes of temperature and polarization power spectra from defects are strictly related. Thus, though no experiment had constrained defects in the BB channel until BICEP2 released their data, the possible defect contribution to it has already been limited by the temperature constraints.

In Fig 1 we show spectra obtained using field theory simulations [10, 11]. There it can be seen how the constraints derived in the temperature channel affect the possible contribution of each type of defect in B-modes. The most dramatic effect is suffered by AH cosmic strings; their contribution is so suppressed by temperature that it is highly unlikely that they could be a primary source of B-modes. On the other hand, limits imposed on textures and SL strings are not so drastic.

Usually the possible defect contribution is encoded in two related parameters. The first one is $G\mu$, where G is Newton's constant and μ is the string tension. It encapsulates the most general

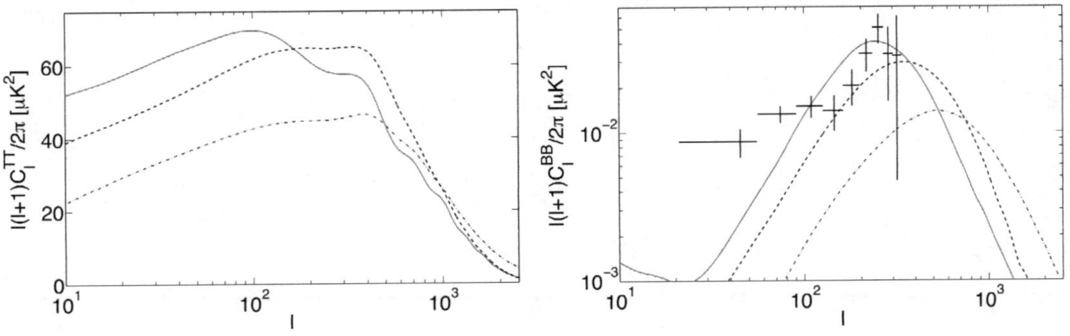

Figure 1. Defect spectra normalised to the 95% upper limits obtained using *Planck* + WMAP (EE and TE) + High-ℓ (SPT+ACT) data [8]. Different lines correspond to textures (solid red line), SL strings (dashed black line), and AH strings (dot-dashed blue line). Normalizations correspond to the values given in the leftmost columns of Table 3.

Dataset	BICEP2 (only *BB*)						
Model (ΛCDM+)	Defects			Defects + r			r
Param	AH	SL	TX	AH	SL	TX	-
r	-	-	-	$0.14^{+0.04}_{-0.06}$	$0.14^{+0.04}_{-0.06}$	$0.14^{+0.04}_{-0.06}$	$0.21^{+0.04}_{-0.05}$
$10^{12}(G\mu)^2$	$0.40^{+0.07}_{-0.08}$	$1.73^{+0.29}_{-0.32}$	$0.86^{+0.14}_{-0.16}$	$0.20^{+0.08}_{-0.09}$	$0.87^{+0.34}_{-0.39}$	$0.43^{+0.17}_{-0.20}$	-
$-\ln\mathcal{L}_{\max}$	8.1	7.4	6.8	1.6	1.6	1.8	4.3

Table 1. Parameter estimations and best-fit likelihood values for various cosmological models, fitting for the BICEP2 data. Only the B-mode is used for these estimations.

properties of the defect network, since μ is strictly linked to the symmetry breaking energy scale[1]. The second is f_{10}, is used only in CMB contexts and measures the relative contribution of defects at multipole $\ell = 10$ in the temperature power spectrum. Although there is no direct analytic correspondence between these two parameters, roughly $f_{10} \propto (G\mu)^2$ for $f_{10} \ll 1$.

3. Fitting BICEP2 data using defects and inflationary tensor modes

The main question faced by this work is whether defects can account for the whole signal measured by the BICEP2 experiment. In the left panel of Fig. 2 we have included BICEP2 points as well as the best-fit gravitational wave spectrum (solid black line) and the possible contributions of AH cosmic strings considering different normalizations. This figure shows qualitatively that a model containing only defects give a poor fit to the data points. If one tries to fit high-ℓ points, the points at low multipoles are underestimated; and vice versa, fitting low multipoles implies overestimating the signal at high-ℓs. The analysis of textures and SL strings leads to similar conclusions.

We promoted our basic and qualitative statements into quantitative results via a Monte Carlo analysis. A summary of the analysis can be found in Table 1. It can be clearly seen that none of the defects is able to give a comparable fit to the one provided by inflationary tensor modes.

[1] For textures the string tension parameter does not make much sense. In that case we use $G\mu = \pi\eta^2$, where η is the vacuum expectation value, and they are usually parametrized by $\epsilon = 4\pi G\eta^2$

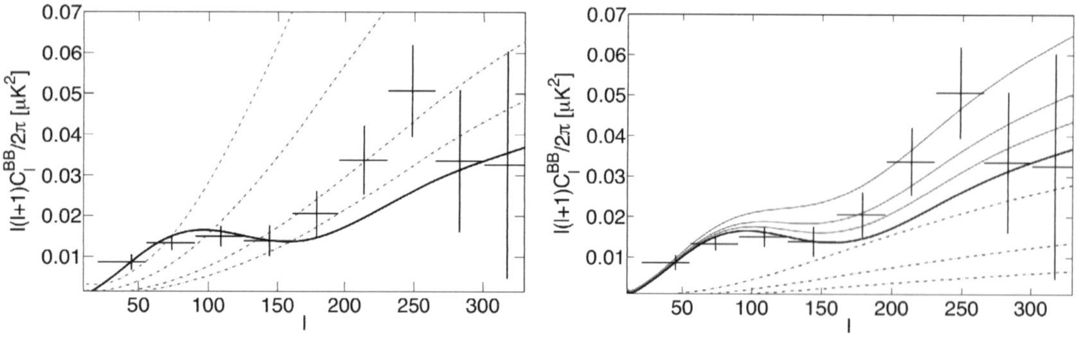

Figure 2. Comparison of different B-mode spectra and BICEP2 *BB* data. The black curve in both panels represents the best-fit gravitational wave case. In the left panel green-dotted curves show AH string contribution at different normalizations ($f_{10} = 0.3$, 0.15, 0.06, and 0.03). In the right panel, in turn, we add a contribution from strings (blue dashed) to the inflationary tensor prediction (total spectra in grey). From bottom to top the string fractions are 0.015, 0.03, 0.04 (highlighted in red), and 0.06. All spectra contain lensing contribution.

Having discarded the role of cosmic defects as primary source of B-modes, at this point we change the perspective and explore how they could assist, as a secondary player, the primary contribution coming from inflationary tensor modes. Defects peak at smaller scales, *i.e.* contribute more significantly at higher multipoles than gravitational waves (see right panel of Fig. 1), hence a mixture of both ingredients could improve the fitting. Results can be found in Table 1, where the maximum likelihoods show how the fit is improved, although it should be noted that this model takes into account two extra contributions.

4. ... and dust

It has become increasingly apparent that the measurements of BICEP2 might have a non-negligible contribution from polarized dust foregrounds [4, 5, 6]. Such an astrophysical effect could have been underestimated in the region where BICEP2 made its measurements and would acquire more importance than previously expected.

Following the recipe given in those works and assuming a power-law profile for the dust power spectrum, we extend our previous analysis including dust into the statistical analysis. The dust spectrum has been characterized in the following way:

$$C_\ell^{BB,\mathrm{dust}} = A_{\mathrm{dust}}\ell^{-2.3} \tag{1}$$

where A_{dust} is used as the normalization parameter that controls the dust contribution[2,3].

Dust foreground B-mode spectrum, in contrast to the defect case, has more importance at low-ℓs (see Fig. 3). Therefore, dust apparently is more in competition with gravitational waves than with defects. The results are shown in Table 2. The first thing we notice is that dust alone

[2] Another parametrisation of dust is used in the literature, given by Δ^2_{BB}, which is related to ours via

$$\Delta^2_{BB,\mathrm{dust},\ell} = \frac{\ell^2}{2\pi}C_\ell = \frac{A_{\mathrm{dust}}}{2\pi}\ell^{-0.3}$$

[3] While this work was in preparation the Planck collaboration submitted a paper [6] where they updated the initial dust spectrum to $C_l^{BB} \propto \ell^{-2.45}$. We tested our results and they do not change significantly with this new power-law.

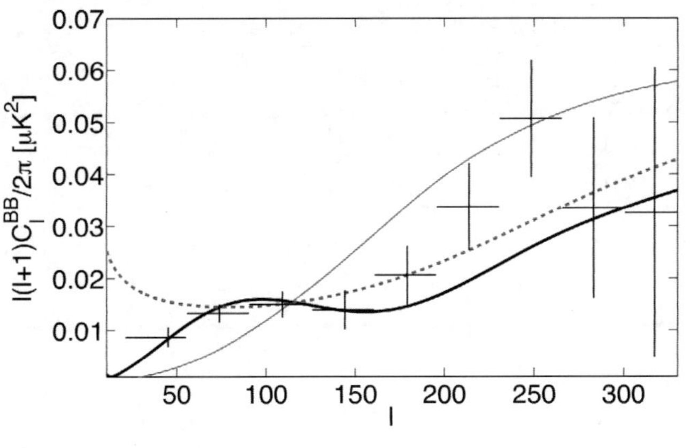

Figure 3. BICEP2 *BB* data points and B-mode spectra of tensors (thick solid black line), dust (thick dashed green line) and AH strings (thin solid red line) plus the corresponding lensing contribution. The spectra are obtained using values given in Tables 1 and 2.

does a very good job, and even more, it is able to improve the fit given by inflationary tensor modes, confirming what previous works suggested [4, 5].

The two rightmost columns of Table 2 show that a combination of gravitational waves and dust is *not* preferred over a model with dust alone. Moreover, analyzing the best-fit point, one could see that data prefers a model with $r \approx 0$. As happened in the previous analysis, defect inclusion improves the fitting, since they assist at higher multipoles.

Dataset	BICEP2 (only *BB*)				
Model (ΛCDM+)	Defects + dust			r + dust	dust
Param	AH	SL	TX	-	-
r	-	-	-	< 0.22	-
$10^{12}(G\mu)^2$	$0.17^{+0.08}_{-0.10}$	$0.74^{+0.40}_{-0.40}$	$0.37^{+0.16}_{-0.24}$	-	-
$A_{\rm dust}$ [μK^2]	$0.20^{+0.06}_{-0.08}$	$0.20^{+0.06}_{-0.08}$	$0.19^{+0.06}_{-0.09}$	$0.19^{+0.10}_{-0.10}$	$0.30^{+0.06}_{-0.07}$
$-\ln \mathcal{L}_{\rm max}$	1.7	1.7	1.8	3.3	3.3

Table 2. Parameter estimations and best-fit likelihood values for different cosmological models, fitting for the BICEP2 data. This is similar to Table 1, but in this case a dust model is included.

5. New constraints

We observe that dust is the globally-preferred ingredient by the data [2]. That is the reason why we derive our latest constraints for defects using a model containing defects and dust contribution. In order to obtain the most accurate constraints, we consider a wider CMB dataset, which not only contains B-mode data by BICEP2, but also includes other channels: Planck (TT), WMAP (TE and EE) and high-ℓ SPT and ACT (TT).

In Table 3 a comparison of the current constraints and the constraints provided by the Planck collaboration is shown. Constraints of our work have been derived for a ΛCDM+defects+dust model and using the full-CMB dataset [2], whereas the Planck collaboration do it for a ΛCDM+defects using the same dataset as us except for the BICEP2 [8].

The inclusion of the B-modes measured by BICEP2 does not change drastically the defect constraints, though their inclusion has slightly tighten them.

	Planck col.		This work	
at < 95% C.L.	$G\mu$	f_{10}	$G\mu$	f_{10}
Abelian Higgs strings	3.2×10^{-7}	0.024	2.7×10^{-7}	0.019
Semilocal strings	11×10^{-7}	0.041	9.8×10^{-7}	0.031
Textures	11×10^{-7}	0.054	7.3×10^{-7}	0.026

Table 3. 95% confidence limits for Gμ and f_{10} derived by the Planck collaboration and this work. We use a combined model with dust and defects to obtain current constraints.

6. Conclusions

Having investigated the possibility of defects being responsible of the recently detected B-mode polarization by the BICEP2 collaboration, we find qualitative and quantitative evidence to assert that defects on their own are a poor fit to the signal. Hence, the need of an additional contribution (gravitational waves or polarized dust) is manifested.

Nevertheless, defects can help those primary contributions to lift the spectrum at high multipoles. The analysis shows that in both cases, when a defect contribution is added to gravitational waves or dust, the overall fit is improved.

On the other hand, dust contamination is in general preferred by the data over a gravitational wave source. This is the reason why we choose dust as the primary player in the defect constraint derivation. We calculate new constraints using a full CMB dataset (Planck, WP, High-ℓ and BICEP2). New constraints, in Table 3, are tighter than the ones provided by the Planck collaboration, evidencing the importance of current B-mode experiments in constraining defect models.

Acknowledgments

JL and JU acknowledge support from the Basque Government (IT-559-10), the University of the Basque Country UPV/EHU (EHUA 12/11), MINECO (FPA2012-34456) and Consolider Ingenio (CPAN CSD2007-00042 and EPI CSD2010-00064). DD and MK acknowledge financial support from the Swiss National Science Foundation. MH and ARL acknowledge support from the Science and Technology Facilities Council (grant numbers ST/J000477/1, ST/K006606/1, and ST/L000644/1).

References

[1] Lizarraga J, Urrestilla J, Daverio D, Hindmarsh M, Kunz M and Liddle A R 2014 *Phys.Rev.Lett.* **112** 171301 (*Preprint* 1403.4924)
[2] Lizarraga J, Urrestilla J, Daverio D, Hindmarsh M, Kunz M *et al.* 2014 *Phys.Rev.* **D90** 103504 (*Preprint* 1408.4126)
[3] Ade P *et al.* (BICEP2 Collaboration) 2014 *Phys. Rev. Lett.* **112**(24) 241101 (*Preprint* 1403.3985)
[4] Mortonson M J and Seljak U 2014 *JCAP* **1410** 035 (*Preprint* 1405.5857)
[5] Flauger R, Hill J C and Spergel D N 2014 *JCAP* **1408** 039 (*Preprint* 1405.7351)
[6] Adam R *et al.* (Planck Collaboration) 2014 (*Preprint* 1409.5738)
[7] Urrestilla J, Bevis N, Hindmarsh M and Kunz M 2011 *JCAP* **1112** 021 (*Preprint* 1108.2730)
[8] Ade P *et al.* (Planck Collaboration) 2014 *Astron.Astrophys.* **571** A16 (*Preprint* 1303.5076)
[9] Lizarraga J, Sendra I and Urrestilla J 2012 *Phys.Rev.* **D86** 123014 (*Preprint* 1207.6266)
[10] Bevis N, Hindmarsh M, Kunz M and Urrestilla J 2010 *Phys.Rev.* **D82** 065004 (*Preprint* 1005.2663)
[11] Urrestilla J, Bevis N, Hindmarsh M, Kunz M and Liddle A R 2008 *JCAP* **0807** 010 (*Preprint* 0711.1842)

Spanish Relativity Meeting (ERE 2014): almost 100 years after Einstein's revolution IOP Publishing
Journal of Physics: Conference Series **600** (2015) 012026 doi:10.1088/1742-6596/600/1/012026

Anisotropic Power-law Inflation: A counter example to the cosmic no-hair conjecture

Jiro Soda
Department of Physics, Kobe University, Kobe 657-8501, Japan
E-mail: jiro@phys.sci.kobe-u.ac.jp

Abstract. It is widely believed that anisotropy in the expansion of the universe will decay exponentially fast during inflation. This is often referred to as the cosmic no-hair conjecture. However, we find a counter example to the cosmic no-hair conjecture in the context of supergravity. As a demonstration, we present an exact anisotropic power-law inflationary solution which is an attractor in the phase space. We emphasize that anisotropic inflation is quite generic in the presence of anisotropic sources which couple with an inflaton.

1. Introduction

As is well known, the event horizon of black holes hides the initial conditions of the collapsed matter other than mass, charge, and angular momenta, which is named the black hole no-hair theorem [1]. The similar thing should happen for inflation because of the cosmological event horizon, namely, any initial conditions should go away beyond the cosmological event horizon. In fact, in the presence of the cosmological constant, there is a cosmic no-hair theorem proved by Wald [2]. Even for a general accelerating universe driven by a scalar field, it is legitimate to expect that the anisotropy decays exponentially first in the presence of the cosmological event horizon. This prejudice is often referred to as the cosmic no-hair conjecture.

Historically, there have been challenges to the cosmic no-hair conjecture [3]. Unfortunately, it turned out that these models suffer from either the instability, or a fine tuning problem, or a naturalness problem [4]. Recently, however, we have succeeded in finding stable anisotropic inflationary solutions in the context of supergravity. More precisely, we have shown that, in the presence of a gauge field coupled with an inflaton, there could be small anisotropy in the expansion rate which never decays during inflation. Since anisotropic inflation is an attractor, this can be regarded as a counter example to the cosmic no-hair conjecture [5, 6]. Moreover, primordial fluctuations created quantum mechanically during inflation also exhibit statistical anisotropy [7]. Indeed, from the point of view of precision cosmology, it is important to explore the role of gauge fields in inflation [8].

It is well known that the supergravity models can be constrained by comparing predictions of inflation with cosmological observations. For example, the tilt of the spectrum gives interesting information of the superpotential $W(\phi^i)$ and the Kaler potential $K(\phi^i, \bar{\phi}^i)$ which are functionals of complex scalar fields. Here, a bar represents a complex conjugate. More concretely, the bosonic part of the action of supergravity reads

$$S = \int d^4x\sqrt{-g}\left[\frac{M_p^2}{2}R + K_{i\bar{j}}\partial^\mu\phi^i\partial_\mu\bar{\phi}^{\bar{j}} - e^K K^{i\bar{j}}D_iWD_{\bar{j}}\bar{W} - f_{ab}(\phi)F^a_{\mu\nu}F^{b\mu\nu}\right] \tag{1}$$

where we have defined Kaler metric $K_{i\bar{j}} = \partial^2 K/\partial\phi^i\partial\bar{\phi}^{\bar{j}}$ and a covariant derivative $D_iW = \partial W/\partial\phi^i + \partial K/\partial\phi^i W$. Here, M_p represents the reduced Planck mass. We note that the gauge kinetic function $f_{ab}(\phi)$ in front of the gauge kinetic term $F^a_{\mu\nu}F^{b\mu\nu}$ could be nontrivial functions of the scalar fields. Curiously, so far, the gauge field part has been neglected in most of discussion of inflation. The reason is partially due to the cosmic no-hair conjecture which states that the anisotropy, curvature, and any matter will be diluted once inflation commences. However, as we will see soon, this is not true when we look at percent level fine structures of inflationary scenarios. We find that gauge fields play an important role in the early universe. In fact, we will show that there exists anisotropic inflation and the degree of the anisotropy we found is smaller than the slow roll parameter. Apparently, the expansion is almost isotropic. Nevertheless, anisotropic inflation provides a qualitatively new picture of the universe.

In this paper, as a demonstration, we consider the cases where the functional form of the potential and the gauge kinetic functions has an exponential dependence on the inflaton. In particular, we present exact anisotropic inflationary solutions. Moreover, we argue anisotropic inflation becomes an attractor in the phase space.

The organization of the paper is as follows. In section 2, we introduce inflationary models where the gauge field couples with an inflaton and obtain exact solutions which contain anisotropic power-law inflationary solutions. In section 3, we show that the anisotropic inflation is an attractor in the phase space. Thus, generic trajectories converge to anisotropic inflation. This implies that the cosmic no-hair conjecture does not hold in general. The final section is devoted to the conclusion.

2. Exact Anisotropic Power-law Inflationary Solutions

In this section, we consider a simple model with exponential potential and gauge kinetic functions and then find exact power-law inflationary solutions. In addition to a well known isotropic power-law solution, we find an anisotropic power-law inflationary solution.

We consider the following action for the metric $g_{\mu\nu}$, the inflaton field ϕ and the gauge field A_μ coupled with ϕ:

$$S = \int d^4x\sqrt{-g}\left[\frac{M_p^2}{2}R - \frac{1}{2}\left(\partial_\mu\phi\right)\left(\partial^\mu\phi\right) - V(\phi) - \frac{1}{4}f^2(\phi)F_{\mu\nu}F^{\mu\nu}\right], \tag{2}$$

where g is the determinant of the metric, R is the Ricci scalar, respectively. The field strength of the gauge field is defined by $F_{\mu\nu} = \partial_\mu A_\nu - \partial_\nu A_\mu$. Motivated by the dimensional reduction of higher dimensional theory such as string theory, we assume the exponential potential and the exponential gauge kinetic functions

$$V(\phi) = V_0 e^{\lambda\frac{\phi}{M_p}}, \qquad f(\phi) = f_0 e^{\rho\frac{\phi}{M_p}}. \tag{3}$$

In principle, the parameters V_0, f_0, λ, and ρ can be determined once the compactification scheme is specified. However, we leave those free hereafter.

Using the gauge invariance, we can choose the gauge $A_0 = 0$. Without loosing the generality, we can take the x-axis in the direction of the gauge field. Hence, we have the homogeneous fields of the form $A_\mu = (\ 0,\ v(t),\ 0,\ 0\ )$ and $\phi = \phi(t)$. As there exists the rotational symmetry in the y-z plane, we take the metric to be

$$ds^2 = -dt^2 + e^{2\alpha(t)}\left[\ e^{-4\sigma(t)}dx^2 + e^{2\sigma(t)}\left(dy^2 + dz^2\right)\ \right], \tag{4}$$

where the cosmic time t is used. Here, e^{α} is an isotropic scale factor and σ represents a deviation from the isotropy. It is easy to solve the equation for the gauge field as

$$\dot{v} = f^{-2}(\phi)e^{-\alpha-4\sigma}p_A, \tag{5}$$

where p_A denotes a constant of integration. Using Eq.(5), we obtain equations

$$\dot{\alpha}^2 = \dot{\sigma}^2 + \frac{1}{3M_p^2}\left[\frac{1}{2}\dot{\phi}^2 + V(\phi) + \frac{p_A^2}{2}f^{-2}(\phi)e^{-4\alpha-4\sigma}\right], \tag{6}$$

$$\ddot{\alpha} = -3\dot{\alpha}^2 + \frac{1}{M_p^2}V(\phi) + \frac{p_A^2}{6M_p^2}f^{-2}(\phi)e^{-4\alpha-4\sigma}, \tag{7}$$

$$\ddot{\sigma} = -3\dot{\alpha}\dot{\sigma} + \frac{p_A^2}{3M_p^2}f^{-2}(\phi)e^{-4\alpha-4\sigma}, \tag{8}$$

$$\ddot{\phi} = -3\dot{\alpha}\dot{\phi} - V'(\phi) + p_A^2 f^{-3}(\phi)f'(\phi)e^{-4\alpha-4\sigma}. \tag{9}$$

Here, an overdot and a prime denote the derivative with respect to the cosmic time t and ϕ, respectively.

In the absence of gauge fields, it is known that there exists the power-law inflationary solution. Therefore, let us first seek the power-law solutions by assuming

$$\alpha = \zeta \log t, \qquad \sigma = \eta \log t, \qquad \frac{\phi}{M_p} = \xi \log t + \phi_0. \tag{10}$$

Apparently, for a trivial gauge field $p_A = 0$, we have the isotropic power-law solution

$$\zeta = \frac{2}{\lambda^2}, \qquad \eta = 0, \qquad \xi = -\frac{2}{\lambda}, \qquad \frac{V_0}{M_p^2}e^{\lambda\phi_0} = \frac{2(6-\lambda^2)}{\lambda^4}. \tag{11}$$

In this case, we have the spacetime

$$ds^2 = -dt^2 + t^{4/\lambda^2}\left(dx^2 + dy^2 + dz^2\right). \tag{12}$$

Hence, we need $\lambda \ll 1$ for obtaining the accelerating expansion.

Next, interestingly, we see that there exists the other non-trivial exact solution in spite of the existence of the no-hair theorem [2]. From the hamiltonian constraint equation (6), we set two relations

$$\lambda\xi = -2, \qquad \rho\xi + 2\zeta + 2\eta = 1 \tag{13}$$

to have the same time dependence for each term. The latter relation is necessary only in the non-trivial gauge field case, $p_A \neq 0$. Then, for the amplitudes to balance, we need

$$-\zeta^2 + \eta^2 + \frac{1}{6}\xi^2 + \frac{1}{3}u + \frac{1}{6}w = 0, \tag{14}$$

where we have defined

$$u = \frac{V_0}{M_p^2}e^{\lambda\phi_0}, \qquad w = \frac{p_A^2}{M_p^2}f_0^{-2}e^{-2\rho\phi_0}. \tag{15}$$

The equation for the scale factor (7) under Eq. (13) yields

$$-\zeta + 3\zeta^2 - u - \frac{1}{6}w = 0. \tag{16}$$

Similarly, the equation for the anisotropy (8) gives

$$-\eta + 3\zeta\eta - \frac{1}{3}w = 0 \ . \tag{17}$$

Finally, from the equation for the scalar (9), we obtain

$$-\xi + 3\zeta\xi + \lambda u - \rho w = 0 \ . \tag{18}$$

Using Eqs. (13), (16) and (17), we can solve u and w as

$$u = \frac{9}{2}\zeta^2 - \frac{9}{4}\zeta - \frac{3\rho}{2\lambda}\zeta + \frac{1}{4} + \frac{\rho}{2\lambda} \ , \qquad w = -9\zeta^2 + \frac{15}{2}\zeta + \frac{9\rho}{\lambda}\zeta - \frac{3}{2} - \frac{3\rho}{\lambda} \ . \tag{19}$$

Substituting these results into Eq. (18), we obtain

$$(3\zeta - 1)\left[6\lambda\left(\lambda + 2\rho\right)\zeta - \left(\lambda^2 + 8\rho\lambda + 12\rho^2 + 8\right)\right] = 0 \ . \tag{20}$$

In the case of $\zeta = 1/3$, we have $u = w = 0$. Hence, it is not our desired solution. Thus, we have to choose

$$\zeta = \frac{\lambda^2 + 8\rho\lambda + 12\rho^2 + 8}{6\lambda(\lambda + 2\rho)} \ . \tag{21}$$

Substituting this result into Eq. (16), we obtain

$$\eta = \frac{\lambda^2 + 2\rho\lambda - 4}{3\lambda(\lambda + 2\rho)} \ . \tag{22}$$

From Eq. (13), we have

$$\xi = -\frac{2}{\lambda} \ . \tag{23}$$

Finally, Eq. (19) reduce to

$$u = \frac{(\rho\lambda + 2\rho^2 + 2)(-\lambda^2 + 4\rho\lambda + 12\rho^2 + 8)}{2\lambda^2(\lambda + 2\rho)^2} \ , \quad w = \frac{(\lambda^2 + 2\rho\lambda - 4)(-\lambda^2 + 4\rho\lambda + 12\rho^2 + 8)}{2\lambda^2(\lambda + 2\rho)^2} \ . \tag{24}$$

Note that Eq. (14) is automatically satisfied.

In order to have inflation, we need $\lambda \ll 1$. For these cases, u is always positive. Since w should be also positive, we have the condition $\lambda^2 + 2\rho\lambda > 4$. Hence, ρ must be much larger than one. Now, the spacetime reads

$$ds^2 = -dt^2 + t^{2\zeta - 4\eta}dx^2 + t^{2\zeta + 2\eta}\left(dy^2 + dz^2\right) \ . \tag{25}$$

This describes the anisotropically inflating universe. The average expansion rate is determined by ζ and the average slow roll parameter is given by

$$\epsilon \equiv -\frac{\dot{H}}{H^2} = \frac{6\lambda(\lambda + 2\rho)}{\lambda^2 + 8\rho\lambda + 12\rho^2 + 8} \ , \tag{26}$$

where we have defined $H = \dot{\alpha}$. In the limit $\lambda \ll 1$ and $\rho \gg 1$, this reduces to $\epsilon = \lambda/\rho$. Now, the anisotropy is characterized by the ratio between the anisotropic expansion rate and the isotropic expansion rate

$$\frac{\Sigma}{H} \equiv \frac{\dot{\sigma}}{\dot{\alpha}} = \frac{2(\lambda^2 + 2\rho\lambda - 4)}{\lambda^2 + 8\rho\lambda + 12\rho^2 + 8} = \frac{1}{3}I\epsilon \ , \qquad I = \frac{\lambda^2 + 2\rho\lambda - 4}{\lambda^2 + 2\rho\lambda} \ . \ . \tag{27}$$

Remarkably, the anisotropy is proportional to the slow roll parameter and the parameter I is smaller than 1. Hence, typically, the anisotropy is quite small. Although the anisotropy is always small, it persists during inflation. Therefore, its effect on the cosmological observables is not negligible.

3. Anisotropic Inflation: A counter example to the cosmic non-hair conjecture

Now, we know two exact solutions, isotropic and anisotropic power-law inflationary solutions. These solutions are not general solutions but particular solutions. In order to eamine if the cosmic no-hair conjecture holds or not, we need to know the fate of generic solutions. To this end, the dynamical analysis is useful. Since two exact solutions correspond to two fixed points in the phase space, we did the stability analysis around the fixed points [5]. It turns out that the isotropic fixed point indicated by orange circle in Fig.1 is a saddle point which is an attractor only on the two-dimensional $Z = 0$ plane. Thus, in the absence of gauge fileds, isotropic inflation is an attractor. Namely, the cosmic no-hair holds in this special case. On the other hand, we found that the anisotropic fixed point is stable. Thus, the end point of trajectories around the unstable isotropic power-law inflation must be anisotropic power-law inflation. In Fig.1, we depicted the phase flow in the phase space for $\lambda = 0.1,\ \rho = 50$. We see that the trajectories converge to the anisotropic fixed point indicated by yellow circle.

When the initial conditions are close to the $Z = 0$ plane, the trajectory approaches to the saddle point, but eventually it goes into the attractor. Interestingly, even if we start from isotropic universe, the final state of the universe is anisotropic (green line). This can be regarded as the spontaneous breakdown of the rotational symmetry.

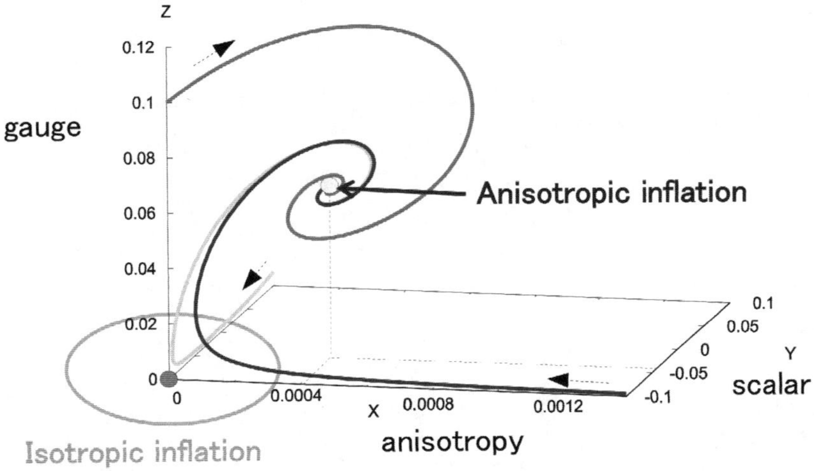

Figure 1. The phase flow in the phase space is depicted for $\lambda = 0.1, \rho = 50$. The axis Z represents the energy density of the gauge field, the axis X describes the anisotropy of the universe, and the axis Y denotes the velocity of the inflaton. The yellow and orange circles indicate the anisotropic and isotropic fixed points respectively. The trajectories converge to the anisotropic fixed point.

As we have shown, clearly, anisotropic inflation gives rise to a counter example to the cosmic no-hair conjecture. This feature is not specific to the exponential potential. Although, in general cases, we do not have exact solutions, qualitative feature is almost the same. Indeed, there are many concrete examples which violate the cosmic no-hair conjecture [9].

We should note that the cosmological constant is assumed in the cosmic no-hair theorem proved by Wald [2]. In the case of inflation, the inflaton can mimic the cosmological constant. Hence, the comic no-hair holds in the conventional cases. However, in the presence of a non-trivial coupling between the inflaton and the gauge field, the cosmic no-hair theorem cannot be applicable anymore [10].

4. Conclusion

We have examined a supergravity model with a gauge kinetic function. It turned out that exact anisotropic power-law inflationary solutions exist when both the potential function for an inflaton and the gauge kinetic function are exponential type. We showed that the degree of the anisotropy is proportional to the slow roll parameter. The slow roll parameter depends both on the potential function and the gauge kinetic function. The dynamical system analysis told us that the anisotropic power-law inflation is an attractor. Therefore, the result we have mentioned in this paper presents a clear counter example to the cosmic no-hair conjecture.

The imprint of anisotropic inflation can be found in the cosmic microwave background radiation [7]. We should note that the statistical anisotropy could be large even when the anisotropy in the expansion is quite small. The result in this paper also has implication in cosmological magnetic fields [6].

4.1. Acknowledgments

I would like to thank Sugumi Kanno and Masa-aki Watanabe for collaborations. This work was supported in part by the Grants-in-Aid for Scientific Research (C) No.25400251 and Grants-in-Aid for Scientific Research on Innovative Areas No.26104708.

References

[1] W. Israel, Phys. Rev. **164**, 1776 (1967); B. Carter, Phys. Rev. Lett. **26**, 331 (1971).
[2] R. M. Wald, Phys. Rev. D **28**, 2118 (1983).
[3] L. H. Ford, Phys. Rev. D **40**, 967 (1989); N. Kaloper, Phys. Rev. D **44**, 2380 (1991); S. Kawai and J. Soda, Phys. Rev. D **59**, 063506 (1999); J. D. Barrow and S. Hervik, Phys. Rev. D **73**, 023007 (2006); L. Ackerman, S. M. Carroll and M. B. Wise, Phys. Rev. D **75**, 083502 (2007) [Erratum-ibid. D **80**, 069901 (2009)]; A. Golovnev, V. Mukhanov and V. Vanchurin, JCAP **0806**, 009 (2008); S. Kanno, M. Kimura, J. Soda and S. Yokoyama, JCAP **0808**, 034 (2008).
[4] B. Himmetoglu, C. R. Contaldi and M. Peloso, Phys. Rev. Lett. **102**, 111301 (2009); A. Golovnev, Phys. Rev. D **81**, 023514 (2010); G. Esposito-Farese, C. Pitrou and J. P. Uzan, Phys. Rev. D **81**, 063519 (2010).
[5] M. a. Watanabe, S. Kanno and J. Soda, Phys. Rev. Lett. **102**, 191302 (2009).
[6] S. Kanno, J. Soda and M. a. Watanabe, JCAP **1012**, 024 (2010).
[7] A. E. Gumrukcuoglu, B. Himmetoglu and M. Peloso, Phys. Rev. D **81**, 063528 (2010); T. R. Dulaney and M. I. Gresham, Phys. Rev. D **81**, 103532 (2010); M. a. Watanabe, S. Kanno and J. Soda, Prog. Theor. Phys. **123**, 1041 (2010); J. Ohashi, J. Soda and S. Tsujikawa, JCAP **1312**, 009 (2013); S. Kanno, J. Soda and M. a. Watanabe, JCAP **0912**, 009 (2009); M. a. Watanabe, S. Kanno and J. Soda, Mon. Not. Roy. Astron. Soc. **412**, L83 (2011); S. Hervik, D. F. Mota and M. Thorsrud, JHEP **1111**, 146 (2011); M. Thorsrud, D. F. Mota and S. Hervik, JHEP **1210**, 066 (2012); R. Emami and H. Firouzjahi, JCAP **1310**, 041 (2013); X. Chen, R. Emami, H. Firouzjahi and Y. Wang, JCAP **1408**, 027 (2014).
[8] J. Soda, Class. Quant. Grav. **29**, 083001 (2012); A. Maleknejad, M. M. Sheikh-Jabbari and J. Soda, Phys. Rept. **528**, 161 (2013).
[9] P. V. Moniz and J. Ward, Class. Quant. Grav. **27**, 235009 (2010); K. Murata and J. Soda, JCAP **1106**, 037 (2011); R. Emami, H. Firouzjahi, S. M. Sadegh Movahed and M. Zarei, JCAP **1102**, 005 (2011); K. Yamamoto, M. a. Watanabe and J. Soda, Class. Quant. Grav. **29**, 145008 (2012); N. Bartolo, S. Matarrese, M. Peloso and A. Ricciardone, JCAP **1308**, 022 (2013); T. Q. Do and W. F. Kao, Phys. Rev. D **84**, 123009 (2011); T. Q. Do, W. F. Kao and I. -C. Lin, Phys. Rev. D **83**, 123002 (2011); S. Bhowmick and S. Mukherji, Mod. Phys. Lett. A **27**, 1250009 (2012); J. Ohashi, J. Soda and S. Tsujikawa, Phys. Rev. D **87**, no. 8, 083520 (2013); J. Ohashi, J. Soda and S. Tsujikawa, Phys. Rev. D **88**, 103517 (2013); R. Emami, H. Firouzjahi and M. Zarei, Phys. Rev. D **90**, 023504 (2014).
[10] A. Maleknejad and M. M. Sheikh-Jabbari, Phys. Rev. D **85**, 123508 (2012).

Spanish Relativity Meeting (ERE 2014): almost 100 years after Einstein's revolution IOP Publishing
Journal of Physics: Conference Series **600** (2015) 012027 doi:10.1088/1742-6596/600/1/012027

Improving on numerical simulations of nonlinear CMB anisotropies

Màrius Josep Fullana i Alfonso[1], Josep Vicent Arnau i Córdoba[2], Robert J. Thacker[3], Hugh M.P. Couchman[4] and Diego P. Sáez Milán[5]

[1] IMM, Universitat Politècnica València, Camí de Vera, s/n 46022 València
[2] DMA, Universitat de València, Dr. Moliner 50, 46100 Burjassot, Spain
[3] DAP, St. Mary's University, Halifax, Nova Scotia, B3H 3C3 Canada
[4] DPA, McMaster University, 1280 Main St. West, Hamilton, Ontario, L8S 4M1 Canada
[5] DAA, Universitat de València, Dr. Moliner 50, 46100 Burjassot, Spain

E-mail: mfullana@mat.upv.es

Abstract. An Adaptative-Particle-Particle-Particle-Mesh code (HYDRA) plus a ray-tracing procedure was used in [1] to perform an exhaustive analysis of the weak lensing anisotropy. Other nonlinear Cosmic Microwave Background anisotropies, such as the Rees-Sciama and the Sunyaev-Zel.dovich effects are also being studied by using the same tools. Here we present some advances in our study of these nonlinear anisotropies. The primary advance is due to the use of better simulations with greater particle densities and appropriate softening, although other parameters have also been adjusted to get better estimates. Thus, we improve on a previous paper [2] where the Rees-Sciama effect was studied with Particle-Mesh simulations. We focus particular attention on the resolution improvement and its consequences.

1. Introduction

For almost a decade, our research team has been working on the computation of Cosmic Microwave Background (CMB) anisotropies using N-body numerical codes (in references [1] to [10], the reader may find a detailed description of the work we have done). The CMB anisotropies computed are those contributing to the CMB angular power spectrum at high ℓ. These anisotropies are mainly produced by nonlinear inhomogeneities placed at large enough distances from the observer; so, these inhomogeneities subtend small angular scales and contribute to large ℓ multipoles. We deal with the anisotropies due to weak lensing (WL), Rees-Sciama (RS) and Sunyaev-Zel'dovich (SZ).

Our initial studies used Particle-Mesh (PM) codes which have low resolution (see [2] to [6]). The resulting calculations were improved by using Adaptative-Particle-Particle-Particle-Mesh (AP3M) codes with more resolution (see [1] and [7] to [10]). We first used the sequential AP3M code HYDRA [7] in dark matter only mode and computed RS and WL CMB anisotropies. Afterwards, a Hydra AP3M parallel code was used to do the same estimates with more resolution and greater boxes [1]. Now, we are moving CMB photons along the simulation boxes of a Hydra AP3M parallel code with baryons [9, 10]. As might be expected, for similar boxes and resolution, the RS and WL CMB anisotropy estimations with baryons give almost the same results as those obtained without baryons [9]. The calculation of the SZ effect is also being performed [10]. Our underlying goal is to compute the RS, WL and SZ effects at the same time, to study the coupling of these three contributions to the CMB anisotropy will be analyzed. The peculiar gravitational potential, its gradients, the electron number density, and other necessary quantities are calculated and used at every time step of the Hydra simulation evolution. The main scope of this paper is to show how the results of some computations improve as the resolution is increased.

2. Mapping

Here we describe the quantities that are computed at every step of the code evolution to calculate CMB anisotropies.

For WL, small unlensed maps of CMB temperature contrasts, $\Delta T/T$, must be constructed to be subsequently deformed by lensing. In order to deform the unlensed maps, the lens deviations corresponding to a set of directions, covering an appropriate region of the sky, must be calculated. These deviations are the quantities:

$$\vec{\delta} = -2 \int_{\lambda_e}^{\lambda_0} W(\lambda) \vec{\nabla}_{\perp} \phi \, d\lambda \ , \tag{1}$$

where $\vec{\nabla}_{\perp}\phi = -\vec{n} \wedge \vec{n} \wedge \vec{\nabla}\phi$ is the transverse gradient of the peculiar gravitational potential, and $W(\lambda) = (\lambda_e - \lambda)/\lambda_e$. The variable λ is

$$\lambda(a) = H_0^{-1} \int_a^1 \frac{db}{(\Omega_{m0} b + \Omega_\Lambda b^4)^{1/2}} \ . \tag{2}$$

H_0, Ω_m, and Ω_Λ being the Hubble constant, and the density parameters of matter and vacuum, respectively. By using these formulas, the temperature contrasts $\Delta T/T$, and the C_ℓ coefficients due to WL may be computed (see [1] for a detailed description of the method).

The RS contribution is:

$$\frac{\Delta T}{T}(\vec{n}) = 2 \int_{\lambda_e}^{\lambda_0} W(\lambda) \, \frac{\partial \phi}{\partial \lambda} \, d\lambda \ , \tag{3}$$

where ϕ is the peculiar gravitational potential, $W(\lambda) = (\lambda_e - \lambda)/\lambda_e$, and λ is given in Eq. (2). (See [2] for details).

The temperature contrast of the SZ thermal effect is:

$$\frac{\Delta T}{T}(\vec{n}) = -2 \, \frac{\sigma_T}{m_e c^2} \int_{l_i}^{l_0} n_e \, k \, (T_e - T_{CMB}) \, dl \ , \tag{4}$$

Finally, for SZ kinematic contribution one has:

$$\frac{\Delta T}{T}(\vec{n}) = - \, \frac{\sigma_T}{c} \int_{l_i}^{l_0} n_e \, v_r \, dl \ , \tag{5}$$

where the subscript e refers to electrons. (See [10] for more details).

3. Algorithm to compute the integrals

The steps we follow to compute the above integrals are now described. Various steps are related with the implementation of our ray-tracing procedure. For a detailed description of this procedure see for instance [4]. Main steps:

(i) Decide the preferred directions.

(ii) Assume the Born approximation, and use the photon step distance Δ_{ps} to determine all the evaluation positions and times on the geodesics. Evaluations are performed inside the simulation volume, from $z_i = 6$ down to the final redshift.

(iii) Associate test particles (without mass) to each of these positions and times.

(iv) For SZ, evaluate the radial component of the peculiar velocity, the number density, and the temperature of the electron gas, in the position and time corresponding to each test particle.

(v) In these positions and times (test particles), evaluate the potential and its gradient (force), using the long-range FFT component and short-range PP correction, as it is done in the HYDRA code.

(vi) During the FFT convolution for the test particles, eliminate contributions from scales larger than $42h^{-1}$ Mpc; namely, remove the signal from wavenumbers $k \leq 0.15h$ Mpc^{-1}. This procedure is required by our ray-tracing method.

(vii) If the evaluation time for a point on a null geodesic lies between two time-steps, calculate a linear interpolation of the physical quantities from the time-steps that straddle the correct time.

4. Hydra Simulations

All the simulations, have been performed in the framework of a concordance model with the following parameters: $h = 0.7$, $\Omega_b = 0.046$, $\Omega_d = 0.233$, $\Omega_\Lambda = 0.721$, $\tau = 0.084$ and $\sigma_8 = 0.817$. The power spectrum of the scalar (adiabatic) energy density perturbations, and the CMB angular power spectrum, necessary for simulations, were calculated using the latest generation of the CMBFAST code. No tensor modes have been considered.

The CMB photons move along specially chosen oblique paths through the simulation box; so, the usual traslations and rotations of boxes are not necessary to avoid periodic effects. In this way, the periodicity inside the box volume ensures that there are no discontinuities in the matter field where photons cross from box to box. We take care to ensure that, for the chosen paths, periodicity effects become negligible. In order to minimize these effects, the photons must cross consecutive boxes through statistically independent regions, which requires: (i) *preferred directions* leading to large enough distances between these regions and, (ii) a suitable cutting to avoid large scale spatial correlations ($k < 0.15h\ Mpc^{-1}$ or spatial scales $r_c > 60h^{-1}Mpc$) between the above distant regions. In [1], a detailed description of the parameters involved in the Hydra code may be found.

5. Results and Discussion

We have focused our attention on the resolution of our codes for CMB anisotropy estimations. Advances on simulations of structure formation have allowed us the design of better codes with more resolution, which led to more accurate results. Our predictions have been compared with previous ones and with observational data (see [1] and [8]). Here, we present some plans to improve on our estimates of the SZ and WL angular power spectra, and some new results relative to the RS effect.

5.1. SZ signal

As it was explained in [10], the computation of the thermal SZ (the largest effect at large enough ℓ) and the kinematic SZ signals could be done in two different ways: (1) the integrands in Eqs. (4) and (5) are computed, for the positions and times of the test particles, in the same way as it is done for real particles in the original Hydra code. This procedure would take a long CPU time, but it is the most coherent one and, (2) we could look for a lower CPU cost by using suitable approaches to estimate the integrands. This second alternative has not led to robust enough codes [10] for the approaches used up to now; nevertheless, new approaches might be designed. The coherent –long CPU time– method (1) is being implemented.

5.2. WL signal

A detailed study of the WL effect was published in [1], where simulations with different mass and space resolutions were considered; so, it was shown that results are better for greater resolutions. Low resolutions lead to Poisson discretization errors producing a wrong power excess at very small angular scales. Our best resolution was reached for simulations of 512^3 particles in boxes of $256h^{-1}$ Mpc. Thus, a signal higher than that obtained in previous estimates was obtained. Actually, for $4000 < \ell < 7000$, we found a signal of 2.0 ± 0.4 μK, which is $\sim 1.4\mu K$ higher than that found elsewhere (see [1] and references cited therein). Although our best resolutions seem to be good enough to avoid discretization errors, new simulations with higher resolutions are desirable to discuss if Poisson errors are really negligible against the high WL signal we estimated.

For resolutions greater than the best one in [1], the presence of baryons should begin to influence the predicted WL power, and simulation errors –in general– may depend on the code structure through various parameters and conditions, On account of these facts, we have tried to improve on our simulations by using a Hydra code with baryons. Two ways may be followed; in the first one, the same code –without baryons– as in [1] is used, but a better resolution is achieved by using 1024^3 (a power of two is required) particles in $256h^{-1}$ Mpc boxes. In the second way, a new Hydra code including baryons is run for 2×640^3 particles evolving in a $200h^{-1}$ Mpc box. Various technical problems must be solved before these simulations can be evolved to low enough redshifts. Further research is in progress with the essential aim of improving on the results of paper [1] and looking for the imprint of baryons in the WL spectra.

5.3. RS signal

Although the Integrated Sachs-Wolfe effect and the RS one may be both calculated by using Eq. (3), according to paper [2], these effects may be properly separated. The RS signal is very small. As can be seen in Fig. 1, this signal reaches a maximum smaller than 1 μK for an ℓ value of the order of 10^3. Similar results were also found in [2]. Although the RS signal is dominated either by primary CMB anisotropies or by the SZ effect, it might be detected, in future, from accurate CMB anisotropy measurements plus complete extended maps of nonlinear structures (clusters, superclusters, and so on). Detection could be achieved by measuring correlations between the distribution of these structures and the CMB anisotropy maps.

Some years ago, the RS angular power spectrum was estimated [2] by using our ray-tracing procedure and a Particle Mesh (PM) N-body code designed by V. Quilis and one of us (DPSM). The main features of this code can be found in [2]. It was adapted by us for CMB applications. Using 512^3 particles in boxes of $512h^{-1}$ Mpc, the spatial resolution of the PM code is only $\sim 350kpc$. In order to obtain better results a greater resolution would be necessary. Hydra codes can be adapted to include our ray tracing method; so, the RS effect may be simulated with much greater resolutions.

The PP part of the Hydra algorithm allows us to get good space and mass resolutions. Neither a too large number of particles nor too small simulation boxes are necessary. In the simulations performed to built up Fig. 1 (hereafter simulations of reference), 256^3 particles evolve inside boxes with a size of $512h^{-1}$ Mpc. The mass of the dark matter particles is 8.8×10^{11} $M_{\odot}$. Moreover, a Plumber softening $S_p = 6h^{-1}$ kpc has been chosen and, consequently, the spatial resolution is $\sim 5S_p = 30h^{-1}$ kpc. This resolution is around ten times greater than that of the PM code used in [2]. These large resolutions are achieved in spite of the fact that: (i) both codes use the same boxes and, (ii) the Hydra code evolves less particles (256^3) than the PM one (512^3). We use the same Hydra code without baryons as in [1], where simulations with the same characteristics led to a good estimate of the WL anisotropy. Finally, the integration step –also called photon step Δ_{ps}– used to perform the integral in Eq. (3) along the background null geodesics is $\Delta_{ps} = 15h^{-1}$ kpc. Moreover, the number of geodesics is 256^2 (a 4 factor less than

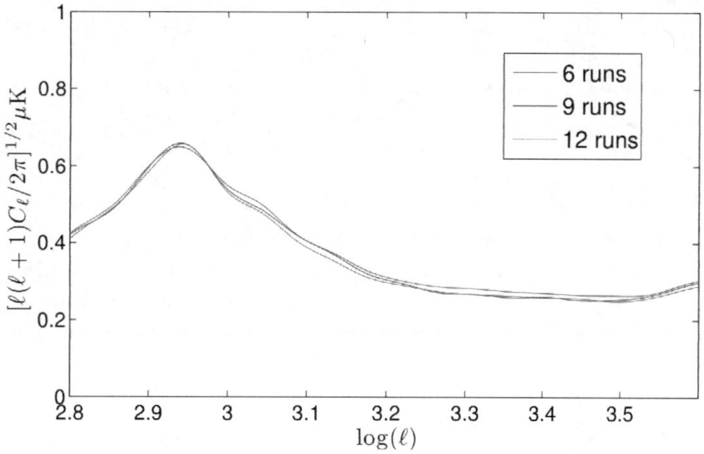

Figure 1. RS power spectrum, in μK, as a function of $\log(\ell)$. Here we plot the mean of 6, 9 and 12 runs estimates. For all cases initial redshift is $z_i = 6$. The number of particles is 256^3 and the box size $512h^{-1}Mpc$.

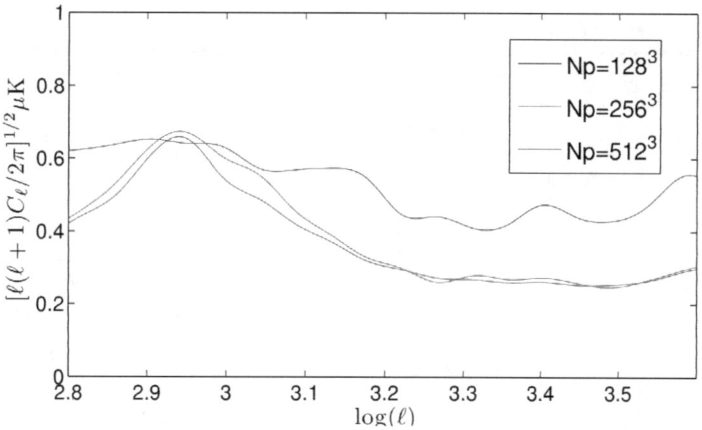

Figure 2. RS power spectrum, in μK, as a function of $\log(\ell)$. Blue, green, and red lines correspond to averages obtained with six simulations of 128^3, 256^3 and 512^3 particles, respectively. For all cases the initial redshift is $z_i = 6$ and the box size $512h^{-1}Mpc$.

that in [2]) and the angular resolution of the RS maps is $1.17'$.

From a unique simulation, it is not possible to obtain an accurate enough RS spectrum; in other words, a unique RS map is not a large enough statistical sample. A certain number of simulations must be run. In Fig. 1, the RS power spectra obtained by averaging over 6, 9 and 12 runs are plotted. The three spectra are very similar and, consequently, six runs are sufficient to get a good enough RS spectrum; however, about 30 runs were necessary in [1] –with a PM code– to get comparable results.

We have studied the impact of mass resolution by using simulations with 128^3, 256^3 and

512^3 particles in boxes with a size of $512h^{-1}$ Mpc. The remaining parameters are assumed to be identical to those of the reference simulations, excepting the masses of the dark matter particles, which are 7.0×10^{12}, and 1.1×10^{11} $M_\odot$ for 128^3 and 512^3 particles, respectively. In Fig. 2, the RS angular power spectra corresponding to the three above simulations are plotted. Evidently, the simulation with 128^3 particles does not have enough resolution, and the resulting signal does not appear adequately resolved. However, the reference simulations (256^3 particles) and those evolving 512^3 particles lead to very similar RS spectra, which we take as suggesting the signal is converging with resolution. The simulations with the best resolution (512^3 particles) give a sligthly higher peak. Since these two spectra have been obtained by averaging six simulations, this small power excess should be real, since according to our discussion about Fig. 1, it is evident that the mentioned excess should not sensibly change as the number of averaged solutions exceeds six.

Acknowledgments

This work has been supported by the Spanish Ministry of *Economía y Competitividad*, MICINN-FEDER project FIS2012-33582. Calculations have been made using the Lluís Vives Computer of Universitat de València. We are greatful to the working team of the Servei d'Informàtica de la Universitat de València for their useful help.

References

[1] Fullana M J, Arnau J V, Thacker R J, Couchman, H M P and Sáez D 2010 *ApJ* **712** 367
[2] Puchades N, Fullana M J, Arnau J V and Sáez D 2006 *MNRAS* **370** 1849
[3] Fullana M J and Sáez D 2006 *Proceedings of the Sixth International Symposium "Frontiers of Fundamental and Computational Physics" (Udine)* ed B G Sidharth et al (Springer, The Netherlands) p 115
[4] Sáez D, Puchades N, Fullana M J and Arnau J V 2006 *Proceedings of the International Conference "CMB and Physics of the Early Universe" (Ischia)* ed G de Zotti (PoS) p058
[5] Sáez D, Puchades N, Fullana M J and Arnau J V 2006 *A century of relativity physics, Proceedings o f the 28th Spanish Relativity Meeting (Oviedo)* ed L Mornas and J Díaz Alonso (AIP Conference Proceedings vol 841 Melville New York) p 594
[6] Fullana M J and Sáez D 2007 *Frontiers of Fundamental and Computational Physics (FFP8) (Madrid)* ed B G Sidhart, A Alfonso-Faus and M J Fullana i Alfonso (AIP Conference Proceedings vol 905 Melville New York) p 13
[7] Fullana M J, Arnau J V and Sáez D 2008 *Frontiers of Fundamental and Computational Physics: 9th International Symposium (Udine and Trieste)* ed B G Sidharth et al (AIP Conference Proceedings vol 1018 Melville New York) p 80
[8] Fullana M J, Arnau J V, Thacker R J, Couchman, H M P and Sáez D 2012 *Frontiers of Fundamental Physics: The Eleventh International Symposium (Paris)* ed J Kouneiher et al (AIP Conference Proceedings vol 1446 Melville New York) p 252
[9] Fullana M J, Arnau J V, Thacker R J, Couchman, H M P and Sáez D 2014 *Frontiers of Fundamental Physics and Physics Education Research 12th edition (Udine)* ed B G Sidharth et al (Springer Proceedings in Physics vol 145 Switzerland) p 188
[10] Fullana M J, Arnau J V, Thacker R J, Couchman, H M P and Sáez D 2014 *Progress in Mathematical Relativity, Gravitation and Cosmology, Proceedings of the Spanish Relativity Meeting ERE2012, (Guimarães)* ed A García-Parrado et al (Springer Proceedings in Mathematics and Statistics vol 60 Springer-Verlag Berlin Heidelberg) p 277

Cosmological constrains on fast transition Unified Dark Matter models

Ruth Lazkoz[1], **Iker Leanizbarrutia**[2], **Vincenzo Salzano**[3]

E-mail: [1]ruth.lazkoz@ehu.es, [2]iker.leanizbarrutia@ehu.com, [3]vincenzo.salzano@ehu.es
[1 2] Department of Theoretical Physics, University of the Basque Country UPV/EHU, P.O. Box 644, 48080 Bilbao, Spain
[3] Institute of Physics, University of Szczecin, Wielkopolska 15, 70-451 Szczecin, Poland

Abstract. In this work we show the results obtained applying a Unified Dark Matter (UDM) model with a fast transition to a set of cosmological data. Two different functions to model the transition are tested, and the feasibility of both models is explored using CMB shift data from Planck [1], Galaxy Clustering data from [2] and [3], and Union2.1 SNe Ia [4]. These new models are also statistically compared with the ΛCDM and quiessence models using Bayes factor through evidence. Bayesian inference does not discard the UDM models in favor of ΛCDM.

1. Introduction

In Unified Dark Matter (UDM) models the role of the dark energy and dark matter are played by a single fluid. The Equation of State $w = p/\rho$ changes through time, usually from a matter like era in the past ($w \simeq 0$) to a cosmological constant form ($w \simeq -1$) or to an effective dark energy content ($w < -1/3$). The best known example is the so-called Chaplygin gas [5][6].

Among the different UDM models, the ones with fast transitions are very interesting. In principle these models can be clearly distinguished from a standard ΛCDM without being ruled out by observational data. Whereas this is not the case for the generalized Chaplygin gas [7]. Thus, UDM models with fast transitions can provide an alternative explanation of the accelerated expansion of the universe [8], as they can fit the experimental data quite well, while they provide interesting and different new features. Besides, fast transition UDM models with scalar fields are also compatible with observational data [9].

A simple set-up parametrizing directly the energy density term [10], instead of the equation of state parameter w_{UDM} [8], is computationally lighter in likelihood calculations because most of its important variables required for the numerical calculations are expressed analytically (less integrals are involved). Here we present two parametrizations for this UDM set-ups with fast transition and constrain them using CMB, Galaxy Clustering and Supernovae Ia data. Then we discuss whether they are statistically favoured or not as compared to ΛCDM.

2. UDM Models

The background geometry considered for the phenomenological UDM models in this case is a flat Friedman-Lemaître-Robertson-Walker (FLRW) metric, $ds^2 = -dt^2 + a^2(t)\delta_{ij}dx_i dx_j$, where $a(t)$ is the scale factor as a function of the cosmic time t. We will consider perfect fluids with fractional densities Ω_i as sources; and taking $8\pi G = c = 1$, the Friedman equation takes the form,

$$E^2(a) = H^2/H_0^2 = \sum_i \Omega_i(a), \tag{1}$$

where $H = \dot{a}/a$ is the Hubble function with the dot denoting differentiation with respect to the cosmic time and H_0 being the Hubble constant today.

Being more specific about the sources, we need to clarify the role and form of our proposed UDM fluid: we want an UDM fluid which exhibit a fast transition from the dark matter stage to a scenario that resembles a ΛCDM. We propose the analytical form

$$\Omega_{UDM} = \Omega_t \left(\frac{a_t}{a}\right)^3 + \Omega_\Lambda \left[1 - \left(\frac{a_t}{a}\right)^3\right] \Theta(a - a_t), \tag{2}$$

being $\Theta(a - a_t)$ a Heaviside function and a_t the value of the scale factor at which the transition happens. We can easily see that for $a < a_t$ the fluid behaves like a pure dark matter fluid, with energy density $\Omega_t \left(\frac{a_t}{a}\right)^3$. For $a > a_t$, the fluid will behave like $(\Omega_t - \Omega_\Lambda)\left(\frac{a_t}{a}\right)^3 + \Omega_\Lambda$, resembling a ΛCDM scenario.

Given the properties of the Heaviside function, one can identify the usual dark matter density Ω_c with the term $\Omega_t a_t^3$. Thus, the total matter component will be $\Omega_m = \Omega_t a_t^3 + \Omega_b$ when the baryonic matter term is also considered. In principle, there would be a degeneracy between the Ω_c and Ω_b terms if we were using only low redshift observational data. As long as we are going to use high redshift CMB data, this degeneracy is broken. Moreover, using CMB data makes necessary the introduction of a radiation term too, $\Omega_r(a)$, which has no influences on the late-time expansion, but is fundamental in the early stages of the Universe history. Resuming, the Friedmann equation 1 can be finally written as

$$E^2(a) = \Omega_c a^{-3} + \Omega_\Lambda \left[1 - \left(\frac{a_t}{a}\right)^3\right] \Theta(a - a_t) + \Omega_b a^{-3} + \Omega_r a^{-4} \,. \tag{3}$$

The last ingredient missing to model this UDM transition is what Heaviside-like functions to consider. We will consider two different functions; the first model for the transition will be :

$$\Theta(a - a_t) = \frac{1}{2} + \frac{1}{\pi} \arctan\left(\beta\pi(a - a_t)\right) \,. \tag{4}$$

The second transition function considered will be:

$$\Theta(a - a_t) = \frac{1}{2}\left[1 + \tanh\left(2\beta(a - a_t)\right)\right] \,. \tag{5}$$

In both cases, the parameter β mainly controls the velocity of the transition, being exactly the value of the first derivative with respect to the scale factor of the transition function, evaluated at the transition point a_t.

3. Observational data

In this section we specify the observational data sets we have used for our analysis, and the analytical expression of the χ^2 we are going to minimize in order to perform our statistical analysis.

3.1. CMB data

CMB data are taken from [1], where distance priors are derived from Planck first release data. The CMB shift parameters are the scaled distance to photon-decoupling surface and angular scale of the sound horizon at the photon-decoupling epoch

$$R \equiv \sqrt{\Omega_m H_0^2}\frac{r(z_*)}{c}\,,\quad l_a \equiv \pi\frac{r(z_*)}{r_s(z_*)}\,, \tag{6}$$

where $r(z_*)$ is the comoving distance

$$r(z_*) = \frac{c}{H_0}\int_0^{z_*}\frac{dz'}{E(z')} \tag{7}$$

and $r_s(z_*)$ is the comoving sound horizon

$$r_s(z_*) \;=\; \frac{c}{H_0}\int_{z_*}^{\infty} dz'\frac{c_s}{E(z')} \equiv \frac{c}{H_0}\int_0^{a_*}\frac{da'}{\sqrt{3(1+\overline{R_b}a')a'^4E^2(a')}}, \tag{8}$$

where $\overline{R_b}a = 3\rho_b/(4\rho_\gamma) = 31500\Omega_b h^2(T_{CMB}/2.7K)^{-4}a$ and $E(a)$ is given by eq. 3. Both shift parameters are evaluated at photon-decoupling epoch z_*. The last shift parameter is the baryon density $\Omega_b h^2$.

The mean values for these shift parameters, their standard deviations and the corresponding normalized covariance matrix are obtained by [1], from where is possible to construct the covariance matrix $\mathbf{C_{CMB}}$.

In order to write the CMB contribution to the χ^2, we first define

$$\mathbf{X_{CMB}} = \begin{pmatrix} l_a - \langle l_a\rangle \\ R - \langle R\rangle \\ \Omega_b h^2 - \langle\Omega_b h^2\rangle \end{pmatrix}\,, \tag{9}$$

and using the inverse of the covariance matrix $\mathbf{C_{CMB}}$, the CMB contribution to the χ^2 term is

$$\chi^2_{CMB} = \mathbf{X^T_{CMB}C^{-1}_{CMB}X_{CMB}}\,. \tag{10}$$

3.2. GC data

The Galaxy Clustering (GC) data we use are the measurements of $H(z)r_s(z_d)/c$ and $D_A(z)/r_s(z_d)$ from the two dimensional two-point correlation function measured at $z = 0.35$ by [2] using a SDSS DR7 Luminous Red Galaxies sample, and at $z = 0.57$ by [3] using the CMASS galaxy sample from BOSS. Here, $H(z)$ is the hubble parameter; $D_A(z)$ is the angular diameter distance

$$D_A(z) = \frac{c}{1+z}\int_0^z\frac{dz'}{H(z')}\,; \tag{11}$$

and the comoving sound horizon $r_s(z_d)$ is evaluated at the drag epoch.

The Galaxy Clustering contribution is calculated independently for each redshift, $\chi^2_{GC} = \chi^2_{GC1} + \chi^2_{GC2}$, where in each point the term is

$$\chi^2_{GCi} = \frac{1}{1-r_i^2}\left(\frac{X1^2_{GCi}}{\sigma^2_{1i}} + \frac{X2^2_{GCi}}{\sigma^2_{2i}} - 2r_i\frac{X1_{GCi}}{\sigma_{1i}}\frac{X2_{GCi}}{\sigma_{2i}}\right)\,, \tag{12}$$

where r_i is the correlation between the two functions at each redshift, and

$$X1_{GCi} = \frac{H(z_i)r_s(z_d)}{c} - \langle\frac{H(z_i)r_s(z_d)}{c}\rangle,\quad X2_{GCi} = \frac{D_A(z_i)}{r_s(z_d)} - \langle\frac{D_A(z_i)}{r_s(z_d)}\rangle\,. \tag{13}$$

3.3. SNe Ia data

The SNe Ia dataset used is the Union2.1 compilation, made of 580 Type Ia Supernovae distributed in the redshift interval $0.015 < z < 1.414$. The dataset provides the distance modulus $\mu(z_i)$ for each supernovae and the full statistical plus systematics covariance matrix. The distance modulus $\mu(z) = 5\log_{10} d_L(z) + \mu_0$ is defined using the dimensionless luminosity distance, which is given by

$$d_L(z) = (1+z)\int_0^z \frac{dz'}{E(z')}, \tag{14}$$

and μ_0 is a nuisance parameter involving the value of the Hubble constant H_0 and the SNe Ia absolute magnitude. We will marginalize over μ_0 (see ref. [11]). Defining the difference vector between the model and the observed magnitudes

$$\mathbf{X_{SN}} = \begin{pmatrix} \mu(z_1) - \mu_{obs}(z_1) \\ \cdots \\ \mu(z_{\mathcal{N}}) - \mu_{obs}(z_{\mathcal{N}}) \end{pmatrix}, \tag{15}$$

and the covariance matrix $\mathbf{C}$ given by [4], we construct the terms $a \equiv \mathbf{X_{SN}}^T \cdot \mathbf{C}^{-1} \cdot \mathbf{X_{SN}}$, $b \equiv \mathbf{X_{SN}}^T \cdot \mathbf{C}^{-1} \cdot \mathbf{1}$, and $d \equiv \mathbf{1}^T \cdot \mathbf{C}^{-1} \cdot \mathbf{1}$, being $\mathbf{1}$ the unitary vector. With these terms, the SNe Ia contribution to the χ^2 marginalizing over μ_0 is

$$\chi^2_{SN} = a + \log\frac{d}{2\pi} - \frac{b^2}{d}. \tag{16}$$

4. Discussion

The statistical analysis will be performed by minimizing the χ^2 function using Markov Chain Monte Carlo (MCMC) Method [12][13][14]. The statistical convergence of each MCMC case has been tested using the method described in [15]. We have tested a fast transition UDM model, with two type of Heaviside-like functions. In order to state the effective statistical weight and validity of our models, we have also analysed the ΛCDM model and the quiessence model as to be compared with.

The priors for the free parameters are: a positive matter density $\Omega_m > 0$ (or $\Omega_c > 0$ for UDM models), proper baryonic bound $0 < \Omega_b < 1$, a positive Hubble function $E(a) > 0$, and $0 < a_t < 1$ because we want the transition have happened. A gaussian prior for the Hubble constant is taken around $H_0 = 100\, h\, km\, s^{-1} Mpc^{-1} = 69\, km\, s^{-1} Mpc^{-1}$. In order to show the same parameters in all the models, the parameter Ω_c is used in both ΛCDM and quiessence models, simply transforming $\Omega_c = \Omega_m - \Omega_b$.

Table 1. Median values for the free parameters using CMB, GC and SN data. The median value for the χ^2_{red} and evidence in favour of ΛCDM is also shown.

Model	$\mathbf{\Omega_c}$	$\mathbf{\Omega_b}$	parameter ♯3	$\chi^2_{\mathbf{red}}$	$\mathbf{lnB_{i\Lambda}}$
arctan($\beta = 500$)	$0.2432^{+0.0067}_{-0.0064}$	$0.04602^{+0.00088}_{-0.00090}$	$a_t = 0.161^{+0.074}_{-0.094}$	0.9525	+0.010
tanh($\beta = 500$)	$0.2433^{+0.0070}_{-0.0065}$	$0.04606^{+0.00081}_{-0.00095}$	$a_t = 0.159^{+0.078}_{-0.094}$	0.9522	−0.015
ΛCDM	$0.246^{+0.010}_{-0.010}$	$0.04696^{+0.00097}_{-0.00097}$	−	0.9472	0
Quiessence	$0.242^{+0.013}_{-0.013}$	$0.0460^{+0.0024}_{-0.0023}$	$w = -1.028^{+0.064}_{-0.067}$	0.9488	−0.330

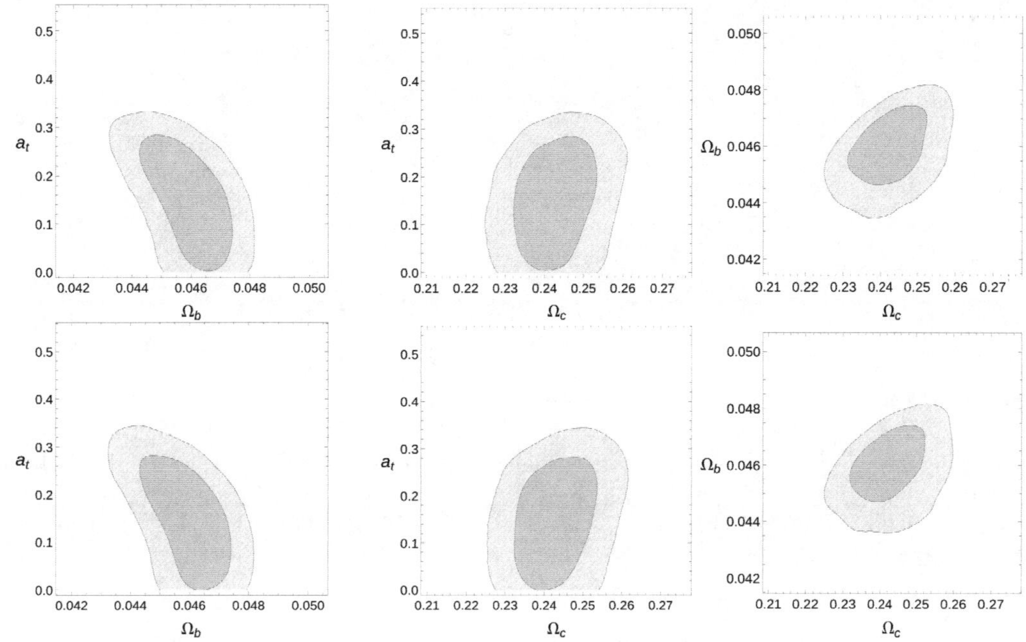

Figure 1. Confidence regions for the arctan (upper row) and tanh (lower row) models, dark grey areas are 1σ region and light grey areas are 2σ region.

Preliminary fits have been carried out using MCMC chains in which the tuning parameter β was left free, but convergence issues occurred. In the end, in order to get reasonable results, it was fixed around a possible minimum. The value of around $\beta = 500$ for both models were hinted by those runs, which is also supported theoretically by [10].

Figures 1 show the constrains of the free parameters of the UDM models, and figure 2 for quiessence and ΛCDM models. The table 1 shows the mean value of the constrained parameters.

A more robust conclusions for model selection can be made with the Bayes factors using the evidence. For the computation of the evidence a nested sampling algorithm has been used [16]. With the resulting evidence, the Bayes factor is obtained comparing the models to the ΛCDM, where the results are shown in the last column of table 1.

All the Bayesian approach shows that the proposed UDM model is worth of consideration, as it can be seen in the table 1 where all the models get a very similar evidence and the difference, according to the so-called "Jeffreys' scale", is inconclusive for all of them. The Arctan model gets even a slightly better evidence than ΛCDM model.

In this work we have analysed two possible functions for the Heaviside function of the fast transition UDM model, both with a quite good results. Due to these results, it could be interesting to analyse deeper other possible Heaviside-like functions in a further research. In any way, the results show that UDM models could be interesting.

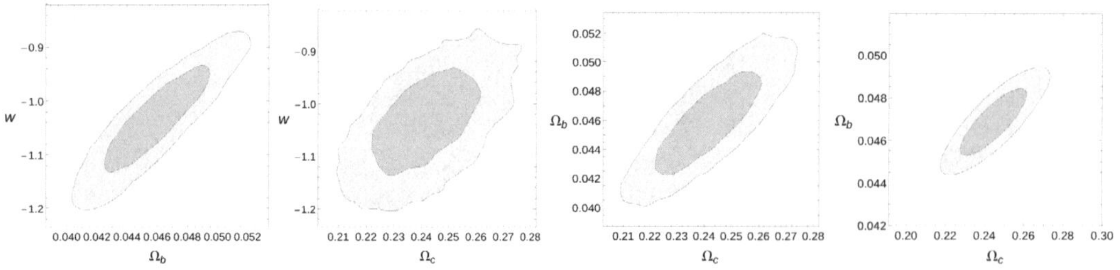

Figure 2. Contour plots for the quiessence (first 3 images) and ΛCDM (last) models, dark grey areas are 1σ region and light grey areas are 2σ region.

Acknowledgments
I.L. acknowledges the University of the Basque Country for a PhD fellowship. The research of V.S. was partially financed by the National Science Center Grant DEC-2012/06/A/ST2/00395. R.L. and V.S. are supported by the Spanish Ministry of Economy and Competitiveness through research projects FIS2010-15492 and Consolider EPI CSD2010-00064. R.L. and I.L. are supported by the University of the Basque Country UPV/EHU under program UFI 11/55 and by the Basque Government through research project GIC12/66.

References
[1] Wang Y and Wang S 2013 *Phys.Rev.* **D88** 043522 (*Preprint* 1304.4514)
[2] Chuang C H and Wang Y 2012 *Mon.Not.Roy.Astron.Soc.* **426** 226 (*Preprint* 1102.2251)
[3] Chuang C H, Prada F, Cuesta A J, Eisenstein D J, Kazin E *et al.* 2013 (*Preprint* 1303.4486)
[4] Suzuki N, Rubin D, Lidman C, Aldering G, Amanullah R *et al.* 2012 *Astrophys.J.* **746** 85 (*Preprint* 1105.3470)
[5] Kamenshchik A Y, Moschella U and Pasquier V 2001 *Phys.Lett.* **B511** 265–268 (*Preprint* gr-qc/0103004)
[6] Bilic N, Tupper G B and Viollier R D 2002 *Phys.Lett.* **B535** 17–21 (*Preprint* astro-ph/0111325)
[7] Bilic N, Lindebaum R J, Tupper G B and Viollier R D 2004 *JCAP* **0411** 008 (*Preprint* astro-ph/0307214)
[8] Piattella O F, Bertacca D, Bruni M and Pietrobon D 2010 *JCAP* **1001** 014 (*Preprint* 0911.2664)
[9] Bertacca D, Bruni M, Piattella O F and Pietrobon D 2011 *JCAP* **1102** 018 (*Preprint* 1011.6669)
[10] Bruni M, Lazkoz R and Rozas-Fernandez A 2013 *Mon.Not.Roy.Astron.Soc.* **431** 2907–2916 (*Preprint* 1210.1880)
[11] Conley A *et al.* (SNLS Collaboration) 2011 *Astrophys.J.Suppl.* **192** 1 (*Preprint* 1104.1443)
[12] Christensen N, Meyer R, Knox L and Luey B 2001 *Class.Quant.Grav.* **18** 2677 (*Preprint* astro-ph/0103134)
[13] Lewis A and Bridle S 2002 *Phys.Rev.* **D66** 103511 (*Preprint* astro-ph/0205436)
[14] Trotta R and Durrer R 2004 (*Preprint* astro-ph/0410115)
[15] Dunkley J, Bucher M, Ferreira P G, Moodley K and Skordis C 2005 *Mon.Not.Roy.Astron.Soc.* **356** 925–936 (*Preprint* astro-ph/0405462)
[16] Mukherjee P, Parkinson D and Liddle A R 2006 *Astrophys.J.* **638** L51–L54 (*Preprint* astro-ph/0508461)

Adiabatic renormalization of the stress-energy tensor for spin-1/2 fields.

Francisco Torrentí

Instituto de Física Teórica UAM/CSIC, Universidad Autónoma de Madrid, Cantoblanco, 28049 Madrid, Spain.

E-mail: f.torrenti@csic.es

Abstract. We review the adiabatic renormalization method for spin-1/2 fields in FLRW spacetimes, emphasizing its similarities and differences with the scalar case. With it, we obtain a generic expression for the renormalized expectation value of the stress-energy tensor and analyze its properties. We particularize this result to de Sitter and radiation-dominated universes, recovering in the second case the equations of cold matter for late times.

1. Introduction

The renormalization of quantum fields in curved spacetime is an extensively studied issue [1, 2]. When one tries to compute the vacuum expectation value of quantities such as the stress-energy tensor, new additional ultraviolet divergences appear which are otherwise absent in Minkowski spacetime. Therefore, one needs to apply a well-motivated renormalization procedure specifically designed to work in curved spacetime.

One of the most useful for FLRW metrics is adiabatic regularization. In this method, the divergent terms are identified and subtracted through an expansion in momenta of the field modes defining the quantum state. By dimensionality, this is equivalent to an expansion in derivatives of the scale factor, hence the *adiabatic* denomination. This method was originally developed for scalar fields [3], and we recently generalized it to deal with spin-1/2 fields [4]. The key difference between both cases is that in the first one, the adiabatic expansion that identifies the divergent terms is of WKB type, while in the second one it is not. In order to check the validity of our approach, we renormalized adiabatically the trace and chiral anomalies and checked that our results were coincident with those obtained with other renormalization procedures.

In [5], we used this construction to obtain a full renormalized expression for the stress-energy tensor of a massive spin-1/2 field. We review here this work. After writing the equations of motion for spin-1/2 and spin-0 fields in a FLRW metric, we show how to obtain the adiabatic expansion of the fermion field modes and the similarities and differences with the well-known scalar case. We then renormalize the stress-energy tensor and analyze some of its properties, including the potential ambiguities associated to the renormalization program. We finally apply this result to two particular examples: de Sitter and radiation-dominated universes. In the second case, we obtain that the created particles obey at late-times the state equation of cold matter.

2. Basic equations

We first review the basic equations for spin-1/2 and spin-0 fields in the FLRW metric $ds^2 = dt^2 - a^2(t)d\vec{x}^2$. The notation used here is explained in more detail in [5].

2.1. Spin-1/2 fields

A spin-1/2 field ψ of mass m propagating in this metric can be written as

$$\psi(\vec{x}, t) = \int d^3\vec{k} \frac{e^{i\vec{k}\vec{x}}}{\sqrt{(2\pi)^3 a^3}} \sum_\lambda \left[B_{\vec{k}\lambda} u_{\vec{k}\lambda}(t) + D^\dagger_{-\vec{k}\lambda} v_{-\vec{k}\lambda}(t) \right] , \tag{1}$$

where $B_{\vec{k}\lambda}$ and $D_{\vec{k}\lambda}$ are destruction operators, and $u_{\vec{k}\lambda}(t)$ and $v_{\vec{k}\lambda}(t)$ are two time-dependent spinors defined as

$$u_{\vec{k}\lambda}(t) = \begin{pmatrix} h^I_k(t)\xi_\lambda(\vec{k}) \\ h^{II}_k(t)\frac{\vec{\sigma}\cdot\vec{k}}{k}\xi_\lambda(\vec{k}) \end{pmatrix} , \quad v_{\vec{k}\lambda}(x) = \begin{pmatrix} -h^{II*}_k(t)\xi_{-\lambda}(\vec{k}) \\ -h^{I*}_k(t)\frac{\vec{\sigma}\cdot\vec{k}}{k}\xi_{-\lambda}(\vec{k}) \end{pmatrix} . \tag{2}$$

Here, $h^I_k(t)$ and $h^{II}_k(t)$ are the field modes that define the quantum state, $\vec{\sigma} = (\sigma_1, \sigma_2, \sigma_3)$ are the usual Pauli matrices, $k \equiv |\vec{k}|$, and ξ_λ are two-component helicity eigenstates with $\lambda = \pm 1$. In this decomposition, the field modes satisfy

$$h^{II}_k = \frac{ia}{k}(\partial_t + im)h^I_k , \qquad h^I_k = \frac{ia}{k}(\partial_t - im)h^{II}_k , \qquad |h^I_k|^2 + |h^{II}_k|^2 = 1 \tag{3}$$

where the first two are the equations of motion and the third one is the normalization condition.

2.2. Spin-0 field

A scalar field of mass m and coupling ξ to the curvature can be written in this metric as

$$\phi(\vec{x}, t) = \int d^3\vec{k} \frac{e^{i\vec{k}\vec{x}}}{\sqrt{(2\pi)^3 a^3}} \left[A_{\vec{k}} f_k(t) + A^\dagger_{-\vec{k}} f^*_k(t) \right] , \tag{4}$$

where $A_{\vec{k}}$ are destruction operators and $f_k(t)$ are the field modes. They satisfy

$$\frac{d^2 f_k}{dt^2} + \left(\frac{k^2}{a^2} + m^2 + \sigma \right) f_k = 0 , \quad f_k \dot{f}^*_k - f^*_k \dot{f}_k = i \tag{5}$$

with $\sigma \equiv (6\xi - 3/4)(\dot{a}/a)^2 + (6\xi - 3/2)\ddot{a}/a$. The first one is the equation of motion and the second one is the corresponding normalization condition.

3. Adiabatic expansion

The adiabatic regularization method is based on an expansion in momenta of the field modes (h^I_k and h^{II}_k for spin-1/2 fields, and f_k for scalars). This construction allows afterwards to identify and subtract the divergent terms from the original quantity. By dimensionality, this is equivalent to an expansion in derivatives of the scale factor, i.e. an adiabatic expansion. In any case, the expansion must recover in the adiabatic limit the solutions

$$f_k(t) \sim \frac{1}{\sqrt{2\omega}} e^{-i\int^t \omega(t')dt'} , \qquad h^I_k(t) \sim \sqrt{\frac{\omega + m}{2\omega}} e^{-i\int^t \omega(t')dt'} \qquad h^{II}_k(t) \sim \sqrt{\frac{\omega - m}{2\omega}} e^{-i\int^t \omega(t')dt'} \tag{6}$$

Journal of Physics: Conference Series **600** (2015) 012029 doi:10.1088/1742-6596/600/1/012029

with $\omega \equiv \sqrt{(k/a(t))^2 + m^2}$. In [4], we found that for spin-1/2 fields, there exists the following self-consistent expansion for the field modes

$$h_k^I(t) \sim \sqrt{\frac{\omega + m}{2\omega}} F(t) e^{-i\int^{t'} W(t')dt'} \quad , \quad h_k^{II}(t) \sim \sqrt{\frac{\omega - m}{2\omega}} G(t) e^{-i\int^{t'} W(t')dt'} \tag{7}$$

where $\Omega(t)$, $F(t)$ and $G(t)$ are time-dependent functions that are expanded adiabatically as

$$W(t) = \sum_{n=0}^{\infty} \omega^{(n)}(t) \ , \quad F(t) = \sum_{n=0}^{\infty} F^{(n)}(t) \ , \quad G(t) = \sum_{n=0}^{\infty} G^{(n)}(t) \ . \tag{8}$$

Here, the superindex (n) in a function indicates that it is of adiabatic order n, which means that it contain n derivatives of the scale factor[1]. Condition (6) imposes that the zeroth order terms are $\omega^{(0)}(t) = \omega$ and $F^{(0)} = G^{(0)} = 1$. The following terms are found by substituting (7) into (3) and solving iteratively the system of equations order by order. The terms obtained in this way contain ambiguities, which in any case do not appear in the final expressions of the renormalized quantities $\langle\bar{\psi}\psi\rangle_{ren}$ and $\langle T_{\mu\nu}\rangle_{ren}$. By simplicity, we can impose at all orders the additional condition $F^{(n)}(m) = G^{(n)}(-m)$ under the change of mass sign, which eliminates all the ambiguities. We obtain and display explicit expressions for $\omega^{(n)}$, $F^{(n)}$ and $G^{(n)}$ up to fourth order in [4].

A natural question to ask is what is the relation of this approach to the one used for scalar fields. For these, the adiabatic expansion of the field modes is of the WKB form [3]

$$f_k(t) = \frac{1}{\sqrt{2\chi(t)}} e^{-i\int^t \chi(t')dt'} \ , \qquad \chi(t) = \sum_{n=0}^{\infty} \chi^{(n)} \ . \tag{9}$$

We have $\chi^{(0)} = \omega$ from (6), and the other terms of the expansion are found by solving iteratively order by order the equation of motion (5) (note that the normalization condition is automatically satisfied). The key difference between (7) and (9) is that, in the first case, the adiabatic expansions of the multiplicative and exponent terms are independent, while in the second case they are not. In fact, one can try to obtain the adiabatic expansion for scalar fields with the more generic ansatz

$$f_k(t) = \frac{1}{\sqrt{2\omega}} H(t) e^{-i\int^t \Omega(t')dt'} \ , \quad \Omega(t) = \sum_{i=0}^{\infty} \Omega^{(i)}(t) \ , \quad H(t) = \sum_{i=0}^{\infty} H^{(i)}(t) \tag{10}$$

which mimics the one used for fermions in (7). Here, $H^{(0)} = 1$ and $\Omega^{(0)} = \omega$. If we substitute (10) into (5) and solve iteratively order by order, we find $H^{(n)} = \sqrt{\omega}(1/\sqrt{\chi})^{(n)}$ and $\Omega^{(n)} = \chi^{(n)}$, rediscovering this way the scalar WKB-type expansion.

To end with, we note that adiabatic regularization can be extended to spin-1 fields [6]. In this case, the adiabatic expansion of the field modes turn out to be WKB again.

4. Renormalization of the stress-energy tensor

We now show how to renormalize the vacuum expectation value of the stress-energy tensor of a spin-1/2 field using the formalism we have explained. As the FLRW universe is homogeneous and spatially isotropic, we only have two independent components: the 00-component (energy density) and the ii-component (pressure)

$$\langle T_{00}\rangle = \frac{1}{2\pi^2 a^3}\int_0^\infty dk k^2 \rho_k(t) \ , \qquad \langle T_{ii}\rangle = \frac{1}{2\pi^2 a}\int_0^\infty dk k^2 p_k(t) \tag{11}$$

[1] For example, $\dot{a}$ is of adiabatic order 1 and $\ddot{a}\dot{a}^2$ is of adiabatic order 4.

where

$$\rho_k = i\left(h_k^I \frac{\partial h_k^{I*}}{\partial t} + h_k^{II}\frac{\partial h_k^{II*}}{\partial t} - h_k^{I*}\frac{\partial h_k^I}{\partial t} - h_k^{II*}\frac{\partial h_k^{II}}{\partial t}\right), \qquad p_k = -\frac{2k}{3a}(h_k^I h_k^{II*} + h_k^{I*}h_k^{II}) . \quad (12)$$

The vacuum state is defined as $B_{\vec{k},\lambda}|0\rangle = D_{\vec{k},\lambda}|0\rangle = 0$. Both quantities in (11) contain quartic, quadratic and logarithmic ultraviolet divergences. To renormalize them, we must first expand their integrands adiabatically

$$\rho_k = \rho_k^{(0)} + \rho_k^{(1)} + \rho_k^{(2)} + \rho_k^{(3)} + \rho_k^{(4)} + \dots , \qquad p_k = p_k^{(0)} + p_k^{(1)} + p_k^{(2)} + p_k^{(3)} + p_k^{(4)} + \dots \quad (13)$$

where here $\rho_k^{(n)}$ and $p_k^{(n)}$ are of nth adiabatic order (note that $\rho_k^{(odd)} = 0 = p_k^{(odd)}$). The different terms of the expansion are found by substituting (7) into (12), and are written explicitly in [5]. After this, we must subtract the terms of the expansion that cause the divergences. As for scalar fields [3, 1, 2], we must subtract up to fourth adiabatic order. Therefore, if we define the adiabatic subtraction terms as

$$\langle T_{00}\rangle_{Ad}^{(n)} \equiv \frac{1}{2\pi^2 a^3}\int_0^\infty dk k^2 \rho_k^{(n)} , \qquad \langle T_{ii}\rangle_{Ad}^{(n)} \equiv \frac{1}{2\pi^2 a}\int_0^\infty dk k^2 p_k^{(n)} , \quad (14)$$

the renormalized stress-energy tensor is

$$\langle T_{\mu\nu}\rangle_{ren} \equiv \langle T_{\mu\nu}\rangle - \langle T_{\mu\nu}\rangle_{Ad}^{(0)} - \langle T_{\mu\nu}\rangle_{Ad}^{(2)} - \langle T_{\mu\nu}\rangle_{Ad}^{(4)} . \quad (15)$$

We now analyze some of its properties.

4.1. Finite fourth-order subtraction terms

One finds that the fourth-order subtraction term can be written as

$$\langle T_{\mu\nu}\rangle_{Ad}^{(4)} = \frac{2}{2880\pi^2}\left[-\frac{1}{2}\,{}^{(1)}H_{\mu\nu} + \frac{11}{2}\,{}^{(3)}H_{\mu\nu}\right] , \quad (16)$$

where ${}^{(1)}H_{\mu\nu}$ and ${}^{(3)}H_{\mu\nu}$ are covariant geometric tensors of fourth adiabatic order [5]. This tensor is mass-independent and finite. Therefore, one could naturally ask if it is really necessary to subtract this term in (15). However, for an arbitrary spacetime, the fourth adiabatic order do contain ultraviolet divergences, they just disappear accidentally for spatially flat FLRW metrics [7]. Therefore, by consistency, we must also subtract in this case the fourth adiabatic order. Note that this is similar to the renormalization of a scalar field with conformal coupling $\xi = 1/6$: in this case the fourth order subtraction term is also finite, but we must subtract it anyway [1].

4.2. Conservation

The renormalized tensor is conserved, $\nabla^\mu \langle T_{\mu\nu}\rangle_{ren} = 0$ as a consequence of the independent conservation laws $\nabla^\mu \langle T_{\mu\nu}\rangle_{Ad}^{(n)} = 0$, for $n = 0, 2, 4$.

4.3. Ambiguities and compatibility with other renormalization procedures

In the curved space renormalization program for the stress-energy tensor, there are in general four types of divergent subtraction terms that generate intrinsic ambiguities: $m^4 g_{\mu\nu}$, $m^2 G_{\mu\nu}$ (where $G_{\mu\nu}$ is the Einstein tensor), ${}^{(1)}H_{\mu\nu}$, and ${}^{(2)}H_{\mu\nu}$ (where ${}^{(2)}H_{\mu\nu}$ is another fourth-order curvature tensor) [2, 8]. We have checked in [5] that our renormalized tensor obeys this property.

In particular, in our FLRW spacetime, ${}^{(2)}H_{\mu\nu}$ is proportional to ${}^{(1)}H_{\mu\nu}$. Therefore, the most generic fourth-order subtraction term is, from (16),

$$\langle T_{\mu\nu}\rangle^{(4)}_{Ad} = \frac{2}{2880\pi^2}\left[c_1\,{}^{(1)}H_{\mu\nu} + \frac{11}{2}\,{}^{(3)}H_{\mu\nu}\right] \tag{17}$$

where c_1 is an arbitrary coefficient. The different renormalization procedures can only potentially differ in the value of this coefficient. As we have seen, adiabatic regularization in particular leads to $c_1 = -1/2$. This ambiguity disappears for spacetime backgrounds for which the tensor ${}^{(1)}H_{\mu\nu}$ vanishes, such as de Sitter and radiation-dominated universes analyzed below. For more generic scale factors, this term rapidly vanishes at late-times.

From (16), the trace anomaly obtained by adiabatic regularization is

$$\langle T^{\mu}_{\mu}\rangle_{ren} = \langle T^{\mu}_{\mu}\rangle^{(4)}_{Ad} = \frac{2}{2880\pi^2}\left[-\frac{11}{2}\left(R_{\mu\nu}R^{\mu\nu} - \frac{1}{3}R^2\right) + 3\Box R\right] . \tag{18}$$

This is in exact agreement with the conformal anomaly for spin-1/2 fields computed by point-splitting, zeta function, and dimensional regularization [2]. However, note that if one considers the general expression (17) instead of (16), the numerical coefficient of $\Box R$ is actually proportional to c_1. Therefore, although we have not obtained the full renormalized stress-energy tensor with all these procedures, we know that they will yield $c_1 = -1/2$.

Finally, we would like to comment some recent results that give insight into the equivalence between adiabatic regularization and Schwinger-DeWitt approaches. In [9], the sixth-order contribution of the renormalized stress-energy tensor for scalar fields $\langle T_{\mu\nu}\rangle^{(6)}_{Ad}$ is computed analytically. This is a good approximation to the full tensor in the limit of high masses. In [10], this term is computed for spin-0, spin-1/2 and spin-1 fields using the Schwinger-DeWitt approach, obtaining for spin-0 fields the same result as in [9]. We have computed this term for spin-1/2 fields with the adiabatic regularization method described here, obtaining also the same result as in [10].

5. Examples

We now apply the formalism developed in the last sections to two particular scale factors: de Sitter and radiation-dominated universes. One must first solve the equations of motion (3). In practice, one can choose a set of two particular solutions $(h^I_{k,p}, h^{II}_{k,p})$ correctly normalized $(|h^I_{k,p}|^2 + |h^{II}_{k,p}|^2 = 1)$ and then construct the full general solution with the Bogolubov-type rotation

$$h^I_k = E_k h^I_{k,p} + F_k h^{II*}_{k,p} , \qquad h^{II}_k = E_k h^{II}_{k,p} - F_k h^{I*}_{k,p} . \tag{19}$$

Here, E_k and F_k are constants that must obey the following three conditions:

(i) We require $|E_k|^2 + |F_k|^2 = 1$ so that the generic solution is also normalized.
(ii) In the limit $k \to \infty$ the physical solutions must recover (6).
(iii) The constants must not add extra ultraviolet divergences to $\langle T_{\mu\nu}\rangle$, so that the subtraction terms cancel all the divergences.

We now see an example of how this works below.

5.1. de Sitter spacetime

For de Sitter spacetime $a(t) = e^{Ht}$ with H a constant, the general solution to the field equations can be conveniently expressed, using the technique of equation (19), in terms of Hankel functions

$$h^I_{k,p} = i\sqrt{\frac{\pi k}{4H}}e^{\frac{\pi m}{2H} - \frac{Ht}{2}} H^{(1)}_{\frac{1}{2}-i\frac{m}{H}}\left(\frac{k}{H}e^{-Ht}\right) , \qquad h^{II}_{k,p} = \sqrt{\frac{\pi k}{4H}}e^{\frac{\pi m}{2H} - \frac{Ht}{2}} H^{(1)}_{-\frac{1}{2}-i\frac{m}{H}}\left(\frac{k}{H}e^{-Ht}\right) . \tag{20}$$

Condition (ii) leads to $E_k \sim 1$ and $F_k \sim 0$ in the limit $k \to \infty$, and condition (iii) leads to $|E_k|^2 - |F_k|^2 = 1 + O(k^{-8})$. In this case, due to the symmetries of de Sitter spacetime, there exists the preferred quantum state $E_k = 1$, $F_k = 0$. This determines a vacuum for spin one-half fields analogous to the Bunch-Davies vacuum [11] for scalar fields. With this choice, the stress-energy tensor can be computed analytically [4, 5].

5.2. Radiation-dominated Universe

For the scale factor $a(t) = a_0 t^{1/2}$, the spin-1/2 equations of motion (3) admit as a solution

$$h^I_{k,p} = \frac{N}{\sqrt{a(t)}} W_{\kappa,\frac{1}{4}}(i2mt)\ , \quad h^{II}_{k,p} = \frac{Nk}{2ma(t)^{3/2}}\left[W_{\kappa,\frac{1}{4}}(i2mt) + \left(\kappa - \frac{3}{4}\right) W_{\kappa-1,\frac{1}{4}}(i2mt)\right] \quad (21)$$

Here, $W_{\kappa,1/4}$ are Whittaker functions, $\kappa = \frac{1}{4} - ik^2/(a_0^2 2m)$, and $N = (a_0^{1/2}/(2m)^{1/4})e^{-(\pi k^2/4a_0^2 m)}$.

Condition (ii) gives $E_k \sim 1$ and $F_k \sim 0$ for $k \to \infty$, while condition (iii) gives $|E_k|^2 - |F_k|^2 = 1 + O(k^{-5})$ [5]. However, in contrast with Sitter spacetime, the absence of extra symmetries in the radiation-dominated universe does not helps us to fix a natural quantum state. Nevertheless, we can make generic predictions of the behavior of the renormalized stress-energy tensor at early and late times even without specifying the choice of constants. We have [5]:

- For late times ($t >> m^{-1}$): $\langle T^{00}\rangle_{ren} \sim \frac{\rho_{0m}}{a^3}$ and $\langle T^{ii}\rangle_{ren} \sim 0$.
- For early times ($t << m^{-1}$): $\langle T^{00}\rangle_{ren} \sim \frac{\rho_{0r}}{a^4}$ and $\langle T^{ii}\rangle_{ren} \sim \frac{1}{3}\langle T^{00}\rangle_{ren}$.

Here, ρ_{0m} and ρ_{0r} are finite and positive constants. Therefore, at late and early times we get the classical equations of radiation and cold matter respectively.

6. Conclusions

In this work, we have obtained the renormalized expectation value of the stress-energy tensor for spin-1/2 fields within the framework of adiabatic regularization and studied its properties.

Acknowledgments

I am very grateful to A. del Rio, A. Landete, and specially to J. Navarro-Salas, for collaboration in the topics of this work. I thank Leonard Parker for useful comments. I also thank Jerzy Matyjasek for bringing to our attention reference [10]. This work is supported by the Severo Ochoa PhD fellowship SVP-2013-067697, the Research Project of the Spanish MINECO FPA2012-39684-C03-02 and the Centro de Excelencia Severo Ochoa Programme SEV-2012-0249.

References

[1] L. Parker and D. J. Toms, *Quantum field theory in curved spacetime: quantized fields and gravity*, Cambridge University Press, Cambridge, England (2009).
[2] N. D. Birrell and P.C.W. Davies, *Quantum fields in curved space*, Cambridge University Press, Cambridge, England (1982).
[3] L. Parker and S. A. Fulling, *Phys. Rev. D* **9**, 341 (1974); S. A. Fulling and L. Parker, *Ann. Phys.* (N.Y.) **87**, 176 (1974); S. A. Fulling, L. Parker, and B. L. Hu, *Phys. Rev. D* **10**, 3905 (1974).
[4] A. Landete, J. Navarro-Salas and F. Torrenti, *Phys. Rev. D* **88**, 061501(R) (2013); *Phys. Rev. D* **89**, 044030 (2014).
[5] A. del Rio, J. Navarro-Salas and F. Torrenti. *Phys. Rev. D* **90**, 084017 (2014).
[6] L.P. Chimento and A. E. Cossarini, *Phys. Rev. D* **41**, 3101 (1990); I. Agullo, A. Landete and J. Navarro-Salas, arXiv:gr-qc/1409.6406
[7] S. M. Christensen, *Phys. Rev. D* **17**, 946 (1978).
[8] R. M. Wald, *Quantum field theory in curved space-time and black hole thermodynamics*, University of Chicago Press, Chicago, (1994).
[9] A. Kaya and M. Tarman, *J. Cosmol. Astropart. Phys.* 04 (2011) 040.
[10] J. Matyjasek and P. Sadurski, *Phys. Rev. D* **88**, 104015 (2013).
[11] T.S. Bunch and P.C.W. Davies, *Proc. P. Soc. London A* **360**, 117 (1978).

Journal of Physics: Conference Series **600** (2015) 012030 doi:10.1088/1742-6596/600/1/012030

Averaging in cosmology based on Cartan scalars

Petr Kašpar and Otakar Svítek

Institute of Theoretical Physics, Faculty of Mathematics and Physics, Charles University in Prague, V Holesovickach 2, Prague 8, 18000

E-mail: `ota@matfyz.cz`

Abstract. Averaging scheme based on the use of Cartan scalars is presented in the cosmological context. The resulting method is applied to LTB cosmological model and the correlation terms appearing due to averaging are physically explained.

1. Introduction

Homogeneous cosmology is a successful but highly simplified model. The real inhomogeneous universe should be suitably averaged to obtain the corresponding homogeneous model. This is a nontrivial task from two perspectives. First, Einstein equations are nonlinear leading to the appearance of a correlation terms which can mimic various sources (e.g. dark energy). Second, there is no established method for averaging tensors covariantly, this can be circumvented by concentrating on scalars only.

Many approaches to averaging were developed in the past. The most popular approach to averaging is Buchert's method of averaging scalar part of Einstein equations [1], [2]. All of Einstein equations are averaged in the case of Macroscopic Gravity [3], [4], at the same time the Cartan structure equations are averaged. Isometric embedding of a 2-sphere into Euclidian space is used for averaging by Korzyński [5]. In [6] the Weitzenböck connection was used for the definition of correct averaging of tensor fields. Coley [7] investigated averaging of scalar invariants constructed from the Riemann tensor and finite number of its covariant derivatives.

Here we propose to use Cartan scalars to covariantly average both geometry and Einstein equations. Originally, the theory of Cartan scalars was developed to solve the equivalence problem for geometries [8], [9] and therefore it is suitable to unambiguously describe the given spacetime and the process of averaging which is suitably trivial for scalars. We also average the left hand side of Einstein equations (which can be expressed using Cartan scalars) and we propose the procedure for obtaining the correlation term.

Usually, correlation term is defined as a difference between Einstein tensor defined via macroscopic highly symmetric ("averaged") metric and average of Einstein tensor computed using unaveraged metric

$$R^{\mu}{}_{\nu}\left(\overline{g_{\alpha\beta}}\right) - \frac{1}{2}R\left(\overline{g_{\alpha\beta}}\right)\delta^{\mu}{}_{\nu} + C^{\mu}{}_{\nu} = 8\pi T^{\mu}{}_{\nu}\left(\overline{g_{\alpha\beta}}\right)$$

$$C^{\mu}{}_{\nu} = \langle R^{\mu}{}_{\nu}\rangle - \frac{1}{2}\langle R\rangle\,\delta^{\mu}{}_{\nu} - R^{\mu}{}_{\nu}\left(\overline{g_{\alpha\beta}}\right) - \frac{1}{2}R\left(\overline{g_{\alpha\beta}}\right)\delta^{\mu}{}_{\nu}.$$

Journal of Physics: Conference Series **600** (2015) 012030 doi:10.1088/1742-6596/600/1/012030

2. Cartan scalars

They are constructed as tetrad projections of Riemann tensor and the finite number of its covariant derivatives and they completely (locally) specify the geometry of a manifold. Cartan scalars are true scalars on the bundle of frames $F(\mathcal{M})$ and for a fixed tetrad they behave as scalars on the manifold as well. We can use them to obtain the dimension of an isometry group and the corresponding algebra of Killing vectors. And as mentioned in introduction, they form a classical approach to the equivalence problem.

First we review the Cartan-Karlhede approach for construction of Cartan scalars. Let $\mathcal{M}$ be a n-dimensional differentiable manifold with a metric

$$\mathbf{g} = \eta_{ij}\boldsymbol{\omega}^i \otimes \boldsymbol{\omega}^j,$$

Frame $\boldsymbol{\omega}^i$ is fixed up to the generalized rotations

$$\boldsymbol{\omega}^i = \omega^i_\nu(x^\mu, \xi^\Upsilon)\mathbf{d}x^\nu,$$

where ξ^Υ denote the coordinates of a Lorentz group G. Bundle of frames $F(\mathcal{M})$ is locally isomorphic to $\mathcal{U} \times G, \mathcal{U} \subset \mathcal{M}$ and we introduce enlarged exterior derivative $\mathbf{d} = \mathbf{d}_x + \mathbf{d}_\xi$. This gives Cartan structure equeations

$$\mathbf{d}\boldsymbol{\omega}^i = \boldsymbol{\omega}^j \wedge \boldsymbol{\omega}^i{}_j,$$

$$\mathbf{d}\boldsymbol{\omega}^i{}_j = -\boldsymbol{\omega}^i{}_k \wedge \boldsymbol{\omega}^k{}_j + \frac{1}{2}R^i{}_{jkl}\boldsymbol{\omega}^k \wedge \boldsymbol{\omega}^l,$$

with a condition $\eta_{ik}\boldsymbol{\omega}^k{}_j + \eta_{jk}\boldsymbol{\omega}^k{}_i = 0$.

Generation of covariant derivatives of the Riemann tensor is performed by successive application of $\mathbf{d}$

$$\begin{aligned} \mathbf{d}R_{ijkl} &= R_{mjkl}\boldsymbol{\omega}^m_i + R_{imkl}\boldsymbol{\omega}^m_j + R_{ijml}\boldsymbol{\omega}^m_k + R_{ijkm}\boldsymbol{\omega}^m_l + R_{ijkl;m}\boldsymbol{\omega}^m, \\ \mathbf{d}R_{ijkl;n} &= R_{mjkl;n}\boldsymbol{\omega}^m_i + R_{imkl;n}\boldsymbol{\omega}^m_i + \ldots + R_{ijkl;nm}\boldsymbol{\omega}^m, \\ &\;\;. \\ &\;\;. \end{aligned}$$

Now, R^p denotes the set $\left\{R_{ijkm}, R_{ijkm;n_1}, \ldots, R_{ijkm;n_1\ldots n_p}\right\}$ where p is such that R^{p+1} contains no additional functionally independent element (over $F(\mathcal{M})$, with dependence defined using $\mathbf{d}f \sim \mathbf{d}g$). The set R^p characterizes the geometry up to isometry and the elements are Cartan scalars. R^p has to satisfy (nonlinear) constraints so we can form a smaller subset R'^p that can be used to generate the full set.

3. Averaging

Geometry of $\mathcal{M}$ is now characterized by the set R^p of scalars which can be naturally averaged over a given domain $\mathcal{D}$ without any problems related to covariant tensor averaging

$$\langle f \rangle (x) = \frac{1}{V_\mathcal{D}} \int_\mathcal{D} f\,(x + x')d^N x'.$$

New geometry $\langle \mathcal{M} \rangle$ describes a manifold identical as a set but with averaged geometry given by averaged Cartan scalars. However, naively averaging R^p would result in a set potentially violating nonlinear constraints and therefore not generating viable geometry. To overcome this we will use the restricted set R'^p to obtain a new set $\langle R'^p \rangle$ and then we generate the whole set $\langle R^p \rangle$ from $\langle R'^p \rangle$ using constraints. Then there exists a metric tensor $\langle g_{\mu\nu} \rangle$ (or equivalently $\langle \boldsymbol{\omega}^i \rangle$)

giving the same Cartan scalars. Such metric may be constructed using the averaged Cartan scalars.

Averaging generally leads to decrease in the number of independent functions in $\langle R^p \rangle$, enlargement of isotropy group of the new spacetime $\langle \mathcal{M} \rangle$ and a new algebra of Killing vectors.

Left hand side of Einstein equations can be projected on tetrad of the Cartan-Karlhede algorithm to obtain expression in Cartan scalars enabling straightforward averaging.

We have two roads to averaging with Cartan scalars. Direct method described above - which is very difficult. And indirect method which is performed by comparing averaged scalars with scalars of known highly symmetric spacetimes (FRLW etc.), possibly getting conditions for their equivalence and correlation tensor

$$C^{\mu}{}_{\nu} = \langle R^{\mu}{}_{\nu} \rangle - \frac{1}{2} \langle R \rangle \delta^{\mu}{}_{\nu} - R^{\mu}{}_{\nu} (\overline{g_{\alpha\beta}}) - \frac{1}{2} R (\overline{g_{\alpha\beta}}) \delta^{\mu}{}_{\nu}$$

and note that the conserved stress energy tensor has the following form

$$^{(ef)}T^{\mu}_{\ \nu} = T^{\mu}_{\ \nu} (\overline{g_{\alpha\beta}}) - C^{\mu}{}_{\nu}. \tag{1}$$

This technique passes the basic test of correct averaging: constant curvature spacetimes are preserved.

4. Applications - LTB metric

Lemaître-Tolman-Bondi (LTB) metric is an exact spherically symmetric solution with inhomogeneous dust $T_{\mu\nu} = \rho u_{\mu} u_{\nu}$. The line element reads

$$ds^2 = -dt^2 + \frac{(R'(t,r))^2}{1+2E(r)} dr^2 + R^2(t,r)(d\theta^2 + \sin^2(\theta) d\phi^2).$$

Einstein equations reduce to the following

$$R_{,t}^2 = 2E + \frac{2M(r)}{R} + \frac{\Lambda}{3} R^2, \quad 4\pi\rho = \frac{M'}{R' R^2},$$

where function $E(r)$ determines a curvature of spatial section $t = const.$ and $M(r)$ is the gravitational mass contained within the comoving radius r. The first of the above equations can be integrated to obtain

$$\int_0^R \frac{d\tilde{R}}{\sqrt{2E + \frac{2M}{\tilde{R}} + \frac{1}{3}\Lambda \tilde{R}^2}} = t - t_B(r),$$

where $t_B(r)$ is the bang time function meaning the Big Bang is not simultaneous in all points.

4.1. First model

We will consider $E = 0$ and assume the ansatz

$$R(t,r) = A(t,r) \exp \psi(t,r),$$

where $\psi(t,r)$ is a quickly varying function, and $\psi << \psi_{,x} \sim \psi_{,xy} \sim \psi_{,xyz}$, where x, y and z denote time or radial coordinate; additionally $\psi_{,x} >> A(t,r)$ and its derivatives. We consider this tetrad

$$\begin{aligned}
\boldsymbol{\omega}^0 &= \frac{1}{\sqrt{2}}(dt + R_{,r}dr), \\
\boldsymbol{\omega}^1 &= \frac{1}{\sqrt{2}}(dt - R_{,r}dr), \\
\boldsymbol{\omega}^2 &= \frac{1}{\sqrt{2}}(Rd\theta + iR\sin\theta d\phi), \\
\boldsymbol{\omega}^3 &= \frac{1}{\sqrt{2}}(Rd\theta - iR\sin\theta d\phi).
\end{aligned}$$

After following the above described prescription for averaging we obtain following nontrivial zero-order Cartan scalars

$$\psi_2 = -\tfrac{1}{6}(R_{,r})^{-1}R_{,ttr} + \tfrac{1}{6}R^{-1}R_{,t}(R_{,r})^{-1}R_{,tr} + \tfrac{1}{6}R^{-1}R_{,tt} - \tfrac{1}{6}R^{-2}(R_{,t})^2,$$
$$\phi_{00'} = \phi_{22'} = \tfrac{1}{2}R^{-1}R_{,t}(R_{,r})^{-1}R_{,tr} - \tfrac{1}{2}R^{-1}R_{,tt},$$
$$\phi_{11'} = -\tfrac{1}{4}(R_{,r})^{-1}R_{,ttr} + \tfrac{1}{4}R^{-2}(R_{,t})^2,$$
$$\Lambda = \tfrac{1}{12}(R_{,r})^{-1}R_{,ttr} + \tfrac{1}{6}R^{-1}R_{,t}(R_{,r})^{-1}R_{,tr} + \tfrac{1}{6}R^{-1}R_{,tt} + \tfrac{1}{12}R^{-2}(R_{,t})^2.$$

In the leading order all averages are equal to zero except

$$\langle\Lambda\rangle = \frac{1}{2}\langle\psi_{,t}^2\rangle. \tag{2}$$

First order scalars are (in the leading order) all equal to zero. Comparison with known solutions evidently leads to de Sitter space, correlation term is in the form of a positive cosmological constant.

4.2. Onion model

We consider approximate LTB solution by Biswas, Mansouri and Notari - a spacetime with radial shells of overdense and underdense regions. It is given by $E(r) > 0$, $a(t,r) := \frac{R(t,r)}{r}$ with

$$a(t,r) := \left(\frac{6}{\pi}\right)^{1/3} t^{2/3}(1 + Lt^{2/3}\frac{1}{r}\sin^2\pi r)$$

and can be interpreted as a perturbation of the flat dust FRW if we consider L as a small parameter.

Cartan scalar Λ reads

$$\Lambda(r,t,L) = \frac{1}{12}\frac{1}{a^2(a + a_{,r})r}[3a^2a_{,tt} + a^2ra_{,ttr} + 3a(a_{,t})^2$$
$$+2a_{,tt}aa_{,r}r + 2a_{,t}raa_{,tr} + (a_{,t})^2ra_{,r} + a_{,r}rK + 3Ka + aK_{,r}].$$

Function $K(r,L)$ is related to curvature function $E(r)$ by

$$K(r,L) = -\frac{2E(r)}{r^2} = \frac{-L}{\pi r}\sin\pi r\sin\pi r \tag{3}$$

Here we perform spatial averaging only and expand in the powers of L (coefficients A, B, C, D are given by averaging)

$$\langle\Lambda\rangle \approx \frac{A}{t^2} + \frac{B}{t^{4/3}}L + \frac{C}{t^{2/3}}L^2 + \frac{D}{t^0}L^3$$

Individual terms as interpreted with respect to Einstein-de Sitter model have the following meaning (beyond first term they represent correlation terms):

(i) standard dust
(ii) behaves like curvature
(iii) causes acceleration equivalent to $p = -\frac{2}{3}\rho$
(iv) cosmological constant

Other zero order Cartan scalars are at most linear in L. However this means that the interpretation given above should be treated with caution beyond second term.

5. Conclusion

We have seen that using Cartan scalars for averaging provides conceptually straightforward method. However, it presents challenging technical (for direct method) and interpretation problems (for indirect one). The correlation term is just a cosmological constant in the simple model. However it is rather complex in the more realistic Onion model.

Acknowledgments

We would like to thank J. E. Åman and M. Bradley for providing us algebraic program SHEEP which was used for computation of Cartan scalars. P.K. was supported by grant SVV-267301. O.S. was supported by grant GAČR 14-37086G.

References

[1] Buchert T 2000 *Gen. Rel. Grav* **32** 105
[2] Buchert T 2001 *Gen. Rel. Grav* **33** 1381
[3] Zalaletdinov R M 1992 *Gen. Rel. Grav* **24** 1015
[4] Zalaletdinov R M 1993 *Gen. Rel. Grav* **25** 673
[5] Korzyński M 2010 *Class. Quantum Grav.* **27** 105015
[6] Brannlund J, van den Hoogen R and Coley A 2010 *Int.J.Mod.Phys. D* **19** 1915
[7] Coley A 2010 *Class. Quantum Grav.* **27** 245017
[8] Cartan E 1946 *Leçons sur la Geometrie des Espaces de Riemann* (Paris, Gauthier-Villars).
[9] Karlhede A 1980 *Gen. Rel. Grav.* **12** 693
[10] Biswas T, Mansouri R and Notari A 2007 *JCAP* **12** 0712

Cauchy horizon stability and mass inflation with a cosmological constant

João L. Costa(1,2), **Pedro M. Girão**(2), **José Natário**(2) **and Jorge Drumond Silva**(2)

(1) ISCTE - Instituto Universitário de Lisboa, Portugal, (2) Center for Mathematical Analysis, Geometry and Dynamical Systems, Instituto Superior Técnico, Universidade de Lisboa, Portugal.

E-mail: jlca@iscte.pt, pgirao@math.ist.utl.pt, jnatar@math.ist.utl.pt, jsilva@math.ist.utl.pt

Abstract. Motivated by the strong cosmic censorship conjecture, we consider the Einstein-Maxwell-scalar field system with a cosmological constant Λ (of any sign), under spherical symmetry, for characteristic initial conditions, with outgoing data prescribed by a (complete) subextremal Reissner-Nordström black hole event horizon. We study the structure of the future maximal (globally hyperbolic) development, analyze the mass inflation scenarios, identifying, in particular, large choices of parameters for which the Hawking mass remains bounded, and study the existence of regular extensions. We also discuss why our results, although valid for all signs of Λ, only provide evidence for the failure of strong cosmic censorship in the case of a positive cosmological constant.

1. Introduction

The existence and stability of Cauchy horizons is intimately related to the question of global uniqueness for the Einstein equations and, in particular, to the celebrated strong cosmic censorship conjecture. To study this, we consider the following problem:

Problem. Given spherically symmetric characteristic initial data for the Einstein-Maxwell-scalar field system with a cosmological constant Λ, with the data on the outgoing part of the null initial hypersurface prescribed by a (complete) subextremal Reissner-Nordström black hole event horizon, with non-vanishing charge, and the remaining data otherwise free, verify if the corresponding maximal globally hyperbolic development is future inextendible as a suitably regular Lorentzian manifold.

This is a direct extension of the framework of [1] with the introduction of a cosmological constant in the field equations.

Here, we will review some of the results obtained in a recent series of papers [2,3,4] dedicated to the above problem, where we generalize the results of Dafermos concerning the stability of the radius function, for a cosmological constant of any sign. This has the remarkable consequence of allowing continuous extensions of the metric across the Cauchy horizon. We analyze the mass inflation scenarios and, in particular, identify large choices of parameters for which the Hawking mass remains bounded. Then we carefully unveil the consequences of this last fact concerning the existence of regular extensions of the metric beyond the Cauchy horizon. More precisely, we obtain continuous extensions of the metric across the Cauchy horizon with square integrable Christoffel symbols and, under slightly stronger conditions, we construct (non-isometric) extensions which are classical solutions of the Einstein equations. To the best of our knowledge, these are the first results of this kind.

2. Framework

The Einstein-Maxwell-scalar field system with a cosmological constant Λ is given by

$$R_{\mu\nu} - \frac{1}{2}Rg_{\mu\nu} + \Lambda g_{\mu\nu} = 2T_{\mu\nu}, \tag{1}$$

with $T_{\mu\nu}$ obtained from the sum of the stress energy tensor of a massless scalar field ϕ, satisfying the wave equation, and an electromagnetic field F solving the (source free) Maxwell equations.

Imposing the following (spherically symmetric) ansatz

$$g = -\Omega^2(u,v)du\,dv + r^2(u,v)\left(d\theta^2 + \sin^2\theta\, d\varphi^2\right), \tag{2}$$

with u the ingoing and v the outgoing null coordinates, respectively, the electromagnetic field completely decouples:

$$F = -\frac{e\Omega^2(u,v)}{2r^2(u,v)}du \wedge dv. \tag{3}$$

Introducing the *renormalized Hawking mass* ϖ through

$$1 - \frac{2\varpi}{r} + \frac{e^2}{r^2} - \frac{\Lambda}{3}r^2 = -\frac{4\partial_u r \partial_v r}{\Omega^2}, \tag{4}$$

the field equations reduce to

$$\partial_u\partial_v\phi = -\frac{\partial_u r\,\partial_v\phi}{r} - \frac{\partial_v r\,\partial_u\phi}{r}, \tag{5}$$

$$\partial_u\partial_v r = \partial_u r\,\partial_v r\,\frac{\frac{2\varpi}{r^2} - \frac{2e^2}{r^3} - \frac{2\Lambda}{3}r}{1-\frac{2\varpi}{r}+\frac{e^2}{r^2}-\frac{\Lambda}{3}r^2}, \tag{6}$$

$$\partial_u\varpi = \left(1 - \frac{2\varpi}{r} + \frac{e^2}{r^2} - \frac{\Lambda}{3}r^2\right)\frac{(r\partial_u\phi)^2}{2\partial_u r}, \tag{7}$$

$$\partial_v\varpi = \left(1 - \frac{2\varpi}{r} + \frac{e^2}{r^2} - \frac{\Lambda}{3}r^2\right)\frac{(r\partial_v\phi)^2}{2\partial_v r}. \tag{8}$$

These are a wave equation for the scalar field (5), a wave equation for the radial function (6), and equations prescribing the gradient of the mass (7) and (8); from these we can easily derive the remaining unknowns: the metric component Ω from (4) and the Faraday tensor F from (3).

3. The Reissner-Nordström solution and strong cosmic censorship

The first reason to consider the previous Einstein-matter system is that it admits the full Reissner-Nordström (RN) family of solutions, where we include all the anti-de Sitter ($\Lambda < 0$) the asymptotically flat ($\Lambda = 0$) and the de Sitter ($\Lambda > 0$) subfamilies, as particular solutions. This family is obtained by setting $\phi \equiv 0$. Then, the usual mass and charge parameters are given by $M = \varpi$ and $Q = 4\pi e$.

In this context, it is natural to try to derive the black hole solutions of this family from a characteristic initial value problem for the system (5)-(8). To do that in a unified way for all signs of Λ, we can prescribe their corresponding characteristic data on the event horizon: the future maximal globally hyperbolic developments obtained will then coincide with the shaded region in figure 1. But such developments can be continued to the future in a highly non unique

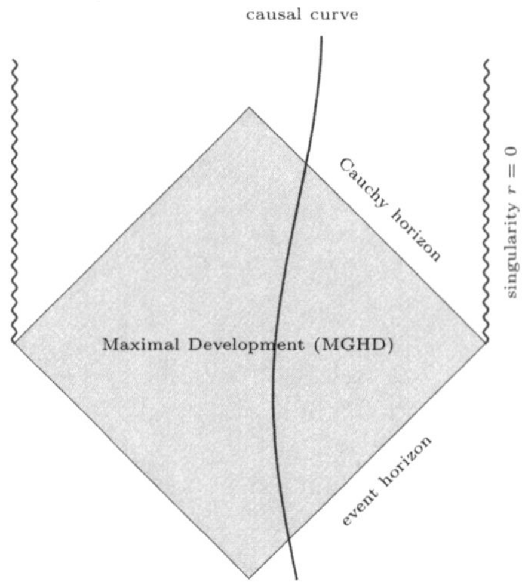

Figure 1. Part of the future of the event horizon of RN black holes.

way, with the metric extending smoothly. This well known fact is quite disturbing since what happens in such extensions is not determined, in any way, by the prescribed initial data. In such extensions, a Cauchy horizon arises as the boundary of the maximal development, signaling the failure of global uniqueness.

This puts into question the deterministic character of general relativity. A heuristic argument devised by Simpson and Penrose [5] suggests that small perturbations might turn the Cauchy horizon into a singularity, via a blue-shift instability, beyond which spacetime cannot be continued in any meaningful way. As a consequence, global uniqueness might still hold as a generic feature of initial value problems for the Einstein equations. These are the core ideas behind the *strong cosmic censorship conjecture* (SCC). Later, Israel and Poisson [6] identified the blow up of the Hawking mass as a source for the expected Cauchy horizon instability, a process known as *mass inflation*.

The previous discussion motivates the problem in the Introduction. Regarding the specific choices made there, the self-gravitating real massless scalar field is the simplest non-pathological matter model with dynamical degrees of freedom in spherical symmetry; moreover it exhibits a wavelike behavior reminiscent of the general Einstein vacuum equations. A non-vanishing charge parameter is needed to exclude the Schwarzschild subfamily, whose solutions do not contain a Cauchy horizon to start with.

We introduce a cosmological constant because it plays a fundamental role in modern physics and, from a purely mathematical point of view, provides rich geometrical structures and dynamical behaviors. But it is not clear why it should matter in the context of the SCC. Note, for instance, that our problem concerns what happens in the interior of a black hole region, whereas the effects of the cosmological constant are specially relevant at large scales. Nonetheless, during the late 90s, there was a considerable amount of activity concerning the SCC with $\Lambda > 0$. The first impressions where that the SCC should fail. But, later on, such claims were dismissed by arguing that the conclusions were based on over-simplified models that did not take into proper account the effects of backscattering (see Section 15.2 of [7] for a detailed discussion).

A full non-linear analysis, for instance as provided by our problem, was clearly in order.

4. Regularity of extensions and Cauchy horizon stability

In the formulation of our problem, a fundamental role is played by the regularity that we allow the extensions to have, beyond the maximal globally hyperbolic development. It turns out that this regularity may be viewed as a measure of the stability of the Cauchy horizon: the more regular the extensions are, the more stable the Cauchy horizon can be thought of. We will now briefly discuss some regularity requirements on the metric.

Inextendibility of the metric in C^2: This is, first of all, motivated by the fact that the Einstein equations are of second order. Also, the blow up of the Kretschmann scalar (a measure of tidal forces) implies this criterion. It is clear that its failure would provide overwhelming evidence against the SCC. Nonetheless, its success would not provide such a compelling argument in favor of the conjecture, since there are plenty of relevant solutions of the Einstein equations whose regularity is well below this threshold. In fact, we have constructed classical solutions which are not C^2.

Inextendibility of the metric in C^0: This corresponds to Christodoulou's choice in his original formulation of the SCC. Its validity would provide overwhelming evidence in favor of the deterministic nature of general relativity.

It turns out that neither of the two previous criteria are enough to capture the full richness of the subject. In fact, in [8], Dafermos proved the existence of (spherically symmetric) extension, with metric in $C^0 \setminus C^1$, for the $\Lambda = 0$ case.

We are thus lead to consider a third type of requirement, based on the possibility of trying to exclude extensions satisfying the field equations.

Inextendibility as a Lorentzian manifold with Christoffel symbols in $L^2_{\rm loc}$: This is enough to ensure inextendibility of the metric, even as a weak solution of the Einstein equations, and it was proposed by Christodoulou in response to Dafermos' results. Chruściel had already proposed a similar criterion with the following conditions imposed directly at the level of the metric: inextendibility with $g \in H^1_{\rm loc}$ and $g^{-1} \in L^\infty_{\rm loc}$. It turns out that, under the genericity conditions alluded to above, this criterion holds for the ($\Lambda = 0$) solutions of Dafermos.

5. Main results

Given a reference subextremal RN solution, from whose event horizon we take the outgoing initial data (recall that the remaining data is fixed freely), we first show that it has a unique maximal development $(\mathcal{M}, g, \phi)$ with the metric (2) and scalar field ϕ defined on $\mathcal{M} = \mathcal{P} \times_r S^2$, where $\mathcal{P}$ is a past set $\mathcal{P} \subset [0, U] \times [0, \infty[$. In our coordinates, the Cauchy horizon of the the reference RN solution is given by $v = +\infty$.

Stability of radius function at the Cauchy horizon. We show that there exists $U > 0$ and $r_0 > 0$ for which

$$r(u, v) > r_0, \text{ for all } (u, v) \in [0, U] \times [0, \infty[\subset \mathcal{P}.$$

So, we see that the symmetry orbits do not collapse to points on $[0, U] \times [0, \infty[$. This is a stability result for the Cauchy horizon. We have that $(\mathcal{M}, g, \phi)$ extends, across the Cauchy horizon $\{v = \infty\}$, to $(\hat{\mathcal{M}}, \hat{g}, \hat{\phi})$, with $\hat{g}$ and $\hat{\phi}$ in C^0.

We now need to introduce the quantity ρ defined as the ratio between the surface gravity of the Cauchy horizon and the surface gravity of the event horizon of the reference solution. Note that $\rho > 1$ and that it provides a measure of how close the initial data is from extremality; the case $\rho = 1$, which we have excluded from beginning, corresponds to the extremal case where the radius of the event horizon and the radius of the Cauchy horizon coincide.

After controlling r, the fundamental quantities that will determine the regularity of extensions are the mass ϖ and the radial derivative of the scalar field along lines of fixed u, $\partial_r(\phi|_{u=c})$. As we will see, our control over these quantities will depend on the decay rate imposed on the free data $\phi(\,\cdot\,, v=0)$ and on how close we are to extremality.

Mass inflation. If $\rho > 2$ and

$$\partial_u\phi(u,0) \geq cu^s, \text{ for some } 0 < s < \frac{\rho}{2} - 1,$$

then

$$\lim_{v\to\infty} \varpi(u,v) = \infty, \text{ for each } 0 < u \leq U.$$

In particular, no C^1 (spherically symmetric) extensions across the Cauchy horizon exist, since ϖ is at the level of first derivatives of the metric. In fact, the Christodoulou-Chruściel inextendibility criterion holds. We stress the fact that we are only considering the existence of spherically symmetric extensions: for instance, when considering general extensions, the blow up of the mass, a priori, only excludes the existence of C^2 extensions, since it leads to the blow up of the Kretschmann scalar. The existence, under these circumstances, of non-spherically symmetric extensions with metric in C^1 is an open problem, to our knowledge.

No mass inflation. If

$$|\partial_u\phi(u,0)| \leq cu^s, \text{ for some } s > \frac{7\rho}{9} - 1,$$

then there exists $C > 0$ such that,

$$|\varpi(u,v)| < C, \text{ for all } (u,v) \in [0,U] \times [0,\infty[.$$

Moreover, the Christodoulou-Chruściel inextendibility criterion fails, i.e. $(\mathcal{M}, g, \phi)$ extends, across the Cauchy horizon, to $(\hat{\mathcal{M}}, \hat{g}, \hat{\phi})$, with $\hat{g}$ and $\hat{\phi}$ in C^0, Christoffel symbols $\hat{\Gamma}$ in L^2_{loc}, and $\hat{\phi}$ in H^1_{loc}.

Bounding $\partial_r(\phi|_{u=c})$. If

$$|\partial_u\phi(u,0)| \leq cu^s, \text{ for some } s > \frac{13\rho}{9} - 1,$$

then, there exists $C > 0$ such that,

$$|\partial_r(\phi|_{u=c})| < C, \text{ for all } (u,v) \in [0,U] \times [0,\infty[.$$

Consequently, $(\mathcal{M}, g, \phi)$ extends (in a non-unique way), across the Cauchy horizon, to a spherically symmetric (classical) solution $(\hat{\mathcal{M}}, \hat{g}, \hat{\phi})$ of the Einstein-Maxwell-scalar field system with cosmological constant Λ, with $\hat{g}$ and $\hat{\phi}$ in C^1. Moreover, the Kretschmann scalar is uniformly bounded. To the best of our knowledge, these are the first results where the generic existence of extensions as solutions of the Einstein equations is established.

6. Remarks about the strong cosmic censorship conjecture

Strictly speaking, our results, which hold for all signs of Λ, do not apply directly to the SCC. The reason for this is that we are considering exact RN data on the initial outgoing hypersurface, with our dynamical degrees of freedom provided by the free data on the ingoing hypersurface. So our results only cover the case where the scalar field has compact support on the event horizon. This scenario is not expected to arise from the gravitational collapse of generic (and appropriate) Cauchy (for $\Lambda \geq 0$) or Cauchy and boundary (for $\Lambda < 0$) initial data, to which a proper formulation of the SCC should refer to.

In fact, for $\Lambda = 0$, the expected generic behavior along the event horizon is described by a Price's law of the form

$$|\partial_v \phi(0, v)| \sim v^{-p},$$

for some $p > 0$ and v a Eddington-Finkelstein coordinate as above. For initial data satisfying this law, in [8], Dafermos has shown that, for all $1 < \rho < \infty$, mass inflation occurs and the Christodoulou-Chruściel criterion holds (see also [7]).

For $\Lambda < 0$, in the general non-symmetric case, the decay along the horizon is expected to be even slower, and the process of gravitational collapse is far less understood. Therefore we excuse ourselves from extrapolating from our results (which, recall, hold when $\Lambda < 0$ as well).

For $\Lambda > 0$ the expected Price law provides "fast" decay

$$|\partial_v \phi(0, v)| \sim e^{-\Delta v}, \tag{9}$$

for some positive Δ, which leads to an exponential approach to the stationary data considered here. So, in the $\Lambda > 0$ case, the *no mass inflation* scenario and some of its remarkable consequences are expected to be true for dynamical data close to extremality $\rho \approx 1$ and satisfying (9). This puts the validity of the SCC, in the $\Lambda > 0$ case, into question.

From the previous discussion it is clear that the stationary data case at the event horizon, discussed here, is the most natural setting to study, in a unified framework, the influence of the sign of Λ in the stability of Cauchy horizons. Our results and techniques show that *differences concerning Cauchy horizon stability and mass inflation when Λ changes sign are due solely to the differences in the decay rates of ϕ along the event horizon, as described by a Price's law.*

[1] Dafermos M 2003 Stability and instability of the Cauchy horizon for the spherically symmetric Einstein-Maxwell-scalar field equations *Ann. Math.* **158** 875–928.

[2] Costa J, Girão P, Natário J and Silva J 2014 On the global uniqueness for the Einstein-Maxwell-scalar field system with a cosmological constant. Part 1. Well posedness and breakdown criterion *Class. Quant. Grav.* **to appear** (*Preprint* gr-qc/1406.7245)

[3] Costa J, Girão P, Natário J and Silva J 2014 On the global uniqueness for the Einstein-Maxwell-scalar field system with a cosmological constant. Part 2. Structure of the solutions and stability of the Cauchy horizon (*Preprint* gr-qc/1406.7253)

[4] Costa J, Girão P, Natário J and Silva J 2014 On the global uniqueness for the Einstein-Maxwell-scalar field system with a cosmological constant. Part 3. Mass inflation and extendibility of the solutions (*Preprint* gr-qc/1406.7261)

[5] Simpson M and Penrose R 1973 Internal instability in a Reissner-Nordström black hole *Int. J. Theor. Phys* **7** 183–197.

[6] Poisson E and Israel W 1989 Inner-horizon instability and mass inflation in black holes *Phys. Rev. Lett* **63** 1663–1666.

[7] Dafermos M 2012 Black holes without spacelike singularities *Comm. Math. Phys.* **online first** (*Preprint* gr-qc/1201.1797)

[8] Dafermos M 2005 The interior of charged black holes and the problem of uniqueness in general relativity *Comm. Pure Appl. Math.* **58** 445–504.

Journal of Physics: Conference Series **600** (2015) 012032 doi:10.1088/1742-6596/600/1/012032

Absence of cosmological constant problem in special relativistic field theory of gravity: one-loop renormalization group

Raúl Carballo-Rubio[1], Carlos Barceló[1] and Luis J Garay[2,3]

[1] Instituto de Astrofísica de Andalucía (IAA-CSIC), Glorieta de la Astronomía, 18008 Granada, Spain

[2] Departamento de Física Teórica II, Universidad Complutense de Madrid, 28040 Madrid, Spain

[3] Instituto de Estructura de la Materia (IEM-CSIC), Serrano 121, 28006 Madrid, Spain

E-mail: raulc@iaa.es

Abstract. There exists a nonlinear theory of gravity which is not structurally equivalent to general relativity and that, in the non-interacting limit, describes a free massless particle with helicity ± 2. We have recently shown that this theory can be understood as the result of self-coupling, in complete parallelism to the well-known case of general relativity. This special relativistic field theory of gravity exhibits a decoupling of vacuum zero-point energies of matter and passes all the known experimental tests in gravitation. It is explicitly demonstrated that there is no flow of the effective cosmological constant under the action of the renormalization group at one-loop level, while simple symmetry arguments show that this would continue to be true for higher-loop corrections. The important lesson is that just mild local assumptions concerning the nature of the particle mediating the gravitational interactions are enough to motivate theories which are free of the cosmological constant problem.

1. Introduction

The cosmological constant problem appears when trying to combine the formalism of quantum field theory with the curved spacetime notions of general relativity. One of the implications of this combination is a problematic radiative instability of the cosmological constant term under the renormalization group [1]. Finding a mechanism which can stabilize the value of the cosmological constant, while keeping intact the low-energy physics as to pass the stringent experimental tests on deviations from general relativity, is what is usually known as the cosmological constant problem [1, 2, 3].[1]

The mainstream school of thought is that this problem would be solved in the framework of a consistent theory of quantum gravity. It is also common the theoretical expectation that the would-be theory of quantum gravity should lead exactly to general relativity in the long-wavelength limit [4]. This motivates the search of different mechanisms which can surpass the cosmological constant problem while keeping the structure of general relativity practically intact. After a long search there is no compelling model and, moreover, they usually involve unknown

[1] Here we do not address the issue of determining the actual value of the cosmological constant by means of first-principles calculations, but to alleviate this tension.

Journal of Physics: Conference Series **600** (2015) 012032 doi:10.1088/1742-6596/600/1/012032

physics [1]. With a totally different spirit, here we argue that there exists a solution to the cosmological constant problem that only rests on the well-settled principles of quantum field theory in flat spacetime, with the proposal that gravity is mediated by a helicity ± 2 graviton, and the equivalence principle as reformulated by Weinberg [5]. That only these mild local assumptions are necessary to propose a solution is the point that we would like to stress in this contribution. As we will see, the key feature is the construction of a nonlinear theory of gravity which exhibits an unbroken, anomaly-free Weyl symmetry at the semiclassical level.

Notation.– We use the metric convention $(-,+,+,+)$. Curvature-related quantities will be defined following Misner-Thorne-Wheeler's convention [6]. When the Minkowski metric η_{ab} is used, it is understood as written in a generic coordinate system. The covariant derivative ∇ is always related to the flat metric and $\mathrm{d}\mathscr{V}$ is the Minkowski volume element.

2. Self-coupling of a helicity ±2 graviton

Within the classification of fundamental interactions in terms of the unitary representations of the Poincaré group [7], gravity is expected to be mediated by a particle corresponding to the massless spin-2 representation. The massless character of the representation means that only the states with helicity ± 2 are physical. This representation has a direct on-shell implementation in field theory (see *e.g.* [8] and references therein). Let us stress that this on-shell description is the only firm statement one can draw from the assumption that gravity is mediated by a helicity ± 2 graviton only, with no admixture of spin 1 or 0. The usual way to proceed is to enlarge the functional space as well as the gauge symmetry, so that the degrees of freedom are kept the same. There exist two of such extensions: the well-known Fierz-Pauli theory and Weyl transverse theory [9].

If we want to describe gravity we need to couple these linear theories to matter. Nowadays it is well understood that this coupling will be through the stress-energy tensor, implying that the resulting theory would be nonlinear [10]. Let us write the resulting action of the self-coupling procedure as

$$\mathscr{A} = \mathscr{A}_0 + \mathscr{A}_{\mathrm{I}}, \tag{1}$$

where $\mathscr{A}_0$ is the free part and $\mathscr{A}_{\mathrm{I}}$ is the self-interacting part we are going to solve for. Given this action, one would be able to obtain its stress-energy tensor. $\mathscr{A}_{\mathrm{I}}$ is then fixed by the requeriment of leading to this stress-energy tensor (obtained by Hilbert's prescription) as the source of the equations of motion. One just needs to expand $\mathscr{A}_{\mathrm{I}} = \sum_{n=1}^{\infty} \lambda^n \mathscr{A}_n$ in terms of the coupling constant λ and compare different orders in the coupling constant λ to obtain the set of iterative equations of the self-coupling problem [11]:

$$\frac{\delta \mathscr{A}_n}{\delta h^{ab}} = \lim_{\gamma \to \eta} \frac{\delta \mathscr{A}_{n-1}}{\delta \gamma^{ab}}, \qquad n \geq 1. \tag{2}$$

Notice that to obtain the right-hand side of these equations one needs to write the partial actions $\mathscr{A}_n$ in terms of an auxiliary metric γ_{ab}. In this procedure different choices can be made in the form of non-minimal couplings to the auxiliary metric [8].

It can be seen that general relativity is a solution to these iterative equations leading, at the lowest order, to Fierz-Pauli theory [8, 12, 13]. In complete parallelism, we have recently shown that Weyl transverse gravity is also a solution to these iterative equations, leading at the lowest order to Weyl transverse theory [14]. Weyl transverse gravity is defined by an action motivated by unimodular gravity:

$$\mathscr{A} = \frac{1}{\lambda^2} \int \mathrm{d}\mathscr{V} R(\hat{g}), \tag{3}$$

where $R(\hat{g})$ has the same functional form as the Ricci scalar of a metric $\hat{g}_{ab}$ whose determinant is constrained by the condition $\det(\hat{g}) = \det(\eta)$. However, for us $\hat{g}_{ab}$ is just a tensor field which lives in a flat background. To make this explicit, let us use the well-known fact [15] that one can express the Ricci scalar in terms of the covariant derivatives associated with the flat metric η_{ab}, denoted by ∇, and integrate by parts to put the action in the form:

$$\mathscr{A} = \frac{1}{4\lambda^2} \int \mathrm{d}\mathscr{V} \left[2\hat{g}_{bj}\delta^a_k\delta^i_c - \hat{g}^{ai}\hat{g}_{bj}\hat{g}_{ck} \right] \nabla_a \hat{g}^{bc} \nabla_i \hat{g}^{jk}. \tag{4}$$

Written in this form, it is clear that it describes a special relativistic field theory, albeit nonlinear. This form of the action is the natural one in the framework of the self-coupling problem. Starting with (3) is just convenient to make contact with our knowledge of general relativity but, indeed, one first obtains the action (4), which can be optionally supplemented with an irrelevant surface term to give (3); see [8] for an analogue discussion in the case of general relativity. The covariant notation we are using makes it clear that this theory is invariant under general coordinate transformations. Moreover, by construction this action is invariant under transverse diffeomorphisms, whose infinitesimal form is

$$\delta_\xi \hat{g}^{ab} = \mathcal{L}_\xi \hat{g}^{ab}, \qquad \nabla_a \xi^a = 0. \tag{5}$$

The field $\hat{g}_{ab}$ is given by

$$\hat{g}_{ab} = \kappa^{-1/4} g_{ab}, \tag{6}$$

where $\kappa := \det(g)/\det(\eta)$, so that the action is invariant under Weyl transformations, whose infinitesimal version is

$$\delta_\omega g_{ab} = \omega\, g_{ab}. \tag{7}$$

As a result, there is no distinction from the perspective of self-coupling between general relativity and the theory presented here. Both nonlinear theories lead to the kind of particle one expects to mediate the gravitational interaction in the non-interacting limit and can be explicitly constructed by self-coupling this particle. The only distinction between them is their internal symmetry group. In previous analyses such as [12, 16] it was always implicitly assumed that the gauge symmetry characteristic of a helicity ± 2 graviton is that of Fierz-Pauli theory. For example in Appendix B of [16], which is nowadays considered as the definite reference as far as the uniqueness of the construction leading to general relativity is concerned, the gauge transformations of Fierz-Pauli theory are extensively used. If one accepts this internal symmetry structure, then general relativity is the only consistent nonlinear theory one would find which preserves the notion of gauge invariance [8]. Nevertheless, this is an assumption which is not necessary from the perspective of special relativistic field theory and, moreover, it is ultimately motivated by the knowledge of the symmetry group of general relativity.

To complete the construction we need to consider the inclusion of matter. The integration of the matter part of the iterative equations is straightforward, although the result is not unique. But we can appeal to two principles to resolve its non-uniqueness. The first one is the quantum version of the equivalence principle shown as a consequence of Poincaré invariance in [5]: the coupling of the field which describes gravity to matter and to itself must be governed by the same coupling constant λ. The second principle rests in the observation that conformal invariance is not a symmetry of matter in our world. A safe way to proceed is then to consider that matter fields are not affected by the Weyl transformations (7) (in the language of [17], matter fields are inert under Weyl transformations). Both principles imply that the resulting action of matter is obtained by replacing η_{ab} with the composite field $\hat{g}_{ab}$. Non-minimal couplings to $\hat{g}_{ab}$ are allowed by construction. Notice that no constraints are imposed on the matter sector.

3. A low-energy solution to the cosmological constant problem

When expressed in terms of the composite variable $\hat{g}_{ab}$, the nonlinear theory of gravity we are considering here has the form of general relativity plus matter, but it is a theory formulated in flat spacetime and the determinant of the metric is fixed by $\det(\hat{g}) = \det(\eta)$. On these grounds it can be understood as an extension of unimodular gravity which is invariant under general coordinate transformations. It is well-known that in unimodular gravity vacuum zero-point energies of matter do not gravitate. Here this is also true: the condition on the determinant implies that the corresponding (regulated) contribution in the effective action would have the form

$$\int \mathrm{d}\mathscr{V} E_{\mathrm{vac}}. \tag{8}$$

These energies act as a mere constant shift in the effective action, as we are used to in any special relativistic quantum field theory which does not contain gravity. Technically, the contributions of vacuum bubbles cancel out of correlation functions as a result of the linked-cluster theorem. This does not happen when considering general relativity (as an effective quantum field theory [18]) instead, leading to the cosmological constant problem [1, 3].

The mechanism behind this result is clearly the invariance of the theory under Weyl transformations of the gravitational field, whose infinitesimal version is given by (7).[2] This symmetry forbids radiative corrections which would otherwise couple to the determinant of the field g_{ab}. Is is not difficult to show this at one-loop level through *e.g.* the effective action scheme [19, 20, 21]. Given a theory of matter with an arbitrary combination of matter fields with different spin (0, 1/2 and 1), minimally coupled to $\hat{g}_{ab}$ as we have argued above, the heat kernel expansion permits to write the regulated effective action in terms of a cutoff κ so that one can take account of the necessary counterterms. In doing this we find three kinds of contributions. The first one is a zeroth-order contribution similar to (8). This contribution is irrelevant as we know from standard quantum field theories in flat spacetime such as electrodynamics [19]. The next contribution has the same functional form than (3), leading to a renormalization of the coupling constant λ. If we call λ_0 the bare coupling constant (whose square is proportional to the bare gravitational constant G_0) then one can read from the one-loop action the equation

$$\frac{1}{\lambda^2} = \frac{1}{\lambda_0^2} + C_1\kappa^2 + C_2 \log\left(\frac{\kappa}{C_3}\right), \tag{9}$$

where C_1, C_2 and C_3 are constants with convenient physical dimensions and whose value depends on the particle content of the matter sector [20]. There exist other contributions which involve quadratic expresions in the "Ricci" tensor $R_{ab}(\hat{g})$ but we are going to ignore these, which also appear in general relativity and are expected to lead to higher-derivative deviations from the second-order field equations at high energies. Thus in essence we have only one physically relevant renormalization group equation, given by (9). The invariance of the theory under Weyl transformations implies, in contrast with general relativity, the absence of any renormalization group equation describing the running of the cosmological constant sector. Most importantly, it seems perfectly possible to realize this symmetry in a non-anomalous way at the semiclassical level: the background structure permits to construct Euclidean path integral measures for matter which are invariant both under transverse diffeomorphisms as well as Weyl transformations (with inert matter). These measures would be defined in terms of scalar products in which the corresponding evolution operators are self-adjoint [21]. The extra structure permits the

[2] Indeed, the extension of this symmetry to matter in the form of the usual conformal transformations was one of the first options considered (and discarded) in the literature to deal with the cosmological constant problem [1]. Previous formulations failed to notice however that, if matter fields are not affected under these transformations, one can ensure both radiative stability of the cosmological constant sector while having non-zero mass terms for matter.

construction of scalar products invariant under the internal symmetries. For example, for a real scalar field one would have simply

$$(\phi_1, \phi_2) = \int \mathrm{d}\mathscr{V} \phi_1 \phi_2. \tag{10}$$

The same symmetry arguments imply that the theory does not contain any bare cosmological constant, *i.e.* Weyl symmetry forbids the occurrence of such a constant in the action. However, this theory describe gravity in a way which is compatible with all the known experiments in gravitation, as the corresponding field equations contain solutions for arbitrary values of the cosmological constant. It is easy to see that the field equations are, by construction, traceless. In particular, in the gauge $\det(g) = \det(\eta)$ they reduce to the trace-free Einstein equations:

$$R_{ab} - \frac{1}{4} R g_{ab} = \lambda^2 \left(T_{ab} - \frac{1}{4} T g_{ab} \right). \tag{11}$$

As a consequence, any solution of the Einstein field equations for arbitrary values of the cosmological constant is imported to this theory. In principle one could expect additional solutions in which the Ricci scalar and the trace of the stress-energy momentum tensor are not tied up through a constant but a general function, but this is not the case.

This can be seen by means of the following argument. Given the fact that the canonical stress-energy tensor of gravity and matter is conserved under solutions [15], we recover the Einstein field equations with a phenomenological integration constant, unrelated to zero-point energies of matter, playing the role of a cosmological constant; this is shown in Sec. VI-A of [8]. The arguments above show that this effective cosmological constant is not renormalized by radiative corrections: its value is protected by local symmetries. The resulting field equations include potential energies in the matter sector as, even if gravity is not directly coupled to these terms by construction, they inevitably appear in the definition of the canonical stress-energy tensor. Notice that the conservation of the canonical stress-energy tensor is not an additional equation on its own, but a consequence of the translational symmetry of the theory and the field equations (11). This is in contrast with the situation in usual formulations of unimodular gravity (see the corresponding discussion in [8] and references therein).

Radiative corrections are intrinsically a feature of the semiclassical theory in which matter fields are quantized, so that the form of the matter action is fundamental to address any question concerning them. In Weyl transverse gravity, Weyl invariance dictates a fairly different coupling between gravity and matter fields with respect to the situation in general relativity. Although it should be clear that any argument trying to read the form of these contributions from the form of the classical equations of motion (maybe in some specific gauge) alone is misleading, there still exist (somewhat inexplicably) confusion in the literature in this respect. A well-known procedure to obtain the form of these radiative corrections leads to a definite answer which settles down the issue of radiative stability unambiguosly, as dictated by the form of the symmetries of the gravitational action.

In sumary, Weyl transverse gravity is a promising candidate to reconcile the classical and quantum-mechanical aspects of the cosmological constant problem. Any theory of emergent gravity which is described by means of this theory in the infrared regime would surpass this problem: on the one hand, Weyl invariance implies that there is no bare cosmological constant at the level of the action nor radiative corrections to it when semiclassical effects are taken into account; on the other hand, one recovers the same predictions than in classical general relativity at the purely classical level. As a result one can construct classical solutions to the gravitational field equations whose cosmological constant is guaranteed to be stable (indeed, unchanged) under radiative corrections due to the protection of the underlying symmetries. And, maybe surprisingly, all the necessary input to obtain this is deceptively simple: that gravity is mediated by gravitons.

Acknowledgments
Financial support was provided by the Spanish MICINN through Projects No. FIS2011-30145-C03-01 and FIS2011-30145-C03-02 (with FEDER contribution), and by the Junta de Andalucía through Project No. FQM219. R. C-R. acknowledges support from CSIC through the JAE-predoc program, cofunded by FSE.

[1] C. P. Burgess. The Cosmological Constant Problem: Why it's hard to get Dark Energy from Micro-physics. *ArXiv e-prints*, September 2013.

[2] Steven Weinberg. The cosmological constant problem. *Rev. Mod. Phys.*, 61:1–23, Jan 1989.

[3] J. Martin. Everything you always wanted to know about the cosmological constant problem (but were afraid to ask). *Comptes Rendus Physique*, 13:566–665, July 2012.

[4] S. Carlip. Challenges for emergent gravity. *Studies in History and Philosophy of Science Part B: Studies in History and Philosophy of Modern Physics*, (0):–, 2013.

[5] S. Weinberg. Photons and Gravitons in S-Matrix Theory: Derivation of Charge Conservation and Equality of Gravitational and Inertial Mass. *Physical Review*, 135:1049–1056, August 1964.

[6] C.W. Misner, K.S. Thorne, and J.A. Wheeler. *Gravitation*. Number pt. 3 in Gravitation. W. H. Freeman, 1973.

[7] E. Wigner. On unitary representations of the inhomogeneous lorentz group. *Annals of Mathematics*, 40(1):pp. 149–204, 1939.

[8] Carlos Barceló, Raúl Carballo-Rubio, and Luis J. Garay. Unimodular gravity and general relativity from graviton self-interactions. *Phys.Rev.*, D89:124019, 2014.

[9] E. Álvarez, D. Blas, J. Garriga, and E. Verdaguer. Transverse Fierz Pauli symmetry. *Nuclear Physics B*, 756:148–170, November 2006.

[10] T. Ortín. *Gravity and Strings*. Cambridge Monographs on Mathematical Physics. Cambridge University Press, 2007.

[11] S. N. Gupta. Gravitation and Electromagnetism. *Physical Review*, 96:1683–1685, December 1954.

[12] S. Deser. Self-interaction and gauge invariance. *General Relativity and Gravitation*, 1:9–18, March 1970.

[13] L. M. Butcher, M. Hobson, and A. Lasenby. Bootstrapping gravity: A consistent approach to energy-momentum self-coupling. *Physical Review D*, 80(8):084014, October 2009.

[14] C. Barceló, R. Carballo-Rubio, and L. J. Garay. Absence of cosmological constant problem in special relativistic field theory of gravity. *ArXiv e-prints*, June 2014.

[15] N. Rosen. General relativity and flat space. i. *Phys. Rev.*, 57:147–150, Jan 1940.

[16] D. G. Boulware and S. Deser. Classical general relativity derived from quantum gravity. *Annals of Physics*, 89:193–240, January 1975.

[17] Enrique Alvarez and Roberto Vidal. Weyl transverse gravity (WTDiff) and the cosmological constant. *Phys.Rev.*, D81:084057, 2010.

[18] C. P. Burgess. Quantum Gravity in Everyday Life: General Relativity as an Effective Field Theory. *Living Reviews in Relativity*, 7:5, April 2004.

[19] D. V. Vassilevich. Heat kernel expansion: user's manual. *Phys. Rep.*, 388:279–360, December 2003.

[20] M. Visser. Sakharov's Induced Gravity. *Modern Physics Letters A*, 17:977–991, 2002.

[21] V. Mukhanov and S. Winitzki. *Introduction to Quantum Effects in Gravity*. Cambridge University Press, 2007.

Spanish Relativity Meeting (ERE 2014): almost 100 years after Einstein's revolution IOP Publishing
Journal of Physics: Conference Series **600** (2015) 012033 doi:10.1088/1742-6596/600/1/012033

Do transient white holes have a place in Nature?

Carlos Barceló[1], Raúl Carballo-Rubio[1], Luis J. Garay[2,3], and Gil Jannes[4]

[1] Instituto de Astrofísica de Andalucía (IAA-CSIC), Glorieta de la Astronomía, 18008 Granada, Spain
[2] Departamento de Física Teórica II, Universidad Complutense de Madrid, 28040 Madrid, Spain
[3] Instituto de Estructura de la Materia (IEM-CSIC), Serrano 121, 28006 Madrid, Spain
[4] Modelling & Numerical Simulation Group, Universidad Carlos III de Madrid, Avda. de la Universidad 30, 28911 Leganés, Spain

E-mail: carlos@iaa.es, raulc@iaa.es, luisj.garay@ucm.es, gil.jannes@uc3m.es

Abstract. The white-hole sector of Kruskal's solution is almost never used in physical applications. However, it can provide a radically different take on the gravitational collapse process, avoiding the problems appearing within the standard paradigm. In this contribution we will try to draw attention to some bouncing geometries that make a democratic usage of the black and white sectors of Kruskal's solution. We will argue that this type of behaviour could be perfectly natural in some approaches to the next physical level beyond classical General Relativity.

1. Theoretical vs astrophysical views on black holes

In general informative publications as well as in the scientific literature the term black hole is used with no distinction nor qualification to represent both, characteristic configurations of theoretical General Relativity and astrophysical objects possessing certain features. However, these two notions need not coincide. Strictly speaking, a General Relativity black hole is a geometry with a region causally disconnected from infinity: the region inside the event horizon. At present we dare saying that most specialists think that an event horizon is a too-idealized notion and that it will have no place in a next-level theory replacing General Relativity, *e.g.* when quantum effects are taken into account. It is largely believed that these strict black holes will be substituted by regularly evaporating black holes and, correspondingly, event horizons substituted by long-lived trapping horizons [1].

On the other side, an astrophysical black hole is nothing but a very dark and compact region containing such a large amount of mass, measured by its gravitational influence in the surrounding matter, that standard General Relativity does not offer other possibility for it than being a black hole. However, one has to take into account that the very notion of event horizon makes it impossible that any astrophysical observation could verify their existence in any of the candidate dark and compact regions. For this reason, before entering into wild philosophical speculations about the implications of the existence of event horizons, it appears more reasonable to propose alternative and verifiable (at some level) hypothesis regarding the nature of the objects hidden in this dark and compact regions.

As already mentioned, the most common hypothesis is that these objects are regularly evaporating black holes. However, although this hypothesis is in principle verifiable, in practice it seems out of reach for the human kind in any foreseeable future. This is what constitutes the lifetime problem of evaporating black holes [2].

2. The lifetime problem of evaporating black holes

Hawking's original calculation [3] concluded that black holes should evaporate quantum mechanically, but at an extremely slow rate if only they contain any realistic initial mass: a stellar-mass black hole would evaporate completely in 10^{67} years! Hawking initially believed that the final evaporation would be accompanied by a loss of information [4]. At present there are several scenarios showing that the evaporation can proceed in a perfectly regular way with no final loss of information (see *e.g.* [5, 6, 1]). However, these scenarios maintain the slowly-evaporating-(now trapping instead of event)-horizon characteristic.

The gigantic lifetime of the trapping horizon makes it difficult to believe on the verifiability of the actual presence of a trapping horizon in any reasonable time expand. This is seldom presented in the literature as a problem, more as a "such is life" fact. However, in our view it constitutes a strong motivation to seek for the viability of an alternative scenario devoided of the lifetime problem.

3. Newtonian gravity vs General Relativity

General Relativity can be thought of as a theory that corrects Newtonian gravity with new effects. Till the sixties of the previous century General Relativity was considered, mostly, a post-Newtonian theory. In most physical situations General Relativity predicts small deviations from the Newtonian behaviour. Even today, there are not direct experimental evidences of the working of General Relativity in regimes far apart from their Newtonian counterparts. However, the similarity between Newtonian gravity and General Relativity breaks up when event horizons are at stake.

Let us compare how one would describe what happen in the head-on collision of two equal balls of matter in Newtonian Gravity and General Relativity. Initially we put the two balls very far apart and with certain initial velocity towards each other. Then, we set them free subject to the rules of gravity. First of all in Newtonian gravity one can imagine the balls of matter to be as small as desired, even as point particles. In the approaching phase their relative velocity increases at the expense of the gravitational potential. Then at some point they collide with each other. In this colliding phase one naturally invokes other physics, the collision itself cannot be understood with gravitational physics alone, that is without further hypothesis. The simplest situations one can imagine is one in which the collision is elastic as if they were hard-balls, or one in which they are transparent to each other, meaning that while traversing each other they do not feel their respective gravity fields. In both cases the colliding phase will be followed by an escaping phase in which the balls will be climbing the gravitational potential. This escaping phase will be the perfectly time-symmetric version of the approaching phase.

The first new feature of General Relativity is that when the two balls start approaching each other there is some dissipation in the form of gravitational-waves emission. Leaving aside this issue (of crucial importance, but not for the point we want to make in this article) the collision process proceeds in a similar way to that in Newtonian gravity ... *except if the collision is so energetic that a horizon forms.* If the initial kinetic energy of the balls is sufficiently high, the collision will generate an event horizon so that the approaching phase will not be followed by a time-symmetric escaping phase: once the event horizon is formed there is no possible escaping phase, the very collision will end up with the formation of a singularity. This time asymmetry [7] separates definitively General Relativity from Newtonian gravity.

Is this separation inescapable? It is puzzling to realize that the time-symmetric version of the approaching phase exists as a classical General Relativity solution even in the cases that involve horizons. These configurations represent lumps of matter emerging from a singularity and escaping from each other, or using the standard terminology, they represent white holes. If only the regularization of the classical singularity allowed to connect the two time-symmetric branches this final separation would not happen.

Given the lifetime problem and the time asymmetry of classical General Relativity, and of most of the scenarios incorporating some quantum effects, it is interesting to seek whether there is space for an alternative scenario devoided of these two characteristics.

4. An alternative scenario

In a series of works [8, 9, 2] the authors have proposed an alternative scenario to the gravitational collapse process and the evaporating black hole paradigm. The alternative scenario has been worked out only for the simplest situation of a spherically symmetric collapse. However, we believe that this situation displays in the clearest possible way the new conceptual notions involved.

The logic of our works is the following. First, we realize that it is possible to imagine a regularization of the classical singularity that would appear in a spherically symmetric collapse so that it gives place to a time symmetric configuration. In geometrical terms the resulting geometry is nothing but the smooth continuation of a black hole spacetime, through its future singularity, to a white hole spacetime, both integrated in a single universe (a single asymptotic region). Then, on the one hand, we reflect about what would be the most salient consequences of this scenario in case it was actually at work in Nature. Most importantly, we argue that the scenario should leave distinct observable traces. On the other hand, we reflect about what characteristics should have a next level gravitational theory (beyond classical General Relativity) to be able to accommodate a regularization of the proposed type.

4.1. The geometries: Two salient features

In [2] we present a family of geometries representing the time-symmetric continuation of the spherically symmetric collapse of a homogeneous ball of dust initially at rest in a finite radius $r_{\rm i}$ (Oppenheimer-Snyner collapse [10]). The geometries are composed of i) a past region corresponding exactly to the Oppenheimer-Snyner spacetime, *i.e.* an external Schwarzschild region and an internal collapsing Friedman-Roberson-Walker region; ii) a future region corresponding exactly to the time reversal of the previous geometry; and iii) an intermediate region smoothly interpolating between the past and future regions. The precise interpolating function will depend on details of the underlying theory, which we do not know at this stage. However, here we are concerned with the generic and robust features of these geometries, independent of those details. The main features of these geometries can be read directly from their representation in Figure 1 (for more detailed arguments and an algebraic description of the geometries see [2]).

Two salient features of these geometries are:

- The total time inverted in the collapse and reestablishment of the initial configuration is extremely short as seen by external observers. Essentially this time is twice the proper time to form the classical singularity in the Oppenheimer-Snyder collapse. For figures taken from a initial neutron-star-like object this bouncing time will be of the order of 10ths of millisecond.
- The central grey triangular region in Figure 1 represents the transient region that is not a solution of standard General Relativity. Below and above this triangle the geometry is perfectly Einsteinian. What is remarkable is that in these geometries the non-standard

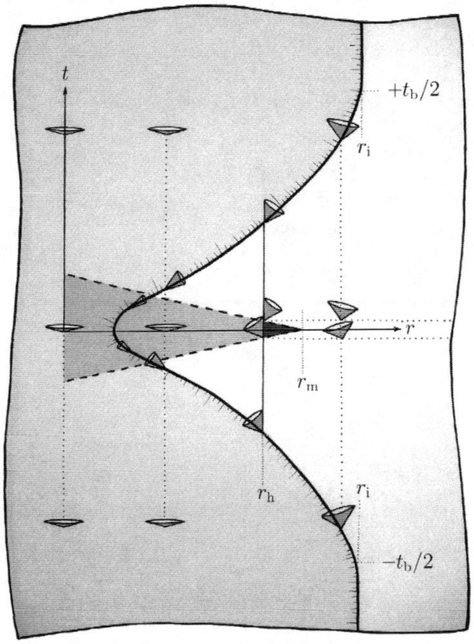

Figure 1. The figure represents the collapse and time-symmetric bounce of a stellar object in our proposal (the thick line). The past thick dashed line from $r = 0$ to r_m marks the boundary where the non-standard gravitational effects start to occur. In all the external white region the metric is Schwarzschild. In the region between the two thick dashed lines (which extends outside the stellar matter itself) the metric is not Schwarzschild, including the small dark grey triangle outside the Schwarzschild radius r_h. The drawing tries to capture the general features of any interpolating geometry. The slope of the almost Minkowskian cones close to the origin has been taken larger than the usual 45 degrees to cope with a convenient and explicit time-symmetric drawing.

region always extends beyond the classical gravitational radius (represented by the small dark-grey triangle in Figure 1). An observer sufficiently close to the gravitational radius will realize that something strange (in General Relativity terms) is going on even before the matter starts to reappear from beyond this radius. Obviously, this radius will not longer act as an event horizon.

The conformal diagram of these geometries is presented in Figure 2. Here you can see that the spacetime associated with this time-symmetric bounce is causally equivalent to Minkowski spacetime. However, the geometries posses transient regions containing outer trapped surfaces and inner trapped surfaces (respectively, the bottom and top down-gray regions in Figure 2). There are no long-lived trapping horizons of any sort.

A geometry causally equivalent to this one has also been presented in [11]. However, as opposed to our proposal, the geometry in that paper is such that the total duration of the bounce as seen by external observers is extremely long, so it does not solve the lifetime problem.

Journal of Physics: Conference Series **600** (2015) 012033 doi:10.1088/1742-6596/600/1/012033

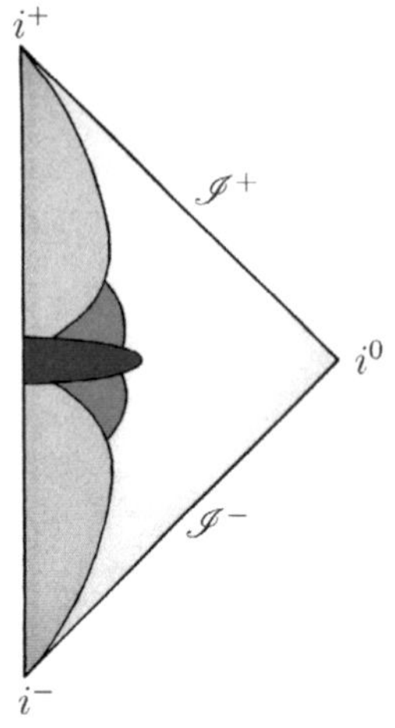

Figure 2. The figure represents the Penrose diagram of the proposed geometry. Globally it has the same causality as Minkowski spacetime. Locally it has some peculiarities. The dark grey region represents a non-standard gravitational field, while the up and down grey regions are respectively regions with inner and outer trapped surfaces. The light grey regions on the left-hand side are those filled by matter.

4.2. A verifiable scenario

If there exists a regularization of the classical standard General Relativity behaviour of the form we have described, the collapse process itself would not constitute the final stage of collapse in stellar physics. One would immediately be impelled to wonder about what would happen after the bounce. The search for new states of equilibrium, on the one hand, and the understanding of the transient collapse process itself, on the other, become entirely distinct issues.

In an ideal situation, perfectly spherically symmetric and without dissipation, the collapsing body would enter into a never-ending cycle of contracting and expanding phases. In a realistic situation though, one expects that the system will dissipate at least quantum mechanically while searching for new equilibrium configurations. However, now these new equilibrium configurations need not conform with the standard image of an evaporating black hole. For instance they could be objects close to their gravitational radius but with no long-lived horizons whatsoever (what we call generically black stars [12]).

The observable features of our scenario must be separated into two categories, those associated with the transient phase (the collapse and relaxation period) and those associated with possible new states of equilibrium different from evaporating black holes.

The transient phase should leave some traces, for instance, in the physics of Gamma Ray Bursts (GRBs) and its coincident gravitational-waves emission. One would expect some signatures associated with a reverberant collapse. In the collapsar model of GRBs however (see *e.g.* [13]) the emission zone is supposed to be very far from the collapsed core. This means that the connection between the processes at the core and those at the external wind shells could be very far from direct. The signatures should be clearer in the Gravitational Waves band. The last part of the gravitational-waves signal should clearly distinguish between the standard direct relaxation towards a black hole and the presence of some bounces.

Regarding the possible existence of black stars in place of evaporating black holes, here let us just say that these objects, having no trapping horizons, would be at least susceptible to complete astrophysical exploration.

4.3. A necessary underlying hypothesis

The past thick dashed line in Fig. 1 marks the boundary where non-standard gravitational effects start to happen. It is born at $r = 0$ and travels outwards even though the light cones are pointing inwards. This signal, should it exist, cannot follow the causality associated with the gravitational light cones. Rather it must follow an underlying causality that is explored only when Planck energies are at stake. This background causality should be trivial in the sense of containing no horizons whatsoever, the simplest example one can think of being a Minkowskian structure. This is the only hypothesis we need for our proposal to make sense. Otherwise general-relativistic light cones could not suffer such a dramatic turn.

The underlying causality we talk about could be nothing but the causality produced by the rest of the matter in the universe (*i.e* not considering the very matter undergoing the high-energy collision). This causality will be naturally (almost) Minkowskian in what concerns a localized process of gravitational collapse. Therefore, we are far from claiming that there need existing a background geometry unrelated to the matter content of the universe.

The previous hypothesis finds natural connections with some conceptualizations of gravity coming from particle physics and condensed matter physics. On the one hand, our proposal resonates with Rosen's reformulation of General Relativity as a nonlinear theory on a flat Minkowski background [14]. This reformulation indeed goes further than the standard formulation of General Relativity in the sense that it is a convenient effective framework to describe the switching-off of gravity at high energies. Rosen's reformulation can be understood as the long-wavelength limit of a nonlinear theory of gravitons (see [15] and references therein). It is still an open possibility that an ultraviolet completion of such a theory would exhibit asymptotic freedom (as its QCD cousin). On the other hand, similar ideas also appear when thinking of gravity as an emergent notion in a condensed matter framework [8] (see also [16]). The nonlinear theory of gravity describes in that case the behaviour of collective degrees of freedom. There, it is reasonable to think that the first quantum gravitational effect is that, above some Planckian energy scale, the collective degrees of freedom corresponding to gravity are diluted, leaving a Minkowskian background for the matter excitations.

Acknowledgments

Financial support was provided by the Spanish MICINN through the projects FIS2011-30145-C03-01 and FIS2011-30145-C03-02 (with FEDER contribution), and by the Junta de Andalucía through the project FQM219. R. C-R. acknowledges support from CSIC through the JAE-predoc program, cofunded by FSE.

References

[1] SW Hawking. Information Preservation and Weather Forecasting for Black Holes. *ArXiv e-prints*, 2014.
[2] C Barceló, R Carballo-Rubio, LJ Garay, and G Jannes. The lifetime problem of evaporating black holes: Mutiny or resignation. *Class. Quantum Grav.*, 2014.
[3] SW Hawking. Particle Creation by Black Holes. *Commun. Math. Phys.*, 43:199–220, 1975.
[4] SW Hawking. Breakdown of Predictability in Gravitational Collapse. *Phys. Rev.*, D14:2460–2473, 1976.
[5] L Susskind. *The Black Hole War: My Battle with Stephen Hawking to Make the World Safe for Quantum Mechanics.* Little, Brown, 2008.
[6] A Ashtekar and M Bojowald. Black hole evaporation: A Paradigm. *Class. Quant. Grav.*, 22:3349–3362, 2005.
[7] R Penrose. Singularities and time-asymmetry. In S.W. Hawking and W. Israel, editors, *General Relativity: An Einstein Centenary Survey.* Oxford University Press, Cambridge University Press, 1979.
[8] C Barceló, LJ Garay, and G Jannes. Quantum Non-Gravity and Stellar Collapse. *Found. Phys.*, 41:1532–1541, 2011.
[9] C Barceló, R Carballo-Rubio, and LJ Garay. Mutiny at the white-hole district. *Int. J. Mod. Phys. D*, 2014.

Journal of Physics: Conference Series **600** (2015) 012033 doi:10.1088/1742-6596/600/1/012033

[10] JR Oppenheimer and H Snyder. On continued gravitational contraction. *Phys. Rev.*, 56:455–459, 1939.

[11] HM Haggard and C Rovelli. Black hole fireworks: quantum-gravity effects outside the horizon spark black to white hole tunneling. *ArXiv e-prints*, 2014.

[12] M Visser, C Barceló, S Liberati, and S Sonego. Small, dark, and heavy: But is it a black hole? *PoS*, BHGRS:010, 2008.

[13] A MacFadyen and SE Woosley. Collapsars: Gamma-ray bursts and explosions in 'failed supernovae'. *Astrophys. J.*, 524:262, 1999.

[14] N Rosen. General relativity and flat space. i. *Phys. Rev.*, 57:147–150, 1940.

[15] C Barceló, R Carballo-Rubio, and LJ Garay. Unimodular gravity and general relativity from graviton self-interactions. *Phys. Rev.*, D89:124019, 2014.

[16] G Jannes. Condensed matter lessons about the origin of time. *ArXiv e-prints*, 2009.

Journal of Physics: Conference Series **600** (2015) 012034 doi:10.1088/1742-6596/600/1/012034

On evaporation of Randall-Sundrum black holes

P I Dyadina[1], **B N Latosh**[2] **and S O Alexeyev**[1]

1 Sternberg Astronomical Institute, Lomonosov Moscow State University, Universitetsky Prospekt, 13, Moscow 119991, Russia

2 Physics Department, Institute of Natural Sciences, Ural Federal University. 19 Mira street, 620002, Ekaterinburg Russian Federation.

E-mail: guldur.anwo@gmail.com, latosh.boris@gmail.com alexeyev@sai.msu.ru

Abstract. We investigate the possibilities of experimental search for new physics which is predicted by Randall-Sundrum II gravity model. For this purpose we obtained the law of evaporation for Dadhich-Rezania black hole solution and black hole solution by S. Abdolrahimi et al. within RS II. Thus we look for cosmological tests of these theory via a study of gamma-ray bursts from primordial black holes radiation.

1. Introduction

The theory of general relativity (GR) is a huge step in the understanding of our Universe as it allows to go beyond the Newtonian picture of the world.

But our understanding of the fundamental laws still has several shortcomings. Already in 1930s the problem of galactic rotation curves appeared [1, 2]. Moreover the accelerated expansion of the universe (i.e., dark energy) was found from cosmological observations recently [3] and we still can't merge gravity with quantum mechanics. So it became obvious that GR should be modified.

Brane world models are one of the ways to expand GR. One of the first brane world model was proposed by Shaposhnikov and Rubakov [4]. In brane world scenario matter is localized on three-dimensional surface in multidimensional space. There are several brane world models and the Randall-Sundrum scenario [5], [6] is a rather common one. In the Randall-Sundrum model matter and three fundamental interactions are localized on the three-dimensional surface in five-dimensional spacetime (bulk) and gravity can propagate into the bulk. In the RS model extra dimension is not compact. The RS scenario includes two models. Model RSI contains two branes. It was created to solve the hierarchy problem of particle physics. In RSII model there is only one brane in the bulk. We use RSII because in this model possible difference between the behavior of gravity GR appears only because of the extra dimension. We research Dadhich-Rezania solution [7], [8] and S. Abdolrahimi, C. Cattoën, D. N. Page, S. Yaghoobpour-Tari solution (ACPY) [9] within RSII model.

2. Models

Dadhich-Rezania solution is one of the first black hole solutions. It is exact asymptotically flat solution describing an uncharged, nonrotating black hole:

$$ds^2 = -\left(1 - \left(\frac{2M}{r} + \left(\frac{q}{\tilde{M}_{Pl}}\right)\frac{1}{r^2}\right)\right)dt^2 + \frac{dr^2}{1 - \left(\frac{2M}{r} + \left(\frac{q}{\tilde{M}_{Pl}}\right)\frac{1}{r^2}\right)} + r^2(d\theta^2 + \sin^2\theta d\varphi^2)\ , \quad (1)$$

where $\tilde{M}_{Pl}$ is the multidimensional Planck mass, M is the black hole mass and q is the tidal charge.

ACPY solution is asymptotically AdS solution. It was obtained by numerical simulation. To search for the experimental consequences of modified gravity theories we used the effects of black hole evaporation.

$$ds^2 = \left[1 - \frac{1}{(-\Lambda r^2)}\frac{r-2M}{r-1.5M}\left(F - r\frac{dF}{dr}\right)\right]\left(1 - \frac{2M}{r}\right)^{-1}dr^2 - \left(1 - \frac{2M}{r}\right)dt^2 + \left[r^2 + \frac{1}{(-\Lambda)}F\right]d\Omega^2\ ,(2)$$

$$F = \ 1 - 1.062\left(\frac{2M}{r}\right) + 0.554\left(\frac{2M}{r}\right)^2 - 0.120\left(\frac{2M}{r}\right)^3 + ... + 2.900\left(\frac{2M}{r}\right)^{11}. \quad (3)$$

The function $F(r)$ describes the perturbation caused by the bulk.

In this work we consider the effects of black hole evaporation [10, 11] to search the experimental consequences of modified gravity theories.

3. Black hole evaporation

In general relativity the effects of evaporation of the black holes with stellar masses are negligible. In extended gravity models typical lifetimes of black holes can differ from GR and the consequences of the evaporation can be observable. We can look for the effects of black hole evaporation on primordial black holes [12]. Primordial black holes are supposed to appear in the early universe due to density fluctuations with masses much smaller than the solar mass. The effects of evaporation play a significant role for them as the evaporation rate is inversely propotional to the black hole mass. The final stage of evaporation should be accompanied by the emission of high energy particles including gamma particles with energies of the TeV order. Events with similar energies occur at large distances from us ($z \leq 9.4$ [13]) and are called gamma ray bursts (GRB). Many GRB are identified with supernovae. But the nature of the most distant GRB remains a mystery. The last stage of the evaporation of primordial black holes may be the decision of this problem. Extended gravity models predict the existense different masses black holes [14], [15, 16, 17, 18], [19] which evaporate completely during the lifetime of the universe. Therefore the cosmological data can be used to search for events that can be explosion of primordial black holes in the final stages of evaporation.

Using data from the telescope Fermi LAT T. N. Ukwatta, J. H. MacGibbon and others got the closest distance at which the telescope is able to detect the evaporation of primordial black holes [20].

$$d \simeq 0.04\left(\frac{\Omega}{\mathrm{sr}}\right)^{-0.5}\left(\frac{E}{\mathrm{GeV}}\right)^{0.7}\left(\frac{T}{\mathrm{TeV}}\right)^{0.8}\ \mathrm{pc}, \quad (4)$$

where d is the distance,Ω is the angular resolution of the telescope, E is the energy range of the telescope, T is the temperature of the black hole. After spending the same reverse procedure and using a telescope detected gamma-ray bursts, we can conclude that the detectable value of mass of primordial black holes should differ from the value predicted by general relativity (with a mass of 10^{15} g), more than five orders of magnitude.

Journal of Physics: Conference Series **600** (2015) 012034 doi:10.1088/1742-6596/600/1/012034

4. Method

To determine the initial mass of the primordial black hole evaporating in our days, we need to measure its temperature. The original Hawking method contains complex calculations and is not very convenient. We used a modification of the method by S. Shankaranarayanan, T. Padmanabhan, and K. Srinivasan [21]. This method was tested on Schwarzschild and Reissner-Nordstrm black holes and gave the same result as the Hawking method. For this method we use the quasi-classical approximation of the wave function of a particle near the the black hole and consider particle tunneling through the potential barrier of the event horizon. Thus we get the temperature of the black hole.

In this method we use several approximations. As rotating black holes lose angular moment faster than the mass so we can consider nonrotating black holes. The emitted particles can be considered massless and spinless. In 1976 Page showed that black holes in GR emit massless particles with zero spin [22, 23, 24]. For extended theories of gravity we can consider the emission of such particles because we determine the temperature of a black hole on their distribution. The temperature of a black hole does not depend on the type of particles. So we use the wave function without spin part for the quantum description of a massive particle in the black hole field. We also assume that there is no radiation into the bulk and that the lifetime of virtual pairs is negligible because radiation occurs due to the process of creation and annihilation of virtual particles. This method is not applicable for rotating black holes. And it is not applicable for quantum black holes whose lifetimes are comparable with lifetimes of pairs of virtual particles.

We can neglect the matter surrounding black hole because the evaporation effects are important for low-mass black holes and such black holes are not able to keep any significant amount of matter in orbit. Pairs of virtual particles appear near a black hole horizon permanently and we can consider the tunneling of one particle of the pair under the horizon of the black hole. The lifetime of virtual particles is extremely small. So we can assume instantaneous tunneling.

Now we consider the metric in the following form

$$ds^2 = A(r)dt^2 - \frac{dr^2}{B(r)} - f(r)\left(d\theta^2 + \sin^2\theta d\varphi^2\right) , \tag{5}$$

Within this method the temperature of black hole depends only on the geometric parameters of the black hole:

$$T = -\frac{1}{4\pi \times res\left(\frac{1}{\sqrt{AB}}\right)} . \tag{6}$$

Using described method, we investigate Dadhich-Rezania solution and ACPY solution.

5. Results

Let's start with the solution which was found Naresh Dadhich, Roy Maartens, Philippos Papadopoulos, and Vahid Rezania [7]. This solution is similar to the Reissner-Nordstrm metric. But it contains a so called with tidal charge instead of the electric charge. For this solution, the rate of the mass loss is:

$$-\frac{dM}{dt} = \frac{1}{216}\,\frac{k_B}{\pi^3 M^2} + \mathcal{O}\left(M^{-6}\right). \tag{7}$$

The leading term in (7) cannot produce the needed 5-order difference. The initial mass of the Dadhich-Rezania black hole that evaporates completely during the lifetime of the Universe proves this fact:

$$M_{\rm DR} = 5.3 \times 10^{14}\text{g}. \tag{8}$$

As the obtained difference is much less than the cutoff threshold the "tidal charge" influence is vanishing and cannot have experimentally verifiable consequences.

S. Abdolrahimi, C. Cattoën, D. N. Page and S. Yaghoobpour-Tari have recently received a large black hole solution [9]. It describes the astrophysical black holes in RSII as a perturbation solution of Einstein's equations within the approach AdS_5-CFT_4 in five-dimensional space - time with negative cosmological constant. Evaporation law for the metric Abdolrahimi - Page has the form:

$$-\frac{dM}{dt} \simeq \frac{1}{256}\frac{k_B}{\pi^3 M^2} \tag{9}$$

This evaporation law completely coincides with the evaporation law for Schwarzschild black hole solution. The initial mass of the ACPY black hole also coincides with initial mass of the Schwarzschild black hole and equals

$$M_{\mathrm{DR}} = 5.0 \times 10^{14}\mathrm{g}. \tag{10}$$

6. Conclusions

Our results allow to conclude that the accuracy of the current cosmological data is still insufficient in order to distinguish the general theory of relativity from RSII by analyzing the evaporation of primordial black holes.

References

[1] Zwicky F 1933 *Helvetica Physica Acta* **6** 110
[2] Oort J H 1932 *Bull. Astron. Inst. Netherlands* **6** 249
[3] Kowalski M, Rubin D, Aldering G, Agostinho R J, Amadon A, Amanullah R, Balland C, Barbary K, Blanc G, Challis P J and et al 2008 *Astrophys. J.* **686** 749
[4] Rubakov V and Shaposhnikov M 1983 *Phys. Lett.* B **125** 136
[5] Randall L and Sundrum R 1999 *Phys. Rev.* Lett. **83** 3370
[6] Randall L and Sundrum R 1999 *Phys. Rev.* Lett. **83** 4690
[7] Dadhich N, Maartens R, Papadopoulos P and Rezania V 2000 *Phys. Lett.* B **487** 1
[8] Alexeyev S O and Starodubtseva D A 2010 *JETP* **111** 576
[9] Abdolrahimi S, Cattoën C, Page D N and Yaghoobpour-Tari S 2013 *J. Cosmology Astropart. Phys.* JCAP06(2013)039
[10] Hawking S W 1974 *Nature* **248** 30
[11] Hawking S W 1975 *Comm. Math. Phys.* **43** 199
[12] Carr B J, Kohri K, Sendouda Y and Yokoyama J 2010 *Phys. Rev.* D **81** 104019
[13] Cucchiara R A, Levan A J, Fox D B, Tanvir N R, Ukwatta T N, Berger E, Kruhler T, Yoldas A K, Wu X F, Toma K and et al 2011 *Astrophys. J.* **736** 7
[14] MacGibbon J H and Carr B J 1991 *Astrophys. J.* **371** 447
[15] Alexeyev S O and Pomazanov M V 1997 *Phys. Rev.* D **55** 2110
[16] Alexeyev S O and Sazhin M V 1998 *Gen. Relativ. Grav.* **8** 1187
[17] Alexeyev S O, Barrau A and Rannu K A 2009 *Phys. Rev.* D **79** 067503
[18] Alexeyev S O and Rannu K A 2012 *JETP* **114** 406
[19] Alexeyev S, Barrau A, Boudoul G, Khovanskaya O and Sazhin M 2002 *Class. Quant. Grav.* **19** 4431
[20] Ukwatta T N, MacGibbon J H, Parke W C, Dhuga K S, Eskandarian A, Gehrels N, Maximon L and Morris D C 2009 *AIP Conf. Proc.* **1133** 440
[21] Shankaranarayanan S, Padmanabhan T and Srinivasan K 2002 *Class. Quant. Grav.* **19** 2671
[22] Page D N 1976 *Phys. Rev.* D **13** 198
[23] Page D N 1976 *Phys. Rev.* D **14** 3260
[24] Page D N 1977 *Phys. Rev.* D **16** 2402

Journal of Physics: Conference Series **600** (2015) 012035 doi:10.1088/1742-6596/600/1/012035

Excavations at the gravitationally collapsed site: Recent findings[1]

Suprit Singh
Inter-University Centre for Astronomy and Astrophysics, Ganeshkhind, Pune 411 007 India
E-mail: suprit@iucaa.ernet.in

Abstract. Hawking effect was digged out of the gravitationally collapsed site forty years back when it was realised that quantum effects at the horizon could propagate outward to infinity giving rise to a thermal flux at late-times. However, the situation regarding non-asymptotic observers was never completely clear. Also, recently a debate has sprung in the community as to what would infalling observers perceive while crossing the horizon. We set out to settle this question and more at least semi-classically in the articles [7] and [8] with a fresh approach. We introduce a new set of coordinates that are regular everywhere, consider the adiabatic expansion of detector response and its link to the trajectory-dependent 'effective' temperature/s and also local invariant observables, energy density and flux, built from renormalized stress energy tensor. This paper is a concise summary of the new procedure and the results obtained thereof.

1. Introduction

Hawking famously showed [1] that black holes emit a stationary thermal flux with temperature $T_H = 1/8\pi M$ which is perceived by the asymptotic observers at late-times. But that is not all that happens in the system that undergoes a gravitational collapse. For example, in a typical collapse scenario (see Fig. 1 (Left)), the usual Hawking effect is understood due to the 'peeling' of modes that travel close to the horizon just before the collapse and escape to infinity with an infinite redshift. This is depicted by ray number 1 and the corresponding static observer encounters a temperature with the standard Tolman shift which becomes unity in the asymptotic limit. However, we can have a similar null ray numbered 2 here, which would be just inside the horizon and hits the singularity at some finite time. The infallers (radial or otherwise) would encounter this situation. The third possibility is of the ray 3 that is inside the collapsing star all the time and here we would have comoving observers (the inside spacetime being Friedmann for dust collapse). Thus, we can have many different observers/frames which *should* see the effect of collapse on the quantum vacuum differently and there is a rich structure beneath waiting to be explored.

On a related note, a class of observers termed the *infallers* have recently sprung a hot debate in the community on what would happen *at the horizon crossing* à la the Hawking effect. It is usually argued [3, 4] on the basis of equivalence principle that the infalling observer should see a quantum vacuum state while crossing the horizon and nothing special. However, this line of thought has been questioned lately by AMPS [2] claiming on the quantum-information-theoretic grounds that the infalling observers will encounter a *firewall* at the horizon. Amidst

[1] Based on the talk 'Black holes: Up, close and personal', I gave at ERE 2014, Valencia, Spain.

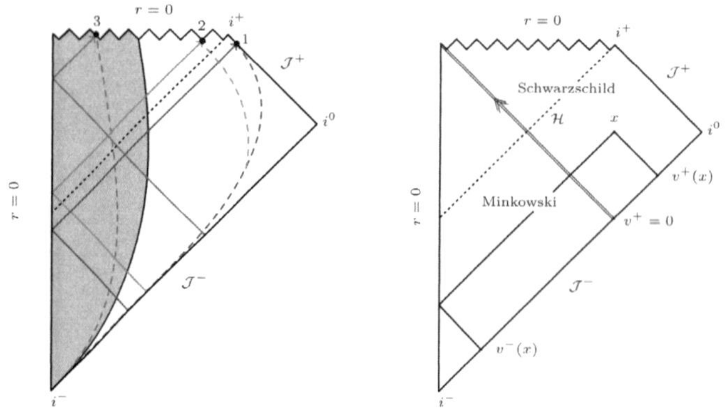

Figure 1. Left. Penrose diagram for a generic gravitational collapse scenario showing three different null rays and observers. The static observers outside correspond to reaching point 1, point 2 corresponds to the radial infallers while those inside the collapsing body reach point 3 presenting different aspects of the Hawking phenomenon. **Right.** Construction of globally defined null coordinates. Any point x on the spacetime can be ascribed two coordinates (v^+, v^-) which correspond to one ray coming directly from $\mathcal{J}^-$ and the other traced back to $\mathcal{J}^-$ after a reflection from the origin. Here, we take Vaidya null collapse where the thin null shell lies at $v^+ = 0$ separating the Minkowski interior from the Schwarzchild exterior.

these speculations which are not completely understood at this time, we seek to know what happens at least *semi-classically*, relying on the well-established grounds.

Working on these lines, Barbado et al [5] introduced the notion of an 'effective' temperature as a measure of the response of an Unruh-Dewitt detector [3, 6] moving on a given trajectory. They found that the effective temperature increases for a detector dropped from infinity on a radially infalling geodesic as the infall progresses and reaches the temperature *four* times the Hawking temperature at the horizon. As much as this result is intriguing, it should be noted their analysis does not give any direct link with the detector response nor does it provide any understanding to how thermal or non-thermal the effect is. Another point of concern is the choice of vacuum which they choose to be some new unconventional state in extended Schwarzchild geometry. We might think the peculiar result could be an artefact of this contraption. But more importantly, the use of both the ingoing and outgoing Eddington-Finkelstein coordinates together is not suitable if we wish to study the interior of the black hole horizon.

With this retrospection of the cards on the table, we now proceed to bring in a new deck [7,8] with the following assumptions and characteristics: (a) We introduce a new set of globally defined coordinates which are regular everywhere. (b) Working with the massless scalar field, the vacuum state in a collapse scenario is a natural and uniquely defined in-vacuum on $\mathcal{J}^-$. (c) We define and study a suitably defined time-dependent 'temperature' function also relating it explicitly to UDW response functions by an *adiabatic expansion* and finally (d) We look at the observables – energy density and fluxes – built from the renormalized stress energy tensor (RSET) which explicitly depend on observer's kinematics for an alternate description. This also confirms the results of the effective temperature approach.

This paper is meant to give a concise description of the formalism and results obtained in the previous works [7,8] which relate with the collapse of a thin null shell (ingoing Vaidya geometry) and to provide a preview of the results (relating to time-like dust collapse) which are a work in progress [9].

2. The new procedure

Without further ado, we shall present the new procedure and its application to the thin null shell collapse. As remarked earlier, it begins with the introduction of the *new* coordinates.

(1) **Coordinatization:** Any point x on the spacetime manifold in the collapse scenario (Fig. 1 (Right)) can be ascribed two coordinates (v^+, v^-) which correspond to one ray coming directly from $\mathcal{J}^-$ and the other traced back to $\mathcal{J}^-$ after a reflection from the origin. This construction is completely general and valid for any type of collapse. For the Vaidya collapse where the thin null shell lies at $v^+ = 0$ separating the inside Minkowski from the Schwarzchild exterior, this coordinatization can be computed (by ray tracing) in terms of the Eddington-Finkelstein ingoing coordinates to give $v^+ = v$ and

$$v^-(v,r) = \begin{cases} v - 2r & \text{for } v < 0 \\ -2r_s\left[1 + W\left(\delta\, e^{\delta - \kappa v}\right)\right] & \text{for } v \geq 0, \end{cases} \quad (1)$$

where $\delta \equiv r/r_s - 1$ and $W(z)$ is the Lambert W-function. The metric in these coordinates is given (with $a \equiv 1 + \kappa v^-$) by,

$$ds^2 = -C(v^+, v^-)dv^+dv^- + r^2 d\Omega^2; \quad C(v^+, v^-) = 1 + \left(-1 + \frac{a-1}{a}\frac{W(-ae^{-a+\kappa v^+})}{1 + W(-ae^{-a+\kappa v^+})}\right)\Theta(v^+). \quad (2)$$

(2) **Field setup:** The minimally coupled massless scalar field in the s-wave approximation at x is given by

$$\frac{\phi(x)}{r} \sim \lim_{r\to\infty} \frac{\phi(v_+(x), r) - \phi(v_-(x), r)}{r}, \quad (3)$$

that is, superposition of two spherical waves one impinging directly and one that comes after a reflection from $r = 0$. This also gives the interpretation to the coordinates (v^+, v^-) as being tied to the eikonal approximation.

(3) **Detector response:** The response of a detector [10] moving on a general trajectory $\gamma(\tau)$ can be written as

$$\dot{\mathcal{R}}(\Omega) = 2\,\mathrm{Re}\int_0^\infty ds\, e^{-i\Omega s} G(\gamma(\tau), \gamma(\tau - s)) \quad (4)$$

where the Wightmann function, $G(x, y)$ takes the logarithmic form [11],

$$G(x,y) \propto \ln\Big(\big(v_+(x) - v_+(y) - i0\big)\big(v_-(x) - v_-(y) - i0\big)\Big) \quad (5)$$

for s-waves. This splits the response into two: $\mathcal{R} = \mathcal{R}_+ + \mathcal{R}_-$ decoupling the ingoing and outgoing modes giving *two temperatures*! The response function in its generality is difficult to compute analytically. However, if $v(\tau)$ is any solution of the differential equation,

$$\ddot{v}(\tau) = -k(\tau)\dot{v}(\tau) \quad (6)$$

with $k(\tau)$ as the slowly varying function, then we can consider the adiabatic expansion of the response functional, which gives,

$$\dot{\mathcal{R}}(\tau, \Omega; k = 2\pi T\,] = \frac{2\pi}{\Omega(e^{\Omega/T} - 1)}\left(1 + \frac{\dot{T}(\tau)}{T(\tau)^2}\frac{\Omega}{8\pi T(\tau)}\frac{e^{\Omega/2T(\tau)}}{\sinh^2(\Omega/2T(\tau))} + \cdots\right). \quad (7)$$

That is, all observers, asymptotic or non-asymptotic will perceive a *quasi-thermal* spectrum at

$$T_\pm = \frac{1}{2\pi}\left|\frac{\ddot{v}^\pm}{\dot{v}^\pm}\right| \quad (8)$$

which is related to the 'peeling' condition. It becomes proper thermal temperature in the UV limit ($\Omega \to \infty$) or under adiabatic condition, $\eta_\pm \equiv \left|\dot{T}_\pm/T_\pm^2\right| \ll 1$.

(4) **Energy density and Fluxes:** For an alternate prescription, we can compute the RSET of the quantum field in $(1+1)$ dimensions from

$$\langle T_{uu(vv)}\rangle = -\frac{1}{12\pi}C^{1/2}\partial^2_{u(v)}C^{-1/2}; \qquad \langle T_{uv}\rangle = \frac{1}{24\pi}\partial_u\partial_v \ln C. \tag{9}$$

For s-waves, the $(1+3)$ dimensional counterpart is usually considered [12] by multiplying the above expressions with the inverse square distance $(1/4\pi r^2)$. Given the velocity of the observer u^a and a vector n^a such that $n_a u^a = 0$, we can construct two local invariant observables *energy density* and *flux* as $U = \langle T_{ab}\rangle u^a u^b$ and $F = \langle T_{ab}\rangle u^a n^b$ respectively and which depend explicitly on the observer's kinematics. These have the advantage over detector response in being local (not dependent on the past trajectory) and exact (no need of adiabatic expansion).

3. Findings

With the above procedure, the usual Hawking effect for an asymptotic observer at rest relative to the hole can be recovered as follows. For this observer, $v^-(\gamma(\tau))$ satisfies

$$-\frac{\ddot{v}^-(\gamma(\tau))}{\dot{v}^-(\gamma(\tau))} = \kappa \tag{10}$$

and subsequently

$$v^-(\tau) - v^-(\tau - s) \propto e^{-\kappa(\tau-s)/2}\sinh(\kappa s/2) \tag{11}$$

which recovers *thermal spectrum* at $T_- = T_H = \kappa/2\pi$

$$\dot{\mathcal{R}}_-(\Omega) \propto \left(\Omega(e^{2\pi\Omega/\kappa} - 1)\right)^{-1}. \tag{12}$$

while $T_+ = 0$ in this case. For a non-asymptotic static observer at some finite radius $r > r_s$ from the hole, the temperature gets scaled by the redshift factor, $z_r = (1 - r_s/r)^{-1/2}$. While many other trajectories can be considered, we shall look at the results for the radially infalling observer characterized by energy per unit rest mass, E or the initial radius from where the infall begins, r_i. The analysis of the effective temperatures here then shows a dependence on E and is presented in Table 1 (these results are also seen to hold in the case of time-like collapse [9]). The non-stationarity of the regime shows in having two temperatures now. The temperature T_- presents itself with a doppler shift for $E \neq 1$ and the result in the near-horizon limit matches with ref. [5] for $E = 1$, i.e., for the detector dropped in from infinity. The surprise is that in the near-horizon region T_+ (incoming modes) dominates over T_- (outgoing modes) for $E \to 0$ (see

Eff. Temperatures	Asymptotic	At horizon crossing
T_-	$T_H\left(E + \sqrt{E^2-1}\right)$	$4ET_H$
T_+	0	$T_H/2E$

Table 1. Effective temperatures for a detector on a radially infalling trajectory specified by energy per unit rest mass, E in the asymptotic and near-horizon limits.

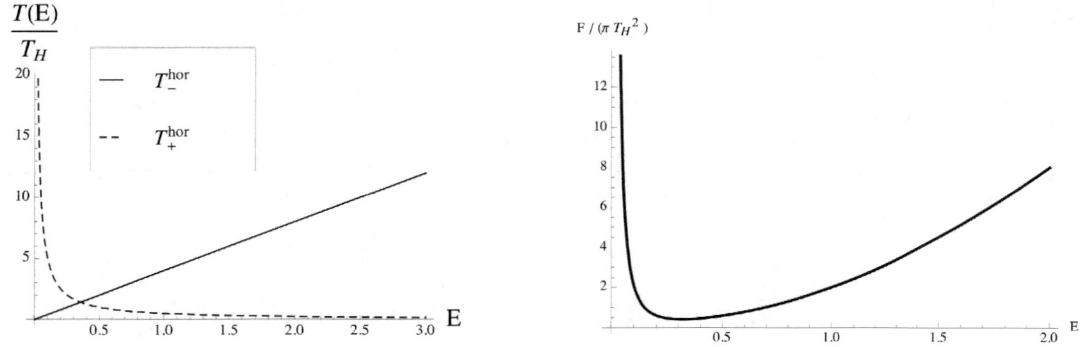

Figure 2. Left. We have the plot of $T_{\pm}$ in the near horizon limit which shows dependence on energy and the dominance of ingoing mode (T_{+}) over outgoing (T_{-}) for $E < 1/\sqrt{8}$. **Right.** We have the flux in the near horizon region confirming the effective temperature results.

Fig 2 Left) which implies a duality between the asymptotic and the near-horizon observer in that the near-horizon observer sees a dominant and high radiation from the *sky* [13]. However, it is to be noted that the adiabaticity in these cases is $\mathcal{O}(1)$ and hence the temperatures can only be interpreted in the UV limit.

From another angle, we can compute the energy density and fluxes for various observers. For the static observers in the late-time limit, we have,

$$U = \frac{\pi T_H^2}{12}\left(1 - \frac{2\,r_s^4}{r^4}\right)\left(1 - \frac{r_s}{r}\right)^{-1}; \quad F = \frac{\pi T_H^2}{12}\left(1 - \frac{r_s}{r}\right)^{-1}. \tag{13}$$

These expressions reduce to the Hawking energy density and flux for the asymptotic observers with $r \to \infty$ and otherwise the temperature is given by the Tolman blueshift factor. The *onset* of thermality depends on the location of the observer. It is earlier for the near-horizon observers and late for those away from the horizon. The flux diverges at the horizon. The energy density presents a contrast that in the near-horizon region it reaches a maximum positive value, then decreases and becomes negative (Fig. 3). This corresponds to negative energy fluxes directed into the black hole which have been discussed before (see ref. [14]).

For the radially infalling observer, the energy density and flux increases (see Fig. 3) as the infall occurs and is *finite* and *regular* at the horizon for the infall beginning far from the horizon and diverges only at the eventual singularity. At the horizon, these observables are given by,

$$U_H = \pi T_H^2\left(\frac{2}{3} - \frac{1}{48E^2} + 2E^2\right); F_H = \pi T_H^2\left(\frac{1}{48E^2} + 2E^2\right) \tag{14}$$

which show agreement with the effective temperatures such that for $E \to 0$, that is, if the infall occurs starts close to the horizon, a high flux is perceived at the horizon crossing (see Fig. 2 Right).

But why the clicking? We can show by considering artificial black holes which are flat at the horizon, that the effective temperature due to outgoing modes T_{-} vanishes at the horizon and also the flux contribution corresponding to it. However, the effect due the ingoing modes remains. This shows that curvature has something to do with the detector response and the flux. When the outgoing modes do not feel the curvature at the horizon, we see no effect but the ingoing modes surely encounter some curvature while travelling to the horizon and account for the observed effect. The varying Riemann curvature (as the infall progresses) gives the clicking at the horizon.

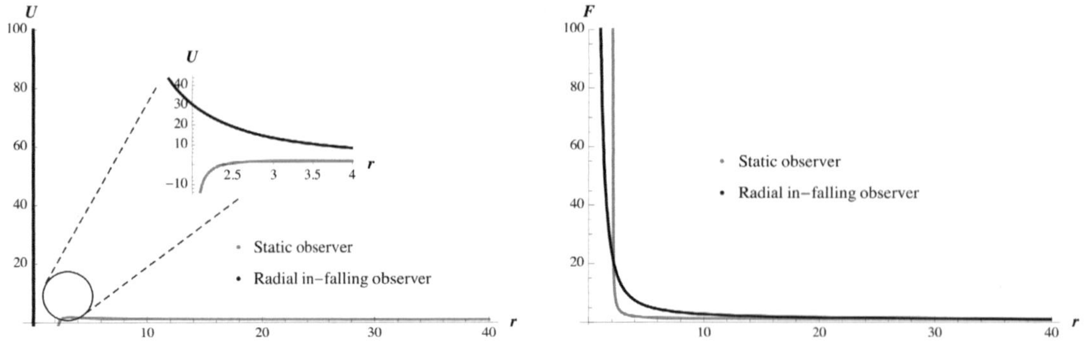

Figure 3. The plots show the increase in energy density (left) and flux (right) as the infall occurs in comparison with those perceived by the static observers at the same radii. Both the observables are finite at the horizon crossing (we have taken $r_s = 2$ or $M = 1$ here) and diverge only at the eventual singularity. For the static observers, as we go near the horizon, the flux diverges at the horizon due to infinite acceleration. The energy density, however, reaches a maximum and then decrease to become negative and diverges negatively at the horizon.

4. Conclusions

We see that using a new set of tools and procedure unravels some new and surprising results in the age old problem of quantum fields in gravitational collapse. The response of an UDW detector can be computed using the adiabatic expansion and hence we see not one but *two* temperatures attributed to ingoing and outgoing modes. This formalism can be applied to many different trajectories. In particular, we see for the infalling radial observers, in the near-horizon limit, the temperature due to ingoing modes dominates over outgoing modes for *slowly* moving detectors. A look at the energy density and flux confirms this. So the punchline is that *if you wish to drop into the black hole, start far off.* Also, now we know, at least semi-classically, that there *is* clicking at the horizon and we need to dig in further for more insights.

Acknowledgements

I thank Thanu Padmanabhan and Dawood Kothawala for various discussions, Matteo Smerlak and Sumanta Chakraborty for the collaborations [7] and [9] respectively. I also thank IUCAA and the organisers of ERE 2014 for the financial support that allowed me to attend the meeting.

References

[1] S.W. Hawking, *Nature (London)* **248**, 30 (1974).
[2] A. Almheiri, D. Marolf, J. Polchinski and J. Sully, *JHEP* 1302, 062 (2013) [arXiv:1207.3123].
[3] W.G. Unruh, *Phys. Rev. D* **14**, 870 (1976).
[4] D. Singleton and S. Wilburn, *Phys. Rev. Lett.* **107**, 081102 (2011).
[5] L.C. Barbado, C. Barcelo and L.J. Garay, *Class. Quant. Grav.* **28**, 125021 (2011).
[6] B. S. DeWitt, "Quantum gravity: the new synthesis," in *General relativity: an Einstein centenary survey*, S. W. Hawking and W. Israel, eds., pp. 680–745 Cambridge University Press, 1979.
[7] M. Smerlak and S. Singh, *Phys. Rev D* **88**, 104023 (2013) [arXiv:1304.2858].
[8] S. Singh and S. Chakraborty, *Phys. Rev. D* **90** 024011 (2014) [arXiv:1404.0684].
[9] S. Chakraborty, S. Singh and T. Padmanabhan, work in progress.
[10] N. D. Birrell and P. C. W. Davies, *Quantum Fields in Curved Space* (Cambridge University Press, 1982).
[11] A. D. Helfer, *Rept. Prog. Phys.* **66**, 943 (2003).
[12] R. Brout et al, *Phys. Rept.* **260**, 329 (1995) [arXiv:0710.4345].
[13] M. Smerlak, Int. J. Mod. Phys. D Vol. 22, No. 12, 1342019 (2013).
[14] L. H. Ford and T. A. Roman, *Phys. Rev. D* **48**, 776 (1993) [arXiv:gr-qc/9303038].

Journal of Physics: Conference Series **600** (2015) 012036 doi:10.1088/1742-6596/600/1/012036

Semiclassical energy conditions and wormholes

Prado Martín-Moruno[1,2]

[1] Centro de Astronomia e Astrofísica da Universidade de Lisboa, Campo Grande, Edifício C8, 1749-016 Lisboa, Portugal

[2]Instituto de Astrofísica e Ciências do Espaço, Universidade de Lisboa, OAL, Tapada da Ajuda, PT1349 018 Lisboa, Portugal.

E-mail: pmmoruno@fc.ul.pt

Abstract. We consider the nonlinear energy conditions and their quantum extensions. These new energy conditions behave much better than the usual pointwise energy conditions in the presence of semiclassical quantum effects. Analogous quantum extensions for the linear energy conditions are not always satisfied as one enters the quantum realm, but they can be used to constrain the violation of the classical conditions. Thus, the existence of wormholes supported by a fluid which violates the null energy condition in a controlled way is of particular interest.

1. Introduction

As is well known the energy conditions are assumptions made on the material content of the spacetime to extract generic characteristics of the geometry through the equations of the dynamics (see, for example, references [1, 2]). On one hand, one could assume a *reasonable behaviour of matter in a given theory of gravity*. A particular example of this approach is the strong energy condition (SEC) which assumes that gravity is always attractive in General Relativity. That is, requiring the convergence of timelike geodesics and assuming Einstein equation, one gets that the SEC is the requirement of having a stress-energy tensor which satisfies

$$\mathrm{SEC}:\ T^a{}_b\, V_a\, V^b - \frac{1}{2}T\, V^a V_a \geq 0, \tag{1}$$

for any timelike vector V^a. For a stress-energy tensor that has only one timelike eigenvector (type I in the classification presented in reference [1]) and, therefore, can be written in an orthonormal basis as $T^{\hat{a}}{}_{\hat{b}} = \mathrm{diag}\,(-\rho,\, p_1,\, p_2,\, p_3)$, the SEC implies

$$\mathrm{type\,I}:\ \rho + p_i \geq 0\ \&\ \rho + \sum p_i \geq 0. \tag{2}$$

On the other hand, one can also formulate energy conditions by assuming that the *material content has some reasonable properties*. The dominant energy condition (DEC) assumes that the energy density measured by any observer (any timelike vector V^a) is non-negative and propagates in a causal way. That is, defining the flux vector $F^a = -T^a{}_b V^b$ we have

$$\mathrm{DEC}:\ T^a{}_b\, V_a\, V^b \geq 0\ \&\ F^a F_a \leq 0\ \Rightarrow\ \mathrm{type\,I}:\ \rho \geq |p_i|. \tag{3}$$

The weak energy condition (WEC) only requires the energy density measured by any observer to be non-negative, implying

$$\mathrm{WEC}:\ T^a{}_b\, V_a\, V^b \geq 0\ \Rightarrow\ \mathrm{type\,I}:\ \rho \geq 0\ \&\ \rho + p_i \geq 0. \tag{4}$$

Finally, the null energy condition (NEC) can be seen as a minimum requirement on the stress-energy tensor. This is

$$\text{NEC}: \; T^a{}_b \, k_a \, k^b \geq 0 \;\Rightarrow\; \text{type I}: \; \rho + p_i \geq 0, \tag{5}$$

where k^a is any null vector. As it is implied by both the WEC (which is a consequence of the DEC) and the SEC, a hypothetical material violating this condition would not satisfy any of the aforementioned conditions.

These energy conditions have been proven to be of great utility to extract conclusions about generic geometries [1, 2]. Moreover, they have also been used to argue that some solutions of the Einstein equations with potential pathological characteristics should be discarded, since they are supported by a matter content violating some of these conditions. That is the case for wormholes [2], which are shortcuts between different regions of one universe or connections between different universes. For traversable wormholes the metric can be given in the Morris-Thorne form [3]

$$ds^2 = -e^{2\Phi(r)}\, dt^2 + \frac{dr^2}{1 - b(r)/r} + r^2 \left(d\theta^2 + \sin^2\theta \, d\phi^2\right) \tag{6}$$

where $r_0 \leq r \leq \infty$, r_0 is the radius of the throat, $r_0 = b(r_0) \leq b(r)$, $\Phi(r) < \infty$ for any value of r to avoid horizons, and $b'(r) < b(r)/r$ close to the throat for it to "flare-out". Considering the Einstein equations it can be seen that one would need the existence of exotic matter close to the throat to keep it open, which exoticity resides precisely on the fact that it violates the NEC since it has

$$\rho + p_r < 0. \tag{7}$$

Thus, if the energy conditions were universal, wormholes would not exist in a universe described by general relativity. (This is not necessarily the case when considering alternative theories of gravity [4].)

Nevertheless, some doubts have been raised about this universality. It is clear that the SEC has to be significantly violated in cosmological scenarios, both during inflation and right now. Moreover, the SEC is also violated by semiclassical quantum effects that have been measured in the laboratory. Therefore, this condition is almost abolished. On the other hand, the other pointwise energy conditions, which are independent of the theory of gravity, are also known to be violated, for example, by the renormalized stress-energy tensor in the Casimir vacuum (bounded violations) and in the Schwarzschild spacetime (unbounded violations for the Boulware and Unruh vacua [5, 6, 7]).

2. New pointwise energy conditions

Given the situation presented in the previous section, the main approach that has been developed in the literature consists in averaging those energy conditions along timelike or null curves, or in using some quantum inequalities to restrict the magnitude and duration of negative energies [8]. However, we will follow a different approach, considering other pointwise energy conditions.

2.1. Nonlinear energy conditions

The first condition we present was first considered in reference [9] when obtaining entropy bounds for uncollapsed systems, and was latter studied in deeper detail in reference [10]. This is the flux energy condition (FEC) which assumes that the energy density measured by any observer propagates in a causal way. That is, it states [9, 10]

$$\text{FEC}: \; F^a F_a \leq 0 \;\Rightarrow\; \text{type I}: \; \rho^2 \geq p_i^2. \tag{8}$$

Therefore, by their very definition, the WEC plus the FEC is equal to the DEC. As some encouraging results were obtained for this condition in relation to the Boulware vacuum of a Schwarzschild black hole (as we will summarize in the next section), in reference [11] other nonlinear energy conditions were formulated. These are the determinant energy condition (DETEC) and the trace-of-square energy condition (TOSEC), which are given by [11]

$$\mathrm{DETEC:}\ \det\left(T^{ab}\right) \geq 0 \ \Rightarrow\ \mathrm{type\,I:}\ \rho \cdot \prod p_i \geq 0, \tag{9}$$

and

$$\mathrm{TOSEC:}\ T^{ab}\,T_{ab} \geq 0 \ \Rightarrow\ \mathrm{type\,I:}\ \rho^2 + \sum p_i^2 \geq 0, \tag{10}$$

respectively. Therefore, the TOSEC is always satisfied if the stress-energy tensor is type I. Indeed, it can only be violated for type IV stress-energy tensors [11]. Although the physical interpretation of these conditions is not so clear as in the case of the FEC, we present them for completeness.

2.2. Semiclassical energy conditions

Following a different point of view, one can note that the violations of the energy conditions which have been measured in the laboratory are not arbitrarily large. Therefore, one could formulate new pointwise energy conditions based on considering bounded violations of the energy conditions for the renormalized stress-energy tensor in semiclassical scenarios, as it has been done in references [10, 11].

In the first place, we consider the quantum flux energy condition (QFEC) which states that the energy density measured by any observer either propagates in a causal way or does not propagate "too much". That is, this condition assumes that either the flux vector is non-spacelike or at least its norm is bounded by above by a bound that depends on the characteristics of the system: number of fields (N), system 4-velocity (U^a), and a characteristic distance (L). A possible formulation is [10]

$$\mathrm{QFEC:}\ F^a F_a \leq \zeta\,(\hbar N/L^4)^2\,(U_a\,V^a)^2 \ \Rightarrow\ \mathrm{type\,I:}\ \rho^2 - p_i^2 \geq -\zeta\,(\hbar N/L^4)^2, \tag{11}$$

where ζ is a parameter of order unity. In the second place, similar quantum generalizations of the other nonlinear conditions have been formulated [11]. Moreover, one can also consider similar quantum extensions for the linear energy conditions. For example, the quantum weak energy condition (QWEC) assumes that the energy density measured by any observer should not be "excessively negative", that is [11]

$$\mathrm{QWEC:}\ T^a{}_b\,V_a V^b \geq -\zeta\hbar N/L^4\,(U_a\,V^a)^2 \ \Rightarrow\ \mathrm{type\,I:}\ \rho + p_i \geq -\zeta\,\hbar N/L^4. \tag{12}$$

3. Quantum vacuum states and energy conditions

Now, let us summarize the behaviour of the renormalized stress-energy tensor of some vacuum states with respect to the energy conditions (see reference [11] for further details). As it is included in table 1, the quantum extensions of the linear and nonlinear energy conditions are satisfied in the Casimir vacuum taking the distance between the parallel plates as the characteristic distance of the system. Moreover, the DETEC and TOSEC are also satisfied.

Table 1. Casimir spacetime. v and s denote that the condition is violated and satisfied, respectively.

WEC	QWEC	FEC	QFEC	DETEC	TOSEC
v	s	v	s	s	s

On the other hand, we consider the renormalized stress-energy tensor of a massless conformally coupled scalar field theory in a Schwarzschild geometry obtained taking Page's analytic approximation [12] for the Hartle–Hawking vacuum, combining that approximation with the results of Brown and

Ottewill [13, 6] for the Bolware vacuum, and Visser's semi-analytical approximation for the Unruh vacuum [7]. The results are summarized in table 2. As the Hartle–Hawking vacuum entails only finite quantities, one could expect a better behaviour for the quantum extensions of the energy conditions, and this is the case taking the characteristic distance to be the black hole radius (which is a natural assumption). It should be noted that the quantum nonlinear energy conditions are satisfied in all the cases. One can wonder why the TOSEC is violated in the Unruh vacuum since it is fulfilled by any type I stress-energy tensor. The reason is because this renormalized stress-energy tensor is of type IV in the asymptotic region [11].

Table 2. Schwarzschild geometry: Massless conformally coupled scalar. v and s denote that the condition is violated and satisfied, respectively.

	WEC	QWEC	FEC	QFEC	DETEC	QDETEC	TOSEC	QTOSEC
Hartle–Hawking	v	s	v	s	v	s	s	s
Boulware	v	v	s	s	v	s	s	s
Unruh	v	v	v	s	v	s	v	s

4. Wormholes satisfying the QWEC

As we have already pointed out in order to have a wormhole geometry one would need some amount of exotic matter. Due to the violation of the NEC by this material, it has been considered how to minimize the quantity of exotic matter needed to support a wormhole by considering thin-shells to construct the tunnel itself (see [14] and references therein) or to surround the mouth of a traversable wormhole matching it to a spacetime where no exotic matter is present [15]. It must be noted that although the NEC can be violated by semiclassical quantum effects, unbounded violations are only associated to some black hole vacua (where the stress-energy tensor itself diverges). Therefore, a different (and compatible) approach to reduce the exoticity of these geometries has been followed in reference [16], that is to minimize the violation of the NEC considering that only small bounded violations of this energy condition are possible. Thus, one would minimize not only the quantity of exotic stuff to be used but also its exoticity.

4.1. Equation of state minimally violating the WEC

The equation of state that we will take as a starting point has been first considered in a cosmological setting when investigating how a small constant deviation from a cosmological constant would change the future of the universe to what has been called *the little sibling of the big rip* [17]. That equation of state is

$$p_{\mathrm{DE}} + \rho_{\mathrm{DE}} = -\frac{A}{3}. \tag{13}$$

As $\rho_{\mathrm{DE}} \geq 0$, the QWEC would be satisfied for $A \ll \hbar H_0^4$ [17].

For wormhole geometries an anisotropic version of this equation of state has been considered in reference [16]. This is

$$p_r(r) + \rho_r(r) = -\frac{A}{8\pi}, \tag{14}$$

with $\rho(r) \geq 0$ and $A \ll 1/r_0^2$. Although there cannot be asymptotically flat solutions with this equation of state (since it has to be also satisfied in the asymptotic limit), this kind of solutions has been constructed considering thin-shells. It has been proven that the resulting construction is stable and satisfies the QWEC, minimizing also the quantity of material violating the WEC [16]. On the other hand, one can also consider an inhomogeneous version of this equation of state such as $\rho + p_r \to -\frac{A}{8\pi}$ when $r \to r_0$ and $\rho + p_r \to 0$ when $r \to \infty$, which is not only compatible with asymptotic flatness (at least in principle), but minimizes even more the violation of the WEC without the introduction of thin-shells. In reference [16] the following equation has been considered:

$$\rho(r) + p_r(r) = -\frac{A}{8\pi}\left(\frac{r_0}{r}\right)^\alpha \tag{15}$$

where $\rho(r) \geq 0$ and $A \ll 1/r_0^2$.

4.2. Particular solution

Considering equation of state (15) with $\alpha > 2$ and assuming a constant shift function $\Phi(r) = \Phi_0$, one can obtain the shape function $b(r)$ and the transverse pressure through the Einstein equations. The metric describing this geometry is [16]

$$\mathrm{d}s^2 = -e^{2\Phi_0}\mathrm{d}t^2 + \frac{\alpha-2}{A\,r_0^2}\frac{\mathrm{d}r^2}{1-(r_0/r)^{\alpha-2}} + r^2\mathrm{d}\Omega_{(2)}^2. \tag{16}$$

Thus, close to $r_0 \lesssim r$ we have $b'(r) < b(r)/r$, that is a wormhole throat. In the asymptotic limit the geometry tends to

$$d\bar{s}^2 = -d\bar{t}^2 + d\bar{r}^2 + \frac{A\,r_0^2}{\alpha-2}\bar{r}^2 d\Omega_{(2)}^2, \tag{17}$$

where $\bar{r}^2 = \frac{(\alpha-2)r^2}{A\,r_0^2}$ and $\bar{t} = e^{\Phi_0}t$. The area of the sphere of radius $\bar{r}$ is $4\pi(1-\Delta)\bar{r}^2$, with $\Delta = 1 - \frac{A\,r_0^2}{\alpha-2}$. Thus, this geometry describes a space with a deficit of solid angle [18] in the asymptotic limit if $\alpha > 3$ and if $2 < \alpha < 3$ & $Ar_0^2 < \alpha - 2$. This solution may, therefore, be interpreted as a wormhole carrying a global monopole [16].

The energy density and pressures of the material content can be obtained using the Einstein equations. These are [16]

$$\rho = \frac{1}{8\pi(\alpha-2)r^2}\left\{\alpha - 2 - Ar_0^2\left[1 + (\alpha-3)\left(\frac{r_0}{r}\right)^{\alpha-2}\right]\right\} \tag{18}$$

$$p_r = -\frac{1}{8\pi(\alpha-2)r^2}\left\{\alpha - 2 - Ar_0^2\left[1 - \left(\frac{r_0}{r}\right)^{\alpha-2}\right]\right\} \tag{19}$$

$$p_t = \frac{A}{16\pi}\left(\frac{r_0}{r}\right)^\alpha > 0. \tag{20}$$

Therefore, $p_t(r) \geq 0$. As $\rho(r) \geq 0$ for $\alpha > 3$ and $2 < \alpha < 3$ & $Ar_0^2 < \alpha - 2$, we would have $\rho(r) + p_t(r) > 0$ and, considering also equation (15), the QWEC is satisfied through the whole space.

5. Discussion

We have reviewed how the quantum generalizations of the energy conditions behave better than their classical counterparts. Moreover, the nonlinear energy conditions are satisfied in more situations than the usual energy conditions. The QFEC is satisfied for classical situations and quantum vacuum states.

Regarding the applicability of these new energy conditions, the FEC has been used to obtain some interesting entropy bounds for uncollapsed systems [9]. As the QFEC is satisfied when Hawking evaporation is present, one should not expect this condition to be compatible with a derivation of the second law of black hole physics. Although its utility for other theoretical problems is completely unexplored for the moment, we can already argue that some steps in its application to generic spacetimes would be particularly tricky due to both the nonlinear character of this condition and its non-vanishing

r. h. s. It has already been pointed out [11] that one could think that two quantum field theories satisfying the QFEC may interfere in a destructive way under some circumstances due to the nonlinear character of this condition and produce a final situation where it is violated, although an explicit example of such behaviour have not been presented yet.

Inspired by the results of the Casimir vacuum, one may think that the quantum linear energy conditions could be enough for describing semiclassical phenomena, giving a preferred status to the Hawking–Hartle vacuum for the black hole space. In this case, it would not only be interesting to study the possible existence of wormhole geometries satisfying the QWEC (as it has been done in reference [16]), but also to explore the utility of this condition to deduce characteristics of a generic spacetime where it is satisfied. Moreover, in the case that this utility may be proven, it would be of particular interest to check if the mentioned wormhole geometries could avoid pathologies (as to be used as hypothetical time-machines) as long as they continue satisfying the QWEC.

Acknowledgments

I want to thank Matt Visser for enlightening discussions and co-authorship of references [10, 11], and Mariam Bouhmadi-López and Francisco S. N. Lobo for collaboration on reference [16], papers in which this talk was mainly based on. This work was supported by FCT (Portugal) through the projects PTDC/FIS/111032/2009 and EXPL/FIS-AST/1608/2013.

References

[1] Hawking S W and Ellis G F R 1972 *The large scale structure of spacetime* (England: Cambridge University press).
[2] Visser M 1996 *Lorentzian wormholes: from Einstein to Hawking* (New York: AIP Press, now Springer-Verlag).
[3] Morris M S and Thorne K S (1988) *Am. J. Phys.* **56** 395.
[4] Lobo F S N 2008 *Class. Quant. Grav.* **25** 175006. Lobo F S N and Oliveira M A, *Phys. Rev.* D **80** 104012.
[5] Visser M 1996 *Phys. Rev.* D **54** 5103.
[6] Visser M 1996 *Phys. Rev.* D **54** 5116.
[7] Visser M 1997 *Phys. Rev.* D **56** 936.
[8] Ford L H, Pfenning M J and Roman T A 1998 *Phys. Rev.* D **57** 4839. Borde A, Ford L H and Roman T A 2002 *Phys. Rev.* D **65** 084002. Ford L H and Roman T A 1999 *Phys. Rev.* D **60** 104018. Fewster C J and Smith C J 2008 *Annales Henri Poincare* **9** 425.
[9] Abreu G, Barceló C and Visser M 2011 *J.High Energy Physics* JHEP12(2011)092.
[10] Martín-Moruno P and Visser M 2013 *Phys. Rev.* D **88** 061701.
[11] Martín-Moruno P and Visser P 2013 *J.High Energy Physics* JHEP09(2013)050.
[12] Page D N 1982 *Phys. Rev.* D **25** 1499.
[13] Brown M R and Ottewill A C 1985 *Phys. Rev.* D **31** 2514.
[14] García N M, Lobo F S N and Visser M 2012 *Phys. Rev.* D **86** 044026.
[15] Lobo F S N and Crawford P 2005 *Class. Quant. Grav.* **22** 4869.
[16] Bouhmadi-López M, Lobo F S N and Martín-Moruno P 2014 *J. Cosmol. Astropart. Phys.* JCAP11(2014)007.
[17] Bouhmadi-López M, Errahmani A, Martín-Moruno P, Ouali T and Tavakoli Y 2014 *Preprint* gr-qc/1407.2446.
[18] Barriola M and Vilenkin A 1989 *Phys. Rev. Lett.* **63** 341.

A perspective on Black Hole Horizons from the Quantum Charged Particle

José Luis Jaramillo
Université de Bretagne Occidentale, Brest, France
E-mail: jose-luis.jaramillo@univ-brest.fr

Abstract. Black hole apparent horizons possess a natural notion of stability, whose spectral characterization can be related to the problem of the stationary quantum charged particle. Such mathematical relation leads to an "analyticity conjecture" on the dependence of the spectral properties on a complex "fine-structure-constant" parameter, that can reduce the study of the spectrum of the (non-selfadjoint) MOTS-stability operator to that of the (selfadjoint) Hamiltonian of the quantum charged particle. Moreover, this perspective might open an avenue to the spinorial treatment of apparent horizon (MOTS-)stability and to the introduction of semiclassical tools to explore some of the qualitative aspects of this black hole spectral problem.

1. The analogy: black hole apparent horizons and quantum charged particles

Given a black hole apparent horizon section $\mathcal{S}$, the systematic study of the spectral problem of the so-called stability operator $L_\mathcal{S}$ [1] of *marginally outer trapped surfaces* (MOTS), namely

$$L_\mathcal{S}\psi \equiv \left[-\Delta + 2\Omega^a D_a - \left(|\Omega|^2 - D_a\Omega^a - \frac{1}{2}R_\mathcal{S} + G_{ab}k^a\ell^b\right)\right]\psi = \lambda\psi \ , \tag{1}$$

has been proposed in [2] as a methodology to explore aspects of the black hole horizon geometry, possibly leading to insights into the black hole stability/instability problem (paraphrased, following Kac's spectral discussion [3], as "can one hear the stability of a black hole horizon?").

The general resolution of the MOTS-spectral problem in (1) represents a challenging task. In this sense, any hint relating this problem to a better known and controlled problem is of clear relevance. This is precisely the context of Ref. [4], where a relation is presented between the MOTS-spectral problem and the study of the stationary states of a spin-0 quantum charged particle moving in the presence of electric and magnetic fields.

To make this statement more precise, we start by briefly describing the terms in the MOTS-stability operator $L_\mathcal{S}$. Let us consider a n-dimensional spacetime with metric g_{ab}, associated Levi-Civita connection ∇_a and Einstein curvature tensor G_{ab}. Let $\mathcal{S}$ be a codimension-2 spacelike closed (compact without boundary) surface with induced metric q_{ab}. $L_\mathcal{S}$ contains both intrinsic and extrinsic geometry elements. The intrinsic geometry ones associated with q_{ab} are D_a, $\Delta = D^aD_a$ and $R_\mathcal{S}$, respectively the Levi-Civita connection, the scalar Laplacian operator and the Ricci scalar on $\mathcal{S}$. Let us also introduce the volume form $\epsilon_{a_1\ldots a_{n-2}}$ with associated measure dS. Regarding the extrinsic geometry, we introduce first two null vectors ℓ^a and k^a spanning the normal bundle $T^\perp\mathcal{S}$, normalized as $k^a\ell_a = -1$. Then the Hájiček form $\Omega_a = -k^b q_a{}^c\nabla_c\ell_b$

encodes part of the extrinsic geometry of $\mathcal{S}$, in particular providing a connection in the normal bundle. Physically, Ω_a represents an angular momentum density through the (Komar) expression $J = \int_{\mathcal{S}} \Omega_a \phi^a dS$, where ϕ^a is an axial Killing vector on $\mathcal{S}$. The term $G_{ab}k^a\ell^b$ is related to ambient spacetime dynamics through the Einstein equations $G_{ab} + \Lambda g_{ab} = 8\pi T_{ab}$. Defining the (outgoing) expansion on $\mathcal{S}$ as $\theta^{(\ell)} = q^{ab}\nabla_a \ell_b$, the MOTS condition on $\mathcal{S}$ is characterized as $\theta^{(\ell)} = 0$, whereas the MOTS-stability operator is defined [1] through the variation of this expansion $L_{\mathcal{S}}\psi \equiv \delta_{-\psi k}\theta^{(\ell)}$, where ψ is a function on $\mathcal{S}$ characterizing the surface deformation along k^a. Note that $L_{\mathcal{S}}$ is (formally) non-selfadjoint, namely due to the $2\,\Omega^a D_a\psi$ term.

Given these elements, and as shown in [4], the MOTS-stability operator is related to the Hamiltonian of the quantum charged particle through the identifications

$$\Omega_a \leftrightarrow \frac{ie}{\hbar c} A_a \ , \quad R_{\mathcal{S}} \leftrightarrow \frac{4me}{\hbar^2}\phi \ , \quad G_{ab}k^a\ell^b \leftrightarrow -\frac{2m}{\hbar^2} V \ , \tag{2}$$

where A_a and ϕ correspond to the vector and scalar potentials of given magnetic and electric fields on $\mathcal{S}$, whereas V represents an external mechanical potential. Indeed, $L_{\mathcal{S}}$ becomes

$$\begin{aligned} \frac{\hbar^2}{2m} L_{\mathcal{S}} \leftrightarrow \hat{H} &= -\frac{\hbar^2}{2m}\Delta + \frac{i\hbar e}{mc} A^a D_a + \frac{i\hbar e}{2mc} D_a A^a + \frac{e^2}{2mc^2} A_a A^a + e\phi + V \\ &= \frac{1}{2m}\left(-i\hbar D - \frac{e}{c}A\right)^2 + e\phi + V \ , \end{aligned} \tag{3}$$

where $\hat{H}$ is the Hamiltonian of the spin-0 quantum charged particle if the formal parameter $\hbar$ is interpreted as the Planck constant (over 2π) and m and e as the mass and charge of the particle. The main points in [4] are, first, to bring attention to the fact that the derivative and Hájiček terms in (1) can be combined as a perfect square into a single connection term, as follows

$$L_{\mathcal{S}}\psi = \left[-(D-\Omega)^2 + \frac{1}{2}R_{\mathcal{S}} - G_{ab}k^a\ell^b \right]\psi \ . \tag{4}$$

This underlines the role of $(D_a - \Omega_a)$ as the relevant connection in the problem (actually a connection in the normal bundle $T^{\perp}\mathcal{S}$). And secondly, pushing forward this simple remark, Ref. [4] establishes a structural connection between the MOTS-stability and quantum charged particle problems beyond the purely formal analogy embodied in the correspondences (2). Specifically, both problems possess and Abelian gauge symmetry defined by the transformations

$$\begin{array}{rrcll} \text{Quantum Charged Particle:} & A_a & \rightarrow & A_a - D_a\sigma \ , & \psi \rightarrow e^{ie\sigma/(c\hbar)}\psi \\ \text{MOTS:} & \Omega_a & \rightarrow & \Omega_a - D_a\sigma \ , & \psi \rightarrow e^{-\sigma}\psi \ . \end{array} \tag{5}$$

The first one defines the compact $U(1)$-gauge symmetry of electromagnetism, whereas the second one defines a non-compact $\mathbb{R}^+$-gauge symmetry corresponding to the invariance of the MOTS-geometry under a rescaling of the null vectors (namely a boost-transformation in the normal direction): $\ell'^a = e^{-\sigma}\ell^a, k^a = e^{\sigma}k^a$. In particular, this identifies the Hájiček form as the gauge potential of this $\mathbb{R}^+$-abelian symmetry. Moreover, as apparent in (4) this gauge potential Ω_a (and therefore the black hole angular momentum) is introduced into the MOTS-problem via a *minimal coupling mechanism*, i.e. by substitution $D_a \rightarrow (D_a - \Omega_a)$ in the non-rotating problem. This is exactly the mechanism for switching on magnetic (and electric) fields in the quantum particle. We can summarize these gauge symmetries considerations in the following lemma [4]:

Lemma 1 *(MOTS-gauge transformations).*
Under the null normal rescaling $\ell'^a = f\ell^a$, $k'^a = f^{-1}k^a$, *with* $f > 0$*:*

i) The expansion and Hájiček form transform as: $\theta^{(\ell')} = f\theta^{(\ell)}$ *and* $\Omega'_a = \Omega_a + D_a(\ln f)$.

ii) The MOTS-stability operator transforms covariantly: $(L_\mathcal{S})'\psi = fL_\mathcal{S}(f^{-1}\psi)$, *where* $(L_\mathcal{S})'\psi \equiv \delta_{\psi(-k')}\theta^{(\ell')}$.

iii) The MOTS-eigenvalue problem is invariant under the additional eigenfunction transformation, $\psi' = f\psi$. *That is,* $L_\mathcal{S}\psi = \lambda\psi$ *goes to* $(L_\mathcal{S})'\psi' = \lambda\psi'$.

It is interesting to note that the gauge-connection $D^\Omega_a = D_a - \Omega_a$ is intimately related to the "covariant derivative" introduced in the GHP formalism [5] to define a notion of "Fermi-Walker" transport of a vector along the surface $\mathcal{S}$. In particular, a vector γk^a is "Fermi-transported" along $\mathcal{S}$ if $(D_a - \Omega_a)\gamma = 0$ (such transport is in general non-integrable on $\mathcal{S}$ and therefore dependent on the path). More generally the GHP formalism, based on a choice of two null directions at each spacetime point and naturally adapted to the study of codimension-2 surfaces, could offer insights into the MOTS-stability problem through its concepts of spin- and boost-weighted quantities (the latter directly connected with the notion of MOTS-gauge symmetry).

The presented analogy between the MOTS-stability problem and the quantum charged particle offers the possibility of transferring tools and concepts from one problem to the other. Here we will comment: on the possibility of addressing the MOTS-spectral problem by importing the expertise in the stationary spectrum of the quantum charged particle through an analytic continuation procedure, on the possibility of introducing semi-classical tools into the study of black horizon geometry and, finally, on an approach to a spinorial formulation of MOTS-stability by mimicking Pauli's and Dirac's steps in the introduction of spinors in the quantum charged particle. We will conclude with a brainstorming list of possible directions of further study.

2. MOTS-spectrum analyticity in the fine structure constant α

Let us introduce the "fine structure constant" $\alpha \equiv \frac{e^2}{\hbar c}$ and set $\hbar$=m=c=1. We define the operator family $L[\sqrt{\alpha}]$ on the square root of this "fine-structure-constant" complex parameter α, as

$$\begin{aligned} L[\sqrt{\alpha}] &\equiv -\frac{1}{2}(D - i\sqrt{\alpha}\Omega)^2 - \frac{\alpha}{4}R_\mathcal{S} - \frac{1}{2}G_{ab}k^a\ell^b \\ &= -\frac{1}{2}\Delta + i\sqrt{\alpha}(\Omega\cdot D + \frac{1}{2}D\cdot\Omega) + \frac{\alpha}{2}|\Omega|^2 - \frac{\alpha}{4}R_\mathcal{S} - \frac{1}{2}G(k,\ell)\ , \end{aligned} \quad (6)$$

so that the quantum charged particle corresponds to a real positive $\alpha = 1$ (normalized as $e^2 = 1$), whereas (half) the MOTS-stability operator corresponds to a real negative fine structure constant $\alpha = -1$. More precisely, we make branch choices $\hat{H} = L[\sqrt{\alpha} = 1]$ and $L_\mathcal{S}/2 = L[\sqrt{\alpha} = -i]$.

The operator family $L[\sqrt{\alpha}]$ provides an "analytical continuation" in $\sqrt{\alpha}$ from the self-adjoint quantum charged particle Hamiltonian $\hat{H}$ to the non-selfadjoint MOTS-stability operator $L_\mathcal{S}$. This naturally raises the following question: *can we recover the MOTS-spectrum* ($\alpha = -1$) *as an analytic extension of the quantum charged particle spectrum* ($\alpha = 1$) *self-adjoint problem?*. If this is indeed possible, we can perform the following "self-adjoint trick" to reduce the resolution of the non-selfadjoint spectral problem associated with $L_\mathcal{S}$ to a self-adjoint one: first multiply formally Ω_a by $\sqrt{\alpha}$, then rotate $\sqrt{\alpha}$ in the complex plane to $i\sqrt{\alpha}$ to produce a self-adjoint operator whose spectral problem can be explicitly solved using standard techniques, and finally perform a "back rotation" $\sqrt{\alpha} \to \frac{1}{i}\sqrt{\alpha}$ in the explicitly obtained eigenvalues.

Answering the question above in its full generality sets a difficult problem in perturbation theory of linear operators [6]. We formulate the following conjecture [4] as an open problem:

Analyticity Conjecture. *Given an orientable closed surface* $\mathcal{S}$ *and the one-parameter family of operators* $L[\sqrt{\alpha}]$ *defined in (6), (in the complex* $\sqrt{\alpha}$*), the MOTS-spectrum* ($\sqrt{\alpha} = -i$) *can be recovered as an "analytic continuation" of the quantum charged particle spectrum* ($\sqrt{\alpha} = 1$)*.*

In order to provide some support on the validity of this conjecture, we consider an explicit example that we can fully solve and that presents the essential qualitative features to be expected in the generic case. This simple example provides an analogue in our context to the Landau levels for the quantum charged particle in a constant magnetic field in $\mathbb{R}^3$. We take as $\mathcal{S}$ a topological sphere S^2 endowed with the round metric $q_{ab} = r^2(d\theta^2 + \sin^2\theta d\varphi^2)$. On S^2, the Hodge decomposition of the Hájiček form is $\Omega_a = \epsilon_a{}^b D_b\omega + D_a\zeta$. We make the simplest non-trivial choice $\omega = \sqrt{\alpha}\cos\theta, \zeta = 0$ (with $\sqrt{\alpha} \in \mathbb{R}$) that leads to $\Omega = \sqrt{\alpha}\sin^2\theta d\varphi$. Finally, we assume vacuum ($G_{ab} + \Lambda g_{ab} = 8\pi T_{ab} = 0$). With these choices, the relevant terms in (1) are

$$R_{\mathcal{S}} = \frac{2}{r^2} \;,\; 2\Omega^a D_a\psi = \frac{2\sqrt{\alpha}}{r^2}\partial_\varphi\psi \;,\; |\Omega|^2 = \frac{\alpha}{r^2}\sin^2\theta \;,\; D^a\Omega_a = 0 \;,\; G_{ab}k^a\ell^b = \Lambda \;. \tag{7}$$

We insert this in Eq. (1), separate variables as $\Psi = S(\theta)e^{im\varphi}$ and write $\lambda = \lambda_\mathrm{R} + i\lambda_\mathrm{I}$. The imaginary part of the equation produces $\lambda_\mathrm{I} = 2\sqrt{\alpha}m/r^2$, whereas the real part ($x = \cos\theta$) is

$$\frac{d}{dx}\left((1-x^2)\frac{d}{dx}\right)S + \left(((\lambda_\mathrm{R} + \Lambda)r^2 + \alpha - 1) - \alpha x^2 - \frac{m^2}{1-x^2}\right)S = 0 \;. \tag{8}$$

Solutions to this equation are given [7] by the (prolate) spheroidal harmonics $S_{\ell m}(\sqrt{\alpha}, cos\theta)$, with values $\lambda_{\ell m}(\sqrt{\alpha}) = (\lambda_\mathrm{R} + \Lambda)r^2 + \alpha - 1$. The solutions of the MOTS-problem are therefore

$$\lambda = \frac{\lambda_{\ell m}(\sqrt{\alpha}) + 1 - \alpha}{r^2} - \Lambda + i\frac{2\sqrt{\alpha}m}{r^2} \;,\; \psi_{\ell m}(\theta,\phi) = S_{\ell m}(\sqrt{\alpha}, \cos\theta)e^{im\varphi} \;. \tag{9}$$

To study the corresponding quantum charged particle problem, we make $\sqrt{\alpha} \to i\sqrt{\alpha}$ in Ω_a. Denoting the new eigenvalues as $\bar{\lambda}$ and the eigenfunctions by $\bar{\psi}$ and repeating the steps above, we first obtain $\lambda_\mathrm{I} = 0$ (the operator is now self-adjoint), whereas the equation on x is now

$$\frac{d}{dx}\left((1-x^2)\frac{d}{dx}\right)S + \left(((\bar{\lambda}_\mathrm{R} + \Lambda)r^2 + 2\sqrt{\alpha}m - \alpha - 1) + \alpha x^2 - \frac{m^2}{1-x^2}\right)S = 0 \;. \tag{10}$$

This is now the equation for the (oblate) spheroidal harmonics [7] with solutions $\bar{S}_{\ell m}(\sqrt{\alpha}, cos\theta) = S_{\ell m}(i\sqrt{\alpha}, cos\theta)$ and where the corresponding $\bar{\lambda}_{\ell m}(\sqrt{\alpha}) = (\bar{\lambda}_\mathrm{R} + \Lambda)r^2 + 2\sqrt{\alpha}m - \alpha - 1$ satisfy $\bar{\lambda}_{\ell m}(\sqrt{\alpha}) = \lambda_{\ell m}(i\sqrt{\alpha})$ (note that $\lambda_{\ell m}(i\sqrt{\alpha})$ are real numbers [7]). The solution is now

$$\bar{\lambda} = \frac{\lambda_{\ell m}(i\sqrt{\alpha}) + 1 + \alpha - 2\sqrt{\alpha}m}{r^2} - \Lambda \;,\; \bar{\psi}_{\ell m}(\theta,\phi) = S_{\ell m}(i\sqrt{\alpha}, \cos\theta)e^{im\varphi} \;. \tag{11}$$

This is in agreement with the "analyticity conjecture" formulated above, since both eigenvalues and eigenfunctions in the solution (9) to the non-selfadjoint problem are indeed recovered from the solutions of the self-adjoint case (11), when applying back the shift $\sqrt{\alpha} \to \frac{1}{i}\sqrt{\alpha}$.

3. Semi-classical tools in the MOTS-spectral problem. Towards an action principle for MOTS-stability

The possibility of reducing the spectral problem of $L_{\mathcal{S}}$ to that of a selfadjoint operator, admitting in addition the interpretation of a quantum Hamiltonian, opens the possibility of considering a particular semi-classical approach to the study of $L_{\mathcal{S}}$. Inverting the quantization rule $p_i \to -iD_i$ (with $\hbar = 1$), the Hamiltonian $\hat{H}(\sqrt{\alpha}) \equiv L[\sqrt{\alpha}]$ leads to the classical Hamiltonian function

$$H_\mathrm{cl}[\sqrt{\alpha}](x,p) = (p - \sqrt{\alpha}\Omega)^2 + \frac{1}{2}R_{\mathcal{S}} - G(k,\ell) \;. \tag{12}$$

Then, we can use $H_\mathrm{cl}[\sqrt{\alpha}](x,p)$ as the starting point to obtain approximate solutions to the spectral problem of $\hat{H}(\sqrt{\alpha})$ by employing semi-classical tools such as the WKB techniques [8].

Under the assumption of the validity of the "analyticity conjecture", the relevant expressions approximating the MOTS-problem would be then obtained by evaluating $\sqrt{\alpha} \to -i$.

A relevant question is: can we define an action from which the MOTS-stability problem can be derived variationally? The action for a (complex) scalar field coupled to an electromagnetic field provides the answer in the self-adjoint case, but its straightforward application to $L_{\mathcal{S}}$ fails. We have not been able to find a real action from which $L_{\mathcal{S}}$ emerges variationally. On the other hand, considering a complex scalar ψ on $\mathcal{S}$ we can introduce the following complex action

$$S = \int_{\mathcal{S}} dS \left(q^{ab}(D_a + \Omega_a)\psi^*(D_b - \Omega_b)\psi + \left(\frac{1}{2}R_{\mathcal{S}} - G_{ab}k^a\ell^b\right)\psi^*\psi - F_\psi\psi^* - F_{\psi^*}\psi \right) . \quad (13)$$

Variating independently ψ^* and ψ, the corresponding Euler-Lagrange equations are

$$\begin{aligned} L_{\mathcal{S}}\psi &= \left(-(D-\Omega)^2 + \frac{1}{2}R_{\mathcal{S}} - G_{ab}k^a\ell^b\right)\psi = F_\psi \\ (L_{\mathcal{S}})^\dagger\psi^* &= \left(-(D+\Omega)^2 + \frac{1}{2}R_{\mathcal{S}} - G_{ab}k^a\ell^b\right)\psi^* = F_{\psi^*} . \end{aligned} \quad (14)$$

Choosing $F_\psi = \lambda\psi$, with λ a parameter to be determined, we recover the eigenvalue problem (1). One recovers not only the elliptic problem for $L_{\mathcal{S}}$, related to the ingoing variation of the outgoing expansion $L_{\mathcal{S}}\psi = \delta_{-k\psi}\theta^{(\ell)}$, but also the problem $(L_{\mathcal{S}})^\dagger\psi^*$ related to the outgoing variation of the ingoing expansion: $(L_{\mathcal{S}})^\dagger\psi^* = \delta_{-\ell\psi^*}\theta^{(k)} - \kappa^{(\psi^*\ell)}\theta^{(k)}$. This suggests that the relevant object in this context is actually a two-component vector related to second variations of the element of area. "Diagonal" second variations would correspond to Raychaudhuri equations, whereas the "non-diagonal" ones are expressed in terms of the MOTS-stability operator and its adjoint.

4. Spinors and MOTS-stability

Having a spinor (first-order) characterization of MOTS-stability would present various potential interests: i) inner boundary conditions for elliptic problems with a horizon (e.g. in Witten's proof of mass positivity, approaches to Penrose-like inequalities...), ii) reduction of the spectral problem to that of a first-order operator and, more generally, iii) natural setting for studying MOTS-stability when studying spinorial fields propagating on a black hole spacetime background.

The analogy of the MOTS-stability operator with the Hamiltonian of the quantum charged particle provides a natural way for introducing spinors by mimicking Pauli's and Dirac's approaches to particles' spin. Let us first introduce some notation. Considering a tetrad ${e^a}_i$, such that $g_{ab}{e^a}_i{e^b}_j = \eta_{ij}$, assuming a spin-structure we introduce gamma-matrices $\boldsymbol{\gamma}^a = {e^a}_i\boldsymbol{\gamma}^i$ acting on spinors $\boldsymbol{\Psi}$ and satisfying the Clifford algebra $\{\boldsymbol{\gamma}^a, \boldsymbol{\gamma}^b\} = 2g^{ab}\mathbf{1}$. In the Riemannian case of the MOTS $\mathcal{S}$, we consider $\{\boldsymbol{\gamma}^a, \boldsymbol{\gamma}^b\} = 2q^{ab}\mathbf{1}$. The derivative connection on spinors is $\mathbf{D}_a\boldsymbol{\Psi} = \left(D_a + \frac{1}{8}\omega_a^{ij}[\gamma_i, \gamma_j]\right)\boldsymbol{\Psi}$, where ω_a^{ij} is the standard spin-connection associated with ${e^a}_i$.

A straightforward approach to recast MOTS-stability in terms of a first-order condition, would be to take the "square-root" of $L_{\mathcal{S}}$ in the same spirit in which the square root of the Klein-Gordon equation of a field of mass m, namely $((i\hbar)^2\Box + m^2c^2)\Psi = 0$, is the Dirac equation $(i\hbar\gamma^i D_i + mc)\boldsymbol{\Psi} = 0$. However, this does not work for $L_{\mathcal{S}}$, due to the non-constancy of the corresponding "mass terms". As an alternative, we consider the second-order Pauli equation starting from the following key remark: the Laplacian (in Euclidean $\mathbb{R}^3$ and acting on spinors) can be written in two ways as follows ($\boldsymbol{\sigma}^i$ are Pauli matrices, the gamma matrices in this setting)

$$\Delta = D^iD_i = \left(\boldsymbol{\sigma}^iD_i\right)^2 \quad , \quad \{\boldsymbol{\sigma}^i, \boldsymbol{\sigma}^j\} = 2\delta^{ij}\mathbf{1} \text{ (Clifford relations)} . \quad (15)$$

The crucial point is that introducing the magnetic vector potential through minimal coupling actually depends on the starting version we choose for the Laplacian. In particular, the spin-magnetic field coupling term $\frac{\hbar e}{2mc}\boldsymbol{\sigma}^iB_i$, with giromagnetic factor 2, is recovered when performing

$\Delta = \left(\boldsymbol{\sigma}^i D_i\right)^2 \to \left(\boldsymbol{\sigma}^i (D_i - ieA_i)\right)^2$. In the curved MOTS case, Eq. (15) is generalized through (an adaptation of) the Lichnerowicz-Weitzenböck formula (we denote $F^{\Omega}_{ab} = D_a\Omega_b - D_b\Omega_a$)

$$(i\gamma^a(\mathbf{D}_a - \Omega_a))^2 = -(D_a - \Omega)^2 + \frac{1}{4}R_{\mathcal{S}} + \frac{1}{4}[\gamma^a, \gamma^b]F^{\Omega}_{ab} \ . \quad (16)$$

The MOTS-stability operator can then be written as

$$L_{\mathcal{S}} = (i\gamma^a(\mathbf{D}_a - \Omega_a))^2 + \frac{1}{4}R_{\mathcal{S}} - \frac{1}{4}[\gamma^a, \gamma^b]F^{\Omega}_{ab} - G_{ab}k^a\ell^b \ . \quad (17)$$

From the Pauli equation perspective, the F^{Ω}_{ab} term modifies the "spin-magnetic field" coupling term correcting the "giromagnetic factor" and then indicating that, from this point of view, the MOTS is a composite object. More importantly, although this is still a second-order operator, it indicates the relevance of the first-order operator defined from the codimension-2 Sen connection, $i\gamma^a(\mathbf{D}_a - \Omega_a)$, and its related spectral problem. Finally, Pauli's second-order equation can be obtained from the Dirac first-order one by taking the (non-relativistic) limit $c \to 0$. This remark indicates a formal path to define a first-order spinorial operator from which the MOTS-stability operator $L_{\mathcal{S}}$ can be recovered through a limit procedure. This will be presented elsewhere [9].

5. Perspectives

The analogy between stable MOTS and quantum particles (stable MOTS behave as charged particles with "negative fine-structure constant") provides the seed for a research program relying on the transfer of knowledge between black hole and quantum charged particle physics.

We list some possible directions of research: i) self-adjoint "shortcut" to the spectral MOTS-problem through the analyticity conjecture; ii) applications to the MOTS-stability of Kerr; iii) MOTS-spectrum statistics and applications of a "$L_{\mathcal{S}}$-spectral zeta function $\zeta_{L_{\mathcal{S}}}(s)$" constructed by analytic continuation from the $\zeta_{L[\sqrt{\alpha}]}(s)$ of the selfadjoint $L[\sqrt{\alpha}]$; iv) semiclassical (and dynamical-systems) tools in the study of the MOTS-stability operator, in particular to construct approximate explicit expressions for MOTS-stability in Kerr, as well as generic "high-eigenvalue" asymptotics; v) spinor reformulation of MOTS-stability and applications to inner boundary conditions in elliptic problems relevant for the "mass problem" in General Relativity (positivity, Penrose-like inequalities, quasi-local gravitational mass); vi) variational derivation of MOTS-stability from an action functional and relation to Ginzburg-Landau theory on closed Riemannian manifolds; vii) higher-dimensional black hole horizons, with richer topologies and field content through the Hodge-decomposition $\Omega = d\alpha + \delta\beta + \gamma$ (with γ harmonic). Finally, we mention other complementary issues: gauge-invariant expression of the charged particle ground state from Donsker-Varadhan theory [4], possibility of a MOTS "Aharonov-Bohm-like effect", signature of quasi-normal modes/superradiance, "second-quantization"-approach (motivated by a "many-particle" treatment of the Schrödinger equation), study of effective MOTS-deformation "degrees of freedom" and possible statistical-mechanics/thermodynamical properties of MOTS.

Acknowledgments. I thank J.M.M. Senovilla and J.A. Valiente-Kroon for discussions.

[1] Andersson L, Mars M, Simon W 2005 Local existence of dynamical and trapping horizons *Phys. Rev. Lett.* **95**, 111102 ; 2008 Stability of marginally outer trapped surfaces and existence of marginally outer trapped tubes *Adv. Theor. Math. Phys.* **12**, 853–888
[2] Jaramillo JL 2014 A Young-Laplace law for black hole horizons *Phys. Rev.* D**89** 021502
[3] Kac M 1966 Can one hear the shape of a drum? *American Mathematical Monthly* **73**(4), 1–23
[4] Jaramillo JL 2014 Black hole horizons and quantum charged particles, *arXiv:1410.0509* submitted
[5] Geroch R, Held A, Penrose R 1973 A spacetime calculus based on pairs of null directions J. Math. Phys. **14**, 874–881
[6] Kato T 1995 *Perturbation theory for linear operators* (Berlin: Springer-Verlag)
[7] Abramowitz M, Stegun IA 1964 *Handbook of Mathematical Functions* (New York: Dover Publications)
[8] Berry M V, 1983 Semiclassical Mechanics of regular and irregular motion, N. Holland, Amsterdam, 171-271
[9] Jaramillo JL 2015 MOTS stability and spinors *in preparation*

Journal of Physics: Conference Series **600** (2015) 012038 doi:10.1088/1742-6596/600/1/012038

Wormhole in higher-dimensional space-time

Hisa-aki Shinkai[1,2], **Takashi Torii** [3]

[1] Faculty of Information Science and Technology, Osaka Institute of Technology, Hirakata, Osaka 573-0196, Japan

[2] Computational Astrophysics Laboratory, Institute of Physical & Chemical Research (RIKEN), Hirosawa, Wako, Saitama, 351-0198 Japan

[3] Faculty of Engineering, Osaka Institute of Technology, Osaka, Osaka 535-8585, Japan

E-mail: hisaaki.shinkai@oit.ac.jp

E-mail: takashi.torii@oit.ac.jp

Abstract. We introduce our recent studies on wormhole, especially its stability aspect in higher-dimensional space-time both in general relativity and in Gauss-Bonnet gravity. We derived the Ellis-type wormhole solution in n-dimensional general relativity, and found existence of an unstable mode in its linear perturbation analysis. We also evolved it numerically in dual-null coordinate system, and confirmed its instability. The wormhole throat will change into black hole horizons for the input of the (relatively) positive energy, while it will change into inflationary expansion for the (relatively) negative energy input. If we add Gauss-Bonnet terms (higher curvature correction terms in gravity), then wormhole tends to expand (or change to black hole) if the coupling constant α is positive (negative), and such bifurcation of the throat horizon is observed earlier in higher dimension.

1. Introduction

Wormhole is a popular tool for interstellar travel in science fiction. It connects two different space-time points directly, which does not contradict with the Einstein's field theory. Although such an astrophysical object is not yet observed, studying wormhole structure gives us many chances to understand unknown aspects of the gravitational field.

The most famous and the common starting model of wormhole is, we think, the one announced by Morris and Thorne as a "traversable wormhole" [1]. They considered "traversable conditions" for human travel through wormholes responding to Carl Sagan's idea for his novel *Co tact*, and concluded that such a wormhole solution is available if we allow "exotic matter" (negative-energy matter).

The introduction of exotic matter sounds to be unusual for the first time, but such matter appears in quantum field theory and in alternative gravitational theories such as scalar-tensor theories. The Morris-Thorne solution is constructed with a massless Klein-Gordon field whose gravitational coupling takes the opposite sign to normal, which is found in Ellis's earlier work [2], so that we call it Ellis wormhole, hereafter. (See a review e.g. by Visser [3] for earlier works; See also e.g. Lobo [4] for recent works).

Ellis wormhole solution was studied in many contexts. Among them, we focus on its dynamical features in higher-dimensional space-time.

The first numerical simulation on Ellis wormhole in 4-dimensional space-time was reported by one of the authors [5]. They use a dual-null formulation for spherically symmetric space-time

Journal of Physics: Conference Series **600** (2015) 012038 doi:10.1088/1742-6596/600/1/012038

integration, and observed that the wormhole is unstable against Gaussian pulses in either exotic or normal massless Klein-Gordon fields. The wormhole throat suffers a bifurcation of horizons and either explodes to form an inflationary universe or collapses to a black hole, if the total input energy is negative or positive, respectively. These basic behaviors were repeatedly confirmed by other groups [6, 7].

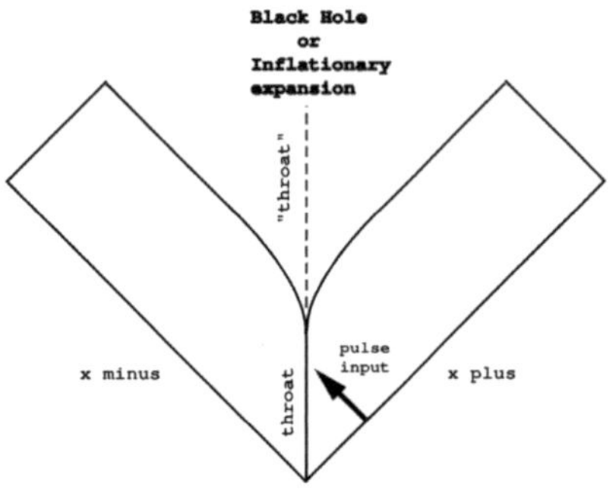

Figure 1. Partial Penrose diagram of the evolved space-time. Suppose we live in the right-side region and input a pulse to an Ellis-wormhole in the middle of the diagram. The wormhole throat suffers a bifurcation of horizons and either explodes to form an inflationary universe or collapses to a black hole, if the total input energy is negative or positive, respectively [5].

The changes of wormhole either to a black hole or an expanding throat support a unified understanding of black holes and traversable wormholes proposed by Hayward [8]. His proposal is that the two are dynamically interconvertible, and that traversable wormholes are understandable as black holes under the existence of negative energy density.

In this article, we introduce our recent studies on the stability in higher-dimensional space-time. The higher-dimensional theories such as string M theories are applied for various unsolved problems in gravitational phenomena and cosmology, and gain new insights into them. We believe that wormholes will also give us new fundamental physical landscapes. We therefore demonstrate wormhole dynamics also in Gauss-Bonnet (GB) gravity, which is one of the modified gravity theory including higher-order corrections of curvatures, one of the string-motivated gravity theories.

The main four contents in this article are: (a) constructing Ellis solutions in higher-dimensional general relativity, (b) stability analysis using a linear perturbation method in n-dimensional general relativity (GR) [9], (c) stability analysis using a numerical evolution method in $5, 6, 7$-dimensional GR, and (d) the similar numerical analysis with the GB coupling.

2. Wormhole solution and its linear perturbation in higher-dimensional GR

2.1. Field e atio s

We start from the n-dimensional Einstein-Klein-Gordon system

$$S = \int_M d^n x \sqrt{-g}\left[\frac{1}{2\kappa_n^2}R - \frac{1}{2}\epsilon(\nabla\phi)^2 - V(\phi)\right], \tag{1}$$

where κ_n^2 is the n-dimensional gravitational constant. The scalar field ϕ is the *or al* (or *ghost*) field if $\epsilon = 1$ (-1).

The metric of the space-time is assumed to be

$$ds^2 = -f(t,r)e^{-2\delta(t,r)}dt^2 + f(t,r)^{-1}dr^2 + R(t,r)^2 h_{ij}dx^i dx^j, \tag{2}$$

where $h_{ij}dx^i dx^j$ represents the line element of a unit $(n-2)$-dimensional constant curvature space with curvature $k = \pm 1, 0$ and volume $_k$. In order to construct a static wormhole solution, we restrict the metric function as $f = f(r)$, $R = R(r)$, $\phi = \phi(r)$, and $\delta = 0$.

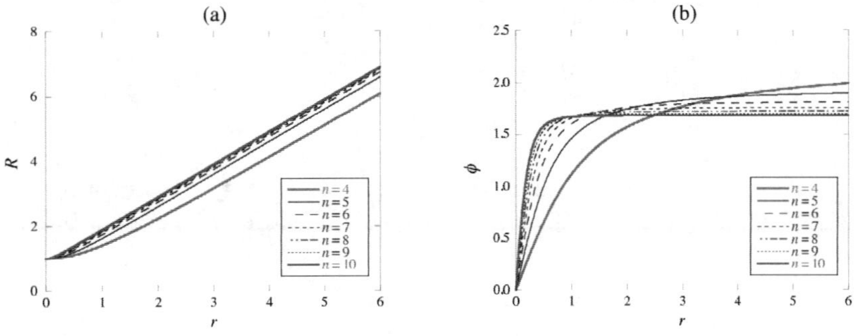

Figure 2. The n-dimensional wormhole solutions; (a) The circumference radius R and (b) the scalar field ϕ are plotted as a function of the radial coordinate r.

For massless ghost scalar field ($V(\phi) = 0$ and $\epsilon = -1$), the Klein-Gordon equation is integrated as $\phi' = C/(fR^{n-2})$, where C is an integration constant, and $' = d/dr$. By locating the throat of the wormhole at $r = 0$, and imposing the reflection symmetry at the throat, the solution of the field equations is obtained only for the spherically symmetric case $k = 1$, as

$$f \equiv 1, \qquad R' = \sqrt{1 - \left(\frac{a}{R}\right)^{2(n-3)}}, \qquad \phi = \frac{\sqrt{(n-2)(n-3)}}{\kappa_n} a^{n-3} \int \frac{1}{R(r)^{n-2}} dr, \tag{3}$$

where a is the radius of the throat, i.e. $R(0) = a$.

The throat of the wormhole has large curvature and the scalar field ϕ becomes steeper as n goes higher. We plotted these behaviors in Figure 2. For $n \to \infty$, the functions have the limiting solution, $R = r + a$ and $\phi = \pi/2$ ($r > 0$).

2.2. E iste ce of sta le ode i li ear ert r atio a al sis

We study linear stability of the solution obtained above. We focus on the "spherical" modes, where the $(n-2)$-dimensional constant curvature space is not perturbed. For metric variables f, δ, R and for scalar field ϕ, we assume the perturbed functions in the form of $x(t,r) = x_0(r) + x_1(r)e^{i\omega t}$, where x_0 denotes the static solution above. Following [7], we obtain the master equation for gauge-invariant variable. After some transformations of variable, we reach the single Schrödinger-type master equation, which potential does not diverge at the throat,

$$-\Psi_1'' + W(r)\Psi_1 = \omega^2\Psi_1, \quad \text{where} \quad W(r) = -\frac{1}{4R_0^2}\Big[\frac{3(n-2)^2}{R_0^{2(n-3)}} - (n-4)(n-6)\Big]. \tag{4}$$

We search eigenfunctions $\Psi_1(r)$ of eq. (4), and find that in any dimension n there exists one negative eigenvalue for ω^2, which implies that the solution is unstable. We find large negative ω^2 for higher n, which indicates the timescale of instability becomes shorter [9].

3. Numerical evolutions of wormhole solutions

3.1. D al ll sche e

We implemented our evolution code [5] for higher-dimensional space-time, and with GB gravity terms. The system we consider is spherical symmetry, and expressed using dual-null coordinate.

The use of dual-null coordinate simplifies the treatment of horizon dynamics and radiation propagation clearly.

The Einstein-Gauss-Bonnet (EGB) action in n-dimensional space-time is described as

$$S = \int_{\mathcal{M}} d^n X \sqrt{-g}\left[\frac{1}{2\kappa^2}\left\{\alpha_{\rm GR}\left(\mathcal{R} - 2\Lambda\right) + \alpha_{\rm GB}\mathcal{L}_{\rm GB}\right\} + \mathcal{L}_{\rm matter}\right], \quad (5)$$

$$\text{where } \mathcal{L}_{\rm GB} = \mathcal{R}^2 - 4\mathcal{R}_{\mu\nu}\mathcal{R}^{\mu\nu} + \mathcal{R}_{\mu\nu\rho\sigma}\mathcal{R}^{\mu\nu\rho\sigma}. \quad (6)$$

The additional term $\mathcal{L}_{\rm GB}$ forms ghost-free combinations [10] and does not give higher derivative equations but an ordinary set of equations with up to the second derivative in spite of the higher curvature combinations.

We adopt the line element

$$ds^2 = -2e^{f(x^+,x^-)}dx^+\,dx^- + r^2(x^+,x^-)h_{ij}dz^i dz^j, \quad (7)$$

where the coordinate (x^+, x^-) are along to null propagation directions.

For writing down the Einstein equations, we introduce the variables

$$\Omega = \frac{1}{r}, \qquad \vartheta_\pm = (n-2)\partial_\pm r, \qquad \nu_\pm = \partial_\pm f, \qquad p_\pm = \partial_\pm \phi, \quad (8)$$

which are conformal factor, expansions, in-affinities, and scalar momentum, respectively.

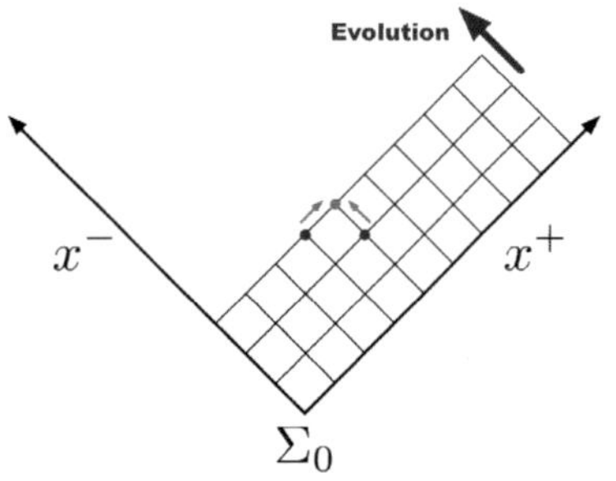

Figure 3. Numerical grid structure. Initial data are given on null hypersurfaces $_\pm$ ($x^\mp = 0$, $x^\pm > 0$) and their intersection S.

We prepare our numerical integration range as drawn in Figure 3. The grid will cover both universes connected by the wormhole throat $x^+ = x^-$. We give initial data on a surface S and the two null hypersurfaces $_\pm$ generated from it. Generally the initial data have to be given as

$$(\Omega, f, \vartheta_\pm, \phi) \quad \text{on } S\text{: } x^+ = x^- = 0 \quad (9)$$

$$(\nu_\pm, p_\pm) \quad \text{on} \quad _\pm\text{: } x^\mp = 0,\ x^\pm > 0. \quad (10)$$

We then evolve the data $u = (\Omega, \vartheta_\pm, f, \nu_\pm, \phi, p_\pm)$ on a constant-x^- slice to the next.

As a virtue of the dual-null scheme, we can follow the wormhole throat or black hole horizons easily. They are both trapping horizons, hypersurfaces where $\vartheta_+ = 0$ or $\vartheta_- = 0$. In order to evaluate the energy, we apply the Misner-Sharp mass in n-dimensional EGB gravity [11].

3.2. Evol tio s of 4 5 a d 6 di e sio al wor hole sol tio s i GR

We checked our numerical code whether it reproduces the static wormhole solution obtained in §2. We next add a perturbation to the scalar momentum on the static wormhole solution in the form of Gaussian pulses from the right-hand universe, $\delta p_+ = c_1 \exp(-c_2(l - c_3)^2)$,where

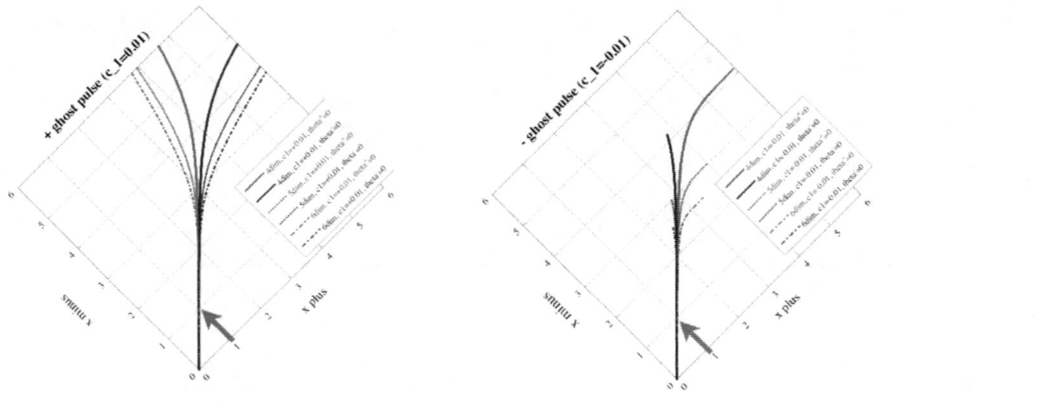

Figure 4. Location of the expansion ϑ_+ (red lines) and ϑ_- (blue lines) for evolutions of a solution in 4, 5, and 6-dimensional General Relativity ($\alpha_{\rm GB} = 0$) as a function of (x_+, x_-). The throat begins expanding if we input negative energy scalar flux (left panel), while the throat turns to be a black hole if we input positive energy scalar flux (right panel).

c_1, c_2, c_3 are parameters. In Figure 4, we plotted the cases with small amplitude $c_1 = \pm 0.01$ and width $c_2 = 3$, and the initial location $c_3 = 1$. That is, the pulse will hit the wormhole throat at $x^+ = x^- = 1$. Positive (or negative) c_1 corresponds to enhancing (or reducing) the supporting ghost field.

Figure 4 shows the results of $n = 4, 5$, and 6 dimensional wormhole solution with above perturbation. The plot shows locations of vanishing expansions, $\vartheta_\pm = 0$, in (x^+, x^-) plane. We see first the wormhole throat locates where $\vartheta_\pm = 0$, but after a small pulse hit it, then the throat (or horizon) split into two ($\vartheta_+ = 0$ and $\vartheta_- = 0$), depending on the signature of the energy of pulse.

If the location of ϑ_+ is outer (in x^+-direction) than that of ϑ_-, then the region $\vartheta_- < x < \vartheta_+$ is judged as a black hole. Otherwise the region $\vartheta_+ < x < \vartheta_-$ can be judged as an expanding throat. The throat begins expanding if we input negative energy scalar flux (left panel in Figure 4), while the throat turns to be a black hole if we input positive energy scalar flux (right panel).

3.3. Evol tio s of wor hole sol tio i EGB gravit

We also construct the initial data with GB terms $\alpha_{\rm GB} \neq 0$, and study their effects to the evolutions. We confirm 4-dimensional EGB equation reproduces the equivalent results with those of GR.

Figure 5 shows the case of 5 and 6-dimensional EGB gravity. The initial data of wormhole on $_\pm = 0$ are obtained numerically, solving the set of equations. The evolutions with $\alpha_{\rm GB} \neq 0$ are quite unstable, and we are hard to keep the static configurations long enough. We see if $\alpha_{\rm GB} < 0$ the throat turns to be a black hole (left panel in Figure 5). On the contrary, if $\alpha_{\rm GB} > 0$, then the throat begins to expand (right two panels).

4. Conclusions and Discussions

We studied the simplest wormhole solutions and their stability in higher-dimensional space-time both in GR and EGB gravity. The space-time is assumed to be static and spherically symmetric, has ghost scalar field, and has reflection symmetry at the throat.

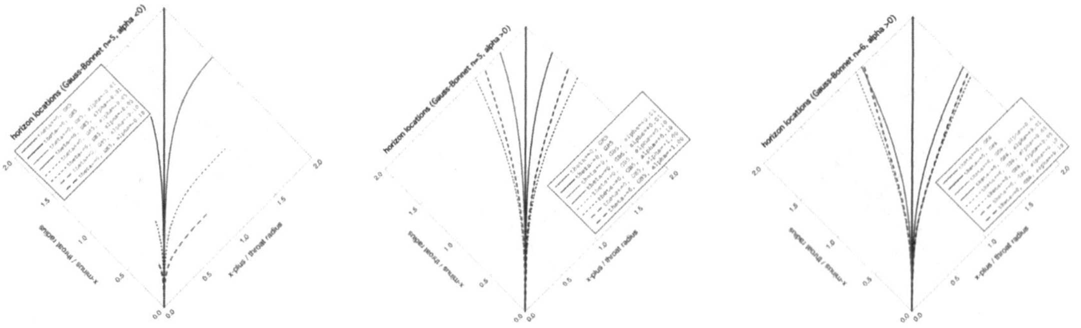

Figure 5. The same with Figure 4, but for Einstein-Gauss-Bonnet gravity. When $\alpha_{\rm GB} < 0$ (left panel), the throat turns to be a black hole, while for $\alpha_{\rm GB} > 0$ the throat begins expanding (middle left panel).

Using the linear perturbation technique, we showed that the solutions have at least one negative mode, which concludes that all wormholes are linearly unstable. The time scale of instability becomes shorter as n becomes large. This prediction is confirmed using numerical evolutions.

Numerical evolutions show that if we put a scalar pulse with small amplitude to the throat, then the wormhole throat bifurcates, its horizon structure changes into a black hole or an expanding throat depending the pulse energy is positive or negative, respectively. We also investigated the effect of the higher-order curvature corrections using Gauss-Bonnet terms, and found that such corrections do not work for stabilization of wormholes.

All the behavior of wormholes may be explained simply with energy balance. In [5], for small perturbations, an existence of critical solution is suggested. The similar behaviors are also observed in our investigations, which will be reported elsewhere.

We guess a wormhole with exotic matter is a disguise to avoid public notice, and does prefer to appear as a black hole or an expanding universe.

Acknowledgments
This work was supported in part by the Grant-in-Aid for Scientific Research Fund of the JSPS (C) No. 25400277. Numerical computations were carried out on SR16000 at YITP in Kyoto University, and on the RIKEN Integrated Cluster of Clusters (RICC).

References

[1] M. S. Morris & K. S. Thorne, Am. J. Phys. **56**, 395 (1988).
[2] H. G. Ellis, J. Math. Phys. **14**, 395 (1973).
[3] M. Visser, *Lorentzian Wormholes* (AIP Press, 1995).
[4] F. S. N. Lobo, in *Classical and Quantum Gravity Research* (Nova Sci. Pub., 2008). [arXiv:0710.4474]
[5] H. Shinkai & S.A. Hayward, Phys. Rev. D **66**, 044005 (2002).
[6] A. Doroshkevich, J. Hansen, I. Novikov, & A. Shatskiy, Int. J. Mod. Phys. D **18**, 1665 (2009)
[7] J. A. Gonzalez, F. S. Guzman & O. Sarbach, Class. Quant. Grav. **26**, 015010, 015011 (2009); Phys. Rev. D **80**, 024023 (2009). O. Sarbach & T. Zannias, Phys. Rev. D **81**, 047502 (2010).
[8] S A Hayward, Int. J. Mod. Phys. D **8**, 373 (1999).
[9] T. Torii & H. Shinkai, Phys. Rev. D **88**, 064027 (2013).
[10] B. Zwiebach, Phys. Lett. B **156**, 315 (1985).
[11] H. Maeda & M. Nozawa, Phys. Rev. D **78**, 024005 (2008).

Numerical Brill-Lindquist initial data with a Schwarzschildean end at spatial infinity

Georgios Doulis[1] and Oliver Rinne[1,2]

[1] Max Planck Institute for Gravitational Physics (Albert Einstein Institute), Am Mühlenberg 1, 14476 Golm, Germany

[2] Department of Mathematics and Computer Science, Freie Universität Berlin, Arnimallee 26, 14195 Berlin, Germany

E-mail: georgios.doulis@aei.mpg.de, oliver.rinne@aei.mpg.de

Abstract. We construct numerically time-symmetric initial data that are Schwarzschildean at spatial infinity and Brill-Lindquist in the interior. The transition between these two data sets takes place along a finite gluing region equipped with an axisymmetric Brill wave metric. The construction is based on an application of Corvino's gluing method using Brill waves due to Giulini and Holzegel. Here, we use a gluing function that includes a simple angular dependence. We also investigate the dependence of the ADM mass of our construction on the details of the gluing procedure.

1. Introduction

The long lasting question of whether it is possible to construct solutions to the vacuum constraint equations that are static in a neighbourhood of space-like infinity and non-static in the interior was answered to the affirmative by Corvino [1]. Therein, Corvino proved that an external Schwarzschild region can be glued smoothly along a finite intermediate annulus to generic asymptotically flat time-symmetric data in the interior. This result paved the way to its generalisations to initial data with stationary Kerr [2] or Kerr-de Sitter [3] ends and static Kottler-de Sitter [4] or Kottler-anti-de Sitter [5] ends. Although the above analytical work provides results about the existence of the data sets considered there, it is not very explicit in the sense that it is not clear how to construct these data explicitly through analytical or numerical methods. The first step, to our knowledge, towards this direction was taken in [6], where the authors proposed an explicit formulation of the gluing procedure for vacuum axisymmetric initial data. Recently, we presented the first attempt to implement this scheme numerically [7]. In the present short contribution, we focus on a setting where the gluing function acquires a simple angular dependence.

The motivation to construct such kind of solutions to the vacuum constraint equations comes from the fact that initial data of static or stationary character close to spatial infinity seem to have a smooth development to null infinity [8]—a result of great importance if one is interested in extracting unambiguously information about the gravitational radiation emitted by isolated self-gravitating systems. In order to evolve the aforementioned initial data, one could use a combination of Cauchy and hyperboloidal evolution in the following sense. For example, in the static case the initial data are Schwarzschild outside the gluing annulus, thus

Journal of Physics: Conference Series **600** (2015) 012039 doi:10.1088/1742-6596/600/1/012039

one can place an artificial boundary there and evolve the data for a short time using a standard Cauchy evolution with exact Schwarzschild boundary conditions being imposed on the artificial boundary. Then, these evolved data can be used as initial conditions for a hyperboloidal evolution [9]. Alternatively, the initial data could be evolved using a Cauchy evolution in the framework of the conformal representation of Einstein's equations [10], where space-like infinity i^0 has been blown up to a cylinder $I = [-1, 1] \times \mathbb{S}^2$ of finite length along the time direction. This approach has been already proven to be successfully numerically implementable, e.g. [11; 12].

2. Mathematical formulation

We intend to construct initial data that in the interior region $0 < r < r_{\rm int}$ consist of two identical non-rotating black holes of mass m lying symmetrically to the origin on the z-axis, $\vec{c} = (0, 0, \frac{d}{2})$, at a moment of time symmetry:[1]

$$g_{\text{B-L}} = \left(1 + \frac{m}{2|\vec{r} - \vec{c}|} + \frac{m}{2|\vec{r} + \vec{c}|}\right)^4 \delta, \tag{1}$$

where $\delta = dr^2 + r^2(d\theta^2 + \sin^2\theta\, d\phi^2)$ denotes the three-dimensional Minkowski line element in spherical polar coordinates. In the intermediate gluing region $r_{\rm int} < r < r_{\rm ext}$ the data are of the general Brill wave form

$$g_{\rm Brill} = \psi^4 \left(e^{2\,q(r,\theta)}(dr^2 + r^2 d\theta^2) + r^2 \sin^2\theta\, d\phi^2\right), \tag{2}$$

and in the exterior region $r > r_{\rm ext}$ they are Schwarzschild,

$$g_{\rm Schw} = \left(1 + \frac{M}{2|\vec{r}|}\right)^4 \delta, \tag{3}$$

with ADM mass M. Following [6], one can glue together the above data sets by defining the conformal factor

$$\psi = \left(1 + \frac{m}{2|\vec{r} - \vec{c}|} + \frac{m}{2|\vec{r} + \vec{c}|}\right) \beta(r, \theta) + (1 - \beta(r, \theta)) \left(1 + \frac{M}{2|\vec{r}|}\right). \tag{4}$$

Whereas in [7] we also considered a θ-independent gluing function β, we focus here on a gluing function with a simple θ-dependence, namely we will assume that

$$\beta(r, \theta) = \frac{1}{2}\left(1 + \tanh\left(\frac{1}{r - r_{\rm int}} + \frac{1}{r - r_{\rm ext}}\right)\right) + \operatorname{sech}\left(\frac{1}{r - r_{\rm int}} + \frac{1}{r - r_{\rm ext}}\right)\sin^2\theta. \tag{5}$$

Notice that $\beta(r \le r_{\rm int}, \theta) = 1$ and $\beta(r \ge r_{\rm ext}, \theta) = 0$.

The Brill wave function $q(r, \theta)$ will be our unknown here and will be specified by Einstein's equations, which in our setting reduce to the vanishing of the Ricci scalar of the Brill wave metric (2), i.e. $R(g_{\rm Brill}) = 0$. After a little bit of algebra the latter reads

$$\frac{\partial^2 q}{\partial r^2} + \frac{1}{r^2}\frac{\partial^2 q}{\partial \theta^2} + \frac{1}{r}\frac{\partial q}{\partial r} = -\frac{4}{\psi}\left(\frac{\partial^2 \psi}{\partial r^2} + \frac{1}{r^2}\frac{\partial^2 \psi}{\partial \theta^2} + \frac{2}{r}\frac{\partial \psi}{\partial r} + \frac{\cot\theta}{r^2}\frac{\partial \psi}{\partial \theta}\right). \tag{6}$$

[1] Here only black holes with non-intersecting horizons will be considered and all cases that a third outer horizon enclosing both black holes appears will be excluded.

Notice that the right-hand side of the elliptic equation (6) is known a priori as the conformal factor (4) and the gluing function (5) are given. The inhomogeneous Poisson equation (6) will be supplemented by the boundary conditions

$$q = 0 \quad \text{and} \quad \frac{\partial q}{\partial \theta} = 0 \qquad \text{at} \quad \theta = 0, \pi, \tag{7a}$$

$$q = 0 \quad \text{and} \quad \frac{\partial^n q}{\partial r^n} = 0 \qquad \text{at} \quad r = r_{\text{int}}, r_{\text{ext}}, \tag{7b}$$

for all $n \in \mathbb{N}$. Conditions (7a) guarantee the absence of any conical singularities on the z-axis, while (7b) ensure that the transition between the different data sets at the boundaries $r_{\text{int}}, r_{\text{ext}}$ of the gluing annulus is smooth.

It is worth mentioning that by inserting into (6) the definition of the conformal factor (4), the former takes the form

$$\frac{\partial^2 q}{\partial r^2} + \frac{1}{r^2}\frac{\partial^2 q}{\partial \theta^2} + \frac{1}{r}\frac{\partial q}{\partial r} = \frac{1}{M + 2\,r + f_0\,\beta}\left(f_1\,\frac{\partial^2 \beta}{\partial r^2} + f_2\,\frac{\partial^2 \beta}{\partial \theta^2} + f_3\,\frac{\partial \beta}{\partial r} + f_4\,\frac{\partial \beta}{\partial \theta}\right),$$

where f_k are functions of r and θ. Notice that the constancy of the gluing function (5) outside the gluing annulus forces the right-hand side of the above equation to vanish. In addition, the boundary conditions (7b) guarantee that q is zero outside the gluing region. Thus, the Poisson equation (6) reduces to the trivial identity $(0 = 0)$ in the exterior of the gluing annulus.

In summary, our goal in the following section will be to solve numerically the elliptic second-order PDE (6) for $q(r, \theta)$ subject to the boundary conditions (7).

3. Numerical implementation

3.1. Numerical scheme

We use pseudo-spectral methods to solve numerically the system (6), (7). First, we perform the transformation $x \mapsto r(x) := \frac{1}{2}(r_{\text{ext}} - r_{\text{int}})\,x + \frac{1}{2}(r_{\text{ext}} + r_{\text{int}})$ on the radial coordinate in order to map the physical domain $(r, \theta) \in [r_{\text{int}}, r_{\text{ext}}] \times [0, \pi]$ to the computational domain $(x, \theta) \in [-1, 1] \times [0, \pi]$, which will enable us to expand the radial dependence of q in Chebyshev polynomials T_k and the θ-dependence in Fourier-sine series. We discretize the computational domain by introducing the collocation points along the angular and the new radial direction

$$\theta_i = \frac{i\,\pi}{L} \qquad \text{and} \qquad x_j = -\cos\left(\frac{j\,\pi}{K}\right) \qquad \text{with} \qquad i = 0, \ldots, L \quad \text{and} \quad j = 0, \ldots, K,$$

where L and K are the numbers of collocation points.

To satisfy the boundary conditions (7) we make the following ansatz for the Brill wave function,

$$q(x, \theta) = B(x) \sin\theta \sum_{k=0}^{K} \sum_{l=1}^{L-1} a_{kl}\, T_k(x)\, \sin(l\,\theta), \tag{8}$$

where K, L as above, the constants a_{kl} are the expansion coefficients in our series expansion, and $B(x)$ is a "bump" function of the form

$$B(x) = \text{sech}\left(\frac{b_1}{x - 1} + \frac{b_2}{x + 1}\right) \quad \text{with} \quad b_1, b_2 \quad \text{constants.} \tag{9}$$

While the presence of $B(x)$ in (8) guarantees that the set of boundary conditions (7b) is satisfied, the expansion in Fourier-sine series and the presence of $\sin\theta$ ensures that our numerical solutions will respect the conditions (7a) on the z-axis.

3.2. Numerical solutions

In order to produce numerical solutions for the system (6), (7) using the method described in Sec. 3.1, one has first to express (6) in terms of the new radial coordinate x and then substitute the ansatz (8) and the definitions (4), (5) into the left and right-hand side, respectively, of the resulting expression. In addition, one has to specify the free parameters entering the definitions of the conformal factor (4) and the gluing function (5), namely the mass m of the individual Brill-Lindquist black holes, their mutual distance d, the mass M of the Schwarzschild region, and the locations of the boundaries $r_{\text{int}}, r_{\text{ext}}$ of the gluing annulus. There are three conditions that constrain the choice of the free parameters: *i*) the mass-to-distance ratio $m/d \leq 0.64$ that prevents the occurence of a third outer horizon enclosing the Brill-Lindquist data, see [13], *ii*) the inequality $r_{\text{int}} > d/2 + m\,d/(2\,d + m)$ that guarantees that the gluing annulus is placed outside of any horizons of the Brill-Lindquist data [7], and *iii*) the integrability condition presented in Sec. 3.3 that constrains the relation between the masses m and M.

In Fig. 1 numerical solutions of the system (6), (7) are presented for a choice that respects the above three conditions, i.e. $m = 2$, $d = 10$, $r_{\text{ext}} = 2\,r_{\text{int}}$, and M takes values in accordance with Tab. 1. We provide solutions for two different locations of the gluing annulus. In Fig. 1(a) the gluing region has been placed very close to the black holes ($r_{\text{int}} = 6$), in Fig. 1(b) far away ($r_{\text{int}} = 500$). The numerically computed values of q drop with the distance of the gluing annulus from the origin. This behaviour was also observed in [7] for solutions corresponding to a θ-independent gluing function.

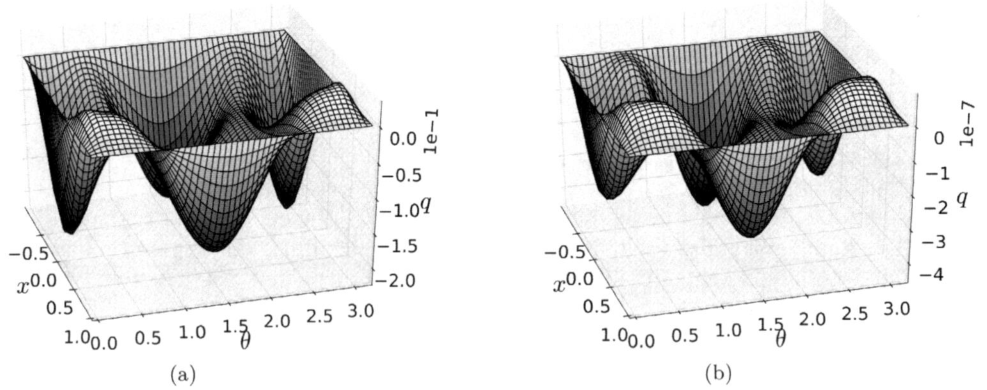

Figure 1. Numerical solutions. The numerical values of the function q for the choice $m = 2$, $d = 10$, $r_{\text{ext}} = 2\,r_{\text{int}}$, and M according to Tab. 1 with the gluing annulus placed at (a) $r_{\text{int}} = 6$ and (b) $r_{\text{int}} = 500$. The values of q decrease with the distance of the gluing annulus from the origin. The resolution used to produce the plots is $K = L = 35$ collocation points.

The accuracy and convergence properties of the solutions presented in Fig. 1 are similar to the ones observed in [7]. Namely, the expansion coefficients in (8) decay to zero exponentially, and their actual values a_{kl} for $k, l > 60$ are of the order of 10^{-14} (numerical roundoff).

3.3. Behaviour of the ADM mass

Here, we will investigate any possible effects that the introduction of the gluing region has on the total ADM mass M of our construction. One expects that the presence of the Brill wave in the gluing region would in general increase the overall ADM mass of the gluing construction.

As was pointed out in [7], the positive mass theorem for time-symmetric, axisymmetric, vacuum gravitational waves proved by Brill [14] can be used in the form

$$M = \int_0^\pi \int_0^\infty \left[\left(\frac{1}{\psi} \frac{\partial \psi}{\partial r} \right)^2 + \left(\frac{1}{r\,\psi} \frac{\partial \psi}{\partial \theta} \right)^2 \right] r^2 \sin\theta \, dr \, d\theta, \qquad (10)$$

not only to determine the relation between the masses m and M involved in our construction, but also to study the behaviour of the ADM mass M on the details of the gluing construction. In the expression above ψ is the conformal factor defined by (4), thus the ADM mass M enters also the right-hand side of (10). In the light of this observation, the relation (10) can be viewed as an integrability condition for the ADM mass in the following sense. For a specific choice of M, we compute numerically the integral quantity on the right-hand side of (10) and denote the result with M_I. We will be interested here in the cases that $M_I = M$ holds, as they correspond to true physical solutions. The case that M_I takes values $M_I < 2m$ corresponds to a reduction of the ADM mass, the case $M_I > 2m$ to an increase.

In Tab. 1 a detailed study of the restrictions that the condition (10) poses on the ADM mass is presented for the choice $m = 2$, $d = 10$, $r_{\rm ext} = 2\,r_{\rm int}$ of the free parameters that was used to generate the solutions of Fig. 1. The closer we place the gluing annulus to the origin, the greater is the increase of the ADM mass. The increase is limited only by the requirement that the gluing region lies outside of the horizons of the Brill-Lindquist data, i.e. $r_{\rm int} > d/2 + m\,d/(2\,d + m)$. Notice that the actual increase of the ADM mass decreases extremely fast to zero with increasing gluing radius $r_{\rm int}$.

$r_{\rm int}$	M_I	$M_I - 2m$	$\lvert\Delta M\rvert$	δM
6	4.08778	0.08778	1.4×10^{-5}	1.5×10^{-5}
10	4.0119	0.0119	3.9×10^{-5}	4.5×10^{-5}
30	4.00017185	0.00017185	3.4×10^{-8}	7.9×10^{-8}
100	4.00000145	1.45×10^{-6}	1.6×10^{-8}	2.2×10^{-8}
500	4.0000000024	2.4×10^{-9}	3.3×10^{-11}	4×10^{-11}

Table 1. Increase of the ADM mass. The increased ADM mass, the actual increase, the violation of the integrability condition $\Delta M = M_I - M$, and the numerical error δM involved in these calculations are presented for different locations of the gluing annulus. Notice that by just quintupling the gluing radius (from 6 to 30), the increase decreases by two orders of magnitude.

In [7] it has been shown that for the θ-independent gluing function

$$\beta(r) = \frac{1}{2}\left(1 + \tanh\left(\frac{1}{r - r_{\rm int}} + \frac{1}{r - r_{\rm ext}}\right)\right),$$

it is possible to reduce the ADM mass in the rather extreme case that the black holes are placed very close to each other and the gluing is performed along a very wide gluing annulus that starts very close to the horizons of the black holes. Here, it was not possible to find a setup that could lead to a reduction of the ADM mass for the choice (5), even for such extreme configurations. It seems that even the simplest inclusion of θ-dependence in the gluing function increases the contribution of the Brill wave to the ADM mass to a degree that makes it impossible to find an arrangement of our data that leads to reduction of the ADM mass.

4. Conclusions

In [7] we presented a numerical implementation of Corvino's gluing method [1] applied to vacuum axisymmetric spacetimes [6]. Here, we focused on a gluing function with a simple

θ-dependence. It was shown numerically that for the choice (5), the vacuum time-symmetric constraint equations (6) admit smooth solutions that are of Schwarzschild type close to spatial infinity, see Fig. 1. We showed that these solutions converge exponentially and depend on the location of the gluing annulus in the expected way: the Brill wave function q decreases with the distance of the gluing annulus from the origin. The latter observation confirms the expectation that in the limiting case that the gluing annulus is placed at infinity q must tend to zero.

Our study of the dependence of the total ADM mass M on the details of the gluing procedure carried out in Sec. 3.3 strongly indicates that for the choice (5) the presence of the gluing annulus tends always to increase the ADM mass. It seems that the Brill wave corresponding to (5) contributes to the overall ADM mass to such a degree that reduction was not possible, even in the extreme setup of [7] where a very wide gluing region starting close to the black hole horizons was chosen.

References

[1] Corvino J 2000 *Commun. Math. Phys.* **214** 137

[2] Corvino J and Schoen R M 2006 *J. Diff. Geom.* **73** 185 (*Preprint* gr-qc/0301071)

[3] Cortier J 2013 *Ann. Henri Poincaré* **14** 1109 (*Preprint* 1202.3688 [gr-qc])

[4] Chruściel P T and Pollack D 2008 *Ann. Henri Poincaré* **9** 639 (*Preprint* 0710.3365 [gr-qc])

[5] Chruściel P T and Delay E 2009 *Comm. Anal. Geom.* **17** 343 (*Preprint* 0711.1557 [gr-qc])

[6] Giulini D and Holzegel G 2005 Corvino's construction using Brill waves *Preprint* gr-qc/0508070

[7] Doulis G and Rinne O 2014 Numerical construction of initial data for Einsteins equations with static extension to space-like infinity *Preprint* 1411.7878 [gr-qc]

[8] Chruściel P T and Delay E 2002 *Class. Quantum Grav.* **19** L71 (*Preprint* gr-qc/0203053)

[9] Rinne O 2010 *Class. Quantum Grav.* **27** 035014 (*Preprint* 0910.0139 [gr-qc])

[10] Friedrich H 1998 *J. Geom. Phys.* **24** 83

[11] Beyer F, Doulis G, Frauendiener J and Whale B 2012 *Class. Quantum Grav.* **29** 245013 (*Preprint* 1207.5854 [gr-qc])

[12] Doulis G and Frauendiener J 2013 *Gen. Relativ. Gravit.* **454** 1365 (*Preprint* 1301.4286 [gr-qc])

[13] Brill D R and Lindquist R W 1963 *Phys. Rev.* **131** 471

[14] Brill D R 1959 *Ann. Phys.* **7** 466

Extended Kerr–Schild spacetimes of any dimension

Tomáš Málek

Institute of Mathematics, Academy of Sciences of the Czech Republic, Žitná 25, 115 67 Praha 1, Czech Republic

E-mail: malek@math.cas.cz

Abstract. We study geometric and algebraic properties of extended Kerr–Schild spacetimes (xKS), i.e. an extension of the Kerr–Schild (KS) ansatz where, in addition to the null KS vector, a spacelike vector field appears in the metric. In contrast to the KS case, it turns out that xKS spacetimes with a geodetic KS vector are not necessarily algebraically special and we obtain, in general, only a necessary condition under which the KS vector is geodetic. However, it is shown that this condition becomes sufficient if we appropriately restrict the geometry of the null and spacelike vector fields. Examples of xKS spacetimes belonging to the Kundt class and also expanding xKS spacetimes, namely the CCLP black hole, are provided and briefly discussed.

1. Kerr–Schild spacetimes

The Einstein field equations are a very complex system of partial differential equations of the 2nd order and finding exact solutions is a considerably non-trivial task especially in dimension $n > 4$. A possible approach to this problem is to assume an appropriate form of the unknown metric and an important example of such a technique is the Kerr–Schild (KS) ansatz [1]

$$g_{ab} = \eta_{ab} - 2\mathcal{H}k_a k_b, \qquad k_a k^a = 0, \tag{1}$$

where η_{ab} is a flat background metric and the KS vector $\boldsymbol{k}$ is null with respect to both g_{ab} and η_{ab}. This choice ensures a simple form of the inverse metric

$$g^{ab} = \eta^{ab} + 2\mathcal{H}k_a k_b \tag{2}$$

and therefore the full metric corresponds exactly to its linear approximation around the flat background. The cosmological constant Λ can be included by taking the corresponding maximally symmetric spacetime as the background metric $\bar{g}_{ab}$

$$g_{ab} = \bar{g}_{ab} - 2\mathcal{H}k_a k_b. \tag{3}$$

The analysis of metrics (3) in [2, 3] shows that if $\boldsymbol{k}$ is geodetic, these metrics are algebraically special[1] with $\boldsymbol{k}$ being the multiple Weyl aligned null direction. It also turns out that non-expanding Einstein KS spacetimes are of the Weyl type N and belong to the Kundt class, on

[1] In the context of the algebraic classification of the Weyl tensor in arbitrary dimension based on the existence of preferred null directions and their multiplicity, see e.g. the recent review [4].

the other hand, expanding Einstein KS spacetimes are of genuine types II or D and the optical matrix[2] $\rho_{ij} \equiv \ell_{a;b} m^a_{(i)} m^b_{(j)}$ satisfies the optical constraint [2, 7]

$$\rho_{ik}\rho_{jk} = \frac{\rho_{lk}\rho_{lk}}{(n-2)\theta}\rho_{(ij)} \tag{4}$$

implying

$$\rho_{ij} = \mathrm{diag}\left(\mathcal{M}_1, \ldots, \mathcal{M}_p, \frac{1}{r}, \ldots, \frac{1}{r}, 0, \ldots, 0\right), \tag{5}$$

where

$$\mathcal{M}_\mu = \frac{1}{r^2 + a_\mu^2}\begin{bmatrix} r & a_\mu \\ -a_\mu & r \end{bmatrix}. \tag{6}$$

Despite the simplicity of the KS ansatz (3), the class of KS metrics contains many physically interesting exact solutions of four-dimensional general relativity and also some of their higher dimensional analogues such as, for instance, the Schwarzschild black hole, the Vaidya radiating star, the Kinnersley photon rocket, the Kerr–(A)dS rotating black hole and type N *pp*-waves, see e.g. [8]. In fact, the KS ansatz has led to the discovery of the rotating black holes in higher dimensional general relativity with a (non-)vanishing cosmological constant [9, 10], respectively, and has been successfully applied also in the context of higher order gravities such as the Gauss–Bonnet theory [11] or quadratic gravity [12, 13].

We are motivated to generalize the KS ansatz for several reasons. More general ansatz could lead to exact solutions of more general Weyl types, e.g. black rings which are of type I_i. Another reason is that although the Kerr–Newman black hole can be cast to the KS form

$$\eta_{ab}\,\mathrm{d}x^a\,\mathrm{d}x^b = -\mathrm{d}t^2 + \mathrm{d}x^2 + \mathrm{d}y^2 + \mathrm{d}z^2 \tag{7}$$

$$k_a\,\mathrm{d}x^a = \mathrm{d}t + \frac{rx+ay}{r^2+a^2}\,\mathrm{d}x + \frac{ry-ax}{r^2+a^2}\,\mathrm{d}y + \frac{z}{r}\,\mathrm{d}z \tag{8}$$

$$\mathcal{H} = -\frac{r^2}{r^4+a^2z^2}\left(Mr - \frac{Q^2}{2}\right), \qquad A = \frac{Qr^3}{r^4+a^2z^2}\boldsymbol{k}, \tag{9}$$

an exact charged rotating black hole solution of higher dimensional Einstein–Maxwell theory is unknown. It is also known that a straightforward generalization of five-dimensional rotating black hole solutions of general relativity in the KS form to the Gauss–Bonnet theory [11] do not represent rotating black holes [14]. Moreover, as will be mentioned later, some already known exact solutions can be cast to an extended KS form.

In the following, we briefly present our main results published in [15].

2. Extended Kerr–Schild spacetimes

Let us consider extended Kerr–Schild (xKS) metrics as an extension of the KS ansatz in the form

$$g_{ab} = \bar{g}_{ab} - 2\mathcal{H}k_ak_b - 2\mathcal{K}k_{(a}m_{b)} \tag{10}$$

involving an additional unit spacelike vector $\boldsymbol{m}$

$$k^ak_a = 0, \qquad k^am_a = 0, \qquad m^am_a = 1, \tag{11}$$

where $\bar{g}_{ab}$ is a maximally symmetric background metric

$$\bar{g}_{ab} = \Omega\eta_{ab}, \qquad \eta_{ab}\,\mathrm{d}x^a\mathrm{d}x^b = -\mathrm{d}t^2 + \mathrm{d}x_1^2 + \ldots + \mathrm{d}x_{n-1}^2 \tag{12}$$

[2] Throughout the paper, we employ the higher dimensional generalization of the Newman–Penrose formalism [5, 6].

with a corresponding conformal factor

$$\Omega_{\rm M}=1, \quad \Omega_{\rm dS}=-\frac{(n-2)(n-1)}{2\Lambda {x_1}^2}, \quad \Omega_{\rm AdS}=\frac{(n-2)(n-1)}{2\Lambda t^2}. \tag{13}$$

The inverse metric can be expressed as

$$g^{ab}=\bar{g}^{ab}+\left(2\mathcal{H}-\mathcal{K}^2\right)k^a k^b+2\mathcal{K}k^{(a}m^{b)}. \tag{14}$$

It is appropriate to identify the vectors $\boldsymbol{k}$, $\boldsymbol{m}$ with the null and spacelike frame vectors

$$\boldsymbol{k}\equiv\boldsymbol{\ell}, \qquad \boldsymbol{m}\equiv\boldsymbol{m}^{(2)} \tag{15}$$

and define indices $\tilde{\imath},\tilde{\jmath}=3,\ldots,n-1$ so that $\boldsymbol{m}$ is excluded in the notation $\boldsymbol{m}^{(\tilde{\imath})}$.

2.1. *Geodeticity of the KS vector* $\boldsymbol{k}$

In the case of KS spacetimes, the null KS vector $\boldsymbol{k}$ is geodetic if and only if the boost weight 2 component of the Ricci tensor $R_{00}=R_{ab}k^a k^b$ vanishes. For the xKS metric (10), R_{00} reads

$$R_{00}=2\mathcal{H}L_{i0}L_{i0}-\frac{1}{2}\mathcal{K}^2 L_{\tilde{\imath}0}L_{\tilde{\imath}0}+\mathcal{K}(2L_{i(i}L_{2)0}+L_{\tilde{\imath}0}M_{\tilde{\imath}0}+\mathrm{D}L_{20})+2\mathrm{D}\mathcal{K}L_{20} \tag{16}$$

and therefore R_{00} vanishes if $\boldsymbol{k}$ is geodetic, but the converse implication does not hold.

If we appropriately restrict the arbitrariness in the choice of the vectors $\boldsymbol{k}$ and $\boldsymbol{m}$

$$k_{[a;b]}m^b=0, \qquad (\zeta m_{[a});_{b]}k^b=0, \tag{17}$$

which can be expressed in terms of the Lie derivative as

$$\pounds_{\boldsymbol{m}}k_a=0, \quad \pounds_{\boldsymbol{k}}(\zeta m_a)=0, \tag{18}$$

the boost weight 2 component of the Ricci tensor reduces to

$$R_{00}=\left(2\mathcal{H}-\frac{1}{2}\mathcal{K}^2\right)L_{\tilde{\imath}0}L_{\tilde{\imath}0}. \tag{19}$$

Therefore, assuming $\mathcal{K}^2\neq 4\mathcal{H}$, the KS vector $\boldsymbol{k}$ is geodetic if and only if R_{00} vanishes.

Note that, in the context of the Einstein field equations the vanishing of the boost weight 2 component of the Ricci tensor is related to the vanishing of the boost weight 2 component of the energy-momentum tensor $R_{00}=\kappa T_{00}$ and the case $R_{00}=0$ not only includes the vacuum case, i.e. Einstein spacetimes

$$R_{ab}=\frac{2\Lambda}{n-2}g_{ab}, \tag{20}$$

but also an aligned Maxwell field

$$R_{ab}=\frac{\kappa}{4}\left(F^c{}_a F_{cb}-\frac{1}{2(n-2)}F^2 g_{ab}\right), \qquad F_{ab}k^b\propto k_a, \tag{21}$$

or aligned pure radiation

$$R_{ab}=\Phi k_a k_b. \tag{22}$$

Note also that the relation (18) is compatible with the optical constraint (4) if $\boldsymbol{m}$ does not correspond to any 2×2 block $\mathcal{M}_\mu$ in the optical matrix (5) implying $\boldsymbol{k}$ and $\boldsymbol{m}$ are surface forming.

2.2. Algebraic types of xKS spacetimes

As already mentioned above, KS spacetimes (3) of any dimension with a geodetic $\boldsymbol{k}$ are algebraically special. On the other hand, xKS spacetimes (10) with a geodetic $\boldsymbol{k}$ are in general of Weyl type I with $\boldsymbol{k}$ being the Weyl aligned null direction. However, in case one assumes the relation (18) between the vectors $\boldsymbol{k}$ and $\boldsymbol{m}$ holds, xKS spacetimes are algebraically special with $R_{0i} = 0$ if and only if the optical matrix ρ_{ij} and the function $\mathcal{K}$ take one of the following forms

$$\rho_{ij}^{(1)} = 0, \quad \mathcal{K}^{(1)} = c_1 r + c_2, \tag{23}$$

$$\rho_{ij}^{(2)} = \mathrm{diag}\left(\frac{1}{r}, 0, \dots, 0\right), \quad \mathcal{K}^{(2)} = c_1 r + \frac{c_2}{r}, \tag{24}$$

$$\rho_{ij}^{(3)} = \frac{1}{1+c_1^2 r^2}\,\mathrm{diag}\left(\begin{bmatrix} \frac{1}{r} & c_1 \\ c_1 & c_1^2 r \end{bmatrix}, 0, \dots, 0\right), \quad \mathcal{K}^{(3)} = \frac{\sqrt{1+c_1^2 r^2}}{c_2 r}, \quad c_1 \neq 0, \tag{25}$$

$$\rho_{ij}^{(4)} = \mathrm{diag}\left(0, \frac{1}{r}, \frac{1}{r+c_2}, \dots, \frac{1}{r+c_p}, 0, \dots, 0\right), \quad \mathrm{rank}\,\rho_{ij}^{(4)} \geq 1, \quad \mathcal{K}^{(4)} = c_1, \tag{26}$$

$$\rho_{ij}^{(5)} = \mathrm{diag}\left(\frac{1}{r}, \frac{1}{r+c_2}, \dots, \frac{1}{r+c_p}, 0, \dots, 0\right), \quad \mathrm{rank}\,\rho_{ij}^{(5)} \geq 2, \quad \mathcal{K}^{(5)} = c_1 r, \tag{27}$$

$$\rho_{ij}^{(6)} = \mathrm{diag}\left(\frac{1}{r}, \mathcal{M}, \dots, \mathcal{M}\right), \quad \mathcal{K}^{(6)} = c_1 r + \frac{c_2}{r}, \quad (c_1 \neq 0) \wedge (c_2 \neq 0), \tag{28}$$

$$\mathcal{M} = \begin{bmatrix} s & A \\ -A & s \end{bmatrix}, \quad s = \frac{r}{r^2 + \frac{c_2}{c_1}}, \quad A = \sqrt{\frac{c_2}{c_1}}\frac{1}{r^2 + \frac{c_2}{c_1}}, \tag{29}$$

where r is an affine parameter along the null geodesics $\boldsymbol{k}$ and c_i are arbitrary scalar functions independent of r.

3. Kundt xKS spacetimes

It turns out that for Kundt xKS metrics the following statements are equivalent:

(i) the spacetime is algebraically special,

(ii) the boost weight 1 components of the Ricci tensor $R_{0i} \equiv R_{ab}k^a m^b_{(i)} = 0$ vanish,

(iii) the function $\mathcal{K}$ and the Ricci rotation coefficients $M_{\tilde{\imath}0}$ take one of the forms

$$\mathcal{K} = d\sqrt{(r+b)^2 + \mu_{\tilde{\jmath}}\mu_{\tilde{\jmath}}}, \quad M_{\tilde{\imath}0} = \frac{\mu_{\tilde{\imath}}}{(r+b)^2 + \mu_{\tilde{\jmath}}\mu_{\tilde{\jmath}}}, \tag{30}$$

or

$$\mathcal{K} = fr + e, \quad M_{\tilde{\imath}0} = 0, \tag{31}$$

where r is an affine parameter along the null non-expanding, non-shearing, and non-twisting geodesics $\boldsymbol{k}$ and b, d, e, f, $\mu_{\tilde{\imath}}$ are arbitrary scalar functions independent of r.

3.1. Examples of Kundt xKS spacetimes

The most straightforward examples of Kundt xKS spacetimes are metrics with vanishing scalar invariants (VSI) [16, 17]

$$\mathrm{d}s^2 = 2\mathrm{d}u\,\mathrm{d}r + \delta_{ij}\,\mathrm{d}x^i\,\mathrm{d}x^j + 2H(u,r,x^k)\,\mathrm{d}u^2 + 2W_i(u,r,x^k)\,\mathrm{d}u\,\mathrm{d}x^i \tag{32}$$

which are of the Weyl type III and clearly admit the xKS form (10) with the flat background $\bar{g}_{ab}\,\mathrm{d}x^a\,\mathrm{d}x^b = 2\mathrm{d}u\,\mathrm{d}r + \delta_{ij}\,\mathrm{d}x^i\,\mathrm{d}x^j$ where

$$\mathcal{H} = -H, \quad k_a\,\mathrm{d}x^a = \mathrm{d}u, \quad \mathcal{K} = -\sqrt{W_i W_i}, \quad m_a\,\mathrm{d}x^a = \frac{W_i\,\mathrm{d}x^i}{\sqrt{W_j W_j}}. \tag{33}$$

Table 1. The relation of the class of higher dimensional Ricci-flat *pp*-waves to the classes of KS, xKS, and VSI spacetimes depending on the Weyl types.

Weyl type	KS	xKS	VSI
N	✓	✓	✓
III	×	✓	✓
II	×	only CSI	×

The class of *pp*-wave spacetimes is geometrically defined as metrics admitting a covariantly constant null vector field which can be written in the form [18]

$$\mathrm{d}s^2 = 2\mathrm{d}u\left[\mathrm{d}r + H(u,x^k)\,\mathrm{d}u + W_i(u,x^k)\,\mathrm{d}x^i\right] + g_{ij}(u,x^k)\,\mathrm{d}x^i\,\mathrm{d}x^j \tag{34}$$

It can be shown that Einstein *pp*-waves are Ricci-flat and of the Weyl type II. Type N Ricci-flat *pp*-waves admit the KS form. Type III Ricci-flat *pp*-waves are VSI and therefore belong to the class of xKS spacetimes. Type II Ricci-flat *pp*-waves can be cast to the xKS form only if they have constant scalar invariants (CSI). The situation is summarized in table 1.

Note that four dimensional Ricci-flat *pp*-waves are of the Weyl type N, belong to the VSI class and take the KS form.

4. Example of expanding xKS spacetimes

The Chong–Cvetič–Lü–Pope charged rotating black hole [19] can be cast to the xKS form [20]

$$\begin{aligned}\bar{g}_{ab}\,\mathrm{d}x^a\,\mathrm{d}x^b =& -\left(1-\lambda r^2\right)\frac{\Delta}{\Xi_a\Xi_b}\,\mathrm{d}t^2 - 2\mathrm{d}r\left(\frac{\Delta}{\Xi_a\Xi_b}\,\mathrm{d}t - \frac{a\sin^2\theta}{\Xi_a}\,\mathrm{d}\phi - \frac{b\cos^2\theta}{\Xi_b}\,\mathrm{d}\psi\right)\\ &+\frac{\rho^2}{\Delta}\,\mathrm{d}\theta^2 + \frac{\left(r^2+a^2\right)\sin^2\theta}{\Xi_a}\,\mathrm{d}\phi^2 + \frac{\left(r^2+b^2\right)\cos^2\theta}{\Xi_b}\,\mathrm{d}\psi^2,\end{aligned} \tag{35}$$

$$k_a\,\mathrm{d}x^a = -\frac{\Delta}{\Xi_a\Xi_b}\,\mathrm{d}t + \frac{a\sin^2\theta}{\Xi_a}\,\mathrm{d}\phi + \frac{b\cos^2\theta}{\Xi_b}\,\mathrm{d}\psi, \tag{36}$$

$$\hat{m}_a\,\mathrm{d}x^a = \lambda ab\frac{\Delta}{\Xi_a\Xi_b}\,\mathrm{d}t + \frac{b\sin^2\theta}{\Xi_a}\,\mathrm{d}\phi + \frac{a\cos^2\theta}{\Xi_b}\,\mathrm{d}\psi, \tag{37}$$

$$\mathcal{H} = -\frac{M}{\rho^2} + \frac{Q^2}{2\rho^4}, \qquad \mathcal{K} = -\frac{Q\nu}{r\rho^2}, \qquad A = -\frac{\sqrt{3}Q}{2\rho}\boldsymbol{k} \tag{38}$$

where a and b are spins, M and Q is mass and charge, respectively, $\rho^2 = r^2+\nu^2$, $\Delta = 1+\lambda\nu^2$, $\Xi_a = 1+\lambda a^2$, $\Xi_b = 1+\lambda b^2$, and $\nu = \sqrt{a^2\cos^2\theta + b^2\sin^2\theta}$. It is a solution of 5D minimal gauged supergravity which is equivalent to the Einstein–Maxwell–Chern–Simons theory with the Chern–Simons coefficient $\chi = 1$ and $\Lambda < 0$

$$R_{ab} - \frac{1}{2}Rg_{ab} + \Lambda g_{ab} = 2F_{ac}F_b{}^c - \frac{1}{2}F_{cd}F^{cd}g_{ab}, \qquad \nabla_b F^{ab} + \frac{\chi}{2\sqrt{3}\sqrt{-g}}\epsilon^{abcde}F_{bc}F_{de} = 0. \tag{39}$$

It can be shown that the CCLP black hole is of the Weyl type I_i and the vectors $\boldsymbol{k}$ and $\boldsymbol{m}$ satisfy the relation (18). Interestingly, the optical matrix takes the same form as for the 5D Kerr–(A)dS black hole

$$\rho_{ij} = \begin{pmatrix} \frac{1}{r} & 0 & 0 \\ 0 & \frac{r}{\rho^2} & \frac{\nu}{\rho^2} \\ 0 & -\frac{\nu}{\rho^2} & \frac{r}{\rho^2} \end{pmatrix} \tag{40}$$

and therefore the optical constraint (4) is met and the vectors $\boldsymbol{k}$ and $\boldsymbol{m}$ are surface-forming. In the case of the uncharged ($Q = 0$) and static ($\nu = 0$) limit, the metric reduces to the KS form (3) and is of the Weyl type D.

5. Conclusion

We believe that the xKS form (10) may lead to the discovery of new solutions of general relativity in higher dimensions in vacuum and also in the presence of matter fields aligned with the KS vector $\boldsymbol{k}$, such as aligned Maxwell field. Using the xKS ansatz, one could also obtain new vacuum solutions of more general theories of gravity, for instance, the Gauss–Bonnet theory or Lovelock gravities of higher order. We hope that the results of our analysis of xKS spacetimes [15] will be useful for finding such new solutions in a subsequent work.

Acknowledgments

The author acknowledges the support of the Albert Einstein Center for gravitation and astrophysics, Czech Science Foundation 14-37086G.

References

[1] Kerr R P and Schild A 1965 *Proc. Symp. Appl. Math.* **17** 199–209
[2] Ortaggio M, Pravda V and Pravdová A 2009 *Class. Quant. Grav.* **26** 025008 (*Preprint* 0808.2165)
[3] Málek T and Pravda V 2011 *Class. Quant. Grav.* **28** 125011 (*Preprint* 1009.1727)
[4] Ortaggio M, Pravda V and Pravdová A 2013 *Class. Quant. Grav.* **30** 013001 (*Preprint* 1211.7289)
[5] Pravda V, Pravdová A, Coley A and Milson R 2004 *Class. Quant. Grav.* **21** 2873–2898 (*Preprint* gr-qc/0401013)
[6] Ortaggio M, Pravda V and Pravdová A 2007 *Class. Quant. Grav.* **24** 1657–1664 (*Preprint* gr-qc/0701150)
[7] Ortaggio M, Pravda V, Pravdová A and Reall H 2012 *Class. Quant. Grav.* **29** 205002 (*Preprint* 1205.1119)
[8] Málek T 2012 *General relativity in higher dimensions* Ph.D. thesis Faculty of Mathematics and Physics, Charles University in Prague (*Preprint* 1204.0291)
[9] Myers R C and Perry M J 1986 *Annals Phys.* **172** 304
[10] Gibbons G, Lü H, Page D N and Pope C 2005 *J. Geom. Phys.* **53** 49–73 (*Preprint* hep-th/0404008)
[11] Anabalon A, Deruelle N, Morisawa Y, Oliva J, Sasaki M, Tempo D and Troncoso R 2009 *Class. Quant. Grav.* **26** 065002 (*Preprint* 0812.3194)
[12] Güllü I, Gürses M, Şişman T Ç and Tekin B 2011 *Phys. Rev.* **D83** 084015 (*Preprint* 1102.1921)
[13] Gürses M, Şişman T Ç and Tekin B 2012 *Phys. Rev.* **D86** 024009 (*Preprint* 1204.2215)
[14] Anabalon A, Deruelle N, Tempo D and Troncoso R 2011 *Int. J. Mod. Phys.* **D20** 639–647 (*Preprint* 1009.3030)
[15] Málek T 2014 *Class. Quant. Grav.* **31** 185013 (*Preprint* 1401.1060)
[16] Coley A, Milson R, Pravda V and Pravdová A 2004 *Class. Quant. Grav.* **21** 5519–5542 (*Preprint* gr-qc/0410070)
[17] Coley A, Fuster A, Hervik S and Pelavas N 2006 *Class. Quant. Grav.* **23** 7431–7444 (*Preprint* gr-qc/0611019)
[18] Brinkmann H W 1925 *Math. Ann.* **94** 119–145
[19] Chong Z W, Cvetič M, Lü H and Pope C 2005 *Phys. Rev. Lett.* **95** 161301 (*Preprint* hep-th/0506029)
[20] Aliev A N and Çiftçi D K 2009 *Phys. Rev.* **D79** 044004 (*Preprint* 0811.3948)

Journal of Physics: Conference Series **600** (2015) 012041 doi:10.1088/1742-6596/600/1/012041

Non-Riemannian geometry: towards new avenues for the physics of modified gravity

Gonzalo J. Olmo[1,2] **and D. Rubiera-Garcia**[3,2]

[1] Departamento de Física Teórica and IFIC, Centro Mixto Universidad de Valencia - CSIC. Universidad de Valencia, Burjassot-46100, Valencia, Spain

[2] Departamento de Física, Universidade Federal da Paraíba, 58051-900 João Pessoa, Paraíba, Brazil

[3] Center for Field Theory and Particle Physics and Department of Physics, Fudan University, 220 Handan Road, 200433 Shanghai, China

E-mail: gonzalo.olmo@csic.es, drubiera@fudan.edu.cn

Abstract. Less explored than their metric (Riemannian) counterparts, metric-affine (or Palatini) theories bring an unexpected phenomenology for gravitational physics beyond General Relativity. Lessons of crystalline structures, where the presence of defects in their microstructure requires the use of non-Riemannian geometry for the proper description of their properties in the macroscopic continuum level, are discussed. In this analogy, concepts such as wormholes and geons play a fundamental role. Applications of the metric-affine formalism developed by the authors in the last three years are reviewed.

1. Introduction and motivation

Despite the tremendous experimental success of General Relativity (GR) [1], it is widely assumed that at the Planck scale, $l_P\sqrt{\equiv \hbar G/c^3} \simeq 10^{-35}m$, it will be superseded by a quantum theory of gravity. At this scale, dramatic changes are expected to arise. Among the proposals in the literature one finds, for example, those of non-commutative geometry, quantized geometry, or higher-dimensional scenarios (e.g. string theory). It is natural to wonder what kind of frameworks could be suitable for an effective description in the low-energy limit of such a theory. In this sense, since the number of physically accessible observables in this regime might be limited and observations have a finite precision, at the end of the day one might only have access to some sector of the whole theory. Effective descriptions through classical models might be necessary and, indeed, they could serve as a guide for developing the underlying fundamental theories.

GR predicts the existence of black holes (BH), which entail singularities in the space-time [2], though the very definition of singularities remains still obscure. At this point one could either live with the existence of those singularities, postulate that quantum gravity somehow regularizes the metric, or look for self-consistent descriptions of the notion of body within GR. The latter path was taken by Wheeler [3], who formulated the concept of geon, namely, self-supported gravitational electromagnetic entities. Combined with the notion of multiply-connected topology, Misner and Wheeler [4] showed that through a wormhole a sourceless electromagnetic field can create mass and charge. Self-gravitating geons and non-Euclidean topologies are thus classical concepts with non-trivial consequences for quantum gravity. They indeed play a role within a larger picture: that of space-time foam [5]. If topology change

Journal of Physics: Conference Series **600** (2015) 012041 doi:10.1088/1742-6596/600/1/012041

can occur dynamically in Nature, then the smoothness of Minkowski space would disappear at Planckian scales, where not only geometry, but also topology, may fluctuate in all possible manners. Quantum fluctuations would lead then to creation and annihilation of wormholes. This raises a question: what kind of framework should be used to represent a scenario where a macroscopic, geometrical, continuum description arises from a discrete, microscopic structure?.

Crystalline structures and, in particular, Bravais crystals, provide valuable lessons to address the above question. The microscopic structure of crystals is represented by a discrete net-like arrangement of atoms, whose respective separation is given by a lattice constant. However, at the macroscopic level the crystal is naturally described in terms of differential geometry, resulting from a continuizing procedure in which the lattice spacing is taken to zero, while keeping the point mass unchanged. At each atom of the microscopic network one can define a triad of vectors (crystallographic directions); moving along these vectors we jump from atom to atom. Distances can be measured by step counting along crystallographic directions, $ds^2 = g_{ij}dx^i dx^j$. A connection $\Gamma^\lambda_{\mu\nu}$ is used to transport vectors and define geodesics. For ideal and perfect crystals (the latter being just a deformation of the former) the macroscopic geometry can be described by a Riemannian structure, where the connection is compatible with the metric, $\nabla^\Gamma g_{\alpha\beta} = 0$.

It turns out, however, that all known ordered structures contain defects. These are defined as a breakdown of the regular network pattern. They can be of different types, e.g., point-like (vacancies and intersticials), line-like (dislocations/disclinations) or two-dimensional (textures). Such defects are very important for the macroscopic description of materials, since they largely determine collective properties such as viscosity or plasticity. Here we shall deal only with point defects. Intersticials corresponds to atoms that have abandoned their equilibrium positions, while vacancies are the empty places left behind. At the location of the defects the step-counting procedure breaks down. The situation becomes non-metric, $\nabla^\Gamma g_{\alpha\beta} \neq 0$. This defected crystal can be also continuized to the macroscopic level, provided that, in addition, the density of defects is kept finite. The new description turns out to be given in terms of a metric-affine geometry [6]. To overcome the presence of defects, a new metric, $h_{\mu\nu}$, that ignores intersticials and fills vacancies, allows to step-count in the defected crystal, $\nabla^\Gamma h_{\alpha\beta} = 0$. The deformation relating $g_{\mu\nu}$ and $h_{\mu\nu}$ depends on the kind and density of defects. Similarly, Cartan's torsion, $S^\alpha_{\beta\gamma} = \Gamma^\alpha_{\beta\gamma} - \Gamma^\alpha_{\gamma\beta}$, represents the continuum version of crystal dislocation. In summary, systems with defects require a non-Riemannian description.

2. Main framework of metric-affine approach.

Let us consider a gravitational action of the form

$$S = \frac{1}{2\kappa^2}\int d^4x\sqrt{-g}f(g_{\mu\nu}, R^\alpha{}_{\beta\mu\nu}) + S_m(g_{\mu\nu}, \Psi) \ , \tag{1}$$

where $\kappa^2 = 8\pi G$ represents Newton's constant, f is a given function of objects constructed with the Riemann tensor $R^\alpha{}_{\beta\mu\nu}$ and its contractions with the space-time metric $g_{\mu\nu}$ (where g is its determinant), and S_m is the matter action with Ψ_m representing collectively the matter fields. Note that the connection $\Gamma^\lambda_{\mu\nu}$ does not couple to the matter sector. As we are working in the metric-affine (or Palatini) approach, the metric $g_{\mu\nu}$ and the affine connection $\Gamma^\lambda_{\mu\nu}$ are taken as independent objects and not constrained a priori. The dynamics of such theories is, in general, inequivalent to that of the standard metric (Riemannian) approach [7].

Typically the function f in (1) will be a combination of curvature invariants such as $R = g_{\mu\nu}R^{\mu\nu}$, $Q = R_{\mu\nu}R^{\mu\nu}$ and $K = R^\alpha{}_{\beta\gamma\delta}R_\alpha{}^{\beta\gamma\delta}$ ($R_{\mu\nu} \equiv R^\alpha{}_{\mu\alpha\nu}$). In this section we focus on the family of theories $f(R,Q)$. To implement the Palatini approach we perform independent variation of the action (1) with respect to metric and connection, which yields [8]

Journal of Physics: Conference Series **600** (2015) 012041 doi:10.1088/1742-6596/600/1/012041

$$f_R R_{\mu\nu} - \frac{f}{2} g_{\mu\nu} + 2 f_Q R_{\mu\alpha} R^\alpha{}_\nu = \kappa^2 T_{\mu\nu} \tag{2}$$

$$\nabla^\Gamma_\beta \left[\sqrt{-g}\,(f_R g^{\mu\nu} + 2 f_Q R^{\mu\nu})\right] = 0 , \tag{3}$$

where $f_R \equiv df/dR$, $T_{\mu\nu}$ is the energy-momentum tensor, and vanishing torsion, $\Gamma^\lambda_{[\mu\nu]} = 0$, has been imposed [9]. Tracing in (2) with the metric $g^{\mu\nu}$ one obtains that $R = R(T_\mu{}^\nu)$ and $Q = Q(T_\mu{}^\nu)$ only depend on the matter sources. This allows us to introduce a new rank-two tensor, $h_{\mu\nu}$, satisfying $\nabla^\Gamma_\beta(\sqrt{-h}h^{\mu\nu}) = 0$. From Eq.(3) the relation between $g_{\mu\nu}$ and $h_{\mu\nu}$ becomes

$$h^{\mu\nu} = \frac{g^{\mu\alpha}\Sigma_\alpha{}^\nu}{\sqrt{\det \hat{\Sigma}}} , \quad h_{\mu\nu} = \left(\sqrt{\det \hat{\Sigma}}\right) \Sigma^{-1}{}_\mu{}^\alpha g_{\alpha\nu} , \tag{4}$$

where the matrix $\Sigma_\mu{}^\nu = f_R \delta^\nu_\mu + 2 f_Q R_\mu{}^\nu$ contains the relative deformation between them. Simple algebraic transformations allow us to rewrite (2) as

$$R_\mu{}^\nu(h) = \frac{1}{\sqrt{\det \Sigma}} \left(\frac{f}{2} \delta^\nu_\mu + \kappa^2 T_\mu{}^\nu \right) \tag{5}$$

This constitutes a second-order system of differential field equations which in vacuum, $T_\mu{}^\nu = 0$, reduce to those of GR plus a cosmological constant term (this term can be also generated through a non-linear electrodynamics model [10]). Thus no extra propagating degrees of freedom arise. Note that since $g_{\mu\nu}$ is algebraically related to $h_{\mu\nu}$ via the matter sources only, $g_{\mu\nu}$ satisfies second-order equations as well. In the case of $f(R)$ theories ($f_Q = 0$) only the trace of the energy-momentum tensor, $T = g_{\mu\nu} T^{\mu\nu}$, contributes to the modifications to the GR structure and, consequently, theories with non-vanishing trace, like the case of Born-Infeld non-linear electrodynamics [11], must be considered [12].

In summary, in Palatini $f(R,Q)$ gravity, the matter built up a Riemannian structure associated to $h_{\mu\nu}$. The structure for $g_{\mu\nu}$, however, is non-Riemannian, and the role of the deformation matrix $\Sigma_\mu{}^\nu$ is akin to that of point defects in crystalline structures. Once specific choices of gravity $f(R,Q)$ and matter S_m are made, this framework provides a full solution to a given problem. Applications are now possible.

3. Applications

3.1. A realization of Wheeler's geon

Consider a matter source of the form

$$S_m = -\frac{1}{16\pi} \int d^4x \sqrt{-g} F_{\mu\nu} F^{\mu\nu} \tag{6}$$

corresponding to a Maxwell field $F_{\mu\nu} = \partial_\mu A_\nu - \partial_\nu A_\mu$. The gravity Lagrangian is taken to be $f(R,Q) = R + l_P^2(aR^2 + Q)$, where a is a dimensionless constant. Let us consider static spherically symmetric solutions. Solving the field equations (5), these solutions can be cast, in ingoing Eddington-Finkelstein coordinates, as [13]

$$ds^2 = -A(x)dv^2 + \frac{2}{\sigma_+} dv dx + r^2(x) d\Omega^2 , \tag{7}$$

with the definitions

$$A(x) = \frac{1}{\sigma_+}\left[1 - \frac{r_S}{r}\frac{(1+\delta_1 G(r))}{\sigma_-^{1/2}}\right] \tag{8}$$

$$\delta_1 = \frac{1}{2r_S}\sqrt{\frac{r_q^3}{l_P}} \; ; \; \sigma_\pm = 1 \pm \frac{r_c^4}{r^4(x)} \; ; \; r^2(x) = \frac{x^2 + \sqrt{x^4 + 4r_c^4}}{2} , \tag{9}$$

where $r_S = 2M$ is Schwarzschild radius and $r_c = \sqrt{r_q l_P}$, with $r_q = 2Gq^2$ a charge scale. A few Planck scale units away from the surface $x = 0$ one has $r^2(x) \approx x^2$ and the geometry reduces to the standard Reissner-Nordström (RN) solution of GR. However, as the surface $x = 0$ (or $r = r_c$) is approached, which replaces the point-like singularity of the RN solution, important non-perturbative modifications occur. For arbitrary values of the mass and charge curvature divergences arise on a sphere of radius $A = 4\pi r_c^2(x)$ (though they are much milder than in the RN case), but they disappear when a certain charge-to-mass ratio is satisfied. The surface $x = 0$ ($r = r_c$) represents the mouth of the wormhole, a tunnel to another region of space-time. The presence of such a wormhole allows to define an electric charge in terms of lines of electric flux flowing through $x = 0$, though no point-like charges are present in the system. These facts allow to consistently interpret these solutions as explicit realizations of Wheeler's geon. Their existence may have important phenomenological consequences for the dark matter problem [14], for objects at particle accelerators [15] and for the quantum foam picture [16].

3.2. Dynamical formation

Consider now an ingoing flux of particles carrying mass and charge, $T_{\mu\nu}^{flux} = \rho_{in} k_\mu k_\nu$, where k_μ is a null vector, $k_\mu k^\mu = 0$, and ρ_{in} is the flux density. Upon interaction with this flux, the static geonic solutions above get replaced by a generalization of the Bonnor-Vaidya solution of GR [17] of the form [18]

$$ds^2 = -\left[\frac{1}{\sigma_+}\left(1 - \frac{1+\delta_1(v)G(z)}{\delta_2(v)z\sigma_-^{1/2}}\right) - \frac{2l_P^2\kappa^2\rho_{in}}{\sigma_-(1-\frac{2r_c^4}{r^4})}\right]dv^2 + \frac{2}{\sigma_+}dvdx + r^2(x,v)d\Omega^2 \,, \qquad (10)$$

where the time-dependent $M(v)$ and $q(v)$ functions reflect the change in mass and charge due to the action of the ingoing flux. Assuming a charged perturbation in a compact interval $[v_i, v_f]$, the metric (10) describes either the process of formation (if the initial state v_i is Minkowski space) or the change in the geometry of the wormhole whose throat has an area $A = 4\pi r_c^2(v_f)$. Apparently the first process implies a topology change, though this may not be the case if one assumes that the exact case $q = 0$ is never physically realized and that only the limit $q \to 0$ is physically meaningful.

3.3. Born-Infeld gravity and its extensions

A different example of Palatini gravity is the proposal by Deser and Gibbons [19], given by the Lagrangian density

$$S_{BI} = \frac{1}{\kappa^2\epsilon}\int d^4x\left[\sqrt{-|g_{\mu\nu} + \epsilon R_{\mu\nu}(\Gamma)|} - \lambda\sqrt{-|g_{\mu\nu}|}\right] + S_m(g_{\mu\nu}, \psi_m) \qquad (11)$$

dubbed, due to its resemblance with the non-linear electromagnetic theory, Born-Infeld (BI) gravity. When expanded in series of the (length-squared) parameter $\epsilon \ll 1$, the Lagrangian (11) reduces, at first order, to the quadratic $f(R, Q)$ Lagrangian studied above. The representation of the field equations (5) of a $f(R, Q)$ theory is also valid for this Lagrangian formulated in the Palatini approach. In addition, for an electromagnetic field, all the subsequent orders in the above expansion exactly cancel out due to the symmetries of the energy-momentum tensor or the matter. As a consequence, BI gravity also supports (for $\epsilon < 0$) geon-like solutions [20].

The Lagrangian density (11) admits an alternative representation as $L_{BI} = \sqrt{\det \hat{\Omega}}$, where the matrix $\hat{\Omega}$ is defined as $\Omega^\alpha{}_\nu = g^{\alpha\beta}q_{\beta\nu}$, where $q_{\mu\nu} \equiv g_{\mu\nu} + \epsilon R_{\mu\nu}(\Gamma)$. This representation suggests that BI gravity is just a particular case of a larger family of Lagrangians, given by $f(A)$, with $A = \det(\hat{\Omega})$. In particular, the family $f = A^{n/2}$ has been recently shown [21] to support bouncing cosmological solutions for different values of n and different equations of state, which confirms the robustness of this kind of theories to avoid the Big Bang singularity as well.

3.4. Higher-dimensional solutions
One major motivation to extend metric-affine theories to extra dimensions is provided by potential applications within the AdS/CFT correspondence. The simplest case is that of $f(R)$ gravity in five space-time dimensions. The field equations (5), with $f_Q = 0$, are valid also in the higher-dimensional case [22]. This means that the field equations of Palatini gravity are second-order in any dimension, without any need to constrain a priori the gravity Lagrangian or to consider ad hoc approaches such as those of quasi-topological gravities [23]. Note that in arbitrary space-time dimension, n, the relation between $h_{\mu\nu}$ and $g_{\mu\nu}$ becomes now

$$h_{\mu\nu} = f_R^{\frac{2}{n-2}} g_{\mu\nu} \; ; \; h^{\mu\nu} = f_R^{\frac{2}{2-n}} g^{\mu\nu}, \tag{12}$$

so they are conformally related and depending on the number of space-time dimensions. In addition, $\det(\Sigma) = f_R^{\frac{4}{n-2}}$. For an electromagnetic field, solutions for a given $f(R)$ model can be obtained and expanded in series in the region of interest, with the result that, though curvature divergences appear, similar discussion as in the four-dimensional case brings that wormhole solutions may exist [22]. The analytic tractability of this framework, and the second-order character of the field equations, could therefore used to investigate concrete proposals within the AdS/CFT scenario. Further extensions to other theories of gravity such as $f(R,Q)$ or Born-Infeld gravity are also possible.

3.5. Thick branes
To reconciliate the apparent contradiction between the addition of extra dimensions with the fact that Nature is four-dimensional, scenarios where the extra dimension is very small (Kaluza-Klein models) or where it is infinitely large has been introduced. One example of the latter is the Randall-Sundrum scenarios, specifically the RS2 one [24], where one assumes that the four space-time dimensional world is embedded in an AdS5 warped geometry, with a single extra spatial dimension of infinite extension. Adding scalar fields to this scenario gives rises to thick branes, in which the warp factor depends on the dynamics of the scalar field, and can be used to control the splitting of the brane. We consider a standard scalar field, $L_S = \frac{1}{2}g^{\mu\nu}\partial_\mu\phi\partial_\nu\phi + 2V(\phi))$ and a space-time metric $ds^2 = e^{2A(y)}g_{ab}dx^a dx^b + dy^2$, where $A(y)$ is the warp factor. A first-order formalism can be implemented to solve the corresponding Palatini field equations, through the introduction of a superpotential $W(\phi)$ such that $\theta = -\frac{1}{3}f_R^{1/3}W$, where we have defined $\theta \equiv A_y + \frac{1}{3}\frac{f_{R,y}}{f_R}$. The corresponding first-order equations are thus obtained as [25]

$$\phi_y = \frac{f_R^{4/3}}{\kappa^2}W_\phi \; ; \; A_y = -\frac{1}{3}f_R^{1/3}\left(W + \frac{1}{\kappa^2}f_{R,\phi}W_\phi\right) . \tag{13}$$

and the curvature reads $\frac{R}{20} = \frac{f_R^{2/3}}{3}\left(\frac{2}{5\kappa^2}f_R W_\phi^2 - \frac{1}{3}W^2\right)$. Specifying a pair $(W(\phi), f_R(\phi))$ these equations allow to obtain the warp factor $A(y)$. It has been shown [25], both perturbatively and with an exact solution, that a general property of these Palatini branes is that the warp factor tends to localize extra dimension, a feature not previously observed in this kind of models.

4. Final remarks
Metric-affine theories have an unexpected high-energy phenomenology and might contribute to shed light on the approaches to quantum gravity. The lessons from crystalline structures show

that a macroscopic, continuum description in terms of differential geometry may arise from a discrete microscopic structure. Since the latter contain defects, this supports the view that perfect (Riemannian) structures might be unstable, and that a description in terms of metric-affine (non-Riemannian) structures is physically more sensible. Insights on how these lessons can be useful to unveil properties of the microscopic structure of space-time have been already obtained in [26], though many open questions remain and further research is needed. As a final lesson, we point out that determining whether the structure of space-time is Riemannian or not, is a question to be answered by experiment, not by tradition or convention.

Acknowledgments

G.J.O. is supported by the Spanish grant FIS2011-29813-C02-02, the Consolider Program CPANPHY-1205388, the JAE-doc program of the Spanish Research Council (CSIC), and the i-LINK0780 grant of CSIC. D.R.-G. is supported by the NSFC (Chinese agency) grant No. 11305038, the Shanghai Municipal Education Commission grant for Innovative Programs No. 14ZZ001, the Thousand Young Talents Program, and Fudan University. The authors also acknowledge funding support of CNPq (Brazilian agency) project No. 301137/2014-5. The authors are grateful to Dionisio Bazeia, Eduardo Guendelman, Francisco S. N. Lobo, Laercio Losano, Jesus Martinez-Asencio, Roberto Menezes, Sergei Odintsov, Helios Sanchis-Alepuz and Mahary Vasihoun for their collaboration in several parts of this framework.

References

[1] C. M. Will, Living Rev. Rel. **17**, 4 (2014).
[2] B. Carter, Phys. Rev. Lett. **26**, 331 (1971); R. Penrose, Riv. Nuovo Cim. Numero Speciale **1**, 252 (1969); Gen. Relativ. Gravit. **34**, 1141 (2002); S. W. Hawking, G. F. R. Ellis, *The Large Scale Structure of Space-Time* (Cambridge University Press, Cambridge, 1973).
[3] J. A. Wheeler, Phys. Rev. **97**, 511 (1955).
[4] C. W. Misner and J. A. Wheeler, Annals Phys. **2**, 525 (1957).
[5] S. W. Hawking, Nucl. Phys. B **144**, 349 (1978).
[6] E. Kröner, Z. angew. Math. Mech. **66**, 5 (1986); Int. J. Theor. Phys. **29**, 1219 (1990).
[7] M. Borunda, B. Janssen, and M. Bastero-Gil, JCAP **0811**, 008 (2008).
[8] G. J. Olmo Int. J. Mod. Phys. D **20**, 413 (2011).
[9] G. J. Olmo and D. Rubiera-Garcia, Phys. Rev. D **88**, 084030 (2013).
[10] E. Guendelman, G. J. Olmo, D. Rubiera-Garcia, and M. Vasihoun, Phys. Lett. B **726**, 870 (2013).
[11] M. Born and L. Infeld, Proc. Roy. Soc. London. A **144**, 425 (1934).
[12] G. J. Olmo and D. Rubiera-Garcia, Phys. Rev. D **84**, 124059 (2011).
[13] G. J. Olmo and D. Rubiera-Garcia, Phys. Rev. D **86**, 044014 (2012); Int. J. Mod. Phys. D **21**, 1250067 (2012); Eur. Phys. J. C **72**, 2098 (2012).
[14] F. S. N. Lobo, G. J. Olmo, and D. Rubiera-Garcia, JCAP **1307**, 011 (2013).
[15] G. J. Olmo and D. Rubiera-Garcia, JCAP **1402**, 010 (2014).
[16] F. S. N. Lobo, Gonzalo J. Olmo, and D. Rubiera-Garcia, Eur. Phys. J. C. **74**, 2924 (2014).
[17] W. B. Bonnor and P. C. Vaidya, Gen. Rel. Grav. **1**, 127 (1970).
[18] F. S. N. Lobo, J. Martinez-Asencio, G. J. Olmo and D. Rubiera-Garcia, Phys. Rev. D **90**, 024033 (2014); Phys. Lett. B **731**, 163 (2014).
[19] S. Deser and G. W. Gibbons, Class. Quant. Grav. **15**, L35 (1998); M. Bañados and P. G. Ferreira, Phys. Rev. Lett. **105**, 011101 (2010).
[20] G. J. Olmo, D. Rubiera-Garcia, and H. Sanchis-Alepuz, Eur. Phys. J. C **74**, 2804 (2014).
[21] S. D. Odintsov, G. J. Olmo and D. Rubiera-Garcia, Phys. Rev. D **90**, 044003 (2014).
[22] D. Bazeia, L. Losano, G. J. Olmo, and D. Rubiera-Garcia, Phys. Rev. D **90**, 044011 (2014).
[23] R. C. Myers and B. Robinson, JHEP **1008**, 067 (2010); W. G. Brenna and R. B. Mann, Phys. Rev. D **86**, 064035 (2012).
[24] L. Randall and R. Sundrum, Phys. Rev. Lett. **83**, 4690 (1999).
[25] D. Bazeia, L. Losano, R. Menezes, G. J. Olmo, and D. Rubiera-Garcia, arXiv:1411.0897 [hep-th].
[26] F. S. Lobo, G. J. Olmo, and D. Rubiera-Garcia, arXiv:1412.4499 [hep-th].

Geometric aspects of charged black holes in Palatini theories

Gonzalo J. Olmo[1,2], **D. Rubiera-Garcia**[3] **and Antonio Sanchez-Puente**[1]

[1] Departamento de Física Teórica and IFIC, Centro Mixto Universidad de Valencia - CSIC. Universidad de Valencia, Burjassot-46100, Valencia, Spain

[2] Departamento de Física, Universidade Federal da Paraíba, 58051-900 João Pessoa, Paraíba, Brazil

[3] Center for Field Theory and Particle Physics and Department of Physics, Fudan University, 220 Handan Road, 200433 Shanghai, China

E-mail: gonzalo.olmo@csic.es, drubiera@fudan.edu.cn, asanchez@ific.uv.es

Abstract. Charged black holes in gravity theories in the Palatini formalism present a number of unique properties. Their innermost structure is topologically nontrivial, representing a wormhole supported by a sourceless electric flux. For certain values of their effective mass and charge curvature divergences may be absent, and their event horizon may also disappear yielding a remnant. We give an overview of the mathematical derivation of these solutions and discuss their geodesic structure and other geometric properties.

1. Introduction

In General Relativity (GR), the curvature of the space-time is calculated using the Levi-Civita connection of the metric (Riemannian formalism). A natural way to extend GR is to allow the connection to be independent of the metric (Palatini formalism). In this way, the variation of the action will give raise to two sets of equations: one set coming from the variation with respect to the metric, and another set coming from the variation with respect to the independent connection. In exchange for the greater number of equations, these new systems turn out to be linear in the derivatives of the metric and connection, which contrasts with the second order derivatives of the metric in the usual Riemannian approach. Within the Palatini formalism, this property is preserved for Lagrangians nonlinear in curvature scalars, which allows us to consider higher curvature corrections to the gravity action without worrying about perturbative instabilities like ghosts.

We study charged black holes in this context, focusing our attention on an action well studied in the Riemannian formalism,namely, $S = \int \left\{R + l_p^2(aR^2 + R_{\mu\nu}R^{\mu\nu})\right\} \sqrt{-g}\mathrm{d}^4x$, where $l_P^2 = \hbar G/c^3$ is Planck's length squared and a a dimensionless constant. The solution for a charged black hole with this action à la Palatini was first found in [1]. This black hole solution is very similar to the Reissner-Nordström black hole of GR at scales larger than a few Planck lengths away from the center. In the Palatini solution, curvature divergences do not arise at the center but on a sphere of area $4\pi r_c^2$. However, for a certain charge-to-mass ratio curvature divergences at r_c disappear. With a convenient choice of coordinates, one finds

Journal of Physics: Conference Series **600** (2015) 012042 doi:10.1088/1742-6596/600/1/012042

that for these configurations the metric is completely regular and the curvature invariants are finite at r_c. The geometry is that of a *wormhole*, the radius r_c being the minimum radius where the wormhole throat lies. Electric charge in this geometry arises as a topological effect resulting from the electric flux flowing through the wormhole mouth. No point-like charges are needed to generate the field. This geometry arises naturally in the Palatini context and provides a nice realization of the concept of *geon* introduced by J. A. Wheeler back in 1955 [2]. Given that the wormhole is a topologically non-trivial object, one can show that even configurations with curvature divergences at $r = r_c$ are indeed solutions with two sides connected at r_c. In this talk, we try to improve our understanding of the wormhole structure in both regular and divergent configurations by exploring the geodesic structure of these space-times and constructing Euclidean embeddings as a tool to better visualize the geometry.

2. Solution for the metric

We start from the action:

$$S = \frac{1}{2\kappa^2}\int f(R,Q)\sqrt{-g}\mathrm{d}^4x + S_m \tag{1}$$

Where g is the determinant of the space-time metric $g_{\mu\nu}$, $R = g^{\mu\nu}R_{\mu\nu}$, $Q = R_{\mu\nu}R_{\alpha\beta}g^{\alpha\mu}g^{\beta\nu}$, and the matter is coupled only to the metric and not the connection. The variation of this action with respect to metric and connection (which are independent from each other) gives raise to two sets of equations (vanishing torsion)

$$(\partial_R f)R_{\mu\nu} - \frac{f}{2}g_{\mu\nu} + 2(\partial_Q f)R_{\mu\alpha}R_{\beta\nu}g^{\alpha\beta} = \kappa^2 T_{\mu\nu} \tag{2}$$

$$\nabla_\beta\left[\sqrt{-g}((\partial_R f)g^{\mu\nu} + 2(\partial_Q f)R^{\mu\nu})\right] = 0 \tag{3}$$

The details to solve the equations given $f(R,Q)$ and S_m are given in [1]. A quick summary goes as follows: Eq. (2) can be manipulated raising one index with the metric to show that $P_\mu{}^\nu \equiv R_{\mu\alpha}g^{\alpha\nu}$ can be expressed as an algebraic function of $T_\mu{}^\nu$. Being that relation a matrix equation, there might be multiple solutions. However, agreement with GR in the low curvature regime should select just one of them. Tracing over the indices of $P_\mu{}^\nu$, one finds $R = P_\mu{}^\mu$, whereas $Q = P_\mu{}^\alpha P_\alpha{}^\mu$. With these expressions, it is possible to obtain the functional form of R and Q in terms of the matter fields and, in the case of an electric field with spherical symmetry, in terms of r. Now, with $P_\mu{}^\nu$ written in terms of the matter fields, we can write (3) in the form $\nabla_\beta\left[\sqrt{-g}g^{\mu\alpha}((\partial_R f)\delta_\alpha{}^\nu + 2(\partial_Q f)P_\alpha{}^\nu\right] = 0$. Here the connection appears linearly and can be solved using elementary algebraic manipulations. A simple way to see this, consists on defining an auxiliary metric $h_{\mu\nu}$ such that the connection equation becomes $\nabla_\beta\left[\sqrt{-h}h^{\mu\nu}\right] = 0$. This implies that $\Gamma^\alpha_{\mu\nu}$ is the Levi-Civita connection of $h_{\mu\nu}$, with $h_{\mu\nu}$ and $g_{\mu\nu}$ related by

$$h^{\mu\nu} = \frac{g^{\mu\alpha}\Sigma_\alpha{}^\nu}{\sqrt{\det\hat{\Sigma}}}\ , \qquad h_{\mu\nu} = \left(\sqrt{\det\hat{\Sigma}}\right)\Sigma^{-1}{}_\mu{}^\alpha g_{\alpha\nu}\ , \tag{4}$$

where $\Sigma_\alpha{}^\nu = (\partial_R f\delta^\nu_\alpha + 2\partial_Q f P_\alpha{}^\nu)$. With this notation, it is possible to write eq. (2) only in terms of $h_{\mu\nu}$ (and its derivatives), R and Q as functions of r, and $T^\mu{}_\nu$. The resulting equation can be integrated to obtain $h_{\mu\nu}$, which allows to compute $g_{\mu\nu}$ using the relations (4).

Before showing the metric solution for the particular case of $f(R,Q) = R + l_p^2(aR^2 + Q)$, and the $T^\mu{}_\nu$ of electrovacuum fields, let us define some constants and factors that will appear: integration of the the equations will give a constant r_S that is identified with the Schwarzschild radius of the black hole. The number of charges of the black hole together with the Planck length gives two characteristic lengths $r_q = N_q l_P$ (where N_q is the number of charges) and $r_c = \sqrt{r_q l_P}$. With these we can define a dimensionless parameter δ_1 that will characterize the behaviour of the black hole near the wormhole throat, and two factors $\sigma_\pm(r)$ that represent the non-zero components of the matrix $\Sigma_\alpha{}^\nu$ introduced above

$$\delta_1 = \frac{1}{2}\frac{r_q^2}{r_c r_s} \qquad \sigma_\pm(r) = 1 \pm \frac{r_c^4}{r^4} \; . \tag{5}$$

With this notation, the metric takes the form

$$ds^2 = -A(r)dt^2 + \frac{1}{A(r)\sigma_+^2}dx^2 + r^2(x)d\Omega^2 \; , \tag{6}$$

where the area function $r^2(x)$ and the radial coordinate x are related as follows

$$x^2 = r^2\sigma_- \; , \; r^2 = \frac{x^2 + \sqrt{x^4 + 4r_c^4}}{2} \; . \tag{7}$$

The behaviour of $A(r)$ in the limit of large and small r, respectively, is [1, 3, 4, 5]

$$\begin{array}{ll} A(r) \approx 1 - \frac{r_S}{r} + \frac{r_q^2}{2r^2} + O\left(\frac{r_c^4}{r^4}\right) & \text{for} \quad r \gg r_c \\ A(r) \approx \frac{N_q}{4N_c}\frac{(\delta_1-\delta_c)}{\delta_1\delta_c}\sqrt{\frac{r_c}{r-r_c}} + \frac{N_c-N_q}{2N_c} + O\left(\sqrt{r-r_c}\right) & \text{for} \quad r \sim r_c \end{array} \tag{8}$$

Here $\delta_c \simeq 0.572069$ and $N_c \simeq 16.55$. The rapid decay of the corrections for $r \gg r_c$ puts forward that at distances slightly larger than r_c the metric is essentially the same as in the Reissner-Nordström black hole of GR. For stellar size objects, no difference with respect to GR could be seen outside the horizon or even a few Planck lengths away from the center. However, as $r \to r_c$, the behavior changes radically and becomes strongly dependent on δ_1. If δ_1 is greater than δ_c, there is a time-like curvature divergence. If δ_1 is smaller than δ_c, the divergence is space-like. For $\delta_1 = \delta_c$ the geometry becomes completely smooth. Moreover, if the number of charges of the black hole is less than a critical number, $N_q < N_c$ there will be no horizon hiding the hypersurface $r = r_c$.

Fig. 1 illustrates why the radial function $r(x)$ cannot be used as a valid coordinate. When $x = 0$, the function $r(x)$ has a minimum and ceases to be monotonic, which prevents it from mapping the whole domain of the solution. It is important to note that the relation between x and $r(x)$ is independent of the charge-to-mass ratio. In the regular case, $\delta_1 = \delta_c$, nothing prevents extending the domain of x to the negative axis, i.e., $x \in]-\infty, +\infty[$. Given that the wormhole is a topological structure, thus insensitive to metric perturbations, the domain of x must always be the same. This is further supported by the properties of the electric flux across the wormhole throat, which is independent of the charge-to-mass ratio. Moreover, the ratio of the electric flux by the area of the wormhole turns out to be a universal constant, independent of the charge and mass of the solutions. This further reinforces the topological character of the solution and the existence of *two sides* regardless of the existence of curvature (metric) divergences at the throat.

3. Euclidean embeddings

The geometry changes from being smooth to having divergences if δ_1 moves infinitesimally away from the value δ_c. It seems shocking that such a small change can lead from something regular to something as pathological as curvature divergences are thought to be. In order to better understand the differences between such configurations, it is useful to look at the Euclidean embedding of a bidimensional slice of the spacetime ($t = constant, \theta = \pi/2$). The induced metric in this slice is

$$\mathrm{d}l^2 = \frac{1}{A\sigma_-}\mathrm{d}r^2 + r^2\mathrm{d}\varphi^2 \; , \tag{9}$$

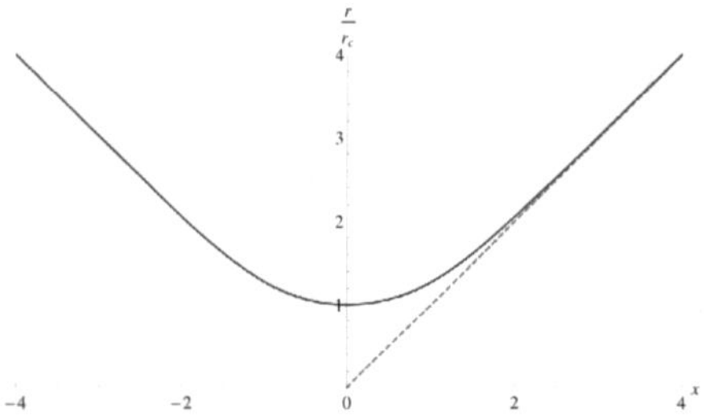

Figure 1: r as a function of x, in units of r_c

while the metric of the three-dimensional Euclidean space with cylindrical symmetry in which we want to embed this slice has the form [6]

$$dl^2 = d\xi^2 + dr^2 + r^2 d\varphi^2. \tag{10}$$

To find the embedding one just needs to restrict ξ to be a function $\xi(r)$ so that the line element in eq. (9) is recovered. Fig. 2 shows $\xi(r)$ for two particular cases with $\delta_1 = \delta_c$ and $\delta_1 > \delta_c$ (the latter representing the analogous of a naked singularity in GR). In the regular case the wormhole is smooth, whereas in the $\delta_1 > \delta_c$ case, there is a kind of vertex at the wormhole throat. It is worth noting that in the $\delta_1 > \delta_c$ case $r(\xi) \simeq \xi^{\frac{4}{3}}$ is both continuous *and* differentiable everywhere. It is the second derivative of $r(\xi)$ that tends to infinity as the wormhole throat is approached, causing the divergence of curvature scalars. The embeddings are shown in Fig. 3.

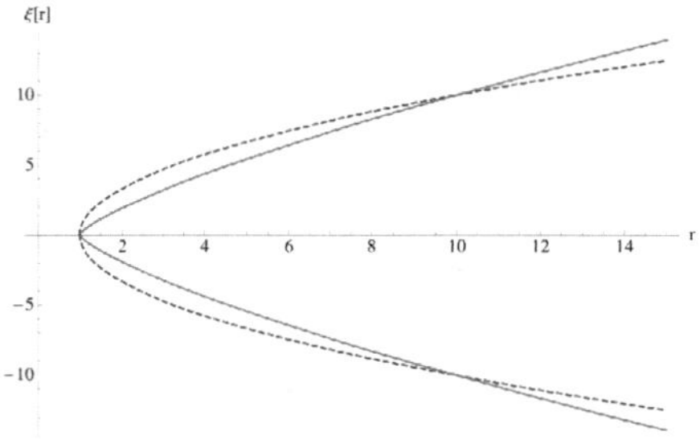

Figure 2: $\xi(r)$ for the regular case (blue dotted line) and the singular case (red line)

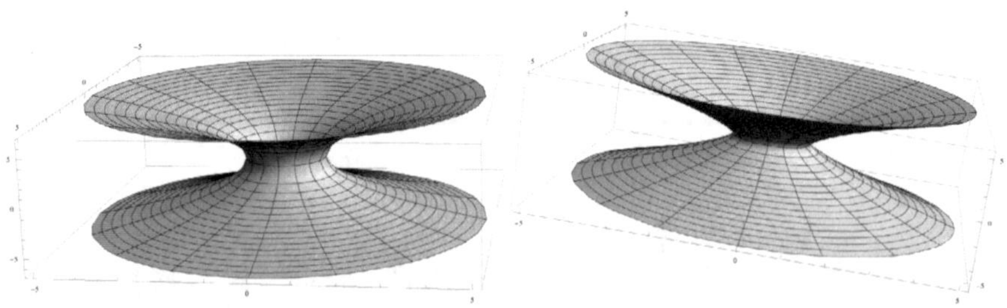

Figure 3: Embedding of a bidimensional slice of the spacetime in 3D euclidean space fo the regular and the singular solution

4. Geodesics

Geodesics are curves whose tangent vector is parallel transported along itself. This definition generalizes the concept of "straight lines" of Euclidean geometry to curved geometry. According to the Einstein Equivalence Principle, freely falling objects follow geodesics. A geodesic $x^\mu(\lambda)$ satisfies the following equation:

$$\frac{\mathrm{d}^2 x^\mu}{\mathrm{d}\lambda^2} + \Gamma^\mu_{\alpha\beta} \frac{\mathrm{d}x^\alpha}{\mathrm{d}\lambda} \frac{\mathrm{d}x^\beta}{\mathrm{d}\lambda} = 0 \ , \tag{11}$$

In the Riemannian formalism, the affine connection that appears in the geodesic equation is the Levi-Civita connection of the metric. However, in the Palatini formalism, the independent connection could also be used to define geodesic paths. If the theory is constructed in such a way that the connection does not couple to the matter fields (as in our case), the geodesics of freely falling objects (or of light rays) are the ones derived from the Levi-Civita connection of the metric.

Since the geometry we are dealing with is very symmetric, it is not necessary to solve the geodesic equation. Time translation symmetry gives us a conserved quantity along the geodesic $E = A\frac{\mathrm{d}t}{\mathrm{d}\lambda}$, and spherical symmetry gives us another conserved quantity, $L = r^2\frac{\mathrm{d}\varphi}{\mathrm{d}\lambda}$. The radial component can be obtaining normalizing the tangent vector to $\kappa = 0$ or $\kappa = 1$ depending if it is a null or a time-like geodesic:

$$\frac{1}{\sigma_+(x)}\frac{dx}{d\lambda} = \sqrt{E^2 - \underbrace{A(x)\left(\kappa + \frac{L^2}{r^2(x)}\right)}_{V(x)}} \tag{12}$$

The motion in the radial direction is analog to the motion of a particle in a one dimensional potential $V(x)$ with energy E^2. In this case, the energy is squared and cannot take negative values. Therefore, wherever the potential $V(x)$ is negative, $\frac{dx}{d\lambda}$ is necessarily different from 0 and the particle is forced to cross that region. The regions where $V(x)$ is negative correspond to the regions where $A(x)$ is negative (inside horizons) and the x coordinate becomes time-like (the t coordinate becomes space-like). The zeros of the potential correspond to the regions where $A(x)$ vanishes, which occurs at the black hole horizons.

Radial null geodesics are characterized by $\kappa = 0$, $L = 0$. The potential is null in this case and insensitive to the details of $A(x)$. In this case, light rays will follow their path crossing the wormhole unimpeded.

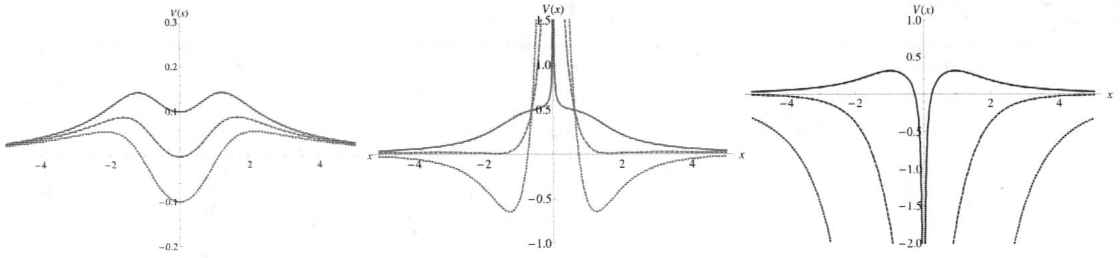

(a) $\delta_1 = \delta_c$ (b) $\delta_1 > \delta_c$ (c) $\delta_1 < \delta_c$

Figure 4: Potential for null geodesics with angular momentum. In each plot, the potential is shown for 3 different values for the number of charges N_c (but the same δ_1)

The potential for null geodesics with angular momentum is shown in Fig. 4. For large values of r, the geodesics feel a repulsive potential that goes like L/r^2. Near the wormhole, the behaviour of the potential depends on the value of δ_1. If $\delta_1 < \delta_c$, the potential will reach a maximum and then go to $-\infty$ as the geodesic approaches the wormhole. All geodesics with energy greater than the maximum will cross to the other side of the wormhole. If $\delta_1 = \delta_c$ the potential will have a finite value at the wormhole throat and reach a maximum before that. The potential at the wormhole throat is a minimum, and there will be stable light orbits if the value of the potential is positive (which happens when $N_q < N_c$). If $\delta_1 > \delta_c$, the potential will go to $+\infty$ as the geodesic approaches the wormhole throat. As a consequence, all geodesics will be repelled and none will cross the wormhole.

The potential for radial time-like geodesics is shown in Fig. 5. In this case, for large values of r, the geodesics feel an attractive potential that goes like $1 - r_S/r$. The behavior near the wormhole depends on δ_1 and is similar to the previous case. For $\delta_1 < \delta_c$, the potential goes to $-\infty$ at the wormhole throat and the geodesics will cross to the other side. For $\delta_1 = \delta_c$, the potential will be finite, and there will be stable orbits at the throat if $N_q < N_c$. If $\delta_1 > \delta_c$, the potential will diverge to $+\infty$ and the geodesics will bounce back.

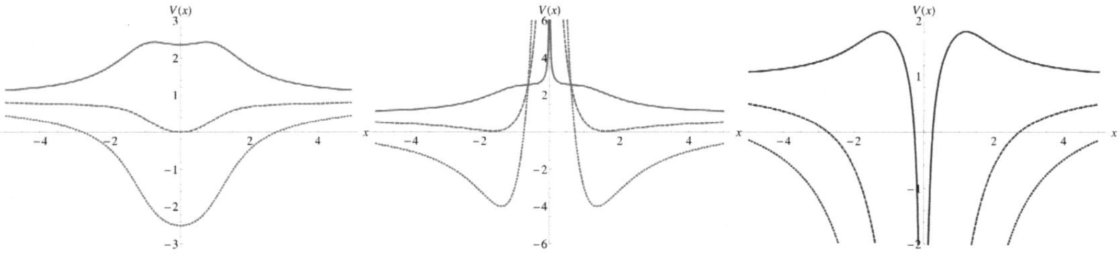

(a) $\delta_1 = \delta_c$ (b) $\delta_1 > \delta_c$ (c) $\delta_1 < \delta_c$

Figure 5: Potential for radial time-like geodesics. In each plot, the potential is shown for 3 different values of the number of charges N_c (but the same δ_1)

5. Concluding Remarks

We have shown that charged black holes for a $f(R, Q)$ gravity theory in the Palatini formalism possess a wormhole structure. This wormhole may have curvature divergences at its throat depending on the value of its charge-to-mass ratio. Using embeddings in Euclidean space, curvature divergences can be understood as a vertex at the wormhole throat. Null and time-like geodesics are shown to be able to cross the wormhole.

Acknowledgments

Work supported by the Spanish grant FIS2011-29813-C02-02, the Consolider Program CPANPHY-1205388, the JAE-doc program of CSIC, the i-LINK0780 grant of CSIC, the NSFC (Chinese agency) grant No. 11305038, the Shanghai Municipal Education Commission grant for Innovative Programs No. 14ZZ001, the Thousand Young Talents Program, Fudan University, and by CNPq (Brazilian agency) project No. 301137/2014-5.

References

[1] G. J. Olmo and D. Rubiera-Garcia, Phys. Rev. D **86**, 044014 (2012).
[2] J. A. Wheeler, Phys. Rev. **97**, 511 (1955).
[3] F. S. N. Lobo, G. J. Olmo and D. Rubiera-Garcia, JCAP **1307**, 011 (2013).
[4] F. S. N. Lobo, J. Martinez-Asencio, G. J. Olmo and D. Rubiera-Garcia, Phys. Rev. D **90**, 024033 (2014).
[5] F. S. N. Lobo, J. Martinez-Asencio, G. J. Olmo and D. Rubiera-Garcia, Phys. Lett. B **731**, 163 (2014).
[6] C. W. Misner, K. S. Thorne, and J. A. Wheeler, *Gravitation and cosmology* (W. H. Freeman, 1973).

On the dynamical content of MAGs

Vincenzo Vitagliano

CENTRA, Departamento de Física, Instituto Superior Técnico, Universidade de Lisboa - UL, Av. Rovisco Pais 1, 1049 Lisboa, Portugal

E-mail: vincenzo.vitagliano@ist.utl.pt

Abstract. Adopting a procedure borrowed from the effective field theory prescriptions, we study the dynamics of metric-affine theories of increasing order, that in the complete version include invariants built from curvature, nonmetricity and torsion. We show that even including terms obtained from nonmetricity and torsion to the second order density Lagrangian, the connection lacks dynamics and acts as an auxiliary field that can be algebraically eliminated, resulting in some extra interactions between metric and matter fields.

Introduction. The intriguing choice to treat alternative theories of gravity by means of the Palatini approach, namely elevating the affine connection to the role of independent variable, contains the seed of some interesting (usually under-explored) generalizations of General Relativity, the metric-affine theories of gravity. The peculiar aspect of these theories is to provide a natural way for matter fields to be coupled to the independent connection through the covariant derivative built from the connection itself.

The simplest prototypical version of a metric-affine theory of gravity is the Einstein-Cartan-Sciama-Kibble (ECSK) theory [1, 2]. In ECSK theory, it is still possible to show that the independent (and not necessarily symmetric) connection can be algebraically eliminated in favour of the metric and its derivatives, plus matter fields [3]. This means that the connection is not propagating in the spacetime, but it is indissolubly confined inside matter configurations.

Interestingly, once the independent connection is introduced as a new variable, nothing prevents us from adding further scalar invariants in the Einstein-Cartan action. The action describing a metric-affine theory is assumed to be a suitable limit, at a certain order, of a some fundamental theory; one can then follow the standard approach of effective field theory, and consider all the operators having the dimension of the specific order of the approximation. In the present case of ECSK, these operators are all the possible second order (namely, all the operators of the same dimension of the scalar curvature) invariants that can be formulated starting from the structures (torsion and nonmetricity) induced by the non-Riemannian nature of the actual spacetime.

Generalized metric-affine theories of gravity have been recently studied from different perspectives. Apart from the huge amount of work done in Einstein-Cartan gravity, progress was also made for what concerns the higher order versions of these theories [4, 5], also attracted by the tantalizing possibility of relaxing the theoretical puzzles associated with the recent discovery of accelerated expansion. The dynamical content of a metric-affine theory of gravity was already analyzed in [6] including the set of second order corrections built with torsion tensor. Here we are going to extend the previous result to the most general second order metric-affine theory of gravity, including quadratic operators in nonmetricity and torsion [7]. Note that, as it will be showed on the basis of a dimensional analysis, this will not include the quadratic Ricci terms, whose presence would introduce *brevi manu* new dynamical degrees of freedom.

The general action. We refer the reader to [7] for the notation and the conventions used also here. In a Lagrangian formulation of a metric-affine theory of gravity, the general action will be of the form

$$\mathcal{S} = \int d^4x\sqrt{-g}[\mathcal{L}_G(g_{\mu\nu}, \Gamma^{\gamma}{}_{\alpha\beta}) + \mathcal{L}_M(g_{\mu\nu}, \Gamma^{\gamma}{}_{\alpha\beta}, \psi)]\,, \tag{1}$$

where g is the determinant of the metric, $\mathcal{L}_G(g_{\mu\nu}, \Gamma^{\gamma}{}_{\alpha\beta})$ and $\mathcal{L}_M(g_{\mu\nu}, \Gamma^{\gamma}{}_{\alpha\beta}, \psi)$ are respectively the gravitational and the matter Lagrangian density (where we have made explicit the dependence of the matter from the connection), and ψ is a convenient way to refer collectively to the matter fields included in the theory under scrutiny.

We want to study the dynamics of the most general lowest order theory associated with the gravitational Lagrangian $\mathcal{L}_G(g_{\mu\nu}, \Gamma^{\gamma}{}_{\alpha\beta})$. Our prescription for constructing such general action is based on power counting of the dimensions of the gravitational terms. In natural units, the metric tensor is adimensional, the connection has the dimension of the inverse of a length and consequently the Ricci tensor has dimension [length]$^{-2}$. Since the total action must be dimensionless, the Lagrangian density must have dimension [length]$^{-4}$. We can think the Lagrangian density as the product of a geometrical scalar invariant times an opportune overall constant, in the form of a power of a length L_P, to adjust the total dimension. For the Einstein-Cartan action, for example, the correct Lagrangian density is $\mathcal{L}_G^{EC} = \mathcal{R}/(16\pi L_P^2)$.

Discarding for simplicity a cosmological constant term, it is not possible to build a Lagrangian density whose geometrical factor has dimension [length]$^{-1}$. This is trivially seen for the Ricci tensor case that is already of dimension [length]$^{-2}$. Nonmetricity and torsion tensor are also of dimension [length]$^{-1}$, but it is not possible to form a scalar invariant from just one rank-three tensor saturated with the metric (rank two); for such reason they will appear only as quadratic terms, hence only as terms of order [length]$^{-2}$, that is at the same order of the Ricci scalar $\mathcal{R}$.

The most general gravitational action with a Lagrangian density of dimension [length]$^{-2}$ is

$$\mathcal{S}_G = \frac{1}{16\pi L_P^2}\int d^4x\sqrt{-g}\left(\mathcal{R} + \sum_i a_i Q_{(i)}^2 + \sum_i b_i Q_{(i)} * S_{(i)} + \sum_i c_i S_{(i)}^2\right)\,, \tag{2}$$

where the last three terms are a symbolic representation of all the possible independent contractions that can be obtained from nonmetricity Q and torsion tensor S (the symbol "*" in (2) indicates the tensorial product between torsion and nonmetricity). Using the symmetries of these two (Q is symmetric with respect to its last two indices, while S is antisymmetric with respect to its first two indices), eleven different combinations are found (see [7]).

It is interesting to note that the full Lagrangian density is free of terms obtained by the covariant derivative of nonmetricity and torsion. The reason is easily understood. The independent connection can be decomposed in the Levi-Civita connection of the metric plus a combination of terms in Q and S called the distortion tensor

$$K_{\alpha\beta}{}^{\gamma} = \frac{1}{2}(-Q_{\alpha}{}^{\gamma}{}_{\beta} + Q^{\gamma}{}_{\beta\alpha} - Q_{\beta\alpha}{}^{\gamma}) + S_{\alpha\beta}{}^{\gamma} - S_{\beta}{}^{\gamma}{}_{\alpha} + S^{\gamma}{}_{\alpha\beta}\,. \tag{3}$$

Using this last condition, one can always think to write the covariant derivative with respect to the independent Γ as a Riemannian part (namely a covariant derivative with respect to the metric) plus another one encoding the non-Riemannian structures of the spacetime; the Riemannian derivative leads to a total divergence resulting in a surface term, the second part is instead constituted of contractions of the distortion tensor with Q and S, i.e. (quadratic) combinations of torsion and nonmetricity, already included in the general action above.

The dynamical content of the independent connection. We are mostly interested in solving the Palatini field equation for the connection and to re-express it as a function of matter fields plus Christoffel symbols of the metric. Without loss of generality, the global structure of this field equation is

$$\frac{1}{\sqrt{-g}}[\overset{\Gamma}{\nabla}_\lambda\left(\sqrt{-g}g^{\mu\nu}\right)-\overset{\Gamma}{\nabla}_\sigma\left(\sqrt{-g}g^{\sigma\mu}\right)\delta^\nu_\lambda]+\sum_j \alpha_j\,Q^{(j)}+\sum_j \beta_j\,S^{(j)}=(8\pi L_P^2)\Delta^{\mu\nu}{}_\lambda\,, \tag{4}$$

where

$$\begin{aligned}\sum_j \alpha_j\,Q^{(j)} &= \alpha_1\delta^\mu{}_\lambda Q^\nu{}_\alpha{}^\alpha+\alpha_2 g^{\mu\nu}Q_\alpha{}^\alpha{}_\lambda+\alpha_3\delta^\nu{}_\lambda Q^\alpha{}_\alpha{}^\mu+\alpha_4 Q^{\mu\nu}{}_\lambda+\alpha_5 Q_\lambda{}^{\nu\mu}+\alpha_6 Q^{\nu\mu}{}_\lambda+\\ &\quad+\alpha_7\delta^\mu{}_\lambda Q^\alpha{}_\alpha{}^\nu+\alpha_8 g^{\mu\nu}Q_{\lambda\alpha}{}^\alpha+\alpha_9\delta^\nu{}_\lambda Q^\mu{}_\alpha{}^\alpha\,, \end{aligned} \tag{5}$$

$$\sum_j \beta_j\,S^{(j)} = \beta_1 g^{\mu\nu}S_{\lambda\alpha}{}^\alpha+\beta_2\delta^\nu{}_\lambda S^\mu{}_\alpha{}^\alpha+\beta_3\delta^\mu{}_\lambda S^\nu{}_\alpha{}^\alpha+\beta_4 S^\nu{}_\lambda{}^\mu+\beta_5 S^{\nu\mu}{}_\lambda+\beta_6 S^\mu{}_\lambda{}^\nu\,, \tag{6}$$

and α_j and β_j are some linear combinations of the primitive coefficients a_i, b_i and c_i used in equation (2). The right hand side of (4) is given by the so-called hypermomentum tensor $\Delta^{\mu\nu}{}_\lambda \equiv -\frac{2}{\sqrt{-g}}\frac{\delta S_M(g,\Gamma,\psi)}{\delta\Gamma^\lambda{}_{\mu\nu}}$, that is, the tensor describing the intrinsic properties of matter as spin angular momentum, shear and dilation current [8]. Also note that

$$\frac{1}{\sqrt{-g}}[\overset{\Gamma}{\nabla}_\lambda\left(\sqrt{-g}g^{\mu\nu}\right)-\overset{\Gamma}{\nabla}_\sigma\left(\sqrt{-g}g^{\sigma\mu}\right)\delta^\nu_\lambda]=\delta^\nu{}_\lambda Q_\alpha{}^{\alpha\mu}+\frac{1}{2}g^{\mu\nu}Q_{\lambda\alpha}{}^\alpha-Q_\lambda{}^{\mu\nu}-\frac{1}{2}\delta^\nu{}_\lambda Q^\mu{}_\beta{}^\beta\,, \tag{7}$$

that is, we can reformulate the Palatini equation just in terms of an expression linear in the torsion tensor and in the nonmetricity tensor, that we can symbolically rewrite as

$$\sum_j \tilde{\alpha}_j\,Q^{(j)}+\sum_j \beta_j\,S^{(j)}=(8\pi L_P^2)\Delta^{\mu\nu}{}_\lambda\,, \tag{8}$$

with $\tilde{\alpha}_3=\alpha_3+1$, $\tilde{\alpha}_5=\alpha_5-1$, $\tilde{\alpha}_8=\alpha_8+\frac{1}{2}$, $\tilde{\alpha}_9=\alpha_9-\frac{1}{2}$ and $\tilde{\alpha}_{j\neq 3,5,8,9}=\alpha_j$. We can contract this equation in three different independent ways: with the metric $g_{\mu\nu}$, with δ^λ_μ and with δ^λ_ν, resulting in a simple linear system whose solution can be written as follows

$$\begin{aligned} Q^{\alpha\rho}{}_\alpha &= (8\pi L_P^2)(A_1\Delta_\alpha{}^{\alpha\rho}+B_1\Delta^{\alpha\rho}{}_\alpha+C_1\Delta^{\rho\alpha}{}_\alpha)\,,\\ Q^{\rho\alpha}{}_\alpha &= (8\pi L_P^2)(A_2\Delta_\alpha{}^{\alpha\rho}+B_2\Delta^{\alpha\rho}{}_\alpha+C_2\Delta^{\rho\alpha}{}_\alpha)\,,\\ S^{\rho\alpha}{}_\alpha &= (8\pi L_P^2)(A_3\Delta_\alpha{}^{\alpha\rho}+B_3\Delta^{\alpha\rho}{}_\alpha+C_3\Delta^{\rho\alpha}{}_\alpha)\,. \end{aligned} \tag{9}$$

where A_i, B_i and C_i are some elementary, but rather lenghty, expressions of the coefficients $\tilde{\alpha}_i$ and β_i. We can now use these three equations to substitute the corresponding terms in (8); their contribution is fully determined by the matter content of the theory, so we can move them on the right hand side, where they are collectively denoted as "$[f(\text{traces of }\Delta)]^{\mu\nu}{}_\lambda$". What remains is the equation

$$\tilde{\alpha}_4 Q^{\mu\nu}{}_\lambda+\tilde{\alpha}_5 Q_\lambda{}^{\nu\mu}+\tilde{\alpha}_6 Q^{\nu\mu}{}_\lambda+\beta_4 S^\nu{}_\lambda{}^\mu+\beta_5 S^{\nu\mu}{}_\lambda+\beta_6 S^\mu{}_\lambda{}^\nu=(8\pi L_P^2)\Delta^{\mu\nu}{}_\lambda+[f(\text{traces of }\Delta)]^{\mu\nu}{}_\lambda. \tag{10}$$

The antisymmetric part with respect to $\{\mu\nu\}$ pairs of indices of the previous equation gives an equation to express the torsion tensor in terms of the antisymmetric part of nonmetricity tensor plus terms in hypermomenta

$$\beta_5 S^{\nu\mu}{}_\lambda+(\beta_6-\beta_4)S^{[\mu}{}_\lambda{}^{\nu]}=(\tilde{\alpha}_6-\tilde{\alpha}_4)Q^{[\mu\nu]}{}_\lambda+(8\pi L_P^2)\Delta^{[\mu\nu]}_\lambda+[f(\text{traces of }\Delta)]^{\mu\nu}{}_\lambda\equiv\Theta^{\mu\nu}{}_\lambda\,, \tag{11}$$

that can be solved considering a suitable combination of the three different permutations of the indices $(\mu\nu\lambda) \to (\lambda\mu\nu)$ and $(\mu\nu\lambda) \to (\lambda\nu\mu)$. At the end, we get:

$$\begin{aligned} S_{\mu\nu\lambda} &= \frac{2\beta_5\Theta_{\mu\nu\lambda} - \beta_4(\Theta_{\mu\nu\lambda} - \Theta_{\lambda\mu\nu} + \Theta_{\lambda\nu\mu}) + \beta_6(\Theta_{\mu\nu\lambda} - \Theta_{\lambda\mu\nu} + \Theta_{\lambda\nu\mu})}{(\beta_4 + 2\beta_5 - \beta_6)(\beta_4 - \beta_5 - \beta_6)} \\ &= [\widehat{f}(\Delta)]_{[\mu\nu]\lambda} - \frac{2(\tilde{\alpha}_6 - \tilde{\alpha}_4)}{\beta_4 + 2\beta_5 - \beta_6} Q_{[\mu\nu]\lambda} \,. \end{aligned} \tag{12}$$

Since we are not interested in the exact form of the contribution of matter to the torsion tensor, we have here defined another tensor $[\widehat{f}(\Delta)]_{\mu\nu\lambda}$ that includes all the contributions coming from the hypermomenta in $\Theta_{\mu\nu\lambda}$. Note that the tensor $\Theta_{\mu\nu\lambda}$, and hence the torsion tensor $S_{\mu\nu\lambda}$, is linear in nonmetricity $Q_{\mu\nu\lambda}$. Using this expression in equation (10) to eliminate the torsion, we can rewrite it in the form

$$\xi_1 Q_{\mu\nu\lambda} + \xi_2 Q_{\lambda\nu\mu} + \xi_3 Q_{\nu\mu\lambda} = [\overline{f}(\Delta)_{\mu\nu\lambda}] \,, \tag{13}$$

where $\xi_i \equiv \xi_i(\tilde{\alpha}_j, \beta_k)$ are some coefficients determined by the equations (10) and (12) and $[\overline{f}(\Delta)]_{\mu\nu\lambda}$ is defined, in analogy to f and $\hat{f}$, as the collective contribution from matter to the right hand side of the expression; equation (13) can be now solved with respect to $Q_{\mu\nu\lambda}$ adding and subtracting the further two equations obtained permuting the indices $(\mu\nu\lambda) \to (\lambda\mu\nu)$ and $(\mu\nu\lambda) \to (\nu\mu\lambda)$. Having expressed nonmetricity in terms of just matter fields, we can reuse it in the equation for torsion to have another expression using just the matter fields. At the end, the total connection, that can be written in terms of Christoffel symbols of the metric plus distortion (where we recall that the distortion tensor (3) is a combination of nonmetricity and torsion), is hence reduced to a (not trivial) expression of metric with its derivatives and of matter fields under the guise of the hypermomenta combination.

An important point to stress is the following. Given a matter Lagrangian that does contain at most linear terms in the covariant derivative, and hence linear terms in the connection, our demonstration shows the lack of dynamics, and the consequent reduction to an auxiliary field, of $\Gamma^\lambda{}_{\mu\nu}$. On the other hand, some specific and exotic forms of matter can evade this fulfillment. Anyway, equations of motion of matter fields are required to be at most of second order, which forces the matter Lagrangian to contain only linear derivatives. Therefore, even in the most convoluted case, the hypermomentum tensor will be algebraic in the connection and the latter can be still eliminated at the component level.

What is the physical consequence of the lack of dynamics of the connection? Once it has been shown that the connection can be algebraically written in terms of derivatives of the metric and matter fields, it is clear that we can substitute all the terms explicitly dependent on connection (or torsion, or nonmetricity) in the field equation obtained varying with respect to the metric. Due to the extreme length of the equation, we will omit to write it here completely. It is anyway clear that, varying the general action (2) with respect to the metric, we will basically obtain terms that are quadratic in the connection (torsion/nonmetricity), and hence that will carry extra contributions of the kind "(hypermomentum)^2" to the effective stress energy tensor. This is similar to what happens in Einstein-Cartan theory as shown in [3], with the main difference that now, because of the presence of nonmetricity terms, the field equations will contain new terms of different nature, coupling matter fields (in the form of hypermomenta) to Christoffel symbols.

Higher orders terms. It is an easy task to argue that scalar Lagrangian corrections of the order $[\text{length}]^{-2n}$, with integer n, are the only ones that can be written starting from our elementary geometrical objects. This is essentially due to the fact that the only quantities carrying dimension $[\text{length}]^{-1}$ are odd-rank objects (torsion, nonmetricity and covariant derivative), and they cannot be trivially saturated with the (even-rank) metric. For such reason, the next-order invariants are all the possible terms with dimension $[\text{length}]^{-4}$. It goes beyond the scope of this paper to enumerate all the possible invariants of the fourth order (just for comparison, it should be taken into account that in a spacetime with torsion and vanishing nonmetricity there are 151 independent scalar invariants [9]). It is anyway possible to show that these terms are inevitably introducing further degrees of freedom, even assuming the simplifying hypothesis of matter fields not coupled to the connection. It is what for example occurs for the generalized Palatini theories considered in [10], whose specific choice of a Lagrangian density of the form $\mathcal{R} + \mathcal{R}_{\mu\nu}\mathcal{R}_{\kappa\lambda}(ag^{\mu\kappa}g^{\nu\lambda} + bg^{\mu\lambda}g^{\nu\kappa})$ has been shown to be equivalent to Einstein gravity plus a (dynamical) Proca vector field.

A rather peculiar case is the metric-affine version of $f(\mathcal{R})$ theories of gravity [5, 11]. The Ricci scalar $\mathcal{R}$ is invariant under the projective transformation

$$\Gamma^{\rho}{}_{\mu\nu} \to \Gamma^{\rho}{}_{\mu\nu} + \delta^{\rho}{}_{\mu}\xi_{\nu}\,, \tag{14}$$

where ξ_ν is an arbitrary covariant vector field. Consequently also any function of $\mathcal{R}$ will respect the same symmetry. While this issue is not a problem when matter does not couple to the connection, for a metric-affine theory this feature can lead to inconsistent field equations. In general the matter Lagrangian is not projective invariant, neither it is reasonable to restrict the matter content to those fields that are fulfilling this property. To circumvent the problem, it is necessary to break the projective invariance of the gravitational sector, fixing the four degrees of freedom of the transformation (related to the four components of the vector field ξ_ν) by a Lagrange multiplier. Note that the projective invariance is automatically broken if distortion-squared terms and, *a fortiori*, higher order curvature invariants are added to the action; such theories have been shown already to carry further dynamics.

Since the number of degrees of freedom to be fixed is four, and since the projective transformation suggests that the goal of breaking projective invariance should be achieved by constraining the connection, it is reasonable to propose an additional term in the gravitational Lagrangian involving a contraction of either nonmetricity tensor or torsion tensor. As already shown in [12], the term of the form $A^\mu Q_\mu \equiv A^\mu g^{\alpha\beta}Q_{\mu\alpha\beta}$, previously proposed in [13] is not suitable for a generic $f(\mathcal{R})$ metric-affine theories, since it requires the function of the Ricci scalar to reduce to the Einstein-Cartan term and the theory results to have no solutions of the field equations whenever the $f(\mathcal{R})$ is non-linear. Interestingly, the result is still valid even if we try to fix the four degrees of freedom through the other independent contraction of the nonmetricity tensor, namely adding to the Lagrangian density the Lagrange multiplier the $B^\mu \widetilde{Q}_\mu \equiv B^\mu g^{\alpha\beta}Q_{\alpha\beta\mu}$, as it can be easily proved: in a torsion-less theory without matter fields, the independent Levi-Civita connection is written as [7]

$$\Gamma^{\gamma}{}_{\alpha\beta} = \{^{\gamma}_{\alpha\beta}\} + \frac{1}{2}\left(-Q_{\alpha}{}^{\gamma}{}_{\beta} + Q^{\gamma}{}_{\beta\alpha} - Q_{\beta\alpha}{}^{\gamma}\right)\,; \tag{15}$$

on the other hand, the field equation of the connection reduces to the usual $f(\mathcal{R})$-Palatini equation[1], that can be solved with respect to the connection to give

$$\Gamma^{\gamma}{}_{\alpha\beta} = \{^{\gamma}_{\alpha\beta}\} + \frac{1}{2f'(\mathcal{R})}\left(2\partial_{(\alpha}f'(\mathcal{R})\delta^{\gamma}_{\beta)} - g^{\gamma\sigma}g_{\alpha\beta}\partial_{\sigma}f'(\mathcal{R})\right). \tag{16}$$

Equating (15) and (16) gives a condition expressing the contribution to nonmetricity coming from the gravitational sector of the theory

$$-Q_{\alpha}{}^{\gamma}{}_{\beta} + Q^{\gamma}{}_{\beta\alpha} - Q_{\beta\alpha}{}^{\gamma} = \frac{1}{f'(\mathcal{R})}\left(2\partial_{(\alpha}f'(\mathcal{R})\delta^{\gamma}_{\beta)} - g^{\gamma\sigma}g_{\alpha\beta}\partial_{\sigma}f'(\mathcal{R})\right); \tag{17}$$

[1] The condition about the absence of matter fields implies also the vanishing of the Lagrange multipliers A_μ and B_μ that can be proven to be proportional to the trace of the hypermomentum tensor.

we can find two independent expressions by contracting respectively α and β indices or α and γ in the last equation; the resulting conditions being

$$
\begin{aligned}
Q_\gamma - 2\widetilde{Q}_\gamma &= -\frac{2}{f'(\mathcal{R})}\partial_\gamma f'(\mathcal{R}) \,, \\
-Q_\beta &= \frac{4}{f'(\mathcal{R})}\partial_\beta f'(\mathcal{R}) \,.
\end{aligned}
\tag{18}
$$

It is now clear that both the constraint $Q_\mu = 0$ or $\widetilde{Q}_\mu = 0$ lead to the same result $Q_\mu = \widetilde{Q}_\mu = \partial_\mu f'(\mathcal{R}) = 0$, that makes the theory obviously inconsistent since it forces the function $f(\mathcal{R})$ to be at most linear.

A viable alternative [5, 6] is the theory obtained constraining the trace of the torsion tensor $S_{\mu\rho}{}^{\rho}$ through the term $C^\mu S_\mu \equiv C^\mu S_{\mu\rho}{}^{\rho}$

$$
\mathcal{S} = \frac{1}{16\pi L_P^2}\int d^4x\sqrt{-g}(f(\mathcal{R}) + C^\mu S_\mu) + \mathcal{S}_M(g_{\mu\nu}, \Gamma^\gamma{}_{\alpha\beta}, \psi) \,, \tag{19}
$$

whose field equations are written as

$$
\begin{aligned}
f'(\mathcal{R})\mathcal{R}_{(\mu\nu)} - \frac{1}{2}f(\mathcal{R})g_{\mu\nu} &= (8\pi L_P^2)T_{\mu\nu} \,, \\
-\overset{\Gamma}{\nabla}_\lambda \left(\sqrt{-g}f'g^{\mu\nu}\right) + \overset{\Gamma}{\nabla}_\sigma \left(\sqrt{-g}f'g^{\sigma\mu}\right)\delta^\nu{}_\lambda + 2\sqrt{-g}f'S^\mu{}_\lambda{}^\nu &= (8\pi L_P^2)\sqrt{-g}\left(\Delta^{\mu\nu}{}_\lambda - \frac{2}{3}\Delta^{\sigma[\nu}{}_\sigma\delta^{\mu]}{}_\lambda\right), \\
S_{\alpha\mu}{}^{\alpha} &= 0 \,.
\end{aligned}
\tag{20}
$$

Note that this choice is fully consistent: if we require vacuum solutions, the second of equations (20) simply reduces to the two conditions $S_{\mu\nu\rho} = 0$ and $\overset{\Gamma}{\nabla}_\lambda \left(\sqrt{-g}f'(\mathcal{R})g^{\mu\nu}\right) = 0$, that are the usual field equations found in Palatini-$f(\mathcal{R})$ theories of gravity. On the other side, this version of $f(\mathcal{R})$-metric-affine theories has the further feature to avoid propagation of torsion waves in vacuum. In fact (modulo the dependence of the matter Lagrangian on the covariant derivative, that must be at most linear), taking the antisymmetric part of second equation in (20) with respect to μ and ν, and adding suitable permutations of the obtained expression, we can show that torsion tensor is

$$
S_{\mu\nu}{}^{\lambda} = \frac{8\pi L_P^2}{f'(\mathcal{R})}g^{\rho\lambda}(\Delta_{[\rho\mu]\nu} + \Delta_{[\nu\rho]\mu} - \Delta_{[\mu\nu]\rho}) \,, \tag{21}
$$

namely, the torsion tensor is algebraically defined by the antisymmetric part of the hypermomentum tensor, $\Delta^{[\mu\nu]}{}_\lambda$.

The Author is pleased to thank FCT - Portugal for the support through the grant SFRH/BPD/77678/2011.

References

[1] D. W. Sciama, Rev. Mod. Phys. **36**, 463 (1964) [Erratum-ibid. **36**, 1103 (1964)].
[2] T. W. B. Kibble, J. Math. Phys. **2**, 212 (1961).
[3] F. W. Hehl, P. Von Der Heyde, G. D. Kerlick and J. M. Nester, Rev. Mod. Phys. **48**, 393 (1976).
[4] F. W. Hehl, J. D. McCrea, E. W. Mielke and Y. Ne'eman, Phys. Rept. **258**, 1 (1995).
[5] T. P. Sotiriou and S. Liberati, Annals Phys. **322**, 935 (2007).
[6] V. Vitagliano, T. P. Sotiriou and S. Liberati, Annals Phys. **326**, 1259 (2011) [Erratum-ibid. **329**, 186 (2013)].
[7] V. Vitagliano, Class. Quant. Grav. **31**, 045006 (2014).
[8] F. W. Hehl, G. D. Kerlick and P. Von Der Heyde, Z. Naturforsch. **31A**, 111 (1976).
[9] S. M. Christensen, J. Phys. A **13**, 3001 (1980).
[10] V. Vitagliano, T. P. Sotiriou and S. Liberati, Phys. Rev. D **82**, 084007 (2010).
[11] T. P. Sotiriou and V. Faraoni, Rev. Mod. Phys. **82**, 451 (2010).
[12] T. P. Sotiriou, arXiv:0710.4438 [gr-qc].
[13] F. W. Hehl, E. A. Lord and L. L. Smalley, Gen. Rel. Grav. **13**, 1037 (1981).

Spanish Relativity Meeting (ERE 2014): almost 100 years after Einstein's revolution IOP Publishing
Journal of Physics: Conference Series **600** (2015) 012044 doi:10.1088/1742-6596/600/1/012044

Testing vector-tensor gravity with current cosmological observations

Roberto Dale[1] **and Diego Sáez**[2]

[1] Departament d'estadística, matemàtiques i informàtica, Universitat Miguel Hernández, 03202, Elx, Alacant, Spain

[2] Departamento de Astronomía y Astrofísica, Universidad de Valencia, 46100, Burjassot, Valencia, Spain

E-mail: rdale@umh.es

Abstract.
A certain vector-tensor theory of gravitation (VT) has been recently applied to cosmology (Phys. Rev. D, 89, 2014, 044035). It leads to encouraging results. The zero order energy density of the vector field accounts for the cosmological constant. It has been recently proved that the VT vector field cannot play the role of the electromagnetic field. The evolution of the scalar perturbations is different in VT and general relativity. Tensor fluctuations evolve in the same way in both theories. Here, the VT evolution equations of the scalar modes are appropriately written, and the initial conditions at high redshift – for numerical integration– are given. The codes COSMOMC and CAMB are modified for applications to VT cosmology and, then, by using the new version of COSMOMC, statistical methods (Markov chains) allow us to fit theoretical VT predictions with current cosmological data obtained in various experiments. Previous encouraging fits were based on WMAP7 data, SN Ia observations, and a minimal model involving seven cosmological parameters. Here a similar preliminary study based on PLANCK data and updated SN Ia observations is presented.

1. Introduction and Generalities

In any vector-tensor theory, there are two fields, the metric $g_{\mu\nu}$, and a four-vector A^{μ}. Various theories of this type have been studied (see [1], [2]). From these fields, we may define: The antisymmetric tensor $F_{\mu\nu} = \nabla_{\mu} A_{\nu} - \nabla_{\nu} A_{\mu}$, where the symbol ∇ stands for a covariant derivative, and quantities R and g, which are the scalar curvature and the determinant of the matrix $g_{\mu\nu}$ formed by the covariant components of the metric, respectively.

Here, we are concerned with a cosmological model, which is based on a certain vector-tensor theory of gravitation. This theory is hereafter referred to as VT (see [3]). The basic VT equations may be derived from the following action [4]:

$$I = \int \left[\frac{R}{16\pi G} + (\frac{\gamma}{2} - \varepsilon) F^{\mu\nu} F_{\mu\nu} + \gamma \, (\nabla_{\mu} A^{\mu})^2 - \rho(1+\epsilon) \right] \sqrt{-g} \, d^4x \ , \tag{1}$$

where the dimensionless parameters γ and ε must satisfy the inequality $2\varepsilon > \gamma$, to avoid ghosts, and the relation $\gamma > 0$ to have a positive cosmological energy density for the vector field A^{μ}. Quantity ρ is the conserved energy density of an isentropic perfect fluid, ϵ is its internal energy density [see [5] and [6] for details], and G is the constant of gravitation.

The VT basic equations may be found in [4]. Here, these equations are directly written in the cosmological case. We write the zero order (background universe) equations, plus the first order corrections due to scalar perturbations of $g_{\mu\nu}$, A^{μ}, and the energy momentum tensor of the cosmological fluid $T^{\mu\nu}$. Vector perturbations do not contribute at first order, and there are no new tensor perturbations associate to A^{μ}, which implies that the tensor modes evolve as in general relativity (GR).

Hereafter, our signature is (–,+,+,+), Greek (Latin) indices run from 0 to 3 (1 to 3), units are chosen in such a way that the speed of light is $c = 1$, the present value of the scale factor is $a = 1$ (flat universe), the coordinate and conformal times are t and τ, respectively, and finally, whatever quantity S may be, S_B stands for its background value and $\dot{S}$ is its derivative with respect to the conformal time.

The background universe is assumed to be flat, homogeneous, and isotropic. It contains matter and radiation. In this flat background, the metric has the Robertson-Walker form, and the covariant components of the VT vector field are $[A_{0B}(\tau), 0, 0, 0]$. Then, at zero order (no perturbations), we get

$$3\frac{\dot{a}^2}{a^2} = 8\pi G a^2(\rho_B + \rho_B^A) \tag{2}$$

$$-2\frac{\ddot{a}}{a} + \frac{\dot{a}^2}{a^2} = 8\pi G a^2(P_B + P_B^A) \tag{3}$$

$$D_B \equiv (\nabla \cdot A)_B = -\frac{1}{a^2}[\dot{A}_{0B} + 2\frac{\dot{a}}{a}A_{0B}] = constant\ , \tag{4}$$

where ρ_B and P_B are the background energy density and pressure of the cosmological fluid (baryons, dark matter, massless neutrinos and radiation), $\nabla \cdot A \equiv \Delta_{\mu}A^{\mu}$, and ρ_B^A (P_B^A) is the energy density (pressure) of the field A^{μ}. These quantities may be easily calculated from the A^{μ} energy momentum tensor [4] to get

$$\rho_B^A = -P_B^A = \gamma D_B^2 = constant\ , \tag{5}$$

which proves that ρ_B^A plays the role of dark energy with the vacuum equation of state $W = P_B^A/\rho_B^A = -1$ [3]. The resulting cosmology involves all the free parameters and equations of the standard ΛCDM model, plus Eq. (4), which may be numerically solved for appropriate initial conditions (see below) to get function $A_{0B}(\tau)$. Parameters γ and ϵ and the sign of A_{0B} kept arbitrary at zero order (background) and also at first order (see [4]).

Let us now include scalar perturbations in VT [4]. It is done as in GR. The metric and the energy momentum tensor of the cosmological fluid are perturbed as in [7]. All the physical quantities are expanded in terms of the scalar harmonics $Q^{(0)}$, which are plane waves in the flat case; e.g., the photon energy density is $\rho_{\gamma} = \rho_{\gamma B}[1 + \delta_{\gamma}(k,\tau)Q^{(0)}]$, k being the comoving wavenumber and $\delta_{\gamma}(k,\tau)$ the first order relative perturbation associated to ρ_{γ} in momentum space, where the perturbed cosmological equations are treated. The metric expansion involves the functions $h(k,\tau)$ and $\eta(k,\tau)$. See [7, 4] to get a full list of the perturbation functions in GR. In VT, a new function is associated to the expansion of the field A^{μ}. It has been justified [3, 4] that the most appropriate scalar mode, $D^{(0)}$, associated to A^{μ} may be defined as follows:

$$D = \nabla \cdot A = D_B[1 + D^{(0)}(k,\tau)Q^{(0)}] \tag{6}$$

In momentum space, the A^{μ} field equations reduce to the following second order differential equation:

$$\ddot{D}^{(0)} + 2\frac{\dot{a}}{a}\dot{D}^{(0)} + k^2 D^{(0)} = 0\ . \tag{7}$$

Since only the scalar mode $D^{(0)}$ is involved in the last equation, this mode is not coupled to any other scalar mode. Let us now write the equations for h and η since most of them are different from the corresponding equations in standard cosmology based on GR. These equations are

$$k^2\eta - \frac{1}{2}\frac{\dot{a}}{a}\dot{h} = 4\pi G[-a^2\rho_B\delta - 2\gamma D_B(a^2 D_B D^{(0)} + A_{0B}\dot{D}^{(0)})] \tag{8}$$

$$k^2\dot{\eta} = 4\pi G[a^2(\rho_B + P_B)\theta + 2\gamma k^2 A_{0B} D_B D^{(0)}] \tag{9}$$

$$\ddot{h} + 2\frac{\dot{a}}{a}\dot{h} - 2k^2\eta = -24\pi G[a^2 P_B \pi_L - 2\gamma D_B(a^2 D_B D^{(0)} - A_{0B}\dot{D}^{(0)})] \tag{10}$$

$$\ddot{h} + 6\ddot{\eta} + 2\frac{\dot{a}}{a}(\dot{h} + 6\dot{\eta}) - 2k^2\eta = -24\pi G a^2(\rho_B + P_B)\sigma \ . \tag{11}$$

The terms involving γ are characteristic of VT; so, for $\gamma = 0$, these equations become identical to those of GR. The remaining equations have the same form in both GR and VT (see [7, 4]) The complete system of equations must be solved for appropriate initial values of $D^{(0)}$, $\dot{D}^{(0)}$, and any other scalar perturbation.

As it is usual, the initial conditions are calculated at redshift $z = 10^8$, in the radiation dominated era. At this redshift, all the cosmological scales were outside the horizon ($k\tau << 1$), the background functions a and A_{0B} may be assumed to be proportional to powers of τ, and any scalar mode Y –including $D^{(0)}$– may be expanded in the form

$$Y = \sum_{n,m} \beta_{nm} k^n \tau^m \ . \tag{12}$$

The smallness of $k\tau$ for cosmological scales, the existence of growing and decaying terms in Eq. (12), and additional considerations allow us to calculate the n and m values being relevant for each mode. Length but straightforward calculations leads to the VT initial conditions. For the background, we get

$$\tau_{in} = \left(\frac{\dot{a}}{a}\right)^{-1}_{in} , \quad (A_{0B})_{in} = -\frac{D_B}{5(1+z_{in})^2} \Big/ \left(\frac{\dot{a}}{a}\right)_{in} \ . \tag{13}$$

The initial conditions for the scalar modes are identical to those of GR (see [7, 4]), except for the following modes:

$$D^{(0)} = C_1 k^n \ , \qquad \dot{D}^{(0)} = 0 \ , \tag{14}$$

where C_1 is an arbitrary constant, and

$$h = C_2(k\tau)^2 + C_4(k\tau)^4, \ \ \theta_\gamma = \theta_b = -\frac{1}{18}C_2 k^4\tau^3 - \frac{1}{30}C_4 k^6 \tau^5,$$
$$\theta_\nu = -\frac{23 + 4R_\nu}{18(15 + 4R_\nu)}C_2 k^4 \tau^3 - \frac{1}{30}C_4 k^6\tau^5, \tag{15}$$

where constant C_2 is another arbitrary parameter, and $C_4 = [8\pi G C_1 \rho_v / 3(1+z_{in})^2](\dot{a}/a)^2_{in}$. The initial conditions only involve the arbitrary parameters C_1 and C_2, which are fully independent of the parameters γ and ε involved in the action (1). The evolution of the background universe and its linear perturbations do not allow us to calculate γ and ε. Here, these parameters are kept undetermined.

For $C_1 = 0$, the same evolution equations and initial conditions as in standard GR cosmology are recovered and, then, C_2 is the unique normalization constant. For $C_1 \neq 0$, the VT scalar modes deviate from those of the standard ΛCDM model. Deviations depend on $|C_1|$ (see [4]).

2. Numerical results and discussion

In the framework of GR, there are available codes, e.g., CAMB, to calculate the predictions of a given cosmological model (angular CMB power spectra, energy density power spectrum, and so on), and there are other codes, as COSMOMC, which use statistical techniques to fit model predictions and current observational data (CMB PLANCK data, updated data from SN I_a observations, and data from other experiments and galaxy surveys). By using these codes, the cosmological parameters may be either estimated or bounded.

Once evolution equations and initial conditions have been derived for both the background universe and the scalar modes of VT (see section 1), the codes CAMB and COSMOMC may be modified for VT cosmological applications. Modifications must be well planed since the original versions of these codes have been designed to work in standard GR cosmology. Such versions involve various conditions leading to accurate and fast calculations, which are not appropriate for VT applications. We have carefully modified recent versions of CAMB and COSMOMC to work with PLANCK data, and with current data about galaxy surveys, I_a supernovae, and so on. The resulting codes are hereafter denoted CAMB-VT and COSMOMC-VT.

Numerical calculations (performed with CAMB-VT) prove that quantities $T_1 = a\ddot{h} + \dot{a}\dot{h}$, $T_2 = \dot{h}$ and $T_3 = \dot{\eta}$, and other quantities involved in the evolution equations of the CMB photon distribution function (see [7]) evolve in the same way, until small redshifts, in both GR and VT. For redshifts $z \leq 5$ and spatial scales smaller than a few hundreds of megaparsec, some of these VT quantities show a significant separation with respect to those of GR. See, e.g., Fig. (1), where quantity T_1 is represented in terms of z for various k-values. At redshifts smaller than ~ 2, an oscillatory evolution of T_1 arises in VT, whereas no oscillations appear in GR. For too large spatial scales there are no differences between VT and GR. Deviations between both theories are responsible for the differences between the CMB angular power spectra of VT and GR. Very small integration steps are necessary to numerically estimate some integrals, at least, inside the oscillatory intervals associated to small enough scales and low redshifts. CAMB and COSMOMC codes have been modified to consider all the significant scales and also to integrate up to zero redshift.

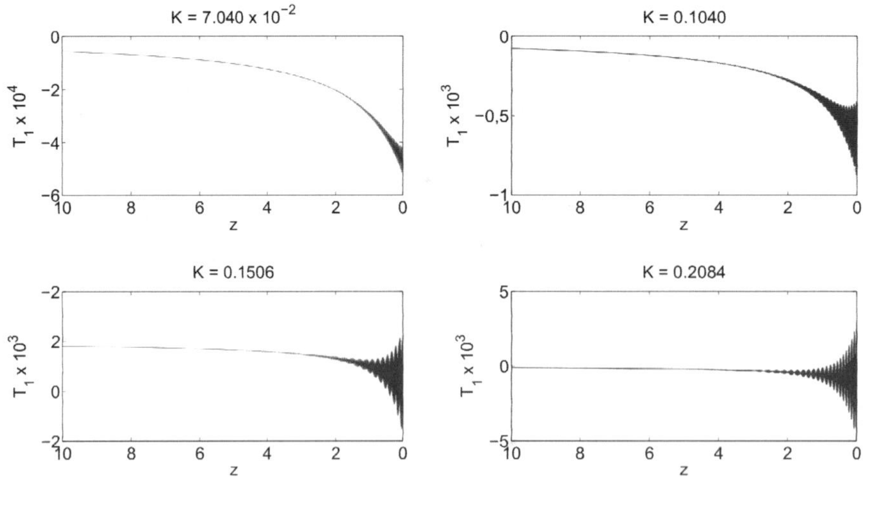

Figure 1. Quantity $T_1 = a\ddot{h} + \dot{a}\dot{h}$ in terms of the redshift z for various scales. The comoving wavenumbers in units of Mpc^{-1} are given in the top of each panel.

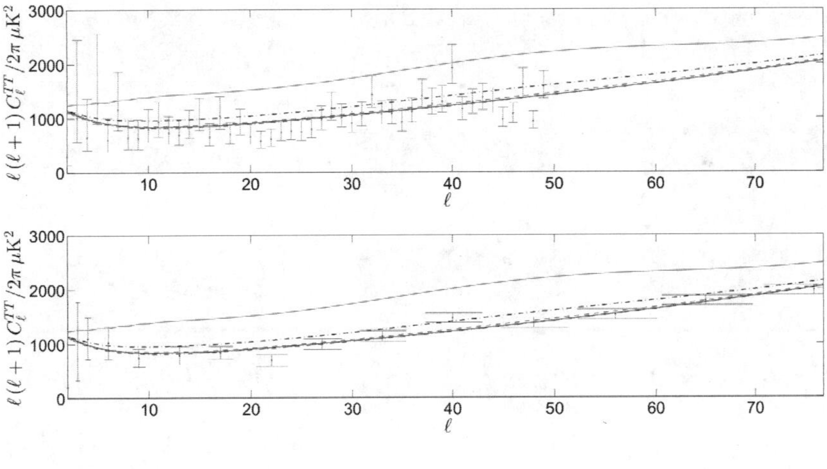

Figure 2. CMB temperature angular power spectra in terms of ℓ. Solid blue line corresponds to the best fit between PLANCK data and the standard ΛCDM model ($C_1 = 0$). The C_1 values of the remaining lines increase from bottom to top. In the top (bottom) panel, dots with error bars are PLANCK (WMAP7) measurements.

By using CAMB-VT, the CMB angular power spectra corresponding to various values of constant C_1 have been calculated. They are shown in Fig. 2 together with the C_ℓ values observed by PLANCK (top panel) and WMAP7 (bottom panel). Since the two C_ℓ samples are rather different just in the region where the VT and GR predicted spectra separate, namely, for ℓ values smaller than ~ 100, the statistical fit –obtained with COSMOMC-VT– could be rather different for PLANCK (here) and WMAP ([4]) data; in particular, for the C_1 parameter. In order to verify these suspicions, we have run the code COSMOMC-VT to fit observations and predictions by using seven parameters: C_1, plus the six usual parameters of standard cosmology used to fit WMAP and PLANCK data. These parameters are $\Omega_b h^2$, $\Omega_{DM} h^2$, τ, n_s, $\log[10^{10} A_s]$, and θ, where Ω_b and Ω_{DM} are the density parameters of baryons and dark matter, respectively, h is the reduced Hubble constant, τ is the reionization optical depth, n_s is the spectral index of the power spectrum of scalar modes, and A_s is the normalization constant of the same spectrum whose form is $P(k) = A_s k^{n_s}$, finally, the parameter θ is defined by the relation $\theta \times 10^{-2} = d_A(z_*)/r_s(z_*)$, where $d_A(z_*)$ is the angular diameter distance at decoupling redshift z_*, and $r_s(z_*)$ is the sound horizon at the same redshift. Moreover, all the calculations have been performed under the following basic assumptions: the background is flat, perturbations are adiabatic, the lensing effect is considered, there are no massive neutrinos, the equation of state of the dark energy is $P = W\rho$ with $W = -1$, vector and tensor modes are negligible, the mean CMB temperature is $T_{CMB} = 2.726$, the effective number of relativistic species is 3.046, and the total number of effectively massless degrees of freedom is $g_* = 10.75$. Only PLANCK and SN Ia data are considered to look for statistically admissible values of the cosmological parameters.

The COSMOMC-VT runs –described above– have given the mean likelihoods displayed in Fig. (3), as well as the best fits presented in Table 1, where we see that the six common parameters of the best GR and VT fits are very similar, and also that the C_1 value of the best VT fit is close to 3×10^{-9}. Similar results were obtained in [4] from the WMAP7 data. In each panel of Fig. (3), there is a red spot on the left hand side. The best fit is inside this spot, but there are other elongated spots where the mean likelihood is smaller (color) than in the main

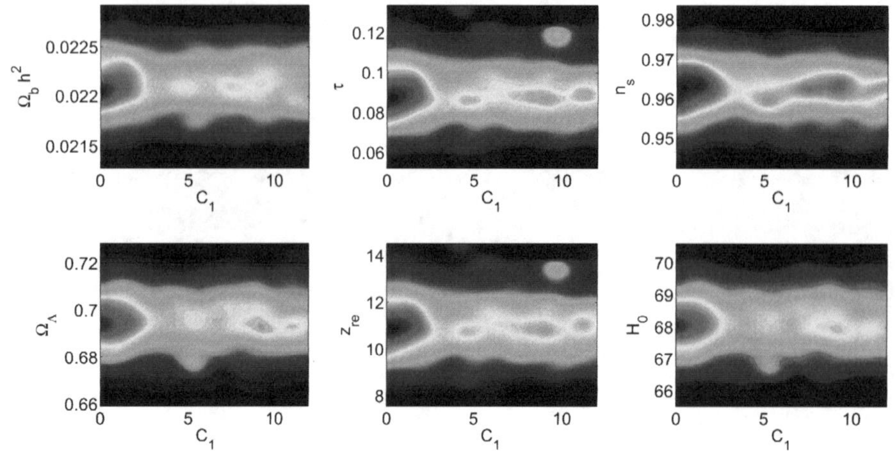

Figure 3. Each panel corresponds to a pair of parameters. In all cases, one of these parameters is C_1. The mean likelihood function is displayed by using a red green yellow blue color bar. The greatest (smallest) values are red (dark blue).

Table 1. Best fits: values of the basic cosmological parameters

THEORY	$C_1 \times 10^{-8}$	$\Omega_b h^2$	$\Omega_{DM} h^2$	τ	n_s	$\log[10^{10} A_s]$	θ
GR	0.0	0.0221	0.119	0.0927	0.963	3.103	1.041
VT	0.2903	0.0220	0.119	0.0909	0.963	3.091	1.041

red spot. More research is necessary to interpret these spots. They could be due to a poor convergence of COSMOMC-VT, but they could also have a real statistical meaning. In this second case, the enlargement of the data set by considering PLANCK and SN Ia observations plus cosmological data from other experiments might lead to interesting conclusions.

Acknowledgments
This work has been supported by the Spanish Ministry of *Economía y Competitividad*, MICINN-FEDER project FIS2012-33582. Calculations have been performed with SIUV (Servicio de Informática de la Universidad de Valencia) computers. We thank the SIUV staff for valuable help.

References

[1] Will C M 1993 *Theory and experiment in gravitational physics* Cambridge University Press, New York
[2] Will C M 2006 *Living Rev. Relativity* **9** 3
[3] Dale R, Morales J A and Sáez D 2009 arXiv:0906.2085[astro-ph.CO]
[4] Dale R and Sáez D 2014 *Phys. Rev. D* **89** 044035
[5] Dale R and Sáez D 2012 *Phys. Rev. D* **85** 124047
[6] Hawking S W and Ellis G F R 1999 *The large scale structure of space-time* Cambridge Monographs on Mathematical Physics (Cambridge University Press, New York).
[7] Ma C P and Bertschinger E 1995 *Astrophys. J.* **455** 7

Journal of Physics: Conference Series **600** (2015) 012045 doi:10.1088/1742-6596/600/1/012045

Astrophysical constraints and insights on extended relativistic gravity

S. Mendoza

Instituto de Astronomía, Universidad Nacional Autónoma de México, A.P. 70-264, Distrito Federal 04510, México

E-mail: sergio@astro.unam.mx

Abstract. It is discussed how observations of gravitational lensing at scales of individual, groups and clusters of galaxies can be understood in terms of non-Newtonian gravitational interactions with a relativistic structure compatible with the Einstein Equivalence Principle. This result is derived on very general grounds without knowing the underlying structure of the gravitational field equations. As such, any developed gravitational theory built to deal with these astrophysical scales (relevant to MOND phenomenology) needs to reproduce those results.

1. Introduction

In the second decade of the 17th century, the observed residuals of Neptune's orbit by were postulated to be due to [1]: (i) the effect of the Sun's gravity, at such a great distance might differ from Newton's description; or (ii) the discrepancies might simply be observational error; or (iii) perhaps Uranus was being pulled, or perturbed, by an as-yet undiscovered planet. Future work by Le Verrier and Adams [2] lead to the discovery of Neptune (the baryonic dark matter perturber of the motion of Uranus). When anomalies on the advance of the perihelium of Mercury were later observed, Le Verrier postulated the existence of an hypothetical unknown planet called Vulcan, closer to the sun than Mercury [cf. 3]. Vulcan was never observed and the relativistic extension made by Einstein when building up his general theory of relativity showed that almost all of the perihelium's advance of Mercury came from relativistic corrections of Newton's theory of gravity [cf. 4]. The important lessons to learn from these stories are that: (i) There could be unseen objects that perturbe gravitational orbits and (ii) there may be cases in which gravity needs to be extended. Note however that, in the previous cases, the anomalies were tiny as compared to the observed dynamics. As opposed to this, the current dark matter/energy hypothesis require huge energy corrections in order to leave Einstein's or Newton's field equations untouched. As such, one may be more tempted to search for non-relativistic and relativistic extensions of gravity.

The flattening of rotation curves in spiral galaxies is usually taken as the basis for either modifying gravity or incorporating unknown dark non-baryonic matter into dynamics. It has always been a complicated issue to distinguish between the path to follow, although it is quite clear that finding dynamical systems with accelerations greater than Milgrom's acceleration constant $a_0 \approx 10^{-10}\mathrm{m/s^2}$ which require dark matter, will contradict all types of MOdified Newtonian Dynamics (MOND) proposals. The absence of high accelerated systems requiring dark matter and the non-experimental detection of the mysterious dark matter particle is a good starting point in order to question the validity of gravity on mass and length scales such that the accelerations induced by the gravitational system on test particles are smaller than a_0.

The flattening of rotation curves in spiral galaxies is best described by the Tully-Fisher relation:

$$v \propto M^{1/4}, \tag{1}$$

where v represents the velocity of a test star about the centre of a galaxy and M is the mass of that galaxy. This relation is only valid in the outer regions of spiral galaxies and it is well known that the inner parts are well described by Kepler's third law, which for circular motions take the form:

$$v \propto \sqrt{M/r}. \tag{2}$$

Globular clusters are traditionally considered perfect Newtonian systems since the velocity dispersion as a function of radius is well modelled with Newtonian gravity. However, recent observations show flattening profiles of these "rotation curves" in the outer parts. As shown by [5, 6], this flattening occurs when the equivalent Newtonian acceleration reaches a value $\approx a_0$. Furthermore, at these scales, the systems obeys Tully-Fisher scaling (1) for a dispersion velocity v.

The simplest gravitational system available in astrophysics is a binary star system. Studies of wide open binaries as a test of Newtonian gravity [7] show that gravity deviates from its Newtonian behaviour when the accelerations $\lesssim a_0$, i.e. for systems with $\sim 1M_{\odot}$, separated by $\sim$ 1pc.

Many astrophysical and cosmological data can either be explained by unknown dark matter/energy entities or by a modification to the theory of gravity [e.g. 8–11]. The fact that the outer parts of globular clusters follow a Tully-Fisher law and that wide open binaries do not follow Kepler's third law are unfeasible to explain with the introduction of dark matter.

As such as an starting point one can think that Kepler's third law is valid at sufficiently small scales, where the gravitational acceleration a of test particles is $a \gtrsim a_0$ and that it gets substituted by the Tully-Fisher law at sufficiently large scales -where $a \lesssim a_0$, making gravity non-scale invariant.

2. The law of gravity

As explained by [12, 13], consider a test particle in circular orbit about a central mass source M. Using equation (2) and centrifugal balance means that the acceleration $a = v^2/r = -GM/r^2$, where $G_{\rm N}$ is Newton's gravitational constant. The minus sign is introduced in order to account for the attractive force of gravity. This calculation was performed by Newton in his *PhilosophiæNaturalis Principia Mathematica* book [14]. If we now assume that there is a scale for which Kepler's third law is substituted by the Tully-Fisher law, then using equation (1) and centrifugal balance it follows that $a = v^2/r = -G_{\rm M}M^{1/2}/r$, where a Modified gravitational constant $G_{\rm M}$ has been introduced. This relation can be transformed into the basic traditional MOND with the definition $G_{\rm M}^2 := a_0 G$ to obtain:

$$a = -\frac{\sqrt{a_0 GM}}{r}. \tag{3}$$

Note that $G_{\rm M}$ and/or a_0 are as fundamental as Newton's gravitational constant since they are applied to different sets of observations at different scales. The purpose of any extended theory of gravity not dealing with dark matter is to extend equation (3) into a proper theory at both relativistic and non-relativistic regimes with a transition from the Newtonian to this modified regime occurring at accelerations $\sim a_0$.

3. Relativistic constraints

As noted by [12], a constraint on the PPN γ parameter can be built following the same procedure as it is done in general relativity [15]. To do so it is only necessary to acknowledge a curved space-time, where the Einstein Equivalence Principle is valid and where particles follow geodesic trajectories. In Schwarzschild spherical coordinates, a point mass source M generating a gravitational field is described by the following line element:

$$\mathrm{d}s^2 = g_{\mu\nu}\mathrm{d}x^\mu \mathrm{d}x^\nu = g_{00}\, c^2\mathrm{d}t^2 + g_{11}\mathrm{d}r^2 - r^2\mathrm{d}\Omega^2, \tag{4}$$

for a metric tensor $g_{\mu\nu}$ and space-time coordinates $x^\alpha = (ct, r, \theta, \varphi)$, where t represents time, r the radial coordinate and the polar and azimuthal angles are given by θ and φ respectively, with an angular displacement $\mathrm{d}\Omega^2 := \mathrm{d}\theta^2 + \sin^2\theta\, \mathrm{d}\varphi^2$ and velocity of light c. The symmetry of the problem means that the unknown metric components g_{00} and g_{11} are functions that depend on the radial coordinate r only. For non-relativistic particles, we must consider the limit where the speed of light $c \to \infty$. In this case, the radial component of the geodesic equation is given by [see e.g. 16]:

$$\frac{1}{c^2}\frac{\mathrm{d}^2 r}{\mathrm{d}t^2} = \frac{1}{2} g^{11} \frac{\partial g_{00}}{\partial r}. \tag{5}$$

In this weak-field slow-motion approximation, a particle bounded to a circular orbit about the mass M has a centrifugal radial acceleration given by:

$$\frac{\mathrm{d}^2 r}{\mathrm{d}t^2} = \frac{v^2}{r}, \tag{6}$$

for a circular tangential velocity v, and so equating this with equation (5) it follows that:

$$\frac{v^2}{c^2 r} = \frac{1}{2}\frac{\partial\, {}^{(2)}g_{00}}{\partial r}. \tag{7}$$

Substitution of equation (3) in the previous relation yields:

$${}^{(2)}g_{00} = \frac{2\phi}{c^2} = -2\left(\frac{v}{c}\right)^2 \ln\left(\frac{r}{r_\star}\right) = -\frac{2G_{\mathrm{M}}M^{1/2}}{c^2}\ln\left(\frac{r}{r_\star}\right), \tag{8}$$

where $r_\star$ is an arbitrary length and ϕ is the Newtonian gravitational potential.

Having obtained the ${}^{(2)}g_{00}$ component, which determines the non-relativistic motion of massive particles, we now proceed to obtain the ${}^{(2)}g_{11}$ component. In the literature it is customary to define a new scalar potential ψ as:

$${}^{(2)}g_{11} = -\frac{2\psi}{c^2}, \tag{9}$$

in complete analogy with the definition of the Newtonian gravitational potential ϕ. The introduction of this potential can be justified considering a more general scenario. Without requiring spherical symmetry, the spatial part of the metric can be written as $g_{ik}\mathrm{d}x^i\mathrm{d}x^k$, with ${}^{(0)}g_{kl} = \delta_{kl}$ being the Minkowskian part. The second order perturbation corrections of g_{kl} could in principle involve other potentials (and not only ϕ or ψ). By a suitable choice of coordinates, one can get rid of the anisotropic contributions at the same perturbation order, which turns g_{kl} into a diagonal form. Given the isotropy of space, there is no preferred direction and so ${}^{(2)}g_{ik} \propto \delta_{ik}$. It is natural to expect that the leading order $\mathcal{O}(2)$ correction must be of the same order of magnitude as the gravitational potential ϕ. Accordingly $g_{kl} = \left(1 + 2\gamma\phi/c^2\right)\delta_{kl}$, where γ is a proportionality constant, and so

$$\mathrm{d}s^2 = g_{00}\mathrm{d}t - \left(1 + 2\gamma\phi/c^2\right)\delta_{kl}\mathrm{d}x^k\mathrm{d}x^l. \tag{10}$$

Since spherical Schwarzschild coordinates are widely used in astrophysical literature, let us calculate the metric component g_{11} in such coordinates.

The conversion to spherical Schwarzschild coordinates is straightforward since:

$$g_{11}\mathrm{d}r^2 + r^2\mathrm{d}\Omega^2 = \left(1 + 2\gamma\phi/c^2\right)\left(\mathrm{d}\tilde{r}^2 + \tilde{r}^2\mathrm{d}\Omega^2\right), \tag{11}$$

for spherical isotropic coordinates $(ct, \tilde{r}, \theta, \varphi)$. Substitution of equation (8) in the previous relation yields:

$$r = \tilde{r}\left[1 - \gamma\left(\frac{G_\mathrm{M}M^{1/2}}{c^2}\right)\ln\left(\frac{r}{r_\star}\right)\right], \tag{12}$$

and so,

$$\mathrm{d}r = \mathrm{d}\tilde{r}\left[1 - \frac{G_\mathrm{M}M^{1/2}}{c^2}\ln\left(\frac{\tilde{r}}{r_\star}\right) - \frac{G_\mathrm{M}M^{1/2}}{c^2}\right], \tag{13}$$

at perturbation order $\mathcal{O}(2)$. This means that:

$$\psi = -\gamma G_\mathrm{M}M^{1/2}, \qquad \text{and so} \qquad {}^{(2)}g_{11} = -\frac{2\gamma G_\mathrm{M}M^{1/2}}{c^2}. \tag{14}$$

4. Weak gravitational lensing at large scales

As pointed out by [17], recent observations have shown that gravitational lensing on individual, groups and clusters of galaxies can be modelled with the standard Schwarzschild solution of general relativity, assuming the existence of a total dark plus baryonic isothermal halo, where the Tully-Fisher law holds for the baryonic matter. As such, the bending angle of light can be calculated using the standard lensing equation The result is that this bending angle does not depend on the impact parameter and scales with the square root of the total baryonic mass. Under a modified theory of gravity scheme with no dark matter component, the bending angle has the same value and so, since the time metric component ${}^{(2)}g_{00}$ is known from the Tully-Fisher law as shown in equation (8), then a reconstruction of the spatial metric component ${}^{(2)}g_{11}$can be made [17].

To be more specific, take a static spherically symmetric total matter distribution M_T, assuming the validity of Einstein's general relativity so that Schwarzschild's metric holds. Therefore $g_{00\mathrm{S}} = -1/g_{11\mathrm{S}}$ and so:

$$g_{00\mathrm{S}} = 1 - \frac{2r_\mathrm{g}}{r} = 1 - \frac{2GM_\mathrm{T}(r)}{c^2 r} = 1 - 2\left(\frac{v}{c}\right)^2. \tag{15}$$

The subscript S identifies the coefficients of the Schwarzschild metric, and $M_\mathrm{T}(r) = v^2 r/G$ refers to the hypothetical isothermal total matter distribution [cf. 18] needed to explain the observed lensing, when assuming general relativity. From this it follows that the dark matter hypothesis provides a self-consistent interpretation of observed phenomenology: the same dark matter halos, which are required to explain the observed rotation curves, have been solved for by analysing extensive lensing observations.

From equation (15) it follows that for isothermal total matter halos under Einstein's general relativity, the metric coefficient $g_{00\mathrm{S}}$ does not depend on the radial coordinate. We can see this by using the empirical Tully-Fisher relation (1) between the velocity and the total baryonic mass in the last identity above. With the aid of the lens equation and making approximations to $\mathcal{O}(2)$ it follows that [17]:

$$g_{11}(r) = -1 - 2\left(\frac{v}{c}\right)^2 = -1 - \frac{2(GM_{\mathrm{b}}a_0)^{1/2}}{c^2} \tag{16}$$

Thus, any metric theory of gravity where g_{11} matches the above expression in the regime where gravitational lenses are observed will accurately reproduce all the observed lensing phenomenology, with M_{b} the total baryonic mass of the object in question (galaxies or group of galaxies), and no hypothetical dark matter assumed to exist. Equations (8) and (16) give empirical mathematical relations for the metric coefficients at perturbation order $\mathcal{O}(2)$ which reproduce all observed rotation velocity and gravitational lensing phenomenology, without the inclusion of any dark matter component.

In terms of the non-relativistic scalar potentials ϕ and ψ defined the above results yield:

$$\phi = -G_{\mathrm{M}}M^{1/2}\ln\left(\frac{r}{r_\star}\right), \quad \psi = -G_{\mathrm{M}}M^{1/2}, \quad \text{and so} \quad \gamma = 1. \tag{17}$$

As shown by [12], this is an extremely important result meaning that the relativistic structure of the underlying theory of gravity at individual, groups and cluster of galaxies scales is compatible with that found in the solar system, where the PPN parameter γ has also a value of one. The difference however, lies on the fact that the gravitational potential ϕ appearing in equation (10) is not the one associated with Kepler's third law, but the one inferred from the Tully-Fisher law.

5. Discussion

The fact that the PPN parameter $\gamma = 1$ serves as a basis for the construction of any extended theory of gravity based on the Tully-Fisher relation and/or MOND. Theories of gravity built to account for MOND phenomenology with $\gamma \neq 1$ are not compatible with weak lensing observations of individual groups and clusters of galaxies and should be disregarded.

The metric extension of gravity by [19], built with the idea of extending MOND to a relativistic regime, is an example of a theory of gravity with $\gamma = 1$. As explained by **(author?)** [13, 20] this proposal can be only applied to systems with a high degree of symmetry. A much general approach to be published elsewhere is being developed in which this requirement is not necessary.

6. Acknowledgements

This work was supported by a DGAPA-UNAM grant (IN111513-3) and the economic support from CONACyT (26344 and grant 101958).

References

[1] Bouvard A 1821 *Tables Astronomiques Publiees Par Le Bureau Des Longitudes De France, Contenant Les Tables De Jupiter, De Saturne Et D'Uranus, Construites D'Apres La Theorie De La Mecanique Celeste* (Bachelier et Huzard)

[2] Kollerstrom N Neptune's discovery: The british case for co-prediction http://www.ucl.ac.uk/sts/nk/neptune/chron.htm accessed: 2014-01-27

[3] Hsu J and Fine D 2005 *100 Years of Gravity and Accelerated Frames: The Deepest Insights of Einstein and Yang-Mills* Advanced series on theoretical physical science (World Scientific) ISBN 9789812563354

[4] Clemence G M 1947 *Rev. Mod. Phys.* **19**(4) 361–364

[5] Hernandez X and Jiménez M A 2012 *ApJ* **750** 9 (*Preprint* 1108.4021)

[6] Hernandez X, Jiménez M A and Allen C 2013 *MNRAS* **428** 3196–3205 (*Preprint* 1206.5024)

[7] Hernandez X, Jiménez M A and Allen C 2012 *European Physical Journal C* **72** 1884 (*Preprint* 1105.1873)

[8] Famaey B and McGaugh S S 2012 *Living Reviews in Relativity* **15** 10 (*Preprint* 1112.3960)

[9] Sotiriou T P and Faraoni V 2010 *Reviews of Modern Physics* **82** 451–497 (*Preprint* 0805.1726)

[10] Capozziello S and de Laurentis M 2011 *Physics Reports* **509** 167–321 (*Preprint* 1108.6266)

[11] Nojiri S and Odintsov S D 2011 *Physics Reports* **505** 59–144 (*Preprint* 1011.0544)

[12] Mendoza S and Olmo G J 2014 *ArXiv e-prints* (*Preprint* 1401.5104)

[13] Mendoza S 2015 *Canadian Journal of Physics* **93(2)** 217–231

[14] Newton I 1729 *The Mathematical Principles of Natural Philosophy* 3rd ed The Mathematical Principles of Natural Philosophy (B. Motte)

[15] Will C 1993 *Theory and Experiment in Gravitational Physics* (Cambridge University Press) ISBN 9780521439732

[16] Landau L and Lifshitz E 1975 *The classical theory of fields* Course of theoretical physics (Butterworth Heinemann) ISBN 9780750627689

[17] Mendoza S, Bernal T, Hernandez X, Hidalgo J C and Torres L A 2013 *MNRAS* **433** 1802–1812 (*Preprint* 1208.6241)

[18] Binney J and Tremaine S 2008 *Galactic Dynamics: Second Edition* (Princeton University Press)

[19] Bernal T, Capozziello S, Hidalgo J C and Mendoza S 2011 *Eur. Phys. J. C* **71** 1794 (*Preprint* 1108.5588)

[20] Mendoza S 2012 *Open Questions in Cosmology* ed Olmo D G J (InTech) ISBN 978-953-51-0880-1 available from: http://www.intechopen.com/books/open-questions-in-cosmology/extending-cosmology-the-metric-approach

Extended models of gravity in SNIa cosmological data using genetic algorithms

O. López-Corona

Instituto de Astronomía, Universidad Nacional Autónoma de México, AP 70-264, Distrito Federal 04510, México

E-mail: olopez@astro.unam.mx

Abstract. In this talk I explained briefly the advantages of using genetic algorithms on any measured data but specially astronomical ones. This kind of algorithms are not only a better computational paradigm, but they also allow for a more profound data treatment enhancing theoretical developments. As an example, I will use the SNIa cosmological data to fit the extended metric theories of gravity of Carranza et al. (2013, 2014) showing that the best parameters combination deviate from theoretical predicted ones by a minimal amount. This means that these kind of gravitational extensions are statistically robust and show that no dark matter and/or energy is required to explain the observations.

1. Introduction

A Genetic Algorithm (GA) is an evolutionary based stochastic search algorithm that in some sense mimic natural evolution. Points in the search space are considered as individuals (solution candidates), which form a population. The particular fitness of an individual is a number, indicating their quality for the problem at hand. As in nature, GAs include a set of fundamental genetic operations that work on the genotype (solution candidate codification). A concrete GA is specified by a particular mutation, recombination, or selection operators [1].

These algorithms operate with a population of individuals $P(t) = x_1^t, ..., x_N^t$, for t iteration, where the fitness of each x_i individual is evaluated according to a set of objective functions $f_j(x_i)$. This objectives functions allows to order from best to worst individuals of the population in a continuum of degrees of adaptation. Then, individuals with higher fitness, recombine their genotypes to form the gene pool of the next generation, in which random mutations are also introduced to produce new variability.

A fundamental advantage of GA versus traditional methods is that GAs solve discrete, nonconvex, discontinuous, and non-smooth problems successfully [2], and thus they have been widely used in Ecology, Natural Resources Management, among other fields, but not so much in astrophysics. Nevertheless they have been recently used by Nesseris (2011) for parameter search in ΛCDM models with SNIa data [3].

When Zwicky applied the virial theorem to the cluster of galaxies COMA, he obtained some results suggesting the existence of additional mass that was not visible. Based on the motions of galaxies near its edge, Zwicky estimated the total mass of the cluster should be about 400 times more than expected. The gravity of the visible galaxies in the cluster was very little to

Journal of Physics: Conference Series **600** (2015) 012046 doi:10.1088/1742-6596/600/1/012046

account for the actually observed galaxy speed, so it should be much more. This is known as the "missing mass problem".

In the 1960s and 1970s, there were measurements of the rotation curves of spiral galaxies with a higher degree of accuracy than anything previously achieved before, that supported Zwicky findings. These results showed that many stars in different orbits of spiral galaxies rotate at about the same angular velocity, implying that their densities were very uniform beyond the location of many of the stars (the galactic bulge). This result suggests that even Newtonian gravity does not apply universally or that, conservatively, over 50% of the mass of galaxies was contained in the relatively dark galactic halo. This non-baryonic interacting matter was termed as Dark Matter [4–6].

A few years later it was discovered that the Universe is undergoing an accelerated expansion [7–9]. This acceleration is usually attributed either to a cosmic fluid with negative pressure dubbed Dark Energy or to an IR modification of gravity. As pointed out by Mendoza (2014), it seems that ideas of introducing occult entities to describe physical phenomena reoccurs from time to time.

Modified theories of gravity have received increased attention lately due to combined motivation coming from high-energy physics, cosmology and astrophysics. Among numerous alternatives to Einsteins theory of gravity, theories which include higher order curvature invariants, and specifically the particular class of $F(R)$ theories, have a long history [14]. (see e.g. Mariana Espinosa talk about the Gravity Apple Tree and the many theories of gravity developed after General Relativity)

Bernal et. al [10] and Mendoza [11] have shown that a useful way to write the gravitational actions S_g is given by:

$$S_g = -\frac{c^3}{16\pi G}\int d^4x\sqrt{-g}\frac{f(\chi)}{L^2} \tag{1}$$

where the dimensionless Ricci scalar ir given by $\chi := L^2 R$, for the Ricci scalar R. In the previous equation, c is the speed of light, G Newton's gravitational constant and g is the determinant of the space-time metricc. Note that the function $f(\chi) = \chi$ recovers general relativity and for the case of $f(\chi) \neq \chi$ the "coupling factor" with dimensions of length L makes that equation dimensionally correct.

As explained in Mendoza (2014) there are two cases that can yield a MONDian behaviour:

$$f(\chi) = \begin{cases} \chi^{\frac{3}{2}} \\ \chi^{-3} \end{cases} \tag{2}$$

both reproduce flat rotational curves and the Tully-Fisher relation and show the fact that the energy-momentum tensor - or the matter Langrangian - is somewhat embedded into the gravitational action. We only on the description based on $f(R,T)$ by Harko et al. and covered by Carranza et al. [13], but also see Mendoza [11] and the proceedings by S. Mendoza in this collection.

It has been shown that the interpretation of the SNIa data depends greatly on the type of parametrization used to perform a data fit. For this reason, choosing a priori a model can thus adversely affect the validity of the fitting method and lead to compromised or misleading results. Then a very important complication in the investigation of the behavior of dark energy or for that matters, its counterpart of extended models of gravity, occurs due to the bias introduced by the parameterizations used [3].

2. Methods

We conceptualized the parameter calibration as an optimization problem and propose to resolve it using both Monte Carlo (MC) techniques and Genetic Algorithms.

As it is well known from Taylor series, any (normal) function may be approximated using a polynomial of the correct order. Of course, even when this is mathematical correct no one would consider that a polynomial is a universal model for every physical phenomenon. In that line of thought, one may fit any data using a model with many free parameters, and even when this model may have a great performance in a statistical sense, it could be at the same time an incorrect model in the physical perspective.

In these sense, an important question is, How much better must a complex (more parametrers) model fit before we say that the extra parameter is necessary? or in a more straight forward, How do we trade off fit with simplicity? That has been the motivation in recent year for new model selection criteria developement in statistics, all of which define simplicity in terms of the paucity of parameters, or the dimension of a model (see Forster and Sober, 1994, for a non-technical introduction). These criteria include Akaikes Information Criterion (AIC) [15, 16], the Bayesian Information Criterion (BIC) [17], and Minimum Description Length (MDL) [18]. They trade off simplicity and fit a little differently, but all of them address the same problem as significance testing: Which of the estimated "curves" from competing models best represent reality? This work has led to a clear understanding of why this form of simplicity is relevant that question [19].

Akaike (Akaike-74,Akaike-85) has shown that choosing the model with the lowest expected information loss (i.e., the model that minimizes the expected Kullback Leibler discrepancy) is asymptotically equivalent to choosing a model M_j, from a set of models $i = 1, 2, ..., k$ that has the lowest AIC value:

$$AIC = -2Log(\mathcal{L}_j) + 2V_j, \tag{3}$$

where $\mathcal{L}_j$, is the maximum likelihood for the candidate model and is determined by adjusting the V_j free parameters in such a way as to maximize the probability that the candidate model has generated the observed data. This equation shows that AIC rewards descriptive accuracy via the maximum likelihood , and penalizes lack of parsimony according to the number of free parameters (note that models with smaller AIC values are to be preferred). In that sense, Akaike extended this paradigm by considering a framework in which the model dimension is also unknown, and must therefore be determined from the data. Thus, Akaike proposed a framework wherein both model estimation and selection could be simultaneously accomplished. For those reasons, AIC is generally regarded as the first, most widely known and used model selection tool.

Taking as objective function the AIC, we performed a Monte Carlo optimization process using an R script. For the GA analysis we used a modified versin of Kumara Sastry code in C++ [20]. For both Monte Carlo and GA, we take averages over 1000 independent sets of parameters each one with 1000 iterations or generation respectively.

3. Results

The results are quite straightforward and show consistently that GA are much more better that the other parameter search techniques used. For example, GA for $\chi^{3/2}$ is of the order of forty percent better than Monte Carlo. In Fig.1 we see data from SNIa, predicted values of the model using theoretical set of parameters in green, values obtained using MC parameter search in blue, and values obtained using GA as search technique in red. This performance was accomplished by desegregating the original data into three subsets and defining an independent AIC objective function for each one, so we make sure that differences in data density over z values doest interfere with parameter estimation.

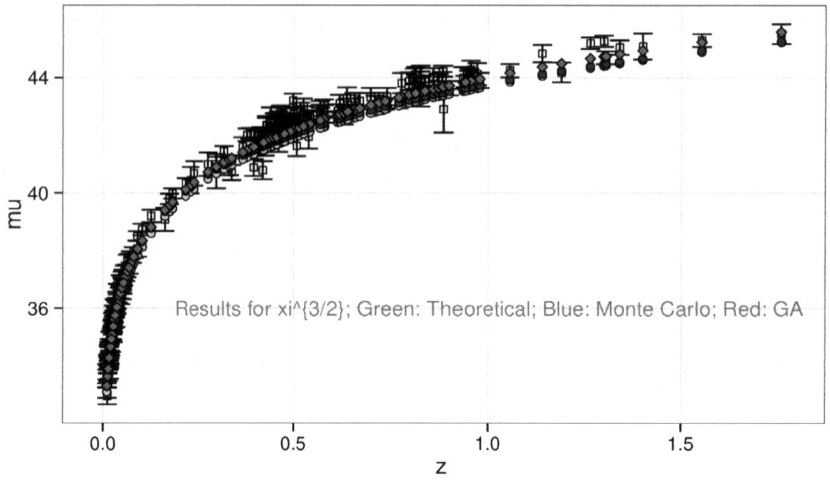

Figure 1. Comparative of optimization techniques for $\chi^{3/2}$ model. Predicted values of the model using theoretical set of parameters in green, values obtained using MC parameter search in blue, and values obtained using GA as search technique in red.

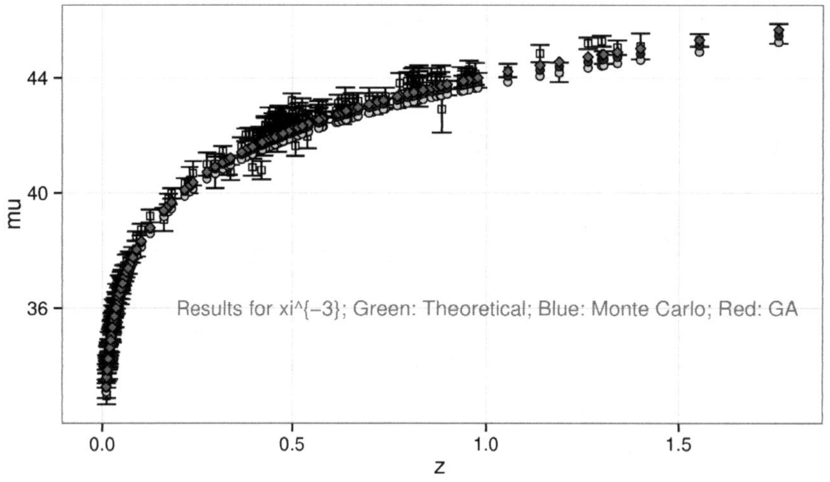

Figure 2. Comparative of optimization techniques for χ^{-3} model. Predicted values of the model using theoretical set of parameters in green, values obtained using MC parameter search in blue, and values obtained using GA as search technique in red.

As shown in Fig.2, χ^{-3} we have the same result, with GA as the best parameter optimization option. It is important to point out that the estimation $b = -3.0001 \pm 0.0015$, implies that the Hubble constant $h = 0.6919 \pm 0.0035$ by the GA is consistent with standard values used in the concordance ΛCDM model.

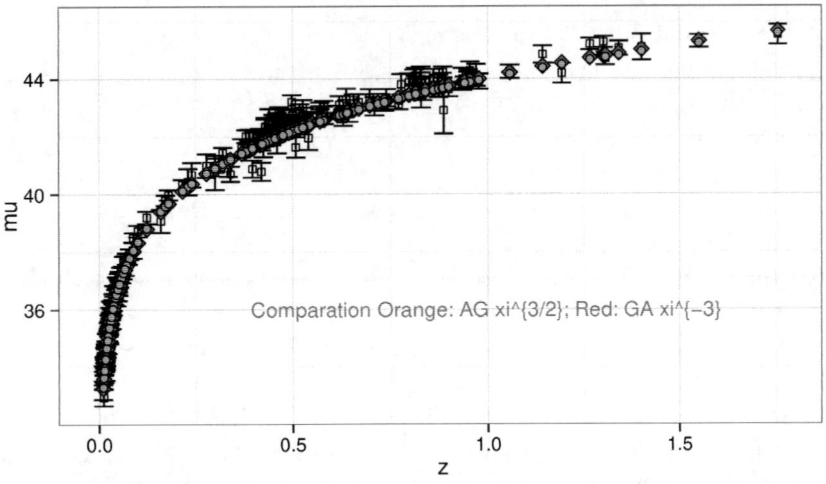

Figure 3. Comparative results of $\chi^{3/2}$ and χ^{-3} model. In this range of redshift z, $\chi^{3/2}$ and χ^{-3} are almost indistinguishable

Comparing χ^{-3} and $\chi^{3/2}$, it turns out that χ^{-3} result better but just for a 1.3% (Fig.3). In this range of z, $\chi^{3/2}$ and χ^{-3} are almost indistinguishable but $\chi^{3/2}$ is base in a punctual concept mass (dust type of models) meanwhile χ^{-3} use a density concept, allowing to apply it to more astrophysical systems than the former.

4. Discussion and Conclusions

Genetic Algorithms (GA) have proven to be an efficient technique for parameter estimation in cosmological data, but as proposed models performance are almost the same with the optimized set of parameters. Which is then the best model? In order to respond this critical question we need to work with data in a far as $z \geq 10$ (perhaps using gamma ray data) where the differences between models are expected to be higher.

Another path to overcome this statistical indistinguishability is to incorporate to the current data driven criteria a physics driven one, for model selection. This new critiria could be based on information-thermodynamical principles; and use a muti-data optimization process (which include different z regimes and observed objects); or a mix of multi-data and multi-criteria optimization process, that would allow us to really exploit the real power of GA by incorporating a multi-objective optimization approach. The point is to have a set of various databases representing different aspects or regimes of a physical phenomena and then estimate the parameters of a model minimizing at the same time

Finally, with this results in mind we are now in position to compare the χ^{-3} model with the standard ΛCDM model.

Acknowledgments

This work was supported by a DGAPA-UNAM posdoctoral fellowship at Instituto de Astronomía and the grant IN111513-3; it was also supported by CONACyT-SNI 62929 (and grant CB-2009-01, #132400)

References

Mitchell, M. (1998). An introduction to genetic algorithms. MIT press.

López-Corona, O., Padilla, P., Escolero, O., Armas, F., García-Arrazola, R. and Esparza, R. (2013), Playing with models and optimization to overcome the tragedy of the commons in groundwater. Complexity, 19: 921. doi: 10.1002/cplx.21462

Savvas Nesseris 2011 J. Phys.: Conf. Ser. 283 012025 doi:10.1088/1742-6596/283/1/012025

Sanders, R. H. (2010). The dark matter problem: a historical perspective. Cambridge University Press.

Bahcall, J. N., Piran, T., Weinberg, S. (Eds.). (2004). Dark matter in the universe. World Scientific.

Trimble, V. (1987). "Existence and nature of dark matter in the universe". Annual Review of Astronomy and Astrophysics 25: 425472.

Riess A. G. et al. [Supernova Search Team Collaboration], 2004, ApJ,607, 665.

Spergel D. N. et al. [WMAP Collaboration], 2007, ApJ.Sup.,170, 377, [arXiv:astro-ph/0603449].

Readhead A. C. S. et al., 2004, ApJ.,609, 498, [arXiv: astro-ph/0402359].

Bernal, T., Capozziello, S., Hidalgo, J. C., Mendoza, S. (2011). Recovering MOND from extended metric theories of gravity. The European Physical Journal C-Particles and Fields, 71(11), 1-7.

Mendoza, S., Bernal, T., Hernandez, X., Hidalgo, J. C., Torres, L. A. (2013). Gravitational lensing with f ()= 3/2 gravity in accordance with astrophysical observations. Monthly Notices of the Royal Astronomical Society, stt752.

Mendoza, S. (2014). MOND as the basis for an extended theory of gravity. Canadian Journal of Physics, (ja).

Carranza, D. A., Mendoza, S., Torres, L. A. (2013). A cosmological dust model with extended f () gravity. The European Physical Journal C, 73(1), 2282.

Sotiriou, T. P., Faraoni, V. (2010). f (R) theories of gravity. Reviews of Modern Physics, 82(1), 451.

Akaike, H. (1974): "A New Look at the Statistical Model Identification." IEEE Transactions on Automatic Control, vol. AC-19: 716-23.

Akaike, H. (1985): "Prediction and Entropy." In A. C. Atkinson and S. E. Fienberg (eds.), A Celebration of Statistics. New York: Springer. 1-24.

Schwarz, Gideon (1978): "Estimating the Dimension of a Model." Annals of Statistics 6: 461-5.

Rissanen, Jorma (1989): Stochastic Complexity in Statistical Inquiry. Singapore: World Books

Forster, Malcolm R. and Elliott Sober (1994): "How to Tell when Simpler, More Unified, or Less Ad Hoc Theories will Provide More Accurate Predictions." British Journal for the Philosophy of Science 45: 1 - 35

Sastry, K. Single and Multiobjective Genetic Algorithm Toolbox in C++; IlliGAL Report No. 2007016, 2007, retrived on Jul 26 2013 from http://www.kumarasastry.com/2007/06/11/single-and-multiobjective-genetic-algorithm-toolbox-in-

Extended Theories of Gravity with Generalized Energy Conditions

J P Mimoso*[1,2], F S N Lobo[1,2] and S Capozziello[3]

[1] Departamento de Física, Faculdade de Ciências da Universidade de Lisboa
Faculdade de Ciências da Universidade de Lisboa, Edifício C8, Campo Grande, P-1749-016 Lisbon, Portugal
[2] Instituto de Astrofísica e Ciências do Espaço, Faculdade de Ciências da Universidade de Lisboa, Edifício C8, Campo Grande, P-1749-016 Lisbon, Portugal
[3] Dipartimento di Fisica, Università di Napoli "Federico II", Napoli, Italy INFN Sez. di Napoli, Compl. Univ. di Monte S. Angelo, Edificio G, Via Cinthia, I-80126, Napoli, Italy.

E-mail: [1]jpmimoso@fc.ul.pt, [2]fslobo@fc.ul.pt, [3]capozzie@na.infn.it

Abstract. We address the problem of the energy conditions in modified gravity taking into account the additional degrees of freedom related to scalar fields and curvature invariants. The latter are usually interpreted as generalized *geometrical fluids* that differ in meaning with respect to the matter fluids generally considered as sources of the field equations. In extended gravity theories the curvature terms are encapsulated in a tensor H^{ab} and a coupling $g(\Psi^i)$ that can be recast as effective Einstein field equations, with corrections to the energy-momentum tensor of matter. The formal validity of standard energy inequalities does not assure basic requirements such as the attractive nature of gravity, so we argue that the energy conditions have to be considered in a wider sense.

1. Introduction

Cosmological observations lead to the introduction of additional ad-hoc concepts like Dark Energy and Dark Matter within the standard Einstein theory. However the evasive nature of these components may be interpreted as a possible signal of a breakdown of GR on large, infrared scales [1]. In such a way, modifications and extensions of GR become a natural alternative if such "dark" elements are not found out. Adopting this viewpoint, several recent works focussed on the cosmological implications of alternative gravity [2, 3, 4] since such models may lead to the explanation of the acceleration effect observed in cosmology [5, 6] and to the explanation of the missing matter puzzle observed at astrophysical scales [7].

When one considers modified theories of gravity, at least for a large class of interesting cases, the generalized field equations can be cast in the following form

$$g(\Psi^i)\ (G_{ab} + H_{ab}) = 8\pi G\, T_{ab}\ , \tag{1}$$

where H_{ab} is an additional geometrical term with regard to GR that encapsulates the geometrical modifications introduced by the modified theory, and $g(\Psi^i)$ is a factor that modifies the coupling with the matter fields in T^{ab}, where Ψ^i generically represents either curvature invariants or other gravitational fields contributing to the dynamics. In the latter expression, as usual,

$G_{ab} \equiv R_{ab} - \frac{1}{2} g_{ab} R$ is the Einstein tensor, defined from the Ricci tensor R_{ab} and its trace $R = R^a{}_a$, the curvature scalar. The field equations govern the interplay between the geometry of the spacetime and the matter content. General Relativity (GR) is recovered for $H_{ab} = 0$ and $g(\Psi^i) = 1$.

In the Einstein field equations, $G_{ab} = 8\pi G\, T_{ab}$, there is a clear separation between the left-hand side that corresponds to the geometry, and the right-hand side where one finds the energy-matter distribution. The underlying idea is that the matter-energy distribution determine the spacetime curvature, and hence how gravity acts, and it follows that any conditions that we impose on T_{ab} immediately translate into corresponding conditions on the Einstein tensor G_{ab} [8]. In this sense, the causal and geodesic structures of space-time are determined by the matter-energy distribution. In this context, the energy conditions guarantee that the causality principle is respected and suitable physical *sources* have to be considered [8]. However, in the extended theories (1) not only does the additional term H_{ab} modify the relation between the geometry of gravity and matter, but also the dynamical coupling conveyed by $g(\Psi^i)$ interferes in that interplay.

In this work, we address the problem of the energy conditions in modified gravity [9]. This issue is extremely delicate since a standard approach is to consider the gravitational field equations as effective Einstein equations. More specifically, the further degrees of freedom carried by these theories [10] can be recast as generalized *geometrical fluids* that have different meanings with respect to the standard matter fluids generally adopted as sources of the field equations [4]. While standard fluids generally obey standard equations of states, these "fictitious" fluids can be related to scalar fields or further gravitational degrees of freedom, as in $f(R)$ gravity. In these cases, the physical properties may be ill-defined and the consequences can be dramatic, since the causal and geodesic structures of the theory could present serious shortcomings as well as the energy-momentum tensor could not be consistent with the Bianchi identities and the conservation laws. Thus, we add a cautionary note of the results obtained in the literature [11]. In this work, we adopt the $(-+++)$ signature and $c = 1$.

2. Energy conditions in modified theories of gravity.

The definition of the energy conditions entails an arbitrary flow which represents a generic observer or a reference frame. In general, we consider a congruence of timelike curves whose tangent 4-vector W^a represents the velocity vector of a family of observers. Thus, the energy conditions emerge directly from the geodesic structure of the space-time. More specifically, consider the *Raychaudhuri equation*, given by [12]

$$\dot{\theta} + \frac{\theta^2}{3} + 2\,(\sigma^2 - \omega^2) - \dot{W}^a{}_{;a} = -R_{ab}\, W^a W^b\,. \tag{2}$$

where $2\sigma^2$ is the square of the *shear tensor*, θ is the *expansion scalar* and ω^2 is the square of the vorticity tensor. This Eq. (2) carries only a geometrical meaning, as the quantities in it are directly derived from the Ricci identities. It is only when we choose a particular theory that we establish a relation between $R_{ab}\, W^a W^b$ in Eq. (2), and the energy-momentum tensor describing matter fields [8]. One may also consider a null congruence k^a and a vanishing vorticity $\omega_{ab} = 0$, which means that, in GR, it is possible to associate the null energy condition with the focusing (attracting) characteristic of the spacetime geometry.

Taking into account the diffeomorphism invariance of the matter action, the covariant conservation of the energy-momentum tensor, $\nabla_a T^{ab} = 0$, is obtained. Thus, from the contracted Bianchi identities, we derive the following conservation law

$$\nabla_b H^{ab} = -\frac{8\pi G}{g^2} T^{ab}\, \nabla_b g\,. \tag{3}$$

Journal of Physics: Conference Series **600** (2015) 012047 doi:10.1088/1742-6596/600/1/012047

The fact that H^{ab} is a geometrical quantity, in the sense that it can be given by geometrical invariants or scalar fields different from ordinary matter fields, implies that the imposition of a specific energy condition on T^{ab} carries an implication for the combination of G_{ab} with H_{ab} and not just for the Einstein tensor. So we cannot obtain a simple geometrical implication from it, as in GR. For instance, if we assume that the strong energy condition, $T_{ab}\,W^aW^b \geq \frac{1}{2}T\,W^aW_a$ holds, in GR it would mean, on the one hand, that $R_{ab}\,W^aW^b \geq 0$ and, on the other hand, given Eq. (2), that the geodesics are focusing, and gravity possesses an attractive character. This is one of the assumptions of the singularity theorems of Hawking and Penrose [8]. However in the modified gravity context under consideration, this condition just states that

$$g(\Psi^i)\,(R_{ab} + H_{ab} - \frac{1}{2}g_{ab}H)\,W^aW_a \geq 0\,, \tag{4}$$

which does not necessarily imply $R_{ab}\,W^aW^b \geq 0$ and hence we cannot straightforwardly conclude that the satisfaction of the strong energy condition (SEC) is synonymous of the attractive nature of gravity in the particular modified theory of gravity under consideration.

In the literature, the term H^{ab} is usually treated as a correction to the energy-momentum tensor, so that the meaning which is attributed to the energy conditions is the satisfaction of a specific inequality using the combined quantity $T^{ab}_{\text{eff}} = T^{ab}/g - H^{ab}$. It is thus misleading to associate this effective energy-momentum tensor to the energy conditions, since they do not emerge only from T^{ab} but from the geometrical quantity H^{ab}, which is considered as an additional energy-momentum tensor. However, if the modified theory of gravity under consideration allows an equivalent description upon an appropriate conformal transformation, it then becomes justified to associate the transformed H^{ab} to the redefined T^{ab} in the conformally transformed Einstein frame. Several generalized theories of gravity can be redefined as GR plus a number of appropriate fields coupled to matter by means of a conformal transformation in the so-called Einstein frame. This is, for instance, the case for scalar-tensor gravity theories, for $f(R)$ gravity, etc [4].

In the original Jordan frame one has a separation between geometrical terms and standard matter terms that can be cast as in (1), where H_{ab} involves a mixture of both the scalar and tensor gravitational fields. Upon a suitable conformal transformation we are able to cast the field equations as $\tilde{G}_{ab} = 8\pi G\,\tilde{T}^{\text{eff}}_{ab}$, where $\tilde{T}^{\text{eff}}_{ab} = \tilde{T}^{M}_{ab} + \tilde{T}^{\varphi}_{ab}$, and it thus makes sense to consider $\tilde{T}^{\text{eff}}_{ab}$ as an effective energy-momentum tensor, where $\tilde{T}^{M}_{ab}$ is the transformed energy-momentum of matter, and $\tilde{T}^{\varphi}_{ab}$ is an energy-momentum tensor for the redefined scalar field φ which is coupled to matter. Then conclusions about the properties of $\tilde{G}_{ab}$ such as whether it focuses geodesics directly from those conditions holding on $\tilde{T}^{\text{eff}}_{ab}$, ignore the fact that H_{ab} originally possesses a geometrical character, and may be too hasty if not supported by the physical analysis of the sources.

If we assume that in this frame the effective energy-momentum tensor $\tilde{T}^{\text{eff}}_{ab}$ satisfies some energy condition, for instance, the null energy condition (NEC), this implies that $\tilde{G}_{ab}$ has to satisfy such a condition. Thus, it is possible to write the Raychaudhuri equation as

$$\frac{\mathrm{d}\tilde{\theta}}{\mathrm{d}v} = -\left[\frac{\tilde{\theta}^2}{3} + 2\tilde{\sigma}^2 + \tilde{R}_{ab}\tilde{k}^a\tilde{k}^b\right]\,, \tag{5}$$

which enables us to conclude on the attractive/repulsive character of the given theory of gravity in the Einstein frame. Reversing the conformal transformation, we can assess, in principle, what happens in the original frame, namely, the Jordan frame. This operation requires to know how the kinematical quantities, present in Eq. (5), transform under a conformal transformation. This means that if $g_{ab} \to \tilde{g}_{ab} = \Omega^2\,g_{ab}$ and $W^a \to \tilde{W}^a = \Omega^{-1}W^a$, we have $\tilde{\nabla}_a\tilde{W}_b = \Omega\,\nabla_a W_b + \Omega\,\gamma^c{}_{ab}W_c + W_b\,\nabla_a\Omega$, where $\gamma^c{}_{ab} = \delta^c_a\partial_b\Omega/\Omega + \delta^c_b\partial_a\Omega/\Omega - g_{ab}\partial^c\Omega/\Omega$.

From this result, it follows that we can pass from the Einstein to the Jordan frame by the following transformations $\tilde{\theta}_{ab} = \Omega\,(\theta_{ab} - \dot{\Omega}\,h_{ab})$, $\tilde{\sigma}_{ab} = \Omega\,\sigma_{ab}$, $\tilde{\omega}_{ab} = \Omega\,\omega_{ab}$, $\tilde{\theta} = \Omega^{-1}\,(\theta - 3\dot{\Omega})$, respectively. Thus, Eq. (5) can finally be written as $\frac{d\tilde{\theta}}{dv} = \frac{\dot{\theta}}{\Omega^2} - \frac{\theta}{\Omega^2}\frac{\dot{\Omega}}{\Omega} - \frac{3}{\Omega}\,(\ln\Omega)^{\cdot\cdot}$. The latter result shows that whereas, in the Einstein frame, the NEC implies the attractive nature of gravity, a similar implication does not necessarily follow in the Jordan frame. In fact, $d\tilde{\theta}/dv \leq 0$ only implies that $\dot{\theta} \leq \frac{\dot{\Omega}}{\Omega}\theta + 3\Omega\,(\ln\Omega)^{\cdot\cdot}$, and thus it depends on the sign of the term on the right-hand side of the inequality. On the other hand, we see that $\tilde{R}_{ab}\tilde{k}^a\tilde{k}^b \geq 0$ does not necessarily entail $R_{ab}k^a k^b \geq 0$. What we do indeed obtain is

$$\left(\Omega^{-2}R_{ab} + 2\nabla_a\nabla_b\ln\Omega + 2\nabla_a\ln\Omega\,\nabla_b\ln\Omega\right)\,k^a k^b \geq 0\,. \tag{6}$$

This emphasizes that if, in one of the conformally related frames, we have attractive gravity (due to the NEC), in the other frame neither the NEC is simultaneously satisfied, nor, in case it is, this means that gravity will be straightforwardly attractive. This fact could be extremely relevant in view of identifying a physical meaning of conformal transformations [4, 13, 14].

3. Example of a modified theory of gravity: Scalar-tensor gravity.

Consider scalar-tensor gravity [15] given by the action

$$S = \frac{1}{16\pi}\int\sqrt{-g}d^4x\,\left[\phi R - \frac{\omega(\phi)}{\phi}\phi_{,a}\phi^{,a} + 2\phi\lambda(\phi)\right] + S_M\,, \tag{7}$$

where S_M is the standard matter part, the gravitational coupling is assumed variable and a self-interaction potential is present. Varying this action with respect to the metric g_{ab} and the scalar field ϕ yields the field equations (1), with H_{ab} given by

$$H_{ab} \;=\; -\frac{\omega(\phi)}{\phi^2}\left[\phi_{;a}\phi_{;b} - \frac{1}{2}g_{ab}\,\phi_{;c}\phi^{;c}\right] - \frac{1}{\phi}\left[\phi_{;ab} - g_{ab}\phi_{;c}{}^{;c}\right] - \lambda(\phi)g_{ab}\,, \tag{8}$$

and $g(\Psi^i) = \phi$, which we shall assume positive, and

$$\Box\phi + \frac{2\phi^2\lambda'(\phi) - 2\phi\lambda(\phi)}{2\omega(\phi)+3} = \frac{1}{2\omega(\phi)+3}\,\left[8\pi GT - \omega'(\phi)\phi_{;c}\phi^{;c}\right]\,, \tag{9}$$

where $T \equiv T^c{}_c$ is the trace of the matter energy-momentum tensor and $G \equiv \frac{2\omega+4}{2\omega+3}$ is the gravitational constant normalized to the Newton value. One also requires the conservation of the matter content $\nabla^a T_{ab} = 0$, to preserve the equivalence principle. The archetype Brans-Dicke theory is characterized by the restriction of $\omega(\phi)$ being a constant, and of $\lambda = \lambda' = 0$.

The above considerations on the energy conditions apply straightforwardly. In particular, Eq. (4) is easily recovered like the other energy conditions. Since we assume $\phi > 0$, we see that the condition $R_{ab}\,W^a\,W^b \geq 0$ becomes

$$(T_{ab} - \frac{1}{2}g_{ab}\,T)\,W^a W^b \geq \phi\,(H_{ab} - \frac{1}{2}g_{ab}\,H)\,W^a W^b. \tag{10}$$

and allows for the focusing of the time-like paths even when a mild violation of the energy condition occurs. Matter may exhibit unusual thermodynamical features, e.g. including negative pressures, and yet gravity remains attractive. Alternatively, repulsive gravity may occur for common matter, i.e., for matter that satisfies all the energy conditions.

In close analogy with the decomposition of the energy-momentum tensor with respect to the vector field W^a [8], one may consider the following the decomposition of the tensor H_{ab} into the parallel and orthogonal components to the time-like vector flow W^a is given by

$$\begin{aligned} H^{ab} &= H_{||}W^aW^b + H_{\perp}h^{ab} + 2\,H_{\perp}^{(a}\,W^{b)} + H_{\perp}^{<ab>} \\ &= \frac{1}{\phi}\left[\tilde{\rho}W^aW^b + \tilde{p}h^{ab} + 2\,\tilde{q}^{(a}\,W^{b)} + \tilde{\pi}^{ab}\right]. \end{aligned} \tag{11}$$

where $H_{||} = H_{ab}\,W^aW^b$, $3H_{\perp} = H_{ab}\,h^{ab}$, $H_{\perp}^{<ab>} = \left(h^{ac}h^{bd} - \frac{1}{3}h^{ab}h^{cd}\right)H_{cd}$, and $H_{\perp}^{a} = W^c\,g\,H_{cd}\,h^{ad}$, where $H_{||}$ and $H_{\perp}$ are scalars, $H_{\perp}^{a}$ is a vector and $H_{\perp}^{<ab>}$ is a projected trace-free symmetric tensor. Thus, the inequality (10) may be written as $(\rho+3p)/\phi - (H_{||}+3H_{\perp}) \geq 0$, where we have used the definitions

$$H_{||} = -\frac{\omega(\phi)}{2\phi^2}\left(3\dot{\phi}^2 - h^{cd}\,\nabla_c\phi\,\nabla_c\phi\right) - \frac{1}{\phi}\,h^{cd}\nabla_c\nabla_d\phi + \lambda(\phi)\,, \tag{12}$$

$$H_{\perp} = -\frac{\omega(\phi)}{3\phi^2}\left(\frac{\dot{\phi}^2}{2} - \frac{1}{2}h^{cd}\,\nabla_c\phi\,\nabla_c\phi\right) - \frac{1}{2\phi}\left(W^aW^b\,\nabla_c\nabla_d\phi - \frac{1}{3}\,h^{cd}\nabla_c\nabla_d\phi\right) - \lambda(\phi). \tag{13}$$

Thus, $\omega(\phi)$ and $\lambda(\phi)$ define whether gravity is attractive or repulsive in the scalar-tensor cosmological models. On the other hand, performing $g_{ab} \to \bar{g}_{ab} = (\phi/\phi_*)\,g_{ab}$, the condition for gravity to be attractive with the redefined Ricci tensor becomes

$$\tilde{R}_{ab}W^aW^b = \frac{4\pi}{\phi_*}\left(\bar{\rho}+3\bar{p}\right) + \frac{8\pi}{\phi_*}\left[\dot{\varphi}^2 - \tilde{V}(\varphi)\right] \geq 0\,, \tag{14}$$

where $\varphi = \int\sqrt{(2\omega+3)/2}\ \mathrm{d}\ln\phi$ is the redefined scalar field, $V(\varphi) = \lambda(\phi(\varphi))/\phi(\varphi)$ is the rescaled potential, and $\bar{\rho} = \rho/\phi^2$, $\bar{p} = p/\phi^2$. The role of the functions $\omega(\phi)$ and $\lambda(\phi)$ underlies the result because the definitions of φ and $V(\varphi)$ depend on them. In addition, in the Einstein frame, the matter and the scalar field are interacting with each other as revealed by the scalar field equation

$$\ddot{\varphi} + \bar{\theta}\dot{\varphi} = -\frac{\partial V(\varphi)}{\partial\varphi} - \frac{\partial\bar{p}(\varphi,\bar{a})}{\partial\varphi}\,. \tag{15}$$

Thus, the dependence of the self-interacting potential $V(\varphi)$, and the coupling $\partial_\varphi\bar{\rho} \propto \alpha(\varphi)a^{-3\gamma}$ is important, where $\alpha = (\sqrt{2\omega+3})^{-1}$, when considering a perfect fluid with $\bar{p} = (\gamma-1)\bar{\rho}$. In a cosmological setting, the interplay of the intervening components such that those which violate the SEC dominate imply that gravity exhibits a transition from being attractive into becoming repulsive. This feature is relevant in view of dark energy.

4. Discussion and conclusions.

We have discussed the formulation and the meaning of the energy conditions in the context of modified theories of gravity. The procedure consists in disentangling the further degrees of freedom that emerges with respect to GR and in grouping them as an effective energy-momentum tensor of the form $T^{ab}/g - H^{ab}$ where $g(\Psi^i)$ is the effective coupling and H^{ab} the contribution due to scalar fields and/or curvature invariants of the given modified theory of gravity. Formally, the weak, null, dominant and strong energy conditions can be rewritten as in GR. Despite of this analogy, their meaning can be totally different with respect to GR since the causal structure, geodesic structure and gravitational interaction may be altered.

A main role in this analysis is played by recasting the theory in the Einstein frame. However, the energy conditions can assume a completely different meaning going back to the Jordan frame and then they could play a crucial role in identifying the physical frame as firstly pointed out in [14]. On the other hand, geometrical implications change in the two frames since optical scalars like σ, θ and ω can give rise to the convergence or divergence of geodesics. This means that the physical meaning of a given extended theory strictly depends on the energy conditions and initial conditions (in relation to the choice of the source [16]). From an observational point of view, this fact could constitute a formidable tool to test the dark components since deviations from standard GR could be put in evidence.

Acknowledgments
SC acknowledges the INFN (*iniziative specifiche* TEONGRAV and QGSKY). FSNL is supported by a Fundação para a Ciência e Tecnologia Investigador FCT Research contract, with reference IF/00859/2012, funded by FCT/MCTES (Portugal). FSNL and JPM thank FCT's grants CERN/FP/123618/201, EXPL/FIS- AST/1608/2013, and PEst-OE/FIS/UI2751/2014.

References

[1] S. Capozziello, V.F. Cardone and A. Troisi, 2006 *JCAP* **0608**, 001; S. Capozziello, V.F. Cardone and A. Troisi, 2007 *MNRAS* **375**, 1423
[2] H. Weyl, 1918 *Math. Zeit.* **2**, 384; W. Pauli, 1919 *Phys. Zeit.* **20**, 457; H. Weyl, 1921 *Raum-Zeit-Materie*, Springer Berlin; R. Bach, 1921 *Math. Zeit.* **9**, 110; C. Lanczos, 1931 *Z. Phys.* **73**, 147
[3] H-J. Schmidt, 2004 *Lectures in Mathematical Cosmology*, arXiv: gr-qc/0407095; S. Nojiri and S.D. Odintsov, 2007 *Int. Jou. Geom. Meth. Mod. Phys.* **4**, 115 S. Capozziello and M. Francaviglia, 2008 *Gen. Rel. Grav.* **40**, 357
[4] S. Capozziello and M. De Laurentis, 2011 *Phys. Rept.* **509**, 167
[5] S. Capozziello, 2002 *Int. Jou. Mod. Phys. D* **11**, 483
[6] S. Nojiri and S.D. Odintsov, 2003 *Phys. Lett. B* **576**, 5; S. Nojiri and S.D. Odintsov, 2003 *Phys. Rev. D* **68**, 123512; D.N. Vollik, 2003 *Phys. Rev. D* **68**, 063510; S.M. Carroll, A. de Felice, V. Duvvuri, D.A. Easson, M. Trodden and M.S. Turner, 2005 *Phys. Rev. D* **71**, 063513; T. Harko, T. S. Koivisto, F. S. N. Lobo and G. J. Olmo, 2012 *Phys. Rev. D* **85**, 084016; S. Capozziello, T. Harko, T. S. Koivisto, F. S. N. Lobo and G. J. Olmo, 2013 *JCAP* **1304**, 011 S. Capozziello, T. Harko, F. S. N. Lobo and G. J. Olmo, 2013 *Int. J. Mod. Phys. D* **22**, 1342006; T. Harko and F. S. N. Lobo, 2010 *Eur. Phys. J. C* **70**, 373 T. Harko, F. S. N. Lobo, S. 'i. Nojiri and S. D. Odintsov, 2011 *Phys. Rev. D* **84**, 024020 T. Harko and F. S. N. Lobo, 2012 *Int. J. Mod. Phys. D* **21**, 1242019 Z. Haghani, T. Harko, F. S. N. Lobo, H. R. Sepangi and S. Shahidi, 2013 *Phys. Rev. D* **88**, 044023 S. D. Odintsov and D. Sáez-Gómez, 2013 *Phys. Lett. B* **725**, 437
[7] S. Capozziello, M. De Laurentis, 2012 *Annalen Phys.* **524**, 545; S. Capozziello, V. F. Cardone and A. Troisi, 2006 *JCAP* **0608**, 001; S. Capozziello, V. F. Cardone and A. Troisi, 2007 *Mon. Not. R. Astron. Soc.* **375**, 1423; A. Borowiec, W. Godlowski and M. Szydlowski, 2007 *Int. J. Geom. Meth. Mod. Phys.* **4** 183; C. F. Martins and P. Salucci, 2007 *Mon. Not. Roy. Astron. Soc.* **381**, 1103; C. G. Boehmer, T. Harko and F. S. N. Lobo, 2008 *Astropart. Phys.* **29**, 386; C. G. Boehmer, T. Harko and F. S. N. Lobo, 2008 *JCAP* **0803**, 024; O. Bertolami, C. G. Boehmer, T. Harko and F. S. N. Lobo, 2007 *Phys. Rev. D* **75**, 104016; S. Capozziello, T. Harko, T. S. Koivisto, F. S. N. Lobo and G. J. Olmo, To appear in Astroparticle Physics, arXiv:1307.0752 [gr-qc]; S. Capozziello, T. Harko, T. S. Koivisto, F. S. N. Lobo and G. J. Olmo, 2013 *JCAP* **07** 024
[8] S. W. Hawking, G. F. R. Ellis, 1973 *The Large scale structure of space-time*, Cambridge University Press, Cambridge; R. M. Wald, 1984 *General Relativity*, The University of Chicago Press, Chicago
[9] S. Capozziello, F. S. N. Lobo and J. P. Mimoso, 2014 *Phys. Lett. B* **730**, 280; ibidem arXiv:1407.7293 [gr-qc].
[10] F.W. Hehl, P. von der Heyde, G.D. Kerlick and J.M. Nester, 1976 *Rev. Mod. Phys.* **48**, 393; F.W. Hehl and J.D. Mccrea, E.W. Mielke and Y. Neeman, 1995 *Phys. Rep.*, 4258, 1;
[11] J. Santos, J. S. Alcaniz, M. J. Reboucas and F. C. Carvalho, 2007 *Phys. Rev. D* **76**, 083513 N. M. Garcia, T. Harko, F. S. N. Lobo and J. P. Mimoso, 2011 *Phys. Rev. D* **83**, 104032 T. Harko, F. S. N. Lobo, M. K. Mak and S. V. Sushkov, 2013 *Phys. Rev. D* **87**, 067504
[12] P. Schneider, J. Ehlers, E.E. Falco, 1992 *Gravitational Lenses*, Springer, Berlin
[13] M. Tsamparlis, A. Paliathanasis, S. Basilakos, S. Capozziello, 2013 *Gen. Rel. Grav.* **45**, 2003
[14] G. Magnano, L. M. Sokolowski, 1994 *Phys.Rev.* **D 50**, 5039
[15] J. D. Barrow and J. P. Mimoso, 1994 *Phys. Rev. D* **50**, 3746 J. P. Mimoso and D. Wands, 1995 *Phys. Rev. D* **51**, 477 J. P. Mimoso and D. Wands, 1995 *Phys. Rev. D* **52**, 5612
[16] S. Capozziello, and S. Vignolo, 2009 *Int. J. Geom. Methods Mod. Phys.*, **6**, 985

Probing $f(R)$ gravity with PLANCK data on cluster pressure profiles

I. De Martino[1,4], M. De Laurentis[2,3,4], F. Atrio-Barandela[1], S. Capozziello[3,4]

[1] Física Teórica, Universidad de Salamanca, 37008 Salamanca, Spain;
[2]Department of Theoretical Physics, Tomsk State Pedagogical University (TSPU), pr. Komsomolsky, 75, Tomsk, 634041, Russia;
[3] Dipartimento di Fisica, Università di Napoli "Federico II";
[4]INFN sez. di Napoli Compl. Univ. di Monte S. Angelo, Edificio G, Via Cinthia, I-80126 - Napoli, Italy.

E-mail: ivan.demartino@usal.es

Abstract. Analytical $f(R)$-gravity models introduce Yukawa-like corrections to the Newtonian potential in the weak field limit. These models can explain the dynamics of galaxies and cluster of galaxies without requiring dark matter. To test the model, we have computed the pressure profile of 579 X-ray galaxy clusters assuming the gas is in hydrostatic equilibrium within the potential well of the modified gravitational potential. We have compared those profiles with the ones measured in the foreground cleaned SMICA released by the Planck Collaboration. Our results show that Extended Theories of Gravity explain the dynamics of self-gravitating systems at cluster scales and represent an alternative to dark matter haloes.

1. Introduction

The Cosmic Microwave Background (CMB) temperature anisotropies measured by the Wilkinson Microwave Anisotropy Probe (WMAP) [2] and the Planck satellites [3, 4, 5, 6] as well as luminosity distances measured of the Supernovae Type Ia (SNeIa) [1], strongly favor the concordance ΛCDM model, where the Universe is dominated by two unknown energy densities: Dark Energy (DE) and Dark Matter (DM). As an alternative, analytical $f(R)$-models modify the description of gravity instead of requiring extra matter components. In these extensions of General Relativity (GR), the Hilbert-Einstein action is replaced with a more general function of the Ricci scalar R [7]. These models have been found to introduce negligible corrections at the scale of the Solar systems [8] so they verify the constraints imposed by the classical test of GR, and the perturbations in self-gravitating systems behave similarly as in GR [9]. Nevertheless, this modified gravity introduces a Yukawa-like correction to the gravitational potential in the Post-Newtonian limit [7, 10]. This modified potential has been used to explain the dynamics of galaxies without requiring DM [11, 12].

Since, the exact functional form of the Lagrangian is unknown, it is important to test potential models using all the available data on self-gravitating systems. Galaxy clusters are reservoirs of hot gas and when CMB photons cross the cluster potential well, they are scattered off by the free electron of the intracluster medium. The effect, known as Sunyaev-Zeldovich (SZ) [13, 14], induces secondary anisotropies that have been extensively measured [15, 16, 5, 17].

Pressure profiles of cluster of galaxies based on the Navarro-Frenk-White (NFW, [18]) DM density profile have been found to be in agreement with SZ [19] and X-ray observations [20]. We have constructed the pressure profiles of clusters of galaxies assuming the gas is in hydrostatic equilibrium within a Newtonian potential with a Yukawa correction and compared them with Planck data. This article is organized as follows: in Sec. 2, we describe the weak field limit of $f(R)$ gravity; in Sec. 3, we summarize the pressure profiles most commonly used in the literature and compare them with the pressure profile for $f(R)$ models; in Sec. 4 we describe the data used in our analysis; in Sec. 5 we present our results and in Sec. 6 we will summarize our main conclusions.

2. Yukawa corrections to the Newtonian potential in $f(R)$-gravity

In Extended Theories of Gravity (ETGs), the field equations are obtained by varying the action

$$\mathcal{A} = \frac{c^4}{16\pi G}\int d^4x\sqrt{-g}f(R) + \mathcal{L}_m\,, \tag{1}$$

resulting in [1]

$$f'(R)R_{\mu\nu} - \frac{f(R)}{2}g_{\mu\nu} - f'(R)_{;\mu\nu} + g_{\mu\nu}\Box_g f'(R) \;=\; 8\pi G T_{\mu\nu}\,.$$

To construct spherically symmetric solutions, we write the metric element as

$$ds^2 = g_{tt}c^2dt^2 - g_{rr}dr^2 - r^2d\Omega, \tag{2}$$

where $d\Omega$ is the solid angle. We restrict our study to $f(R)$-Lagrangians that are expandable in Taylor series around a fixed point R_0

$$f(R) = \sum_n \frac{f^n(R_0)}{n!}(R-R_0)^n \simeq f_0 + f_0'R + \frac{f_0''}{2}R^2 + \ldots\,. \tag{3}$$

In the Post-Newtonian limit of these models the gravitational potential can be written as [7]

$$\Phi_{grav}(r) = \frac{\Phi_N(r)}{(1+\delta)}\left(1+\delta e^{-\frac{r}{L}}\right). \tag{4}$$

In this expression, δ represents the deviation from GR at zero order and the Newtonian limit of GR is recovered at $\delta \to 0$, irrespective of the scale parameter L. The latter is the scale length of the self-gravitating object [7, 10] and its effects are negligible at Solar System scales where GR, is restored [10]. In terms of the coefficients in the Taylor expansion they are given by $\delta = 1 - f_0' = 1$ and $L = [-f_0'/(6f_0'')]^{1/2}$. In this context, the dynamical effects of DM are now given by the modifications of the Newtonian potential.

3. Cluster pressure profiles in $f(R)$ gravity

SZ effect induces two different type of secondary temperature anisotropies on the CMB: a thermal contribution due to the motion of the electrons within the cluster potential well and a kinematic one (KSZ) due to the peculiar motion of the cluster. It is given by

$$\frac{\Delta T}{T_0} = g(\nu)\frac{k_B\sigma_T}{m_e c^2}\int n_e T_e dl, \tag{5}$$

[1] $f'(R) = df(R)/dR$ is the first derivative with respect to the Ricci scalar, $\Box_g = {}_{;\sigma}{}^{;\sigma}$ is the d'Alembertian with covariant derivatives, $T_{\mu\nu} = -2(-g)^{-1/2}\delta(\sqrt{-g}\mathcal{L}_m)/\delta g^{\mu\nu}$ is the matter energy-momentum tensor, T its trace, g the determinant of the metric tensor $g_{\mu\nu}$. Greek indices run from 0 to 3.

Table 1: Parameters of the GNFW corresponding to [20, 15, 24], β for Coma cluster and $f(R)$ is the best fit model to Planck data.

Model	c_{500}	α_a	β_a	γ_a	P_0
Arnaud	1.177	1.051	5.4905	0.3081	$8.403h_{70}^{3/2}$
Planck	1.81	1.33	4.13	0.31	6.41
Sayers	1.18	0.86	3.67	0.67	4.29
β	β	$n_{c,0}/m^{-3}$	r_c/Mpc	T_e/keV	
	2/3	3860.	0.25	6.48	
$f(R)$	δ	L/Mpc	γ		
	-0.98	0.1	1.2		

where T_e is the electron temperature, n_e the electron density and the integration is carried out along the line of sight l. In Eq. (5) k_B is the Boltzmann constant, m_ec^2 the electron annihilation temperature, c the speed of light, ν the frequency of observation, σ_T the Thomson cross section and T_0 the mean temperature of the CMB. Finally, $g(\nu) = x\coth(x/2) - 4$ is the frequency dependence of the TSZ effect, with the reduced frequency $x = h\nu/KT_0$. For individual clusters, only the TSZ anisotropy has been measured.

Eq. (5) shows that in order to compute the TSZ anisotropies we need to estimate the cluster pressure profile, n_eT_e. Several cluster profiles, based on X-ray data and numerical simulations, have appeared in the literature:

• The isothermal β-model [21, 22] specifies the electron density: $n_e(r) = n_{e,0}[1+(r/r_c)^2]^{-3\beta/2}$. From X-ray surface brightness data it has been found that $\beta = 0.6 - 0.8$ [23].

• The universal profile [20] is a phenomenological model based on the NFW DM-profile

$$p(x) \equiv \frac{P_0}{(c_{500}x)^{\gamma_a}[1+(c_{500}x)^{\alpha_a}]^{(\beta_a-\gamma_a)/\alpha_a}}, \tag{6}$$

where x is the radial distance in units of r_{500}, the radius where the average density is 500 times the critical density, and c_{500} is the concentration parameter at r_{500}. Different groups have fit the model parameters $[c_{500}, \alpha_a, \beta_a, \gamma_a, P_0]$ to X-ray or CMB data; their best fit values are given in Table 1.

To compute in $f(R)$ gravity the pressure profile n_eT_e of Eq. (5) we assume that the gas is in hydrostatic equilibrium within the modified potential well of the cluster (without the DM contribution)

$$\frac{dP(r)}{dr} = -\rho(r)\frac{d\Phi_{grav}(r)}{dr}, \tag{7}$$

and the physical state of the gas is described by a polytropic equation of state

$$P(r) \propto \rho^{\gamma}(r). \tag{8}$$

Eqs. (7) and (8), together with the continuity equation

$$\frac{dM(r)}{dr} = 4\pi\rho(r), \tag{9}$$

and the cluster gravitational potential given by eq. (4) form a close system that can be solved numerically to obtain the pressure profiles of any given cluster as a function of two gravitational parameters (δ, L) and the polytropic index γ. Fig. 1 we represent the different profiles integrated along the line of sight with the parameters given in Table 1. The profiles are particularized for the parameters of the Coma cluster.

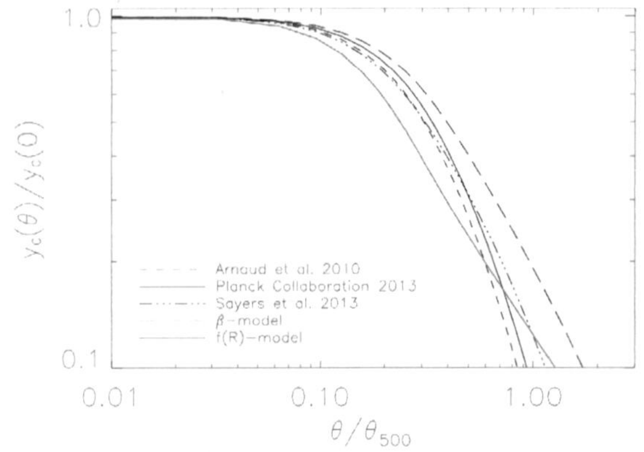

Figure 1: Pressure profiles integrated along the line of sight for the Coma cluster ($z = 0.023$). We plot three GNFW profiles (dashed, solid and dash-dotted lines), one $\beta = 2/3$ model (long dashed line) and a $f(R)$ model (red solid line). The parameters of each model are given in Table 1.

4. Data.

We use Planck data and a X-ray selected cluster catalog to constrain the pressure profiles of clusters of galaxies.

• Our cluster catalog is fully described in [25] and lists position, redshift, X-ray flux, X-ray luminosity. The X-ray electron temperature is derived from the $L_X - T_X$ relation of [26]. A total of 579 clusters are located outside the minimal Planck mask; the region where the CMB data is less affected by galactic foreground residuals.

• The Planck Collaboration released in March 2013 nine maps of the CMB temperature anisotropies spanning a frequency range from 30 to 845 GHz. Since foreground cleaned maps at all frequencies were not made available, we performed our analysis on a foreground cleaned map known as SMICA. This map has been constructed using a component separation method by combining the data at all frequencies [27]. The SMICA map has $5'$ resolution.

For all except the brightest clusters, the underlying CMB anisotropies have large amplitude than the TSZ contributions. Then, to compare the expected profile with the data, we stack the signal of all the clusters in our sample. The anisotropy is averaged in rings of width $\theta_{500}/2$, where θ_{500} is the angular scale subtended by the r_{500}. For each data point an error bar is obtained by evaluating 1,000 times the average profiles at 579 random positions in the SMICA map. To avoid overlapping real and simulated clusters we removed a disc of $80'$ around each of the clusters in our sample.

Using eqs. (4), (7), (8) and (9), we constructed the pressure profile of all clusters in the data as a function of three parameters: (δ, L, γ). We integrated the profiles along the line of sight and convolved them with a Gaussian beam with the same resolution of the SMICA map. We considered two parameterizations. In (A) $L = \zeta r_{500}$ differs from cluster to cluster but scales linearly with r_{500}; in this parametrization we assume that $\zeta = [0.1, 4]$. In (B) L has the same value for all the clusters; it varies in the range $L = [0.1, 20]$Mpc, from the scale of a typical core radius to the mean cluster separation. Since if $\delta < -1$ the potential is repulsive and if $\delta = -1$ the potential diverges, we took $\delta = [-0.99, 1.0]$. Finally, the polytropic index γ was set to vary in the range $\gamma = [1.0, 1.6]$, that corresponds to an isothermal and adiabatic monoatomic gas, respectively. We took 30 equally spaced steps in all intervals.

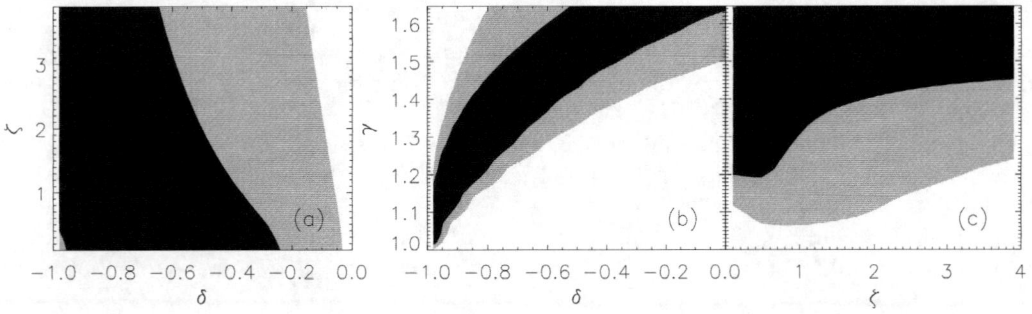

Figure 2: Confidence contours for pairs of parameters of Model A. Contours are at the 68% and 95% confidence level.

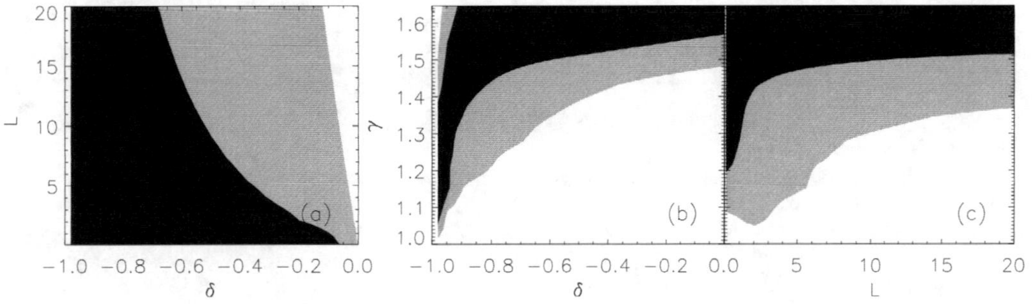

Figure 3: Same as in Fig. 2 for Model B.

5. Results and discussion

We computed the likelihood function $\log \mathcal{L} = -\chi^2/2$ as

$$\chi^2(p) = \Sigma_{i,j=0}^{N}(y(p, x_i) - d(x_i))C_{ij}^{-1}(y(p, x_j) - d(x_j)) \tag{10}$$

where $N = 7$ is the number of data points, and $p = (\delta, L, \gamma)$. In eq. (10), $d(x_i)$ is the data and $C_{i,j}$ is the correlation function between the average temperature anisotropy on discs and rings in bins of size θ_{500}. The process was computed on 579 random positions outside the cluster locations and repeated 1,000 times.

In Figs. 2 and 3, we computed the 2D contours at the 68% and 95% confidence levels of the marginalized likelihoods of pairs of parameters of Model A and Model B, respectively. Since the contours are not closed, we can only quote the following upper limits at the same level of confidence: $\delta < -0.46, -0.10$, $\zeta < 2.5, 3.7$ and $\gamma > 1.35, 1.12$ for Model A and $\delta < -0.43, -0.08$, $L < 12, 19$ Mpc and $\gamma > 1.45, 1.2$ for Model B. Notice that L is similar in both A and B and γ dominated by the physical boundary imposed.

Although model parameters are weakly constrained by the data, the value $\delta = 0$ is excluded at more than 95% confidence level. Since $\delta \simeq 0$ corresponds to the standard Newtonian potential without DM, our result indicates that a self-gravitating gas, without DM or modified gravity, does not fit the data.

6. Conclusions

We have constrained the analytical $f(R)$ model of gravity by comparing the pressure profile of clusters of galaxies with CMB data. We have demonstrated that these models can accurately fit cluster profiles without a potential well dominated by DM. In this context, $f(R)$ gravity can be considered a viable alternative to DM. Our results also showed that a profile made of gas, without a modified potential or without significant DM contributions could not fit the observed profiles.

Our constraints on model parameters are rather weak since we only used the data on the cluster profile but we did not use the frequency dependence of the TSZ effect, the reason being that the Planck Collaboration has not released foreground cleaned maps at each frequency. Then, our results could be substantially improved when including this frequency information, opening the possibility of constraining - ruling out the modified Newtonian potential at this scale.

References

[1] Suzuki N et al 2012 *ApJ* **746**, 85
[2] Hinshaw G et al 2013 *ApJS* **208**, 19
[3] Planck Collaboration. "Planck 2013 results. XV. CMB power spectra and likelihood", 2014 *A&A* **571**, 15
[4] Planck Collaboration. "Planck 2013 results. XVI. Cosmological parameters", 2014 *A&A* **571**, 16
[5] Planck Collaboration. "Planck 2013 results. XX. Cosmology from SunyaevZeldovich cluster counts", 2014 *A&A* **571**, 20
[6] Planck Collaboration. "Planck 2013 results. XXI. Cosmology with the all-sky Planck Compton parameter y-map", 2014 *A&A* **571**, 21
[7] Capozziello S and De Laurentis M 2011 *Physics Reports* **509**, 167
[8] Berry C P L and Gair J L 2011 *Phys Rev D* **83**, 104022
[9] Capozziello S De Laurentis M de Martino I Formisano M and Odintsov S 2012 *Phys Rev D* **85**, 044022
[10] Capozziello S and De Laurentis M 2012 *Ann. Phys.* **524**, 1
[11] Cardone V F and Capozziello S 2011 *MNRAS* **414**, 1301
[12] Napolitano N R Capozziello S Romanowsky A J Capaccioli M and Tortora C 2012 *ApJ* **748**, 87
[13] Sunyaev R A and Zeldovich Y B 1972 *Comments on Astrophys. Space Phys.* **4** 173
[14] Sunyaev R A and Zeldovich Y B 1980 *MNRAS* **190**, 413
[15] Planck Collaboration. "Planck intermediate results V: Pressure profiles of galaxy clusters from the Planck Survey", 2013 *A&A* **550**, 131
[16] Planck Collaboration. "Planck intermediate results X: Physics of the hot gas in the Coma cluster", 2013 *A&A* **554**, 140
[17] Planck Collaboration. "Planck 2013 results. XXIX. Planck catalogue of Sunyaev-Zeldovich sources", 2014 *A&A* **571**, 29
[18] Navarro J F Frenk C S and White S D M 1997 *ApJ* **490**, 493
[19] Atrio-Barandela F Kashlinsky A Kocevski D and Ebeling H 2008 *ApJ* **675**, L57
[20] Arnaud M Pratt G W Piffaretti R Bhringer H Croston J H and Pointecouteau E 2010 *A&A* **517**, 92
[21] Cavaliere A and Fusco-Femiano R 1976 *A&A* **49**, 137
[22] Cavaliere A and Fusco-Femiano R 1978 *A&A* **70**, 677
[23] Jones C and Forman W 1984 *ApJ* **276**, 38
[24] Sayers J et al 2013 *ApJ* **768**, 177
[25] Kocevski D and Ebeling H 2006 *ApJ* **645**, 1043
[26] White D A Jones C and Forman W 1997 *MNRAS* **292**, 419
[27] Planck Collaboration. "Planck 2013 results. XII. Component separation", 2014 *A&A* **571**, A12

Post-Newtonian limits for Randall-Sundrum model

K A Rannu, P I Dyadina and S O Alexeyev

Sternberg Astronomical Institute, Lomonosov Moscow State University, Universitetsky Prospekt, 13, Moscow 119991, Russia

E-mail: rannu@xray.sai.msu.ru, guldur.anwo@gmail.com, alexeyev@sai.msu.ru

Abstract. Parametrized post-Newtonian formalism is one of the standard methods of experimental check of gravity theories and limiting the model parameters. In this work we consider the PPN-paramertization of Randall-Sundrum II black hole solutions and show that these solutions are in good agreement with the GR predictions and the observation results.

1. Introduction

Black hole solution is a basic one for any theory of gravity. It describes the compact object that a very massive star at the end of its life cycle collapses into and it also features the curvature of the spacetime produced by the presence of matter specific for the considered gravity model. Any extended theory of gravity should be consistent with the predictions of GR and the observations' results therefore the existence of black holes and their properties are important indicators of the theory's viability. In this work we consider the weak-field limit of the Randall-Sundrum solutions found by Figueras and Wiseman and by Abdolrahimi, Cattoën, Page and Yaghoobpour-Tari to make certain that they agree with GR. We also look for the opportunity to distinguish the Randall-Sundrum model experimentally via these solutions.

2. Parameterized post-Newtonian formalism

Parametrized post-Newtonian limit (PPN) was originally constructed for comparing different metric theories of gravity with each other [1, 2, 3, 4, 5, 6]. Using the parametrized post-Newtonian formalism the metric can be represented as the perturbative expansion around Minkowski spacetime $\eta_{\mu\nu}$ [5]:

$$\begin{gathered} g_{\mu\nu} = \eta_{\mu\nu} + h_{\mu\nu}, \\ h_{00} \sim 1/r^2, \quad h_{oj} \sim 1/r^3, \quad h_{ij} \sim 1/r. \end{gathered} \tag{1}$$

PPN needs several requirements to be fulfilled:

- weak field limit,
- asymptotically flat spacetime,
- small velocities of matter the motion of which obeys the hydrodynamical equations for the prefect fluid.

Such parametrization allows to distinguish metric theories from each other on a number of properties such as the measure of space curvature produced by unit mass, the non-linearity in gravitational superposition, the existence of preferred location and frame effects and the violation of conservation laws of energy, momentum and angular momentum. These properties are expressed through the post-Newtonian parameters that are measured experimentally with high precision [6].

The units in which $G = c = 1$ are used here.

3. Randall-Sundrum gravity

The Randall-Sundrum model considers four-dimensional branes with tension embedded into a five-dimensional spacetime (bulk) that is assumed to have an AdS_5 geometry [7, 8]. All the matter and the three fundamental interactions (electromagnetic, strong nuclear and weak nuclear) are localized on the brane except gravity which can propagate into the bulk along the additional dimension. This extra dimension is allowed to be noncompact and even infinite. Randall-Sundrum I (RSI) model considers two branes with different properties helping to solve the hierarchy problem [7]. Moving the second brane to the infinity led to the Randall-Sundrum II (RSII) model with one brane [8].

First black hole solutions were obtained by Chambling, Hawking and Reall [9] and by Dadhich, Maartens, Papadopoulos and Rezania [10] in 2000. A remarkable conjecture was then made that static black holes cannot exist in RSII for a radius much greater than the AdS length ℓ [11, 12, 13]. By using the numerical methods [14, 15] black holes in 5D RSII with a radius up to $\sim 0.2\ell$ and for 6D up to $\sim 2.0\ell$ were constructed [16, 17, 18]. However, by using the same methods, it has subsequently been argued that even very small RSII static black holes do not exist [19, 20]. In 2011 Figueras and Wiseman numerically constructed large RSII static black holes with radius up to $\sim 20\ell$ close to the associated AdS_5-CFT_4 solution [21]. Soon after them in 2013 Abdolrahimi, Cattoën, Page and Yaghoobpour-Tari found the infinite-mass black hole solution [21] by a different numerical method (ACPY solution). This solution agrees well with the Figueras-Wiseman one [21] and thus adds further evidence for the existence of large RSII black holes, despite the doubts expressed by previous works.

4. Results

4.1. Figueras-Wiseman solution

To obtain the post-Newtonian parametrization of the Figueras-Wiseman solution the following field equations were used:

$$G_{\mu\nu} = 8\pi G_4 T^{brane}_{\mu\nu} + \epsilon^2 \left\{16\pi G_4 \langle T^{CFT}_{\mu\nu}[g]\rangle + a_{\mu\nu}[g] + \log\epsilon\, b_{\mu\nu}[g]\right\} + O(\epsilon^4 \log\epsilon), \tag{2}$$

where G_4 is the usual four-dimenstional graviational constant, $T^{brane}_{\mu\nu}$ is the stress-enegry tensor of the matter localized on the brane, the additional stress-enegry terms $\langle T^{CFT}_{\mu\nu}[g]\rangle$, $a_{\mu\nu}[g]$ and $b_{\mu\nu}[g]$ describe the influence of the extra dimension. They depend on metric only and are expressed via the Fefferman-Graham expansion [22, 23]; ϵ is a small perturbation parameter indicating the deviation of the brane position from the equilibrium state $z = 0$, z is the coordinate along the extra dimension.

The additional term in the post-Newtonian expansion of the Figueras-Wiseman solution calculated in this paper is

$$\delta h^{\mathrm{FW}}_{00} = \frac{121}{27}\frac{\epsilon^2}{\ell^2}\frac{M^2}{r^2}, \tag{3}$$

where M is the mass of the central object. In the considered case it equals the solar mass.

The obtained value (3) lies within the 1PN limit (1) and points at a potentially observable effect. In Randall-Sundrum model gravity is allowed propagate into the bulk along the extra dimension therefore the effect described by (3) most likely leads to the negative nonlinearity in gravitational superposition. In other words the resulting gravitational field produced by two or more massive objects can be less than the direct vector sum of their contributions. The parameterized post-Newtonian (PPN) parameter β is responsible for such an effect [5, 6, 24]. Therefore the result (3) should be expressed as follows:

$$\beta = 1 - \frac{\epsilon^2}{\ell^2}\,\frac{121}{108}\,M^2. \tag{4}$$

The constraint on the PPN parameter β obtained from analysis of the lunar laser ranging data [24] is $|\beta - 1| \leq 1.1\times 10^{-4}$. The admitted region of the AdS length is limited by the results of the Newton's law test $l < 10^{-5}$ m [25]. Therefore the upper limit on the value of ϵ is:

$$\epsilon \leq 5.7\times 10^{-47} \ll m_{Pl}. \tag{5}$$

Originally the parameter ϵ was assumed to be negligibly small and the vanishing value found in (5) implies that in fact $\epsilon = 0$. Thus the Figueras-Wiseman four-dimensional black hole solution is not only self-consistent but well consistent with the solar system constraints as well. Therefore this solution is indistinguishable form GR in the 1PN limit after all.

4.2. ACPY solution

The ACPY solution [26] is asymptotically conformal to the Schwarzschild metric and includes a negative five-dimensional cosmological constant Λ_5:

$$\begin{aligned} ds^2 = & - \left(1-\frac{2M}{r}\right)dt^2 + \left[1 - \frac{1}{-\Lambda_5 r^2}\,\frac{r-2M}{r-1.5M}\left(F - r\,\frac{dF}{dr}\right)\right]\left(1-\frac{2M}{r}\right)^{-1} dr^2 + \\ & + \left[r^2 + \frac{F}{-\Lambda_5}\right] d\Omega^2 \\ F = & \ 1 - 1.1241\left(\frac{2M}{r}\right) + 1.956\left(\frac{2M}{r}\right)^2 - 9.961\left(\frac{2M}{r}\right)^3 + \ \ldots\ + 2.900\left(\frac{2M}{r}\right)^{11}. \end{aligned} \tag{6}$$

The function $F(r)$ describes the perturbation caused by the bulk. The best fit for it was obtained in [26].

The field equations induced on the brane were derived by Sasaki, Shiromizu and Maeda [27]:

$$G_{\mu\nu} = -\,\Lambda_4 g_{\mu\nu} + \frac{8\pi}{M^2_{\rm Pl4}}\,T_{\mu\nu} + \frac{8\pi}{M^3_{\rm Pl5}}\,S_{\mu\nu} - \mathcal{E}_{\mu\nu}, \tag{7}$$

where Λ_4 is usual four-dimensional cosmological constant, $g_{\mu\nu}$ is the metric on the brane, $T_{\mu\nu}$ is the stress-energy tensor of the matter localized on the brane, $S_{\mu\nu}$ is the local quadratic stress-energy correction, $\mathcal{E}_{\mu\nu}$ is the four-dimensional projection of the five-dimensional Weyl tensor. $M_{\rm Pl4}$ is usual four-dimensional Planck mass and $M_{\rm Pl5}$ is the fundamental five-dimensional Planck mass which is typically much less than the effective Planck mass on the brane.

The induced metric on the brane is asymptotically flat, the bulk is an anti-de-Sitter spacetime as in the original Randall-Sundrum scenario [8], then $\mathcal{E}_{\mu\nu} = 0$ [28]. Therefore the correction term due to the contribution from ACPY topology (6) that follows from (7) has the form

$$\delta h_{00}^{\rm ACPY} = \frac{l^2 M^2}{96}\,\frac{1}{r^4} + \mathcal{O}(r^{-5}). \tag{8}$$

According to (1) the expansion term of the 1PN-order should be proportional to r^{-2}. The correction (8) contains the next perturbation order which lies beyond 1PN. Therefore the obtained contribution (8) cannot be observed in the solar system experiments as well. This conclusion on the Randall-Sundrum model predictions confirms the result for the Figueras-Wiseman solution.

5. Conclusions

Consideration of the post-Newtonian expansion of the Figueras-Wiseman solution [21] reveals such possible effect as a negative nonlinearity of gravitational superposition (4). It naturally results from the theory itself because gravity is allowed to propagate to the extra dimension in Randall-Sundrum model. However the breaking of gravitational superposition turns out to depend on a negligibly small parameter (5) thus the predictions of the Figueras-Wiseman solution fully agrees with GR and the present observations. This effect may influence the strong field regime (close binary systems, black holes) as a consequence of curvature growth. So the next step could be the search of such features of the Randall-Sundrum model in the strong field limit. Fortunately this investigation is admissible as the large stable black hole solutions for RSII black holes have been found [21, 26].

The consideration of the other black hole solution by Abdolrahimi, Page et al. [26] shows that the terms describing the bulk influence (8) greatly exceed the limits of the post-Newtonian approximation. As a result both recent large Randall-Sundrum black holes solutions seems to be well consistent with GR at the solar system scale. For obtaining the limits on its parameters other tests are needed.

Acknowledgments

The work was supported by Federal Agency on Science and Innovations of Russian Federation, state contract 02.740.11.0575. S.A. also was supported by individual grants from Dmitry Zimin Foundation "Dynasty". Authors would like to thank Dr. M. Smolyakov and Dr. D. Levkov for useful discussions on the subject of this work.

References

[1] Thorn K S and Will C M 1971 *Astrophys. J.* **163** 595
[2] Will C M 1971 *Astrophys. J.* **163** 611
[3] Will C M and Nordtvedt K J 1972 *Astrophys. J.* **177** 757
[4] Nordtvedt K J and Will C M 1972 *Astrophys. J.* **177** 775
[5] Will C M 1981 *Theory and experiment in gravitational physics* (Cambridge: Cambridge University Press)
[6] Will C M 2014 *Living Rev. Rel.* **17** 4
[7] Randall L and Sundrum R 1999 *Phys. Rev.* Lett. **83** 3370
[8] Randall L and Sundrum R 1999 *Phys. Rev.* Lett. **83** 4690
[9] Chamblin A, Hawking S W and Reall H S 2000 *Phys. Rev.* D **61** 065007
[10] Dadhich N, Maartens R, Papadopoulos P and Rezania V 2000 *Phys. Lett.* B **487** 1
[11] Tanaka T 2002 *Prog. Theor. Phys. Suppl.* **148** 307
[12] Emparan R, Fabbri A and Kaloper N 2002 *J. High Energy Phys.* JHEP08(2002)043
[13] Emparan R, Garcia-Bellido J and Kaloper N 2003 *J. High Energy Phys.* JHEP01(2003)079
[14] Wiseman T 2003 *Phys. Rev.* D **65** 124007
[15] Wiseman T 2003 *Class. Quant. Grav.* **20** 1137
[16] Kudoh H, Tanaka T and Nakamura T 2003 *Phys. Rev.* D **68** 024035
[17] Kudoh H 2003 *Prog. Theor. Phys.* **110** 1059
[18] Kudoh H 2004 *Phys. Rev.* D **69** 104019
[19] Yoshino H 2009 *J. High Energy Phys.* JHEP01(2009)068
[20] Kleihaus B, Kunz J, Radu E and Senkbeil D 2011 *Phys. Rev.* D **63** 104050
[21] Figueras P and Wiseman T 2011 *Phys. Rev.* Lett. **107** 081101
[22] de Haro S, Skenderis K and Solodukhin S N 2001 *Commun. Math. Phys.* **217** 595
[23] de Haro S, Skenderis K and Solodukhin S N 2001 *Class. Quant. Grav.* **18** 3171
[24] Turyshev S G 2009 *Phys. Usp.* **52** 1
[25] Kapner D J, Cook T S, Adelberger E G, Gundlach J H, Heckel B R, Hoyle C D and Swanson H E 2007 *Phys. Rev.* Lett. **98** 021101
[26] Abdolrahimi S and Catto C
[27] Sasaki M, Shiromizu T and Maeda K I 2000 *Phys. Rev.* D **62** 024008
[28] Maartens R 2000 *Phys. Rev.* D **62** 084023

Spanish Relativity Meeting (ERE 2014): almost 100 years after Einstein's revolution IOP Publishing
Journal of Physics: Conference Series **600** (2015) 012050 doi:10.1088/1742-6596/600/1/012050

The gravity apple tree

Mariana Espinosa Aldama

Posgrado en Filosofa de la Ciencia, Universidad Nacional Autónoma de México, Distrito Federal 04510, México

E-mail: eamariana@yahoo.com

Abstract.
The *gravity apple tree* is a genealogical tree of the gravitation theories developed during the past century. The graphic representation is full of information such as guides in heuristic principles, names of main proponents, dates and references for original articles (See under **Supplementary Data** for the graphic representation). This visual presentation and its particular classification allows a quick synthetic view for a plurality of theories, many of them well validated in the Solar System domain. Its diachronic structure organizes information in a shape of a tree following similarities through a formal concept analysis. It can be used for educational purposes or as a tool for philosophical discussion.

1. Introduction

The quest for the fundamental principles of the physical world its at an exciting moment. Obscure new phenomena, –commonly explained as dark matter, dark energy and inflation– has pushed into a crisis the standard paradigm of gravitation: General Relativity and the Concordance Cosmological Model. The plurality of theories available (some of them well validated) is a testimony to this critical condition. Nowadays, most professionals are ultra specialized, so is quite a job for the layperson to form a synthetic view of the hypothesis, concepts and theories being explored as well as their fruits. The *gravity apple tree* is a genealogical graphic representation of gravitational theories that illustrates the bulk of knowledge on this matter over the past 100 years.

This structured map is full of information. Theories, authors and dates are presented in green. Models for dark matter and dark energy are presented in orange and principles followed are in violet. Some selected principles are written in the middle of branches as they allow to distinguish groups of theories. The timeline grows upwards and radially according to the main article publication dates, included in the bibliography for completeness.

The theoretical framework of the tree is based on the work of philosopher Mario Bunge [1], who points out that foundations of physics are not rigid as building blocks, as such would turn our science into dogma. However, their provisional character does not render them nonexistent or less fundamental. Acknowledging the organic character of physics as a system of mutually dependent parts, I will focus on a hierarchical image and an axiomatic approach as it allows one to design this particular visual representation. I have applied a formal concept analysis that provides diagrams linked by a similarity criteria, which I then draw as tree branches [See 2]. The classification of theories is based on [3–8].

1.1. Background
The common early theoretical background is represented as grass, and is that of continuum mechanics. The formal background consists of logic, semantics, algebra, topology, analysis, and manifold geometry. The material background consists of protophysics, mainly the theory of local time and the general systems theory, and some elements of physical geometry [See 1].

The *gravity apple tree* has its base at the end of the 17th century, with the publication of Newton's *Principia* [9], represented by an apple. This theory, remained (and still does) as a paradigm for some 200 years, until 1905, when Special Relativity Theory joins the big explosion of physical knowledge (along with Quantum Theory and Maxwell's electrodynamics). Ten main branches emerge when different principles are followed. I will briefly explain them by groups, according to the agenda on gravitation.

2. Unification of Special Relativity and Gravitation
The Special theory of relativity has two mayor limitations. The first one is that it is restricted to non accelerated reference frames. The second one is that it does not include Newton's law in a self-consistent way. In Newton's theory gravitational force acts instantaneously; but according to the relativity theory, nothing can travel faster than light, which has a finite constant value [See 10, 11]. The birth of the first three branches, identified as the Physical Field branch, the Scalar Lorentz Covariat branch and the General Relativity branch (which keeps growing as the main trunk of the tree), represent the quest to solve this problems.

2.1. Physical Field Theories
Henri Poincaré (1904), Hermann Minkowski (1908) and Willem de Sitter (1905) considered gravitation should propagate with the velocity of light in a retarded way, and gravity to be a physical field, very much like the Maxwell-Faraday electromagnetic field with action at a distance in a flat four dimensional space-time [See 12].Their investigations failed to specify a unique result and the deviations for planetary motions computed by Minkowki were so small that they did not allow to make a firm decision about the law of gravity in the relativistic regime [See e.g. 13, 14].

The interest to develop the relativistic field theories after the second world war was in connection with the interactions between gravity and the meson, electromagnetic and pair fields. This was done mainly in two ways, with its own quantum approaches [15–20][See 14, 21].

2.2. Scalar Lorentz Covariant theories
To avoid action at a distance, the next natural approach was to consider a field law for gravitation, retaining the Laplacian scalar potential of gravity, extending the Poisson equation through a time derivative term. This dynamical scalar field, in contrast with a vector field, has no direction in space. Gunnar Nordstrom [22] propose a metric $\mathbf{g}$ constructed from a flat background metric η and a scalar dynamical field ϕ: $\mathrm{g} = \mathrm{f}(\phi)\eta$. Einstein and Fokker [23] applied the new methods of absolute differential calculus of Ricci and Levi-Civita to Nordstrom's theory, leading to a theory with kinematical effects where gravitational fields would slow clocks and alter the lengths of rods, concluding that it is actually only conformal to a Minkowsky space-time with the gravitational potential to be the conformal factor, and the presence of a gravitational field coincides with deviations of space-time from flatness [See 13, 22, 24–28].

2.3. General Relativity (GR)
Einstein was in a race for discovery. After an initial paper in 1907, from 1912 untill 1916 when he finally published the complete theory of General Relativity (GR), he published some 25 articles related to his progress. The development of the theory can be divided into tree parts: (i) The

understanding of the Strong Equivalence Principle (1907-1908); (ii) The choice of a Riemannian geometry and of the appropriate metric tensor field for the relativistic Newtonian potential (1912-1913); (iii) The search for the correct field equations for the metric tensor field (1913-19015). While working on the patent office, in 1907, Einstein suddenly realized that a falling man would not feel the difference between gravity and an accelerating force acting upon him. The equality between acceleration and gravity meant an equality of inertial mass and the gravitational mass. In Newton's theory, inertial mass and gravitational mass, even though seem equivalent, are not supposed to be the same thing. Inertial mass is a property of an isolated body in an empty universe, while gravitational mass is a property of a system of bodies. Einstein's strong version of the Equivalence Principle guided his research for a generalization of the relativity theory. The next step in his quest was to find a way to express such equivalent condition in a formal mathematical way where physics laws were the same in accelerated and rotating reference frames. Einstein struggled for some 5 years until his university friend, mathematician Marcel Grossmann, suggested the use of the non-euclidean geometry of Bernhard Riemann. He also adopted differential calculus and the notion of a metric tensor, developed also by Ricci and Levi-Civita in 1900. The metric tensor is a set of 10 functions of two tangent vectors defined on a manifold which produces a real number. A Riemannian manifold gives a positive value to every non-zero vector. Metric tensors are used to define and calculate the length of curves in the manifold. The Riemannian manifold and the metric tensor form a metric space with a distance function. Einstein and Grossmann's (1913) proposal ran into three serious problems: (i) It violated the principle of causality; (ii) the motion equation did not reduce to Poisson's equation for Newtonian gravity; (iii) the field equations violated the conservation of energy. In the summer of 1915, mathematician David Hilbert invited Einstein to give a series of lectures at the University of Gottingen. In november 1915 Hilbert sent a postcard to Einstein in Berlin, explaining he had derived the correct field equations by a variational principle using a Lagrangian function that was invariant under coordinate transformations. Einstein rushed to derive his equations the hard way and publish his results in december 1915. He published a polished complete review with the whole theory in march of 1916, which is the date we consider in the *gravity apple tree* [See 10, 29–31].

2.4. Prior Geometry

Scalar Lorentz Covariant theories enter in a much wider group of theories that still contemplate a flat non-dynamical space-time as background η [See 32–37]These theories with prior geometry are a subgroup of the metric theories, which is represented by the right and top side branches. Metric theories postulate that matter follows geodesics responding only to the metric $\mathbf{g}$ and not to any extra field. Extra fields can contribute to the metric's curvature, and can be generated by matter, but they cannot interact with it directly. Therefore, $\mathbf{g}$ is the only field that enters the equations of motion. Metric theories were developed during the 50s 60s and 70s, when It became clear the need for some neutral parameters that allowed comparison between theories and with experiments in the Solar System. Sir Arthur Eddington (1922) formulated the first post-newtonian parametrization (PPN). Ken Nordtvedt (1968, 1969), Clifford Will (1971) and Wei-Tou Ni (1972) have developed 9 new parameters that answer questions such as how much space curvature is produced by unit rest mass? Or how much gravity is produced by unit of kinetic energy, of gravitational potential energy, internal energy or by pressure?

3. For a classical unification

The idea to unifying gravity and electromagnetism was first put forward by Poincaré in his lecture on "New conceptions of matter" (1912). Four branches emerged from this unifying ideas exploring, mainly, mathematical alternatives to treat symmetries and measuring objects such as the affine connection and the metric tensor, obtaining much general geometries, and theories with gauge invariance, torsion, non-metricity and extra dimensions [38–49] [See 13, 14, 50–54].

4. Generalizations of GR

Metric theories with additional dynamical fields were suggested separately by five groups of scientists, represented by: Willy Scherrer (1941)[55] a german mathematician working at the university of Bern; Pascual Jordan (1947) [56], in Princeton; Yves Thiry (1948) [57, 58] in Paris; and Robert H. Dicke and his Phd student Carl H. Brans (1961), also at Princeton. Scherrer's approach came form "wave mechanics"; Jordan's and Thiry's group projected Kaluza's five dimensional space-time back into the fourth dimension, associating ϕ with a variable gravitational "constant" G, according with Eddington's and Dirac's Large Numbers Hypothesis [59, 60]. This hypothesis states that the strength of gravity is inversely proportional to the age of the universe, and its mass being proportional to the square of the universe's age implies a a particular cosmology. [See 61]. Dicke's and Brans' theory considered strongly Mach's Principle, which basically states that "the inertial forces observed locally in an accelerated laboratory may be interpreted as gravitational effects having their origin in distant matter accelerated relative to the laboratory" [62] [See 62, 63].This theory expects to prove only a weak version of the Equivalence Principle. The formulation presented by [64, 65] is a generalization of the scalar-tensor theories; Jacob Bekenstein (1977) [66] proposed a Variable Mass Theory (VMT) where rest masses of elementary particles are allowed to vary in space-time via the scalar field, while Barker (1978) [67] proposed a constant G. An even greater generalization is that of Hordenski (1974) [68]. See also [69–71] for vector-tensor theories.

5. Towards quantum gravity

There are two main ways to approach the quantum field theory. One, supposes a fictitious background space, separating the gravitational field into the sum of two components: the background and a quantum field. This would be the approach taken by String Theory [72] which departs from Kaluza's-Klein Theory. The second one is to reconstruct the Quantum Field Theory with no background space, where the space itself is quantified. Charles Misner proposed that calculations could be performed by summing over all possible space-times, applying the ADM formalism (named after its developers: Richard Arnowitt, Stanley Deser and Charles Misner) [73]; and along John Wheeler [74] thought space-time must had a foam-like structure at very small scales . Wheeler and Bryce Dewitt (1967) [75] proposed a wave function over geometries which expresses the probability of having one space-time geometry rather than another. Loop quantum gravity is a mathematical description that uses a novel formulation of GR due to Abhay Ashtekar [76] where the Ashtekar connection field replaces Einstein's gravitational potential in terms of elementary quantum excitations called loops [See 77].

6. New Phenomena

The top set of branches is related to the unexplained new phenomena: (i) The homogeneity of the early universe as interpreted by the measurements of the Cosmic Microwave Background radiation (CMB) and its supposed transition from a curved into a flat structured universe, which are commonly explained by invoking an inflationary period. (ii) The present accelerated expansion of the universe, called dark energy. (iii) The dark matter problem, which refers to the discrepancies between the predictions of Newton's laws and the actual dynamics observed in astronomical systems, as well as the failure of general relativity to explain the observed gravitational lenses and the structure formation, amongst other anomalies. Besides the top set of branches that includes the standard model LCDM, the F(R) theories, and the *MONDian* theories, we can find other alternative theories at the extremes of the main branches, such as John Moffat's Modified Gravity (MOG)[31, 78] better related to Brans-Dicke theory or Mannheim's Conformal Gravity, more related to Wey'ls geometry [See 7, 79–81].

6.1. Dark Matter

A first natural hypothesis to the flat rotational galactic curves was to maintain Newton's law valid, as well as Einstein's GR, and suppose the existence of undetected baryonic matter. This approach produced a number of models for dark entities such as baryonic MACHO's which have been ruled out, since the amount of baryons needed does not match with the early nucleosynthesis model, the observed abundances of elements and the baryon acoustic oscillations. Other alternatives are scalar fields, axions, neutrinos, massive neutrinos and other non-baryonic particles with seemingly ad hoc properties. Hot, warm or cold dark matter models, depending on velocity, are tested on computer simulations of cosmic and galactic structure expecting to tune the right properties that match the observations [See e.g. 82–86] The standard model has given more weight to the very precise data obtained from the CMB and in some sense, has worked its way down from cosmology towards galactic domains, having here most of its problems to matching theory and data. Nevertheless, only direct detection and manipulation of these particles will prove the theory right, since it seems to be almost unfalsifiable [See 81, 87, 88].

6.2. The MONDian Branch

In contrast to the standard model, the top right sided branch emerged from a phenomenological theory that gave priority to galactic phenomena and supposes the possibility that Newton's laws may not be valid at some point. This branch follows the heuristic insights of Mordehai Milgrom's Modified Newtonian Dynamics (MOND) where modifications of the inertial or gravitational laws are solved by the addition of fields [See e.g. 89–96]. The guiding line for *MONDian*theories is the observation that the dynamics depends on the acceleration regime that a particular system experiences, which depends on the surface density of the system. Below a critical acceleration known as Milgrom's fundamental acceleration $\sim 10^{-10}\mathrm{m/s^2}$ [See 97], systems are affected in such a way that the gravitational acceleration a test particle experiences at such sufficiently large distance is greater than its Newtonian value [See 98]. Most *MONDian* theories do not fulfill the strong version of the equivalence principle; therefore, it is conjectured that gravitational dynamics of the systems are to be affected by the field of their host system. This external field effect has not been observed [See 6, 99, 100].

6.3. F(R)

The top middle branch, which explores higher order invariants in the Einstein-Hilbert's action, has two clear periods of growth. Only the second one appears in the *gravity apple tree* for reasons of space and clarity. Nevertheless, it is interesting to note that as early as 1919, Hermann Weyl and Arthur Eddington (1923) explored functions of curvature, mainly as a scientific mathematical curiosity, without any theoretical or experimental reason, but that of a better understanding. In this same spirit, Einstein explored what is now known as the Palatini variation of the action, where the metric and the connection are considered independent variables, and so one can vary the action with respect to both of them. Even though both variational principles lead to the same equations when the Lagrangian is linear in R, this is not true for functions of higher order. [See 101]. Motivations started coming in the 1960s when Utiyama showed that if the Einstein-Hilbert action had higher order curvature invariants with respect to the Ricci scalar, it could be renormalizable. These corrections seemed to be only relevant for very strong gravity (near black holes) or at Planck scales as in the early universe. It is in 1980 when Starobinsky[102] applies this option to model a curvature driven inflation scenario avoiding singularities. [See 103].

6.4. Extended Gravity (EG)

More recently, extensions to the Lagrangian have been added in order to explain galactic anomalies without the hypothesis of dark matter. The proposal of Hernández and Mendoza has qualitatively described the dynamics of all sort of symmetric galactic systems [104–106], including dwarf spheroidal galaxies, as well as gravitational lenses and extended binary star systems [C.f. 107, 108]. Extended gravity is also compatible with the current acceleration of the universe and provides a good solution for the initial values problem [109–111]. This approach relates to the F(R) theories since it tries to preserve the same principles of GR such as Einstein's Equivalence Principle, the Principle of Relativity, the geodesic motion of particles trough space-time, but incorporates as basic empirical principles Milgrom's acceleration constant and a surface density scale. EG does not necessarily predict the external field effect, though finer comparisons await a more finished theory including a generalized poisson equation [See 112, 113].

7. Remarks

The *gravity apple tree* is a visual tool that facilitates reflexions about foundations on physics, components of theories, and the image of science. With its aid we can easily illustrate progress and relations between theories; distinguish with different colors metric from non-metric theories, or use circles to make valid theories stand out from unviable ones; we can use arrows to show how theories are recycled or merged, and it can be easily illustrated how new objectives and new phenomena have revitalized most of the paths explored.

This type of visualizations are not intended to confirm facts, but as a learning tool. The image of the science in question projected by the *gravity apple tree* may be misleading in some senses. For example, it does not [yet] reflect how popular theories are. It does not include all the theories, and for any unintended omissions I apologize. Nevertheless, the image of such plurality of theories is shocking and invites to some philosophical reflections. Plurality of theories is a natural scientific way, which is to explore all the possible options available, mostly, just for scientific spirit. But the image of science provided by such plurality and the fact that many of these theories can be proved viable in the Solar System takes us to the natural philosophical question of how to rely on the truthfulness of our theories from a realistic and monistic point of view. This point of view generally considers this kind of plurality as a crisis and expects a revolution to happen soon. The key toward the next revolution is not just a new great hypothesis, or the look for the best method, since, as we have seen, approaches and methods are plural as well. Crucial experiments may serve to discard some theories, but scientists are ingenious and will look for ways to go around such problems to save their theories as much as possible. Simplicity of concepts will always be preferred, as well as the power to unify lower level principles. Nicholas Maxwell [114] claims that a strategy that always succeeds is to devise neutral observational terms between competing theories, which can be used to describe phenomena that constitute crucial experiments. Nature is the inevitable dictator. It constrains our theories through measurements of phenomena, observations and experiments. In this sense, the PPN have been with out a doubt the best way to prove a theory valid in the Solar System. Such kind of formalisms are to be considered for new theories in the low acceleration domain. For example, according to [115] it is possible to calculate the γ PPN in the *MONDian* regime, just where the third law of Kepler is substituted by the Tully-Fisher law. Amazingly, observations of individual galaxies, galaxy groups and galaxy clusters all imply that $\gamma \sim 1$, much in coincidence with the observed value in the Solar System where the third law applies. PPN type formalisms are being currently developed towards comparing theories for the cosmological scenarios. These might be called the Post Friedmanian Parameters. Many questions have been answered the past century, and much about curved spaces, acceleration and inertia. The actual question for gravitation is how matter actually curves space-time. This can only be known by astronomical observations.

8. Acknowledgements

The author acknowledges support granted by the Universidad Nacional Autónoma de México and by CONACyT (357910).

References

[1] Bunge M 1967 *Foundations of physics* Springer tracts in natural philosophy (Springer-Verlag) URL http://books.google.com.mx/books?id=8pPQAAAAMAAJ

[2] Belohlavek R 2008 URL http://phoenix.inf.upol.cz/esf/ucebni/formal.pdf

[3] Thorne K S, Lee D L and Lightman A P 1973 *Phys. Rev. D* **7**(12) 3563–3578 URL http://link.aps.org/doi/10.1103/PhysRevD.7.3563

[4] Will C 1993 *Theory and Experiment in Gravitational Physics* (Cambridge University Press) ISBN 9780521439732 URL http://books.google.com.mx/books?id=BhnUITA7sDIC

[5] Gronwald F and Hehl F W 1996 *ArXiv General Relativity and Quantum Cosmology e-prints* (*Preprint* gr-qc/9602013)

[6] Famaey B and McGaugh S S 2012 *Living Reviews in Relativity* **15** URL http://www.livingreviews.org/lrr-2012-10

[7] Mannheim P D 2006 *Progress in Particle and Nuclear Physics* **56** 340–445 (*Preprint* astro-ph/0505266)

[8] Wikipedia 2014 Alternatives to general relativity — wikipedia, the free encyclopedia [Online; accessed 6-November-2014] URL \url{http://en.wikipedia.org/w/index.php?title=Alternatives_to_general_relativity&oldid=632239596}

[9] Newton I 1687 *Philosophiae naturalis principia mathematica* (J. Societatis Regiae ac Typis J. Streater) URL http://books.google.com.mx/books?id=-dVKAQAAIAAJ

[10] Misner CW Thorne K and Wheeler J 1973 *Gravitation* (*Gravitation* no parte 3) (W. H. Freeman) ISBN 9780716703440 URL http://books.google.com.mx/books?id=w4Gigq3tY1kC

[11] Norton J D 1993 *Reports on Progress in Physics* **56** 791 URL http://stacks.iop.org/0034-4885/56/i=7/a=001

[12] Poincaré H 2007 *The Genesis of General Relativity* (*Boston Studies in the Philosophy of Science* vol 250) ed Janssen M, Norton J, Renn J, Sauer T and Stachel J (Springer Netherlands) pp 1179–1198 ISBN 978-1-4020-3999-7 URL http://dx.doi.org/10.1007/978-1-4020-4000-9_19

[13] Norton J 1992 *Archive for History of Exact Sciences* **45** 17–94 ISSN 0003-9519 URL http://dx.doi.org/10.1007/BF00375886

[14] Baryshev Y V 1999 *ArXiv General Relativity and Quantum Cosmology e-prints* (*Preprint* gr-qc/9912003)

[15] Birkhoff G D 1943 *Proceedings of the National Academy of Sciences* **29** 231–239 (*Preprint* http://www.pnas.org/content/29/8/231.full.pdf+html) URL http://www.pnas.org/content/29/8/231.short

[16] Birkhoff G D 1944 *Proceedings of the National Academy of Sciences* **30** 324–334 (*Preprint* http://www.pnas.org/content/30/10/324.full.pdf+html) URL http://www.pnas.org/content/30/10/324.short

[17] Moshinsky M 1950 *Phys. Rev.* **80**(4) 514–519 URL http://link.aps.org/doi/10.1103/PhysRev.80.514

[18] Kalman G 1961 *Phys. Rev.* **123**(1) 384–390 URL http://link.aps.org/doi/10.1103/PhysRev.123.384

[19] Logunov A and Mestvirishvili M 1984 *Theoretical and Mathematical Physics* **61** 1170–1183 ISSN 0040-5779 URL http://dx.doi.org/10.1007/BF01035001

[20] Logunov A A and Mestvirishvili M A 1986 *Foundations of Physics* **16** 1–26

[21] Galindo S, Cabrera C and Cervantes-Cota J L 2013 *Revista mexicana de fÃsica E* **59** 101 – 114 ISSN 1870-3542 URL http://www.scielo.org.mx/scielo.php?script=sci_arttext&pid=S1870-35422013000200008&nrm=iso

[22] Nordstrom G 1912 *Physikalische. Zeitschrift* **13**(1126)

[23] Einstein Albert; Fokker A D 1914 *Annalen der Physik* **44** 321–328

[24] Littlewood D E 1953 *Mathematical Proceedings of the Cambridge Philosophical Society* **49**(01) 90–96 ISSN 1469-8064 URL http://journals.cambridge.org/article_S0305004100028085

[25] Bergmann O 1956 *American Journal of Physics* **24** 38–42

[26] Whitrow G and Morduch G 1965 *Vistas in Astronomy* **6** 1 – 67 ISSN 0083-6656 URL http://www.sciencedirect.com/science/article/pii/0083665665900024

[27] Page C and Tupper B O J 1968 *MNRAS* **138** 67

[28] Yilmaz H 1971 *Phys. Rev. Lett.* **27**(20) 1399–1402 URL http://link.aps.org/doi/10.1103/PhysRevLett.27.1399

[29] Einstein A 1916 *Annalen der Physik* **354** 769–822 ISSN 1521-3889 URL http://dx.doi.org/10.1002/andp.19163540702

[30] Stachel J 1999 *Journal of Astrophysics and Astronomy* **20** 91–101 ISSN 0250-6335 URL http://dx.doi.org/10.1007/BF02702345

[31] Moffat J W 2005 *J. Cosmology Astropart. Phys.* **5** 003 (*Preprint* astro-ph/0412195)

[32] Deser S and Laurent B 1968 *Annals of Physics* **50** 76 – 101 ISSN 0003-4916 URL http://www.sciencedirect.com/science/article/pii/0003491668903175

[33] Bollini C, Giambiagi J and Tiomno J 1970 *Lettere al Nuovo Cimento* **3** 65–70 ISSN 0375-930X URL http://dx.doi.org/10.1007/BF02755901

[34] Ni W T 1972 *ApJ* **176** 769

[35] Ni W T 1973 *Phys. Rev. D* **7**(10) 2880–2883 URL http://link.aps.org/doi/10.1103/PhysRevD.7.2880

[36] Lightman A P and Lee D L 1973 *Phys. Rev. D* **8**(10) 3293–3302 URL http://link.aps.org/doi/10.1103/PhysRevD.8.3293

[37] Rosen N 1973 *General Relativity and Gravitation* **4** 435–447

[38] Weyl H 1919 *Annalen der Physik* **364** 101–133 ISSN 1521-3889 URL http://dx.doi.org/10.1002/andp.19193641002

[39] Utiyama R 1956 *Phys. Rev.* **101**(5) 1597–1607 URL http://link.aps.org/doi/10.1103/PhysRev.101.1597

[40] Cartan E 1923 *Ann. c. Norm.* **40** 325–412

[41] Cartan E 1924 *Ann. c. Norm.* **41** 1–25

[42] Cartan E 1925 *Ann. c. Norm.* **42** 17–88

[43] Kibble T W B 1961 *Journal of Mathematical Physics* **2** 212–221 URL http://scitation.aip.org/content/aip/journal/jmp/2/2/10.1063/1.1703702

[44] Sciama D W 1964 *Rev. Mod. Phys.* **36**(1) 463–469 URL http://link.aps.org/doi/10.1103/RevModPhys.36.463

[45] Moffat J W 1995 *Physics Letters B* **355** 447–452 (*Preprint* gr-qc/9411006)

[46] Hehl F W, von der Heyde P, Kerlick G D and Nester J M 1976 *Rev. Mod. Phys.* **48**(3) 393–416 URL http://link.aps.org/doi/10.1103/RevModPhys.48.393

[47] Einstein A 1928 *Session Report of the Prussian academy of Sciences* **354** 224–227

[48] Kaluza T 1921 *Sitzungsberichte der Preussischen Akademie der Wissenschaften* **Juli-Dec** 966–972

[49] Klein O 1926 *Zeitschrift fr Physik* **37** 895–906 ISSN 0044-3328 URL http://dx.doi.org/10.1007/BF01397481

[50] Demopoulos W, Frappier M and Bub J 2012 *Studies in History and Philosophy of Science Part B: Studies in History and Philosophy of Modern Physics* **43** 221 – 225 ISSN 1355-2198 URL http://www.sciencedirect.com/science/article/pii/S1355219812000147

[51] Hacyan S 2007 *ArXiv e-prints* (*Preprint* 0712.2222)

[52] Kastrup H A 2008 *Annalen der Physik* **520** 631–690 (*Preprint* 0808.2730)

[53] Arcos H and Pereira J 2004 *International Journal of Modern Physics D* **13** 2193–2240

[54] Romero C, Fonseca-Neto J B and Pucheu M L 2012 *Classical and Quantum Gravity* **29** 155015 URL http://stacks.iop.org/0264-9381/29/i=15/a=155015

[55] Scherrer W 1941 *Helv. Phys.* **14** 81 84 (*Preprint* gr-qc/0412038)

[56] Jordan P and Müller C 1947 *Z. Naturforsch.* **2a** 1–2

[57] Thiry Y 1948 *C. R. Hebd. Seanc. Acad. Sci.* **226** 216–218

[58] Thiry Y and de Paris U 1951 *Etude mathématique des équations d'une théorie unitaire à quinze variables de champ* (Université de Paris) URL http://books.google.com.mx/books?id=W08AHAAACAAJ

[59] Dirac P A 1937 *Nature* **139** 323

[60] Dirac P A M 1938 *Royal Society of London Proceedings Series A* **165** 199–208

[61] Barrow J 2002 *The Constants of Nature: From Alpha to Omega–the Numbers that Encode the Deepest Secrets of the Universe* (Pantheon Books) ISBN 9780375422218 URL http://books.google.com.mx/books?id=MJXuAAAAMAAJ

[62] Brans C and Dicke R H 1961 *Phys. Rev.* **124**(3) 925–935 URL http://link.aps.org/doi/10.1103/PhysRev.124.925

[63] Goenner H 2012 *General Relativity and Gravitation* **44** 2077–2097 ISSN 0001-7701 URL http://dx.doi.org/10.1007/s10714-012-1378-8

[64] Wagoner R V 1970 *Phys. Rev. D* **1**(12) 3209–3216 URL http://link.aps.org/doi/10.1103/PhysRevD.1.3209

[65] Bergmann P 1968 *International Journal of Theoretical Physics* **1** 25–36 ISSN 0020-7748 URL http://dx.doi.org/10.1007/BF00668828

[66] Bekenstein J D 1977 *Phys. Rev. D* **15**(6) 1458–1468 URL http://link.aps.org/doi/10.1103/PhysRevD.15.1458

[67] Barker B M 1978 *ApJ* **219** 5–11

[68] Horndeski G 1974 *International Journal of Theoretical Physics* **10** 363–384 ISSN 0020-7748 URL http://dx.doi.org/10.1007/BF01807638

[69] Will C M and Nordtvedt Jr K 1972 *ApJ* **177** 757

[70] Nordtvedt Jr K 1970 *ApJ* **161** 1059

[71] Hellings R W and Nordtvedt K 1973 *Phys. Rev. D* **7**(12) 3593–3602 URL http://link.aps.org/doi/10.1103/PhysRevD.7.3593

[72] Bergshoeff E, Sezgin E and Townsend P K 1987 *Physics Letters B* **189** 75–78

[73] Arnowitt R, Deser S and Misner C W 2008 *General Relativity and Gravitation* **40** 1997–2027 (*Preprint* gr-qc/0405109)

[74] Wheeler J A 1955 *Phys. Rev.* **97**(2) 511–536 URL http://link.aps.org/doi/10.1103/PhysRev.97.511

[75] Dewitt B S 1967 *Physical Review* **160** 1113–1148

[76] Ashtekar A 1987 *Physical Review D* **36** 1587–1602

[77] Rovelli C 2003 *Physics World*

[78] Moffat J W 2006 *J. Cosmology Astropart. Phys.* **3** 004 (*Preprint* gr-qc/0506021)

[79] Magueijo J 2003 *Faster Than the Speed of Light: The Story of a Scientific Speculation* (Perseus Book Group) ISBN 9780738205250 URL http://books.google.com.mx/books?id=RwobAQAAIAAJ

[80] Moffat J 2008 *Reinventing Gravity: A Physicist Goes Beyond Einstein* (Dundurn) ISBN 9780887622823 URL http://books.google.ca/books?id=JQg-NwAACAAJ

[81] Capozziello S, Consiglio L, De Laurentis M, De Rosa G and Di Donato C 2011 *ArXiv e-prints* (*Preprint* 1110.5026)

[82] White S, Frenk C S and Davis M 1983 *ApJL* **274** L1–L5

[83] Ostriker J P and Peebles P J E 1973 *ApJ* **186** 467–480

[84] Bond J, Efstathiou G and Silk J 1980 *Phys. Rev. Lett.* **45** 1980–1984

[85] Blumenthal G R, Faber S M, Primack J R and Rees M J 1984 *Nature* **311** 517–525

[86] Lovell M R, Eke V, Frenk C S, Gao L, Jenkins A, Theuns T, Wang J, White S D M, Boyarsky A and Ruchayskiy O 2012 *MNRAS* **420** 2318–2324 (*Preprint* 1104.2929)

[87] Sanders R 2010 *The Dark Matter Problem: A Historical Perspective* (Cambridge University Press) ISBN 9781139485739 URL http://books.google.com.mx/books?id=RpuAoqSOWQIC

[88] Peebles P J E 2013 *ArXiv e-prints* (*Preprint* 1305.6859)

[89] Milgrom M 1983 *ApJ* **270** 365–370

[90] Milgrom M 1983 *ApJ* **270** 371–389

[91] Milgrom M 2010 *MNRAS* **403** 886–895 (*Preprint* 0911.5464)

[92] Bekenstein J D 2004 *Physical Review D* **70** 083509 (*Preprint* astro-ph/0403694)

[93] Bekenstein J and Milgrom M 1984 *ApJ* **286** 7–14

[94] Sanders R H 1997 *ApJ* **480** 492 (*Preprint* astro-ph/9612099)

[95] Skordis C 2008 *Physical Review D* **77** 123502 (*Preprint* 0801.1985)

[96] Zlosnik T G, Ferreira P G and Starkman G D 2007 *Physical Review D* **75** 044017 (*Preprint* astro-ph/0607411)

[97] Bernal T, Capozziello S, Cristofano G and de Laurentis M 2011 *Modern Physics Letters A* **26** 2677–2687 (*Preprint* 1110.2580)

[98] Bruneton J P and Esposito-Farèse G 2007 *Physical Review D* **76** 124012

[99] Blanchet L and Novak J 2011 *MNRAS* **412** 2530–2542 (*Preprint* 1010.1349)

[100] Lüghausen F, Famaey B and Kroupa P 2014 *MNRAS* **441** 2497–2507 (*Preprint* 1404.4049)

[101] Sotiriou T P and Faraoni V 2010 *Reviews of Modern Physics* **82** 451–497 (*Preprint* 0805.1726)

[102] Starobinsky A A 1980 *Physics Letters B* **91** 99–102

[103] Capozziello S, Cardone V F and Troisi A 2006 *J. Cosmology Astropart. Phys.* **8** 001 (*Preprint* astro-ph/0602349)

[104] Mendoza S, Hernandez X, Hidalgo J C and Bernal T 2011 *MNRAS* **411** 226–234 (*Preprint* 1006.5037)

[105] Hidalgo J C, Mendoza S, Hernandez X, Bernal T, Jimenez M A and Allen C 2012 *American Institute of Physics Conference Series* (*American Institute of Physics Conference Series* vol 1458) ed Beltrán Jiménez J, Ruiz Cembranos J A, Dobado A, López Maroto A and De la Cruz Dombriz A pp 427–430 (*Preprint* 1202.4189)

[106] Mendoza S 0 *Canadian Journal of Physics* **0** null (*Preprint* http://dx.doi.org/10.1139/cjp-2014-0208) URL http://dx.doi.org/10.1139/cjp-2014-0208

[107] Mendoza S 2012 *ArXiv e-prints* (*Preprint* 1208.3408)

[108] Hernandez X, Jiménez M A and Allen C 2012 *European Physical Journal C* **72** 1884 (*Preprint* 1105.1873)

[109] Mendoza S, Bernal T, Hernandez X, Hidalgo J C and Torres L A 2013 *Monthly Notices of the Royal Astronomical Society* URL http://mnras.oxfordjournals.org/content/early/2013/06/17/mnras.stt752.abstract

[110] Hernandez X and Jimenez M A 2013 *ArXiv e-prints* (*Preprint* 1307.0777)

[111] Carranza D A, Mendoza S and Torres L A 2013 *European Physical Journal C* **73** 2282 (*Preprint* 1208.2502)

[112] Capozziello S and De Laurentis M 2013 *ArXiv e-prints* (*Preprint* 1307.4523)

[113] Harko T, Lobo F S N, Nojiri S and Odintsov S D 2011 *Physical Review D* **84** 024020 (*Preprint* 1104.2669)

[114] Maxwell N 2014 *Journal for General Philosophy of Science* **45** 133–149 ISSN 0925-4560 URL http://dx.doi.org/10.1007/s10838-013-9225-5

[115] Mendoza S and Olmo G J 2014 *ArXiv e-prints* (*Preprint* 1401.5104)

[116] White S, Frenk C S and Davis M 1983 *ApJ* **271** 417–430

[117] Witten L 1962 *Gravitation: An Introduction to Current Research* (Wiley) URL http://books.google.com.mx/books?id=3llAAAAAIAAJ

[118] Brans C H 1961 *Mach's principle and a varying gravitational constant* Ph.D. thesis Princeton University

[119] Mannheim P D and Kazanas D 1989 *ApJ* **342** 635–638

Spanish Relativity Meeting (ERE 2014): almost 100 years after Einstein's revolution IOP Publishing
Journal of Physics: Conference Series **600** (2015) 012051 doi:10.1088/1742-6596/600/1/012051

Symmetry operators and conserved currents

Thomas Bäckdahl

The School of Mathematics, University of Edinburgh, James Clerk Maxwell Building, Peter Guthrie Tait Road, Edinburgh EH9 3FD, UK

E-mail: t.backdahl@ed.ac.uk

Abstract. In these proceedings we present recent progress concerning existence of symmetry operators for the conformal scalar wave equation, the Dirac-Weyl equation and the source-free Maxwell equation. Furthermore, we consider conserved currents for the source-free Maxwell equations.

1. Introduction

In these proceedings we consider *symmetry operators*, i.e. linear differential operators which take solutions of a differential equation to solutions. These are versatile tools for the analysis of fields on spacetimes with special structure. A well-known example of a symmetry operator for the scalar wave equation is provided by the Lie derivative along a Killing vector field. However, in the Kerr spacetime, there is also another symmetry operator that arises from a Killing spinor rather than a Killing vector. This was an essential tool in a proof of decay of scalar waves on the Kerr background by Andersson and Blue [4].

As the symmetry operators are so useful, it is natural to ask when they exist. Here we will consider what conditions on a spacetime are necessary and sufficient for existence of symmetry operators for the conformal wave equation, the Dirac-Weyl equation, and the Maxwell equation, i.e. for massless test fields of spins 0, 1/2 and 1. We will investigate how the conditions for the symmetry operators for the different field equations are related, and how they are related to existence of conserved currents. This exposition is mainly based on the paper [3] about symmetry operators, but also ideas from [2] about conserved currents are presented.

2. Preliminaries

2.1. Spinors

Throughout, we will work on an orientable and time orientable globally hyperbolic spacetime with metric g_{ab} satisfying Einstein's field equations. We will use the $(+---)$ sign convention and extensively use spinor notation from [6].

2.2. Irreducible decompositions

Any spinor can be decomposed so it is written entirely in terms of the spin metric and symmetric spinors formed from the original spinor by taking traces and symmetrizing the remaining indices. For instance, a general valence $(3,0)$ spinor is decomposed as

$$T_{ABC} = T_{(ABC)} - \tfrac{1}{6}T^{D}{}_{CD}\epsilon_{AB} - \tfrac{1}{6}T^{D}{}_{DC}\epsilon_{AB} - \tfrac{1}{6}T^{D}{}_{BD}\epsilon_{AC} - \tfrac{1}{6}T^{D}{}_{DB}\epsilon_{AC} - \tfrac{1}{2}T_{A}{}^{D}{}_{D}\epsilon_{BC}.$$

Journal of Physics: Conference Series **600** (2015) 012051 doi:10.1088/1742-6596/600/1/012051

We call this an *irreducible decomposition.* This is an important simplification because now we *only need to work with symmetric spinors.* From an algebraic point of view this is of great importance when one wants to simplify complicated spinorial expressions and tries to write them in a canonical form. Most calculations for the research presented here were done using the *xAct* suite for *Mathematica.* In fact I have written the *SymManipulator* package which is a part of *xAct*, and includes an efficient algorithm to do irreducible decompositions of arbitrary spinor expressions.

As we have seen, one can work with only symmetric spinors, but a covariant derivative of a symmetric spinor is not necessarily symmetric, so we need to do an irreducible decomposition of that too. This gives us four natural differential operators that maps symmetric spinors to symmetric spinors. For instance, consider the irreducible decomposition of the covariant derivative of a symmetric valence $(3, 2)$ spinor.

$$\nabla_A{}^{A'}T_{BCD}{}^{B'C'} = (\mathscr{T}_{3,2}T)_{ABCD}{}^{A'B'C'} - \tfrac{2}{3}\bar{\epsilon}^{A'(B'}(\mathscr{C}_{3,2}T)_{ABCD}{}^{C')} \\ - \tfrac{3}{4}\epsilon_{A(B}(\mathscr{C}^{\dagger}_{3,2}T)_{CD)}{}^{A'B'C'} + \tfrac{1}{2}\epsilon_{A(B}\bar{\epsilon}^{A'(B'}(\mathscr{D}_{3,2}T)_{CD)}{}^{C')},$$

where

$$\begin{aligned}(\mathscr{D}_{3,2}T)_{AB}{}^{A'} &= \nabla^{CB'}T_{ABC}{}^{A'}{}_{B'},\\ (\mathscr{C}_{3,2}T)_{ABCD}{}^{A'} &= \nabla_{(A}{}^{B'}T_{BCD)}{}^{A'}{}_{B'},\\ (\mathscr{C}^{\dagger}_{3,2}T)_{AB}{}^{A'B'C'} &= \nabla^{C(A'}T_{ABC}{}^{B'C')},\\ (\mathscr{T}_{3,2}T)_{ABCD}{}^{A'B'C'} &= \nabla_{(A}{}^{(A'}T_{BCD)}{}^{B'C')}.\end{aligned}$$

These operators generalizes to arbitrary valence as long as enough indices exist to form the index contractions, so we get $\mathscr{D}_{k,l}$, $\mathscr{C}_{k,l}$, $\mathscr{C}^{\dagger}_{k,l}$ and $\mathscr{T}_{k,l}$ for symmetric valence (k, l) spinors. More details about these operators including commutator relations can be found in [3].

Definition 2.1. *A symmetric valence (k,l) spinor $L_{A\dots F}{}^{A'\dots F'}$ is called a valence (k,l)* Killing spinor *if $(\mathscr{T}_{k,l}L)_{A\dots G}{}^{A'\dots G'} = 0$.*

3. Symmetry operators

With our notation the field equations we would like to study now takes the forms:

- The conformal wave equation (Spin-0): $(\square + 4\Lambda)\phi = 0$.
- The Dirac-Weyl equation (Spin-1/2): $(\mathscr{C}^{\dagger}_{1,0}\phi)_{A'} = 0$.
- The Maxwell equation (Spin-1): $(\mathscr{C}^{\dagger}_{2,0}\phi)_{AA'} = 0$.

As mentioned in the introduction, what we mean by a symmetry operator is a linear differential operator that maps a solution of a differential equation to a solution of the differential equation. However, we can slightly extend this notion so that an operator that map to solutions of the complex conjugate version of the differential equation instead is also considered to be a symmetry operator. As the complex conjugate of the Dirac-Weyl equation is $(\mathscr{C}_{0,1}\bar{\phi})_A = 0$ and similarly for the Maxwell equation, we can formulate the two different kinds of symmetry operators for spin-1/2 and spin-1 as:

- First kind: $\ker\mathscr{C}^{\dagger} \to \ker\mathscr{C}^{\dagger}$. (Preserves parity)
- Second kind: $\ker\mathscr{C}^{\dagger} \to \ker\mathscr{C}$. (Changes parity)

The conformal wave equation is real so we do not have a notion of second kind symmetry operators for that case. In the paper [3], we study both the first and second kind symmetry operators up to second order, but here we will restrict ourselves to the first kind of operators.

To give clear statements for existence of symmetry operators, we introduce the following operators that depends on the curvature.

Definition 3.1. *For a valence* $(2,2)$ *Killing spinor* $L_{AB}{}^{A'B'}$, *we define*

$$\begin{aligned}(\mathcal{O}^{(0)}_{2,2}L)_{A}{}^{A'} \equiv{}& \tfrac{1}{3}\Psi_{ABCD}(\mathscr{C}_{2,2}L)^{BCDA'} + \tfrac{1}{3}\bar{\Psi}^{A'}{}_{B'C'D'}(\mathscr{C}^{\dagger}_{2,2}L)_{A}{}^{B'C'D'}\\ &+ L^{BCA'B'}(\mathscr{C}_{2,2}\Phi)_{ABCB'} + L_{A}{}^{BB'C'}(\mathscr{C}^{\dagger}_{2,2}\Phi)_{B}{}^{A'}{}_{B'C'},\\ (\mathcal{O}^{(1)}_{2,2}L)_{AB}{}^{A'B'} \equiv{}& L^{CDA'B'}\Psi_{ABCD} - L_{AB}{}^{C'D'}\bar{\Psi}^{A'B'}{}_{C'D'}.\end{aligned}$$

It turns out that the conditions for existence of symmetry operators involves existence of valence $(2,2)$ Killing spinors related to the curvature through the operators in definition 3.1.

Definition 3.2. *Let* $L_{AB}{}^{A'B'}$ *be a Killing spinor of valence* $(2,2)$. *Define the conditions*

(A0) There is a function Q *such that* $(\mathcal{O}^{(0)}_{2,2}L)_{A}{}^{A'} = (\mathscr{T}_{0,0}Q)_{A}{}^{A'}$.

(A1) There is a vector field $P_{A}{}^{A'}$ *such that* $(\mathcal{O}^{(1)}_{2,2}L)_{AB}{}^{A'B'} = (\mathscr{T}_{1,1}P)_{AB}{}^{A'B'}$.

The conditions for existence of symmetry operators up to second order for the conformal wave equation was already settled by Michel, Radoux & Šilhan [5], but here we present it in our notation. In the paper [3], we derived conditions for existence of symmetry operators of the first kind up to second order for the Dirac-Weyl and Maxwell equations.

Theorem 3.3 ([5], [3]). *(i) The conformal wave equation has a symmetry operator* $\phi \to \chi$, *with order less or equal to two, if and only if there is a valence* $(2,2)$ *Killing spinor satisfying condition A0.*

(ii) There exists a symmetry operator of the first kind for the Dirac-Weyl equation $\phi_A \to \chi_A$, *with order less or equal to two, if and only if there is a valence* $(2,2)$ *Killing spinor satisfying condition A0 and condition A1.*

(iii) There exists a symmetry operator of the first kind $\phi_{AB} \to \chi_{AB}$ *for the Maxwell equation, with order less or equal to two, if and only if there is a valence* $(2,2)$ *Killing spinor satisfying condition A1.*

To see how one can construct valence $(2,2)$ Killing spinors that satisfies these conditions, we observe that if they are built from conformal Killing vectors or valence $(2,0)$ Killing spinors, they naturally satisfy both conditions A0 and A1.

Proposition 3.4 ([3]). *Let* $\xi^{AA'}$ *and* $\zeta^{AA'}$ *be (not necessarily distinct) conformal Killing vectors and let* κ_{AB} *be a Killing spinor of valence* $(2,0)$.

(i) The symmetric spinor $\xi_{(A}{}^{(A'}\zeta_{B)}{}^{B')}$ *is a Killing spinor of valence* $(2,2)$, *which admits solutions to the auxiliary conditions A0 and A1.*

(ii) The symmetric spinor $\kappa_{AB}\bar{\kappa}_{A'B'}$ *is also a Killing spinor of valence* $(2,2)$, *which admits solutions to the auxiliary conditions A0 and A1.*

It is worth to note that the conditions A0 and A1 are not equivalent to each other. In fact in [3] we give an example of a spacetime and a valence $(2,2)$ Killing spinor that satisfies the A1 condition, but not the A0 condition.

3.1. Maxwell equation

Although we have found expressions for all the symmetry operators in the paper [3], we will here restrict ourselves to symmetry operators of the first kind for the Maxwell equation.

Theorem 3.5 ([3]). *There exists a symmetry operator of the first kind up to order 2 for the Maxwell equation iff*

$$(\mathscr{T}_{2,2}L)_{ABC}{}^{A'B'C'} = 0, \qquad (\mathscr{T}_{1,1}P)_{AB}{}^{A'B'} = -\tfrac{2}{3}(\mathcal{O}^{(1)}_{2,2}L)_{AB}{}^{A'B'}, \qquad (\mathscr{T}_{0,0}Q)_{BA'} = 0.$$

The symmetry operator takes the form

$$\begin{aligned} \phi_{AB} &\to \chi_{AB} = Q\phi_{AB} + (\mathscr{C}_{1,1}A)_{AB}, \\ A_{AA'} &= -L^{BC}{}_{A'}{}^{B'}(\mathscr{T}_{2,0}\phi)_{ABCB'} + \tfrac{1}{3}\phi^{BC}(\mathscr{C}_{2,2}L)_{ABCA'} - \tfrac{4}{9}\phi_{AB}(\mathscr{D}_{2,2}L)^{B}{}_{A'} - P^{B}{}_{A'}\phi_{AB}. \end{aligned} \tag{3.1}$$

We also note that the potential satisfies $(\mathscr{C}^\dagger_{1,1}A)_{A'B'} = 0$. Furthermore, note that (3.1) is the general form of a first order operator $\ker\mathscr{C}^\dagger_{2,0} \to \ker\mathscr{C}^\dagger_{1,1}$ up to addition of a gradient.

4. Conserved currents

In this section we will consider conserved currents for the Maxwell field, and see how these are connected with the symmetry operators. Here, any divergence-free vector field formed as a concomitant of the Maxwell field will be called a *conserved current.* Typically, we will study the flux of a conserved currents through a timelike hypersurface, which is then a *conserved energy.* A special class of conserved currents are the *trivial currents*, which should not be confused with the zero current.

Definition 4.1. *A* trivial current $\tilde{J}^{AA'}$ *is a field that can be written as*

$$\tilde{J}_{AA'} = (\mathscr{C}^\dagger_{2,0}S)_{AA'} + (\mathscr{C}_{0,2}T)_{AA'}$$

for some symmetric spinor fields S_{AB} and $T_{A'B'}$.

The flux through a hypersurface of a trivial current $\tilde{J}^{AA'}$ is given by a pure boundary term. We will study *equivalence classes* of currents up to trivial currents. For most applications it is natural to study currents which are bilinear in ϕ_{AB} and $\bar{\phi}_{A'B'}$. We can write this class of currents in a simple form.

Lemma 4.2 ([2]). *Assume that $J^{AA'} \in \ker\mathscr{D}_{1,1}$ is a conserved current that is a differential operator that is bilinear in ϕ_{AB} and $\bar{\phi}_{A'B'}$, where $(\mathscr{C}^\dagger_{2,0}\phi)_{AA'} = 0$, then it can be written as*

$$J^{AA'} = A^{A}{}_{B'}\bar{\phi}^{A'B'} + \tilde{J}^{AA'},$$

where $A_{AA'}$ satisfies $(\mathscr{C}^\dagger_{1,1}A)_{A'B'} = 0$ and $\tilde{J}^{AA'}$ is a trivial current.

Remark 4.3. *Observe that $\phi_{AB} \to (\mathscr{C}_{1,1}A)_{AB}$ is a symmetry operator of the first kind for the Maxwell equation.*

Remark 4.4. *The general first order $A_{AA'}$ is given by* (3.1)*. Hence, the conditions for existence of a first order conserved current of this form are the same as for a second order symmetry operator of the first kind.*

[1] L. Andersson, T. Bäckdahl, and P. Blue. A new energy-momentum tensor for Maxwell fields on the Kerr spacetime. In preparation.
[2] L. Andersson, T. Bäckdahl, and P. Blue. Conserved currents. In preparation.
[3] L. Andersson, T. Bäckdahl, and P. Blue. Second order symmetry operators. *Classical and Quantum Gravity*, 31(13):135015, July 2014. arXiv.org:1402.6252.
[4] L. Andersson and P. Blue. Hidden symmetries and decay for the wave equation on the Kerr spacetime. Aug. 2009. arXiv.org:0908.2265.
[5] Michel J P, Radoux F and Šilhan J 2014 *SIGMA* **10**, 016, 26 pages doi:10.3842/SIGMA.2014.016 arXiv.org:1308.1046
[6] R. Penrose and W. Rindler. *Spinors and Space-time I & II.* Cambridge Monographs on Mathematical Physics. Cambridge University Press, Cambridge, 1986.

Journal of Physics: Conference Series **600** (2015) 012052 doi:10.1088/1742-6596/600/1/012052

Asymptotic properties of gravitational and electromagnetic fields in higher dimensions

Marcello Ortaggio, Alena Pravdová

Institute of Mathematics, Academy of Sciences of the Czech Republic
Žitná 25, 115 67 Prague 1, Czech Republic

E-mail: ortaggio@math.cas.cz, pravdova@math.cas.cz

Abstract.
We summarize the fall-off of electromagnetic and gravitational fields in $n > 5$ dimensional Ricci-flat spacetimes along an asympotically expanding non-singular geodesic null congruence.

1. Introduction

Under suitable assumptions, the well-known peeling-off property characterizes the behavior of the gravitational and electromagnetic fields at null infinity (see, e.g., [1, 2] and references therein). It has been observed [3] that the Weyl tensor peels off differently in $n > 4$ dimensions. Here, we summarize our recent results [4, 5] on the leading-order behavior of gravitational and electromagnetic fields in higher dimensions. Ref. [4] partly recovers the results of [3] but uses a different method and different assumptions. We restrict to Ricci-flat spacetimes with suitable properties at null infinity (a cosmological constant can be included [4,5]), formulated in terms of a geodesic null vector field $\boldsymbol{\ell} = \partial_r$ (r is an affine parameter) and of the Weyl tensor, using a "null" frame [6] based on two null vectors $\boldsymbol{m}_{(0)} = \boldsymbol{\ell}$, $\boldsymbol{m}_{(1)} = \boldsymbol{n}$ and $n-2$ orthonormal spacelike vectors $\boldsymbol{m}_{(i)}$ $(i, j, \cdots = 2, \dots, n-1)$. First, we assume that the optical matrix $\rho_{ij} = \ell_{a;b} m^a_{(i)} m^b_{(j)}$ is *asymptotically non-singular and expanding* [4,5] (this includes asymptotically flat spacetimes [3] but also holds more generally – see [7] in four dimensions). Furthermore, we assume that the boost-weight (b.w.) $+2$ Weyl components $\Omega_{ij} \equiv C_{0i0j} = C_{abcd}\ell^a m^b_{(i)} \ell^c m^d_{(j)}$ fall off as

$$\Omega_{ij} = O(r^{-\nu}) \qquad (\nu > 2). \tag{1}$$

Again, this is satisfied in asymptotically flat spacetimes [3] (e.g., $\Omega_{ij} = O(r^{-5})$ in the 4D spacetimes of [7]). Under the above conditions, one is able to determine how the Maxwell and Weyl tensors fall off as $r \to \infty$, as we summarize in sections 2 and 3. However, as an intermediate step, one also needs the r-dependence of the Ricci rotation coefficients and of the derivative operators [6], which is given in [4] (it follows from the Ricci identities [8], also using the commutators [9] and the Bianchi identities [10]). For example, $\rho_{ij} = \frac{\delta_{ij}}{r} + \dots$. For brevity, in this paper, we discuss only results in $n > 5$ dimensions – the case $n \geq 5$ is studied in [4, 5].

2. Electromagnetic field

We start from the simpler case of *test* Maxwell fields in the background of an n-dimensional Ricci-flat spacetime satisfying the assumptions of section 1 [5]. The gravitational field (Weyl tensor) can be treated similarly, however, resulting in a larger number of possible cases (section 3).

In the frame of section 1, we assume that for $r \to \infty$ the Maxwell components have a *power-like* behavior described by

$$F_{0i} = O(r^{\alpha}), \qquad F_{01} = O(r^{\beta}), \qquad F_{ij} = O(r^{\gamma}), \qquad F_{1i} = O(r^{\delta}). \tag{2}$$

The empty-space Maxwell equations $F^{a}{}_{b;a} = 0 = F_{[ab;c]}$ (see [5,11] for their GHP and NP form) determine the possible values of α, β, γ and δ. We assume that if a generic component f behaves as $f = O(r^{-\zeta})$ then $\partial_r f = O(r^{-\zeta-1})$ and $\partial_A f = O(r^{-\zeta})$. As it turns out, α can be chosen arbitrarily, giving raise to two main cases, $\alpha \geq -2$ or $\alpha < -2$. In the latter, one needs to choose whether $\gamma \geq -2$ or $\gamma < -2$, and then specify more precisely the value of α, as we detail.

2.1. *Case $\alpha \geq -2$.*

In this case, all components fall off at the same speed, i.e.,

$$F_{0i} = O(r^{\alpha}), \qquad F_{01} = O(r^{\alpha}), \qquad F_{ij} = O(r^{\alpha}), \qquad F_{1i} = O(r^{\alpha}). \tag{3}$$

The electromagnetic field does not peel. This describes, e.g., a uniform magnetic field permeating asymptotically flat black holes [12] (or black rings [13] if $n = 5$ is included, cf. [5]).

2.2. *Case $\alpha < -2$.*

Generically, we have

$$F_{0i} = O(r^{\alpha}), \tag{4}$$

$$F_{01} = o(r^{-2}), \qquad F_{ij} = O(r^{-2}), \tag{5}$$

$$F_{1i} = O(r^{-2}). \tag{6}$$

The above behavior includes the special case when $\boldsymbol{\ell}$ is an aligned null direction of the Maxwell field, i.e., $F_{0i} = 0$ (in the formal limit $\alpha \to -\infty$). The leading term is of type II. Examples can be obtained as a "linearized" Maxwell field limit of certain full Einstein-Maxwell solutions given in [14] for even n. Several subcases are possible when $\gamma < -2$.

2.2.1. *Subcase (a): $\gamma < -2$ with $1 - \frac{n}{2} \leq \alpha < -2$.*

In this case, one has the same results as in section 2.1 above. This subcase does not exist for $n = 6$.

2.2.2. *Subcase (b): $\gamma < -2$ with $-\frac{n}{2} \leq \alpha < 1 - \frac{n}{2}$.*

Here, we have

$$F_{0i} = O(r^{\alpha}), \tag{7}$$

$$F_{01} = O(r^{\alpha}), \qquad F_{ij} = O(r^{\alpha}), \tag{8}$$

$$F_{1i} = O(r^{1-n/2}). \tag{9}$$

The leading term falls off as $1/r^{\frac{n}{2}-1}$ and is of type N. This is characteristic of radiative fields (note that $T_{11} \propto F_{1i}F_{1i} \sim 1/r^{n-2}$ and the energy flux along $\boldsymbol{\ell}$ can be directly related to the energy loss, at least in the case of asymptotically flat spacetimes – cf. [15–17] for $n = 4$). As opposed to the well-known four-dimensional case, here, $\boldsymbol{\ell}$ cannot be aligned with F_{ab} if radiation is present (since $\alpha \geq -\frac{n}{2}$). In the case $\alpha = -\frac{n}{2}$, if one assumes that F_{1i} has a power-like behavior also at the subleading order, from the Maxwell equations, one finds $F_{1i} = F^{(0)}_{1i} r^{1-\frac{n}{2}} + O(r^{-n/2})$, which gives the peeling-off behavior

$$F_{ab} = \frac{N_{ab}}{r^{\frac{n}{2}-1}} + \frac{G_{ab}}{r^{\frac{n}{2}}} + \ldots \qquad \left(\alpha = -\frac{n}{2}\right). \tag{10}$$

The subleading term is algebraically general, which is qualitatively different from the 4D case [1,2,16,17]. This resembles the behavior of the Weyl tensor of higher dimensional asymptotically flat spacetimes [3]. See [5] for a possible different peeling-off in five dimensions.

2.2.3. Subcase (c): $\gamma < -2$ with $2 - n \leq \alpha < -\frac{n}{2}$. The same results as in section 2.1 apply.

2.2.4. Subcase (d): $\gamma < -2$ with $\alpha < 2 - n$. We have

$$F_{0i} = O(r^{\alpha}), \tag{11}$$
$$F_{01} = O(r^{2-n}), \qquad F_{ij} = o(r^{2-n}), \tag{12}$$
$$F_{1i} = O(r^{2-n}). \tag{13}$$

The leading term is of type II and falls off as $1/r^{n-2}$ (it is purely electric in the subcase $F_{1i} = o(r^{2-n})$). This behavior includes the Coulomb field of a weakly charged asymptotically flat black hole [12, 18] (or black ring [13] if $n = 5$ is included [5]). In the special subcase $F_{01} = o(r^{2-n})$, the same results as in section 2.1 again apply (for example, for $n = 5$ and $\alpha = -4$, this is the case of the weak-field limit of the 5D dipole black rings of [19]).

Let us observe that in all cases, type N fields for which $\boldsymbol{\ell}$ is aligned are not permitted [11, 20].

2.3. The case of p-forms

The above results for a 2-form F_{ab} can be extended easily [5] to p-form fields satisfying the generalized Maxwell equations (given in [11] in the GHP notation). In *even* dimensions, the special case $p = n/2$ (including $n = 4$, $p = 2$) has unique properties. It peels off as

$$F_{a_1\ldots a_p} = \frac{N_{a_1\ldots a_p}}{r^{\frac{n}{2}-1}} + \frac{II_{a_1\ldots a_p}}{r^{\frac{n}{2}}} + \ldots \qquad \left(p = \frac{n}{2}\right). \tag{14}$$

The (radiative) leading term is of type N and falls off as $1/r^{\frac{n}{2}-1}$. In contrast to the case $p = 2$ discussed above (or, in fact, any other $p \neq n/2$), Maxwell fields of type N aligned with $\boldsymbol{\ell}$ are now permitted [5] and the peeling (14) applies also in the presence of a cosmological constant [5]. Corresponding solutions of the *full Einstein-Maxwell equations* have recently been obtained [21].

3. Gravitational field

The method to be used for the Weyl tensor [4] is essentially similar, now $-\nu$ playing the role that α played above. Instead of the Maxwell equations, one has to integrate the system "Bianchi-Ricci-commutators". However, there is now extra freedom in the choice of possible boundary conditions. In particular, three possible choices for the behavior of b.w. $+1$ components Ψ_{ijk} are possible (cases (i), (ii) and (iii) below). Once the fall-off of Ω_{ij} and Ψ_{ijk} has been specified, the next step is to determine the fall-off of the b.w. 0 components Φ_{ijkl}

$$\Phi_{ijkl} = O(r^{\beta_c}). \tag{15}$$

The parameter β_c can then be used to label various possible subcases, which we now present.

3.1. Case (i): $\Omega_{ij} = O(r^{-\nu})$, $\Psi_{ijk} = O(r^{-\nu})$

In all cases given here, we have (this will not be repeated every time below)

$$\Omega_{ij} = O(r^{-\nu}) \quad (\nu > 2), \qquad \Psi_{ijk} = O(r^{-\nu}). \tag{16}$$

3.1.1. Subcase (A): $\beta_c = -2$. In this case, necessarily $\beta_c > -\nu$ and we have the following possible behaviors, depending on how ν is chosen (cf. [4] for a few further special subcases):

A1:

$$\begin{aligned} &\Phi_{ijkl} = O(r^{-2}), \qquad \Phi^S_{ij} = o(r^{-2}), \qquad \Phi^A_{ij} = o(r^{-2}) \qquad (2 < \nu \leq 3),\\ &\Psi'_{ijk} = O(r^{-2}),\\ &\Omega'_{ij} = O(r^{\sigma}) \qquad (-2 \leq \sigma < -1); \end{aligned} \tag{17}$$

A2:

$$\begin{aligned} &\Phi_{ijkl} = O(r^{-2}), \quad \Phi^S_{ij} = O(r^{-3}), \quad \Phi = O(r^{-\nu}), \quad \Phi^A_{ij} = O(r^{-3}) \quad (3 < \nu < 4),\\ &\Psi'_{ijk} = O(r^{-2}), \quad \Psi'_i = O(r^{-3}),\\ &\Omega'_{ij} = O(r^{-2}); \end{aligned} \tag{18}$$

A3:

$$\begin{aligned} &\Phi_{ijkl} = O(r^{-2}), \qquad \Phi^S_{ij} = O(r^{-3}), \qquad \Phi = O(r^{-4}), \qquad \Phi^A_{ij} = O(r^{-3}) \qquad (\nu \geq 4),\\ &\Psi'_{ijk} = O(r^{-2}), \quad \Psi'_i = O(r^{-3}),\\ &\Omega'_{ij} = O(r^{-2}), \end{aligned} \tag{19}$$

with the further restrictions $\Phi^S_{ij} = O(r^{1-\nu})$ for $4 \leq \nu < 5$ and $\Phi^S_{ij} = O(r^{-4})$ for $\nu \geq 5$;

A4:

$$\begin{aligned} &\Phi_{ijkl} = O(r^{-2}), \quad \Phi^S_{ij} = O(r^{1-\nu}), \quad \Phi = O(r^{-\nu}), \quad \Phi^A_{ij} = O(r^{-\nu}) \quad (\nu \geq 4,\, \nu \neq n),\\ &\Psi'_{ijk} = O(r^{-2}), \quad \Psi'_i = O(r^{1-\nu}),\\ &\Omega'_{ij} = O(r^{-2}); \end{aligned} \tag{20}$$

A5:

$$\begin{aligned} &\Phi_{ijkl} = O(r^{-2}), \qquad \Phi^S_{ij} = O(r^{1-n}), \qquad \Phi^A_{ij} = O(r^{-n}) \qquad (\nu \geq n),\\ &\Psi'_{ijk} = O(r^{-2}), \quad \Psi'_i = O(r^{1-n}),\\ &\Omega'_{ij} = O(r^{-2}). \end{aligned} \tag{21}$$

None of the above five cases can describe asymptotically flat spacetimes, cf. [3]. In cases A2–A5, the leading term falls off as $1/r^2$ at infinity and it is of type II(abd). In cases A3–A5, $\boldsymbol{\ell}$ can be a multiple WAND. Examples in case A5 are Robinson-Trautman spacetime [22].

When $\beta_c < -2$, its precise value depends on the value of ν so that we have to consider the following possible cases.

3.1.2. Subcase (B): $\beta_c < -2$ with $\frac{n}{2} < \nu \leq 1 + \frac{n}{2}$. In this case, $\beta_c = -\frac{n}{2}$ and we have

$$\begin{aligned} &\Phi_{ijkl} = O(r^{-n/2}), \qquad \Phi = O(r^{-\nu}), \qquad \Phi^A_{ij} = O(r^{-\nu}) \qquad \left(\frac{n}{2} < \nu \leq 1 + \frac{n}{2}\right),\\ &\Psi'_{ijk} = O(r^{-n/2}),\\ &\Omega'_{ij} = O(r^{1-n/2}). \end{aligned} \tag{22}$$

Here, $\boldsymbol{\ell}$ cannot be a WAND. The leading term at infinity falls off as $1/r^{n/2-1}$ and it is of type N. This includes *radiative spacetimes* [3] that are asymptotically flat in the Bondi definition [23,24]. If one takes for b.w. +2 components $\nu = 1+\frac{n}{2}$ and additionally *assumes* that $\Omega_{ij} = \Omega_{ij}^{(0)} r^{-n/2-1} + \Omega_{ij}^{(1)} r^{-n/2-2} + o(r^{-n/2-2})$, then one finds [4] the peeling-off behavior

$$C_{abcd} = \frac{N_{abcd}}{r^{n/2-1}} + \frac{II_{abcd}}{r^{n/2}} + o(r^{-n/2}). \tag{23}$$

This agrees with [3] for asymptotically flat spacetimes. See [3, 4] for special properties of the case $n = 5$. When $\beta_c < -2$ but ν is not in the range $\frac{n}{2} < \nu \leq 1+\frac{n}{2}$ one has the following subcases (B*) and (C).

3.1.3. Subcase (B): $\beta_c < -2$ with $2 < \nu \leq \frac{n}{2}$ or $1+\frac{n}{2} < \nu \leq n-1$.* In this case, $\beta_c = -\nu$ and we have (cf. section IV A 5 of [4])

$$\begin{aligned} &\Phi_{ijkl} = O(r^{-\nu}), && \Phi_{ij}^{A} = O(r^{-\nu}), \\ &\Psi'_{ijk} = O(r^{-2}) \quad \text{if } 2 < \nu \leq 3, && \Psi'_{ijk} = O(r^{-\nu}) \quad \text{if } \nu > 3, \\ &\Omega'_{ij} = o(r^{1-\nu}) \quad \text{if } \nu \neq \frac{n}{2}, && \Omega'_{ij} = O(r^{1-n/2}) \quad \text{if } \nu = \frac{n}{2}. \end{aligned} \tag{24}$$

Here, $\boldsymbol{\ell}$ cannot be a WAND.

3.1.4. Subcase (C): $\beta_c < -2$ with $\nu > n-1$. In this case, $\beta_c = 1-n$ and we have

$$\begin{aligned} &\Phi_{ijkl} = O(r^{1-n}), && \Phi_{ij}^{A} = o(r^{1-n}) \qquad (\nu > n-1), \\ &\Psi'_{ijk} = O(r^{1-n}), && \\ &\Omega'_{ij} = o(r^{2-n}), && \end{aligned} \tag{25}$$

with $\Phi_{ij}^{A} = O(r^{-\nu})$ for $n-1 < \nu < n$ and $\Phi_{ij}^{A} = O(r^{-n})$ for $\nu \geq n$. Here, $\boldsymbol{\ell}$ can become a multiple WAND, cf. [25]. This includes asymptotically flat spacetimes in the case of *vanishing radiation* [3], such as those for which $\boldsymbol{\ell}$ is a multiple WAND [25], e.g., the Schwarzschild-Tangherlini metric and Kerr-Schild spacetimes [26] with a non-degenerate Kerr-Schild vector.

3.2. Case (ii): $\Omega_{ij} = o(r^{-n})$, $\Psi_{ijk} = O(r^{-n})$

3.2.1. Subcase $\beta_c = -2$. Generically, one has

$$\begin{aligned} &\Omega_{ij} = o(r^{-n}), && && \\ &\Psi_{ijk} = O(r^{-n}), && && \\ &\Phi_{ijkl} = O(r^{-2}), && \Phi_{ij}^{S} = O(r^{-4}), && \Phi_{ij}^{A} = O(r^{-3}), \\ &\Psi'_{ijk} = O(r^{-2}), && \Psi'_{i} = O(r^{-3}), && \\ &\Omega'_{ij} = O(r^{-2}). && && \end{aligned} \tag{26}$$

For $\Psi_{ijk}^{(n)} = 0$, this case reduces to (19) (with $\nu > n$). See [4] for possible subcases.

3.2.2. Subcase $\beta_c = 1-n$. When $\beta_c < -2$ then necessarily $\beta_c = 1-n$ and generically, one has

$$\begin{aligned} &\Omega_{ij} = o(r^{-n}), && \\ &\Psi_{ijk} = O(r^{-n}), && \\ &\Phi_{ijkl} = O(r^{1-n}), && \Phi_{ij}^{A} = O(r^{-n}), \\ &\Psi'_{ijk} = O(r^{1-n}), && \Psi'_{i} = O(r^{1-n}), \\ &\Omega'_{ij} = o(r^{2-n}). && \end{aligned} \tag{27}$$

This includes asymptotically flat spacetimes in the case of vanishing radiation [3]. For $\Psi^{(n)}_{ijk} = 0$, this case reduces to (25) (with $\nu > n$).

3.3. Case (iii): $\Omega_{ij} = o(r^{-3})$, $\Psi_{ijk} = O(r^{-3})$

This case cannot represent asymptotically flat spacetimes [3]. Generically, $\beta_c = -2$ and

$$\begin{aligned} &\Omega_{ij} = O(r^{-\nu}) && (\nu > 3), \\ &\Psi_{ijk} = O(r^{-3}), && \Psi_i = o(r^{-3}), \\ &\Phi_{ijkl} = O(r^{-2}), && \Phi^S_{ij} = O(r^{-3}), \qquad \Phi = o(r^{-3}), \qquad \Phi^A_{ij} = O(r^{-3}), \\ &\Psi'_{ijk} = O(r^{-2}), && \Psi'_i = O(r^{-3}), \\ &\Omega'_{ij} = O(r^{-2}), \end{aligned} \tag{28}$$

where $\Psi_i = O(r^{-\nu})$, $\Phi = O(r^{-\nu})$ for $3 < \nu \leq 4$ while $\Psi_i = O(r^{-4})$, $\Phi = O(r^{-4})$ for $\nu > 4$. Here, $\boldsymbol{\ell}$ can be a single WAND and the asymptotically leading term is of type II(abd). For $\Psi^{(3)}_{ijk} = 0$, this case reduces for $3 < \nu < 4$ to (18) (with $\nu > n$), for $4 \leq \nu \leq n$ to (19) and for $\nu > n$ to (26). If $\beta_c < -2$ then $\Phi_{ijkl} = O(r^{-3})$ and the leading term at infinity becomes of type III(a).

Acknowledgments

The authors acknowledge support from research plan RVO: 67985840 and research grant GAČR 13-10042S.

References

[1] Newman E T and Tod K P 1980 *General Relativity and Gravitation: One Hundred Years After the Birth of Albert Einstein* vol 2 ed Held A (London and New York: Plenum Press) pp 1–36
[2] Penrose R and Rindler W 1986 *Spinors and Space-Time* vol 2 (Cambridge: Cambridge University Press)
[3] Godazgar M and Reall H S 2012 *Phys. Rev.* D **85** 084021
[4] Ortaggio M and Pravdová A 2014 *Phys. Rev.* D **90** 104011
[5] Ortaggio M 2014 *Phys. Rev.* D **90** 124020
[6] Ortaggio M, Pravda V and Pravdová A 2013 *Class. Quantum Grav.* **30** 013001
[7] Newman E T and Penrose R 1962 *J. Math. Phys.* **3** 566–578 see also E. Newman and R. Penrose (1963), Errata, *J. Math. Phys.* 4:998.
[8] Ortaggio M, Pravda V and Pravdová A 2007 *Class. Quantum Grav.* **24** 1657–1664
[9] Coley A, Milson R, Pravda V and Pravdová A 2004 *Class. Quantum Grav.* **21** L35–L41
[10] Pravda V, Pravdová A, Coley A and Milson R 2004 *Class. Quantum Grav.* **21** 2873–2897 see also V. Pravda, A. Pravdová, A. Coley and R. Milson *Class. Quantum Grav.* **24** (2007) 1691 (corrigendum).
[11] Durkee M, Pravda V, Pravdová A and Reall H S 2010 *Class. Quantum Grav.* **27** 215010
[12] Aliev A N and Frolov V P 2004 *Phys. Rev.* D **69** 084022
[13] Ortaggio M and Pravda V 2006 *JHEP* **12** 054
[14] Ortaggio M, Podolský J and Žofka M 2008 *Class. Quantum Grav.* **25** 025006
[15] Penrose R 1963 *Phys. Rev. Lett.* **10** 66–68
[16] Exton A, Newman E and Penrose R 1969 *J. Math. Phys.* **10** 1566–1570
[17] van der Burg M 1969 *Proc. R. Soc.* A **310** 221–230
[18] Aliev A N 2006 *Phys. Rev.* D **74** 024011
[19] Emparan R 2004 *JHEP* **03** 064
[20] Ortaggio M *Proceedings of the XVII SIGRAV Conference (Torino, September 4–7, 2006)* (*Preprint* `gr-qc/0701036`) URL `http://www.sigrav.org/Material/Torino2006/Ortaggio.pdf`
[21] Ortaggio M, Podolský J and Žofka M 2014 (*Preprint* `arXiv:1411.1943`)
[22] Podolský J and Ortaggio M 2006 *Class. Quantum Grav.* **23** 5785–5797
[23] Tanabe K, Tanahashi N and Shiromizu T 2010 *J. Math. Phys.* **51** 062502
[24] Tanabe K, Kinoshita S and Shiromizu T 2011 *Phys. Rev.* D **84** 044055
[25] Ortaggio M, Pravda V and Pravdová A 2009 *Phys. Rev.* D **80** 084041
[26] Ortaggio M, Pravda V and Pravdová A 2009 *Class. Quantum Grav.* **26** 025008

Journal of Physics: Conference Series **600** (2015) 012053 doi:10.1088/1742-6596/600/1/012053

Inertial forces in General Relativity

L. Filipe O. Costa, José Natário

CAMGSD — Instituto Superior Técnico, Universidade de Lisboa, Portugal

E-mail: lfpcosta@math.ist.utl.pt, jnatar@math.ist.utl.pt

Abstract. An exact equation describing the inertial forces for arbitrary orthonormal frames on arbitrary spacetimes is presented. It manifests that the so-called gravitomagnetic field consists of a combination of two effects of independent origin: the vorticity of the observer congruence, plus the angular velocity of rotation, relative to Fermi-Walker transport, of the triad of spatial axes that each observer "carries" with it. Such formulation encompasses different gravitomagnetic fields in the literature. The formalism is applied to notable reference frames in Kerr spacetime, and their inertial forces shown to have familiar Newtonian analogues.

1. Introduction

Inertial forces are fictitious forces that arise in the description of the motion of test particles when the reference frame is not inertial. That is always the case in a curved spacetime, since no globally inertial frames exist therein. The inertial forces have been defined in different ways in the literature (see [1] and Refs. therein), from the best known linearized theory approaches, to exact formulations (e.g. [2, 3, 1]). Herein we will follow the exact approach in [1], to which we refer for more details and notation/conventions; it embodies many of the approaches in the literature as special cases. The problem may the stated as follows. Consider a congruence of observers of 4-velocity u^α, and a test particle of worldline $z^\alpha(\tau)$ and 4-velocity $dz^\alpha/d\tau = U^\alpha$. Let $h^\alpha_{\ \beta} \equiv u^\alpha u_\beta + \delta^\alpha_{\ \beta}$ be the projector orthogonal to u^α (the "space" projector). The spatial projection of the particle's 4-velocity, $U^{\langle\alpha\rangle} \equiv h^\alpha_{\ \beta}U^\beta$, gives a notion[1] of velocity of the particle *relative to the observers*. For simplicity, we consider only particles in geodesic motion. It is the variation of $U^{\langle\alpha\rangle}$ along $z^\alpha(\tau)$, that one casts as *inertial forces*. To determine it one must specify the reference frame. The observers' 4-velocity $\mathbf{u}$ defines the time axis; but one still needs to specify the spatial axes, in particular their transport law along the observer congruence (see Fig. 1 of [1]), which is where the different approaches mainly differ. If we take an orthonormal frame (a "tetrad") $\mathbf{e}_{\hat\alpha}$, with $\mathbf{e}_{\hat 0} = \mathbf{u}$, the general form for the transport law reads

$$\nabla_{\mathbf{u}}\mathbf{e}_{\hat\beta} = \Omega^{\hat\alpha}_{\ \hat\beta}\mathbf{e}_{\hat\alpha}; \quad \Omega^{\alpha\beta} = 2u^{[\alpha}a^{\beta]} + \epsilon^{\alpha\beta}_{\ \ \nu\mu}\Omega^\mu u^\nu \ , \tag{1}$$

where $a^\alpha \equiv \nabla_{\mathbf{u}}u^\alpha$, and Ω^α the angular velocity of rotation of the spatial triad $\mathbf{e}_{\hat i}$ relative to Fermi-Walker transport. We will leave Ω^α *arbitrary* for now.

In order to measure the rate of change of $U^{\langle\alpha\rangle}$ with respect to such frame, we need a connection $\tilde\nabla$ (i.e., a *covariant derivative*) for spatial vectors with the following properties: i) with respect

[1] More precisely, one can decompose $U^\alpha = \gamma(u^\alpha + v^\alpha)$, where $\gamma = -U^\alpha u_\alpha = \sqrt{1 - v^\alpha v_\alpha}$ is the Lorentz factor, and $v^\alpha = U^{\langle\alpha\rangle}/\gamma$. v^α is the usual 3-velocity of the particle relative to the observers; in a coordinate system $\{t, x^i\}$ where the observers are at rest ($u^i = 0$), one has $v^i = dx^i/dt$.

to it, the spatial triad vectors $\mathbf{e}_{\hat{i}}$ are constant along the observer congruence: $\tilde{\nabla}_{\mathbf{u}}\mathbf{e}_{\hat{i}} = 0$. That is, $\tilde{\nabla}_{\mathbf{u}}$ should become an ordinary time derivative ($\partial_{\mathbf{u}}$) *in the prescribed frame*[2]. ii) Along the directions orthogonal to u^α, $\tilde{\nabla}$ should equal the projected Levi-Civita spacetime connection, $\tilde{\nabla}_{\mathbf{X}}Z^\alpha = h^\alpha_{\ \beta}\nabla_{\mathbf{X}}Z^\beta$, for all X^α and Z^α orthogonal to u^α. This is to correct for the variation of the axes $\mathbf{e}_{\hat{i}}$ in the directions orthogonal to u^α, which are not related to inertial forces (e.g. the trivial variation from point to point of the $\mathbf{e}_{\hat{i}}$ of a non-rectangular coordinate system in flat spacetime). This is achieved by the connection $\tilde{\nabla}$ whose action on a spatial vector X^α reads

$$\tilde{\nabla}_\alpha X^\beta = (h^u)^\beta_{\ \gamma}\nabla_\alpha X^\gamma + u_\alpha \epsilon^\beta_{\ \delta\gamma\lambda}u^\gamma X^\delta \Omega^\lambda \; . \tag{2}$$

The first term $(h^u)^\beta_{\ \gamma}\nabla_\alpha X^\gamma \equiv \nabla^\perp_\alpha X^\beta$ is called [2] the "Fermi-Walker connection", since $\nabla^\perp_{\mathbf{u}}X^\alpha$ is the Fermi-Walker derivative of X^α along the congruence. The second term adjusts the connection to the transport law chosen for the $\mathbf{e}_{\hat{i}}$. It is easily seen that $\tilde{\nabla}$ preserves the spatial metric $h_{\alpha\beta}$.

Differentiating $U^{\langle\alpha\rangle}$, with respect to $\tilde{\nabla}$, along the particle's worldline (of tangent U^α) yields

$$\tilde{\nabla}_{\mathbf{U}}U^{\langle\alpha\rangle} = -\gamma\nabla_{\mathbf{U}}u^\beta + \epsilon^\alpha_{\ \beta\gamma\delta}u^\delta U^\beta \Omega^\gamma \equiv F^\alpha_{\rm GEM} \tag{3}$$

which is the *inertial* (or "gravitoelectromagnetic") "force". It consists of two terms of distinct origin: a first term that depends only on the variation of the observers' 4-velocity u^α along the particle's worldline, and a second *independent* term, that arises from the transport law for the spatial triads $\mathbf{e}_{\hat{i}}$. Decomposing $\nabla_\beta u_\alpha$ in the congruence's kinematics

$$\nabla_\beta u_\alpha = -a_\alpha u_\beta - \epsilon_{\alpha\beta\gamma\delta}\omega^\gamma u^\delta + K_{(\alpha\beta)}; \qquad K_{(\alpha\beta)} = h_\alpha^{\ \mu}h_\beta^{\ \nu}\nabla_\mu u_\nu = \sigma_{\alpha\beta} + \theta h_{\alpha\beta} \; , \tag{4}$$

where $\omega^\alpha \equiv \epsilon^{\alpha\beta\gamma\delta}\nabla_\beta u_\gamma u_\delta/2$ is the vorticity of the observer congruence, and $K_{(\alpha\beta)}$ the shear/expansion tensor ($\theta \equiv$ expansion, $\sigma_{\alpha\beta} \equiv$ traceless shear), we get

$$F^\alpha_{\rm GEM} = \gamma\left[\gamma G^\alpha + \epsilon^\alpha_{\ \beta\gamma\delta}u^\delta U^\beta H^\gamma - K^{(\alpha\beta)}U_\beta\right] , \tag{5}$$

where

$$G^\alpha = -\nabla_{\mathbf{u}}u^\alpha \; ; \qquad H^\alpha = \omega^\alpha + \Omega^\alpha \tag{6}$$

are, respectively, the "gravitoelectric" and "gravitomagnetic" fields, that play analogous roles to the electric and magnetic fields in the Lorentz force. H^α thus consists of two parts of independent origin: the vorticity ω^α of the observers, plus the angular velocity Ω^α of rotation of the spatial triads relative to Fermi-Walker transport. Ω^α has a physical interpretation in terms of the "precession" of gyroscopes. According to the Mathisson-Papapetrou equations, no torque is exerted on a gyroscope (taken as a *pole-dipole* particle) in a gravitational field; its spin vector (under the Mathisson-Pirani spin condition [4]) undergoes Fermi-Walker transport, $DS^\alpha/d\tau = S_\nu a^\nu U^\alpha$. In a comoving tetrad $\mathbf{e}_{\hat{\alpha}}$ (where $U^{\hat{\alpha}} = \delta^{\hat{\alpha}}_{\hat{0}}$, $S^{\hat{0}} = 0$), this is

$$\frac{DS^{\hat{i}}}{d\tau} = 0 \Leftrightarrow \frac{dS^{\hat{i}}}{d\tau} = -\Gamma^{\hat{i}}_{\hat{0}\hat{k}}S^{\hat{k}} = \left(\vec{S}\times\vec{\Omega}\right)^{\hat{i}} . \tag{7}$$

The inverted commas are because the gyroscopes, in rigor, do not precess (it is the frame that rotates); their rotation axes actually determine the local *compass of inertia*; and Fermi-Walker transport mathematically the non-rotating frame carried by an accelerated observer [1, 4, 7].

Ω^α is up until now arbitrary; different choices are made in the literature. The natural choice would be to choose spatial axes pointing to fixed neighboring observers; in general, however, that

[2] Just like $\nabla_{\mathbf{u}} = \partial_{\mathbf{u}}$ in an inertial frame, or $\nabla^\perp_{\mathbf{u}} = \partial_{\mathbf{u}}$ in a Fermi-Walker transported frame.

is not compatible with orthonormal axes. Let η^α be a connecting vector between the observers' worldlines, $\mathcal{L}_{\mathbf{u}}\eta^\alpha = 0$, and $Y^\alpha = h^\alpha{}_\beta\eta^\beta$ its space projection. Y^α evolves in the tetrad as [1]

$$\dot{Y}_{\hat{i}} = \left[\sigma_{\hat{i}\hat{j}} + \frac{1}{3}\theta\delta_{\hat{i}\hat{j}} + \epsilon_{\hat{i}\hat{k}\hat{j}}(\omega^{\hat{k}} - \Omega^{\hat{k}})\right] Y^{\hat{j}} . \quad (8)$$

If the congruence is rigid ($\sigma_{\hat{i}\hat{j}} = \theta = 0$), and one chooses spatial triads $\mathbf{e}_{\hat{i}}$ *co-rotating* with the observers, which amounts to choosing $\Omega^\alpha = \omega^\alpha$, then Y^α is constant in the tetrad, $\dot{Y}_{\hat{i}} = 0$. If the congruence is not rigid but only expands ($\theta \neq 0$, $\sigma_{\hat{i}\hat{j}} = 0$), then $\dot{Y}_{\hat{i}} = \theta Y_{\hat{i}}/3$, i.e., Y^α, albeit not constant, has a fixed direction on the tetrad, so the $\mathbf{e}_{\hat{i}}$ still point to fixed neighboring observers. If $\sigma_{\hat{i}\hat{j}} \neq 0$, then that is not possible, as the connecting vectors cannot remain orthogonal; the choice $\Omega^\alpha = \omega^\alpha$ means in this case that the triads $\mathbf{e}_{\hat{i}}$ co-rotate with the observers, but without undergoing their shearing effects. This is as much as an orthonormal frame can adapt to the observer congruence; for this reason we dub it *the congruence adapted frame*. It is argued in [5] to be the closest generalization of the Newtonian concept of reference frame.

Other transport laws for the spatial frame (with $\Omega^\alpha \neq \omega^\alpha$) are used in the literature, as they can be suitable for some applications; an example are the so-called "locally non rotating frames" (e.g. [6, 7, 3]), defined in stationary axisymmetric spacetimes, and where the observer congruence are the zero angular momentum observers (whose vorticity is zero, $\omega^\alpha = 0$), and the triads $\mathbf{e}_{\hat{i}}$ are tied to the background symmetries (implying in general that $\Omega^\alpha \neq 0$). Another choice [2] is a Fermi-Walker transported spatial frame ($\Omega^\alpha = 0$). These reference frames might seem strange at first; in the next section we will discuss them in the Kerr spacetime, and show that each of them has actually familiar Newtonian analogues.

2. Rigid observer congruences

2.1. Rigidly rotating frames

Rigidly rotating frames are reference frames *adapted* to a rigidly rotating congruence of observers. The simplest case is the rigidly rotating frame in flat spacetime. Let α = constant be their angular velocity. The metric is obtained from the Minkowski metric in cylindrical coordinates $\{t, r, \varphi, z\}$ by the transformation $\varphi = \phi + \alpha t$. A spatial orthonormal triad $\mathbf{e}_{\hat{i}}$ adapted to the observers is obtained by taking (using $h_{\alpha\beta}$, see Sec. 1) the normalized projection, orthogonal to $\mathbf{u}$, of the coordinate basis vectors $\{\partial_r, \partial_\phi, \partial_z\}$. The $\mathbf{e}_{\hat{i}}$ point to fixed observers (since the ∂_i are connecting vectors), and (in agreement with Eq. (8)) rotate relative to Fermi-Walker transport with an angular velocity that matches the congruence's vorticity, $\vec{\Omega} = \vec{\omega} = \alpha/(1 - \alpha^2 r^2)\partial_z$. The inertial forces in this frame are obtained by setting $K_{(\alpha\beta)} = 0$, $H^\alpha = 2\omega^\alpha$, in Eq. (5),

$$\vec{F}_{\rm GEM} = \gamma\left[\gamma\vec{G} + 2\vec{U} \times \vec{\omega}\right] \quad (9)$$

with $\vec{G} = \alpha^2 r/(1 - \alpha^2 r^2)\partial_r$. These are the inertial forces felt in a merry go round; $\gamma^2\vec{G}$ and $2\gamma\vec{U} \times \vec{\omega}$ are relativistic versions of the well known centrifugal and Coriolis forces.

2.1.1. Frame adapted to the "static" observers in Kerr spacetime Take the Kerr metric in Boyer-Lindquist coordinates. The so-called "static" observers are the observers at rest in that coordinate system, whose worldlines are tangent to the time Killing vector field, $\mathbf{u} \propto \partial/\partial t$. This is a *rigid* observer congruence, and a spatial orthonormal triad $\mathbf{e}_{\hat{i}}$ adapted to it is obtained by projecting the coordinate basis vectors $\{\partial_r, \partial_\theta, \partial_\phi\}$ in the hyperplanes orthogonal to $\mathbf{u}$, and normalizing. The triad $\mathbf{e}_{\hat{i}}$ points to fixed observers (as the ∂_i are connecting vectors), and rotate with respect to Fermi-Walker transport with angular velocity $\vec{\Omega} = \vec{\omega}$, in agreement with Eq. (8). Since the congruence is rigid, and the triads are $\vec{e}_{\hat{i}}$ locked to neighboring observers, this is a *rigid frame*. The vorticity $\vec{\omega}$ decreases as r increases, vanishing at infinity where the $\mathbf{e}_{\hat{i}}$ become

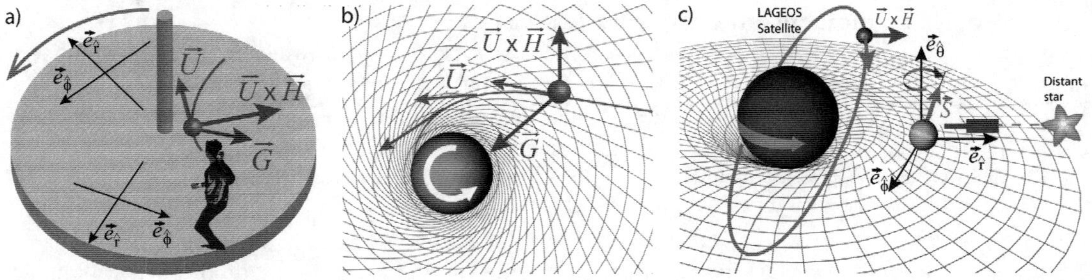

Figure 1. Two examples of rigidly rotating frames: a) a merry go round and b)-c) the frame adapted to the "static" observers in the Kerr spacetime. The inertial forces have the same form (9). $\vec{G}$ has opposite directions (centrifugal in a)); in Kerr $\vec{\omega} \neq 0$ for finite r, whilst $\lim_{r\to\infty} \vec{\omega} = 0$ (manifestation of frame-dragging). The axes at infinity are inertial ("star fixed"); since the frame is rigid, measuring the precession of gyroscopes (e.g. Gravity-Probe B) or deflection of test particles (e.g. the LAGEOS satellite) at any point relative to this frame, amounts to measuring them with respect to the distant stars.

inertial axes. This manifests the frame-dragging effect (absent in the merry go round of Fig. 1a)): the frame is inertial at infinity, and rotating close to the black hole (relative to the local compass of inertia, see Sec. 1) whilst, at the same time, being rigid.

A gyroscope whose center of mass is at rest in this frame is seen to precess as (cf. Eq. (7)) $d\vec{S}/d\tau = \vec{S} \times \vec{H}/2$. Since the frame is rigid, the triads $\mathbf{e}_{\hat{i}}$ are locked to the inertial axes at infinity; experimentally this amounts to anchor the axes to telescopes pointing at the distant stars (Fig. 1c)). This means that, at any point, $-\vec{H}/2$ is in fact the precession with respect to an inertial frame at infinity. It is known as the *Lense-Thirring* gyroscope precession, and has been measured, in the gravitational field of the Earth, by the Gravity Probe B mission [9].

The expression for the inertial forces is formally similar to the one in a merry go round, Eq. (9), only now $\vec{G} = -\vec{a}$ and $\vec{\omega} = \vec{\Omega}$ are minus the acceleration and the vorticity of the static observers in Kerr spacetime (for explicit expressions for $\vec{a}$ and $\vec{\Omega}$, see Eqs. (107)-(108) and (110) of [7]). The gravitomagnetic force, second term of Eq. (9), leads to a precession of the orbits of particles in the field of a rotating source, which has been detected by Laser Ranging to the LAGEOS satellites [10], and is under further scrutiny by the ongoing LARES mission [11].

2.2. Non-adapted frame — the "Fermi-Walker" gravitomagnetic field

Consider now a reference frame composed of the same congruence of static observers in Kerr spacetime, but now carrying spatial triads $\mathbf{e}_{\hat{i}}$ that undergo Fermi-Walker transport along the congruence; i.e., triads attached to the spinning axis of local guiding gyroscopes, which is given by the condition $\Omega^\alpha = 0$. These triads are not *adapted* to the congruence: the congruence has vorticity ($\omega^\alpha \neq 0$), thus we have observers that are rotating, but carrying with them spatial axes that do not rotate. The gravitomagnetic field reduces to the observer's vorticity, $\vec{H} = \vec{\omega} + \vec{\Omega} = \vec{\omega}$ and is known as the "Fermi-Walker gravitomagnetic field" [2]; it is one half of the $\vec{H}$ of the start-fixed rigid frame of Sec. 2.1.1. The inertial forces in such frame are thus

$$\vec{F}_{\rm GEM} = \gamma \left[\gamma \vec{G} + \vec{U} \times \vec{\omega} \right] . \tag{10}$$

These are the inertial forces involved in the "hidden momentum" of a spinning particle [4]. Such reference frame might seem strange, but it actually has a familiar Newtonian analogue: a family of rigidly rotating observers in flat spacetime, carrying a system of axes that remain fixed to the basis vectors of the Cartesian inertial frame (see Fig. 2b) of [1]), which is the situation in a Ferris wheel, for observers sitting in the cabins.

3. Zero angular momentum observers (ZAMOs)

The quantity $u_\phi = \mathbf{u} \cdot \partial_\phi$ is identified with the observer's "orbital" angular momentum. Zero angular momentum observers (ZAMOs) are stationary observers for which $u_\phi = 0$. They have angular velocity $\dot{\phi}_{\rm ZAMO} = -g_{0\phi}/g_{\phi\phi}$, and 4-velocity $u^\alpha_{\rm ZAMO} = u^0_{\rm ZAMO}(1, 0, \dot{\phi}_{\rm ZAMO}, 0)$, the component $u^0_{\rm ZAMO}$ being determined by the normalization condition $\mathbf{u}^2_{\rm ZAMO} = -1$.

These observers, also called "fiducial" or "Eulerian" [8], have special properties. They have *no vorticity* ($\omega^\alpha = 0$), being orthogonal to the hypersurfaces $t = constant$. This means that they do not rotate relative to their neighbors (are locally irrotational). They also measure zero Sagnac effect around the black hole. That is, consider a circular loop, made of optical fiber, at fixed (r, θ) around the black hole, and let a flash of light be emitted at some point, so that light rays travel along the fiber both in the positive and negative ϕ directions (i..e, both co-rotating and counter-rotating with the black hole); it turns out that the ZAMOs will receive back both light signals *at the same time*. Hence the ZAMOs are not rotating in any sense relative to the spacetime geometry — locally and "globally". However, relative to the fixed stars (i.e., to the static observers), they are seen to be rotating, with angular velocity $\dot{\phi}_{\rm ZAMO} \neq 0$, which manifests frame-dragging. The congruence has zero expansion, $\theta = 0$, but shears, $\sigma_{\alpha\beta} \neq 0$.

Analogy with a free vortex.— The ZAMOs congruence exhibits many formal similarities with the irrotational flow of a free vortex (Fig. 2) well known from fluid dynamics. In a free vortex (in flat spacetime) the fluid has angular velocity $\dot{\phi} = K/r^2$. In spite of being rotating relative to the (globally inertial) star-fixed frame, the vorticity of the flow is zero ($\omega^\alpha = 0$), i.e., it is *locally irrotational*, just like the ZAMOs in Kerr spacetime. Also, similarly to the ZAMOs, the congruence does not expand ($\theta = 0$) but shears ($\sigma_{\alpha\beta} \neq 0$). The qualitative difference is that the observers comoving with the flow have non-zero angular momentum, and thus measure non-zero Sagnac effect along a circular optical fiber centered at $r = 0$. This is because the frame-dragging effect of Kerr spacetime has no parallel in this setup: herein, rotating with respect to the distant stars implies having non-zero angular momentum (unlike the ZAMOs in Kerr, which, for finite r, rotate with respect to the distant stars, whilst having zero angular momentum).

3.1. Congruence adapted frame

Since the ZAMOs have no vorticity, the congruence adapted frame amounts to setting $\Omega^\alpha = \omega^\alpha = 0$, i.e., choosing spatial triads Fermi-Walker transported along the congruence. This means that there is no gravitomagnetic field in this frame, $H^\alpha = \omega^\alpha + \Omega^\alpha = 0$, the inertial forces reduce to the gravitoelectric field plus the shear force, see Fig. 2a),

$$F^\alpha_{\rm GEM} = \gamma \left[\gamma G^\alpha - \sigma^\alpha{}_\beta U^\beta\right] \equiv F^\alpha_{\rm G} + F^\alpha_{\rm shear} \; . \tag{11}$$

This reference frame has a close analogue in fluid dynamics. As is well known, a boat circling a free vortex keeps its tip pointing always in (nearly) the same direction (Fig. 2b)); it is so because the boat behaves approximately as a vorticity meter. In a reference frame *adapted* to observers comoving with the flow of a free vortex (where the spatial axes are tied to vorticity meters), the inertial forces take likewise the form (11).

3.2. Locally "non-rotating" frames (LNR)

The so-called "locally non-rotating frames" [6] (also known as "proper reference frames of the fiducial observers" [8, b]) are frames associated with the ZAMOs, where the spatial triads $\mathbf{e}_{\hat{i}}$ are tied to the background symmetries. These frames are regarded as important for black hole physics, because they are defined everywhere (unlike the star fixed static observers of Sec. 2, that do not exist past the ergosphere). For the Kerr spacetime the $\mathbf{e}_{\hat{i}}$ are obtained by simply normalizing the Boyer-Lindquist coordinate basis $\{\partial_r, \partial_\theta, \partial_\phi\}$ (as it is orthogonal to $\mathbf{u} = \mathbf{u}_{\rm ZAMO}$). These triads clearly are *not* Fermi-Walker transported along the congruence (i.e., $\Omega^\alpha \neq 0$), see

Journal of Physics: Conference Series **600** (2015) 012053 doi:10.1088/1742-6596/600/1/012053

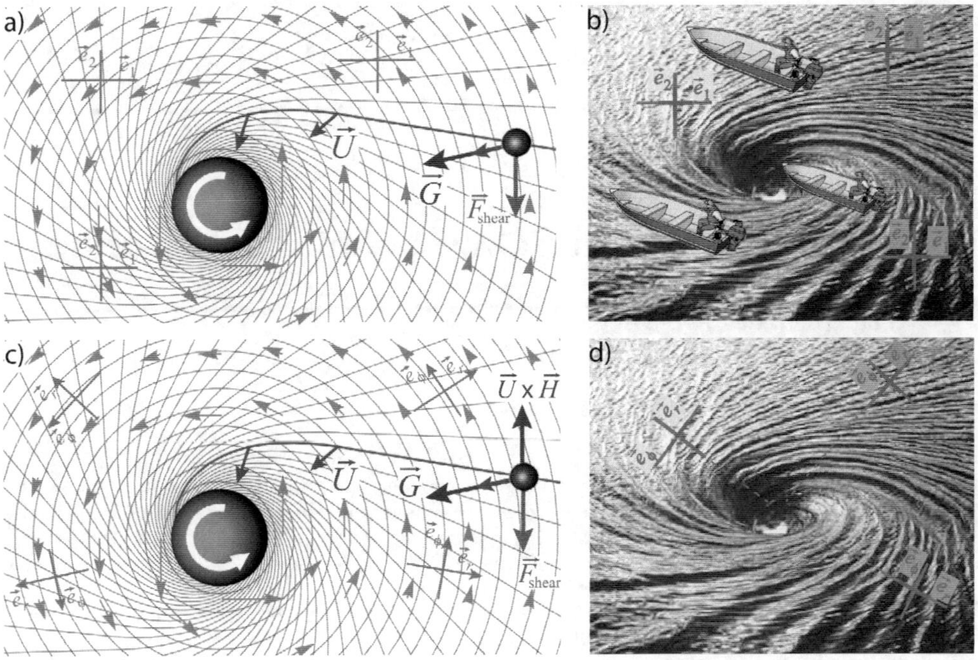

Figure 2. a) Frame adapted to the ZAMOs in Kerr spacetime; gray arrows represent the ZAMOs velocity (relative to the star fixed frame), $\vec{U}$ are the space components of the test particle's velocity *relative* to the ZAMOs, $U^{\langle\alpha\rangle} = h^{\alpha}{}_{\beta}U^{\beta}$. b) Frame adapted to observers comoving with the flow of a free vortex; the spatial frame is set up by tying the axes to vorticity meters (e.g. a boat). c) Locally "non-rotating" frames in Kerr spacetime. d) Frame composed of observers comoving with the flow of a free vortex, carrying spatial triads tied to a cylindrical coordinate system. In a) and b), $H^{\alpha} = \theta = 0$, and the inertial forces reduce to Eq. (11). In c) and d), $H^{\alpha} = \Omega^{\alpha}$, and the inertial forces are given by (12).

Fig. 2c); the explicit expression for $\Omega^{\alpha} \equiv \Omega^{\alpha}_{\rm LNR}$ is given in e.g. Eqs. (73)-(74) of [7]. Hence, the denomination locally "non-rotating" frames is misleading (cf. [7]); indeed the spatial frame rotates relative to the local compass of inertia. This is manifest in the inertial forces; the gravitomagnetic field in this frame is (since $\omega^{\alpha} = 0$) $H^{\alpha}_{\rm LNR} = \Omega^{\alpha}_{\rm LNR}$, and thus

$$\vec{F}_{\rm GEM} = \gamma \left[\gamma\vec{G} + \vec{U} \times \vec{\Omega}_{\rm LNR} - \sigma^{\hat{i}}{}_{\hat{j}}U^{\hat{j}}\vec{e}_{\hat{i}}\right] \equiv \vec{F}_{\rm G} + \vec{F}_{\rm GM} + \vec{F}_{\rm shear} \,, \tag{12}$$

i.e., consist of the gravitoelectric, gravitomagnetic, and shear forces. This reference frame again has an analogue in fluid dynamics: the frame associated to the observers comoving with the flow of a free vortex, and carrying spatial triads tied to the basis vectors of a cylindrical coordinate system, see Fig. 2d); therein, likewise, $H^{\alpha} = \Omega^{\alpha}$ and the inertial forces take the form (12).

[1] Costa L F and Natário J 2014 *Gen. Rel. Grav.* **46** 1792, Section 3
[2] Jantzen R T, Carini P, and Bini D 1992 *Ann. Phys.* **215** 1 *Preprint* gr-qc/0106043
[3] Semerák O 1995 *Il Nuovo Cim.* B **110** 973
[4] Costa L F and Natário J 2014 *Preprint* arXiv:1410.6443.
[5] Massa E 1974 *Gen. Rel. Grav.* **5** 573; Massa E and Zordan C 1975 Meccanica **10** 27
[6] Bardeen J 1970 *Astrophys. J.* **162** 71; Bardeen J *et al* 1972 *Astrophys. J.* **178** 347
[7] Semerák O 1993 *Gen. Rel. Grav.* **25** 1041
[8] Gourgoulhon E gr-qc/0703035; Thorne K *et al* 1986 *Black Holes, the Membrane Paradigm* (Yale U. P)
[9] *Review of Gravity Probe B* 1995 (Nat. Acad. Press); Everitt F *et al* 2011 *Phys. Rev. Lett.* **106** 221101
[10] Ciufolini I and Pavlis E, 2004 *Nature* **431** 958; Ciufolini I, Pavlis E, Peron E 2006 *New Astron.* **11** 527
[11] Ciufolini *et al* 2009 *Space Sci. Rev.* **148**, 71; http://www.lares-mission.com/

Spanish Relativity Meeting (ERE 2014): almost 100 years after Einstein's revolution IOP Publishing
Journal of Physics: Conference Series **600** (2015) 012054 doi:10.1088/1742-6596/600/1/012054

Relativistic positioning in Schwarzschild space-time

Neus Puchades and Diego Sáez

Departamento de Astronomía y Astrofísica, Universidad de Valencia, 46100-Burjassot, Valencia, Spain

E-mail: neus.puchades@uv.es

Abstract. In the Schwarzschild space-time created by an idealized static spherically symmetric Earth, two approaches –based on relativistic positioning– may be used to estimate the user position from the proper times broadcast by four satellites. In the first approach, satellites move in the Schwarzschild space-time and the photons emitted by the satellites follow null geodesics of the Minkowski space-time asymptotic to the Schwarzschild geometry. This assumption leads to positioning errors since the photon world lines are not geodesics of any Minkowski geometry. In the second approach –the most coherent one– satellites and photons move in the Schwarzschild space-time. This approach is a first order one in the dimensionless parameter GM/R (with the speed of light c=1). The two approaches give different inertial coordinates for a given user. The differences are estimated and appropriately represented for users located inside a great region surrounding Earth. The resulting values (errors) are small enough to justify the use of the first approach, which is the simplest and the most manageable one. The satellite evolution mimics that of the GALILEO global navigation satellite system.

1. Introduction

In any Relativistic Positioning System (RPS), which may be based on Minkowski Space-Time (MS-T), Schwarzschild Space-Time (SS-T), or any other space-time, the user to be located must receive -simultaneously- the proper times from four satellites. These proper times are called emission coordinates τ^A, where A runs from 1 to 4 to label the four satellites. The user space-time position is characterized by the inertial coordinates x^α; here and hereafter any Greek index runs from 1 to 4. These inertial coordinates are associated to an almost inertial reference system. From emission to reception, the photons delivering the proper times follow null geodesics. This fact allow us to calculate the inertial coordinates from the emission ones and the satellite world lines (excepting bifurcation cases).

Any Latin index, excepting A, runs from 1 to 3. Quantities G, $M_\oplus$, $R_\oplus$, t and τ stand for the gravitation constant, the Earth mass, the Earth radius, the coordinate time and the proper time, respectively. Units are chosen in such a way that the speed of light is $c = 1$.

The SS-T describes the external gravitational field of an idealized static spherically symmetric Earth. This space-time tends -asymptotically- to a MS-T, in which, the symmetry centre is at rest; hence, the SS-T may be considered as a perturbation of the asymptotic MS-T. The perturbing terms tend to zero as the distance to the symmetry centre tends to infinity.

Let us assume that the emission coordinates τ^A are received from four satellites. From these coordinates, the user position may be found by using two different approaches; in the first one, photons move in the MS-T asymptotic to the SS-T created by Earth, whereas in the second approach, photons move in the proper SS-T and calculations are performed up to first order in $GM_\oplus/R$. In both approaches, satellites move in SS-T. Thus, two different user positions are obtained, whose differences allow us to estimate the positioning E-errors, which are to be compared with the errors due to uncertainties in the satellite world lines (U-errors, see [4]). If the E-errors are small enough as compared with the U-errors, the user position may be calculated by using the first approach instead of the second one, which is more complicated from the numerical point of view.

2. First approach

If the satellite world lines are known in a certain inertial system of reference, the user position can be found, in the same inertial reference, using the following equations:

$$\eta_{\alpha\beta}[x^\alpha - x_A^\alpha(\tau^A)][x^\beta - x_A^\beta(\tau^A)] = 0 \quad (1)$$

where $\eta_{\alpha\beta}$ is the MS-T metric tensor, whose non vanishing components are $\eta_{11} = \eta_{22} = \eta_{33} = 1$ and $\eta_{44} = -1$, and the satellite world lines $x_A^\alpha = x_A^\alpha(\tau^A)$ are written in terms of the proper time parameter. These equations, which relate the emission coordinates τ^A and the inertial ones x^α, are found taking into account that, in this first approach, the photons move along null-geodesics in MS-T from satellite emission to user reception.

An analytical solution of the system of equations (1) was found in [1]. This solution gives the inertial coordinates (x^μ_{M0}) in terms of the emission ones, for photons moving in MS-T. A numerical code based on this solution -hereafter referred as the TX-code- was designed and tested in [2], [3] and [4]. The analytical solution holds for arbitrary satellite world lines.

Moreover, we can numerically calculate the emission coordinates (unknowns) from the inertial ones, by using the Newton-Raphson method [5] to solve the system (1). This is implemented in a numerical code, which is called XT-code.

We assume that the satellite trajectories are circumferences concentric with Earth, which are travelled as it is predicted in SS-T. See [2] and [4], where the equations of the satellite world lines are given in terms of τ. In order to compute E-errors, it may be assumed that satellites strictly follow nominal world lines of the GALILEO constellation.

3. Second approach

In this second approach, the photons move in SS-T. Each photon is emitted from the satellite A at point P_{IA}, whose coordinates $x^\mu_{IA}(\tau^A)$ ($\vec{R}_{IA} \equiv x^i_{IA}$ and $t_{IA} \equiv x^4_{IA}$) have been calculated from τ^A. The user position is then the intersection of four null geodesics that pass through the points P_{IA} and have appropriate propagation directions at these points. The coordinates of the intersection point P_{S0} (user space-time position in SS-T) are denoted x^μ_{S0}. Hereafter, $\vec{R}_{S0} \equiv x^i_{S0}$ and $t_{S0} \equiv x^4_{S0}$.

In the first approach, the photons emitted from the points P_{IA} follow null geodesics in MS-T. Four of these geodesics intersect at user space-time position P_{M0} (with coordinates $\vec{R}_{M0} \equiv x^i_{M0}$ and $t_{M0} \equiv x^4_{M0}$), which may be calculated as shown in the first approach (TX-code).

Moreover, SS-T may be seen as a perturbation of MS-T. Hence, the coordinates of each point on the SS-T geodesic may be written in the form $x^\mu_S = x^\mu_M + \Delta x^\mu_M$; so the SS-T null geodesic may be seen as a perturbation of the MS-T one. Quantities Δx^μ_M are the deviations between both geodesics.

The condition for the intersection of four null geodesics in SS-T is that, at point P_{M0}, the quantities Δx^μ_M ($\Delta\vec{R}_M \equiv \Delta x^i_M$ and $\Delta t_M \equiv \Delta x^4_M$) are identical for the four satellites (see [6]). Lengthy but straightforward calculations (at first order in $GM_\oplus/R$) allow us to impose this condition. So, we obtain the following system of equations:

$$\Delta t_{M0} - \frac{\vec{R}_{M0} - \vec{R}_{IA}}{|\vec{R}_{M0} - \vec{R}_{IA}|}.\Delta\vec{R}_{M0} = 2GM_\oplus \ln\left[\frac{|\vec{r}_{M0}| + |\vec{r}_{IA}| + |\vec{r}_{M0} - \vec{r}_{IA}|}{|\vec{r}_{M0}| + |\vec{r}_{IA}| - |\vec{r}_{M0} - \vec{r}_{IA}|}\right] \quad (2)$$

where $r = |\vec{r}|$ is a radial isotropic coordinate. The relation between the radial coordinates R and r, up to first order in $GM_{\oplus}/R$, is $|\vec{r}| = |\vec{R}|(1 - GM_{\oplus}/|\vec{R}|)$ (see [7]). The quantities $\Delta\vec{R}_{M0}$ and Δt_{M0} are the unknowns. The system (2) is composed of four equations, one equation per satellite.

The determinant of this system of four equations is

$$D = \begin{vmatrix} \frac{x^1_{I1}-x^1_{M0}}{|\vec{R}_{M0}-\vec{R}_{I1}|} & \frac{x^2_{I1}-x^2_{M0}}{|\vec{R}_{M0}-\vec{R}_{I1}|} & \frac{x^3_{I1}-x^3_{M0}}{|\vec{R}_{M0}-\vec{R}_{I1}|} & 1 \\ \frac{x^1_{I2}-x^1_{M0}}{|\vec{R}_{M0}-\vec{R}_{I2}|} & \frac{x^2_{I2}-x^2_{M0}}{|\vec{R}_{M0}-\vec{R}_{I2}|} & \frac{x^3_{I2}-x^3_{M0}}{|\vec{R}_{M0}-\vec{R}_{I2}|} & 1 \\ \frac{x^1_{I3}-x^1_{M0}}{|\vec{R}_{M0}-\vec{R}_{I3}|} & \frac{x^2_{I3}-x^2_{M0}}{|\vec{R}_{M0}-\vec{R}_{I3}|} & \frac{x^3_{I3}-x^3_{M0}}{|\vec{R}_{M0}-\vec{R}_{I3}|} & 1 \\ \frac{x^1_{I4}-x^1_{M0}}{|\vec{R}_{M0}-\vec{R}_{I4}|} & \frac{x^2_{I4}-x^2_{M0}}{|\vec{R}_{M0}-\vec{R}_{I4}|} & \frac{x^3_{I4}-x^3_{M0}}{|\vec{R}_{M0}-\vec{R}_{I4}|} & 1 \end{vmatrix} \tag{3}$$

If this determinant vanishes for the emission coordinates τ^A, there are no solutions of the corresponding system of equations, which means that positioning at the MS-T point P_{M0} [with coordinates (x^i_{M0}, t_{M0})] is not possible.

In [4], it was proved that, for low satellite velocities as those of any GNSS, the relation $V_T \simeq |J|/6$ is satisfied, where J is the Jacobian of the transformation, $\tau^A = \tau^A(x^\alpha)$, calculated at point $(\vec{R}_{M0}, t_{M0})$, and V_T is the volume of the tetrahedron formed by the tips of the four user-satellite unit vectors having their common origin at point $(\vec{R}_{M0}, t_{M0})$ in MS-T. It is important to remark that, for low satellite velocities, the determinant D obtained here is exactly the same as the Jacobian J. Therefore, D is proportional to the volume V_T. See also [8], where this volume is related with the so-called dilution of precision in the framework of Global Positioning System (GPS).

Quantities D, J, and V_T vanish simultaneously. In the points where these three quantities are small, the U-errors and the E-errors become very large; by this reason, these errors are expected to be large in two cases: close to $D = J = V_T = 0$ points, and also at points located far from both $D = 0$ points and satellites. In this last case, the four satellites subtend a small solid angle and the tetrahedron volume V_T is expected to be small (large errors).

4. Numerical Results

Calculations will be performed inside a big sphere, hereafter E-sphere. The center of this sphere is the point E, whose spherical inertial coordinates are $r_E = R_{\oplus}$, $\theta_E = 60°$ and $\phi_E = 30°$. Hence, E is on Earth surface. The radius of this sphere is 10^5 km.

In this paper, as in [2] and [4], the HEALPIx (hierarchical equal area isolatitude pixelization of the sphere) package ([9]) is used to depict appropriate maps. This pixelization was designed to construct and analyze maps of the cosmic microwave background. It is useful to display any scalar quantity depending on the observation direction (pixel). In our maps, there are 3072 pixels.

Given four satellites of the GALILEO constellation and an user position (x^i_{M0}, t_{M0}) in MS-T, the XT-code may be used to get the emission coordinates τ^A, which would be received by the chosen user in MS-T. From these emission coordinates and the satellite world lines (see [2] and [4]), the satellite positions at emission times (x^i_{IA}, t_{IA}) may be obtained and, then, the E-errors at the MS-T chosen point (x^i_{M0}, t_{M0}) may be calculated by solving the system (2) for the unknowns Δx^i_{M0} and Δt_{M0}. Quantities $\Delta_R = \Delta|\vec{R}_{M0}|$ and Δt_{M0} are good estimators of the E-errors.

We have considered 3072 HEALPIx directions, and 10^3 users uniformly distributed along each direction, from point E ($L = 0$) to $L = 10^5$ km. So the E-sphere is covered. The E-errors and the determinant D have been then calculated in the points of this coverage at coordinate time $t = 25$ h.

In Fig. 1, the E-errors corresponding to a HEALPIx particular direction without $D = 0$ points are shown. From this Figure, we can conclude that Δ_R and $|\Delta_t|$ increase with the distance L to E. For the chosen direction, these quantities are smaller than 0.5 m, but they are close to this value nearby the E-sphere boundary.

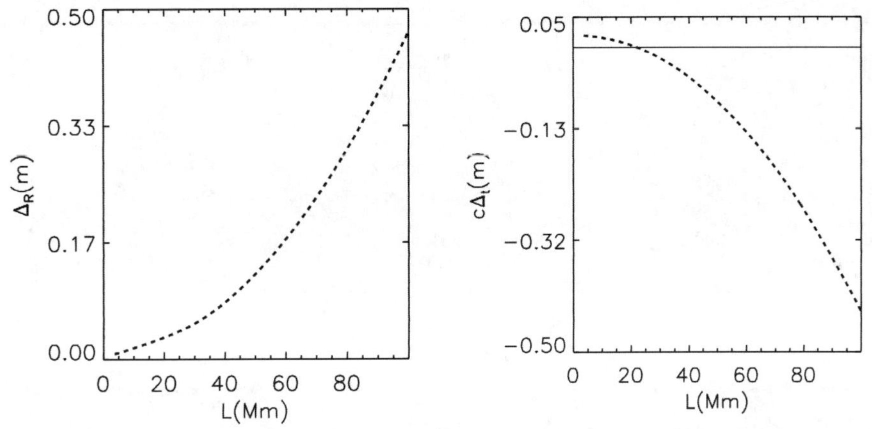

Figure 1. Quantities Δ_R (left) and $c\Delta_t$ (right) are represented along a particular direction from $L = 0$ (point E) to $L = 10^2$ Mm. The determinant D does not vanish along the chosen segment.

In Fig. 2, the distance L_D, from $L = 0$ (point E) to the point where $D = 0$ for the first time is displayed. Grey pixels correspond to HEALPIx directions where the determinant D does not vanish from E to $L = 10^5$ km. Inside a sphere (centered at point E), whose radius is the minimum value of the colour bar (23500 km), the determinant D does not vanish.

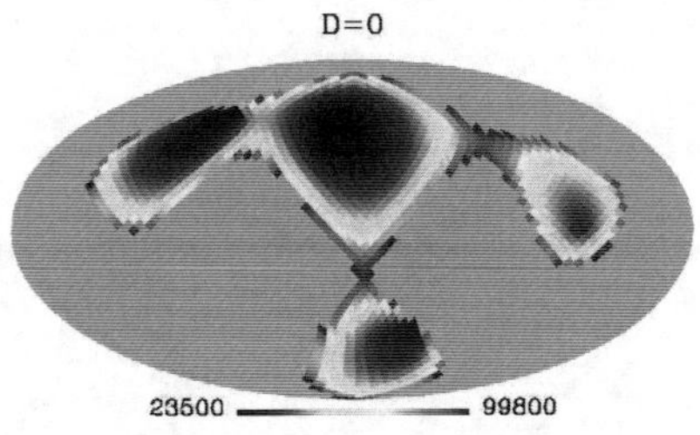

Figure 2. HEALPIx mollwide map. The quantity represented is the distance L_D, in kilometers, from E to the point where D vanishes for the first time.

Journal of Physics: Conference Series **600** (2015) 012054 doi:10.1088/1742-6596/600/1/012054

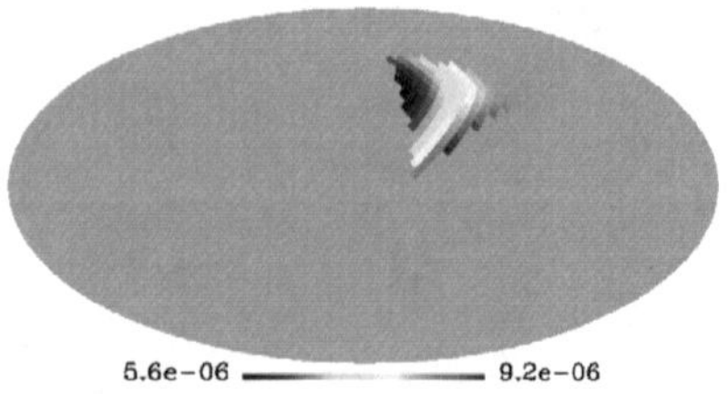

(a) $\rho = R_{\oplus}$

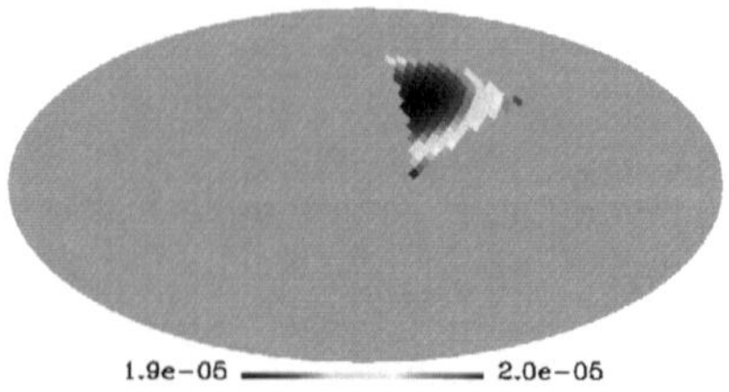

(b) $\rho = R_{\oplus}$

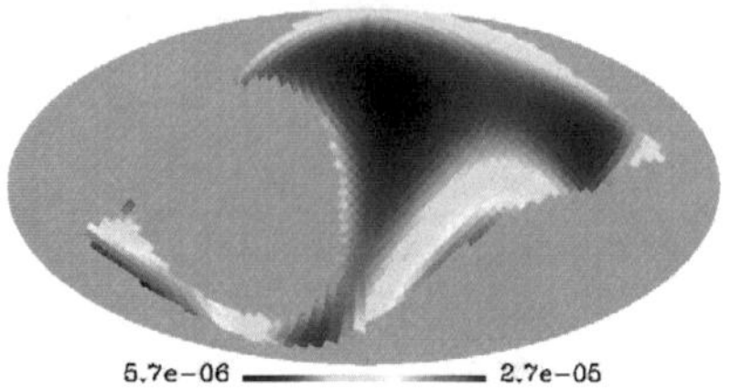

(c) $\rho = 10^4\ km$

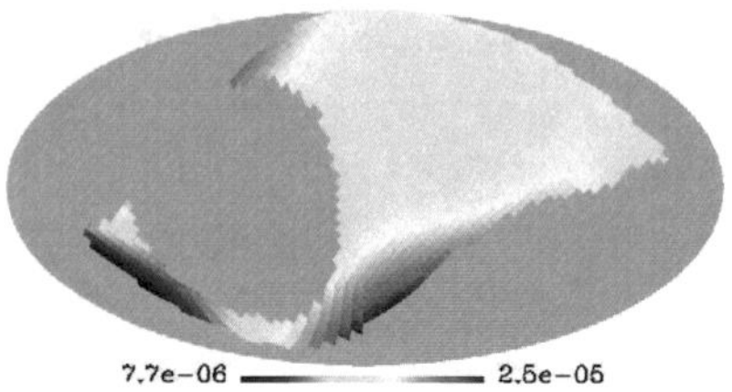

(d) $\rho = 10^4\ km$

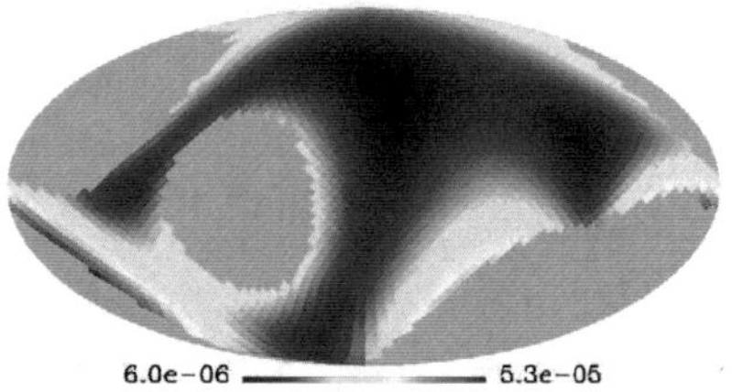

(e) $\rho = 1.5 \times 10^4\ km$

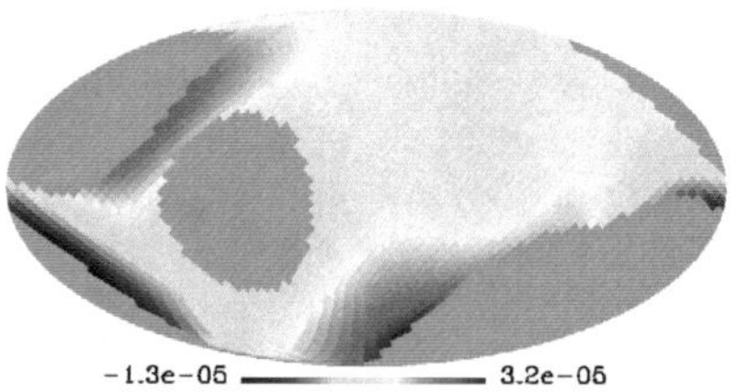

(f) $\rho = 1.5 \times 10^4\ km$

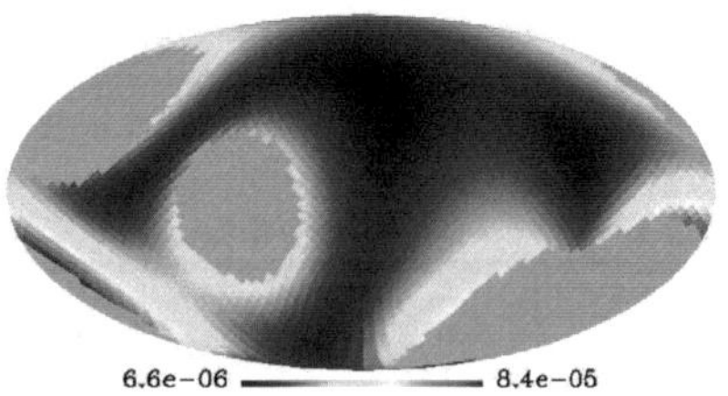

(g) $\rho = 2 \times 10^4\ km$

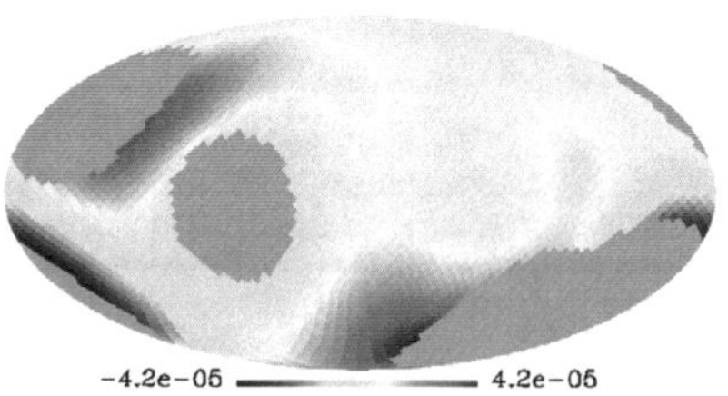

(h) $\rho = 2 \times 10^4\ km$

Figure 3. HEALPIx mollwide maps of Δ_R (left) and $c\Delta_t$ (right), in kilometers, on spherical surfaces concentric with Earth. The radius ρ are given in the bottom of each panel. Users located on the grey pixels cannot see the four satellites at the same time (invisibility zones).

Finally, we have calculated the E-errors (Δ_R and Δ_t) on spherical surfaces concentric with Earth for appropriate radius and $t = 25\ h$. These radius are $\rho = R_{\oplus}$, $\rho = 10^4\ km$,

$\rho = 1.5 \times 10^4$ km and $\rho = 2 \times 10^4$ km. Since the chosen radius are smaller than 23500 km, the determinant D does not vanish on the corresponding spherical surfaces (see Fig. 2).

In Fig. 3, the grey pixels cover invisibility zones, in which, positioning is not possible since the users do not receive, at time $t = 25$ h, the signals from the four satellites. It is due to the fact that one or more satellites are hidden by Earth.

All the spheres, whose E-errors are displayed in Fig. 3, are located inside the satellite orbits in a region where $D \neq 0$. On these spherical surfaces the order of these E-errors is between 10^{-1} cm and 10 cm. The maximum values of the E-errors appear in pixels close to the boundary of the invisibility regions. It is due to the fact that, for these pixels, the photon passes close to Earth surface.

5. Conclusions

Our study has been restricted to regions without $D = 0$ points. One of these regions is surrounding Earth. From Fig. 3 it follows that, in this region, the E-errors are smaller than ~ 10 cm. According to Fig. 2, there are directions (grey pixels) without any $D = 0$ point (from E to the E-sphere boundary). The E-errors have been estimated along many of these directions. One of them has been considered in Fig. 1. From this study, we have concluded that the E-errors increase with the distance L to point E, reaching values of the order of 1 m for some directions (close to 0.5 m in Fig. 1); hence, positioning quality decreases as L grows.

In paper [4], U-errors due to uncertainties in the satellite world lines were calculated. Here, we have estimated the E-errors at the same user space-time positions considered in the U-errors calculations. It has been done to allow comparisons. The order of magnitude of the U-errors, inside the E-sphere, is between 10 m and 100 m. These errors were calculated under the assumption that the amplitude of the deviation in the satellite world lines -with respect to the nominal ones- is 10 m. Then, for the chosen deviation amplitude, we have verified that the E-errors are smaller than the U-errors inside the E-sphere. Taking into account that the U-error values are proportional to the deviation amplitude, if this amplitude would be as smaller as 1 m, in future, the U-errors would be of the same order of magnitude as the E-errors. In this paper, it has been verified that the conditions $D = 0$, $J = 0$ and $V_T = 0$ are equivalent. So, where the U-errors diverge ($J = 0$), the E-errors also diverge ($D = 0$).

Since E-errors are smaller than U-errors inside the E-sphere, we conclude that the approach based on the approximation that photons (satellites) are moving in MS-T(SS-T) can be used. This approach simplifies the calculations.

6. Acknowledgements

This research has been supported by the Spanish Ministry of *Economía y Competitividad*, MICINN-FEDER project FIS2012-33582. We thank B. Coll, J.J. Ferrando, and J.A. Morales-LLadosa for valuable comments.

References

[1] Coll B, Ferrando J J and Morales-LLadosa, J A 2010 *Class. Quantum Grav.* **27** 065013
[2] Puchades N and Sáez D 2012 *Astrophys. Space Sci.* **341** 631
[3] Sáez D and Puchades N 2013 *Acta Futura* **7** 103
[4] Puchades N and Sáez D 2014 *Astrophys. Space Sci.* **307** 352
[5] Press W H et al. 1999 *Numerical recipes in fortran 77: the art scientific computing* Cambridge University Press, New York
[6] Delva P, Kostić U and Čadež A 2011 *Adv. Space Res.* **47** 370
[7] Misner C W, Thorne K H and Wheeler J A 1973 *Gravitation* W H Freeman and Company, New York
[8] Langley R B 1999 *GPS World* **10** 52
[9] Górski K M, Hivon E and Wandelt B D 1999 *Proceedings of the MPA/ESO Conference on Evolution of Large Scale Structure* Banday A J, Sheth R K and Da Costa L (Eds.) Printpartners Ipskamp Enschede p 37-42

Spanish Relativity Meeting (ERE 2014): almost 100 years after Einstein's revolution IOP Publishing
Journal of Physics: Conference Series **600** (2015) 012055 doi:10.1088/1742-6596/600/1/012055

Teaching the gravitational redshift: lessons from the history and philosophy of physics

Robert B Scott

Département de physique, Université de Bretagne Occidentale, Brest, France

E-mail: robert.scott@univ-brest.fr

Abstract. The equivalence principle and the notion of an ideal clock running independently of acceleration suggest that clocks are unaffected by gravity. The apparent contradiction with the gravitational redshift points to a subtlety in general relativity theory. Indeed, early attempts for a clear derivation of the gravitational redshift were fraught with errors and ambiguities, and much confusion endured for the next two decades. This suggests that the subject should be treated carefully in introductory textbooks on relativity theory. I analyze the weaknesses of the presentation in five otherwise excellent modern introductory general relativity books (by Rindler, Schutz, Hobson et al., Weinberg, and Carroll). I also present some analysis from an history and philosophy of physics article, which proves to be a great resource to learn about, anticipate, and clarify problems in teaching the redshift.

1. Introduction

Several authors have made the case that general relativity GR can and should be taught to undergraduate physics students [1–4] and many other articles can be found offering advice in this endeavor [5–10]. Recurring themes are to motivate the mathematics by first introducing the physics [2] and use an interactive approach [10]. Surely another important two components of good teaching are to identify the subtle areas and to present them clearly using careful language. I found that help can be found for both these good-teaching components from an unexpected source, namely the history and philosophy of physics. In this short article I focus on the gravitational redshift as an important example.

The gravitational redshift provides an example of an especially subtle topic. Okun *et al.* [11] note that in the literature there are two interpretations of the gravitational redshift in a static gravitational field: either the photon frequency is modified en route between emitter and receiver (and the clock rate unaffected by gravitational potential) or the clocks at lower potential are slowed down (and the photon unaffected en route). They advocate strongly the clocks-slow-down view, stating that the gravitational redshift should be taught in a way that "centers on the universal modification of the rate of a clock exposed to a gravitational potential" [11][p. 119]. But the situation is subtle and confusing because an important heuristic principle in GR is that the local effects of gravity can always be eliminated with a coordinate transformation. This follows from a version of the principle of equivalence of gravity and inertia that asserts that inertial forces and gravitational forces are one and the same physical effect Møller(1952, §83)[12]. So do clocks slow down? According to Weinberg

Consider a [ideal] clock in an arbitrary gravitational field ... The equivalence principle

> tells us that its rate is unaffected by the gravitational field ..., see Weinberg(1972, §3.5)[13].

The history of physics also alerts us to the subtlety surrounding the gravitational redshift. Indeed, when GR was introduced by Einstein in the first two decades of the XXth century, there was great confusion among the leading physicists of the day especially around the derivation of the gravitational redshift. (If leading physicists could be confused when first confronted with relativity, we should not be surprised when a fresh crop of students are confused too.) This history was reviewed by two philosophers, John Earman and Clark Glymour [14] and makes for both informative and entertaining reading. Furthermore their derivation is admirably clear; we give the full derivation in §3.2.

The gravitational redshift and the equivalence principle are basic concepts and standard material for an introductory GR theory course. Given the subtlety and historical confusion surrounding these issues one would expect surely to find careful discussion of these points in the textbooks and pedagogical literature. Surprisingly I found this not to be the case. I provide a brief survey of the interpretation of the gravitational redshift in my five favourite introductory GR textbooks in §2. In none of them is the tension between ideal clocks/equivalence principle and the gravitational redshift explicitly presented and is at best implicit in two of them [13, 15]. I then turn to the history and philosophy of physics literature and review the lessons from the analysis of the gravitational redshift [14]. First I present the erroneous derivation due to Eddington [16] and show how this was repeated in Weinberg's book [13]. Then I present the admirably clear Earman-Glymour derivation. This reveals in which sense clocks do *not* slow down in a gravitational field. I conclude with a summary in §4.

2. Interpretation of the redshift in introductory textbooks

In §5.1 of his well-known introductory GR textbook [17], Bernard Schutz derives the gravitational redshift first using a conservation of energy argument applied to the setting of the Pound-Rebka-Snider experiment. A very similar argument can be found in advanced textbooks, e.g. §7.2 of ref. [18]. He offers the interpretation:

> ...the gravitational redshift implies that time itself runs slightly faster at the higher altitude than it does on the [surface of] Earth. (Schutz, 2009, § 5.1, p. 113)[17]

Naively one might infer that gravity somehow compresses spacetime. No where is it mentioned that this appears to contradict the idea that the equivalence principle tells us that clock rate is unaffected by the gravitational field.

Similarly, Wolfgang Rindler in his §1.16 of textbook [15] also derives the gravitational redshift first using a conservation of energy argument. He also replaces the gravitational field with a uniform acceleration and finds a Doppler shift for the receiver because of his velocity relative to an inertial frame initially at rest. A similar argument can be found in §7.4 of ref. [18]. Price considers the same calculation but without approximation [19]. For the interpretation Rindler states

> That result (known as the *gravitational frequency shift*) has the important consequence that standard clocks fixed in a *stationary* gravitational field at low potential go slower than clocks fixed at higher potential. (Rindler, 2001, § 1.16, p. 25)[15]

But Rindler continues:

> ...So let a standard clock at some point A of low potential (for example, on the surface of a dense planet) be seen from some point B of higher potential, and let ν_A be **the universal rate at which *all* standard clocks tick**. Let the rate at which the standard A-clock *is seen* to tick at B be ν_B. This is, by [his] Eq. (1.11), less than the

rate ν_A at which the standard clock at B tickes, by the factor on the RHS. But if the clock at A is *seen* to go slow, then it really *does* go slow.
(Rindler, 2001, § 1.16, pp. 25 ... 26, italics in the original, bold added by author)[15]

So Rindler appears to contradict himself by referring to the universal rate at which *all* standard clocks tick, and yet emphasizing that these clocks really do slow down. For static and stationary spacetimes he explains that a 'world-movie' running at constant rate in global coordinate time would display the events slowed down near heavy masses and indeed frozen at the horizon, §9.4 of [15]. My two other favourite introductory GR textbooks [20, 21] avoid an interpretation, as does the advanced book [18]. As we saw from the quote in §1, Weinberg [13] clearly states the clock-rate-gravity-independent view. However, he offers the reader no help in reconciling this with the interpretation of the gravitational redshift nor does he even hint that there is something subtle to be helped with. Surprisingly, he also repeates a fundamental error in deriving the gravitational redshift that we discuss in detail in §3.1.

3. Lessons from the history of physics

The confusion started with an ambiguous derivation of the gravitational redshift by Einstein [22–24] and the confusion was compounded with an erroneous derivation by Eddington [16, 25, 26].

3.1. Erroneous Eddington gravitational redshift derivation

Eddington was concerned with the gravitational redshift of the Sun's spectra received on Earth. He starts his derivation with what Earman and Glymour call the Eddington premise that the proper period $d\tau$ of vibration of a given type of atom (acting as an ideal clock) is independent of its environment. Let A be a point on the surface of the Sun, and B a point on the surface of the Earth. For simplicity we can ignore the relative motions of the source and receiver and focus on the effect of the gravitational potential.

With $(d\tau)_A$ the proper period of a vibrating atom at location A and $(d\tau)_B$ the proper period of the same type of atom at location B, the Eddington premise becomes

$$(d\tau)_A = (d\tau)_B. \quad (1)$$

The proper time period $(d\tau)_A$ is related to the coordinate time interval $(dt)_A$ at some fixed location A (so $dx^i = 0$) via

$$\sqrt{g_{00}(r_A)}(dt)_A = (d\tau)_A, \quad (2)$$

where of course $g_{00}(r_A)$ is the lapse squared or temporal component of the metric tensor at location r_A in the approximately static spacetime. Unfortunately Eddington used his premise Eq. (1) as follows,

$$\sqrt{g_{00}(r_A)}(dt)_A = (d\tau)_A = (d\tau)_B = \sqrt{g_{00}(r_B)}(dt)_B. \quad (3)$$

And for $g_{00}(r_A) < g_{00}(r_B) \Rightarrow (dt)_A > (dt)_B$. The frequency of vibration $\nu = 1/dt$, and, Eddington infers

$$\frac{1}{(dt)_A} = \nu_A < \nu_B = \frac{1}{(dt)_B}. \quad (4)$$

Eddington concludes from Eq. (4) that "the solar atom thus vibrates more slowly, and thus its spectral lines will be displaced towards the red" [16, 25].

The problem with this derivation is that t is coordinate time and hence ν is the coordinate frequency, but observed frequencies are related to $1/d\tau$, the proper time intervals. The result is of course correct in that the observational frequency is redshifted. Two mistakes have cancelled.

Only James Rice queried Eddington on why we observe a redshift if $(d\tau)_A = (d\tau)_B$. Eddington's reply in the *Proceeds of the Royal Society* (1920), and especially in *Observatory* (1920), while confusing, indicates that he has all the elements of the corrected derivation, according to Earman and Glymour[14]. And yet strangely Eddington kept the erroneous redshift derivation in his influential textbook [26]! Sadly Rice repeated the erroneous redshift derivation in *his* textbook [27][p. 287–289]. Importantly the erroneous derivation was repeated consistently in the most influential presentations of GR before 1935, including, Earman and Glymour claim, in all five editions of Weyl's monograph Raum-Zeit-Materie, [28–32]. Recently Harvey and colleagues [33, 34] advocate the use of the definition of the gravitational redshift as presented in the fifth edition of *Raum-Zeit-Materie* [32]. Harvey [34] suggests that this definition has been overlooked because it is only available in German.

It is strange that Earman and Glymour [14] fail to mention that Weinberg repeated the erroneous Eddington derivation in his very authoritative and influential textbook [13][§3.5, pp. 79–80]. Earman and Glymour almost certainly noticed, since they cite [13][§3.5, pp. 84–85] in a different context. But there is no ambiguity that Weinberg repeated the erroneous Eddington derivation above; his notation is clear from the equations. His Eq. (3.5.1) is

$$\frac{dt}{\Delta t} = \left(-g_{\mu\nu} \frac{dx^\mu}{dt} \frac{dx^\nu}{dt} \right)^{-1/2}, \tag{5}$$

which clearly reveals that Δt is the proper time interval, while dt is the coordinate time interval. He considers the case of observering light at point 1 (like our point B above) that came from an atomic transition at point 2 (like our point A above), both points being at rest in a stationary gravitational field. In this case his Eq. (3.5.1) reduces to his Eq. (3.5.2)

$$\frac{dt}{\Delta t} = (-g_{00})^{-1/2}, \tag{6}$$

from which he derives his next two unnumbered equations

$$dt_2 = \Delta t \left(-g_{00}(x_2)\right)^{-1/2} \qquad \text{and} \qquad dt_1 = \Delta t \left(-g_{00}(x_1)\right)^{-1/2}. \tag{7}$$

Taking their ratio dt_1/dt_2 he finds his Eq. (3.5.3)

$$\frac{\nu_2}{\nu_1} = \left(\frac{g_{00}(x_2)}{g_{00}(x_1)} \right)^{1/2}. \tag{8}$$

Just as Eddington did, Weinberg mistakes the coordinate time interval dt with the observable time period. Yet he gets the correct answer by setting the proper time interval Δt to a constant. The text of this section of Weinberg's book is more correct but there is a disconnect with the equations. Weinberg's bibliography for this chapter includes Eddington's and Weyl's books [26, 31], suggesting his presentation was mislead by them.

3.2. Earman-Glymour gravitational redshift derivation

The Earman-Glymour [14] derivation starts with the Eddington premise Eq. (1). A seemingly small but actually very helpful point is their notation; they introduce the notation $(d\tau)_{A,B}$ for the proper time interval between vibrations of atom at A, as measured at B. So then $(d\tau)_{A,A}$ denotes the proper time period of vibrations of the atom on the Sun, as measured on the Sun, while $(d\tau)_{A,B}$ is the proper time interval between makers of the vibration of the atom on the Sun, as measured on Earth. And the Eddington premise in Earman-Glymour notation becomes

$$(d\tau)_{B,B} = (d\tau)_{A,A}. \tag{9}$$

Journal of Physics: Conference Series **600** (2015) 012055 doi:10.1088/1742-6596/600/1/012055

Earman & Glymour are concerned that we'll find this notation "fussy and pedantic"; I find it clarifies a confusing situation. Then, with constant coordinate time interval dt:

$$\frac{(d\tau)_{A,A}}{(d\tau)_{A,B}} = \frac{dt\sqrt{g_{00}(r_A)}}{dt\sqrt{g_{00}(r_B)}} = \frac{\sqrt{g_{00}(r_A)}}{\sqrt{g_{00}(r_B)}}. \tag{10}$$

They note that Max von Laue [35] showed that in a static spacetime Maxwell's equations have solutions proportional to $\exp(i\nu t)$, with constant coordinate frequency ν. This justifies cancelling the coordinate time interval dt in the numerator and denominator in Eq. (10).[1] *Here*, according to Earman & Glymour, is the appropriate place to apply Eddington's premise Eq. (9). We want to compare the interval $(d\tau)_{A,B}$ with a similar atom vibrating on Earth for which $(d\tau)_{B,B} = (d\tau)_{A,A}$. Substituting the Eddington premise into Eq. (10) allows us to relate local observals on Earth:

$$\frac{(d\tau)_{A,A}}{(d\tau)_{A,B}} = \frac{(d\tau)_{B,B}}{(d\tau)_{A,B}} = \frac{\nu_{A,B}}{\nu_{B,B}} = \frac{\sqrt{g_{00}(r_A)}}{\sqrt{g_{00}(r_B)}} < 1, \tag{11}$$

where $\nu \equiv 1/d\tau$ is the proper frequency.

The Eddington premise can be seen as a succinct statement of the clock-rate-gravity-independent view point of §1. And yet, with its careful application, we arrive at the gravitational redshift. This supports the compatibility of the two apparently contradictor two statements of Rindler in §2, and suggests that the redshift is a phenomenon related to message-passing.

4. Conclusion

History repeats itself every year in the classroom; when new students are introduced to subtle concepts they are confused exactly on those points that the leading scientists were confused about when the subject was newly discovered. Indeed the historical analysis of Earman and Glymour [14] warns us that the gravitational redshift is notoriously subtle and confusing.

The gravitational redshift derivation of Earman and Glymour [14] is admirably clear. A key point was their notation; even if a bit cumbersome it pays for itself in clarity. In particular their notation $(d\tau)_{A,B}$ allows one to distinguish between the markers of the beginning and end points of a proper time interval, say a tick of a given atomic clock at event A, as observed at different events B in spacetime.

The statement that "clocks slow down at lower gravitational potential" must be used with caution. In some sense it is true and in another sense it is not. Clarifying exactly in what sense we mean clocks slow down is one of the challenges of teaching introductory GR. It is tempting to play it safe, as do many textbooks [18, 20, 21]. But I think this is a disservice to the most inquisitive students who may become discouraged by the confusion surrounding an interpretation of the gravitational redshift. Recall the wisdom of John A. Wheeler when he said "Sad the week without a paradox, a difficulty, an apparent contradiction! For how can one then make progress?" [37]. We should lead our students to the paradox and then help resolve it.

An important limitation of the Earman-Glymour analysis [14] is that they forgo interpretation of the gravitational redshift, wishing to avoid a "windy warefare" over the Doppler vs. non-Doppler interpretations. But one can and should offer a more detailed interpetation in terms of the structure of spacetime. Furthermore the gravitational time dilation phenomenon has been observed with two clocks at different altitude [38, 39], wherein there is no message-passing. How can this be understood? This will be taken up in our next publication.

[1] Of course one could also appeal at this point to the time independence of the static gravitational field; see [36] and [17][Fig. 5.2].

Acknowledgments
I gratefully acknowledge financial support from a chaire d'excellence provided by the CNRS of France and travel funds from the Laboratoire de physique des océans of the Université de Bretagne Occidentale. I also thank José Luis Jaramillo, Richard Tweed, Eric Poisson, Jose María M. Senovilla and José Navarro-Salas for helpful discussions, encouragement and references. I also thank Wolfgang Rindler and Phil Morrison for helpful comments.

References
[1] Wald R M 2006 *Am. J. Phys.* **74** 471–477
[2] Hartle J B 2006 *Am. J. Phys.* **74** 14–21
[3] Bertschinger E and Taylor E F 2008 *Am. J. Phys.* **76** 103–105
[4] Hobson A 2008 *Am. J. Phys.* **76** 607–607
[5] Silvert W 1973 *Am. J. Phys.* **41** 926–927
[6] Lambourne R 2005 *Eur. J. Phys.* **26** S135
[7] Baez J C and Bunn E F 2005 *Am. J. Phys.* **73** 644–652
[8] Drake S P 2006 *Am. J. Phys.* **74** 22–25
[9] Zimmerman S 2010 *Eur. J. Phys.* **31** 1377
[10] Christensen N and Moore T 2012 *Phys. Today* **65** 41–47
[11] Okun L B, Selivanov K G and Telegdi V L 2000 *Am. J. Phys.* **68** 115–119
[12] Møller C 1952 *The Theory of Relativity* (Oxford Clarendon Press) ISBN 978-1124104133
[13] Weinberg S 1972 *Gravitation and Cosmology: Principles and Applications of the General Theory of Relativity* (New York, NY: John Wiley & Sons)
[14] Earman J and Glymour C 1980 *Studies in History and Philosophy of Science* **11** 175–214
[15] Rindler W 2006 *Relativity: Special, General and Cosmological, 2nd Ed.* (Oxford, UK: Oxford University Press)
[16] Eddington A S 1918 *Report on the Relativity Theory of Gravitation* (Fleetwood Press, London UK)
[17] Schutz B 2009 *A first course in General Relativity* (Cambridge UK: Cambridge University Press)
[18] Misner C, Thorne K and Wheeler J 1973 *Gravitation* (San Francisco: W.H. Freeman)
[19] Price H E 1974 *Am. J. Phys.* **42** 336–337
[20] Hobson M, Efstathiou G and Lasenby A 2006 *General Relativity: An introduction for physicists* (Cambridge University Press, UK: Cambridge)
[21] Carroll S 2004 *Spacetime and Geometry: An introduction to General Relativity* (San Francisco: Addison Wesley) 513 + xiv pp
[22] Einstein A 1907 *Annalen der Physik* **23** 371–384 ISSN 0003-3804
[23] Einstein A 1911 *Annalen der Physik* **35** 898 translation by W. Perrett and G. B. Jeffery, *The Principle of Relativity* (New York: Dover, 1952)
[24] Einstein A 1916 *Annalen der Physik* **49** 720–822 translation by W. Perrett and G. B. Jeffery, *The Principle of Relativity* (New York: Dover, 1952)
[25] Eddington A S 1920 *Report on the Relativity Theory of Gravitation, Second Edition* (Fleetwood Press, London UK)
[26] Eddington A S 1923 *The Mathematical Theory of Relativity* (Chelsa, New York, NY 1975)
[27] Rice J 1923 *Relativity: A systematic treatment of Einstein's Theory* (London, UK: Longmans, Green) 397 + xv pp
[28] Weyl H 1918 *Raum, Zeit, Materie, 1st edition* (Verlag von Julius Springer, Berlin)
[29] Weyl H 1918 *Raum, Zeit, Materie, 2nd edition* (Verlag von Julius Springer, Berlin)
[30] Weyl H 1919 *Raum, Zeit, Materie, 3rd edition* (Verlag von Julius Springer, Berlin)
[31] Weyl H 1921 *Space-Time-Matter, 4th ed.* (Dover, New York) ISBN 978-0486602677
[32] Weyl H 1923 *Raum, Zeit, Materie, 5th edition* (Verlag von Julius Springer, Berlin)
[33] Harvey A, Schucking E L and Surowitz E J 2006 *Am. J. Phys.* **74** 1017–1024
[34] Harvey A 2012 *Eur. J. Phys.* **33** 823
[35] von Laue M 1920 *Phys. Z.* **21** 659–662
[36] Schild A 1967 Lectures on general relativity theory *Relativity Theory And Astrophysics* ed Ehlers J (American Mathematical Society) pp 1–105
[37] Wheeler J A 1996 *At Home in the Universe* (American Institute of Physics, New York) ISBN 978-0883188620
[38] Briatore L and Leschiutta S 1977 *Il Nuovo Cimento B Series 11* **37** 219–231 ISSN 0369-3554
[39] Alley C O 1983 Proper time experiments in gravitational fields with atomic clocks, aircraft, and laser light pulses *Quantum Optics, Experimental Gravity, and Measurement Theory* (*NATO Advanced Science Institutes Series* vol 94) ed Meystre P and Scully M O (Springer US) pp 363–427 ISBN 978-1-4613-3714-0

Journal of Physics: Conference Series **600** (2015) 012056 doi:10.1088/1742-6596/600/1/012056

Constraining the Equation of State of Neutron Stars from Binary Mergers

Kentaro Takami[1], Luciano Rezzolla[2] and Luca Baiotti[3]

[1,2] Institut für Theoretische Physik, Max-von-Laue-Straße 1, 60438 Frankfurt, Germany
[3,] Graduate School of Science, Osaka University, 560-0043 Toyonaka, Japan

E-mail: [1] takami@th.physik.uni-frankfurt.de

Abstract. Determining the equation of state of matter at nuclear density and hence the structure of neutron stars has been a riddle for decades. We show how the imminent detection of gravitational waves from merging neutron star binaries can be used to solve this riddle. Using a large number of accurate numerical-relativity simulations of binaries with nuclear equations of state, we have found that the postmerger emission is characterized by two distinct and robust spectral features. While the high-frequency peak has already been associated with the oscillations of the hypermassive neutron star produced by the merger and depends on the equation of state, a new correlation emerges between the low-frequency peak, related to the merger process, and the compactness of the progenitor stars. More importantly, such a correlation is essentially universal, thus providing a powerful tool to set tight constraints on the equation of state. If the mass of the binary is known from the inspiral signal, the combined use of the two frequency peaks sets four simultaneous constraints to be satisfied. Ideally, even a single detection would be sufficient to select one equation of state over the others. We have tested our approach with simulated data and verified it works well for all the equations of state considered.

1. Introduction

The first direct detection of gravitational waves (GWs) will be made by a series of advanced detectors such as LIGO, Virgo, and KAGRA which will become operational in the next five years. The inspiral and postmerger of neutron-star binaries or neutron-star–black-hole binaries, and binary black holes are expected as the sources of GWs. Population-synthesis models suggest that binary neutron star mergers (BNSs) may be the most promising source and an expected detection rate is $\sim 40\,\mathrm{yr}^{-1}$ [1].

The first evidence that the information of the postmerger signal could be extracted from the corresponding spectrum was provided by Bauswein and Janka [2], who performed a large number of simulations using a smoothed particle hydrodynamics code solving the Einstein field equations assuming conformal flatness, and employing a GW backreaction scheme within a post-Newtonian approximation. In particular, they pointed out the presence of a peak at high-frequency in the spectrum, which corresponds to a fundamental fluid mode with $m = 2$ of the hypermassive neutron star (HMNS) [3], and showed it correlated with the properties of the equation of state (EOS).

By performing a large number of accurate simulations in full general relativity of equal-mass and unequal-mass BNSs with a number of different nuclear EOSs, we have examined the spectral

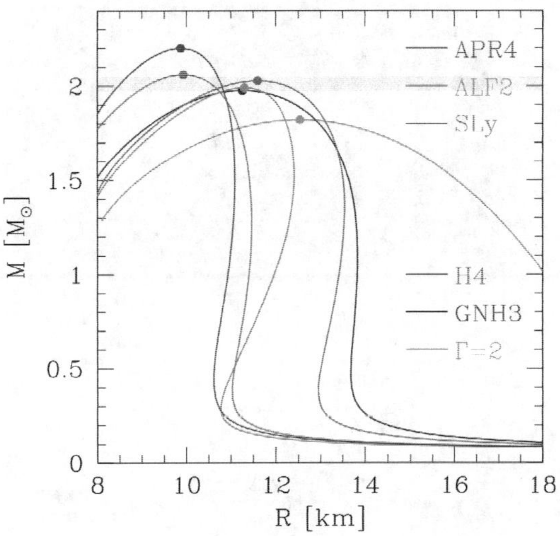

Figure 1. TOV sequences for each EOS considered. A shaded yellow band shows current observational constraint on the observed maximum mass in neutron stars, i.e., $2.01 \pm 0.04 M_{\odot}$ obtained for the pulsar PSR J0348+0432 [6].

properties of the postmerger GW signal. In this proceeding, we summarize our recent results in our papers [4, 5].

2. Numerical Setup

Our results have been obtained in full general relativity solving the Einstein equations with the `McLachlan` code [7, 8]. The solution of the relativistic hydrodynamics equations is obtained using the `Whisky` code [9, 10]. The stars are modeled by using a nuclear EOS and we have considered five different models, i.e., APR4, ALF2, SLy, H4, GNH3. As the corresponding TOV sequences are shown in Fig. 1, all of these EOSs satisfy the current observational constraint on the observed maximum mass in neutron stars, i.e., $2.01 \pm 0.04 M_{\odot}$ obtained for the pulsar PSR J0348+0432 [6]. Rather than using tables, it is more convenient to use n piecewise polytropic approximations to these EOSs [11], expressing the "cold" contribution to pressure and specific internal energy as

$$p_{\rm c} = K_i \rho^{\Gamma_i}, \quad \epsilon_{\rm c} = \epsilon_i + \frac{K_i \rho^{\Gamma_i - 1}}{(\Gamma_i - 1)}, \tag{1}$$

where ρ and K are the rest-mass density and the polytropic constant, respectively. We adapt $n = 4$ for all cases, because it is sufficient to obtain an accurate representation of the different EOSs.

In addition, to model the thermal effects arising from the merger, the cold pressure is augmented through an ideal-fluid EOS, so that the total pressure and specific internal energy are $p = p_{\rm c} + p_{\rm th}$, $\epsilon = \epsilon_{\rm c} + \epsilon_{\rm th}$, with $p_{\rm th} = \rho \epsilon_{\rm th} (\Gamma_{\rm th} - 1)$ [12]. Following Ref. [13], we use $\Gamma_{\rm th} = 2$, because we have verified that our results are not sensitive to this choice [5]. Finally, to span a larger range in stellar compactness and go beyond the one covered by the nuclear EOSs above, we have considered a sixth EOS given by a pure polytrope with $\Gamma = 2$ and $K = 123.6$ in units where $c = G = M_{\odot} = 1$.

For each EOS, we have considered five equal-mass binaries with average gravitational mass

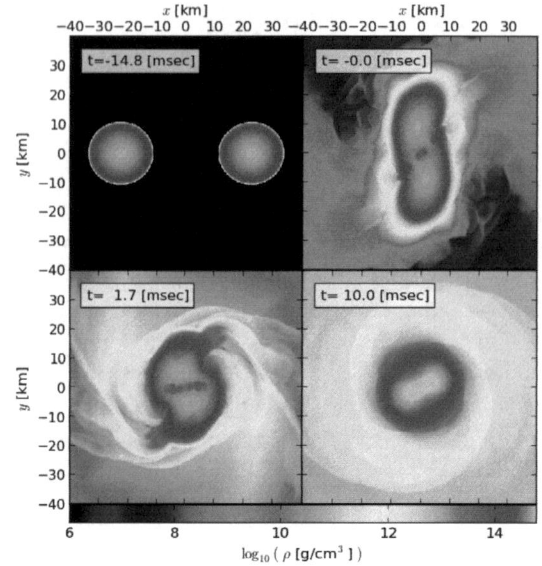

Figure 2. Snapshots of the rest-mass density on the (x, y) plane for the binary with H4 EOS and $\bar{M} = 1.3M_{\odot}$. The panels refer to four characteristic times: the initial time, the time of the merger, the time right after the merger (i.e., at $t = 1.7\,\mathrm{ms}$), and the time with a quasi-stationary core (i.e., at $t = 10.0\,\mathrm{ms}$).

at infinite separation in the range $\bar{M} \equiv (M_1 + M_2)/2 = (1.275 - 1.375)M_{\odot}$ for the APR4 EOS, $(1.225 - 1.325)M_{\odot}$ for the ALF2 EOS, $(1.250 - 1.350)M_{\odot}$ for the GNH3, H4, and SLy EOSs, and $(1.350 - 1.450)M_{\odot}$ for the $\Gamma = 2$ polytrope. We have also considered two unequal-mass binaries for the GNH3 and SLy EOSs having $\bar{M} = 1.300\,M_{\odot}$ and mass ratio $\simeq$0.92. The binaries are modeled as irrotational in quasicircular orbits and computed with the `LORENE` code [14], assuming a conformally flat metric. To increase resolution we have employed a reflection symmetry across the $z = 0$ plane, a π-symmetry condition across the $x = 0$ plane, and a moving-mesh refinement via the `Carpet` driver [15]. We have used six refinement levels, the finest having a resolution of $0.15\,M_{\odot} \simeq 0.221\,\mathrm{km}$, and extracted the GWs near the outer boundary at a distance $R_0 = 500\,M_{\odot} \simeq 738\,\mathrm{km}$.

3. Results

First of all, we show a dynamics of a binary with H4 EOS and $\bar{M} = 1.3M_{\odot}$ in Fig. 2, in which the two stellar cores collide and bounce repeatedly as a result of the strong rotation and very high densities. The figure contains four different panels for the rest-mass density on the (x, y) plane at four characteristic times: the initial time, the time of the merger, the time right after the merger, and the time with a quasi-stationary core.

As pointed out by several authors (e.g., [2, 16]), the power spectral density (PSD) of the postmerger GW signal has three clear peaks. As examples, two binaries with gravitational masses $\bar{M}/M_{\odot} = 1.325$, and APR4 and GNH3 EOSs are presented in Fig. 3. The top panel shows the evolution of the $\ell = m = 2$ plus polarization of the strain aligned at the merger for sources at a polar distance of 50 Mpc. On the other hand, the bottom panel shows the spectral densities $2\tilde{h}(f)f^{1/2}$ for the two EOSs, comparing them with the sensitivity curves of Advanced LIGO and of the Einstein Telescope. The dotted lines refer to the whole time series, while the

Journal of Physics: Conference Series **600** (2015) 012056 doi:10.1088/1742-6596/600/1/012056

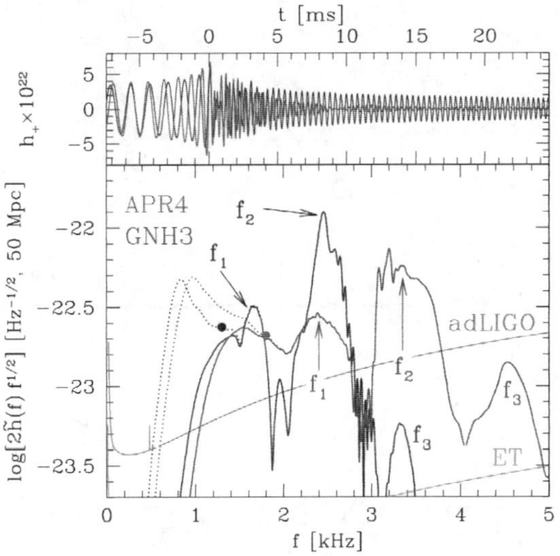

Figure 3. Top panel: evolution of h_+ for representative binaries with the APR4 and GNH3 EOSs (dark-red and blue lines, respectively) for sources at a polar distance of 50 Mpc. Bottom panel: spectral density $2\tilde{h}(f)f^{1/2}$ the merger for the two EOSs and sensitivity curves of Advanced LIGO (green line) and ET (light-blue line); the dotted lines show the power in the inspiral, while the circles mark the contact frequency.

circles mark the "contact frequency" $f_{\rm cont} = \mathcal{C}^{3/2}/(2\pi\bar{M})$ [17], where $\mathcal{C} \equiv \bar{M}/\bar{R}$ is the average compactness, $\bar{R} \equiv (R_1 + R_2)/2$, and $R_{1,2}$ are the radii of the nonrotating stars associated with each binary. In order to evaluate the peak frequencies, we use a fitting procedure for the PSDs with the two peaks [5].

The most interesting and important result of our spectral analysis is that there is a very clear correlation between the low-frequency peak f_1 and the compactness $\mathcal{C}$. The results of the fitting procedure for the low-frequency peak are collected in the left panel of Fig. 4. Also shown as a shaded gray band is the estimate of the total error, which is effectively dominated by the fitting procedure of the PSD. The behavior of the low-frequency peak is remarkably consistent with a cubic polynomial function. The essentially universal behavior of the f_1 frequency with compactness provides a powerful tool to constrain the EOS. On the other hand, the right panel of Fig. 4 shows the behavior of the high-frequency peak f_2 as a function of the average rest-mass density $\left(\bar{M}/\bar{R}^3\right)^{1/2}$. We perform a linear fit for each EOS.

Finally, we discuss how to use them to constrain the EOS. Let us assume that the GW signal from a BNS has been detected and all of the spectral features are clearly identifiable. Using the measured values of f_1 and f_2 we can draw on the $(\bar{M}, \bar{R})$ plane a series of curves given by the relations $\bar{M} = \bar{M}(\bar{R}, f_1)$ (solid gray line) and $\bar{M} = \bar{M}(\bar{R}, f_2; \mathrm{EOS})$ (solid colored lines). This is shown in Fig. 5 for the ALF2 EOS. We can see that the $\bar{M} = \bar{M}(\bar{R}, f_1)$ relation intersects each of the various equilibrium curves (colored dashed lines) at one point (e.g., at $\bar{M} \simeq 1.325\,M_\odot$, $\bar{R} \simeq 12.3$ km for the ALF2 EOS), but also other crossings take place for the other EOSs. However, when using also the relations $\bar{M} = \bar{M}(\bar{R}, f_2; \mathrm{EOS})$, some EOSs can be readily excluded (e.g., APR4, SLy, and GNH3, in our example) and only the ALF2 and H4 EOSs have "near crossings" between the equilibrium-models curves and the frequency-correlations curves. Fortunately, the uncertainty can be removed if the mass of the binary is known from the inspiral signal. In this case, in fact, there will be a horizontal line in the $(\bar{M}, \bar{R})$ plane that will

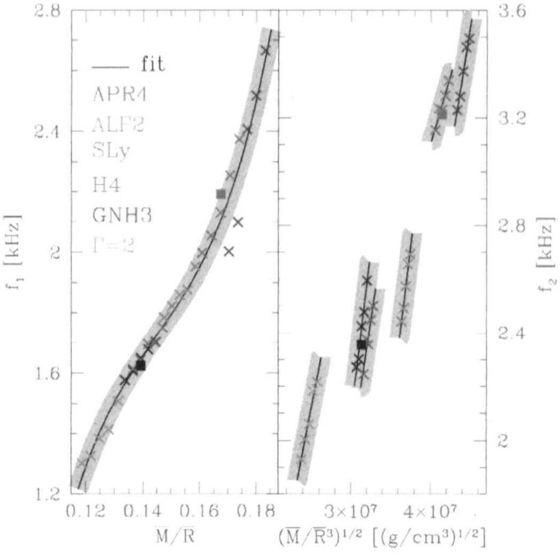

Figure 4. Left panel: the low-frequency peaks as a function of the stellar compactness for the six different EOSs considered, where crosses and filled squares refer to equal-mass and unequal-mass binaries, respectively. The cubic fit is shown by a solid black line, while the gray band is the estimate of the total errors. Right panel: the high-frequency peaks as a function of the average rest-mass density.

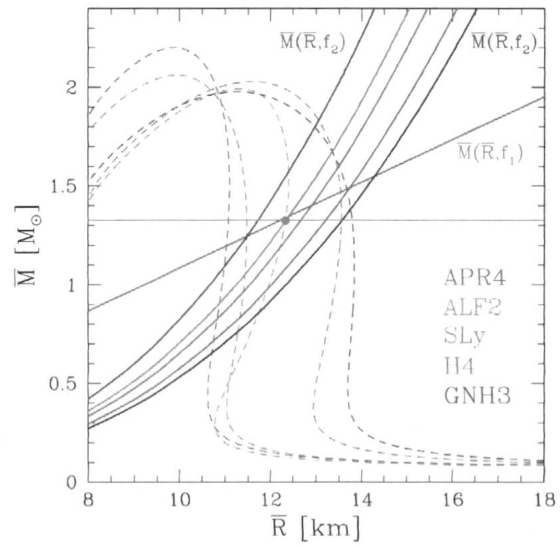

Figure 5. Examples of use of the spectral features to constrain the EOS. Once a detection is made, the relations $\bar{M} = \bar{M}(\bar{R}, f_1)$ and $\bar{M} = \bar{M}(\bar{R}, f_2; \mathrm{EOS})$ (colored solid lines) will cross at one point the curves of equilibrium configurations (colored dashed lines). Information of the mass of the system (horizontal line) will provide a fourth constraint, removing possible degeneracies. The panel refer to the ALF2, but all other EOSs behave in the same way.

break the degeneracy imposing four simultaneous constraints. This is shown with the horizontal light-blue line, which clearly intersects the three curves relative to the ALF2 EOS at one point only (green solid circle). Despite the simplifying assumptions, this method shows that even a single detection of a GW signal with high SNR and from which the mass of the binary can be calculated, would be sufficient to set tight constraints on the EOS. This approach works well for all of other binaries considered.

4. Conclusions

We have carried out a large number of accurate and fully general-relativistic simulations of the inspiral and postmerger of BNSs with nuclear EOSs. This has allowed us to have a comprehensive view of the spectral properties of the complex postmerger GW signal and to highlight the presence of two robust frequency peaks. We have shown for the first time that the low-frequency peaks exhibit a correlation with the stellar compactness that is essentially EOS independent and that can be used to constrain the EOS once the peak is measured. In addition, the combined use of other EOS-dependent correlations from the high-frequency peaks can further constrain the EOS. In principle, if the mass is known from the inspiral and the peaks are clearly measurable, a single detection would be sufficient to set constraints on the EOS.

Acknowledgments

Partial support comes from the DFG grant SFB/Transregio 7 and by "NewCompStar", COST Action MP1304. K.T. is supported by the LOEWE-Program in HIC for FAIR. The simulations were performed on SuperMUC at LRZ-Munich, on Datura at AEI-Potsdam and on LOEWE at CSC-Frankfurt.

References

[1] Abadie J *et al.* 2010 *Class. Quantum Grav.* **27** 173001 (*Preprint* 1003.2480)
[2] Bauswein A and Janka H T 2012 *Phys. Rev. Lett.* **108** 011101 (*Preprint* 1106.1616)
[3] Stergioulas N, Bauswein A, Zagkouris K and Janka H T 2011 *Mon. Not. R. Astron. Soc.* **418** 427–436 (*Preprint* 1105.0368)
[4] Takami K, Rezzolla L and Baiotti L 2014 *Phys. Rev. Lett.* **113** 091104 (*Preprint* 1403.5672)
[5] Takami K, Rezzolla L and Baiotti L 2014 *ArXiv e-prints* (*Preprint* 1412.3240)
[6] Antoniadis J, Freire P C C, Wex N, Tauris T M, Lynch R S, van Kerkwijk M H, Kramer M, Bassa C, Dhillon V S, Driebe T, Hessels J W T, Kaspi V M, Kondratiev V I, Langer N, Marsh T R, McLaughlin M A, Pennucci T T, Ransom S M, Stairs I H, van Leeuwen J, Verbiest J P W and Whelan D G 2013 *Science* **340** 1233232 (*Preprint* 1304.6875)
[7] Brown D, Diener P, Sarbach O, Schnetter E and Tiglio M 2009 *Phys. Rev. D* **79** 044023 (*Preprint* 0809.3533) URL http://dx.doi.org/10.1103/PhysRevD.79.044023
[8] Löffler F, Faber J, Bentivegna E, Bode T, Diener P, Haas R, Hinder I, Mundim B C, Ott C D, Schnetter E, Allen G, Campanelli M and Laguna P 2012 *Class. Quantum Grav.* **29** 115001 (*Preprint* arXiv:1111.3344 [gr-qc])
[9] Baiotti L, Hawke I, Montero P J, Löffler F, Rezzolla L, Stergioulas N, Font J A and Seidel E 2005 *Phys. Rev. D* **71** 024035 (*Preprint* gr-qc/0403029)
[10] Baiotti L, Giacomazzo B and Rezzolla L 2008 *Phys. Rev. D* **78** 084033 (*Preprint* 0804.0594)
[11] Read J S, Lackey B D, Owen B J and Friedman J L 2009 *Phys. Rev. D* **79** 124032
[12] Janka H T, Zwerger T and Mönchmeyer R 1993 *Astron. Astrophys.* **268** 360–368
[13] Bauswein A, Janka H and Oechslin R 2010 *Phys. Rev. D* **82** 084043 (*Preprint* 1006.3315)
[14] Gourgoulhon E, Grandclément P, Taniguchi K, Marck J A and Bonazzola S 2001 *Phys. Rev. D* **63** 064029 (*Preprint* gr-qc/0007028)
[15] Schnetter E, Hawley S H and Hawke I 2004 *Class. Quantum Grav.* **21** 1465–1488 (*Preprint* gr-qc/0310042)
[16] Bauswein A, Janka H T, Hebeler K and Schwenk A 2012 *Phys. Rev. D* **86** 063001 (*Preprint* 1204.1888)
[17] Damour T, Nagar A and Villain L 2012 *Phys. Rev. D* **85**(12) 123007 URL http://link.aps.org/doi/10.1103/PhysRevD.85.123007

Fallback accretion onto magnetized neutron stars and the hidden magnetic field model

A Torres[1], P Cerdá-Durán[1] and J A Font[1,2]

[1]Departamento de Astronomía y Astrofísica, Universitat de València, Dr. Moliner 50, 46100, Burjassot (València), Spain
[2]Observatori Astronòmic, Universitat de València, C/ Catedrático José Beltrán 2, 46980, Paterna (València), Spain

E-mail: `Alejandro.Torres@uv.es`

Abstract. The observation of several neutron stars with relatively low values of the surface magnetic field found in supernova remnants has led in recent years to controversial interpretations. A possible explanation is the slow rotation of the proto-neutron star at birth which is unable to amplify its magnetic field to typical pulsar levels. An alternative possibility, the hidden magnetic field scenario, seems to be favoured over the previous one due to the observation of three low magnetic field magnetars. This scenario considers the accretion of the fallback of the supernova debris onto the neutron star as the responsible for the observed low magnetic field. In this work, we have studied under which conditions the magnetic field of a neutron star can be buried into the crust due to an accreting fluid. We have considered a simplified toy model in general relativity to estimate the balance between the incoming accretion flow an the magnetosphere. We conclude that the burial is possible for values of the surface magnetic field below 10^{13} G. The preliminary results reported in this paper for simplified polytropic models should be confirmed using a more realistic thermodynamical setup.

1. Introduction

The hidden magnetic field scenario proposed by Shabaltas & Lay [1] offers a viable explanation for the unusual features observed in Central Compact Objects (CCOs). CCOs are isolated young neutron stars with no radio emission and located near the center of young supernova remnants (SNR). Nowadays, three such neutron stars (PSR E1207.4-5209, PSR J0821.0-4300, PSR J1852.3-0040) show an inferred magnetic field significantly lower than the common values for neutron stars. PSR E1207.4-5209 in the supernova remnant PKS 1209-51/52 was the first discovered CCO [2, 3] and has been extensively studied. Its period $P = 424.130751(4)$ ms and period derivative $\dot{P} = (9.6 \pm 9.4) \times 10^{-17}$s s^{-1} imply a surface magnetic field strength $B_s < 3.5 \times 10^{11}$ G, and a characteristic age of $\tau_c \equiv P/2\dot{P} > 24$ Myr. PSR J0821.0-4300 in Puppis A [4] is a 112 ms pulsar with $\dot{P} < 8.3 \times 10^{-15}$s s^{-1}. This value of $\dot{P}$ implies a surface magnetic dipole field strength $B_s < 9.8 \times 10^{11}$ G. The characteristic age of PSR J082-4300 is $\tau_c > 220$ kyr. Finally, PSR J1852-0040 in Kes 79 [5] has a period of $P = 105$ ms, a period derivative of $\dot{P} < 7 \times 10^{-14}$s s^{-1}and a surface magnetic field strength of $B_s < 3 \times 10^{12}$ G, and its spin-down age is $\tau > 24$ kyr. In all cases, the difference between the characteristic age of the neutron star and the age of their the remnant indicates that the neutron stars were born spinning at their present periods.

Similar to these observations of CCOs, there also exist three additional observations of so-called low magnetic field magnetars [6], SGR 0418+5729, Swift J1822-1606 and 3XMM J185246.6+003317. SGR 0418+5729 [7] has a period of $P = 9.07838822$ s and a period derivative $\dot{P} = 4 \times 10^{-15}$s s^{-1}, which leads to a surface dipolar magnetic field of $B_s = 6 \times 10^{12}$ G. This measurement confirms SGR 0418+5729 as the lowest magnetic field magnetar known. The inferred spindown age is $\sim$ 550 kyr. Swift J1822-1606 [8] has a period of $P = 8.43772016$ s, a period derivative $\dot{P} = 2.14 \times 10^{-14}$s s^{-1},a surface magnetic field of $B_s = 1.4 \times 10^{13}$ G and spindown age of 6300 kyr. Finally, 3XMM J185246.6+003317 [9] has a period of $P = 11.55871346$ s, a period derivative $\dot{P} < 1.4 \times 10^{-13}$s s^{-1}, which, assuming the classical magneto-dipolar braking model, gives a limit on the dipolar magnetic field of $B_s < 4.1 \times 10^{13}$ G. The spindown age is $>$ 1300 kyr. The spectral characteristics and activity of these objects confirm their identification as magnetars but, as in the case of the CCOs presented before, they have values of the magnetic field significantly lower than the typical values for their class (i.e. $10^{14} - 10^{15}$ G).

The first explanation for the unusual magnetic field found in these objects assumes that the neutron stars were born with a low value of the magnetic field. This idea is based on field amplification models where the magnetic field is amplified by the turbulent dynamos of the proto-neutron star [10, 11]. Therefore, the low values of the magnetic field are due to the fact that the slow rotation at birth of the neutron star does not sufficiently amplify the magnetic field. However, recent studies have shown that even in the absence of rapid rotation magnetic fields in proto-neutron stars can be amplified by other mechanisms such as convection and the standing accretion shock instability (SASI) [12, 13].

The other possible explanation is the hidden magnetic field model. After the supernova explosion, when the neutron star is born, the supernova shock is still traveling outwards through the external layers of the star. When this shock crosses a discontinuity in density, part of it can be reflected and move backwards towards the neutron star. The total mass accreted by the reverse shock in this process is $\sim 10^{-4} - 10^{-1} M_\odot$ in a typical timescale of hours to days [14]. Such high accretion rate can compress the magnetic field of the neutron star which can eventually be buried into the neutron star crust. As a result, the value of the external magnetic field would be significantly lower than the internal 'hidden' magnetic field. When accretion stops, the magnetic field may reemerge after a certain period of time.

There exist several works that have studied the viability of this scenario. On the one hand, the initial works studied the process of reemergence [15, 16, 17] using simplified 1D models and dipolar fields. These works established that the timescale for the magnetic field reemergence is about $1 - 10^7$ kyr, depending on the depth at which the magnetic field is buried. More recent works have confirmed this result. Ho [18] observed similar timescales for the reemergence using a 1D cooling code. Bernal, Lee and Page [19] performed 1D and 2D simulations of a single column of material falling onto a magnetized neutron star and showed how the magnetic field can be buried into the neutron star crust. Viganò and Pons [20] carried out simulations of the evolution of the interior magnetic field during the accretion and magnetic field submergence phase. Our goal is to perform 2D MHD simulations of the accretion phase. In this work we present preliminary results of initial tests aimed at devising the strategy we must follow to perform such simulations and at identifying the potential difficulties. Complete result will be presented elsewhere [21].

2. Pressure balance in an accreting magnetosphere

In this work, we follow the notation defined in [22]. The neutron star magnetosphere refers to the area surrounding the star where the magnetic pressure dominates over the thermal pressure of the accreting fluid. We call magnetopause to the interface between the magnetically dominated area and the thermally dominated area. The space of parameters of the problem includes the

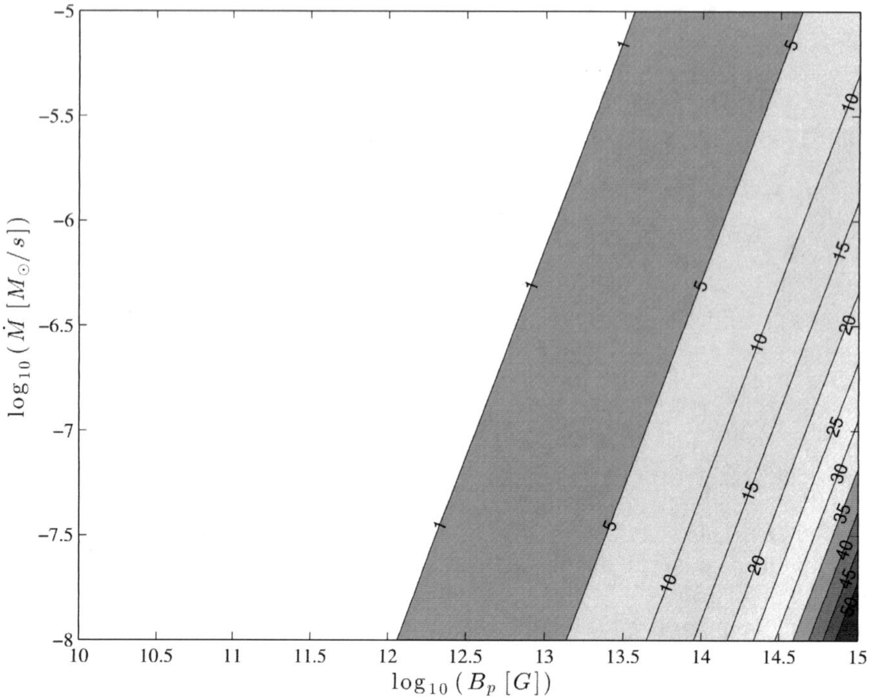

Figure 1. Radial position of the magnetopause for several values of the accretion rate $\dot{M}$ and magnetic field at the pole B_p. The lines indicate isocontours of $r_{\rm mp}/r_{\rm star}$ referred to the radius of the neutron star. Values under 1 are shown in white.

surface magnetic field at the pole of the neutron star, typically in the range $B_p \in [10^{10} - 10^{15}]$ G, and the accretion rate of the infalling material, in the interval $\dot{m} \in [10^{-8} - 10^{-5}] M_\odot/s$. The latter values are derived from the total mass accreted by the reverse shock in a time scale of a few hours.

For our first test we model the accretion fluid with the so-called Michel solution describing the spherical accretion of an unmagnetized relativistic fluid [23]. The magnetosphere is generated by a simple dipole magnetic field and a polytropic equation of state is used to describe the fluid. Our aim is to compute the position of the magnetopause at which the balance between the accreting fluid and the magnetosphere is reached. For this purpose we search for a pressure balance at the equator for the span of parameters. As the velocity of the fluid plays an important role in the form of ram pressure, we include in the expression of the pressure balance an extra term that models the effect of the velocity, namely

$$\frac{B^2}{2} = P_{\rm fluid} + \rho h W^2 v^2, \tag{1}$$

where $\frac{B^2}{2}$ and $P_{\rm fluid}$ are the pressures generated by the magnetic field and the fluid respectively, ρ is the density of the fluid, h is the specific enthalpy, W is the Lorentz factor and v is the velocity. The restriction given by this equation must be satisfied at the magnetopause.

Fig. 1 shows the position of the equilibrium magnetopause ($r_{\rm mp}$) for the span of parameters we are considering. Values of $r_{\rm mp}/r_{\rm star} < 1$ indicate that the pressure equilibrium position is reached inside the star, and therefore the magnetic field can be buried totally. This is the case

Journal of Physics: Conference Series **600** (2015) 012057 doi:10.1088/1742-6596/600/1/012057

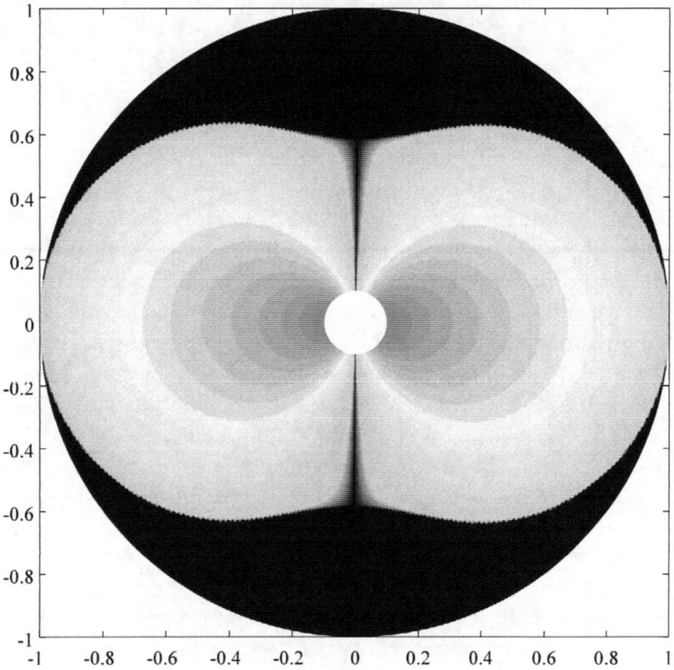

Figure 2. Magnetosphere generated by the multipolar expansion of a dipole proposed by Michel [22]. The white circle at center represents the neutron star, where the magnetic field is not calculated. The scale is normalized to be 1 at the magnetopause.

for magnetic field values below 10^{12} G for all accretion rates considered. In the case of magnetars with $B_p \in [10^{14} - 10^{15}]$ G, only the highest accretion rates considered can compress the magnetic field near the neutron star surface, but not close enough to bury it at the equator. However, it could still be possible to have the magnetic field partially buried in higher latitudes of the star.

3. Equilibrium magnetosphere

In the second test we compute the magnetosphere structure following the procedure proposed in [22]. The dipolar magnetic field is confined by gravitationally trapped plasma external to the star. The magnetic field is modeled as a dipole plus an arbitrary set of multipoles. The vector potential is

$$A_\phi = \sin\theta[r^{-2} - \sum_{n=1}^{\infty} a_n r^n P_n'(\cos\theta)], \tag{2}$$

where P_n' is the derivative with respect to $\cos\theta$ of the nth Legendre polynomial. The field lines are given by lines of constant $f = r\sin\theta A_\phi$ and they are tangent to the surface given by the magnetopause. The goal of this test is to have an idea of the shape of a magnetosphere confined by the accreting fluid before performing the actual numerical simulations.

The shape of Michel's confined magnetosphere is shown in Fig. 2. The radius and the magnetic field values have been normalized for generality. We can see how the magnetosphere approaches zero at the poles as expected for a dipolar magnetic field. This result gives us information of the changes in the geometry of the magnetic field resulting from the accretion we can expect. In a more realistic simulation the initial magnetic field has to be confined by the accretion fluid.

4. Conclusions

The initial results presented here provide support to the idea that the hidden magnetic field scenario is plausible for values of the magnetic field $B_p \in [10^{10} - 10^{13}]$ G, typical in isolated neutron stars, and for the typical accretion rates produced by the supernova reverse shock. In the case of magnetars, the scenario seems to be less plausible as the equatorial point of pressure balance is located far from the neutron star, implying that it could be difficult to bury the magnetic field. We note however that the models we have used in this preliminary work are fairly simple and the results may change accordingly when considering a more realistic scenario. From a technical point of view, the feasibility to carry out the actual simulations is going to be conditioned by the ratio between the magnetic and the thermal pressure. An ideal MHD numerical code must separate the thermal and magnetic contributions to the energy, which, in the scenario under study, is going to be a numerically challenging goal.

The second test reported in this work is key to understand the shape of a confined magnetic field. In future simulations we must compute a force-free solution for the magnetic field that takes into account the boundary conditions imposed by the accreting fluid. In the same way, we must impose very low densities in order to assure a force-free solution in the whole magnetosphere. However, the multipolar expansion of the dipole given by Eq. (2) is not general due to its dependence on the number of multipoles N used to compute the solution, which in our case was not too large as our algorithm shows stability and convergence problems for large values of N. In addition, this expression does not allow to impose arbitrary boundary conditions in order to study different field configurations.

In a future work [21] we plan to use a more realistic equation of state that will allow us to control the temperature and the composition of the accreting fluid. Furthermore, the magnetic field must be computed with increased generality to accommodate different magnetic field configurations and impose distinct magnetopause positions. The incorporation of these new ingredients in a time-dependent MHD numerical code is ongoing and results for a larger number of astrophysical scenarios will be presented elsewhere [21].

Acknowledgements

This work has been supported by the Spanish MINECO (AYA2013-40979-P), by the Generalitat Valenciana (PROMETEOII-2014-069), and by the European Research Council (grant StG-CAMAP-259276).

References

[1] Shabaltas N and Lai D 2012 *ApJ* **748**,148
[2] Helfand D J and Becker R H 1984 *Nature* **307** 215
[3] Zavlin V E, Pavlov G G, Sanwal D and Trumper J 2000 *ApJL* **540** L25
[4] Gottfled E V and Halpern J P 2009 *ApJL* **695** L35
[5] Gottfled E V, Halpern J P and Seward S D 2005 *ApJL* **627** 390-96
[6] Turolla R and Esposito P 2013 *Int. J. Mod. Phys.* D **22** 1330024
[7] Rea N *et al* 2013 *ApJ* **770** 65
[8] Rea N *et al* 2012 *ApJ* **754** 27
[9] Rea N, Viganò D, Israel G L, Pons J A, Torres D F 2014 *ApJL* **781** L17
[10] Thompson C and Duncan R C 1993 *ApJ.* **408** 194
[11] Bonanno A *et al.* 2005 *A&A* **440** 199
[12] Endeve E, Cardall C Y, Budiardja R D, Beck S W and Bejnood A 2012 *ApJ.* **751** 26 (28pp)
[13] Obergaulinger M, Janka H T and Aloy-Torás M A 2014 Accepted for publication in MNRAS
[14] Ugliano M, Janka H T, Marek A and Arcones A 2012 *ApJ* **757** 69

[15] Young E J and Chanmugan G 1995 *ApJ* **442** L53
[16] Muslinov A and Page D 1995 *ApJ* **440** L77
[17] Geppert U Page D and Zannias T 1999 *A&A* **345** 847
[18] Ho W C G 2011 *MNRAS* **414** 2567
[19] Bernal C G, Lee W H and Page D 2010 *RMAA* **46** 309
[20] Viganò D and Pons J A 2012 *MNRAS* **000** 1-6
[21] Torres A, Cerdá-Durán P and Font J A 2015 in preparation.
[22] Michel F C 1977 *ApJ* **214** 261-67
[23] Michel F C 1971 *Ap. Space Sci.* **15** 153

Comparison between the fCCZ4 and BSSN formulations of Einstein equations in spherical polar coordinates

N. Sanchis-Gual[1], P. J. Montero[2], J. A. Font[1,3], E. Müller[2] and T. W. Baumgarte[4]

[1]Departamento de Astronomía y Astrofísica, Universitat de València, Dr. Moliner 50, 46100, Burjassot (València), Spain.
[2]Max-Planck-Institut für Astrophysik, Karl-Schwarzschild-Str. 1, 85748, Garching bei München, Germany.
[3]Observatori Astronòmic, Universitat de València, C/ Catedrático José Beltrán 2, 46980, Paterna (València), Spain.
[4]Department of Physics and Astronomy, Bowdoin College, Brunswick, ME 04011, USA.

E-mail: nicolas.sanchis@uv.es

Abstract. Recently, we generalized a covariant and conformal version of the Z4 system of the Einstein equations using a reference metric approach, that we denote as fCCZ4. We successfully implemented and tested this approach in a 1D code that uses spherical coordinates and assumes spherical symmetry, obtaining from one to three orders of magnitude reduction of the Hamiltonian constraint violations with respect to the BSSN formulation in tests involving neutron star spacetimes. In this work, we show preliminary results obtained with the 3D implementation of the fCCZ4 formulation in a fully 3D code using spherical polar coordinates.

1. Introduction

The Z4 formulation [1] of the Einstein equations is an extension of these equations obtained by adding a four vector ${}^{(4)}Z_\mu$ to the original system. Given a set of initial data which are a solution of the original Einstein equations, the extended equations will yield the same solution provided that at the initial time slice ${}^{(4)}Z_\mu = 0$. Any deviation from the solution of the original Einstein equations will then propagate away. Two conformal and traceless decomposition of the Z4 system have been recently proposed, namely the Z4c formulation [2, 3, 4, 5] and the CCZ4 formulation [6, 7]. Both of them incorporate the constraint damping scheme developed by [8]. This scheme allows to control dynamically the constraint violations by means of constraint damping terms, which are parametrized by two constants κ_1 and κ_2. Both formulations have been extensively tested and results show a reduction of the Hamiltonian constraint violations from one to three orders of magnitude with respect to BSSN [9, 10, 11], the actual values depending on the particular tests considered.

In a recent paper [12] we proposed a generalized covariant and conformal version of the Z4 system of the Einstein equations by adopting a reference metric approach. This approach, which we call fCCZ4 for short, reduces to the CCZ4 formalism in Cartesian coordinates, but is well-suited for curvilinear coordinates as well. In [12] we implemented this fCCZ4 formalism in a 1D

Journal of Physics: Conference Series **600** (2015) 012058 doi:10.1088/1742-6596/600/1/012058

code that employs spherical polar coordinates under the assumption of spherical symmetry using a partially-implicit Runge-Kutta (PIRK) method. This scheme sidesteps the use of traditional regularization procedures when using spherical coordinates and allowed us to evolve both vacuum and non-vacuum spacetimes in spherical symmetry without encountering numerical instabilities. We performed several tests and compared violations of the Hamiltonian constraint in the fCCZ4 with those in the BSSN systems for different choices of the free parameters κ_1 and κ_2. For an optimal choice of parameters and for neutron star spacetimes, the violations of the Hamiltonian constraint are between 1 and 3 orders of magnitude smaller in the fCCZ4 system than in the BSSN formulation. However, for black hole spacetimes the advantages of fCCZ4 over BSSN are less evident.

The next logical step in our program was to implement and test the fCCZ4 formulation in a fully 3D code. This has been accomplished recently and preliminary results are reported in this work. In our research we employ our 3D code `nada`, whose ability to accurately handle numerical evolutions using spherical polar coordinates without any symmetry assumption adopting a covariant form of the BSSN formulation was recently shown in [13] and more recently by [14] who reported the first successful implementation of relativistic hydrodynamics coupled to dynamical spacetimes in spherical polar coordinates with no symmetry assumptions. The new fCCZ4 formulation has been assessed and compared with BSSN by performing two tests in 3D. We anticipated to find the same trend already observed in the spherically symmetric case and that has been indeed the case; that is, we observe a reduction in the violation of the Hamiltonian constraint of several orders of magnitude for fCCZ4 with respect to BSSN also in 3D.

2. The Fully Covariant and Conformal Z4 Formulation

In the Z4 system, the Hamiltonian and momentum constraints result in equations for the four-vector ${}^{(4)}Z_\mu$. In a 3+1 decomposition, these equations can be written as evolution equations for the projection of the ${}^{(4)}Z_\mu$ along the normal n^μ, which, following standard convention, we define as $\Theta \equiv -n_\mu {}^{(4)}Z^\mu = \alpha\, {}^{(4)}Z^0$, and the spatial projection of ${}^{(4)}Z_\mu$, $Z_i \equiv \gamma_i{}^{\mu}\, {}^{(4)}Z_\mu$. Here Z_i denotes a spatial vector whose index can be raised with the (inverse) spatial metric, $Z^i = \gamma^{ij} Z_j$.

By defining

$$\partial_\perp \equiv \partial_t - \mathcal{L}_\beta \tag{1}$$

where $\mathcal{L}_\beta$ denotes the Lie derivative along the shift vector β^i, the fully covariant and conformal Z4 system in a reference-metric approach (fCCZ4) is then given by the following set of evolution equations:

$$\begin{aligned}
\partial_\perp \bar{\gamma}_{ij} &= -\frac{2}{3}\bar{\gamma}_{ij}\bar{D}_k\beta^k - 2\alpha\bar{A}_{ij}, &(2)\\
\partial_\perp \bar{A}_{ij} &= -\frac{2}{3}\bar{A}_{ij}\bar{D}_k\beta^k - 2\alpha\bar{A}_{ik}\bar{A}^k_j + \alpha\bar{A}_{ij}(K-2\Theta) + e^{-4\phi}[-2\alpha\bar{D}_i\bar{D}_j\phi \\
&\quad +4\alpha\bar{D}_i\phi\bar{D}_j\phi + 4\bar{D}_{(i}\alpha\bar{D}_{j)}\phi - \bar{D}_i\bar{D}_j\alpha + \alpha(\bar{R}_{ij} + D_iZ_j + D_jZ_i - 8\pi S_{ij})]^{\rm TF}, &(3)\\
\partial_\perp \phi &= \frac{1}{6}\bar{D}_i\beta^i - \frac{1}{6}\alpha K, &(4)\\
\partial_\perp K &= e^{-4\phi}[\alpha(\bar{R} - 8\bar{D}^i\phi\bar{D}_i\phi - 8\bar{D}^2\phi) - (2\bar{D}^i\alpha\bar{D}_i\phi + \bar{D}^2\alpha)] + \alpha(K^2 - 2\Theta K) + 2\alpha D_iZ^i \\
&\quad -3\alpha\kappa_1(1+\kappa_2)\Theta + 4\pi\alpha(S - 3E), &(5)\\
\partial_\perp \Theta &= \frac{1}{2}\alpha[e^{-4\phi}(\bar{R} - 8\bar{D}^i\phi\bar{D}_i\phi - 8\bar{D}^2\phi) - \bar{A}^{ij}\bar{A}_{ij} + \frac{2}{3}K^2 - 2\Theta K + 2D_iZ^i] - Z^i\partial_i\alpha \\
&\quad -\alpha\kappa_1(2+\kappa_2)\Theta - 8\pi\alpha E, &(6)\\
\partial_\perp \tilde{\Lambda}^i &= \bar{\gamma}^{jk}\hat{D}_j\hat{D}_k\beta^i + \frac{2}{3}\Delta\Gamma^i\bar{D}_j\beta^j + \frac{1}{3}\bar{D}^i\bar{D}_j\beta^j - 2\bar{A}^{jk}(\delta^i{}_j\partial_k\alpha - 6\alpha\delta^i{}_j\partial_k\phi - \alpha\Delta\Gamma^i_{jk})
\end{aligned}$$

$$-\frac{4}{3}\alpha\bar{\gamma}^{ij}\partial_j K + 2\bar{\gamma}^{ki}(\alpha\partial_k\Theta - \Theta\partial_k\alpha - \frac{2}{3}\alpha K Z_k) - 2\alpha\kappa_1\bar{\gamma}^{ij}Z_j - 16\pi\alpha\bar{\gamma}^{ij}S_j. \tag{7}$$

Here the superscript TF denotes the trace-free part of a tensor, κ_1 and κ_2 are the damping coefficients introduced by [8], and $\hat{D}_i$, D_i and $\bar{D}_i$ denote the covariant derivatives built from the connection associated with the reference metric $\hat{\gamma}_{ij}$, the physical metric γ_{ij} and the conformal metric $\bar{\gamma}_{ij}$, respectively. We have also defined

$$\tilde{\Lambda}^i \equiv \Delta\Gamma^i + 2\bar{\gamma}^{ij}Z_j, \tag{8}$$

where

$$\Delta\Gamma^i = \bar{\gamma}^{jk}\Delta\Gamma^i_{jk}. \tag{9}$$

The vector $\tilde{\Lambda}^i$ plays the role of the "conformal connection functions" in the original CCZ4 system; its evolution equation (7) involves the evolution equation for the variables Z_i.

The matter sources E, S_i, S_{ij} and S denote the density, momentum density, stress, and the trace of the stress as observed by a normal observer, respectively:

$$E \equiv n_\mu n_\nu T^{\mu\nu}, \tag{10}$$
$$S_i \equiv -\gamma_{i\mu} n_\nu T^{\mu\nu}, \tag{11}$$
$$S_{ij} \equiv \gamma_{i\mu}\gamma_{j\nu}T^{\mu\nu}, \tag{12}$$
$$S \equiv \gamma^{ij}S_{ij}. \tag{13}$$

In Eq. (3), we compute the Ricci tensor $\bar{R}_{ij}$ associated with $\bar{\gamma}_{ij}$ from

$$\bar{R}_{ij} = -\frac{1}{2}\bar{\gamma}^{kl}\hat{D}_k\hat{D}_l\bar{\gamma}_{ij} + \bar{\gamma}_{(i}\hat{D}_{j)}\Delta\Gamma^k + \Delta\Gamma^k\Delta\Gamma_{(ij)k} + \bar{\gamma}^{kl}(2\Delta\Gamma^m_{k(i)}\Delta\Gamma_{j)ml} + \Delta\Gamma^m_{ik}\Delta\Gamma_{mjl}). \tag{14}$$

Here we compute the $\Delta\Gamma^i$ from their definition (9). Given $\Delta\Gamma^i$, and values for $\tilde{\Lambda}^i$, the vectors Z_i, which are not evolved independently, can be determined from (8).

Unless stated otherwise we fix the gauge freedom by imposing the so-called "non-advective 1+log" condition for the lapse

$$\partial_t\alpha = -2\alpha(K - 2\Theta), \tag{15}$$

and a variation of the "Gamma-driver" condition for the shift vector

$$\partial_t\beta = B^i, \tag{16}$$
$$\partial_t B^i = \frac{3}{4}\partial_t\tilde{\Lambda}^i. \tag{17}$$

Finally, when $\Theta = Z_i = 0$, the evolution equations (2)-(7) imply that the Hamiltonian and momentum constraints hold in the form

$$\mathcal{H} \equiv \frac{2}{3}K^2 - \bar{A}_{ij}\bar{A}^{ij} + e^{-4\phi}(\bar{R} - 8\bar{D}^i\phi\bar{D}_i\phi - 8\bar{D}^2\phi) - 16\pi E = 0, \tag{18}$$
$$\mathcal{M}^i \equiv e^{-4\phi}(\frac{1}{\sqrt{\bar{\gamma}}}\hat{D}_j(\sqrt{\bar{\gamma}}\bar{A}^{ij}) + 6\bar{A}^{ij}\partial_j\phi - \frac{2}{3}\bar{\gamma}^{ij}\partial_j K + \bar{A}^{jk}\Delta\Gamma^i_{jk}) - 8\pi S^i = 0, \tag{19}$$

where $\bar{R}$ is the trace of $\bar{R}_{ij}$. We refer to [12] for a complete list of definitions of the variables used here.

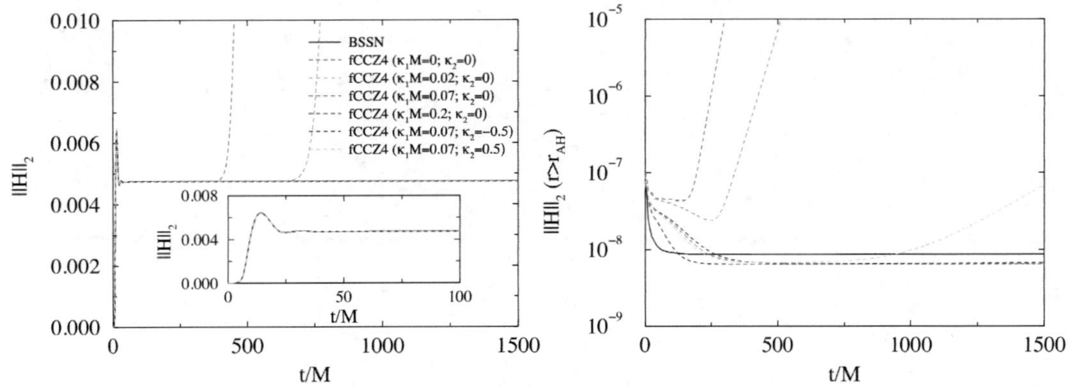

Figure 1. *Left panel:* L2-norm of the Hamiltonian constraint in the single puncture black hole simulation. The inset shows a magnified view of the initial $100M$ in the evolution. *Right panel:* Same quantity but computed outside the apparent horizon.

3. Numerical Results

The Einstein equations coupled to the general relativistic hydrodynamics equations (see [12] for specific details) are solved using both the BSSN and the fCCZ4 formulations. We show results for two 3D tests, namely the propagation of an axisymmetric Teukolsky wave and the evolution of a rotating neutron star in equilibrium, and only one 1D test, the evolution of a Schwarzschild black hole, addressing the interested reader to [12] for additional tests in the case of spherical symmetry. The 1D test corresponds to spherically symmetric initial data that have been evolved using the 1D code mentioned previously. We note that similar results are found when evolving the same initial data using the 3D code `nada`.

3.1. 1D test: Schwarzschild black hole

We evolve a single Schwarzschild black hole given by wormhole initial data and follow the coordinate evolution to the trumpet geometry. In the left panel of Fig. 1 we plot the L2-norm of the Hamiltonian constraint computed in the whole computational domain (including the interior of the apparent horizon (AH)). The largest violation of the Hamiltonian constraint arises from the finite differencing across the puncture. The right panel of Fig. 1 shows that the L2-norm of the Hamiltonian constraint violation computed outside the AH presents some differences between the two formulations which also depends on the values for the damping coefficients. We observe that the numerical evolutions develop instabilities for $\kappa_2 = 0$ and $\kappa_1 M = (0, 0.02)$. Selecting $\kappa_1 M = 0.07$ and $\kappa_2 = 0.5$ leads to an over-damped behavior that is responsible for an exponential growth of the constraint violation at late times. We find that $\kappa_2 = 0$ with $\kappa_1 M = 0.07$ or $\kappa_1 M = 0.2$ give the best results, leading to constraint violations that are comparable to those achieved with BSSN.

The AH mass is defined as $M_{\rm AH} = \sqrt{A/16\pi}$, where A is the proper area of the horizon. For stable black hole evolutions the difference between the AH mass for BSSN and fCCZ4 is less than 0.005% at the end of the simulation ($t = 1875M$), while the error with respect to the initial ADM mass is $\sim 0.7\%$. We note, however, that the black hole mass continues to drift for the fCCZ4 formulation, while it remains constant after an initial transition for the BSSN formulation.

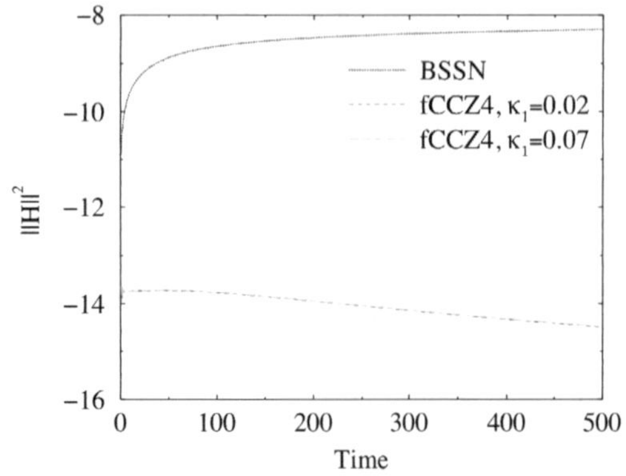

Figure 2. Time evolution of the L2 norm of the Hamiltonian constraint of the Teukolsky wave for BSSN and fCCZ4 for different choices of the damping parameter κ_1, with $\kappa_2 = 0$.

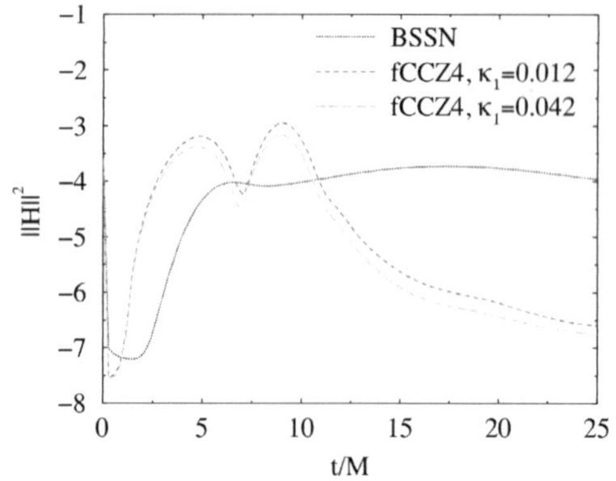

Figure 3. Time evolution of the L2 norm of the Hamiltonian constraint of the rotating neutron star for BSSN and fCCZ4 for different choices of the damping parameter κ_1, with $\kappa_2 = 0$.

3.2. 3D test: Weak gravitational waves

As a first 3D test of our fCCZ4 formulation we consider small-amplitude gravitational waves on a flat Minkowski background. We construct an analytical, linear solution for quadrupolar (l=2) waves from a function

$$F(r,t) = A(r \mp t)e^{-(r \mp t)^2/\lambda^2} \tag{20}$$

where the constant A is related to the amplitude of the wave and λ to its wavelength. For our test we consider the axisymmetric mode $m = 0$, set $\lambda = 1$, and choose an amplitude small enough, $A = 10^{-7}$, to ensure that the test is performed in the linear regime.

In Fig. 2 we show the time evolution of the L2-norm of the Hamiltonian constraint for BSSN and fCCZ4 for two values of the damping parameter κ_1. In [12] we realized that the influence of the parameter κ_2 on the results is not important and can even lead to over-damping effects, so we kept κ_2 equal to zero for most tests. Likewise, we choose $\kappa_2 = 0$ for the present 3D test. Fig. 2 shows that the L2-norm of the Hamiltonian constraint is almost five orders of magnitude smaller for fCCZ4 than for BSSN, as expected. The difference between the two choices of the damping parameter κ_1 is not noticeable in this case.

3.3. 3D test: Rotating neutron star

For our second 3D test we evolve a rapidly rotating relativistic star in equilibrium. The initial data are constructed using the RNS code [15]. In particular, we choose the uniformly rotating BU7 model from [16], a relativistic polytrope with polytropic index $N = 1$ and polytropic constant $K = 100$. This model has a central energy density (in units $G = c = M_\odot = 1$) $\epsilon_c = 1.444 \times 10^{-3}$ and a ratio of polar-to-equatorial coordinate axis $r_p/r_e = 0.65$, resulting in a star with mass $M = 1.666$ and radius $R = 12.3$.

In Fig. 3 we plot the time evolution of the L2-norm of the Hamiltonian constraint for BSSN and fCCZ4 for two different choices of the damping parameter κ_1 and $\kappa_2 = 0$. Again, when using fCCZ4 we observe a reduction in the violation of the Hamiltonian constraint up to almost three orders of magnitude with respect to BSSN. Besides, contrary to the previous test, we find that taking a large enough value of κ_1 leads to slightly smaller constrain violations.

4. Summary

The fCCZ4 formulation of the Einstein equations allows us to write the evolution equations in a fully covariant form suitable for curvilinear coordinate systems. In this paper we have discussed the implementation of the fCCZ4 system in spherical coordinates adopting a PIRK scheme for the time evolution. We have shown that this approach leads to stable evolutions – without regularization of the equations at the coordinate singularities – for both vacuum and non-vacuum spacetimes.

We have tested our approach both in a 1D code and in the 3D code `nada`. Our results for non-vacuum spacetimes show that for optimal choices of the constraint damping parameters of fCCZ4, we obtain a reduction from 1 to 3 orders of magnitude of the violations of the Hamiltonian constraint with respect to BSSN. On the other hand, for black hole spacetimes in the spherically symmetric case, we have seen that fCCZ4 offers little advantages over BSSN. We also note, as discussed in detail in [12], that inadequate choices of the damping parameters may lead to over-damping effects. Of both damping parameters, it is κ_1 the one that plays the most important role in the reduction of the violations of the constraints. Increasing the value of this parameter tends to reduce constraint violations but may also introduce too large damping, thereby making the code unstable and causing it to crash.

The results presented in this work are preliminary. In the near future we plan to fully assess the fCCZ4 formulation in a fully 3D setting by considering more demanding scenarios such as the dynamical evolution of rotating black holes and the gravitational collapse of rotating neutron stars leading to black hole formation, investigating in particular the benefits of the new approach for the extraction of gravitational radiation when employing spherical coordinates.

4.1. Acknowledgments

NSG thanks the Max-Planck-Institut für Astrophysik for its hospitality during the development of part of this project. This work was supported in part by the Spanish MINECO (AYA2013-40979-P), by the Generalitat Valenciana (PROMETEOII-2014-069), by the Deutsche Forschungsgesellschaft (DFG) through its Transregional Center SFB/TR 7 "Gravitational Wave Astronomy", and by NSF grants PHY-1063240 and PHY-1402780 to Bowdoin College.

5. References

[1] Bona C., Ledvinka T., Palenzuela C. and Zacek M., Phys. Rev. D **67**, 104005 (2003).
[2] Bernuzzi S. and Hilditch D., Phys. Rev. D **81**, 084003 (2010).
[3] Ruiz M., Hilditch D. and Bernuzzi S., Phys. Rev. D **83**, 024025 (2011).
[4] Weyhausen A., Bernuzzi S. and Hilditch D., Phys. Rev. D **85**, 024038 (2012).
[5] Hilditch D., Bernuzzi S., Thierfelder M., Cao Z., Tichy W. and Brügmann B., Phys. Rev. D **88**, 084057 (2013).
[6] Alic D., Bona-Casas C., Bona C., Rezzolla L., and Palenzuela C., Phys. Rev. D **85**, 064040 (2012).
[7] Alic D., Kastaun W. and Rezzolla L., Phys. Rev. D **88**, 064049 (2013).
[8] Gundlach C., Martin-Garcia J. M., Calabrese G. and Hinder I., Classical Quantum Gravity **22**, 3767 (2005).
[9] Nakamura T., Oohara K. and Kojima Y., Prog. Theor. Phys. Suppl. **90**, 1 (1987).
[10] Shibata M. and Nakamura T., Phys. Rev. D **52**, 5428 (1995).
[11] Baumgarte T. W. and Shapiro S. L., Phys. Rev. D **59**, 024007 (1998).
[12] Sanchis-Gual N., Montero P. J., Font J. A., Müller E. and Baumgarte T. W. Phys. Rev. D **89**, 104033 (2014).
[13] Baumgarte T. W., Montero P. J., Cordero-Carrión I. and Müller E., Phys. Rev. D **87**, 044026 (2013).
[14] Montero P. J., Baumgarte T. W. and Müller E., Phys. Rev. D **89**, 084043 (2014).
[15] Stergioulas N. and Friedman J. L., ApJ, **444**, 306, (1995).
[16] Dimmelmeier H., Stergioulas N. and Font J. A., Mon. Not. R. Astron. Soc. **368**, 1609-1630, (2006).

Excision technique in constrained formulations of Einstein equations: collapse scenario.

I Cordero-Carrión[1,2], N Vasset[2], J Novak[2] and J L Jaramillo[3]

[1] Department of Applied Mathematics, University of Valencia, C/ Dr. Moliner 50, E-46100 Burjassot, Spain.
[2] Laboratoire Univers et Théories (LUTH), Observatoire de Paris/CNRS/Université Paris Diderot, 5 place Jules Janssen, F-92190 Meudon, France.
[3] Université de Bretagne Occidentale, 6 ave. Le Gorgeu, CS 93837, 29238 Brest Cedex 3, France

E-mail: isabel.cordero@uv.es

Abstract. We present a new excision technique used in constrained formulations of Einstein equations to deal with black hole in numerical simulations. We show the applicability of this scheme in several scenarios. In particular, we present the dynamical evolution of the collapse of a neutron star to a black hole, using the CoCoNuT code and this excision technique.

1. Introduction

When black hole (BH) are present in numerical simulations, one has to handle the physical singularity, where some fields may diverge. One option is the excision technique, where the singularity, together with its neighborhood, is removed from the computational domain. In the constrained formulations of Einstein equations, as in the case of the so-called Fully Constrained formulation (FCF) [1], the elliptic nature of the resulting partial differential equations (PDEs) require boundary conditions at the excision surface which have to be defined and tested.

A geometric approach has been undertaken on the basis of the isolated horizon paradigm [2] or the dynamical "trapping horizon" concept [3]. They have been successfully applied to stationary spacetimes [4], which can then be used as initial data for further dynamical evolutions. When no symmetry is assumed, the apparent horizon (AH) does not have a simple shape in general and we suggest using an arbitrary but nearby sphere inside the AH to define simple and appropriate boundary conditions. We consider here spherically symmetric spacetimes, but we plan to follow similar ideas in more general spacetimes.

We use geometrical units. Greek (latin) indices run from 0 (1) to 3. We adopt the standard convention for the summation over repeated indices. ∂_α denotes partial derivatives.

2. Fully Constrained Formulation

The metric line element in the 3+1 formalism can be written as

$$g_{\mu\nu}dx^\mu dx^\nu = -N^2dt^2 + \gamma_{ij}(dx^i + \beta^i dt)(dx^j + \beta^j dt), \qquad (1)$$

where N is the lapse function, β^i is the shift vector, $\gamma_{ij} = \psi^4\tilde{\gamma}_{ij}$ is the 3-metric onto the spatial hypersurfaces, being ψ the conformal factor and $\tilde{\gamma}_{ij}$ the conformal 3-metric. The difference between the conformal and flat metrics is denoted by h^{ij}. $K_{\mu\nu} = -\frac{1}{2}\mathcal{L}_{\mathbf{n}}\gamma_{\mu\nu}$ is the extrinsic curvature. As gauge conditions, we impose the maximal slicing and the generalized Dirac gauge (see [5] for more details), for which the resulting Einstein equations are formed by a set of elliptic equations for N, β, ψ, and a set of hyperbolic equations for h^{ij} and the components of a conformal traceless extrinsic curvature $\hat{A}^{ij} = \psi^{10}K^{ij}$. In the case of spherical symmetry and with topologically $\mathbb{R}^3$ spatial hypersurfaces, the FCF reduces to the isotropic gauge, where $h^{ij} = 0$. This is not true anymore for more general topologies. In a nonconvex one such as $\mathbb{R}^3 - \mathcal{B}$, where $\mathcal{B}$ is a ball, the expression for $\tilde{\gamma}_{ij}$ involves an extra real function $\omega(t)$; in $\mathbb{R}^3$, $\omega = 0$, but in general this function can be chosen arbitrarily and depends on the given value of the metric at the excision surface. Setting $\omega = 0$ implies $h^{ij} = 0$ and provides a time-independent prescription for $\hat{A}^{ij}$ and a redundant condition in the bulk for the hyperbolic equations; then, we only have to solve the elliptic equations.

3. Excision method

Excision consists in removing from each spatial hypersurface the open interior of a topological sphere. We choose this sphere to be located strictly inside the AH of the modeled BH region, encoded as boundary conditions of the elliptic equations to be solved for. We denote by s^μ the unit outward-directed spacelike vector normal to the spheres, by b the contraction $\beta^i s_i$, and by θ the scalar outward expansion. θ can be expressed in terms of s^i, ψ, and the Levi-Civita connection associated with the conformal 3-metric and could be used as a boundary condition on ψ to compute initial data. θ can be prescribed to be negative to place the excision surface inside the AH. One can choose freely the value of N at the excision surface (differential gauge condition) for the initial data; in a dynamical evolution it has to be consistent with the assumption $h^{ij} = 0$.

We have in mind the BH formation in stellar gravitational collapse simulations. Starting from a simulation for regular data, a trapped region forms at a given time; shortly after this moment, one performs excision inside the trapped region. We choose the excision surface to be located strictly inside the trapped region, where the metric components are determined by the previous evolution and are employed as initial values, and the scalar expansion θ is generically (and on average) negative. Once the initial excised surface has been chosen, one needs to determine a geometrical prescription for the evolution of the excision surface in time. Imposing b to be constant in time at the excised surface provides us with a simple boundary condition for β^i. The trace part of the evolution equations gives a consistent time evolution for ψ, which provides an additional coherent boundary condition for this variable and allows to solve the corresponding elliptic equation. The value of N at the excision surface is given as a consequence of the choice $h^{ij} = 0$ from the evolution equations. During the numerical simulations, we check that $(b - N) \geq 0$. With these boundary conditions, all elliptic equations can be solved in the numerical domain for all times, and no evolution equations are solved in the bulk.

We found in our numerical simulations that the metric exponentially converges to a stationary solution for intrinsic stationary spacetimes. Under some reasonable assumptions (see [5] for more details), one can understand this numerical behavior and check that the coordinates adapt to the stationarity of the spacetime.

4. Numerical results

The excision method already presented for the FCF works well in practice in a numerical code, as we show here in two toy models, the evolution of a Schwarzschild BH and the accretion of a massless scalar field, and also in a full simulation of the collapse of a neutron star to a BH in spherical symmetry.

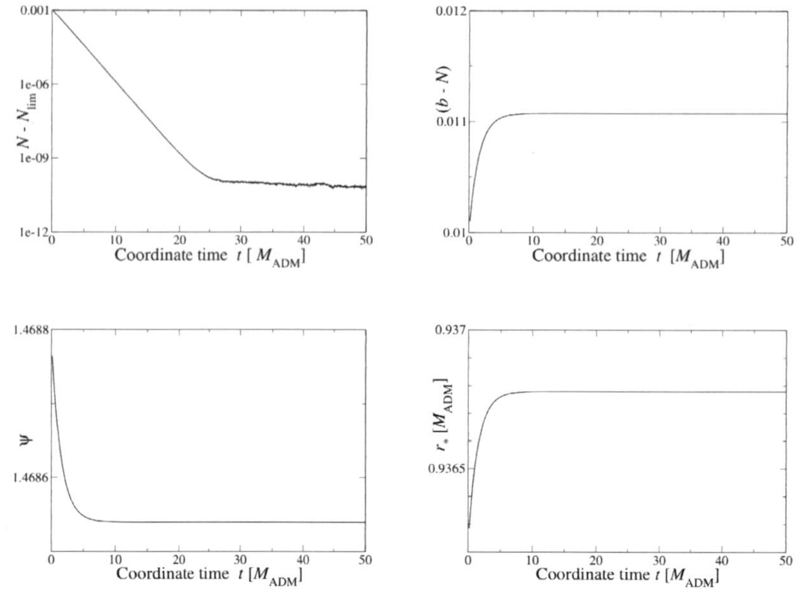

Figure 1. Evolution in terms of coordinate time t of N minus its time asymptotic value $N_{\rm lim} \simeq 0.549$ (left top panel, logarithmic scale); of $(b - N)$ (right top panel); of ψ (left bottom panel), at the excision surface ($r = 0.916\ M_{\rm ADM}$); and of the AH coordinate radius r_* (right bottom panel), for a spherically symmetric vacuum (i.e. Schwarzschild) BH spacetime.

4.1. Setup

We use spherical (polar) coordinates. Since we are imposing $h^{ij} = 0$ and considering spherically symmetric spacetimes, we only need to solve for N, β^r and ψ.

In the first two scenarios we start from an existing BH, with the excision sphere located at the coordinate radius $r = 1$. Therefore, we need to construct initial data at $t = 0$ by solving the elliptic set of equations, with the stress-energy tensor being either zero (vacuum) in the Schwarzschild BH evolution, or given by corresponding expressions in the case of the scalar field accretion (see corresponding subsection below). The initial excision surface is chosen as a sphere with a given (arbitrary) value of the expansion θ, prescribed to be initially negative. This guarantees that the initial excision surface is inside the AH, which then is located using an AH finder. Initial data boundary conditions are needed for the elliptical system and are taken as follows: setting an initial value for θ, a boundary condition for ψ is obtained; the boundary value of N is fixed yielding a Dirichlet condition; a value is prescribed for $(b - N)$, from which one can get a Dirichlet boundary condition for the radial component of the shift vector β^r, the other two components being zero in spherical symmetry. The elliptic system is solved iteratively with the library LORENE [6].

In the third scenario (neutron star collapse to a BH), initial data are obtained solving Einstein equations, coupled to the fluid equilibrium equations in the isotropic gauge, in whole space. The numerical approach is very similar to the isolated BH case, but with no inner boundary condition imposed.

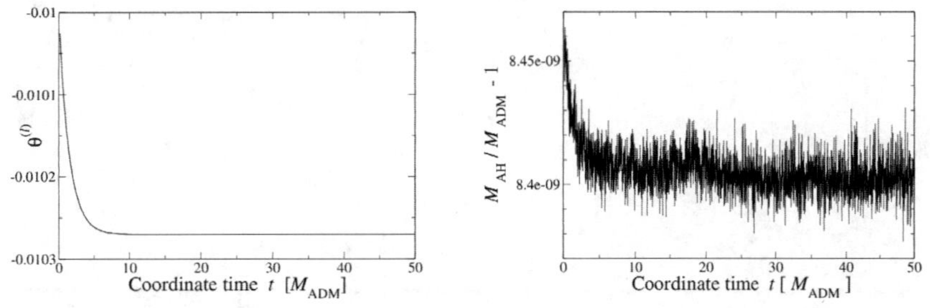

Figure 2. Evolution in terms of coordinate time t of the expansion θ (left panel) and the variation of the AH irreducible mass with respect to the initial ADM mass (right panel), for a spherically symmetric BH spacetime.

4.2. Evolution of a Schwarzschild black hole

A spherically symmetric spacetime is computed for $r \geq 1$, with the previously defined boundary conditions, and the specific values $\theta|_{t=0} = -0.01$, $N|_{t=0,r=1} = 0.55$ and $(b - N)|_{t=0} = 0.01$ for these initial data. Once having solved the corresponding elliptic system, one verifies that this setting induces a nonzero value for the Arnowitt-Deser-Misner (ADM) mass, $M_{\rm ADM} \approx 1.09$. Numerically, it is obtained in isotropic gauge from the asymptotic behavior of the conformal factor (see e.g. [1]). This causes the excision sphere to be located at $r \simeq 0.916\ M_{\rm ADM}$. Using a AH finder, we find in the initial data an AH located at the coordinate radius $r_* \simeq 0.94\ M_{\rm ADM}$. This is evidence that the initial data represent a BH in spherical symmetry. This BH spacetime is then numerically evolved in time. This is obviously only a gauge evolution, since the spacetime is Schwarzschild by construction. The time evolution of the metric variables at the excision surface ($r = 1$) as well as the coordinate radius r_* of the AH, in the interval $0 \leq t \leq 50$, are displayed in Fig. 1. An exponential convergence toward stationary values is observed for all metric quantities, with an explicit behavior shown for the lapse N. From the top right panel of Fig. 1, one can check that the difference $(b - N)$ remains positive. In left panel of Fig. 2 the time evolution of the expansion θ is displayed, which is decreasing during the simulation and always remains negative, ensuring that the excision surface always remains inside the AH.

The scheme appears stable (we have run it for $t \sim 1000$) and, in order to check its accuracy, we have monitored the variation of the AH irreducible mass, $M_{\rm AH}$, determined by the AH finder. The conservation of $M_{\rm AH}$ is displayed in right panel of Fig. 2. The conservation of the ADM mass has also been checked, with quantitatively very similar results to the conservation of the AH surface. Finally, a second-order convergence of these conserved quantities has been obtained numerically while decreasing the time step, in agreement with the implemented second-order Adams-Bahforth scheme. The exponential convergence obtained for the chosen initial data, given by the specific values of $\theta|_{t=0}$, $N|_{t=0,r=1}$ and $(b - N)|_{t=0}$, is independent of these initial data. A similar behavior is found for different initial values.

4.3. Accretion of a massless scalar field

In this case we evolve a BH spacetime with energy content in the form of a minimally coupled massless scalar field in spherical symmetry. The massless Klein-Gordon equation, or (simply) the wave scalar equation, is given by $\nabla^\mu \nabla_\mu \phi = 0$, where ∇ is the Levi-Civita connection associated with the spacetime metric. The stress-energy tensor associated with this scalar field is given by

$$T_{\mu\nu} = \nabla_\mu \phi \nabla_\nu \phi - \frac{1}{2} \gamma_{\mu\nu} \nabla_\rho \phi \nabla^\rho \phi. \tag{2}$$

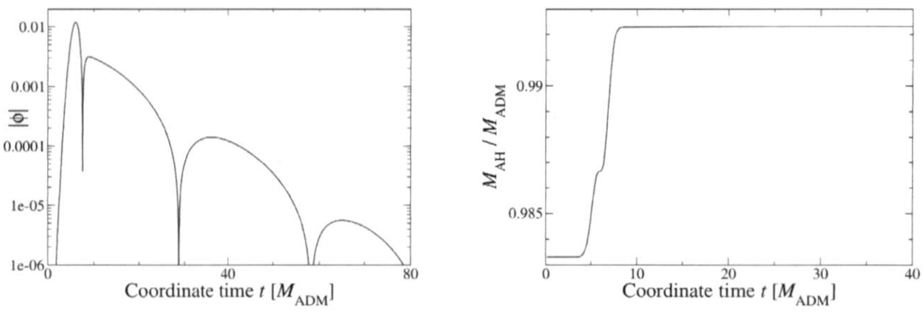

Figure 3. Evolution in terms of the coordinate time t of the absolute value of the accreted scalar field, $|\phi|$, at the excision surface, $\phi|_{r=1}$ (left panel), and of the ratio between the AH irreducible mass and the ADM one, for the case of the accretion of a massless scalar field.

The wave equation can be rewritten as a first-order system in space and time, by introducing auxiliary variables, and solved using a second-order Adams-Bashforth scheme (see more details in [5]). No boundary condition is needed for the field at the excision surface, as all the characteristics are directed out of the computational domain as long as the excision surface is spacelike, which is verified in our case with $(b - N) > 0$ in our numerical simulations.

Initial data are given by a Gaussian profile for the scalar field outside the excision surface. In the evolution of these initial data a fraction of the scalar field is radiated away, while the other part is accreted onto the BH; its time evolution at the excision surface is given in left panel of Fig. 3. The metric quantities follow a similar evolution with respect to the vacuum case and, once the scalar field has been accreted to the BH, they settle rapidly to stationary values. Right panel of Fig. 3 shows the evolution of the AH mass as a function of the coordinate time. As expected, the AH grows in time while accreting energy from the scalar field, before reaching a stationary limit. This limit does not represent all of the ADM mass of the spacetime, as part of this asymptotic mass is still contained in the scalar field traveling to higher radii. The accuracy of the code has been checked by monitoring the variation of the ADM mass, which conservation is obtained up to the level 10^{-6}. The excision boundary conditions allow us to study the growth of a spherically symmetric accreting BH in a stable and accurate way.

4.4. Collapse of a neutron star to a black hole

We here describe a simulation in spherical symmetry, starting from an unstable static neutron star, up to the formation of a BH and its accretion of all matter into the horizon. Excision is switched on during the simulation, after the AH is formed. We have modified the code CoCoNuT [7], so that it uses the excision technique described above. This code solves the general-relativistic Euler equations, with Einstein equations in isotropic gauge. There is no need for boundary conditions for the Euler equations at the excision surface, as all the characteristics point out of the numerical integration domain. The initial data consists of a static neutron star, computed in isotropic gauge with a polytropic equation of state.

The collapse proceeds until the formation of an AH. Excision is switched on when 85% of the total baryon mass has entered the horizon (changing this value has little influence). The excision radius is defined as a fraction of the AH radius r_* between 0.9 and 0.98 (changing this value does not influence the physical results). Density profiles at various time steps around the time when excision is started are given in left panel of Fig. 4. The evolution of the BH irreducible mass is shown in right panel of Fig. 4. The whole simulation is stable. The main error in the

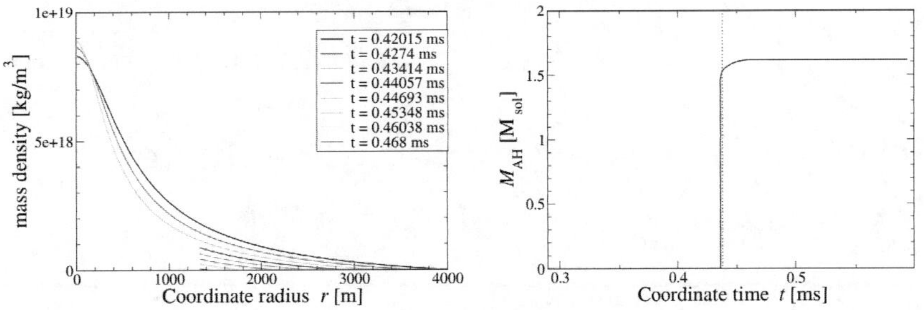

Figure 4. Radial profiles of the rest-mass density, for different moments around the start of the excision in the simulation of the collapse of a neutron star to a BH (left panel), where excision starts at $t = 0.43788$ ms, and evolution in terms of coordinate time t of the AH irreducible mass for the collapse of a neutron star to a BH (right panel), where the vertical dotted line indicates the moment when excision is turned on.

conservation of the ADM mass comes from the numerical solution of the Euler equations. The relative change in the ADM mass after excision has been started is of the order of 10^{-8}, similar to previous tests.

5. Conclusions

In this work we have presented a new excision technique for the dynamical evolution of spherically symmetric spacetimes in the FCF in order to numerically simulate systems forming a BH, like the stellar collapse to a BH. In this case, we have to deal with boundary conditions for elliptic-type PDEs. We have checked the practical applicability of this approach in several cases, showing stable and accurate numerical results. We plan to extend this approach to spacetimes with less symmetries in forthcoming studies.

Acknowledgments

This work has been partially funded by the SN2NS project ANR-10-BLAN-0503. This work was also supported by the Grants No. AYA2010-21097-C03-01 of the Spanish MICINN and No. AYA2013-40979-P of the Spanish MINECO.

References

[1] Bonazzola S, Gourgoulhon E, Grandclément P and Novak J 2006 *Phys. Rev.* D **74** 084028
[2] Cook G B and Pfeiffer H P 2004 *Phys. Rev.* D **70** 104016
Jaramillo J L, Gourgoulhon E and Marugán G A 2004 *Phys. Rev.* D **70** 124036
Dain S, Jaramillo J L and Krishnan B 2005 *Phys. Rev.* D **71** 064003
Gourgoulhon E and Jaramillo J L 2006 *Phys. Rep.* **423** 159
Caudill M, Cook G B, Grigsby J D and Pfeiffer H P 2006 *Phys. Rev.* D **74** 064011
Jaramillo J L, Ansorg M and Limousin F 2007 *Phys. Rev.* D **75** 024019
Jaramillo J L 2009 *Phys. Rev.* D **79** 087506
[3] Jaramillo J L, Gourgoulhon E, Cordero-Carrión I and Ibáñez J M 2008 *Phys. Rev.* D **77** 047501
Jaramillo J L 2011 *Int. J. Mod. Phys.* D **20** 2169–2204
[4] Vasset N, Novak J and Jaramillo J L 2009 *Phys. Rev.* D **79** 124010
[5] Cordero-Carrión I, Vasset N, Novak J and Jaramillo J L 2014 *Phys. Rev.* D **90** 044062
[6] http://www.lorene.obspm.fr
[7] Dimmelmeier H, Novak J, Font J A, Ibáñez J M and Müller E 2005 *Phys. Rev.* D **71** 064023

Spectral approach to axisymmetric evolution of Einstein's equations

Christian Schell and Oliver Rinne

Max Planck Institute for Gravitational Physics (Albert Einstein Institute), Am Mühlenberg 1, 14476 Golm, Germany and
Department of Mathematics and Computer Science, Freie Universität Berlin, Arnimallee 26, 14195 Berlin, Germany

E-mail: christian.schell@aei.mpg.de, oliver.rinne@aei.mpg.de

Abstract. We present a new formulation of Einstein's equations for an axisymmetric spacetime with vanishing twist in vacuum. We propose a fully constrained scheme and use spherical polar coordinates. A general problem for this choice is the occurrence of coordinate singularities on the axis of symmetry and at the origin. Spherical harmonics are manifestly regular on the axis and hence take care of that issue automatically. In addition a spectral approach has computational advantages when the equations are implemented. Therefore we spectrally decompose all the variables in the appropriate harmonics. A central point in the formulation is the gauge choice. One of our results is that the commonly used maximal-isothermal gauge turns out to be incompatible with tensor harmonic expansions, and we introduce a new gauge that is better suited. We also address the regularisation of the coordinate singularity at the origin.

1. Introduction

In this paper we consider the vacuum Einstein equations under the assumption of axisymmetry, i.e. there is an everywhere spacelike Killing vector field with closed orbits, ∂_φ in spherical polar coordinates $(t, r, \vartheta, \varphi)$ adapted to the symmetry. As an initial step we further assume that ∂_φ is hypersurface-orthogonal (or twist-free), so there is no rotation.

The motivation for considering this situation comes from the obvious fact that due to the reduced dimensionality, it is much less computationally demanding than the case without symmetries. On the other hand we are not oversimplifying too much in the sense that our situation shares important features and properties of the full theory such as the existence of gravitational waves. In fact it was shown in [1] that it is not possible to assume any further reasonable symmetry in the given situation when demanding gravitational radiation. Due to the so-called Birkhoff theorem (for interesting historic remarks see [2]), vacuum spherical symmetry is non-dynamical and thus not of interest to us.

We focus on isolated systems here, i.e. we assume spacetime is asymptotically flat. This provides us with appropriate fall-off conditions that can be used at an artificial boundary far away from the situation of interest.

There are many interesting applications we have in mind. These include the collapse of gravitational waves, in particular critical phenomena. Up to date there are, to the best of our knowledge, only two successful implementations finding critical phenomena for the vacuum axisymmetric Einstein equations [3, 4]. These studies obtained rather different results for different initial data, and further investigations are needed. Another interesting question in mathematical relativity is the one of stability of solutions such as black holes. Our setup will allow us to study the evolution of perturbations compatible with our symmetry assumptions.

Our aim here is to derive a fully constrained formulation in spherical polar coordinates which may be implemented by making use of a spectral approach for the angular part. A well-known fact is that only six of the ten Einstein equations are of dynamical character, the other four are constraints. An approach that is often applied to numerical investigations of Einstein's equations is to solve the constraints only on an initial hypersurface and then to evolve the system according to the remaining evolution equations. This gives rise to the so-called free evolution. Analytically the constraints remain preserved, which justifies the approach. Numerically, on the other hand, there may exist constraint-violating modes, which cause instabilities. Therefore our approach consists in enforcing the constraints at each timestep. By construction there are no constraint-violating modes then. Such formulations are called fully constrained, for previous publications see e.g. [5, 6] for the case without symmetries and [7] for the case of axisymmetry. A disadvantage of solving the constraints is that they are of elliptic nature and hence computationally much more involved. We will introduce ideas how to save computational cost at other points in the formulation.

This brings us to the choice of a coordinate system. At least for the implementation it is very common to introduce a fixed coordinate system. In contrast to many previous formulations for similar situations which use cylindrical polar coordinates, we choose spherical polar coordinates $(t, r, \vartheta, \varphi)$. Besides the motivation from astrophysics, where many objects have an approximate spherical shape, the main reason is a mathematical one. If spacetime has a topology $\mathcal{M}^2 \times S^2$ then spherical harmonics $Y_{\ell m}(\vartheta, \varphi)$ form an appropriate system for the spectral expansion on the sphere S^2. Another nice property is that fall-off conditions to be imposed at the outer boundary are usually given as an expansion in inverse powers of r. For the choice of spherical coordinates and spectral expansion in general relativity see for example [8, 9, 10].

In these proceedings we mainly focus on conceptual issues of the formulation, in particular the gauge choice. In section 2 we introduce our spectral expansion. It will be applied in section 3, where we describe the derivation of the nonlinear equations, further their linearisation and regularisation. In particular we focus on the gauge choice and investigate two possibilities in detail. In the last chapter we briefly summarize and give an outlook on work in progress.

For us general relativity is given by the Einstein equations on a four-dimensional Lorentzian manifold $(\mathcal{M}^4, g)$ which is metric compatible, torsion-free and globally hyperbolic. Indices i, j run from 1 to 3.

2. Spectral approach

A general problem when using spherical polar coordinates is the occurrence of coordinate singularities at the origin ($r = 0$) and the axis of symmetry ($\vartheta = 0, \pi$). As an example consider the flat Laplacian in spherical coordinates,

$$\triangle = \partial_r^2 + \frac{2}{r}\partial_r + \frac{1}{r^2}\left(\partial_\vartheta^2 + \frac{\cos\vartheta}{\sin\vartheta}\partial_\vartheta\right). \tag{1}$$

In fact many equations contain operators that have some similarity with the Laplacian, which is why we will take it as a model to illustrate our ideas.

Since spherical harmonics are regular on the axis, they take care of that issue automatically. Because of the assumed twist-free axisymmetry, all quantities are φ-independent. This implies in particular that the spherical harmonics reduce to the $m = 0$ harmonics

$$Y := Y_\ell(\vartheta) = \sqrt{\frac{2\ell+1}{4\pi}}P_\ell(\cos\vartheta) \tag{2}$$

Journal of Physics: Conference Series **600** (2015) 012060 doi:10.1088/1742-6596/600/1/012060

with Legendre polynomials $P_\ell(\cos\vartheta)$. We shall omit the index ℓ if it is clear from the context. However general relativity is a tensor theory. We need to know the particular behaviour of each component of a scalar function, a vector and a symmetric two-tensor [11, 12]. Let $\hat{g}_{AB}$ be the round metric on the unit sphere and $\hat{\nabla}$ its covariant derivative $(A, B = \vartheta, \varphi)$. The general even parity quantities are $Y_A := \partial_A Y$ and $Y_{AB} := [\hat{\nabla}_A \hat{\nabla}_B Y]^{\mathrm{tf}}$, the odd parity quantities play no role because of hypersurface-orthogonality. In the case of twist-free axisymmetry the relevant basis harmonics are given by

$$Y = {}_0Y, \tag{3a}$$

$$Y_\vartheta = -\frac{1}{2}\sqrt{\ell(\ell+1)}\left({}_1Y - {}_{-1}Y\right), \tag{3b}$$

$$Y_{\vartheta\vartheta} = -\left[\frac{\cos\vartheta}{\sin\vartheta}Y_\vartheta + \frac{\ell(\ell+1)}{2}Y\right] = \frac{1}{4}\sqrt{(\ell-1)\ell(\ell+1)(\ell+2)}\left({}_2Y + {}_{-2}Y\right), \tag{3c}$$

where we have used the properties of the Legendre functions to eliminate the second ϑ-derivatives in (3c). For completeness we have also given the expressions in terms of the spin-weighted harmonics ${}_sY$ [12]. In the following we will refer to Y, Y_ϑ and $Y_{\vartheta\vartheta}$ as the scalar, vector and tensor harmonics, respectively. Fields expanding in those harmonics will be called scalar, vector and tensor quantities.

We explicitly give the expansion of some components of the spatial metric γ_{ij} needed in the following:

$$\gamma_{rr} = HY, \tag{4a}$$

$$\gamma_{\vartheta\vartheta} = r^2\left(K - \frac{\ell(\ell+1)}{2}G\right)Y - r^2\frac{\cos\vartheta}{\sin\vartheta}GY_\vartheta = r^2\left(KY + GY_{\vartheta\vartheta}\right), \tag{4b}$$

$$\gamma_{\varphi\varphi} = r^2\sin^2\vartheta\left(K + \frac{\ell(\ell+1)}{2}G\right)Y + r^2\cos\vartheta\sin\vartheta GY_\vartheta = r^2\sin^2\vartheta\left(KY - GY_{\vartheta\vartheta}\right), \tag{4c}$$

where H, K and G are functions of t and r only and a sum over ℓ is implied.

3. Formulation

Our starting point is the usual 3+1 decomposition of general relativity $(\mathcal{M}^4, g) \mapsto (\Sigma^3, \gamma, K)$ [13, 14]. Here Σ^3 is a level set of three dimensional spacelike hypersurfaces and γ and K are their first and second fundamental forms. The evolution takes place along the timelike vector field t (recall that $(\mathcal{M}^4, g)$ is globally hyperbolic by assumption). The line element is given by

$$\mathrm{d}s^2 = -\alpha^2\mathrm{d}t^2 + \gamma_{ij}\left(\mathrm{d}x^i + \beta^i\mathrm{d}t\right)\left(\mathrm{d}x^j + \beta^j\mathrm{d}t\right), \tag{5}$$

where α is the lapse function and β^i the components of the shift vector. We obtain six evolution equations each for γ_{ij} and K_{ij}, a Hamiltonian constraint and three momentum constraints.

Now, in this setting, twist-free axisymmetry means that all variables are φ-independent, $\gamma_{r\varphi} = \gamma_{\vartheta\varphi} = 0$ and $\beta^\varphi = 0$. These identities are preserved under time evolution. It follows that also $K_{r\varphi}$ and $K_{\vartheta\varphi}$ have to vanish and the φ-momentum constraint is identically satisfied.

The diffeomorphism invariance of general relativity is encoded in the lapse and shift. To fix the gauge we have to choose a slicing condition and two spatial gauge conditions.

3.1. Choice of a gauge

An evident choice is the so-called maximal-isothermal gauge, see [15] for a review. This is a combination of maximal slicing, $\mathrm{tr}K = 0 = \partial_t(\mathrm{tr}K)$, and the quasi-isotropic condition. The latter one consists of the diagonal gauge,

$$\gamma_{r\vartheta} = 0 = \partial_t\gamma_{r\vartheta}, \tag{6}$$

and a condition that separates the remaining φ-part of the spatial metric by relating the other components as

$$\gamma_{\vartheta\vartheta} = r^2 \gamma_{rr}, \tag{7}$$

which is also preserved in time, $\partial_t \left(\gamma_{\vartheta\vartheta} - r^2\gamma_{rr}\right) = 0$. This is a widely used gauge in axisymmetric simulations [3, 16, 17, 7] and also analytically well studied [15]. As we will see later in section 3.2 after the linearisation and expansion in spherical harmonics, the maximal-isothermal gauge is unfortunately *not* an appropriate gauge for our purposes. In order to find a well-suited condition we decide to keep maximal slicing and the diagonal gauge as before but to come up with a new condition for the diagonal components of metric, namely

$$\gamma_{\vartheta\vartheta} = r^4 \sin^2\vartheta \; \gamma^{\varphi\varphi} (\gamma_{rr})^2. \tag{8}$$

This gauge should also be preserved in time, $\partial_t \left(\gamma_{\vartheta\vartheta} - r^4\sin^2\vartheta\; \gamma^{\varphi\varphi}(\gamma_{rr})^2\right) = 0$. Note that the nonlinear condition (8) relates all the remaining components of the spatial metric. We will show in section 3.2 that this is indeed an appropriate gauge for our purposes. The three preservation equations in t for the gauge choices give us further elliptic equations to be solved at each time step in the fully constrained system.

Having the new gauge condition at hand we can now follow the usual procedure in the 3+1 formulation of general relativity to derive the nonlinear constraints and evolution equations. For obvious reasons one should choose variables in such a way that their linearisation has a convenient expansion in scalar, vector and tensor harmonics in the sense explained at the end of section 2. We note that it is sometimes useful to add appropriate multiples of the constraints (which vanish for a solution to Einstein's equations) to some of the equations such that the linearisation of the equations also expand in a definite way.

We have eight variables, namely the lapse function, two components of the shift vector, two components of the spatial metric and three components of the extrinsic curvature. On the other hand we have six elliptic equations: the Hamiltonian and two momentum constraints, the preservation of maximal slicing, the diagonal gauge condition (6) and the newly introduced gauge condition (8). Since we are looking for a fully constrained formulation, we will explicitly solve all of these elliptic equations in our scheme. Furthermore we have five evolution equations, two for the remaining components of the spatial metric γ_{ij} and three for the extrinsic curvature K_{ij}. Thus the system is overdetermined. In the following we will concentrate on those two evolution equations for the components of γ_{ij} and K_{ij} that expand in the linearisation in tensor harmonics. Besides being reasonable to choose two canonically conjugated variables as evolved fields, we expect the tensor quantities to carry the gravitational wave degrees of freedom (at least in linearised theory). The other evolution equations may be used for consistency checks but will not be considered in the remainder of this paper.

3.2. Linearisation

Having derived the equations, we next linearise them about a flat background spacetime. This means we expand all quantities in the form $f = f|_{\text{flat}} + \epsilon\tilde{f}$ and just keep terms of linear order in ϵ, ignoring higher-order terms. We obtain two evolution equations and six constraints, all dependent on (t, r, ϑ). As expected we are faced with singularities both on the axis and at the origin. One should think of the Laplacian in (1) as a model operator.

On the linearised level we expand all variables in the corresponding spherical harmonics as given at the end of section 2. E.g. for a variable $\tilde{f}$ that expands in scalar harmonics, we have

$$\tilde{f}(t, r, \vartheta) = \sum_{\ell} \hat{f}_\ell(t, r) Y_\ell(\vartheta). \tag{9}$$

Applying the expansion of the metric coefficients (4) to the quasi-isotropic condition (7), one finds

$$r^2\left(H - K + \frac{\ell(\ell+1)}{2}G\right)Y + r^2\frac{\cos\vartheta}{\sin\vartheta}GY_\vartheta = 0, \tag{10}$$

which implies $G = 0$ and hence $H = K$. Thus only one degree of freedom for the spatial metric remains. Therefore the only situation that is compatible with this choice is the one of spherical symmetry.

On the other hand, linearising (8) about flat space leads to the condition

$$\gamma_{\vartheta\vartheta} = 2r^2\gamma_{rr} - \frac{\gamma_{\varphi\varphi}}{\sin^2\vartheta} \tag{11}$$

and hence, again by using (4),

$$2r^2(K - H)Y = 0. \tag{12}$$

Therefore $H = K$ and G arbitrary are two remaining degrees of freedom, which shows that these conditions are indeed well suited.

The ϑ-dependence is completely absorbed in the spherical harmonics. Thus the interesting part is now contained in $\hat{f}_\ell$ in (9), which depends on (t, r) only. On the linear level one expects a decoupling of all the different ℓ-modes. This is indeed the case provided one considers the "correct" nonlinear equations as explained in the previous subsection 3.1, adding appropriate multiples of the constraints. Therefore we obtain, for each ℓ-mode, a $(1+1)$-dimensional system of equations in (t, r). The equations are still formally singular at the origin $r = 0$.

3.3. Regularisation

In order to regularise the just obtained equations at the origin $r = 0$, we follow a procedure proposed e.g. in [18, 19]. Let us again take the Laplace operator (1) as an example. After expansion in (scalar) spherical harmonics, the Laplace equation reads

$$\partial_r^2\hat{f}_\ell + \frac{2}{r}\partial_r\hat{f}_\ell - \frac{\ell(\ell+1)}{r^2}\hat{f}_\ell = 0. \tag{13}$$

This equation can be solved explicitly. Its solution is a superposition of a singular part proportional to $r^{-(\ell+1)}$ and a regular part proportional to r^ℓ. Since we only consider solutions that are smooth at $r = 0$, we require the integration constant of the part proportional to $r^{-(\ell+1)}$ to vanish. If we isolate the leading-order behaviour by setting $\hat{f}_\ell(t, r) =: r^\ell\bar{f}_\ell(t, r)$, we can continue to work with the barred quantities, which expand, close to the origin $r = 0$, in even power series in r. Therefore (13) is now manifestly regular,

$$\partial_r^2\bar{f}_\ell + \frac{2}{r}(\ell+1)\partial_r\bar{f}_\ell = 0. \tag{14}$$

For our set of equations we have to pull out factors of either r^ℓ, $r^{\ell+1}$ or $r^{\ell+2}$ to obtain similar results, but indeed, the system can be completely regularised in this way.

A further nice property is that, with a little bit of rearranging, one obtains a hierarchy of equations. For the implementation we evolve, from one time step to the next, the two tensor variables by using the evolution equations. Here we remark that, when taking the second time derivative of the tensor component of γ_{ij} and using the evolution equation for the corresponding component of K_{ij}, the principal part of the equations is just the ordinary wave equation. Then, on the new time level, we solve successively the constraints to obtain the remaining variables on that time slice before stepping to the next slice. Given the tensor part of γ_{ij}, the Hamiltonian constraint gives us the scalar component of γ_{ij}, while the preservation of maximal slicing determines the lapse function. Given the tensor contribution to K_{ij}, the two momentum constraints allow us to determine the remaining (scalar and vector) components of the extrinsic curvature. Finally the preservation of the spatial gauge conditions determines the components of the shift vector.

4. Conclusion

We have formulated Einstein's equations for an isolated system in twist-free axisymmetry. Key features of our formulation are that the scheme is fully constrained and uses a spherical polar coordinate system. In its linearisation about a flat background, a spectral expansion in spherical harmonics may be used and the resulting equations are fully regularisable at the origin. A central result presented in this paper is the fact that the well-understood and frequently applied maximal-isothermal gauge is not compatible with tensor spherical harmonic expansions. Instead we proposed another gauge condition which is well-suited. Currently we are in the process of coding the linearised system and finding a proper way to include the non-linearities. Possible directions for further studies include a deeper analysis of the properties of the system, the inclusion of a non-vanishing twist to allow for rotating spacetimes and, ultimately, the application of the code to physically and mathematically interesting situations such as those mentioned in section 1.

Acknowledgments

This research is supported by grant RI 2246/2 of the German Research Foundation (DFG) and a Heisenberg Fellowship to OR. CS also acknowledges support from the International Max Planck Research School (IMPRS) for Geometric Analysis, Gravitation and String Theory.

References

[1] Bičák J and Pravdová A 1998 *J. Math. Phys.* **39** 6011–6039
[2] Johansen N V and Ravndal F 2006 *Gen. Relativ. Gravit.* **38** 537–540
[3] Abrahams A M and Evans C R 1993 *Phys. Rev. Lett.* **70** 2980–2983
[4] Sorkin E 2011 *Class. Quantum Grav.* **28** 025011
[5] Bonazzola S, Gourgoulhon E, Grandclément P and Novak J 2004 *Phys. Rev.* D **70** 104007
[6] Cordero-Carrión I, Ibáñez J M, Gourgoulhon E, Jaramillo J L and Novak J 2008 *Phys. Rev.* D **77** 084007
[7] Rinne O 2008 *Class. Quantum Grav.* **25** 135009
[8] Bonazzola S, Gourgoulhon E and Marck J A 1999 *J. Comp. Appl. Math.* **109** 433–473
[9] Csizmadia P, László A and Rácz I 2013 *Class. Quantum Grav.* **30** 015010
[10] Szilágyi B 2014 *Int.J.Mod.Phys.* **D23** 1430014
[11] Sarbach O and Tiglio M 2001 *Phys. Rev.* D **64** 084016
[12] Rinne O 2009 *Class. Quantum Grav.* **26** 048003
[13] York J W 1979 *Sources of gravitational radiation* ed by Smarr L L (Cambridge University Press) pp 83–126
[14] Gourgoulhon E 2012 *3+1 Formalism in General Relativity* (Springer-Verlag)
[15] Dain S 2011 *J. Phys.: Conf. Series* **314** 012015
[16] Garfinkle D and Duncan G C 2001 *Phys. Rev. D* **63** 044011
[17] Choptuik M W, Hirschmann E W, Liebling S L and Pretorius F 2003 *Class. Quantum Grav.* **20** 1857–1878
[18] Grandclément P and Novak J 2009 *Living Rev. Relativity* **12** 1 cited October 11, 2014
[19] Gundlach C, Martín-García J M and Garfinkle D 2013 *Class. Quantum Grav.* **30** 145003

Journal of Physics: Conference Series **600** (2015) 012061 doi:10.1088/1742-6596/600/1/012061

Unconstrained hyperboloidal evolution of black holes in spherical symmetry with GBSSN and Z4c

Alex Vañó-Viñuales and Sascha Husa

Universitat de les Illes Balears and Institut d'Estudis Espacials de Catalunya, Cra. de Valldemossa km.7.5, 07122 Palma de Mallorca, Spain

E-mail: alex.vano@uib.es

Abstract. We consider unconstrained evolution schemes for the hyperboloidal initial value problem in numerical relativity as a promising candidate for the optimally efficient numerical treatment of radiating compact objects. Here, spherical symmetry already poses nontrivial problems and constitutes an important first step to regularize the resulting singular PDEs. We evolve the Einstein equations in their generalized BSSN and Z4 formulations coupled to a massless self-gravitating scalar field. Stable numerical evolutions are achieved for black hole initial data, and critically rely on the construction of appropriate gauge conditions.

1. Introduction

For isolated systems, their energy loss and radiation are in general only well defined at null infinity ($\mathscr{I}^+$). This is where the observers of astrophysical events are located [1, 2, 3] and where we ideally want to extract radiation signals from numerical simulations. Following the approach of Penrose [4, 5], we conformally compactify spacetime: we rescale the physical metric $\tilde{g}_{\mu\nu}$ by a conformal factor that vanishes at $\mathscr{I}^+$:

$$g_{\mu\nu} \equiv \Omega^2 \tilde{g}_{\mu\nu} \ . \tag{1}$$

The equations of motion in terms of the rescaled metric $g_{\mu\nu}$ diverge at $\mathscr{I}^+$:

$$G_{\mu\nu}[g] = 8\pi \, T_{\mu\nu} - \frac{2}{\Omega}\left(\nabla_\mu\nabla_\nu\Omega - g_{\mu\nu}\nabla^\gamma\nabla_\gamma\Omega\right) - \frac{3}{\Omega^2} g_{\mu\nu}(\nabla_\gamma\Omega)\nabla^\gamma\Omega \ . \tag{2}$$

The character of an initial value formulation depends on how we foliate spacetime, unfortunately all choices which reach future null infinity come with significant technical problems. Characteristic foliations offer many simplifications [6], but are also prone to develop caustics in the strong field region. Cauchy-characteristic matching, where data on spacelike slices are matched to characteristic slices which reach $\mathscr{I}^+$ [7, 6, 8, 9, 10], poses considerable difficulties to find a stable algorithm for the matching procedure. A smooth way of combining spacelike foliations in the strong field region and also reaching future null infinity is offered by the hyperboloidal initial value approach, pioneered by Friedrich [11, 12, 13]. Here one evolves along hyperboloidal slices, i.e. spacelike slices that reach null infinity. In this approach regularizing the singular equations (2) in a way that is not just numerically stable, but also avoids instabilities arising from the continuum equations, has proven to be difficult, in particular for hyperbolic free

Journal of Physics: Conference Series **600** (2015) 012061 doi:10.1088/1742-6596/600/1/012061

evolution schemes. For the use of elliptic-hyperbolic systems see [14, 15, 16]. In this paper we consider the already nontrivial hyperboloidal initial value problem in spherical symmetry. We have recently presented stable evolutions of regular initial data which do not form black holes [17], here we use our code to evolve a scalar field interacting with a black hole as a problem to test how our algorithm handles black holes, and we demonstrate that our code can track the expected power law tails.

2. Hyperboloidal foliations

Starting with the Schwarzschild metric

$$d\tilde{s}^2 = -A(\tilde{r})d\tilde{t}^2 + B(\tilde{r})d\tilde{r}^2 + \tilde{r}^2 d\sigma^2, \quad d\sigma^2 \equiv d\theta^2 + \sin^2\theta d\phi^2, \tag{3}$$

where $A(\tilde{r}) = B(\tilde{r})^{-1} = 1 - \frac{2M}{\tilde{r}}$ and the parameter M denotes the mass of the black hole, we first transform the Schwarzschild time coordinate $\tilde{t}$ to a new time coordinate t whose constant values determine the hyperboloidal slices. We then compactify the radial coordinate and conformally rescale the line element according to (1):

$$t = \tilde{t} - h(\tilde{r}), \qquad \tilde{r} = \frac{r}{\bar{\Omega}} \qquad \text{and} \qquad d\bar{s}^2 = \Omega^2 d\tilde{s}^2, \tag{4}$$

where $h(\tilde{r})$ is the height function (compare e.g. [18]), Ω is the conformal factor used to rescale the metric, and $\bar{\Omega}$ the compactifying factor to rescale the radial coordinate. The resulting line element is

$$d\bar{s}^2 = -A\Omega^2 dt^2 + \frac{\Omega^2}{\bar{\Omega}^2}\left[-2Ah'(\bar{\Omega} - r\bar{\Omega}')dtdr + (B - Ah'^2)\frac{(\bar{\Omega} - r\bar{\Omega}')^2}{\bar{\Omega}^2}dr^2 + r^2 d\sigma^2\right]. \tag{5}$$

Here A and B are now functions of $r/\bar{\Omega}$, and the spatial derivative of the height function is

$$h'(\tilde{r}) = \frac{dh}{d\tilde{r}} = -\frac{\frac{K_{CMC}\tilde{r}^3}{3} + C_{CMC}}{\left(1 - \frac{2M}{\tilde{r}}\right)\sqrt{\left(\frac{K_{CMC}\tilde{r}^3}{3} + C_{CMC}\right)^2 + \left(1 - \frac{2M}{\tilde{r}}\right)\tilde{r}^4}}. \tag{6}$$

The parameter K_{CMC} is the trace of the extrinsic curvature that labels the Constant-Mean-Curvature (CMC) foliation and C_{CMC} is an integration constant. For given values of M and K_{CMC}, there is only one possible choice for C_{CMC} that will give us trumpet initial data [19]. We also obtain the smallest value of the Schwarzschild radius $\tilde{r} = R_0$ that the hyperboloidal foliation can reach, and where the trumpet is located at infinite proper distance from the apparent horizon. Two examples of hyperboloidal CMC foliations of Schwarzschild are shown in figure 1; the larger the value of $|K_{CMC}|$, the closer the hyperboloidal slices are to characteristic ones.

For the conformal factor Ω that rescales the metric we follow Zenginoğlu [20, 21, 22, 23] and use a conformal factor fixed in time, whose expression is

$$\Omega = \frac{r_{\mathscr{I}}^2 - r^2}{6\, r_{\mathscr{I}}}|K_{CMC}|\,, \tag{7}$$

where $r_{\mathscr{I}}$ is the position of null infinity, which in the following we will set to unity.

The value of $\bar{\Omega}$ is determined numerically by imposing a conformally flat spatial metric. The behaviour of $\bar{\Omega}$ near $r = 0$ is linear in the compactified radius r, with a slope inversely proportional to the minimal Schwarzschild radius, R_0. In figure 2 we show Ω, $\bar{\Omega}$ and its approximation at the origin as a function of r. Note that at $r = r_{\mathscr{I}}$, Ω equals $\bar{\Omega}$, as at null infinity the spacetime is asymptotically flat.

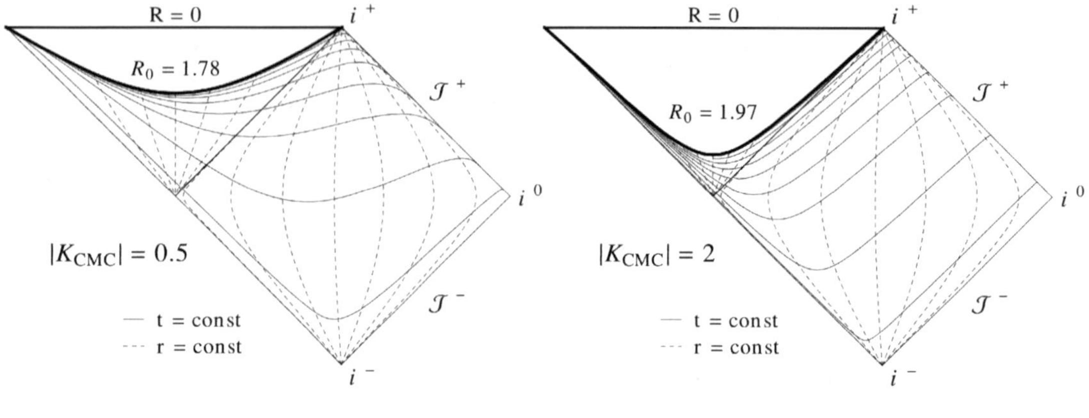

Figure 1. Carter-Penrose diagrams showing foliations for two values of K_{CMC}. The thick line corresponds to the smallest value of the Schwarzschild radial coordinate, R_0, along the slice.

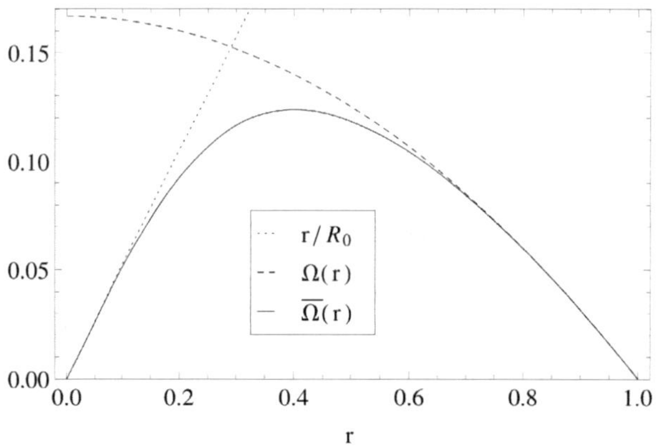

Figure 2. Factors Ω and $\bar{\Omega}$ for $M = 1$ and $K_{CMC} = -1$.

3. Conformally compactified equations

We use either the generalized BSSN formulation [24, 25, 26] or a similar conformal version of the Z4 formulation [27, 28, 29], the Z4c equations [30, 31], in their spherically symmetric reduction. The actual equations used in the simulations can be found in Appendix C of [17]. The massless scalar field satisfies the wave equation in the physical spacetime, and with respect to the unphysical rescaled metric it satisfies

$$g^{\mu\nu}\nabla_\mu\nabla_\nu\Phi - 2g^{\mu\nu}\nabla_\mu\Phi\frac{\nabla_\nu\Omega}{\Omega} = 0. \tag{8}$$

Following Zenginoğlu [32, 22], we use a fixed conformal factor Ω (7), and a scri-fixing gauge where we demand that $(\partial/\partial t)^a = \alpha n^a + \beta^a$ becomes null at $\mathscr{I}^+$. Since $\mathscr{I}$ is a null surface, our condition will be satisfied if $(\partial/\partial t)^a$ is parallel to $\nabla^a\Omega$ at $\mathscr{I}$, and we obtain $\partial_t\Omega|_{\mathscr{I}} = 0$, consistent with the value of Ω not being evolved in time. The time vector $(\partial/\partial t)^a$ is null at $\mathscr{I}^+$ if $-\alpha^2 + \beta^a\beta_a|_{\mathscr{I}} = 0$, which in our variables becomes $-\alpha^2 + \chi^{-1}\gamma_{rr}{\beta^r}^2\Big|_{\mathscr{I}} = 0$. In our simulations this is achieved by making sure that the values of α and β^r stay fixed at $\mathscr{I}^+$ during the evolution.

The shift can either be fixed throughout the evolution or evolved with the rest of the variables. For black hole initial data we choose the second option. As evolution equation we chose a generalized Gamma-driver shift condition:

$$\dot{\beta}^r = \beta^r \beta^{r\prime} + \frac{3}{4}\mu B^r + L_0 - \frac{\xi_{\beta^r}}{\Omega}\beta^r, \tag{9}$$

$$\dot{B}^r = \beta^r B^{r\prime} - \eta B^r + \lambda\left(\dot{\Lambda}^r - \beta^r \Lambda^{r\prime}\right). \tag{10}$$

A source function L_0 calculated from the flat spacetime values was added, as well as a damping term $(-\beta^r/\Omega)$ that makes sure that the value of β^r stays fixed at $\mathscr{I}^+$. The parameters λ and μ have to be chosen carefully, as they determine the eigenspeeds. Appropriate values for our simulations are: $\xi_{\beta^r} = 5$, $\lambda = 3/4$, $\mu = 0.15 + 4(1-r^2)$ and $\eta = 0.1$.

As slicing condition we choose a generalized Bona-Massó equation that takes the form

$$\dot{\alpha} = \beta^r \alpha' - f(\alpha)\,(K - K_0) + L_0, \tag{11}$$

where we have the freedom to choose the two functions K_0 and L_0. The presence of K_0 is necessary to prevent the equation from growing exponentially, as the extrinsic curvature K is negative in its stationary solution. The source function L_0 is calculated from flat spacetime initial data on the hyperboloidal foliation. Due to their properties, we are interested in having the 1+log slicing condition around the origin and the harmonic one near $\mathscr{I}^+$. For this we can make $f(\alpha)$ change smoothly from 2α to α^2. For the numerical results shown below we have used

$$\dot{\alpha} = \beta^r \alpha' - \alpha^2\,(K - K_0) + L_0 + (\alpha_0 - \alpha),$$

where α_0 are the initial data corresponding to the Schwarzschild lapse.

4. Numerical test: power law tails of the scalar field

The equations were implemented in a spherically symmetric code that uses the method of lines with a 4th order Runge-Kutta time integrator and 4th order finite differences. We add Kreiss-Oliger dissipation [33] and evolve on a staggered grid, which does not include $\mathscr{I}^+$, however we obtain 4th order convergence for values interpolated to $\mathscr{I}^+$ as seen in Fig. 4.

We have modified trumpet initial data (5) with either a perturbation in the initial lapse (gauge waves), or coupled to the massless scalar field with initial data

$$\Phi_0 = A e^{-\frac{(r^2-c^2)^2}{4\sigma^4}}. \tag{12}$$

A scalar field perturbation of a Schwarzschild black hole is expected to decay at late times with a power-law tail of the form [34]

$$\lim_{t\to+\infty} \Phi(t,r) \propto t^p. \tag{13}$$

For spherical scalar perturbations the decay rate p has been calculated analytically as $p = -3$ along timelike surfaces [34] and $p = -2$ along null surfaces ($\mathscr{I}^+$) [35, 36]. In our simulations we found a value of $p \sim -2.08 \pm 0.09$ for the scalar field on $\mathscr{I}^+$. In figure 4 we show the behaviour in time of the rescaled scalar field Φ/Ω after an initial perturbation of amplitude 10^{-4} at some selected values of the radial coordinate, as well as extrapolated to $\mathscr{I}^+$.

The convergence results are shown in figure 4. The values of Φ/Ω are extrapolated to $\mathscr{I}^+$ before calculating the differences presented in the plot. We see good convergence up to $t \approx 100$, when the convergence order starts to decrease due to loss of accuracy.

We note that some tuning of the gauge conditions is necessary to suppress continuum instabilities in the evolution systems. Some algebraic modifications of the evolution variables which also suppress continuum instabilities are discussed in [17].

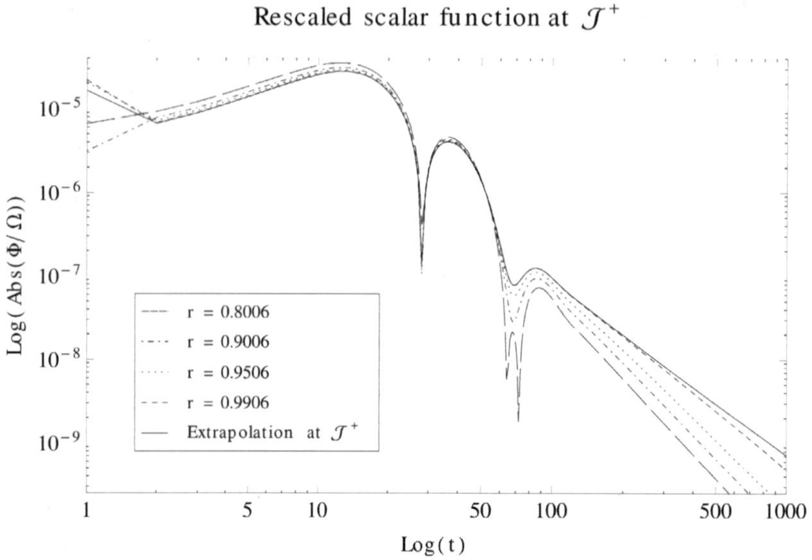

Figure 3. Values of the rescaled scalar field at some values of r and extrapolated to $r = r_{\mathscr{I}}$ as a function of time. The exponent of the tails is $p \sim -3$ up to $r \sim 0.95$ and $p \sim -2$ at $\mathscr{I}^+$.

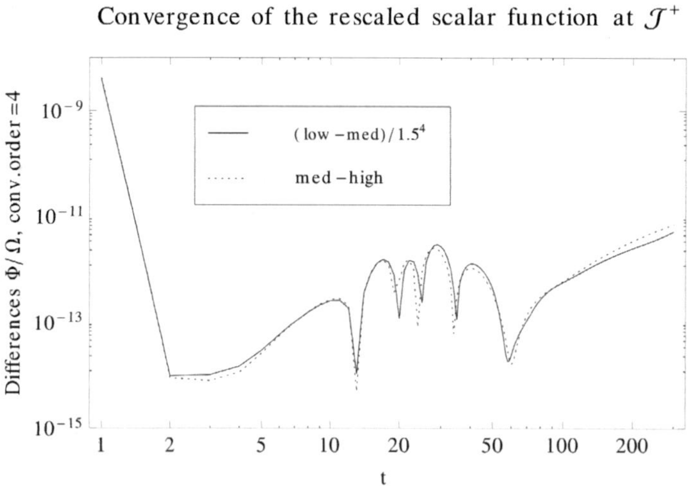

Figure 4. Convergence of the rescaled scalar field at $\mathscr{I}^+$.

5. Conclusions

We have evolved nonlinear spherical perturbations of a Schwarzschild black hole using a hyperboloidal free evolution approach based on conformally compactified versions of the generalized BSSN and Z4c evolutions systems for the Einstein equations. The present paper extends results we recently obtained for regular initial data [17] to the presence of a black hole. Our initial data correspond to nonlinear perturbations of initial data for black hole trumpet geometries, which we have calculated on the compactified hyperboloidal slices. We obtain stable numerical evolutions of the scalar field coupled to the Einstein equations, and show 4th order convergence for the scalar wave signal at $\mathscr{I}^+$. The gauge conditions play a key role in suppressing continuum instabilities in the evolution, and require some degree of tuning. Future work will explore more general initial data, gauge conditions, and different versions of the numerical algorithms.

Acknowledgments
AV was supported by AP2010-1697, AV and SH were supported by Spanish MINECO grants FPA2010-16495, FPA2013-41042-P and CSD2009-00064, European Union FEDER funds, and the Conselleria d'Economia i Competitivitat del Govern de les Illes Balears.

References
[1] Barack L 1999 *Phys.Rev.* **D59** 044016 (*Preprint* gr-qc/9811027)
[2] Leaver E W 1986 *J.Math.Phys.* **27** 1238
[3] Leaver E W 1986 *Phys. Rev. D* **34**(2) 384–408
[4] Penrose R 1963 *Phys. Rev. Lett.* **10**(2) 66–68
[5] Penrose R 1965 *Proc.Roy.Soc.Lond.* **A284** 159
[6] Winicour J 2009 *Living Reviews in Relativity* **12** 3 (*Preprint* 0810.1903)
[7] Bishop N T 1993 *Class. Quantum Grav.* **10** 333–341
[8] Reisswig C, Bishop N, Pollney D and Szilagyi B 2009 *Phys.Rev.Lett.* **103** 221101 (*Preprint* 0907.2637)
[9] Reisswig C, Bishop N, Pollney D and Szilagyi B 2010 *Class.Quant.Grav.* **27** 075014 (*Preprint* 0912.1285)
[10] Taylor N W, Boyle M, Reisswig C, Scheel M A, Chu T *et al.* 2013 *Phys.Rev.* **D88** 124010 (*Preprint* 1309.3605)
[11] Friedrich H 1983 *Comm. Math. Phys.* **91** 445–472
[12] Frauendiener J 2004 *Living Reviews in Relativity* **7**
[13] Friedrich H 2002 *Lect. Notes Phys.* **604** (*Preprint* gr-qc/0304003)
[14] Andersson L 2002 Construction of hyperboloidal initial data *The Conformal Structure of Space-Time* (*Lecture Notes in Physics* vol 604) ed Frauendiener J and Friedrich H (Springer Berlin Heidelberg) pp 183–194 ISBN 978-3-540-44280-6
[15] Rinne O 2010 *Class.Quant.Grav.* **27** 035014 (*Preprint* 0910.0139)
[16] Rinne O and Moncrief V 2013 *Class.Quant.Grav.* **30** 095009 (*Preprint* 1301.6174)
[17] Vañó-Viñuales A, Husa S and Hilditch D 2014 (*Preprint* 1412.3827)
[18] Malec E and Murchadha N O 2003 *Phys.Rev.* **D68** 124019 (*Preprint* gr-qc/0307046)
[19] Hannam M, Husa S, Pollney D, Bruegmann B and O'Murchadha N 2007 *Phys.Rev.Lett.* **99** 241102 (*Preprint* gr-qc/0606099)
[20] Zenginoğlu A 2008 *Class.Quant.Grav.* **25** 145002 (*Preprint* 0712.4333)
[21] Zenginoğlu A 2008 *Class.Quant.Grav.* **25** 175013 (*Preprint* 0803.2018)
[22] Zenginoğlu A 2008 *Class.Quant.Grav.* **25** 195025 (*Preprint* 0808.0810)
[23] Zenginoğlu A, Nunez D and Husa S 2009 *Class.Quant.Grav.* **26** 035009 (*Preprint* 0810.1929)
[24] Shibata M and Nakamura T 1995 *Phys. Rev. D* **52**(10) 5428–5444
[25] Baumgarte T W and Shapiro S L 1999 *Phys.Rev.* **D59** 024007 (*Preprint* gr-qc/9810065)
[26] Brown J D 2008 *Class. Quant. Grav.* **25** 205004 (*Preprint* 0705.3845)
[27] Bona C, Ledvinka T, Palenzuela C and Zacek M 2003 *Phys. Rev.* **D67** 104005
[28] Alic D, Bona-Casas C, Bona C, Rezzolla L and Palenzuela C 2012 *Phys.Rev.* **D85** 064040 (*Preprint* 1106.2254)
[29] Sanchis-Gual N, Montero P J, Font J A, Mueller E and Baumgarte T W 2014 (*Preprint* 1403.3653)
[30] Bernuzzi S and Hilditch D 2010 *Phys.Rev.* **D81** 084003 (*Preprint* 0912.2920)
[31] Weyhausen A, Bernuzzi S and Hilditch D 2012 *Phys.Rev.* **D85** 024038 (*Preprint* 1107.5539)
[32] Zenginoğlu A 2007 *A conformal approach to numerical calculations of asymptotically flat spacetimes* Ph.D. thesis (*Preprint* 0711.0873)
[33] Kreiss H and Oliger J 1973 *Methods for the approximate solution of time dependent problems* GARP publications series No. 10 (International Council of Scientific Unions, World Meteorological Organization)
[34] Price R H 1972 *Phys. Rev. D* **5**(10) 2419–2438
[35] Bonnor W B and Rotenberg M A 1966 *Proc. R. Soc. London* **A289** 247
[36] Gundlach C, Price R H and Pullin J 1994 *Phys. Rev. D* **49**(2) 883–889

Journal of Physics: Conference Series **600** (2015) 012062 doi:10.1088/1742-6596/600/1/012062

Spherically Symmetric solutions on a cosmological dynamical background with BSSN equations

J. Rekier[1], **I. Cordero-Carrión**[2,1,3] **and A. Füzfa**[1]

[1] NaXys, University of Namur (UNamur), 61 Rue de Bruxelles, B-5000 Namur, Belgium
[2] Laboratoire Univers et Théories (LUTh), Observatoire de Paris/CNRS/Université Paris Diderot, 5 place Jules Janssen, F-92190 Meudon, France
[3] IFPA, Dep. AGO, Universite de Liège, Bat B5, Sart Tilman B-4000 Liège 1, Belgium

E-mail: jeremy.rekier@unamur.be

Abstract. We give a summary of the work presented in [1]. We expose a numerical method for the study of cosmological problems in spherical symmetry in full General Relativity. The stability of the code close to the origin is made possible through the use of the *Partially Implicit Runge-Kutta* (PIRK) algorithm described in [2]. We demonstrate the stability and convergence properties and give a simple application to the evolution of the *Lemaître-Tolman-Bondi* spacetime. This work is a generalisation of the study given in [3] performed on an asymptotically flat background.

1. Introduction

A definitive theory of the evolution of the Universe should explain how small portions of spacetime can collapse and lead to the formation of structures on an expanding background. Such processes have been studied in many ways which all make simplifying assumptions departing from General Relativity. On the other hand, recent progresses in the field of Numerical Relativity have allowed to solve many problems on Minkowskian background. The first step towards solving the complete collapse of structures is to assume spherical symmetry. This involves the need to regularise operators of the form $1/r^p$ close to the origin of coordinates. The PIRK methods are an easy way to solve the instability problems [2].

In the following sections, we start by recalling the BSSN formalism. This gives us the opportunity to define the variables used throughout this work. We then illustrate the stability and convergence properties of the code by studying the gauge dynamics on a de Sitter background and proceed to reproduce the dynamics of the LTB solution. This describes the expanding spacetime around a spherically distributed dust density profile hence its interest for both cosmology and numerical relativity.

2. Formalism

The metric ansatz is

$$ds^2 = -(\alpha - \beta^2)dt^2 + 2\beta dtdt + \psi^4 a^2(t)(\hat{a}dr^2 + \hat{b}r^2 d\Omega^2). \quad (1)$$

The only change compared to the usual BSSN formulation lies in the presence of the homogeneous scale factor $a(t)$ in the explicit form of the conformal metric. This scalar function describes the

Journal of Physics: Conference Series **600** (2015) 012062 doi:10.1088/1742-6596/600/1/012062

cosmological background. The extrinsic curvature defined as $K_{ij} := -\frac{1}{2}\pounds_{\mathbf{n}}\gamma_{ij}$, with γ_{ij} being the induced 3-metric can be decomposed in trace K and conformally-scaled trace-free part $\hat{A}_{ij}$ as

$$K_{ij} = \frac{1}{3}\gamma_{ij}K + \psi^4 a^2 \hat{A}_{ij}. \tag{2}$$

In spherical symmetry, $\hat{A}_{ij}$ has only two non zero components, $A_a := \hat{A}^r_r$ and $A_b := \hat{A}^\theta_\theta$. Since $\hat{A}_{ij}$ must be traceless, one further has $A_a + 2A_b = 0$. The last dynamical variable of the BSSN formalism is

$$\hat{\Delta}^r = \frac{1}{\hat{a}}\left[\frac{\partial_r \hat{a}}{2\hat{a}} - \frac{\partial_r \hat{b}}{\hat{b}} - \frac{2}{r}\left(1 - \frac{\hat{a}}{\hat{b}}\right)\right]. \tag{3}$$

Throughout this work, we have used $\beta = 0$. We thus omit this variables in the evolution equations given below.

$$\begin{aligned}
\partial_t \hat{a} &= -2\alpha\hat{a}A_a, \\
\partial_t \hat{b} &= -2\alpha\hat{b}A_b, \\
\partial_t \psi &= -\frac{1}{6}\alpha\psi K - \frac{1}{2}\frac{\dot{a}}{a}\psi, \\
\partial_t K &= -\nabla^2\alpha + \alpha(A_a^2 + 2A_b^2 + \frac{1}{3}K^2) + 4\pi\alpha(E + S_a + 2S_b), \\
\partial_t A_a &= -\left(\nabla^r\nabla_r\alpha - \frac{1}{3}\nabla^2\alpha\right) + \alpha\left(R^r_r - \frac{1}{3}R\right) + \alpha K A_a - \frac{16}{3}\pi\alpha(S_a - S_b) \\
\partial_t \hat{\Delta}^r &= -\frac{2}{\hat{a}}(A_a\partial_r\alpha + \alpha\partial_r A_a) + 2\alpha\left(A_a\hat{\Delta}^r - \frac{2}{r\hat{b}}(A_a - A_b)\right) \\
&+ \frac{\xi\alpha}{\hat{a}}\left[\partial_r A_a - \frac{2}{3}\partial_r K + 6A_a\frac{\partial_r\psi}{\psi} + (A_a - A_b)\left(\frac{2}{r} + \frac{\partial_r\hat{b}}{\hat{b}}\right) - 8\pi j_r\right],
\end{aligned} \tag{4}$$

The strong hyperbolicity of these equations is ensured by choosing $\xi > 1/2$. Following [3], we take $\xi = 2$ which is the standard choice . The functions E, j_r, S_a and S_b are built from the various projections of the stress-energy tensor (see [1]). The Hamiltonian and Momentum constraints read

$$\mathcal{H} \equiv R - (A_a^2 + 2A_b^2) + \frac{2}{3}K^2 - 16\pi E = 0, \tag{5}$$

$$\begin{aligned}
\mathcal{M}^r \equiv &\ \partial_r A_a - \frac{2}{3}\partial_r K + 6A_a\frac{\partial_r\psi}{\psi} \\
&+ (A_a - A_b)\left(\frac{2}{r} + \frac{\partial_r\hat{b}}{\hat{b}}\right) - 8\pi j_r = 0.
\end{aligned} \tag{6}$$

The dynamics of the background Universe is described by the usual Friedmann and acceleration equations

$$\frac{1}{\alpha_{\mathrm{bkg}}^2}\left(\frac{\dot{a}}{a}\right)^2 = \frac{8\pi}{3}\rho_{\mathrm{bkg}}, \tag{7}$$

$$\frac{1}{\alpha_{\mathrm{bkg}}^2}\frac{\ddot{a}}{a} - \frac{\dot{a}}{a}\frac{\dot{\alpha}_{\mathrm{bkg}}}{\alpha_{\mathrm{bkg}}} = -\frac{8\pi}{6}(\rho_{\mathrm{bkg}} + 3p_{\mathrm{bkg}}). \tag{8}$$

The homogeneous background density, ρ_{bkg}, and pressure, p_{bkg}, are functions of time. α_{bkg} is the homogeneous lapse function.

The adjustment of boundary conditions is important in this work as the background is dynamical. We use a generalised form of dissipative boundary conditions

$$\partial_t f = \partial_t f_{\mathtt{bkg}} - v\,\partial_r f - \frac{v}{r}(f - f_{\mathtt{bkg}}), \tag{9}$$

with $f_{\mathtt{bkg}}$, the homogeneous background value of the field f and v, its characteristic velocity.

The Numerical solution of the above equations are obtained by using the PIRK evolution scheme (see [2] for details) along with a fourth order finite difference discretisation of the spatial derivative operators. No numerical dissipation is needed.

3. Gauge Dynamics

We investigate the stability and convergence properties of the code with the study of pure gauge dynamics. We look at the evolution of a lapse pulse on a De Sitter Universe background with a constant energy density mimicking a cosmological constant with $\Lambda = 8\pi\rho_{\mathtt{bkg}}$ and $p_{\mathtt{bkg}} = -\rho_{\mathtt{bkg}}$. The initial lapse pulse is

$$\alpha(t=0) = \alpha^0_{\mathtt{bkg}} + \frac{\alpha_0 r^2}{1+r^2}\left[e^{-(r-r_0)^2} + e^{-(r+r_0)^2}\right]. \tag{10}$$

Fixing $\alpha^0_{\mathtt{bkg}} = 1$, the Friedmann equation gives $E = \rho_{\mathtt{bkg}} = \frac{3}{8\pi}H_0^2$. The initial conditions

$$\begin{aligned} &\hat{a}(t=0) = \hat{b}(t=0) = \psi(t=0) = 1, \\ &K = 3H_0, \quad A_a = 0. \end{aligned} \tag{11}$$

then solve satisfy the constraint equations. The evolution is performed in the harmonic slicing with zero shift:

$$\partial_t \alpha = -\alpha^2 K. \tag{12}$$

We set and $\alpha_0 = 0.01$ and $H_0 = 0.01$, that is well in the exponential regime of the cosmic expansion for time scales up to $t \sim 10$. The *Courant-Friedrichs-Lewy* (CFL) factor for simulations in this section is $\Delta t/\Delta r = 0.25$. The spatial resolution is $\Delta r = 0.05$. Fig. 1 shows the evolution of the lapse function. This should be compared with Fig. 2 showing the evolution of its asymptotic cosmological value.

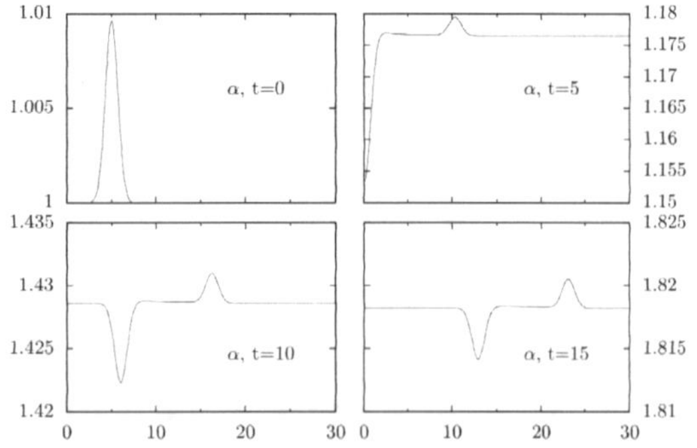

Figure 1. Evolution of a pure gauge pulse in time on a de Sitter background. The asymptotic value of α gets rescaled during the evolution.

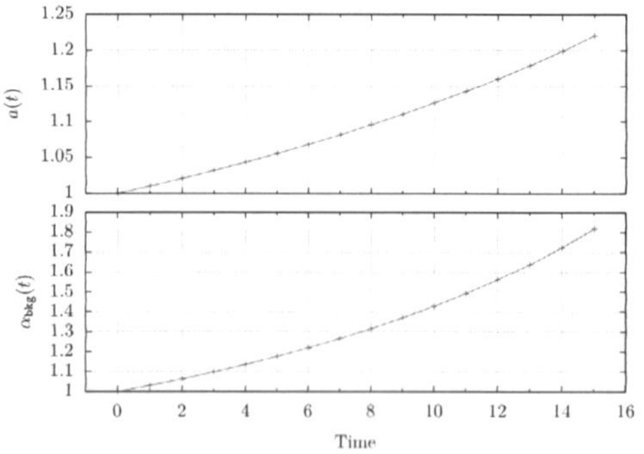

Figure 2. Evolution of the scale factor $a(t)$ (upper panel) and background lapse function $\alpha_{\mathtt{bkg}}$ (lower panel).

The convergence and stability of the code are illustrated on Fig. 3 showing the Hamiltonian constraint violation for different values of the resolution. The rescaling of the error demonstrates the second-order convergence of the method. Similarly to the results of [3] the hamiltonian constraint violation is maximum at the centre of coordinates. Yet the error remains bounded and the central value of all fields remains controlled.

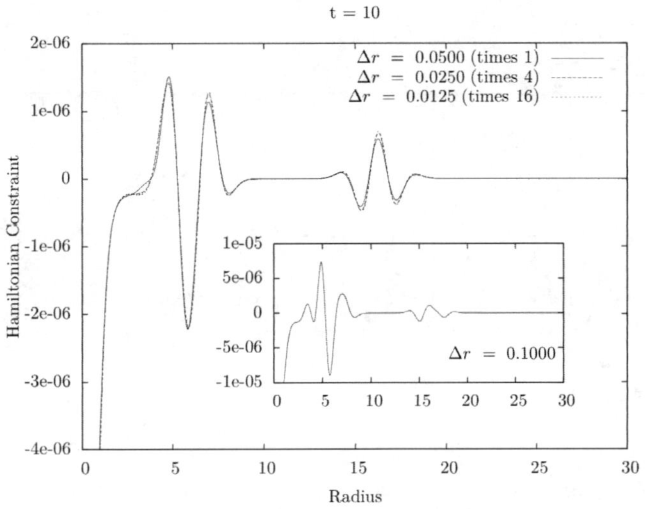

Figure 3. Value of the hamiltonian constraint in pure gauge dynamics for 3 different resolutions at $t = 10$. The rescaling of the curves shows the good agreement with the expected second-order convergence of the numerical method. The inner panel shows that the convergence regime is not yet reached for $\Delta r = 0.1$. These results were obtained for $H_0 = 0.01$ and $\Delta t/\Delta r = 0.25$.

4. Lemaître-Tolman-Bondi Solution

The LTB solution describes a spherically symmetric dynamical spacetime filled with dust. The analytical expression of its metric is usually given in the geodesic gauge $\alpha = 1$. The line element is

$$ds^2 = -dt^2 + \frac{a_\parallel^2(t,r)}{1 + 2E_{\mathtt{ltb}}(r)} + a_\perp^2(t,r) r^2 d\Omega^2, \tag{13}$$

with $a_\parallel = \partial_r(r\, a_\perp)$ and where $E_{\mathtt{ltb}}(r)$ is a free function. The evolution equations for the metric component are generalisation of the Friedmann equations. Those are given in [1] along with the evolution equation for an initial distribution of matter. Reproducing the LTB solution starts with solving the initial data for such initial distribution. We set

$$\begin{aligned} \hat{a}(t=0) = \hat{b}(t=0) = 1, &\ \ K(t=0) = -3H_0, \\ A_a(t=0) = A_b(t=0) = 0&, \\ E(r, t=0) = [1 + \delta_m(r)]\, \rho^0_{\mathtt{bkg}}&, \end{aligned} \tag{14}$$

with $\rho^0_{\mathtt{bkg}} = \rho_{\mathtt{bkg}}(t=0)$ and $\delta_m(r)$ some profile in the constraint equations. Upon using the Friedmann equation, the Hamiltonian constraint reduces to

$$\partial_r^2 \psi + \frac{2}{r}\partial_r \psi = 16\pi\, \rho^0_{\mathtt{bkg}}\, \delta^0_m(r)\, a_0^2\, \psi^5. \tag{15}$$

This can be solved as a boundary value problem. The boundary conditions are $\partial_r \psi|_{r=0} = 0$ and $\psi \to 1$ for $r \to \infty$. Once the solution is found, this is used to write the initial values of $a_\perp$ and $a_\parallel$ and their time derivatives.

Fig. 4 shows the good agreement between the solutions obtained with the LTB equations and the evolution using BSSN variables at early times. The maximum of the relative difference between both solution is of the order $\sim 10^{-5}$. The agreement stays equally valid at later time. Fig. 5 shows the comparison between the background scale factor and the central expansion of spacetime. The results presented here were obtained with $H_0 = 0.1$, $\Delta r = 0.1$. The CFL factor is $\Delta t/\Delta r = 0.5$.

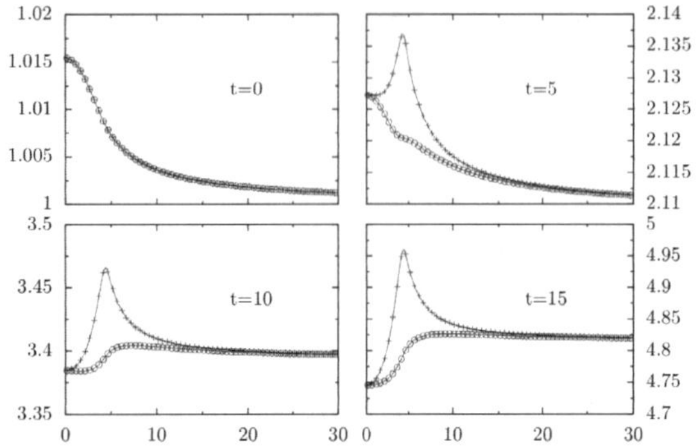

Figure 4. Metric components γ^{rr} (top curve) and $\gamma^{\theta\theta}/r^2$ (bottom curve). The plain lines is the result of the LTB equations. The dots are the result of the BSSN equations.

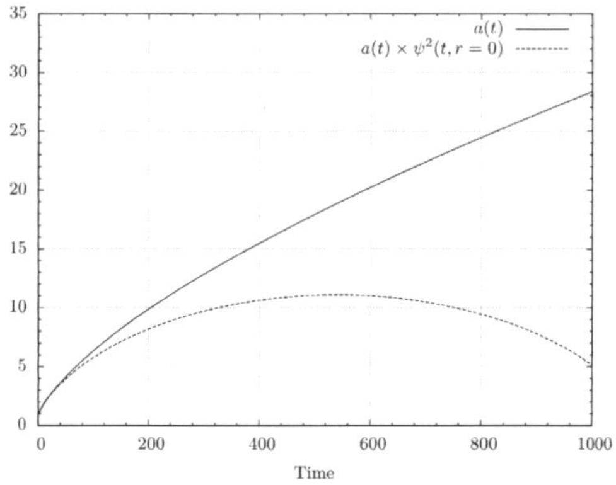

Figure 5. Evolution of the background scale factor (top curve) compared with the central expansion factor (bottom line).

5. Conclusion

We have shown how to apply the BSSN formalism to deal with the evolution of spherically symmetric solutions on a cosmological background. The validity of the numerical method has been proven by studying pure gauge dynamics and we have applied it to study the evolution of a simple dust distribution thus recovering the usual LTB solution. Further directions include the study of spacetime less restrictive than LTB. The application of the method here described in the case of a background Universe filled with a scalar field of Dark Energy is underway as of the moment of the writing of the present proceeding.

Acknowledgments

The authors warmly thank Dr. P. Montero. J.R. is supported by a FRS-FNRS (Belgian Fund for Scientific Research) Research Fellowship. I. C.-C. & A.F. have been partially supported by the ARC convention No. 11/15-040. This work was also supported by the SN2NS project ANR-10-BLAN-0502, the Fonds de la Recherche Scientifique - FNRS under grant 4.4501.05, the ERC Starting Grant CAMAP-259276, and the Grant No. AYA2010-21097-C03-01 of the Spanish MICINN. Computations were performed at the "plate-forme technologique en calcul intensif" (PTCI) of the University of Namur, Belgium, with the financial support of the FRS-FNRS (convention No. 2.4617.07. and 2.5020.11).

References

[1] Rekier J, Cordero-Carrion I and Fuzfa A 2014 (*Preprint* `1409.3476`) URL `http://arxiv.org/abs/1409.3476`
[2] Cordero-Carrión I and Cerdá-Durán P 2012 25 (*Preprint* `1211.5930`) URL `http://arxiv.org/abs/1211.5930`
[3] Montero P J and Cordero-Carrión I 2012 *Phys. Rev. D* **85** 124037 (*Preprint* `arXiv:1204.5377`)

Journal of Physics: Conference Series **600** (2015) 012063 doi:10.1088/1742-6596/600/1/012063

Regularization of geodesics in static spherically symmetric Kerr-Schild spacetimes

Pablo Galindo

Dept. de Geometría y Topología, Universidad de Granada, Campus de Fuentenueva s/n, 18071 Granada, Spain.

E-mail: `pablogsal@correo.ugr.es`

Marc Mars

Inst. de Física Fundamental y Matemáticas (IUFFyM), Universidad de Salamanca, Plaza de la Merced s/n 37008 Salamanca, Spain.

E-mail: `Marc@usal.es`

Abstract. We describe a method to analyze causal geodesics in static and spherically symmetric spacetimes of Kerr-Schild form which, in particular, allows for a detailed study of the geodesics in the vicinity of the central singularity by means of a regularization procedure based on a generalization of the McGehee regularization for the motion of Newtonian point particles moving in a power-law potential. The McGehee regularization was used by Belbruno and Pretorius [1] to perform a dynamical system regularization of the central singularity of the motion of massless test particles in the Schwarzschild spacetime. Our generalization allows us to consider causal (timelike or null) geodesics in any static and spherically symmetric spacetime of Kerr-Schild form. As an example, we apply these results to causal geodesics in the Schwarzschild and Reissner-Nordström spacetimes.

1. Introduction

The motion of freely falling particles in static spherically symmetric spacetimes constitutes a basic topic in General Relativity, which has been studied from several perspectives. In this paper we use dynamical system methods to analyze causal geodesics in static spherically symmetric spacetimes admitting a Kerr-Schild structure (which, in this context, is a very mild restriction and includes virtually all cases of interest). We are thus able to analyze the behaviour of geodesics across Killing horizons and reach the singularity at $r = 0$, when one is present. We are particularly interested in studying the geodesics in the vicinity of the central singularity. To that aim we generalize the McGehee regularization [4] of power-law Newtonian potentials, already used by Belbruno and Pretorius [1] to study null geodesics in the Schwarzschild spacetime, to a general Newtonian potential. We also show that the dynamics described by the Hamiltonian of causal geodesics in a static, spherically symmetric spacetimes of Kerr-Schild form is equivalent to the dynamics of a Newtonian point particle under the action of a suitable central potential. The regularization at the center introduces a collision manifold which is invariant under the flow and which allows for a detailed description of the approach of geodesics to the singularity. The condition that the geodesics be future causal translates into the existence of excluded regions of the phase space portrait. The topological modification of the phase space coming from the excluded regions is intimately linked to the Penrose structure of the maximal extension of a Kerr-Schild patch of the spacetime under consideration, thus allowing us to describe the global behaviour of geodesics in the extended spacetime by means of a single two-dimensional phase space. We illustrate this by studying causal geodesics in the Schwarzchild and Reisner-Norsdtröm spacetimes. Further details on the topics addressed in this paper can be found in [2].

Journal of Physics: Conference Series **600** (2015) 012063 doi:10.1088/1742-6596/600/1/012063

2. Geodesic equations for a general stationary Kerr-Schild metric

Throughout this paper, we will consider spacetimes $\{\mathcal{M} = \mathbb{R} \times (\mathbb{R}^3 \setminus \mathcal{C}), g\}$ where $\mathcal{C} \subset \mathbb{R}^3$ is a closed subset such that $\mathcal{M}$ is connected and g is a Lorentzian metric of Kerr-Schild form [3]. More specifically, let $\{x^\alpha\} = \{T, x^i\}$ ($\alpha, \beta, \cdots = 0,1,2,4$ and $i, j, \cdots = 1,2,3$) be Cartesian coordinates on $\mathbb{R} \times \mathbb{R}^3$ and endow $\mathcal{M}$ with the Minkowski metric $\eta = -dT^2 + \delta_{ij}dx^i dx^j$. Let $\boldsymbol{K}$ be a smooth one-form on $\mathcal{M}$ which is null with respect to the metric η and $h : \mathcal{M} \longrightarrow \mathbb{R}$ a smooth function. The metric g being of Kerr-Schild form means that it takes the form

$$g = \eta + h\boldsymbol{K} \otimes \boldsymbol{K}.$$

In addition we will assume $\mathcal{C}$ to be rotationally symmetric and g to be static and spherically symmetric

$$g = -dT^2 + d\vec{x} \cdot d\vec{x} + h(r)(dr - \sigma dT) \otimes (dr - \sigma dT),$$

where $r = \sqrt{\vec{x} \cdot \vec{x}}$ and $\sigma = \pm 1$. The integrable Killing vector is ∂_T, which we note is timelike, null or spacelike depending on whether $h < 1$, $h = 1$ or $h > 1$. We choose the time orientation so that $\sigma K_\alpha := (dr - \sigma dT)$ is future directed. The function h takes the form $h(r) = \frac{2M}{r}$ in the Schwarzschild spacetime and $h(r) = \frac{2Mr+Q^2}{r^2}$ in the Reissner-Nordström spacetime. In any spacetime $(\mathcal{M}, g)$, affinely parametrized geodesics are the solutions of the Hamilton equations of the Hamiltonian

$$H = \frac{1}{2}(g^{-1})^{\alpha\beta} p_\alpha p_\beta$$

defined on the cotangent bundle of $\mathcal{M}$. In our case, and by the use of the Killing conserved quantities, the following Hamiltonian describing the spatial part of the geodesic flow arises naturally

$$H' := H + \frac{1}{2}E^2 = \frac{1}{2}\left(\vec{p}^2 - h\left(\vec{K} \cdot \vec{p} - E\widehat{K}\right)^2\right), \tag{1}$$

where we have written $K = \{\hat{K}, \vec{K}\}$, $\boldsymbol{p} = \{\widehat{p}, \vec{p}\}$, $E := -\boldsymbol{p}(dT)$ and dot means scalar product with δ_{ij}. H' is now defined on the cotangent bundle of $\mathbb{R}^3 \setminus \mathcal{C}$. In addition to the Hamilton equation of H', the geodesics also satisfy the following ODE describing the temporal part of the geodesic flow

$$\left(1 - h\vec{K}^2\right)\frac{dT}{ds} + h\widehat{K}\vec{K} \cdot \frac{d\vec{x}}{ds} = E, \tag{2}$$

and which is simply the coordinate form of $g(u, \partial_T) = -E$, where u is the geodesic velocity vector $u(s) := (\dot{T}(s), \dot{\vec{x}}(s))$. The Hamiltonian H' is still complicated. The following theorem [2] allows us to replace it by a simpler one.

Theorem 1. *Let $x'(s)$ and $\tilde{x}(s)$ be trajectories of the respective Hamiltonians*

$$H' = \frac{1}{2}\vec{p}^2 - \frac{h(|\vec{x}|)}{2}\left(\frac{\vec{x} \cdot \vec{p}}{|\vec{x}|} - \sigma E\right)^2, \qquad E \in \mathbb{R}$$

$$\tilde{H} = \frac{1}{2}\vec{p}^2 - \frac{h(|\vec{x}|)}{2}\left(\frac{L^2}{|\vec{x}|^2} + \mu\right), \qquad L, \mu \in \mathbb{R}.$$

Then $x'(s) = \tilde{x}(s)$ if and only if their respective initial values satisfy

$$\tilde{x}_0 = x'_0, \qquad \tilde{p}_0 = p'_0 - h(|x'_0|)\left(\frac{x'_0 \cdot p'_0}{|x'_0|} - \sigma E\right)\frac{x'_0}{|x'_0|},$$

$$|x'_0 \times p'_0| = |L|, \qquad H'(x'_0, p'_0) = \frac{1}{2}\left(E^2 - \mu\right).$$

Moreover, in that case, $\tilde{H}(\tilde{x}_0, \tilde{p}_0) = \frac{1}{2}\left(E^2 - \mu\right)$.

Note that the value of μ in the theorem is arbitrary. However, when making contact to the spacetime geodesics, the relevant values are $\mu = 1$ for timelike geodesics, $\mu = 0$ for null geodesics and $\mu = -1$ for spacelike geodesics, cf. (1). It is interesting that the Hamiltonian $\tilde{H}$ is independent of σ, so that it can describe the geodesics in $(\mathcal{M}, g)$ both for the case when $\boldsymbol{K}$ is future directed (plus sign) or past directed (negative sign). Moreover, the Hamiltonian $\tilde{H}$ is a standard Hamiltonian in Newtonian mechanics for a point particle in a central potential. This is a substantial simplification over the original problem of solving the geodesic equations in a stationary and spherically symmetric spacetime of Kerr-Schild form, because we can exploit all the information known for trajectories of point particles in Newtonian mechanics under the influence of a radial potential. Hence the equation for the trajectories are

$$\begin{aligned} \ddot{\vec{x}} &= -\frac{\partial V(|\vec{x}|)}{\partial \vec{x}} = \frac{\partial}{\partial \vec{x}}\left[\frac{h(|\vec{x}|)}{2}\left(\frac{L^2}{|\vec{x}|^2} + \mu\right)\right], \\ \vec{L} &= \vec{x} \times \dot{\vec{x}}, \end{aligned}$$

which need to be supplemented with the energy conservation stated at the end of Theorem 1,

$$\frac{1}{2}|\dot{\vec{x}}|^2 - \frac{h(|\vec{x}|)}{2}\left(\frac{L^2}{|\vec{x}|^2} + \mu\right) = \frac{1}{2}\left(E^2 - \mu\right) := \epsilon$$

and with (2), which becomes

$$E = (1-h)\dot{T} + h\sigma\frac{\vec{x}\cdot\dot{\vec{x}}}{|\vec{x}|}.$$

Since we are interested in future directed causal geodesics we need to find the restrictions on the initial data which guarantee this. This is given in the following proposition, cf. [2]

Proposition 2. *A geodesic starting at a point* $(t_0, \vec{x}_0 \neq 0)$ *is future causal iff* $\dot{\vec{x}}_0$ *satisfies*

$$\begin{aligned} &\text{if } h_0 > 1, && \sigma\dot{r}_0 \in [a_0, \infty) \\ &\text{if } h_0 < 1, && \dot{r}_0 = (-\infty, \infty) \\ &\text{if } h_0 = 1, && \sigma\dot{r}_0 \in [0, \infty) \qquad \text{with} \quad \dot{r}_0 = 0 \Longrightarrow \mu = L = 0 \end{aligned}$$

where $r_0 := |\vec{x}_0|$, $\dot{r}_0 := \frac{\vec{x}_0 \cdot \dot{\vec{x}}_0}{|\vec{x}_0|}$, $h_0 := h(r_0)$ *and* $a_0 := \sqrt{(|1-h_0|)\left(\frac{L^2}{r_0^2} + \mu\right)} \geq 0$.

These restrictions imply the existence of excluded regions in the phase space:

Corollary 3. *The excluded regions in the phase space correspond to* $h_0 \geq 1$ *and* $\epsilon < -\frac{\mu}{2}$, *independently of the sign of* σ *and of the function* $h(|\vec{x}|)$ *in the Kerr-Schild metric.*

Having obtained the geodesic equations for all values of σ we need a method that allows us to regularize the singularity located at $r = 0$. The method that we develop in the following theorem, see [2], is called "McGehee regularization" because it provides a generalization of the original approach by McGehee in [4]. This procedure will allow us to obtain information of the geodesics at the vicinity of the singularity

Theorem 4. *Let* $\mathcal{N}$ *be an open annulus in* $\mathbb{C}$ *and* $V : \mathcal{N} \to \mathbb{R}$ *be a radially symmetric function* $V(x) = V(|x|)$. *Assume that* $V(|x|)$ *is* C^1 *as a function of* $|x|$ *and define* $\nabla = \partial_{x^1} + i\partial_{x^2}$ *where* $x = x^1 + ix^2$, $x^1, x^2 \in \mathbb{R}$. *Then the dynamical system*

$$\begin{aligned} \dot{x} &= y, \\ \dot{y} &= -\nabla V(|x|) := \Lambda(|x|)x, \end{aligned} \tag{3}$$

on $\mathcal{N} \times \mathbb{C}$ is equivalent to the system

$$r' = ru \tag{4}$$
$$u' = r^{-2(\beta+1)}\left(L^2 + r^4\Lambda(r)\right) - \beta u^2 \tag{5}$$
$$\theta' = Lr^{-(\beta+1)} \tag{6}$$

where β is an arbitrary constant. This system also admits the following two constants of motion

$$L = vr^{\beta+1}$$
$$\epsilon = \frac{1}{2}r^{2\beta}\left(u^2 + v^2\right) + V(r).$$

The coordinates $\{r, \theta, u, v\}$ take values in $r \in (a, b) \subset \mathbb{R}^+$, $\theta \in \mathbb{S}^1$ and $u, v \in \mathbb{R}$. The coordinate change is defined by

$$\left.\begin{array}{ll} x &= re^{i\theta} \\ y &= r^{\beta}(u + iv)e^{i\theta} \end{array}\right\} \qquad d\tau = r^{1-\beta}ds,$$

where τ is the flow parameter in (3) and s is the flow parameter in (4)-(6).

The optimal choice of β for a detailed study of the dynamical system (4)-(6) at $r = 0$ is $\beta = \min\{-1, -\frac{\gamma}{2}\}$ [2] with γ selected in such a way that $|x|^\gamma V(|x|)$ admits a C^1 extension to $|x| = 0$ and $\lim_{|x|\to 0}|x|^\gamma V(|x|) \neq 0$. Indeed, a larger value of β is not capable of regularizing the system at $r = 0$. On the other hand, a smaller value of β overkills the singularity. This has the effect that the invariant submanifold $\{r = 0\}$ (which is called *the collision manifold*) has $u = 0$ as a single fixed point, which is moreover always non-hyperbolic. Thus, all details of the phase space structure of the dynamical system at $\{r = 0\}$ are lost by such a choice of β. We will see below an example of this behavior when considering the Schwarzschild limit of the dynamical system describing causal geodesics in the Reissner-Nordström spacetime.

3. The Schwarzschild dynamical system

Let us analyze briefly the Schwarzschild dynamical system. The function $h(r)$ takes the form $h(r) = \frac{2M}{r}$. From Theorems 1 and 4 with the optimal value of $\beta = -\frac{3}{2}$ we derive the geodesic equations

$$r' = ru, \tag{7}$$
$$u' = r\left(L^2 - \mu Mr\right) - 3L^2M + \frac{3}{2}u^2, \tag{8}$$
$$\theta' = L\sqrt{r}. \tag{9}$$

3.1. The collision manifold

The submanifold $r = 0$ is clearly invariant under the flow. Since $r = 0$ corresponds to the spacetime singularity, this submanifold is called *collision manifold*. It can be described globally by the coordinates $\{u, \theta\}$ so its topology is $\mathbb{R} \times \mathbb{S}^1$. The dynamical system (7)-(9) restricted to the collision manifold reads

$$u' = \frac{3}{2}u^2 - 3L^2M,$$
$$\theta' = 0.$$

This system has two lines of critical points: one line of stable points at $(\theta, u) = (\theta_0, -\sqrt{2ML^2})$ and one line of unstable nodes at $(\theta, u) = (\theta_0, \sqrt{2ML^2})$, where $\theta_0 \in \mathbb{S}^1$ is an arbitrary value. For each value of θ_0, there is a trajectory extending from $u = -\infty$ and approaching $u = -\sqrt{2ML^2}$ as its future limit point, a trajectory from $u = -\sqrt{2ML^2}$ to $u = \sqrt{2ML^2}$ and a trajectory having $u = \sqrt{2ML^2}$ as its past limit point and extending to $u = +\infty$, all of them with $\theta = \theta_0$.

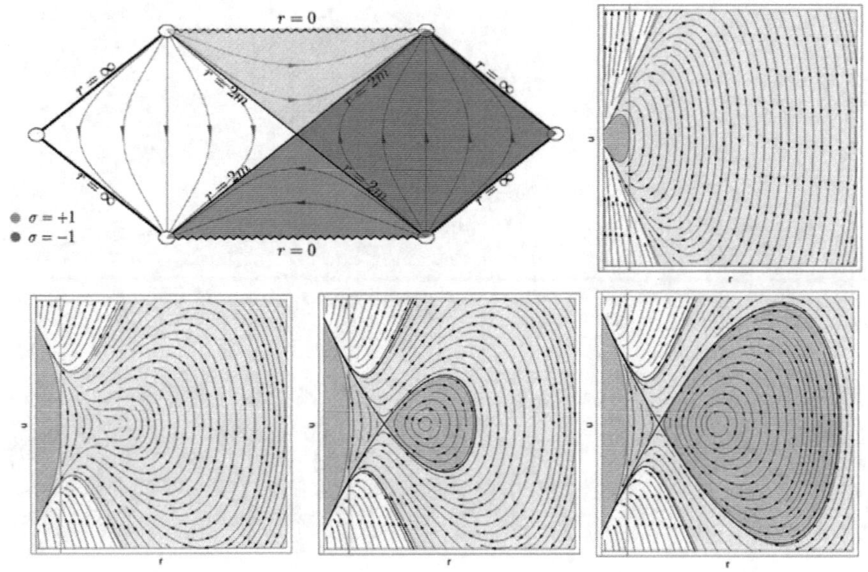

Figure 1: Phase space for timelike particles with increasing value of L from left to right and top to bottom. The dark zone correspond to the forbidden region given by $\epsilon < \frac{\mu}{2}$ and the red line corresponds to the horizon. Different colors are used to distinguish different behaviours for the flow. The upper left image is the Penrose diagram for the Kruskal spacetime where the Kerr-Schild patches have been highlighted.

3.2. The general flow

We are going to center our attention in the phase portrait for timelike geodesics ($\mu = 1$), as this shows more interesting phenomena than in the null case. The fixed points are

$$(r = 0, u = \pm\sqrt{2ML^2}), \qquad (r = r_\pm(M, |L|) := \frac{L^2 \pm |L|\sqrt{L^2 - 12M^2}}{2M}, u = 0),$$

the second pair under the additional condition $L^2 \geq 12M^2$. For $L^2 > 12M^2$ all fixed points are hyperbolic, with $(r = r_+(M, |L|), u = 0)$ being a center (purely imaginary eigenvalues) and $(r = r_-(M, |L|), u = 0)$ being a saddle. When $L^2 = 12M^2$, there is a bifurcation in the phase space, which can be visualized in the transition between the second and third plots in Fig. 1. We thus recover easily all well-known results for geodesics in Schwarzschild outside the horizon. The approach here, however, is perfectly regular both across the horizon at $r = 2M$ and even at the singularity $r = 0$. Moreover, it allows us to treat all points in the Kruskal spacetime with a single dynamical system.

4. The Reissner-Nordström dynamical system

Next, we to analyze the Reissner-Nordström dynamical system. The function $h(r)$ is now $h(r) = \frac{2Mr+Q^2}{r^2}$. From Theorems 1 and 4 with the optimal value of $\beta = -2 < \beta_{\text{Schwarzschild}}$ we derive the geodesic equations

$$r' = ru, \tag{10}$$

$$\theta' = L\,r, \tag{11}$$

$$u' = r\left(r\left(L^2 - \mu Mr + \mu Q^2\right) - 3L^2M\right) + 2\left(L^2Q^2 + u^2\right). \tag{12}$$

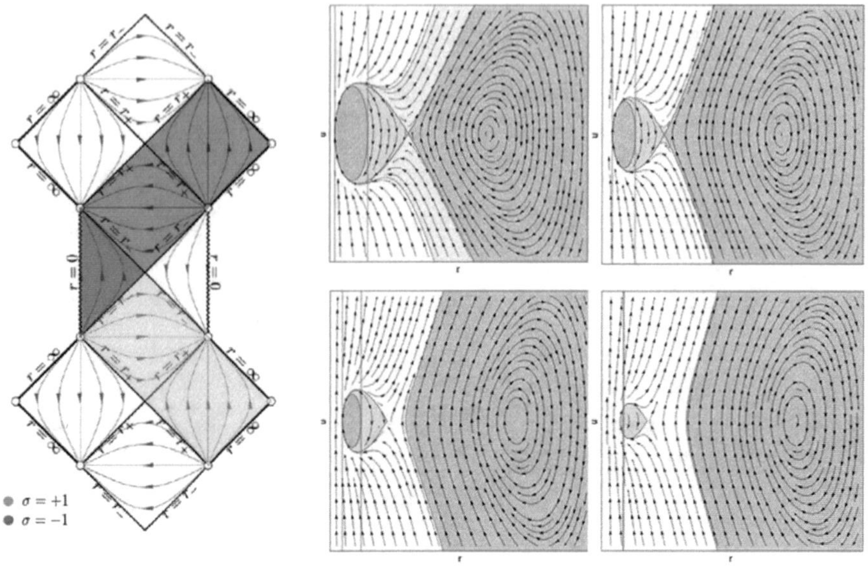

Figure 2: Phase space for timelike particles with fixed value of L increasing value of Q from left to right and top to bottom. The dark zone correspond to the forbidden region and the red line corresponds to the horizons. Different colors are used to distinguish different behaviours for the flow. The left image is the Penrose diagram for the Kruskal spacetime where the Kerr-Schild patches have been highlighted.

Unlike the Schwarzschild collision manifold, the Reissner-Nordström collision manifold is not reachable by timelike geodesics, which reflects the fact that the singularity for charged black holes is repulsive. More details on the approach of causal geodesics to the collision manifold can be found in [2]. Concerning the general flow for timelike geodesics, the phase diagram has now three critical points and the excluded region shows an interesting behaviour: for $0 < |Q| < M$ it detaches from the line $r = 0$ and moves across the phase diagram, diminishing for larger values of Q and vanishing when $|Q| = M$. As geodesics can now encircle the excluded region, the variation ranges of Lemma 2 implies (see [2] for details) that any geodesic travelling from $r = r_0 > r_+ \to r_+ \to r_- \to r_1$ and back to $r_- \to r_+$ must have changed the Kerr-Schild patch along the way (by changing the value of σ).

It is interesting to note that inserting $Q = 0$ in system (10)-(12) the Schwarzschild case is not recovered. This is because the value of the parameter β adapted to Reissner-Nordström is different to that of Schwarzschild. Thus, in the Schwarzschild subcase of Reissner-Nordström we have overkilled the singularity and the fixed points that previously existed at $r = 0$, $u = u_\pm$ have both collapsed to $u = 0$. This collapse can be detected directly on the Reissner-Nordström phase space because the fixed point $\{u = 0, \theta = \theta_0\}$ is no longer hyperbolic when $Q = 0$.

5. References

[1] Belbruno, E., and Pretorius, F. A dynamical system's approach to Schwarzschild null geodesics. *Classical and Quantum Gravity 28*, 19 (2011), 195007.

[2] Galindo, P., and Mars, M. Mcgehee regularization of general so (3)-invariant potentials and applications to stationary and spherically symmetric spacetimes. *arXiv preprint arXiv:1405.7627* (2014).

[3] Kerr, R. P., and Schild, A. A new class of vacuum solutions of the Einstein field equations. *Atti del Congregno Sulla Relativita Generale: Galileo Centenario* (1965).

[4] McGehee, R. Double collisions for a classical particle system with nongravitational interactions. *Commentarii Mathematici Helvetici 56*, 1 (1981), 524–557.

Spanish Relativity Meeting (ERE 2014): almost 100 years after Einstein's revolution IOP Publishing
Journal of Physics: Conference Series **600** (2015) 012064 doi:10.1088/1742-6596/600/1/012064

An upper bound on the number of Killing-Yano tensors

Tsuyoshi Houri
Department of Physics, Kobe University, Kobe, Japan

Abstract. We discuss a simple method for computing an upper bound on the number of Killing-Yano tensors for a given metric. The method is applied to some metrics in four and five dimensions.

1. Introduction

Spacetime symmetry has played an important role in general relativity. When our spacetime admits isometry described by Killing vector fields, if one calculates a physical quantity at a point of spacetime one can obtain the quantity at another point of spacetime without solving equations of motion. Similarly, Killing-Yano symmetry described by Killing-Yano tensors [1] is called hidden symmetry of spacetimes because due to the symmetry spacetimes can possess remarkable properties, e.g., separability of Hamilton-Jacobi for geodesics, Klein-Gordon and Dirac equations. However, it is not always easy to find Killing-Yano tensors for a given metric as we have to partial differential equations obtained from the Killing equations. In this article, we discuss a simple method to find Killing-Yano tensors, as well as Killing vector fields, for a given metric. The detailed calculation can be found in [2].

2. Killing vector fields

Let us consider Killing vector fields on a spacetime $(M, g_{\mu\nu})$ in n dimensions. The equation defining a Killing vector field ξ^μ is given by

$$\nabla_\mu \xi_\nu + \nabla_\nu \xi_\mu = 0\,, \tag{1}$$

where ∇ is the Levi-Civita connection. Introducing a 2-form $L_{\mu\nu} = L_{[\mu\nu]}$, the equation can be decomposed into two equations

$$\nabla_\mu \xi_\nu = L_{\mu\nu}\,, \tag{2}$$

$$L_{\mu\nu} = \nabla_{[\mu}\xi_{\nu]}\,. \tag{3}$$

Covariantly differentiating eq. (1), we also obtain the equation

$$\nabla_\mu L_{\nu\rho} = -R_{\nu\rho\mu}{}^{\sigma}\xi_\sigma\,. \tag{4}$$

The point here is that if we start with eqs. (2) and (4), we obtain eq. (3). Since eqs. (2) and (4) are packaged into the form

$$\mathcal{D}_\mu \begin{pmatrix} \xi_\nu \\ L_{\nu\rho} \end{pmatrix} \equiv \begin{pmatrix} \nabla_\mu \xi_\nu - L_{\mu\nu} \\ \nabla_\mu L_{\nu\rho} + R_{\nu\rho\mu}{}^{\sigma}\xi_\sigma \end{pmatrix} = 0\,, \tag{5}$$

we find that eq. (5) is equivalent to the Killing equation (1). Eq. (5) implies that one can introduce a connection $\mathcal{D}$ on the vector bundle $E^1 \equiv T^*M \oplus \Lambda^2 T^*M$ whose parallel sections $\hat{\xi}_A = (\xi_\mu, L_{\mu\nu})$ are one-to-one corresponding to Killing vector fields,

$$\mathcal{D}_\mu \hat{\xi}_A = 0\,. \tag{6}$$

Since the number of parallel sections is bound by the rank of E^1, the maximum number of Killing vector fields is given by $n(n+1)/2$.

Calculating the curvature of the Killing connection, we obtain some conditions for the parallel sections. Since we have eq. (6), the parallel sections $\hat{\xi}_A$ satisfy the curvature condition

$$\mathcal{R}_{\mu\nu A}{}^{B}\hat{\xi}_B \equiv (\mathcal{D}_\mu \mathcal{D}_\nu - \mathcal{D}_\nu \mathcal{D}_\mu)\hat{\xi}_A = 0\,. \tag{7}$$

Since prallel sections satisfy the curvature condition (7), the number of the parallel sections is bound by the number of solutions of the curvature condition. The condition is no longer partial differential equations but linear algebraic equations for $\hat{\xi}_A$, which enables us to solve the curvature conditions. Thus, we can easily compute an upper bound on the number of Killing vector fields for any metric.

3. Killing-Yano tensors

Let $(M, \boldsymbol{g})$ be an n-dimensional spacetime and ∇ be the Levi-Civita connection. We work in a local orthonormal frame of TM denoted by $\{\boldsymbol{X}_a\}$ and its dual frame of T^*M denoted by $\{\boldsymbol{e}^a\}$. They satisfy $\boldsymbol{X}_a \lrcorner\, \boldsymbol{e}^b = \delta_a^b$ where $\lrcorner$ is the inner product. The Latin indices $a, b, \ldots$ range from 1 to n.

A rank-p KY tensor $\boldsymbol{k}$ is defined as a p-form satisfying the equation

$$\nabla_{\boldsymbol{X}} \boldsymbol{k} = \frac{1}{p+1} \boldsymbol{X} \lrcorner\, d\boldsymbol{k}\,, \tag{8}$$

for any vector field $\boldsymbol{X}$. Covariantly differentiating (8), we obtain

$$\nabla_{\boldsymbol{X}}(d\boldsymbol{k}) = \frac{p+1}{p} R^+(\boldsymbol{X})\boldsymbol{k}\,, \tag{9}$$

where

$$R^+(\boldsymbol{X}) \equiv \boldsymbol{e}^a \wedge R(\boldsymbol{X}, \boldsymbol{X}_a)\,, \tag{10}$$

with the Riemann curvature

$$R(\boldsymbol{X}, \boldsymbol{Y}) \equiv \nabla_{\boldsymbol{X}}\nabla_{\boldsymbol{Y}} - \nabla_{\boldsymbol{Y}}\nabla_{\boldsymbol{X}} - \nabla_{[\boldsymbol{X},\boldsymbol{Y}]}\,. \tag{11}$$

Combining eqs. (8) and (9), one can introduce a connection $\mathcal{D}$ on the vector bundle $E^p(M) \equiv \Lambda^p T^*M \oplus \Lambda^{p+1} T^*M$ [3],

$$\mathcal{D}_{\boldsymbol{X}} \begin{pmatrix} \boldsymbol{\omega} \\ \boldsymbol{\eta} \end{pmatrix} \equiv \left[\nabla_{\boldsymbol{X}} - \begin{pmatrix} 0 & \frac{1}{p+1}\boldsymbol{X}\lrcorner \\ \frac{p+1}{p} R^+(\boldsymbol{X}) & 0 \end{pmatrix} \right] \begin{pmatrix} \boldsymbol{\omega} \\ \boldsymbol{\eta} \end{pmatrix}, \tag{12}$$

where $\boldsymbol{\omega}$ is a p-form and $\boldsymbol{\eta}$ is a $(p+1)$-form. If a section $\hat{\boldsymbol{\omega}} = (\boldsymbol{\omega}, \boldsymbol{\eta})$ of $E^p(M)$ is given by a KY p-form $\boldsymbol{\omega} = \boldsymbol{k}$ and its exterior derivative $\boldsymbol{\eta} = d\boldsymbol{k}$, then it satsifies the parallel equation

$$\mathcal{D}_{\boldsymbol{X}} \hat{\boldsymbol{\omega}} = 0\,. \tag{13}$$

4D metrics	dim $KY^p(M)$ $p=1$	$p=2$	$p=3$
Maximally symmetric space	10	10	5
Plebanski-Demianski	2	0	0
Carter Kerr	2	1	0
Schwarzschild	4	1	0
Friedmann-Lemaître-Robertson-Walker	6	4	1

Table 1. The numbers of rank-p KY tensors on some physical spacetimes in four dimensions.

Conversely, if $\hat{\omega} = (\boldsymbol{\omega}, \boldsymbol{\eta})$ is a parallel section of $E^p(M)$, $\boldsymbol{\omega}$ is a KY p-form and $\boldsymbol{\eta}$ is its exterior derivative, $\boldsymbol{\eta} = d\boldsymbol{\omega}$. It follows that KY p-forms on M are in one-to-one correspondence with parallel sections of $E^p(M)$. Hence, the maximum number of KY p-forms is bound by the rank of $E^p(M)$ [3], which is given by

$$\text{rank } E^p(M) = \begin{pmatrix} n \\ p \end{pmatrix} + \begin{pmatrix} n \\ p+1 \end{pmatrix} = \begin{pmatrix} n+1 \\ p+1 \end{pmatrix} . \tag{14}$$

The equality is attained if a spacetime is maximally symmetric. Note that when we take $p = 1$, eqs. (8) and (9) are equivalent to eqs. (2) and (3). Eqs. (12) and (13) correspond to eqs. (5) and (6), respectively. The maximum number (14) becomes $n(n+1)/2$ for $p = 1$.

We calculate the curvature of the Killing connection (12) by

$$\mathcal{R}(\boldsymbol{X}, \boldsymbol{Y}) \equiv \mathcal{D}_{\boldsymbol{X}}\mathcal{D}_{\boldsymbol{Y}} - \mathcal{D}_{\boldsymbol{Y}}\mathcal{D}_{\boldsymbol{X}} - \mathcal{D}_{[\boldsymbol{X},\boldsymbol{Y}]} , \tag{15}$$

on the vector bundle $E^p(M)$. Then, the Killing curvature is given by

$$\mathcal{R}(\boldsymbol{X}, \boldsymbol{Y}) = \begin{pmatrix} N_{11}(\boldsymbol{X}, \boldsymbol{Y}) & 0 \\ N_{21}(\boldsymbol{X}, \boldsymbol{Y}) & N_{22}(\boldsymbol{X}, \boldsymbol{Y}) \end{pmatrix} . \tag{16}$$

The entries are given by

$$\begin{aligned} N_{11}(\boldsymbol{X}, \boldsymbol{Y}) &= R(\boldsymbol{X}, \boldsymbol{Y}) + \frac{1}{p}\Big\{\boldsymbol{X} \lrcorner R^+(\boldsymbol{Y}) - \boldsymbol{Y} \lrcorner R^+(\boldsymbol{X})\Big\} , \\ N_{21}(\boldsymbol{X}, \boldsymbol{Y}) &= -\frac{p+1}{p}\Big\{(\nabla_{\boldsymbol{X}} R)^+(\boldsymbol{Y}) - (\nabla_{\boldsymbol{Y}} R)^+(\boldsymbol{X})\Big\} , \\ N_{22}(\boldsymbol{X}, \boldsymbol{Y}) &= R(\boldsymbol{X}, \boldsymbol{Y}) + \frac{1}{p}\Big\{R^+(\boldsymbol{X})(\boldsymbol{Y} \lrcorner \bullet) - R^+(\boldsymbol{Y})(\boldsymbol{X} \lrcorner \bullet)\Big\} , \end{aligned} \tag{17}$$

where

$$(\nabla_{\boldsymbol{X}} R)^+(\boldsymbol{Y}) = \boldsymbol{e}^a \wedge (\nabla_{\boldsymbol{X}} R)(\boldsymbol{Y}, \boldsymbol{X}_a) . \tag{18}$$

	dim $KY^p(M)$			
5D metrics	$p = 1$	$p = 2$	$p = 3$	$p = 4$
Maximally symmetric space	15	20	15	6
Myers-Perry	3	0	1	0
Emparan-Reall	3	0	0	0
Kerr string	3	1	0	1

Table 2. The numbers of rank-p KY tensors for Myers-Perry, Emaparan-Reall and Kerr string metrics in five dimensions.

4. Applications

By use of the method, we can compute the upper bounds on the number of Killing-Yano tensors (including Killing vector fields as rank-1 Killing-Yano tensors) for various metrics. The point is that since the curvature conditions are linear algebraic equations, one can easily solve them for any metic. Another point is that the method can be applied to any metric in any dimensions written in any coordinates. Furthermore, by use of the solutions of the curvature conditions, it is possible to integrate the Killing equations. As a result, we can obtain the precise number of Killing-Yano tensors for a given metric. The results for various metrics in four and five dimensions are listed in Tables 1 & 2.

5. Summary

This article has illustrated an idea for counting the number of Killing-Yano tensors. Using the fact that rank-p Killing-Yano tensors are in one-to-one correspondence with parallel sections of the vector bundle $E^p \equiv \Lambda^p T^*M \oplus \Lambda^{p+1} T^*M$, we have seen that the maximum number of rank-p Killing-Yano tensors is given by the rank of E^p. We have also obtained curvature conditions. Using the curvature conditions, we have actually computed the precise numbers of Killing-Yano tensors for various metrics in four and five dimensions.

One possible extension is to see curvature conditions for conformal Killing-Yano tensors. It was shown [3] that one can introduce a Killing connection on the vector bundle $E^p(M) \equiv \Lambda^p T^*M \oplus \Lambda^{p+1} T^*M \oplus \Lambda^{p-1} T^*M \oplus \Lambda^p T^*M$ whose prallel sections are one-to-one corresponding to rank-p conformal Killing-Yano tensors. Similar to this article, one could calculate the curvature of the Killing connection.

Finally, I would like to draw the reader's attention to a package of the computational software, *Mathe atica*, which automatically solves the curvature conditions and returns the upper bound on the number of Killing-Yano tensors for a given metric [4].

Acknowledgments

I would like to thank the organisers for giving me an opportunity of the contribution to the Spanish Relativity Meeting 2014 (ERE2014). This work was supported by the JSPS Grant-in-Aid for Scientific Research No. 26·1237.

References

[1] Yano K 1952 *Ann. Math.* **55** 328-347
[2] Houri T and Yasui Y 2014 arXiv:1410.1023
[3] Semmelmann U 2002 math/0206117
[4] The package is available at the URL, http://www.research.kobe-u.ac.jp/fsci-pacos/KY_upperbound/

Spanish Relativity Meeting (ERE 2014): almost 100 years after Einstein's revolution IOP Publishing
Journal of Physics: Conference Series **600** (2015) 012065 doi:10.1088/1742-6596/600/1/012065

Type N universal spacetimes

S. Hervik[1], **V. Pravda**[2], **A. Pravdová**[2]

[1] Faculty of Science and Technology, University of Stavanger, N-4036 Stavanger, Norway;
[2] Institute of Mathematics, Academy of Sciences of the Czech Republic
Žitná 25, 115 67 Prague 1, Czech Republic

E-mail: sigbjorn.hervik@uis.no, pravda@math.cas.cz, pravdova@math.cas.cz

Abstract. Universal spacetimes are vacuum solutions to all theories of gravity with the Lagrangian $L = L(g_{ab}, R_{abcd}, \nabla_{a_1} R_{bcde}, \dots, \nabla_{a_1 \dots a_p} R_{bcde})$. Well known examples of universal spacetimes are plane waves which are of the Weyl type N. Here, we discuss recent results on necessary and sufficient conditions for all Weyl type N spacetimes in arbitrary dimension and we conclude that a type N spacetime is universal if and only if it is an Einstein Kundt spacetime. We also summarize the main points of the proof of this result.

1. Introduction

Due to the diffeomorphism invariance, correction terms in perturbative quantum gravity which are added to the Einstein action consist of curvature invariants constructed from the Riemann tensor and its covariant derivatives. The resulting modified field equations are in general very complicated. In the case where only correction terms quadratic in the Riemann tensor are added to the Einstein-Hilbert action - the so called quadratic gravity,

$$S = \int \mathrm{d}^n x \sqrt{-g} \left(\frac{1}{\kappa} (R - 2\Lambda_0) + \alpha R^2 + \beta R_{ab}^2 + \gamma \left(R_{abcd}^2 - 4R_{ab}^2 + R^2 \right) \right), \tag{1}$$

the field equations read [1]

$$\begin{aligned} &\frac{1}{\kappa}\left(R_{ab} - \frac{1}{2}Rg_{ab} + \Lambda_0 g_{ab}\right) + 2\alpha R\left(R_{ab} - \frac{1}{4}Rg_{ab}\right) + (2\alpha + \beta)\left(g_{ab}\Box - \nabla_a\nabla_b\right)R \\ &+2\gamma\left(RR_{ab} - 2R_{acbd}R^{cd} + R_{acde}R_b{}^{cde} - 2R_{ac}R_b{}^c - \frac{1}{4}g_{ab}\left(R_{cdef}^2 - 4R_{cd}^2 + R^2\right)\right) \\ &+\beta\Box\left(R_{ab} - \frac{1}{2}Rg_{ab}\right) + 2\beta\left(R_{acbd} - \frac{1}{4}g_{ab}R_{cd}\right)R^{cd} = 0. \end{aligned} \tag{2}$$

Interestingly, there exist vacuum solutions of Einstein's gravity (with possibly non-zero cosmological constant) that are "immune" to these corrections, i.e they are vacuum solutions to both theories, the Einstein gravity and the quadratic gravity.

For instance, it has been shown in [2] that for type N[1] Einstein spacetimes, all terms in (2) are proportional to the metric and thus in arbitrary dimension *all Weyl type N Einstein*

[1] Type N spacetimes in the algebraic classification of tensors introduced in [3], recently reviewed in [4] and briefly discussed in section 2.

Journal of Physics: Conference Series **600** (2015) 012065 doi:10.1088/1742-6596/600/1/012065

spacetimes with appropriately chosen effective cosmological constant Λ *are exact solutions of quadratic gravity* (2).

It is a natural question to ask whether some of the type N Einstein spacetimes solve also more general theories than (2). Here, we will be interested in the so called universal spacetimes first introduced in [5].

Definition 1.1. A metric is called *universal* if all conserved symmetric rank-2 tensors constructed from the metric, the Riemann tensor and its covariant derivatives of arbitrary order[2] are multiples of the metric.

Vacuum field equations of modified gravities obtained by varying a diffeomorphism invariant Lagrangian with respect to the metric are conserved, rank-2, and symmetric and therefore universal spacetimes solve vacuum equations of all theories with the Lagrangian being a polynomial curvature invariant in the form

$$L = L(g_{ab}, R_{abcd}, \nabla_{a_1} R_{bcde}, \dots, \nabla_{a_1 \dots a_p} R_{bcde}). \tag{3}$$

Obviously, all universal spacetimes are Einstein (i.e. $R_{ab} = (R/n) g_{ab}$), where n is the dimension of the spacetime. So far, necessary and sufficient geometrical conditions for universality are unknown. However, when considering only type N spacetimes, such conditions can be found[3].

Theorem 1.2. *A type N spacetime is universal if and only if it is an Einstein Kundt spacetime.*

Note that special examples of Ricci-flat type N Kundt spacetimes are plane waves that were identified as vacuum solutions of all theories given by (3) already in [8] and [9].

So far, we did not properly define type N spacetimes in arbitrary dimension and Kundt spacetimes. These terms together with other basic definitions will be briefly reviewed in the next section. In section 3, we outline the key points of the necessity and sufficiency parts of the proof of the theorem 1.2 and in section 4, we give explicit examples of type N universal metrics.

2. Preliminaries

We employ the algebraic classification of the Weyl tensor [3] and higher dimensional generalizations of the Newman-Penrose [10, 11] and the Geroch-Held-Penrose formalisms [12]. We will follow the notation of [4, 12]. Here, let us give basic definitions just very briefly and refer to [4] for more information.

We will work in a null frame with two null vectors $\boldsymbol{\ell}$ and $\boldsymbol{n}$ and $n-2$ spacelike vectors $\boldsymbol{m}^{(i)}$ obeying

$$\ell^a \ell_a = n^a n_a = 0, \qquad \ell^a n_a = 1, \qquad m^{(i)a} m^{(j)}_a = \delta_{ij}, \tag{4}$$

where coordinate indices $a, b, \dots$ and frame indices $i, j, \dots$ take values from 0 to $n-1$ and 2 to $n-1$, respectively.

We say that a quantity q has a boost weight b if it transforms as

$$\hat{q} = \lambda^{\mathrm{b}} q \tag{5}$$

under boosts

$$\hat{\boldsymbol{\ell}} = \lambda \boldsymbol{\ell}, \qquad \hat{\boldsymbol{n}} = \lambda^{-1} \boldsymbol{n}, \qquad \hat{\boldsymbol{m}}^{(i)} = \boldsymbol{m}^{(i)}. \tag{6}$$

[2] We consider only symmetric rank-2 tensors constructed as contractions of *polynomials* from the metric, the Riemann tensor and its covariant derivatives of arbitrary order, however, most of our results hold also for analytic functions of such invariants.

[3] The necessary conditions for universality were found in [6], sufficient conditions were also first proven in [6], however, without a proof the result was stated already in [5] and brief general comments about some points of the proof were also included by one of us in v2 of [7].

By definition [3, 4], type N spacetimes are spacetimes for which the Weyl tensor (in an appropriately chosen frame (4)) admits only components of boost weight -2 and thus can be expressed as

$$C_{abcd} = 4\Omega'_{ij}\, \ell_{\{a} m^{(i)}_{b} \ell_{c} m^{(j)}_{d\}}, \tag{7}$$

where Ω'_{ij} is symmetric and traceless and for an arbitrary tensor T_{abcd}

$$T_{\{abcd\}} \equiv \frac{1}{2}(T_{[ab][cd]} + T_{[cd][ab]}), \tag{8}$$

so that $C_{abcd} = C_{\{abcd\}}$.

It can be shown [10] that for type N Einstein spacetimes, the multiple WAND[4] $\boldsymbol{\ell}$ is geodetic. If we choose an affine parameterization we can express the covariant derivative of $\boldsymbol{\ell}$ as

$$\ell_{a;b} = L_{11}\ell_a\ell_b + L_{1i}\ell_a m^{(i)}_b + \tau_i m^{(i)}_a \ell_b + \rho_{ij} m^{(i)}_a m^{(j)}_b. \tag{9}$$

Optical scalars of $\boldsymbol{\ell}$, shear σ^2, expansion θ and twist ω^2 can be expressed as

$$\sigma^2 = \ell_{(a;b)}\ell^{(a;b)} - \tfrac{1}{n-2}\left(\ell^a{}_{;a}\right)^2, \qquad \theta = \tfrac{1}{n-2}\ell^a{}_{;a}, \qquad \omega^2 = \ell_{[a;b]}\ell^{a;b}. \tag{10}$$

Now, we are ready to define Kundt spacetimes.

Definition 2.1. *Kundt* spacetimes are spacetimes admitting a null geodetic congruence $\boldsymbol{\ell}$ with vanishing shear, expansion and twist.

Kundt metrics in higher dimensions were introduced in [13, 14].

3. Main points of the proof of theorem 1.2

Let us briefly mention the main points of the proof [6] of the theorem 1.2.

3.1. Sufficiency

First, let us discuss the proof of the sufficiency part of the theorem 1.2 , i.e. the proof of the statement that *all Einstein type N Kundt spacetimes are universal.* Thus, we want to show that in this case, all rank-2 tensors constructed from the Weyl[5] tensor and its covariant derivatives are proportional to the metric (in fact, they vanish).

For rank-2 tensors constructed from the Weyl tensor only (without covariant derivatives), the proof is very simple. Any rank-2 tensor has only terms of boost weight $\geq$ -2 and the type N Weyl tensor admits only boost weight -2 terms. Therefore, all rank-2 tensors constructed from the type N Weyl tensor which are quadratic or of a higher order in the Weyl tensor vanish and, due to the tracelessness of the Weyl tensor, rank-2 tensors linear in the Weyl tensor vanish as well. Thus, it is not possible to construct a non-vanishing rank-2 tensor from the type N Weyl tensor.

For covariant derivatives of the Weyl tensor, the proof is more involved. The key point is

Proposition 3.1. *For type N Einstein Kundt spacetimes, the boost order of* $\nabla^{(k)}C$ *(a covariant derivative of an arbitrary order of the Weyl tensor) with respect to the multiple WAND is at most* -2.

[4] Weyl aligned null direction [3].

[5] Obviously, for Einstein spacetimes, the Ricci tensor is proportional to the metric and its covariant derivatives vanish.

The proof of the above proposition [6] using balanced scalar approach introduced in [15] is rather technical, and it relies on the special form of the Bianchi and Ricci identities for this class of spacetimes.

A direct consequence is

Lemma 3.2. *For type N Einstein Kundt spacetimes, rank-2 tensors constructed from $\nabla^{(k)}C$, which are quadratic or of higher order in $\nabla^{(k)}C$, vanish.*

Using the expression for the commutator of covariant derivatives, the above results and the Bianchi identities, one can generalize the above lemma also to the case of rank-2 tensors constructed from $\nabla^{(k)}C$, which are linear in $\nabla^{(k)}C$ (see [6]). This completes the sufficient part of the proof of the theorem 1.2.

3.2. Necessity

The proof of the necessity part of the theorem 1.2, i.e. the statement that all type N universal spacetimes are Einstein and Kundt is based on another result of [6] that will be discussed in more detail elsewhere in this volume

Theorem 3.3. *A universal spacetime is necessarily a CSI spacetime.*

CSI (constant curvature invariant) spacetimes are spacetimes for which all curvature invariants constructed from the metric, the Riemann tensor and its derivatives of arbitrary order are constant, see e.g. [14].

Let us study the simplest non-trivial curvature invariant for type N spacetimes [16]

$$I_N \equiv C^{a_1 b_1 a_2 b_2 ; c_1 c_2} C_{a_1 d_1 a_2 d_2 ; c_1 c_2} C^{e_1 d_1 e_2 d_2 ; f_1 f_2} C_{e_1 b_1 e_2 b_2 ; f_1 f_2}. \tag{11}$$

In terms of higher dimensional GHP quantities, it can be shown [15] that I_N is proportional (via a numerical constant) to

$$I_N \propto \left[(\Omega'_{22})^2 + (\Omega'_{23})^2\right]^2 (S^2 + A^2)^4, \tag{12}$$

where S and A are closely related to the optical scalars (see [6]). The invariant above is non-constant unless the type N Einstein spacetime is Kundt [15] and thus, in this class of spacetimes, only Kundt spacetimes are CSI.

From theorem 3.3, it follows that type N universal spacetimes are Kundt.

4. Explicit examples of universal type N Kundt metrics

By theorem 1.2, all type N Einstein Kundt metrics are universal. In four dimensions, all type N Einstein Kundt metrics can be expressed as [17]

$$\mathrm{d}s^2 = 2\frac{Q^2}{P^2}\mathrm{d}u\mathrm{d}v + \left(2k\frac{Q^2}{P^2}v^2 + \frac{(Q^2)_{,u}}{P^2}v - \frac{Q}{P}H\right)\mathrm{d}u^2 + \frac{1}{P^2}\left(\mathrm{d}x^2 + \mathrm{d}y^2\right), \tag{13}$$

where

$$P = 1 + \frac{\Lambda}{12}(x^2+y^2), \quad k = \frac{\Lambda}{6}\alpha(u)^2 + \frac{1}{2}\left(\beta(u)^2 + \gamma(u)^2\right),$$
$$Q = \left(1 - \frac{\Lambda}{12}(x^2+y^2)\right)\alpha(u) + \beta(u)x + \gamma(u)y, \quad H = 2f_{1,x} - \frac{\Lambda}{3P}(xf_1 + yf_2),$$

where $\alpha(u), \beta(u), \gamma(u)$ are free functions (see [18] for the canonical forms) and $f_1 = f_1(u,x,y)$ and $f_2 = f_2(u,x,y)$ obey $f_{1,x} = f_{2,y}$, $f_{1,y} = -f_{2,x}$.

Higher dimensional examples of type N universal metrics can be obtained by warping (13). Another higher-dimensional example is (A)dS-wave [6]

$$\mathrm{d}s^2 = e^{-pw}\left(2\mathrm{d}u\mathrm{d}v + H(u,w,x^M)\mathrm{d}u^2 + \delta_{MN}\mathrm{d}x^M\mathrm{d}x^N\right) + \mathrm{d}w^2, \tag{14}$$

with p being a constant and H obeying $H_{,KL}\,\delta^{KL} + \left(H_{,ww} - \frac{n-1}{2}pH_{,w}\right)\mathrm{e}^{-pw} = 0$. Further explicit examples can be found in [6].

Acknowledgments
V.P. and A.P. acknowledge support from research plan RVO: 67985840 and research grant GAČR 13-10042S.

References

[1] I. Gullu and B. Tekin. Massive higher derivative gravity in D-dimensional anti-de Sitter spacetimes. *Phys. Rev.*, D80:064033, 2009.
[2] T. Málek and V. Pravda. Type III and N solutions to quadratic gravity. *Phys. Rev. D*, 84:024047, 2011.
[3] A. Coley, R. Milson, V. Pravda, and A. Pravdová. Classification of the Weyl tensor in higher dimensions. *Class. Quantum Grav.*, 21:L35–L41, 2004.
[4] M. Ortaggio, V. Pravda, and A. Pravdová. Algebraic classification of higher dimensional spacetimes based on null alignment. *Class. Quantum Grav.*, 30:013001, 2013.
[5] A.A. Coley, G.W. Gibbons, S. Hervik, and C.N. Pope. Metrics with vanishing quantum corrections. *Class. Quantum Grav.*, 25:145017, 2008.
[6] S. Hervik, V. Pravda and A. Pravdová, Type III and N universal spacetimes. *Class. Quantum Grav.*, 31:215005, 2014.
[7] M. Gurses, T. C. Sisman, B. Tekin and S. Hervik, AdS-Wave Solutions of f(Riemann) Theories. Phys. Rev. Lett. 111:101101, 2013 [arXiv:1305.1565v2].
[8] D. Amati and C. Klimčik. Nonperturbative computation of the Weyl anomaly for a class of nontrivial backgrounds. *Phys. Lett.* B, 219:443–447, 1989.
[9] G. T. Horowitz and A. R. Steif. Spacetime singularities in string theory. *Phys. Rev. Lett.*, 64:260–263, 1990.
[10] V. Pravda, A. Pravdová, A. Coley, and R. Milson. Bianchi identities in higher dimensions. *Class. Quantum Grav.*, 21:2873–2897, 2004.
[11] M. Ortaggio, V. Pravda, and A. Pravdová. Ricci identities in higher dimensions. *Class. Quantum Grav.*, 24:1657–1664, 2007.
[12] M. Durkee, V. Pravda, A. Pravdová, and H. S. Reall. Generalization of the Geroch-Held-Penrose formalism to higher dimensions. *Class. Quantum Grav.*, 27:215010, 2010.
[13] A. Coley, R. Milson, N. Pelavas, V. Pravda, A. Pravdová, and R. Zalaletdinov. Generalizations of *pp*-wave spacetimes in higher dimensions. *Phys. Rev.* D, 67:104020, 2003.
[14] A. Coley, S. Hervik, and N. Pelavas. On spacetimes with constant scalar invariants. *Class. Quantum Grav.*, 23:3053–3074, 2006.
[15] A. Coley, R. Milson, V. Pravda, and A. Pravdová. Vanishing scalar invariant spacetimes in higher dimensions. *Class. Quantum Grav.*, 21:5519–5542, 2004.
[16] J. Bičák and V. Pravda. Curvature invariants in type-N spacetimes. *Class. Quantum Grav.*, 15:1539–1555, 1998.
[17] I. Ozsváth, I. Robinson and K. Rózga. Plane-fronted gravitational and electromagnetic waves in spaces with cosmological constant. *J. Math. Phys.*, 26:1755-1761, 1985.
[18] J. B. Griffiths and J. Podolský. Exact Space-Times in Einstein's General Relativity. Cambridge University Press, Cambridge, 2009.

Type III and II universal spacetimes

S. Hervik[1], **V. Pravda**[2], **A. Pravdová**[2]

[1] Faculty of Science and Technology, University of Stavanger, N-4036 Stavanger, Norway;
[2] Institute of Mathematics, Academy of Sciences of the Czech Republic
Žitná 25, 115 67 Prague 1, Czech Republic

E-mail: sigbjorn.hervik@uis.no, pravda@math.cas.cz, pravdova@math.cas.cz

Abstract. We briefly summarize our recent results on universal spacetimes. We show that universal spacetimes are necessarily CSI, i.e. for these spacetimes, all curvature invariants constructed from the Riemann tensor and its derivatives are constant. Then, we focus on type III universal spacetimes and discuss a proof of universality for a class of type III Kundt spacetimes. We also mention explicit examples of type III and II universal spacetimes.

1. Introduction

In the contribution [1] in this volume, we have introduced universal spacetimes obeying the following definition [2]

Definition 1.1. A metric is called *universal* if all conserved symmetric rank-2 tensors constructed from the metric, the Riemann tensor and its covariant derivatives of arbitrary order are multiples of the metric.

We have argued that universal metrics solve the vacuum equations of *all* theories of gravity with the Lagrangian of the form

$$L = L(g_{ab}, R_{abcd}, \nabla_{a_1} R_{bcde}, \dots, \nabla_{a_1 \dots a_p} R_{bcde}). \tag{1}$$

In general, it seems too difficult to study universal spacetimes in arbitrary spacetime dimension in full generality. However, employing the algebraic classification of the Weyl tensor [3] (see also [4] for a recent review) leads, in particular for types N and III, to considerable simplification of the problem and allows us to prove universality for various classes of metrics.

In [1], we summarized the main results of [5] for type N universal spacetimes. Here, let us briefly present further results on universal spacetimes (for details see [5]).

First, in section 2, we describe the main points of the proof of the following theorem.

Theorem 1.2. *A universal spacetime is necessarily a CSI spacetime.*

CSI (constant curvature invariant) spacetimes are spacetimes that have all curvature invariants constructed from the metric, the Riemann tensor and its derivatives of arbitrary order constant, see e.g. [6].

Then, we discuss type III spacetimes. Although we expect that type III universal spacetimes are necessarily Kundt, in contrast to the type N case, we cannot use Theorem 1.2 to prove this statement in full generality and in [5], we provide a proof only in the "generic" case. Thus, in section 3, we focus on the sufficient conditions for universality. We identify a universal subclass of type III Kundt metrics[1]

Theorem 1.3. *Type III, $\tau_i = 0$ Einstein Kundt spacetimes obeying*

$$C_{acde}C_b{}^{cde} = 0 \tag{2}$$

are universal.

Finally, in section 4, we present explicit examples of type III and type II universal metrics.

2. Universal spacetimes are CSI

Let us here briefly summarize the main points of the proof of the theorem 1.2.

Let us consider the Lagrangian (1) containing covariant derivatives of the Riemann tensor up to a fixed p. In [7], it has been shown that by varying the action, one arrives at the field equations

$$-T^{ab} = \frac{\partial L}{\partial g_{ab}} + E^a{}_{cde}R^{bcde} + 2\nabla_c\nabla_d E^{acdb} + \frac{1}{2}g^{ab}L, \tag{3}$$
$$E^{bcde} = \frac{\partial L}{\partial R_{bcde}} - \nabla_{a_1}\frac{\partial L}{\partial \nabla_{a_1}R_{bcde}} + \cdots + (-1)^p \nabla_{(a_1}\cdots\nabla_{a_p)}\frac{\partial L}{\partial \nabla_{(a_1}\cdots\nabla_{a_p)}R_{bcde}},$$

where T^{ab} is the associated conserved tensor.

Now, let us consider any polynomial invariant I and let us assume that it contains derivatives of the Riemann tensor of orders at most p. By [7], we can assume it is of the form

$$I = I[g_{ab}, R_{abcd}, \nabla_{a_1}R_{bcde}, ..., \nabla_{(a_1...a_p)}R_{bcde}].$$

Let us consider the (infinite) series of Lagrangians $L = I^n$, $n = 1, 2, 3, ...$ By variation, we get a conserved tensor $T[n]^a{}_b$ for each n. For universal spacetimes, traces of $T[n]^a{}_b$ are constant. By studying the explicit forms of the corresponding expressions [5], one arrives at a conclusion that I has to be a constant and since I is an arbitrary curvature invariant it follows that universal spacetimes are CSI.

3. Type III universal spacetimes

Type III spacetimes by definition [3,4] admit a multiple Weyl aligned null direction, mWAND, $\boldsymbol{\ell}$. Let us complete a frame with another null vector $\boldsymbol{n}$ and $n-2$ spacelike orthonormal vectors $\boldsymbol{m}^{(i)}$ with the only non-vanishing products being $\ell^a n_a = 1$, $m^{(i)a}m^{(j)}_a = \delta_{ij}$ (coordinate indices $a, b, \ldots = 0 \ldots n-1$, frame indices $i, j, \ldots = 2 \ldots n-1$).

In an appropriately chosen frame, the Weyl tensor for type III spacetime reads [3,4]

$$C_{abcd} = 8\Psi'_i\,\ell_{\{a}n_b\ell_c m^{(i)}_{\ d\}} + 4\Psi'_{ijk}\,\ell_{\{a}m^{(i)}_{\ b}m^{(j)}_{\ c}m^{(k)}_{\ d\}} + 4\Omega'_{ij}\,\ell_{\{a}m^{(i)}_{\ b}\ell_c m^{(j)}_{\ d\}}, \tag{4}$$

where the frame components satisfy $\Psi'_i = \Psi'_{jij}$, $\Omega'_{ij} = \Omega'_{ji}$, and $\Omega'_{ii} = 0$ and for an arbitrary tensor T_{abcd}

$$T_{\{abcd\}} \equiv \frac{1}{2}(T_{[ab][cd]} + T_{[cd][ab]}). \tag{5}$$

As discussed in [5], for type III Einstein spacetimes, the symmetric rank-2 tensor

$$S_{ab} \equiv C_{acde}C_b{}^{cde} \tag{6}$$

is conserved. While for type N, S_{ab} vanishes identically, for type III Einstein spacetimes, it is in general a boost weight -2 tensor proportional to $\ell_a\ell_b$. Therefore, for type III universal spacetimes, we have an additional necessary condition (2).

[1] However, note that other type III Einstein Kundt spacetimes may also exist.

Journal of Physics: Conference Series **600** (2015) 012066 doi:10.1088/1742-6596/600/1/012066

3.1. Main points of the proof of the theorem 1.3

The key intermediate result proven in [5] is that

Proposition 3.1. *For type III Einstein Kundt spacetimes, the boost order of* $\nabla^{(k)}C$ *(a covariant derivative of an arbitrary order of the Weyl tensor) with respect to the multiple WAND is at most* -1.

Proof of this result relies on the precise form of various Bianchi and Ricci identities [8, 9].

A direct consequence of proposition 3.1 is that

Lemma 3.2. *For type III Ricci-flat Kundt spacetimes, a non-vanishing rank-2 tensor constructed from the metric, the Weyl tensor and its covariant derivatives of arbitrary order is at most quadratic in the Weyl tensor and its covariant derivatives.*

We find that for type III Ricci-flat Kundt spacetimes, for which the FKWC basis [10] of rank-2, order-6 Weyl polynomials consists of

$$F_1 \equiv C^{pqrs}C_{pqrs;ab}, \quad F_2 \equiv C^{pqrs}{}_{;a}C_{pqrs;b}, \quad F_3 \equiv C^{pqr}{}_{a;s}C_{pqrb}{}^{;s}, \tag{7}$$

F_1 and F_2 are in general non-vanishing. In our case, F_1 and F_2 are conserved and thus, in general, type III Ricci-flat Kundt spacetimes are not universal. However, both F_1 and F_2 vanish for $\tau_i = 0$. In this case, a rather technical proof [5] allows us to arrive at the theorem 1.3.

4. Kundt spacetimes

Since all universal spacetimes we have discussed belong to the Kundt class, let us briefly discuss Kundt metrics. Kundt spacetimes are spacetimes admitting a null geodetic vector field $\boldsymbol{\ell}$ with vanishing shear, expansion and twist. In appropriately chosen coordinates,

$$\ell_{a;b} = L_{11}\ell_a\ell_b + \tau_i(\ell_a m^{(i)}_b + m^{(i)}_a\ell_b) \tag{8}$$

and the metric reads [6, 11]

$$\mathrm{d}s^2 = 2\mathrm{d}u\left[\mathrm{d}r + H(u,r,x^\gamma)\mathrm{d}u + W_\alpha(u,r,x^\gamma)\mathrm{d}x^\alpha\right] + g_{\alpha\beta}(u,x^\gamma)\mathrm{d}x^\alpha\mathrm{d}x^\beta, \tag{9}$$

with $\alpha, \beta, \gamma = 2 \ldots n-1$.

Since universality implies CSI (theorem 1.2) we restrict ourselves to the *Kundt CSI metrics*, where [6, 12]

$$\begin{aligned} W_\alpha(u,r,x^\gamma) &= rW^{(1)}_\alpha(u,x^\gamma) + W^{(0)}_\alpha(u,x^\gamma), \\ H(u,r,x^\gamma) &= \frac{r^2}{8}\left(a + W^{(1)}_\alpha W^{(1)\alpha}\right) + rH^{(1)}(u,x^\gamma) + H^{(0)}(u,x^\gamma), \end{aligned} \tag{10}$$

$g_{\alpha\beta}(x^\gamma)$ (note that $g_{\alpha\beta,u} = 0$) is the metric of a locally homogeneous space and a is a constant. Note that (10) are necessary but not sufficient conditions for Kundt CSI.

4.1. type III

Note also that for type III and $\tau_i = 0$, the Bianchi identities imply $\Lambda = 0$ and thus type III, $\tau_i = 0$, Kundt universal spacetimes are in fact Ricci-flat and VSI (vanishing scalar invariants) with a metric of the form [13, 14]

$$\mathrm{d}s^2 = 2\mathrm{d}u\left[\mathrm{d}r + H(u,r,x^\gamma)\mathrm{d}u + W_\alpha(u,r,x^\gamma)\mathrm{d}x^\alpha\right] + \delta_{\alpha\beta}\mathrm{d}x^\alpha\mathrm{d}x^\beta, \tag{11}$$

with

$$W_2 = 0, \tag{12}$$
$$W_M(u,r,x^\gamma) = W_M^{(0)}(u,x^\gamma), \tag{13}$$
$$H(u,r,x^\gamma) = rH^{(1)}(u,x^\gamma) + H^{(0)}(u,x^\gamma). \tag{14}$$

Since further constraints on H and W_M follow from the Einstein equations we conclude with an explicit example of a four-dimensional type III, $\tau_i = 0$ Ricci-flat Kundt universal metric (expressed in other coordinates) [15]

$$\mathrm{d}s^2 = 2\mathrm{d}u\mathrm{d}v - x(v+\mathrm{e}^x)\mathrm{d}u^2 + \mathrm{e}^x(\mathrm{d}x^2 + \mathrm{e}^{-2u}\mathrm{d}y^2). \tag{15}$$

4.2. type II

An example of a four-dimensional type II universal Kundt spacetime

$$\mathrm{d}s^2 = 2\mathrm{d}u\mathrm{d}v + \left(-v^2\lambda + H(u,x,y)\right)\mathrm{d}u^2 + \frac{1}{\lambda}(\mathrm{d}x^2 + \sinh^2 x\mathrm{d}y^2), \quad \Box H = 0 \tag{16}$$

has been given already in [2]. As we will discuss elsewhere, for some type II classes of Kundt spacetimes, universality depends on dimension of the spacetime.

Acknowledgments

The authors acknowledge support from research plan RVO: 67985840 and research grant GAČR 13-10042S.

References

[1] S. Hervik, V. Pravda, and A. Pravdová. Type N universal spacetimes. Proceedings of the 2014 Spanish Relativity Meeting, to appear in Journal of Physics: Conference Series

[2] A.A. Coley, G.W. Gibbons, S. Hervik, and C.N. Pope. Metrics with vanishing quantum corrections. *Class. Quantum Grav.*, 25:145017, 2008.

[3] A. Coley, R. Milson, V. Pravda, and A. Pravdová. Classification of the Weyl tensor in higher dimensions. *Class. Quantum Grav.*, 21:L35–L41, 2004.

[4] M. Ortaggio, V. Pravda, and A. Pravdová. Algebraic classification of higher dimensional spacetimes based on null alignment. *Class. Quantum Grav.*, 30:013001, 2013.

[5] S. Hervik, V. Pravda and A. Pravdova, Type III and N universal spacetimes. *Class. Quantum Grav.*, 31:215005, 2014.

[6] A. Coley, S. Hervik, and N. Pelavas. On spacetimes with constant scalar invariants. *Class. Quantum Grav.*, 23:3053–3074, 2006.

[7] V. Iyer and R. M. Wald, Some Properties of Noether Charge and a Proposal for Dynamical Black Hole Entropy *Phys. Rev.* D, 50: 846-864 , 1994.

[8] M. Ortaggio, V. Pravda, and A. Pravdová. Ricci identities in higher dimensions. *Class. Quantum Grav.*, 24:1657–1664, 2007.

[9] M. Durkee, V. Pravda, A. Pravdová, and H. S. Reall. Generalization of the Geroch-Held-Penrose formalism to higher dimensions. *Class. Quantum Grav.*, 27:215010, 2010.

[10] S. A. Fulling, R. C. King, B.G. Wybourne, and C.J. Cummins. Normal forms for tensor polynomials. 1: The Riemann tensor. *Class. Quantum Grav.*, 9:1151–1197, 1992.

[11] A. Coley, R. Milson, N. Pelavas, V. Pravda, A. Pravdová, and R. Zalaletdinov. Generalizations of *pp*–wave spacetimes in higher dimensions. *Phys. Rev.* D, 67:104020, 2003.

[12] A. Coley, S. Hervik, G. O. Papadopoulos, and N. Pelavas. Kundt spacetimes. *Class. Quantum Grav.*, 26:105016, 2009.

[13] A. Coley, R. Milson, V. Pravda, and A. Pravdová. Vanishing scalar invariant spacetimes in higher dimensions. *Class. Quantum Grav.*, 21:5519–5542, 2004.

[14] A. Coley, A. Fuster, S. Hervik, and N. Pelavas. Higher dimensional VSI spacetimes. *Class. Quantum Grav.*, 23:7431–7444, 2006.

[15] A. Z. Petrov. Gravitational field geometry as the geometry of automorphisms. In *Recent Developments in General Relativity*, pages 379–386. Pergamon Press/PWN, Oxford/Warszawa, 1962.

Journal of Physics: Conference Series **600** (2015) 012067 doi:10.1088/1742-6596/600/1/012067

Vacuum Spacetimes with Constant Weyl Eigenvalues

A Barnes

School of Engineering & Applied Science, Aston Umiversity, Birmingham B4 7ET, UK (now retired)

E-mail: Alan.Barnes45678@gmail.com

Abstract. Einstein spacetimes (that is vacuum spacetimes possibly with a non-zero cosmological constant Λ) with constant non-zero Weyl eigenvalues are considered. For type Petrov II & D this assumption allows one to prove that the non-repeated eigenvalue necessarily has the value $2\Lambda/3$ and it turns out that the only possible spacetimes are some Kundt-waves considered by Lewandowski which are type II and a Robinson-Bertotti solution of type D.

For Petrov type I the only solution turns out to be a homogeneous pure vacuum solution found long ago by Petrov using group theoretic methods. These results can be summarised by the statement that the only vacuum spacetimes with constant Weyl eigenvalues are either homogeneous or are Kundt spacetimes. This result is similar to that of Coley et al. who proved their result for **general** spacetimes under the assumption that all scalar invariants constructed from the curvature tensor **and all its derivatives** were constant.

1. Introduction

In this paper special classes of solutions of Einstein's vacuum field equations in general relativity, possibly with a non-zero cosmological constant Λ,

$$R_{ab} = \Lambda g_{ab} \tag{1}$$

are considered. The eigenvalues λ and eigenbivectors V^{ab} of the Weyl tensor satisfy the equation

$$C^{ab}{}_{cd} V^{cd} = 2\lambda V^{ab} \tag{2}$$

where the factor two, which arises due to the contraction over a pair of antisymmetric indices, is included for later convenience. Owing to the duality conditions satisfied by the Weyl tensor there are essentially only three roots of the characteristic equation and these sum to zero due to the trace-free nature of the Weyl tensor (see, for example, chapter 4 of Stephani et al. [1]).

In this paper the class of fields considered are Petrov types I, II & D fields where all Weyl eigenvalues are constant; the conformally flat fields are of constant curvature and need not be considered further whereas for Petrov types III and N fields the assumption of the constancy of the Weyl eigenvalues do not restrict these fields as all the eigenvalues are necessarily zero. For Petrov type I fields it may be assumed that the Weyl eigenvalues are non-zero since Brans [2] showed that pure vacuum Petrov type I fields with a zero Weyl eigenvalue do not exist. His proof generalises trivially to the case when Λ is non-zero.

There are a number of motivations for investigating fields of this sort. Firstly, most known vacuum solutions of the field equations are either algebraically special or admit a group of isometries and so it would be interesting to enlarge the class of known vacuum solutions of Petrov type I. Making simplifying assumptions regarding the scalar curvature invariants seems to be a promising approach which has hitherto been largely neglected. Secondly, in recent years there has been considerable interest in general spacetimes in which all the scalar invariants constructed from the curvature tensor and its covariant derivatives are constant. For example Coley and his collaborators [3] showed that such spacetimes were either homogeneous or are a special class of Kundt-waves [4]. In this paper it is shown that a similar result holds for four dimensional Einstein spaces under the weaker assumption that the scalar curvature invariants constructed from the curvature tensor **alone** are constant and non-zero. It would be interesting to see how far this result could be extended to non-vacuum spacetimes.

To conclude this section some useful results on the Petrov classification are briefly reviewed. The conditions on the Weyl eigenvalues are expressed in the notation of the Newman-Penrose [5] formalism which will be used in the analysis later in the paper (see chapters 3 & 4 of Stephani et al. [1] for more details). For Petrov types I & D there is a complex null tetrad: $\ell^a = (u^a + e_3^a)/\sqrt{2}$, $n^a = (u^a - e_3^a)/\sqrt{2}$, $m^a = (e_1^a + \imath e_2^a)/\sqrt{2}$ in which the NP Weyl tensor components satisfy

$$\Psi_0 = \Psi_4 = (\lambda_2 - \lambda_1)/2 \qquad \Psi_1 = \Psi_3 = 0 \qquad \Psi_2 = -\lambda_3/2, \tag{3}$$

where, the λ_A's are the Weyl eigenvalues and $(e_1^a, e_2^a, e_3^a, u^a)$ is an orthonormal Weyl principal tetrad. For type D we have in addition $\Psi_4 = \Psi_0 = 0$ as $\lambda_1 = \lambda_2$. For Petrov type II the NP Weyl tensor components satisfy

$$\Psi_0 = \Psi_1 = \Psi_3 = 0 \qquad \Psi_4 = -2 \qquad \Psi_2 = -\lambda_3/2. \tag{4}$$

Thus the constancy of the Weyl eigenvalues may be restated as (where in all cases $\Psi_1 = \Psi_3 = 0$):

Type I: Ψ_2 and Ψ_4 are both non-zero constants;

Type D: Ψ_2 is a non-zero constant and $\Psi_0 = \Psi_4 = 0$;

Type II: Ψ_2 is a non-zero constant, $\Psi_0 = 0$ and $\Psi_4 = -2$.

For Petrov type I the choice of a canonical tetrad is not unique; it depends on the choice of numbering of the 3 spacelike Weyl principal vectors e_A^a used to construct the complex null tetrad. As there are six distinct numberings a given Petrov type I field will manifest itself as six different solutions of the Newman-Penrose equations for the spin coefficients and Weyl tensor components Ψ_i. These fall naturally into 3 pairs depending on the choice of the spacelike principal vector e_3^a used to construct the real null vectors ℓ^a and n^a. The two members of each pair have the same value of Ψ_2, but Ψ_4 changes opposite sign under interchange of e_1^a and e_2^a.

The ambiguity in the numbering of the principal vectors also means that type D fields with $\lambda_3 = \lambda_1 = -\lambda_2/2$ or $\lambda_2 = \lambda_3 = -\lambda_1/2$ appear as special cases in the analysis of type I spacetimes and need to be excluded. They are characterised by the conditions $\Psi_4 = \pm 3\Psi_2$ respectively. Similarly the Petrov type I fields where one Weyl eigenvalue is zero appears as a special case in the analysis and can be immediately excluded by Brans' result [2]. The cases with $\lambda_1 = 0$, $\lambda_2 = 0$ and $\lambda_3 = 0$ are characterised by the conditions $\Psi_4 = \pm\Psi_2$ and $\Psi_2 = 0$ respectively.

2. Algebraically Special Spacetimes with Constant Weyl Eigenvalues

In this section the type II and type D fields satisfying the assumptions of §1 are investigated. It is convenient to consider these two cases together; the NP Weyl tensor components satisfy $\Psi_0 = \Psi_1 = \Psi_3 = 0$ and $\Psi_2 (\neq 0)$ is a constant. Thus $D\Psi_2$, $\Delta\Psi_2$, $\delta\Psi_2$, $\bar{\delta}\Psi_2$, DR, ΔR, δR and $\bar{\delta}R$ are all zero and for type II $\Psi_4 \neq 0$ whereas for type D, $\Psi_4 = 0$.

In what follows the equation numbers refer to the Newman-Penrose equations in chapter 7 of Stephani et al. [1]. Using the above restrictions on Ψ_i the Bianchi identities (7.32a,b,e,h) become

$$3\kappa\Psi_2 = 0, \qquad 3\sigma\Psi_2 = 0, \qquad 3\rho\Psi_2 = 0, \qquad 3\tau\Psi_2 = 0. \tag{5}$$

Journal of Physics: Conference Series **600** (2015) 012067 doi:10.1088/1742-6596/600/1/012067

Hence $\kappa = \sigma = \rho = \tau = 0$, and so the spacetime belongs to a special subclass of the Kundt spacetimes [4]. The Ricci identity (7.21q) reduces to $\Psi_2 + R/12 = 0$ and thus $\Psi_2 = -\Lambda/3$ or equivalently $\lambda_3 = 2\Lambda/3$. Thus under the assumption of constancy, *the non-repeated eigenvalue of the Weyl tensor of an Einstein spacetime of type D or II necessarily has the value* $2\Lambda/3$.

These Einstein spaces belong to the subclass of the Kundt spacetimes considered by Lewandowski [6]. In terms of a complex coordinate z and real coordinates u & v, the metric becomes

$$ds^2 = 2P^{-2}dzd\bar{z} - 2du(dv + Wdz + \bar{W}d\bar{z} + Hdu) \tag{6}$$

where $P = P(z, \bar{z}, u)$ and $H = H(z, \bar{z}, u, v)$ are real and $W = W(z, \bar{z}, u)$ is complex; W is independent of v as a consequence of $\tau = 0$. The repeated principal null vector ℓ^a is given by ∂_v. As shown by Lewandowski [6], the metric functions P, H and W may be written as

$$P = 1 + \Lambda z\bar{z}/2 \qquad H = -\Lambda v^2/2 + H_0(z, \bar{z}, u), \qquad W = iL_{,z} \tag{7}$$

where L is a *real* potential satisfying

$$P^2 L_{,z\bar{z}} = -\Lambda L. \tag{8}$$

To complete the identification of the type II & D spacetimes with constant Weyl scalars with those considered by Lewandowski, it is necessary to check that all Lewandowski spacetimes are Petrov type II or D with constant Weyl scalars. Choosing a complex null basis of one-forms:

$$\ell_i dx^i = (H + P^2 W\bar{W})du + dv, \qquad n_i dx^i = du, \qquad m_i dx^i = -P\bar{W}du + P^{-1}dz, \tag{9}$$

a straightforward calculation using the computer algebra system Classi [7] shows that $\Psi_2 = \Lambda/3$ & $\Psi_0 = \Psi_1 = \Psi_3 = 0$. Thus the Weyl tensor is either type II or D and the Weyl eigenvalues are constant. Note that this basis of one-forms is a canonical null tetrad of the Weyl tensor.

The general solution of equation (8) for the potential L is [6]

$$L = \Re(\Lambda P^{-1}\bar{z}f(z, u) - f_{,z}(z, u)) \tag{10}$$

where $f(z, u)$ is an arbitrary function analytic in z. The remaining field equation implies

$$H_{0,z\bar{z}} = \Lambda L_{,z}L_{,\bar{z}} - \Lambda^2 P^{-2} L^2. \tag{11}$$

Given L, this can be integrated to give H_0 up to addition of an arbitrary harmonic function $\Re h_0(z, u)$. It can be seen that the general solution of Petrov type II depends on two arbitrary complex functions $f(z, u)$ and $h_0(z, u)$ analytic in z.

The type D condition $\Psi_4 = 0$ implies, after a somewhat messy calculation (see [8] for more details), that the metric may be written in the form

$$ds^2 = 2(1 + \Lambda z\bar{z}/2)^{-2}dzd\bar{z} - 2dudv - \Lambda v^2 du^2. \tag{12}$$

The metric is decomposable into two 2-spaces of constant curvature and so is a Robinson-Bertotti solution [9, 10]. It is homogenous with a multiply-transitive isometry group of dimension 6.

3. Algebraically General Spacetimes with Constant Weyl Eigenvalues

Now consider Petrov type I fields with constant Weyl eigenvalues. The NP Weyl tensor components satisfy $\Psi_1 = \Psi_3 = 0$ and Ψ_2 & $\Psi_4 (=\Psi_0)$ are non-zero constants. The Bianchi identities (7.32) of [1] become purely algebraic equalities:

$$\begin{aligned}
(4\alpha - \pi)\Psi_4 + 3\kappa\Psi_2 &= 0, & 3\pi\Psi_2 - \kappa\Psi_4 &= 0,\\
(4\gamma - \mu)\Psi_4 + 3\sigma\Psi_2 &= 0, & 3\mu\Psi_2 - \sigma\Psi_4 &= 0,\\
(4\epsilon - \rho)\Psi_4 + 3\lambda\Psi_2 &= 0, & 3\rho\Psi_2 - \lambda\Psi_4 &= 0,\\
(4\beta - \tau)\Psi_4 + 3\nu\Psi_2 &= 0, & 3\tau\Psi_2 - \nu\Psi_4 &= 0.
\end{aligned}$$

Journal of Physics: Conference Series **600** (2015) 012067 doi:10.1088/1742-6596/600/1/012067

Solving these for eight of the spin coefficients in terms κ, σ, ν and λ one obtains:

$$\rho = \psi\lambda, \qquad \tau = \psi\nu, \qquad \mu = \psi\sigma, \qquad \pi = \psi\kappa, \tag{13}$$

$$\alpha = \frac{(\psi^2-1)\kappa}{4\psi}, \qquad \beta = \frac{(\psi^2-1)\nu}{4\psi}, \qquad \epsilon = \frac{(\psi^2-1)\lambda}{4\psi}, \qquad \gamma = \frac{(\psi^2-1)\sigma}{4\psi}. \tag{14}$$

where a subsidiary constant $\psi = \Psi_4/(3\Psi_2)$ has been introduced for later convenience. The following values of ψ are excluded: $\psi \neq 0, \pm 1$ (type D) & $\psi \neq \pm 1/3$ (zero Weyl eigenvalue). These eight spin coefficients may be completely eliminated from the 18 Newman-Penrose Ricci identities (7.21) which thus now only involve derivatives of κ, σ, ν and λ. Here and below equation references of the form (7.xx) refer to Chapter 7 of Stephani et al. [1].

Equations (7.21b & h) are equalities for $D\sigma - \delta\kappa$ and result in the purely algebraic identity:

$$24(\kappa\nu - \sigma\lambda)(1-\psi^2) - 12\Psi_2(1-3\psi^2) - R = 0. \tag{15}$$

Similarly (7.21j & q) are both equalities for $\Delta\lambda - \bar{\delta}\nu$ and also results in (15). A second algebraic expression is obtained from (7.21b, ℓ, f & j):

$$6(\kappa\nu - \sigma\lambda)(1-\psi^2)(1-5\psi^2) - \psi^2(18\psi^2\Psi_2 - 42\Psi_2 + R) = 0. \tag{16}$$

From (15) and (16) the following simpler algebraic relations may be deduced:

$$\Psi_2 = \frac{(\kappa\nu - \sigma\lambda)(9\psi^2 - 1)}{9\psi^2}, \qquad R = 4(\kappa\nu - \sigma\lambda)(3\psi^2+1)^2/3. \tag{17}$$

From these two equations it can be immediately deduced that $\kappa\nu - \sigma\lambda \neq 0$ as $\Psi_2 \neq 0$ and that the pure vacuum case ($R = 0$) is characterised by $\psi = \pm i/\sqrt{3}$. Note that for the pure vacuum case the Weyl eigenvalues are proportional to the three cube roots of -1; for example $\lambda_1 = (-1 + i\sqrt{3})\lambda_3/2$ and $\lambda_2 = (-1 - i\sqrt{3})\lambda_3/2$ for the choice $\psi = +i/\sqrt{3}$.

Equations (7.21a & g) reduce to two linear equations for $D\lambda$ & $\bar{\delta}\kappa$ which may be solved for these two derivatives. Similarly, from the pairs (7.21c & i), (7.21k & m) and (7.21n & p) the derivatives $D\nu, \Delta\kappa, \delta\lambda, \bar{\delta}\sigma, \delta\nu$ and $\Delta\sigma$ are obtained. Thus 8 of the 16 derivatives of κ, σ, ν and λ are now known. In addition to the previous two algebraic equations and the eight equations for single derivatives of κ, σ, ν and λ, five independent Ricci identities involving pairs of derivatives of these spin coefficients remain. A more complete exposition of the calculations here and below may be found in [8].

On applying the commutator $\delta D - D\delta$ to λ and ν, equations for $\kappa\Delta\lambda - \sigma\bar{\delta}\lambda$ & $\kappa\Delta\nu - \sigma\bar{\delta}\nu$ are obtained. Similarly applying the commutator $\bar{\delta}\Delta - \Delta\bar{\delta}$ to σ and κ equations for $\nu D\sigma - \lambda\delta\sigma$ & $\nu D\sigma - \lambda\delta\sigma$ result. The first three of these equations and the five remaining Ricci identities may be solved for the unknown spin coefficient derivatives: $D\kappa, \delta\kappa, D\sigma, \delta\sigma, \Delta\nu, \bar{\delta}\nu, \Delta\lambda$ and $\bar{\delta}\lambda$.

Substituting these in the equation for $\nu D\sigma - \lambda\delta\sigma$ a purely algebraic relation is obtained:

$$\bar{\psi}(1+3\psi^2)(\kappa\nu - \sigma\lambda)^2 = 0. \tag{18}$$

It may be concluded that $\psi = \pm i/\sqrt{3}$ (since $\psi \neq 0$ as type D is excluded and from (17) $\kappa\nu - \sigma\lambda \neq 0$ since $\Psi_2 \neq 0$). Without loss of generality we may choose $\psi = +i/\sqrt{3}$ as the negative sign simply corresponds to interchanging the two Weyl eigenvalues λ_1 and λ_2. Thus, from (17), *the spacetime is a pure vacuum spacetime* ($R = 0$ or equivalently $\Lambda = 0$). To summarise it has been shown that *the only Petrov type I Einstein spacetimes with constant Weyl eigenvalues have* $\Lambda = 0$ *(i.e. are pure vacuum) and the Weyl eigenvalues are proportional to the three cube roots of* -1. Thus, *a fortiori* there are no homogeneous proper Einstein spaces of Petrov type I; this recovers a result of MacCallum & Siklos [11] without using group theoretic methods. In the same paper MacCallum & Siklos showed there are no homogeneous proper Einstein spaces of Petrov type II; a result also confirmed by the analysis in §2 as it is easy to show the Lewandowski metrics are not homogeneous. Note that in the proof of non-existence for Petrov type II the assumption of homogeneity cannot be replaced by the weaker one of constant Weyl eigenvalues.

As shown above $\Lambda = 0$ and it can be assumed *wlog* that $\psi = +i/\sqrt{3}$. As all 16 derivatives of the spin coefficients κ, σ, ν & λ are now known, the remaining commutators may be applied to these 4 spin coefficients and all the resulting derivative terms eliminated to produce eight purely algebraic compatibility relations.

$$2\kappa\sigma\nu + \sigma^2\lambda + \sigma\bar{\kappa}\bar{\nu} - \sigma\bar{\sigma}\bar{\lambda} + i\sqrt{3}\nu^2\lambda = 0, \qquad \lambda^3 + \sigma\bar{\sigma}\bar{\lambda} - \bar{\kappa}\sigma\bar{\nu} - i\sqrt{3}\kappa^2\lambda = 0, \tag{19a}$$

$$2\kappa\sigma\lambda + \kappa^2\nu + \kappa\bar{\sigma}\bar{\lambda} - \kappa\bar{\kappa}\bar{\nu} + i\sqrt{3}\nu\lambda^2 = 0, \qquad \sigma^3 + \bar{\sigma}\lambda\bar{\lambda} - \bar{\kappa}\bar{\nu}\lambda - i\sqrt{3}\sigma\nu^2 = 0, \tag{19b}$$

$$2\kappa\nu\lambda + \sigma\lambda^2 + \bar{\kappa}\bar{\nu}\lambda - \bar{\sigma}\lambda\bar{\lambda} + i\sqrt{3}\kappa^2\sigma = 0, \qquad \nu^3 + \kappa\bar{\kappa}\bar{\nu} - \kappa\bar{\sigma}\bar{\lambda} - i\sqrt{3}\sigma^2\nu = 0, \tag{19c}$$

$$2\sigma\nu\lambda + \kappa\nu^2 + \bar{\sigma}\nu\bar{\lambda} - \bar{\kappa}\nu\bar{\nu} + i\sqrt{3}\kappa\sigma^2 = 0, \qquad \kappa^3 + \bar{\kappa}\nu\bar{\nu} - \bar{\sigma}\nu\bar{\lambda} - i\sqrt{3}\kappa\lambda^2 = 0. \tag{19d}$$

There are three sets of solutions and, in each of them, all the spin coefficients are constants:

$$\sigma = \lambda = 0, \quad \nu = +\bar{\kappa}, \quad \kappa = \pm k(1 + \epsilon_2 i)/\sqrt{2}, \quad \Psi_2 = 8k^2/3. \tag{20a}$$

$$\lambda = \sigma = \epsilon_2\kappa, \quad \nu = -\kappa, \quad \kappa = \pm k(-\sqrt{3} + i)/2, \quad \Psi_2 = 4k^2(-1 + i\sqrt{3})/3. \tag{20b}$$

$$\lambda = \sigma = i\epsilon_2\kappa, \quad \nu = +\kappa, \quad \kappa = \pm k(1 - i\sqrt{3})/2, \quad \Psi_2 = -4k^2(1 + i\sqrt{3})/3. \tag{20c}$$

Here k is an arbitrary positive constant and $\epsilon_2 = \pm 1$. The remaining spin coefficients are

$$\rho = i\lambda/\sqrt{3}, \qquad \tau = i\nu/\sqrt{3}, \qquad \mu = i\sigma/\sqrt{3}, \qquad \pi = i\kappa/\sqrt{3}, \tag{21}$$

$$\alpha = i\kappa/\sqrt{3}, \qquad \beta = i\nu/\sqrt{3}, \qquad \epsilon = i\lambda/\sqrt{3}, \qquad \gamma = i\sigma/\sqrt{3}. \tag{22}$$

As discussed in §1 the choice of canonical tetrad is not unique; it depends on the ordering of the spacelike eigenvectors used to construct the complex null tetrad. Thus there will be three pairs of distinct solutions of the Newman-Penrose equations for the spin coefficients and Weyl tensor components Ψ_i corresponding to the same spacetime. The choice of $\psi = +i/\sqrt{3}$ effectively singles out one member of each pair and so the appearance of three solutions in (20) is to be expected. The fourfold sign ambiguity in each of these solutions is also to be expected as the vectors of the orthonormal principal tetrad $(e_1^a, e_2^a, e_3^a, u^a)$ are only determined up to signs.

Thus it is expected that the three cases in (20) all correspond to the same underlying spacetime. In fact, the spacetime in question is homogeneous with metric:

$$ds^2 = \frac{1}{4k^2}\left(dx^2 + e^{-4x/\sqrt{3}}dy^2 + e^{2x/\sqrt{3}}\cos(2x)(dz^2 - dt^2) - 2e^{2x/\sqrt{3}}\sin(2x)dzdt\right). \tag{23}$$

This metric is originally due to Petrov [12] who derived it using group theoretic methods. The orthonormal tetrad of Weyl principal vectors corresponding to case (a) in (20) is:

$$u_a dx^a = \tfrac{e^{x/\sqrt{3}}}{2k}\big(\cos x\ dt - \sin x\ dz\big), \qquad e_{3a}dx^a = \frac{e^{x/\sqrt{3}}}{2k}\big(\sin x\ dt + \cos x\ dz\big),$$

$$e_{1a}dx^a = \tfrac{1}{2\sqrt{2}k}\big(dx - e^{-2x/\sqrt{3}}dy\big), \qquad e_{2a}dx^a = \frac{1}{2\sqrt{2}k}\big(dx + e^{-2x/\sqrt{3}}dy\big).$$

A straightforward calculation using the computer algebra system Classi [7] shows that

$$\Psi_0 = \Psi_3 = 0, \quad \Psi_0 = \Psi_4 = 8ik^2/\sqrt{3}, \quad \Psi_2 = 8k^2/3.$$

The metric (23) is the only homogeneous Einstein space of Petrov type I [12]. The analysis above shows that it is *the only Einstein space of Petrov type I with constant Weyl eigenvalues.*

4. Conclusions

Vacuum spacetimes possibly with a non-zero cosmological constant Λ (that is Einstein spacetimes) with constant non-zero Weyl eigenvalues have been considered. For type II & D the non-repeated eigenvalue necessarily has the value $2\Lambda/3$ and so algebraically special pure vacuum spacetimes of this type are ruled out. It is then shown that the only possible spacetimes are some Kundt-waves considered by Lewandowski [6] which are type II and a Robinson-Bertotti solution of type D. The Lewandowski solutions depend on an arbitrary complex function $f(z,u)$ analytic in $z = x + iy$ and an arbitrary real function of x, y & u harmonic in x and y. The solutions are not homogeneous and in general, they admit no isometries. Originally the solutions were found by Lewandowski by considering Kundt solutions with a reduced holonomy group; this paper gives them a new characterisation namely the constancy of their Weyl eigenvalues.

For Petrov type I the only solutions in which all three Weyl eigenvalues are constant must have $\Lambda = 0$ (i.e. must be pure vacuum); there are no proper Einstein spaces satisfying this assumption. Thus *a fortiori* there are no homogenous proper Einstein spaces of Petrov type I. This provides an independent proof of the result of MacCallum & Siklos [11] (see also [1] §12.9) which does not use group theoretic methods. The only metric turns out to be the homogeneous pure vacuum solution found long ago by Petrov [12] using group theoretic methods. It is the only homogenous Einstein spacetime of Petrov type I; this paper shows that it can be characterised uniquely by the weaker assumption of the constancy of the Weyl eigenvalues.

The above results can be summarised in an alternative way by the statement that the only vacuum spacetimes with constant Weyl eigenvalues are either homogeneous or are Kundt spacetimes of the Lewandowski class. This result is similar to that of Coley et al. [3] who proved their result for **general** spacetimes under the assumption that all scalar invariants constructed from the curvature tensor **and all its derivatives** were constant. The result in this paper is restricted to Einstein spaces of Petrov types I, II & D only, but subject to the weaker assumption of the constancy of the scalar invariants constructed from the curvature tensor **alone**.

It is somewhat disappointing that all solutions in this paper had previously been found by other methods, given the generality of initial assumptions (namely the constancy of the Weyl eigenvalues). However, the paper does provide new characterisations of the solutions based on conditions that are weaker than those originally used to derive them. In particular the proof makes no assumptions regarding the isometry or holonomy groups of the spacetimes.

Acknowledgements

The extensive calculations in §3 were performed using a Maple package for the NP formalism which was kindly supplied to me by Norbert van den Bergh of the University of Ghent.

References

[1] Stephani H, Kramer D, MacCallum M, Hoenselaers C and Herlt E 2003 *Exact Solutions to Einstein's Field Equations* (Cambridge University Press)
[2] Brans C H 1975 *J. Math. Phys.* **16** 1008
[3] Coley A, Hervik S and Pelavas N 2006 *Class. Quantum Grav.* **23** 3053
[4] Kundt W 1961 *Z. Phys.* **163** 7
[5] Newman E T and Penrose R 1962 *J. Math. Phys.* **3** 566
[6] Lewandowski J 1992 *Class. Quantum Grav.* **9** L147
[7] Åman J E 2002 Classification programs for geometries in general relativity – manual for CLASSI, 4th edition *University of Stockholm Report*
[8] Barnes A 2014 Einstein spacetimes with constant Weyl eigenvalues *Preprint* arXiv:1409.4300
[9] Robinson I 1959 *Bull. Acad. Polon. Sci. Math.* **7** 351
[10] Bertotti B 1959 *Phys. Rev.* **116** 1331
[11] MacCallum M A H and Siklos S T C 1992 *J. Geometry Phys.* **8** 221
[12] Petrov A Z 1962 **In** *Recent Developments in General Relativity* p379 (Pergammon Press–PWN Warsaw, Oxford)

Journal of Physics: Conference Series **600** (2015) 012068 doi:10.1088/1742-6596/600/1/012068

Hamiltonian Reduction of Einstein's Equations without Isometries

Jong Hyuk Yoon
School of Physics, Konkuk University, Seoul 143-701, Korea
E-mail: yoonjh@konkuk.ac.kr

Abstract.
I apply the Hamiltonian reduction procedure to general spacetimes of 4 dimensions with no isometries in the (2+2) formalism and find privileged spacetime coordinates. Physical time is chosen as the *area element* of the two dimensional cross-section of null hypersurfaces. The physical spatial coordinates are defined by *equipotential* surfaces on a given spacelike hypersurface of constant physical time. The physical Hamiltonian is manifestly local and positive-definite in the privileged coordinates. The complete set of Hamilton's equations is presented and it is found that they coincide with the Einstein's equations written in the privileged coordinates. This shows that the Hamiltonian reduction is a self-consistent procedure.

1. Introduction

It is well-known that the true physical degrees of freedom of Einstein's theory of gravitation reside in the conformal metric of the two dimensional spatial cross-section of null hypersurfaces[1, 2]. Using the spacetime diffeomorphism invariance, one may choose certain scalar functions in the phase space of the theory as arbitrarily specifiable spacetime coordinates, thereby eliminating unphysical degrees of freedom. In these privileged coordinates, the theory become constraint-free. This procedure is known as the Hamiltonian reduction[3, 4, 5]. Prof. Kuchař and others applied this Hamiltonian reduction to midi-superspace[6, 7, 8] which admits two commuting Killing vector fields, and showed that Einstein's theory after Hamiltonian reduction was equivalent to cylindrical massless scalar field theory propagating in the 1+1 dimensional Minkowski spacetime.

In this article, we review the Hamiltonian reduction of spacetimes with no isometries in the framework of the (2+2) decomposition[9, 10, 11, 12, 13, 14]. The area element of the two dimensional spatial cross section of null hypersurfaces emerges as the physical time, and the physical radial coordinates are defined by *e i ote tial* surfaces on a given spacelike hypersurface of constant physical time. The reduced physical Hamiltonian turns out to be local and manifestly positive-definite in these privileged coordinates[15]. The momentum constraints are simply the *defi i g* equations of the physical momentum densities; hence, there exist no constraints to solve, and therefore, the theory becomes constraint-free. In addition, the Hamiltonian reduction is self-consistent because Hamilton's equations of motion obtained through the Hamiltonian reduction are *ide tical* to the Ricci-flat equations in the privileged coordinates.

2. The action in the (2+2) formalism

Let us recall that the metric in the (2+2) decomposition[1, 2, 9, 10, 11] of 4 dimensional spacetimes can be written as

$$ds^2 = 2dudv - 2hdu^2 + \tau\rho_{ab}\left(dy^a + A_+^{\ a}du + A_-^{\ a}dv\right) \times \left(dy^b + A_+^{\ b}du + A_-^{\ b}dv\right). \quad (1)$$

The vector fields $\hat{\partial}_\pm$ defined as

$$\hat{\partial}_\pm := \partial_\pm - A_\pm^{\ a}\partial_a, \quad (2)$$

where

$$\partial_+ = \partial/\partial u, \quad \partial_- = \partial/\partial v, \quad \partial_a = \partial/\partial y^a \quad (a = 2, 3), \quad (3)$$

are horizontal vector fields orthogonal to the two-dimensional spacelike surface N_2 generated by ∂_a. The inner products of the horizontal vector fields are given by

$$< \hat{\partial}_+, \hat{\partial}_+ >= -2h, \quad < \hat{\partial}_+, \hat{\partial}_- >= 1, \quad < \hat{\partial}_-, \hat{\partial}_- >= 0, \quad (4)$$

which tells us that $\hat{\partial}_-$ is a null vector field, and that $\hat{\partial}_+$ has a norm $-2h$, which can be either positive, negative, or zero, depending on the sign of h. In this article, we choose the sign $-2h > 0$ so that $v =$ constant is a spacelike hypersurface. The metric on N_2 is $\tau\rho_{ab}$, where τ is the area element of N_2 and ρ_{ab} is the conformal two metric with $\det\rho_{ab} = 1$.

As was shown in [11], the Einstein's equations can be obtained from the variational principle of the following action integral:

$$S = \int dvdud^2y\{\pi_\tau\dot{\tau} + \pi_h\dot{h} + \pi_a\dot{A}_+^{\ a} + \pi^{ab}\dot{\rho}_{ab} - \text{“1”}\cdot C_- - \text{“0”}\cdot C_+ - A_-^{\ a}C_a\}, \quad (5)$$

where the overdot $\dot{}$ means ∂_-, and "1", "0", and $A_-^{\ a}$ are Lagrange multipliers that enforce the constraints $C_- = 0$, $C_+ = 0$, and $C_a = 0$, which are given by

$$\begin{aligned} C_- :=& \frac{1}{2}\pi_h\pi_\tau - \frac{h}{4\tau}\pi_h^2 - \frac{1}{2\tau}\pi_h D_+\tau + \frac{1}{2\tau^2}\rho^{ab}\pi_a\pi_b - \frac{\tau}{8h}\rho^{ab}\rho^{cd}(D_+\rho_{ac})(D_+\rho_{bd}) \\ & - \frac{1}{2h\tau}\rho_{ab}\rho_{cd}\pi^{ac}\pi^{bd} - \frac{1}{2h}\pi^{ac}D_+\rho_{ac} - \tau \mathrm{R}_{(2)} + D_+\pi_h - \partial_a(\tau^{-1}\rho^{ab}\pi_b) = 0, \quad (6) \\ C_+ :=& \pi_\tau D_+\tau + \pi_h D_+ h + \pi^{ab}D_+\rho_{ab} - 2D_+(h\pi_h + D_+\tau) + 2\partial_a(h\tau^{-1}\rho^{ab}\pi_b + \rho^{ab}\partial_b h) = 0, \quad (7) \\ C_a :=& \pi_\tau\partial_a\tau + \pi_h\partial_a h + \pi^{bc}\partial_a\rho_{bc} - 2\partial_b(\rho_{ac}\pi^{bc}) - D_+\pi_a - \partial_a(\tau\pi_\tau) = 0, \quad (8) \end{aligned}$$

respectively. Here, $\mathrm{R}_{(2)}$ is the scalar curvature of N_2, and the diffN_2-covariant derivative[11] of a tensor density q_{ab} with weight s is defined as

$$D_\pm q_{ab} := \partial_\pm q_{ab} - [A_\pm,\ q]_{\mathrm{L}ab} = \partial_\pm q_{ab} - A_\pm^{\ c}\partial_c q_{ab} - q_{cb}\partial_a A_\pm^{\ c} - q_{ac}\partial_b A_\pm^{\ c} - s(\partial_c A_\pm^{\ c})q_{ab}, \quad (9)$$

where $[A_\pm,\ q]_{\mathrm{L}ab}$ is the Lie derivative of q_{ab} along $A_\pm := A_\pm^{\ a}\partial_a$. For instance, the diff$N_2$-covariant derivatives of the area element τ and the conformal metric ρ_{ab}, which are a scalar density and a tensor density with weight 1 and -1 with respect to the diffN_2 transformations, respectively, are given by

$$\begin{aligned} D_\pm\tau &= \partial_\pm\tau - A_\pm^{\ a}\partial_a\tau - (\partial_a A_\pm^{\ a})\tau, \quad (10) \\ D_\pm\rho_{ab} &= \partial_\pm\rho_{ab} - A_\pm^{\ c}\partial_c\rho_{ab} - \rho_{cb}\partial_a A_\pm^{\ c} - \rho_{ac}\partial_b A_\pm^{\ c} + (\partial_c A_\pm^{\ c})\rho_{ab}. \quad (11) \end{aligned}$$

On the other hand, h is a scalar under the diffN_2 transformations, whose diffN_2-covariant derivative is given by

$$D_\pm h = \partial_\pm h - A_\pm^{\ a}\partial_a h, \quad (12)$$

and the diffN_2-covariant field strength $F_{+-}{}^{a}$ is defined as

$$F_{+-}{}^{a} := \partial_+ A_-{}^{a} - \partial_- A_+^{a} - [A_+,\ A_-]^a = \partial_+ A_-{}^{a} - \partial_- A_+^{a} - A_+^{b}\partial_b A_-{}^{a} + A_-{}^{b}\partial_b A_+^{a}. \tag{13}$$

The diffN_2-covariant derivatives of the conjugate momenta π_τ, π_h, π_a, and π^{ab}, which are tensor densities of weights $0, 1, 1$ and 2, respectively, are given by

$$D_\pm \pi_\tau = \partial_\pm \pi_\tau - A_\pm^{a}\partial_a \pi_\tau, \tag{14}$$
$$D_\pm \pi_h = \partial_\pm \pi_h - A_\pm^{c}\partial_c \pi_h - (\partial_c A_\pm^{c})\pi_h, \tag{15}$$
$$D_\pm \pi_a = \partial_\pm \pi_a - A_\pm^{c}\partial_c \pi_a - \pi_c \partial_a A_\pm^{c} - (\partial_c A_\pm^{c})\pi_a, \tag{16}$$
$$D_\pm \pi^{ab} = \partial_\pm \pi^{ab} - A_\pm^{c}\partial_c \pi^{ab} + \pi^{cb}\partial_c A_\pm^{a} + \pi^{ac}\partial_c A_\pm^{b} - 2(\partial_c A_\pm^{c})\pi^{ab}. \tag{17}$$

The conformal two metric ρ_{ab}, being a metric of unit determinant, satisfies the condition

$$\rho^{bc}\partial_\pm \rho_{bc} = \rho^{bc}\partial_a \rho_{bc} = \rho^{bc} D_\pm \rho_{bc} = 0. \tag{18}$$

3. Hamiltonian reduction

Let us define a *ote tial* function R and its conjugate momentum π_R as

$$\partial_+ R := -h\pi_h, \quad \pi_R = -\partial_+ \ln(-h), \tag{19}$$

respectively[13, 14]. The transformation from (h, π_h) to (R, π_R) is a *ca o ical* transformation, as it changes the action integral by total derivatives only. If we impose the constraints $C_+ = 0$ and $C_a = 0$ and choose the Lagrange multiplier $A_-{}^{a} = 0$, then the action in (5) becomes

$$S = \int dv du d^2 y \{\pi_\tau \dot{\tau} + \pi_R \dot{R} + \pi_a \dot{A}_+^{a} + \pi^{ab}\dot{\rho}_{ab} - C_-\} \ + \ \text{total derivatives}, \tag{20}$$

where the constraint $C_- = 0$ in the new variables is given by

$$\begin{aligned} C_- = & -\frac{1}{2h}\pi_\tau \partial_+ R - \frac{1}{4h\tau}(\partial_+ R)^2 + \frac{1}{2h\tau}(D_+\tau)(\partial_+ R) - \frac{1}{h}D_+(\partial_+ R) + \frac{1}{h}(\partial_+ R)\partial_+ \ln(-h) \\ & -\frac{1}{h}(\partial_+ R)A_+^{a}\partial_a \ln(-h) - \tau R^{(2)} + \frac{1}{2\tau^2}\rho^{ab}\pi_a\pi_b - \partial_a(\tau^{-1}\rho^{ab}\pi_b) - \frac{1}{2h\tau}\rho_{ab}\rho_{cd}\pi^{ac}\pi^{bd} \\ & -\frac{\tau}{8h}\rho^{ab}\rho^{cd}(D_+\rho_{ac})(D_+\rho_{bd}) - \frac{1}{2h}\pi^{ac}D_+\rho_{ac} = 0. \end{aligned} \tag{21}$$

The function h still appears in (21), but as h is related to R and π_R by (19), the constraint function C_- given by (21) may be viewed as a function of the new variables $(\tau, Q^I; \pi_\tau, \Pi_I)$, where $Q^I = (R, A_+^{a}, \rho_{ab})$ and $\Pi_I = (\pi_R, \pi_a, \pi^{ab})$. Notice that the first term in (21) is linear in π_τ, and that all the remaining terms are independent of π_τ. Thus, the equation of motion for τ is given by

$$\dot{\tau} = \int du d^2 y \frac{\delta C_-}{\delta \pi_\tau} = -\frac{1}{2h}\partial_+ R. \tag{22}$$

Now, recall that $\tau = \tau(v, u, y^a)$. If we solve this equation for v, then v may be viewed as a function of (τ, u, y^a) and consequently, $(R,\ A_+^{a},\ \rho_{ab})$ may be regarded as functions of (τ, u, y^a). Therefore,

$$\begin{aligned} & \dot{R} = \dot{\tau}\partial_\tau R,\ \dot{A}_+^{a} = \dot{\tau}\partial_\tau A_+^{a},\ \dot{\rho}_{ab} = \dot{\tau}\partial_\tau \rho_{ab}, \\ & C_- = -\left(\frac{2h}{\partial_+ R}\right)\dot{\tau}C_-, \end{aligned} \tag{23}$$

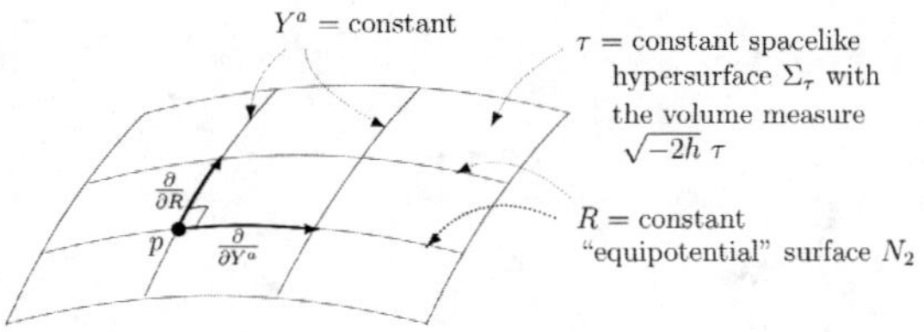

Figure 1. On the $R =$ constant "equipotential" surface N_2 on $_\tau$, Y^a are introduced such that $Y^a =$ constant is normal to N_2 at each point p on $_\tau$. Then the "shift" vector $A_+{}^a$ becomes zero at p.

because $\partial u/\partial v = \partial y^a/\partial v = 0$. Then, the action in (20) becomes

$$S = \int dv du d^2 y \dot{\tau}\{\pi_\tau + \pi_R \partial_\tau R + \pi_a \partial_\tau A_+{}^a + \pi^{ab}\partial_\tau \rho_{ab} + (\frac{2h}{\partial_+ R})C_-\}$$
$$= \int d\tau du d^2 y\{\pi_R \partial_\tau R + \pi_a \partial_\tau A_+{}^a + \pi^{ab}\partial_\tau \rho_{ab} - C_{(1)}\}$$
$$\equiv \int d\tau du d^2 y\{\sum_I \Pi_I \partial_\tau Q^I - C_{(1)}\}, \tag{24}$$

where we replaced $dv\dot{\tau}$ by $d\tau$ in the second line, and $C_{(1)}$ is defined as

$$C_{(1)} = -(\frac{2h}{\partial_+ R})C_- - \pi_\tau. \tag{25}$$

Notice that if we impose the constraint $C_- = 0$, then $C_{(1)}$ reduces to

$$C_{(1)} = -\pi_\tau. \tag{26}$$

The second step in the Hamiltonian reduction consists of *ide tif i g* arbitrarily specifiable coordinates u and y^a as

$$u = R, \quad y^a = Y^a \tag{27}$$

such that the "shift" vector $A_+{}^a$ is zero, i.e.,

$$A_+{}^a = 0. \tag{28}$$

For the class of spacetimes whose spatial topology is either $N_2 \times \mathbb{R}$ or $N_2 \times S^1$, where N_2 is a compact two-dimensional space with the genus g, the condition in (28) can always be met by relabeling N_2 such that $Y^a =$ constant is normal to $u =$ constant at each point on $_\tau$. If we continue to label N_2 this way, then the "shift" $A_+{}^a$ can always be made zero for the class of spacetimes under consideration (see Fig. 1). Thus, in the coordinates $X^A := (\tau, R, Y^a)$, the following coordinate condition

$$\frac{\partial X^A}{\partial X^B} = \delta^A_B \tag{29}$$

holds. For instance, (22) becomes,

$$\dot{\tau} = -\frac{1}{2h} > 0, \tag{30}$$

which means that τ increases monotonically along the out-going null vector field.

Hamilton's equations of motion follow from the variational principle of the action integral in (24):

$$\partial_\tau Q^I = \int_{\Sigma_\tau} dud^2y \frac{\delta C_{(1)}}{\delta \Pi_I}|_{u=R,y^a=Y^a}, \tag{31}$$

$$\partial_\tau \Pi_I = -\int_{\Sigma_\tau} dud^2y \frac{\delta C_{(1)}}{\delta Q^I}|_{u=R,y^a=Y^a}, \tag{32}$$

where $Q^I = (R, A_+^{\ a}, \rho_{ab})$, $\Pi_I = (\pi_R, \pi_a, \pi^{ab})$, and $_\tau$ is a spacelike hypersurface defined by $\tau =$ constant.

4. Main results

In the following, we present the main results of this article[13, 14]. The spacetime metric in these privileged coordinates becomes

$$ds^2 = -4hdRd\tau - 2hdR^2 + \tau\rho_{ab}dY^adY^b. \tag{33}$$

Ei stei 's co strai t e atio s

1. $C_- = 0 \Rightarrow \pi_\tau = -H + 2\partial_R \ln(-h)$ (physical Hamiltonian) (34)
2. $C_+ = 0 \Rightarrow \pi_R = -\pi^{ab}\partial_R\rho_{ab}$ (physical momentum) (35)
3. $C_a = 0 \Rightarrow \tau^{-1}\pi_a = -\pi^{bc}\partial_a\rho_{bc} + 2\partial_b(\pi^{bc}\rho_{ac}) - \tau\partial_a(H + \pi_R)$ (physical momentum) (36)

S er ote tial $\ln(-h)$

4. $\partial_\tau \ln(-h) = H - \tau^{-1}$ (37)
5. $-\partial_R \ln(-h) = \pi_R$ (38)
6. $-\partial_a \ln(-h) = \tau^{-1}\pi_a$ (39)

I tegra ilit co ditio s

7. $\partial_R(\tau^{-1}\pi_a) = \partial_a\pi_R$ (40)
8. $\partial_\tau\pi_R = -\partial_R H$ (41)
9. $\partial_\tau(\tau^{-1}\pi_a) = -\partial_a H$ (42)

In the above equations, H is defined by

$$H = \frac{1}{\tau}\rho_{ab}\rho_{cd}\pi^{ac}\pi^{bd} + \frac{1}{4}\tau\rho^{ab}\rho^{cd}(\partial_R\rho_{ac})(\partial_R\rho_{bd}) + \pi^{ac}\partial_R\rho_{ac} + \frac{1}{2\tau} \geq \frac{1}{2\tau}. \tag{43}$$

To ological ce sorshi

10. $\tau \mathrm{R}_{(2)} = \frac{1}{2}\tau^{-2}\rho^{ab}\pi_a\pi_b - \partial_a(\tau^{-1}\rho^{ab}\pi_b)$ (44)

Ei stei 's evol tio e atio s I

11. $\frac{\partial \pi_a}{\partial \tau} = 2\tau^{-1}\pi_a + (\pi^{bc} + \frac{\tau}{2}\rho^{bd}\rho^{ce}\partial_R\rho_{de})\partial_a\rho_{bc} - \partial_b(2\pi^{bc}\rho_{ac} + \tau\rho^{bc}\partial_R\rho_{ac})$ (45)
12. $\frac{\partial \pi_\tau}{\partial \tau} = \frac{1}{2}\tau^{-2} + \tau^{-2}\rho_{ab}\rho_{cd}\pi^{ac}\pi^{bd} - \frac{1}{4}\rho^{ab}\rho^{cd}(\partial_R\rho_{ac})(\partial_R\rho_{bd}) - 2\tau^{-2}\partial_a(h\rho^{ab}\pi_b)$ (46)

The evolution equations of ρ_{ab} and π^{ab} can be found from the reduced action principle

$$S^* = \int_{\Sigma_\tau} dRd^2Y\{\pi^{ab}\partial_\tau\rho_{ab} - C^*_{(1)}\}, \tag{47}$$

where $C^*_{(1)}$ is the restriction of $C_{(1)}$ to the coordinates $u = R$ and $y^a = Y^a$:

$$\begin{aligned} C^*_{(1)} &:= C_{(1)}|_{u=R, y^a=Y^a} = -\pi_\tau \\ &= H - 2\partial_R \ln(-h). \end{aligned} \tag{48}$$

Ei stei 's evol tio e atio s II

$$13.\ \frac{\partial\rho_{ab}}{\partial\tau} = \int_{\Sigma_\tau} dRd^2Y \frac{\delta C^*_{(1)}}{\delta\pi^{ab}}, \tag{49}$$

$$14.\ \frac{\partial\pi^{ab}}{\partial\tau} = -\int_{\Sigma_\tau} dRd^2Y \frac{\delta C^*_{(1)}}{\delta\rho_{ab}}. \tag{50}$$

It can be shown that, by straightforward calculations, the whole set of equations summarized in Section 4 is *ide tical to the vac Ei stei 's e atio s* $\mathrm{R}_{AB} = 0$ in the privileged coordinates where the metric is given by (33). Thus, the whole procedure of the Hamiltonian reduction is self-consistent, even though the final theory is written in the privileged coordinates.

Finally, I would like to mention that the integral of (44) over a closed two surface N_2 becomes

$$\int_{N_2} d^2Y \tau^{-2}\rho^{ab}\pi_a\pi_b = 16\pi(1-g) \geq 0, \tag{51}$$

where g is the genus of N_2. This identity states that, as long as the out-going null hypersurface forms a congruence of null geodesics which admits a cross section, the spatial topology of that null hypersurface is either a two sphere or a torus. This is a remarkably simple proof of *to ological ce sorshi* , as it does not rely on assumptions such as global hyperbolicity, asymptotic conditions, energy conditions, and so on, either inside or outside the null hypersurface, which are normally assumed in the literature[16, 17, 18].

4.1. Ac owledg e ts
This work is supported by the research program of Konkuk University (2013-A019-0133).

References

[1] Sachs R 1962 *Proc. Roy. Soc.* A **270** 103
[2] d'Inverno R A and Stachel J 1978 *J. Math. Phys.* **19** (12) 2447
[3] Arnowitt R, Deser S and Misner C W 1960 *Phys. Rev.* **117** 1595
[4] Rovelli C and Smolin L 1994 *Phys. Rev. Lett.* **72** 446
[5] Brown J D and Kuchař K V 1995 *Phys. Rev.* D **51** 5600
[6] Kuchař K V 1971 *Phys. Rev.* D **4** 955.
[7] Husain V 1994 *Phys. Rev.* D **50** 6207.
[8] Romano J D and Torre C G 1996 *Phys. Rev.* D **53** 5634
[9] Cho Y M, Park Q H, Soh K S and Yoon J H 1992 *Phys. Lett.* B **286** 251.
[10] Yoon J H 1993 *Phys. Lett.* B **308** 240
[11] Yoon J H 2004 *Phys. Rev.* D **70** 084037
[12] Yoon J H 2014 *Class. Quant. Grav.* **31** 045005
[13] Yoon J H 2014 *J. Korean Phys. Soc.* **54** 192
[14] Yoon J H 2014 *J. Korean Phys. Soc.* **64** 631
[15] Husain V and Pawłowski T 2012 *Phys. Rev. Lett.* **108** 141301
[16] Hawking S W 1972 *Comm. Math. Phys.* **25** 152
[17] Friedman J L, Schleich K and Witt D M 1993 *Phys. Rev. Lett.* **71** 1486
[18] Galloway G J, Schleich K, Witt D M and Woolgar E 1999 *Phys. Rev.* D **60** 104039

Journal of Physics: Conference Series **600** (2015) 012069 doi:10.1088/1742-6596/600/1/012069

Small scale structure of spacetime in presence of a minimal length

Dawood Kothawala
Indian Institute of Technology Madras, Chennai, India
E-mail: tdawood@iitm.ac.in

Abstract. I present some recent results which yields some specific consequences of having a minimal spacetime length L_0 while maintaining Lorentz covariance. In particular: (i) the resultant description of geometry must be in terms of a non-local bi-tensor $q_{ab}(P, p, L_0^2)$ instead of the conventional metric tensor $g_{ab}(p)$, (ii) q_{ab} and g_{ab} are related disformally through the Synge world function, (iii) the resultant non-local d'Alembartian yields Green's functions with a natural covariant short distance cut-off, (iv) the Ricci bi-scalar $R(P, p, L_0^2)$ corresponding to q_{ab} yields a local scalar $\mathcal{S}(P, L_0^2) = R(P, P, L_0^2)$ in the coincidence such that

$$\lim_{L\to 0} \mathcal{S}(P; L) \propto R_{ab}t^a t^b$$

where t^a are arbitrary normalized vectors representing gauge degrees of freedom at each spacetime event; this limit turns out to be independent of the precise details of the short distance modification, (v) the Gibbons-Hawking-York surface term for q_{ab} yields a residual entropy for each spacetime event.

1. Introduction

If the space-time possesses a minimal length, say L_0, which does not violate Lorentz invariance, it has widely been believed that this will result in space-time having some structure at small scales, and also that such a structure would go hand in hand with the short distance regularisation of the propagators which arises from having such a minimal length scale. In recent work, we have arrived at such a characterisation of small scale structure of space-time in terms of the Synge distance function $d(P, p) = \sqrt{\epsilon\sigma^2(P, p)}$. In particular, we have derived a description of geometry in terms of a second rank bi-tensor $q_{ab}(P, p, L_0^2)$ which arises as a consequence of excising a geodesic neighbourhood of size L_0 around any arbitrary space-time event P.

In this talk, I summarise the key results of this work, and in particular highlight those features which are expected to be generic and independent of the precise model of quantum gravity; in fact, I would clarify that the non-perturbative nature of some of these results make them qualitatively similarly to the various anomalies that arise in quantum field theory in curved space-time, and would therefore be independent of how small L_0 is. Here is the plan of the talk:

(i) Minimal length and the small scale structure of space-time

- UV behaviour of propagators and the Synge world function bi-scalar $\Omega(p, P)$.
- Non-local, disformal coupling of the metric to $\Omega(p, P)$: the *qmetric*.

- Anomalous behaviour of space-time curvature – the Cheshire Grin.

$$\lim_{L_0 \to 0} \lim_{p \to P} \neq \lim_{p \to P} \lim_{L_0 \to 0}$$

(ii) Entropy density of space-time from zero point length

- geodesic structure of spacetime
- GHY term & zero-point entropy of space-time

(iii) Discussion and Outlook

Note: All mathematical details of the results presented here can be found in the references in the bibliography. For brevity, the article contains no explicit citations, and all additional references concerning previous related work on this topic can be found in the references listed below, particularly in [1]. The metric signature is $(-, +, +, +, \ldots)$.

2. Minimal length and the small scale structure of space-time

2.1. Background

For a smooth differentiable manifold, the metric near any event has a taylor expansion in the Riemann normal coordinates, given by

$$g_{ab}(x; X) \approx \eta_{ab} - \frac{1}{3} R_{acbd}(X)\ (x - X)^c (x - X)^d + \text{higher order terms}$$

However, such an expansion is questionable at least in 2 cases:

- $d \sim L_0$, when continuous structures like R_{abcd} may not make much sense, and
- near spacetime singularities.

The relevant deformation of geometry which incorporates the fact that space-time might have some structure at small scales can then become non-local. In fact, in a recent paper, it was shown that, if geodesic distances are modified as $\sigma^2 \to \sigma^2 + L_0^2$, then the required deformation of geometry which can yield the modified geodesic function is given by

$$q_{ab}(P, p, L_0^2) = \mathcal{A} g_{ab}(p) - \epsilon \left(\mathcal{A} - \mathcal{A}^{-1}\right) t_a t_b$$

$$\mathcal{A}[\sigma; L_0] = 1 + \frac{L_0^2}{\sigma_g^2}$$

which is:

- disformally coupled to g_{ab}
- a non-local bi-tensor – a 2nd rank tensor at p and a scalar at P
- is singular in the limit $p \to P$

An important consequence of this last characteristic is the fact that, although $\lim_{L_0 \to 0} q_{ab} = g_{ab}$, the *local scalars* constructed out of q_{ab} need not reduce, in the limit $L_0 \to 0$, to their corresponding form in g_{ab}; in fact, this is just what happens! i will now discuss the structure of the two crucial terms in the gravitational action (in units of $\hbar$).

Let $L_0^2 \to 0$ with $p \neq P$ (no surprises here!)	The Strategy	Let $p \to P$ with $L_0^2 \neq 0$ (leads to entropy density functional)
$\sigma^2 \to \sigma^2(P,p)$	(1) Start with the geodesic interval $\sigma^2(P,p)$ for a metric g_{ab}	$\sigma^2 \to 0$
$\sigma^2_{(q)} \to \sigma^2(P,p)$	(2) Incorporate some QG effects by the ansatz $\sigma^2_{(q)}(P,p,L_0^2) = \sigma^2(P,p) + L_0^2$	$\sigma^2_{(q)} \to L_0^2$
$q_{ab}(P,p,L_0^2) \to g_{ab}(P)$	(3) Find the qmetric $q_{ab}(P,p,L_0^2)$ related to $\sigma^2_{(q)}(P,p,L_0^2)$	Diverges as $\frac{L_0^2}{\sigma^2}$
$R(P,p,L_0^2) \to R(P)$	(4) Compute the Ricci biscalar $R(P,p,L_0^2)$ for the $q_{ab}(P,p,L_0^2)$	$R(P,p,L_0^2) \to O(1) R_{ab} t^a t^b + O(L_0^2)$

Table 1. Summary of the results. The strategy adopted in this work is described in the middle column with the logical flow being from top to bottom. We take the coincidence limit to obtain local quantities from bi-tensors; these limits are shown in the right column. As a crosscheck, we have included a left column showing what happens when we set $L_P^2 = 0$. (The results in this column are trivial and as expected.)

$$(16\pi L_p^2)\, S_{\rm grav} = \int_{\mathcal{V}} R[g] \;+\; 2\int_{\partial\mathcal{V}} K[h]$$

and show that

- $\lim_{L_0\to 0}\lim_{\sigma^2\to 0} R[q] \neq R[g]$.
- the surface term $K\sqrt{h}$ is finite in the limit $\sigma^2 \to 0$, and yields *entropy density of spacetime* with a *zero-point term.*
- these results turn out to be independent of precise nature of how geodesic distances are modified, hence presumably also of any specific framework of quantum gravity.

The final results are summarised in Table 1.

The exact form of the Ricci scalar, and many more features of the disformal geometry including the extrinsic curvature on the boundaries of the *equi-geodesic* geodesic neighbourhood, can be found in the references. The exact expressions turn out to be:

$$\frac{1}{16\pi L_P^2}\mathcal{R}(P,L_0) = \alpha\,\mathcal{S}_g \;-\; \frac{\mu^2}{240\pi}\left[\frac{1}{3}\mathcal{S}_{ab}\mathcal{S}^{ab} + \frac{3}{2}\ddot{\mathcal{S}} + \frac{5}{3}\mathcal{S}^2\right]$$

$$\frac{1}{8\pi L_P^2}\lim_{|\sigma^2|\to 0}\left[K\sqrt{h}\right]_q = \mathcal{S}_0 - \frac{\mu^4}{24\pi}\mathcal{S}_g$$

where

- $\mu = L_0/L_P = O(1)$
- $\mathcal{S}_{ab} = R_{aibj}t^i t^j$
- $\mathcal{S}_0 = (3/8\pi)\,\mu^2$
- $\mathcal{S}_g = R_{ab}t^a t^b$

Some important comments on the result:

- Note that there are no terms of the form $1/L_0^2$ in curvature, which one often encounters in *naive* arguments involving vacuum fluctuations of geometry.
- Apart from a cubic divergence which is regularizable in standard point-splitting scheme, there are no divergences despite the manifestly divergent form of the qmetric in the coincidence limit.
- The above two points are a consequence of some miraculous cancellations in the intermediate steps, which remove terms like $1/\sigma^2$ and $1/L_0^2$; this in turn is a consequence of the disformal structure of the qmetric; such terms do appear, for e.g., in quantised conformal fluctuations.

3. Discussion and Future Outlook

In this last section, I discuss the implications of the result and some open issue which need to be addressed by future research.

(i) Relics of the space-time minimal length:

- Our result suggests that non-local and non-analytic effects of a minimal length might leave residues which are independent of L_0.
- In our case, such effects leave their imprints in the form of
  - $\mathcal{S}_0 = 3\mu^2/8\pi$ zero-point entropy density of space-time
  - $\mathcal{S}_g = R_{ab}t^a t^b$ gravitational entropy density of space-time
- Such quantum relics are not unfamiliar in physics; e.g. effects of Lorentz violating regulators at higher energies can generically get dragged to lower energies due to radiative corrections, leaving $O(1)$ residual effects [Collins et. al., Polchinski]; conformal anomaly, $D \to 4$ limit in dimensional regularization [see, for e.g., Birrell & Davies]; non-relativistic relic of the $O(1/c)$ expansion of the relativistic point particle wave function [Padmanabhan et. al.]; nonlocal quantum residue of discreteness of the causal set type [Sorkin].

(ii) Implications for space-time thermodynamics: The above analysis also leads to an unexpected and non-trivial connection between **small scale structure of spacetime** and **spacetime thermodynamics and gravity**. In particular:

- The object $\mathcal{S}_g = R_{ab}t^a t^b$ is well-known to be connected with the Noether charge of diffeomorphism invariance, and hence to entropy of local Rindler horizons; this in turn has been used to derive gravitational dynamics from space-time thermodynamics [Jacobson, Padmanabhan].
- Our result for modified Ricci scalar and surface term in gravitational action suggests that *even classically*, the correct variational principle for gravity must be based on $R_{ab}t^a t^b$ rather than the conventional Einstein-Hilbert lagrangian R.

- Such a thermodynamic variational principle for gravity based on an *entropy functional* was already proposed few years back, in which the degrees of freedom to be varied are arbitrary normalised vectors t^a [Padmanabhan et.al.].

The figure below summarises this connection:

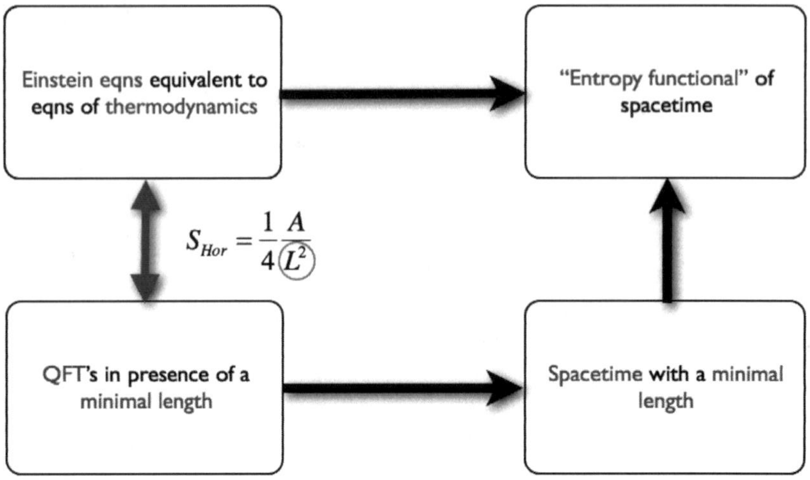

Future Outlook: Some of the future directions which can be explored based on the constructs and ideas described here are as follows:

- Implications for the cosmological constant problem.

  our approach seems to connect the notion of the cosmological constant being a non-local relic of quantum gravity and it's role in the emergent gravity paradigm

- Implications of the non-locality for the high frequency behaviour of Hawking radiation; the trans-planckian issue.

- Short distance structure of the two-point functions & implications for QFT.

- Implications for spacetime singularities

  how does a minimal length scale affect focussing and de-focussing of geodesics?

4. Acknowledgments
I would like to thank the organisers of Spanish Relativity Meeting, ERE-2014, for giving me an opportunity to present this work at the conference in Valencia, Spain.

References
[1] Dawood Kothawala, Phys. Rev. D **88**, 104029 (2013) [arXiv:1307.5618].
[2] Dawood Kothawala, T. Padmanabhan (2014) [arXiv:1405.4967, 1408.3963].
[3] Dawood Kothawala Gen. Rel. Grav. **46**, 1836 (2014) [arXiv:1406.2672].

Journal of Physics: Conference Series **600** (2015) 012070 doi:10.1088/1742-6596/600/1/012070

Regular and chaotic motion in general relativity: The case of an inclined black hole magnetosphere

Ondřej Kopáček and Vladimír Karas

Astronomical Institute, Academy of Sciences, Boční II 1401/1a, CZ-141 31 Prague, Czech Republic

E-mail: kopacek@ig.cas.cz

Abstract.
Dynamics of charged particles in the vicinity of a rotating black hole embedded in the external large-scale magnetic field is numerically investigated. In particular, we consider a non-axisymmetric model in which the asymptotically uniform magnetic field is inclined with respect to the axis of rotation. We study the e ect of inclination onto the prevailing dynamic regime of particle motion, i.e. we ask whether the inclined field allows regular trajectories or if instead, the deterministic chaos dominates the motion. In this contribution we further discuss the role of initial condition, particularly, the initial azimuthal angle. To characterize the measure of chaoticness we compute maximal Lyapunov exponents and employ the method of Recurrence Quantification Analysis.

1. Introduction

In this paper we investigate some aspects of dynamics of ionised matter exposed to the interplay of strong gravitational and electromagnetic fields. In particular, we consider the astrophysically motivated model consisting of a rotating Kerr black hole that is immersed in the asymptotically uniform magnetic field. Previously we have investigated dynamics of charged particles in the axisymmetric case with parallel orientation of the field with respect to the rotation axis [1, 2, 3]. We studied the topology of the particle confinements and located both the regions of mostly regular motion as well as areas dominated by the deterministic chaos.

More recently we have also investigated the generalized scenario in which the magnetic field is arbitrarily inclined with respect to the axis [4]. Breaking the axial symmetry appears to have profound consequences regarding the dynamics of matter. Most importantly, we found that within the given model the stability of regular motion of charged particles depends critically on the perfect alignment of large-scale magnetic field with the rotation axis. Once the field is slightly inclined and the symmetry of the system breaks, the chaotic regime completely dominates the dynamics. We quantified the chaoticness of orbits by the maximal Lyapunov exponent χ. Regarding the trajectories which are regular in the axisymmetric setup we found that inclining the field causes almost immediate onset of chaos. Increasing the inclination angle typically leads to the increase of χ until it saturates. On the other hand, if we consider trajectories which already show distinctive chaotic features in the aligned field, we find that χ is almost indifferent to the increasing inclination. Obliqueness of the field, which may be considered as an extra perturbation of the system, does not enhance chaos in this case.

Naturally, breaking the axial symmetry cancels the conservation of the axial component of the angular momentum L_z and azimuthal angle φ is no longer a cyclic coordinate. Therefore, it becomes dynamically important and, especially, by setting different initial azimuthal angles $\varphi(0)$ we obtain fundamentally distinct trajectories. Nevertheless, we shall only consider small inclinations of the field (typically up to $\approx 6°$). In this context we investigate to which degree the dynamics of test particles may depend on the choice of initial azimuthal coordinate $\varphi(0)$. In particular, we shall explore if different value of $\varphi(0)$ may substantially change the degree of chaoticness.

2. Rotating black hole in the external magnetic field

Kerr metric describing the geometry of the spacetime around the rotating black hole of mass M and spin a may be expressed in Boyer-Lindquist coordinates $x^\mu = (t,\ r,\ \theta,\ \varphi)$ as follows [5]:

$$ds^2 = -\frac{\Delta}{\ }\,[dt - a\sin\theta\, d\varphi]^2 + \frac{\sin^2\theta}{\ }\,[(r^2+a^2)d\varphi - a\, dt]^2 + \frac{\ }{\Delta}\, dr^2 + \ d\theta^2, \tag{1}$$

where

$$\Delta \equiv r^2 - 2Mr + a^2, \qquad \equiv r^2 + a^2\cos^2\theta. \tag{2}$$

Geometrized units are used throughout the paper. Values of basic constants thus equal unity $G = c = k = k_C = 1$.

A test-field solution corresponding to the aligned magnetic field (of the asymptotic strength B_z) was derived by Wald [6]. This solution was later generalized by Bičák and Janiš [7] to describe the field which is arbitrarily inclined with respect to the rotation axis (specified by two independent components B_z and B_x). We may also consider a non-zero electric charge Q of the black hole generating the electromagnetic field of Kerr-Newman black hole, though in the test-field regime (metric remains unaltered by Q). Resulting vector potential of the electromagnetic field $A_\mu = (A_t, A_r, A_\theta, A_\varphi)$ can be given explicitly [4].

We note that in our previous studies [8, 9] we have investigated the structure of the electromagnetic field arising in this setup enriched by an extra ingredient: uniform motion of the black hole in arbitrary direction with respect to (arbitrarily inclined) magnetic field. However, here we consider the inclination of the magnetic field but not the boost of the black hole.

2.1. E atio s of otio

Employing the Hamiltonian formalism we first construct the super-Hamiltonian $\mathcal{H}$ [5]:

$$\mathcal{H} = \tfrac{1}{2}g^{\mu\nu}(\pi_\mu - qA_\mu)(\pi_\nu - qA_\nu), \tag{3}$$

where q is charge of the test particle (of rest mass m), π_μ is the generalized (canonical) momentum, $g^{\mu\nu}$ is the metric tensor, and A_μ denotes the vector potential. The latter is related to the electromagnetic tensor $F_{\mu\nu}$ by $F_{\mu\nu} = A_{\nu,\mu} - A_{\mu,\nu}$.

Hamilton's equations of motion are given as

$$\frac{\mathrm{d}x^\mu}{\mathrm{d}\lambda} \equiv p^\mu = \frac{\partial\mathcal{H}}{\partial\pi_\mu}, \quad \frac{d\pi_\mu}{d\lambda} = -\frac{\partial\mathcal{H}}{\partial x^\mu}, \tag{4}$$

where $\lambda = \tau/m$ is the affine parameter (dimensionless in geometrized units), τ denotes the proper time, and p^μ is the standard kinematical four-momentum for which the first equation reads $p^\mu = \pi^\mu - qA^\mu$ and thus the conserved value of super-Hamiltonian is equal to $\mathcal{H} = -m^2/2$. Moreover, the system is stationary since the Hamiltonian is independent on the coordinate time t. Its conjugate momentum π_t is therefore an integral of motion. Namely, it expresses (negatively taken) energy of the test particle $\pi_t \equiv -E$. The mass of the black hole M is used to scale all quantities which is formally equivalent to setting $M = 1$ in the equations. We also switch to specific quantities $q/m \to q$ and $E/m \to E$ when describing the test particle.

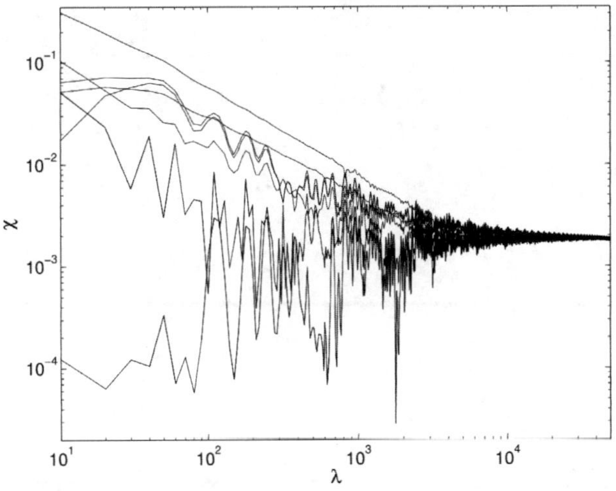

Figure 1. Independence of the asymptotical value of the Lyapunov exponent with respect to the choice of the initial deviation $w(0)$ for a particular choice of the chaotic trajectory. The initial deviation with a single non-zero component (e.g., $w(0) = (0,0,0,1,0,0,0,0)$ pointing in the azimuthal direction) results in wildly oscillating χ while for the 'averaged' initial deviation $w(0) = 1/\sqrt{8}\,(1,1,1,1,1,1,1,1)$ the evolution is rather smooth (uppermost curve in this figure).

2.2. Ma i al L a ov characteristic e o e t

Maximal Lyapunov exponent χ is commonly used as a basic quantitative indicator of chaotic dynamics [10]. Its value directly captures the tendency of nearby orbits to diverge as the system evolves. In other words, it allows us to express how sensitive the given orbit is on the initial condition. High (exponential) sensitivity is a defining property of chaos. Exponent χ is defined as follows [11] [1]

$$\chi \equiv \lim_{\lambda\to\infty} \frac{1}{\lambda} \ln \frac{||w(\lambda)||}{||w(0)||}, \tag{5}$$

where we choose to use standard L^2 (Euclidean) norm to measure the length of deviation vector in the phase space $w(\lambda) = (\delta t, \delta r, \delta\theta, \delta\varphi, \delta\pi_t, \delta\pi_r, \delta\pi_\theta, \delta\pi_\varphi)$.

We employ the usual method of determining the evolution of the deviation $w(\lambda)$ consisting in solving variational equations, which restrict to the linear term in corresponding Taylor expansion [12]. We set the initial deviation as follows $w(0) = 1/\sqrt{8}\,(1,1,1,1,1,1,1,1)$. Theory of Lyapunov spectra guarantees that setting random initial deviation results in the maximal exponent χ with probability one. The set of initial deviations for which we would obtain different Lyapunov exponent has zero measure [11]. In Fig. 1 we illustrate the independence of the asymptotic value of χ on a particular choice of the chaotic orbit.

3. Dynamics of charged particles in the oblique field

We address the question of azimuthal dependence of charged particle dynamics in the inclined magnetic field. The system of test particle in this non-axisymmetric setup has three degrees of freedom and is apparently non-integrable. In general, we expect to find both regular and chaotic orbits in such a system. Our previous analysis [4] shows that most regular orbits found in the axisymmetric case are quickly destroyed once the field inclines even very slightly. We found that a very small inclination angle is sufficient for the deterministic chaos to dominate the phase space completely. We demonstrate this fact on a series of examples and treat separately several

[1] Lyapunov exponent χ is defined as an asymptotic measure (5). In the numerical application we actually compute the quantity usually denoted as *finite time Lyapunov exponent*, which depends on the integration variable, instead of the limit. However, in this text we disregard such distinction and use the term Lyapunov exponent for both quantities as it cannot cause any confusion within the scope of the paper.

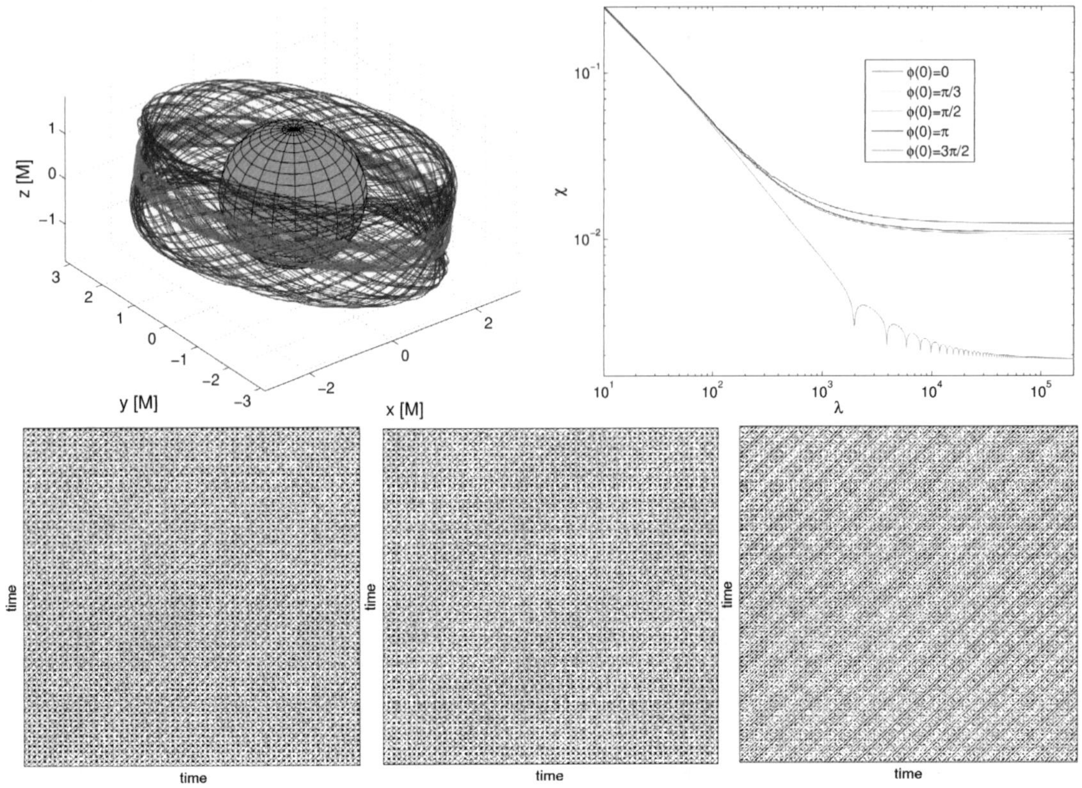

Figure 2. The top left panel shows the three-dimensional view of two chaotic orbits in the oblique magnetic field differing solely in the initial azimuthal angle; $\varphi(0) = 0$ (blue trajectory) and $\varphi(0) = 3\pi/2$ (red orbit). Grey surface marks the horizon of the black hole. In the top right panel we compare Lyapunov exponents of these trajectories which only differ in $\varphi(0)$. Bottom panels compare these trajectories by means of the recurrence plots. The left plot ($\varphi(0) = 0$) shows no differences from the middle one ($\varphi(0) = \pi/3$) while the case $\varphi(0) = 3\pi/2$ (bottom right) produces slightly different pattern.

distinct classes of trajectories. Nevertheless, in [4] we do not discuss the role of initial azimuthal coordinate $\varphi(0)$ which becomes relevant if the axisymmetry is lost.

We shall not present any systematic discussion of the topic here. We rather demonstrate the effect of different initial azimuthal angle on a suitably chosen example. To this end, we pick the chaotic trajectory specified by following values of parameters $a = 0.9$, $E = 1.58$, $qQ = 0$, $qB_z = 1$, $B_x/B_z = 0.1$ and with the initial condition set as follows: $r(0) = 3$, $\theta(0) = \pi/2$, $u^r(0) = 0$ and $\pi_\varphi(0) = 5$. This particular set of values was used in our previous analysis (see Fig. 4 in [4]) in the discussion of the impact of gradually increasing inclination on the trajectory which is regular in the parallel field ($B_x/B_z = 0$). However, here we choose the highest value of inclination permitted for given parameters ($B_x/B_z = 0.1$ corresponding to angle $\approx 6°$) in order to make the possible effect of different $\varphi(0)$ most apparent.

In Fig. 2 we show that most values of $\varphi(0)$ lead to similar results in terms of Lyapunov exponent χ. Indeed, most angles result in a similar evolution as well as the final value of χ,

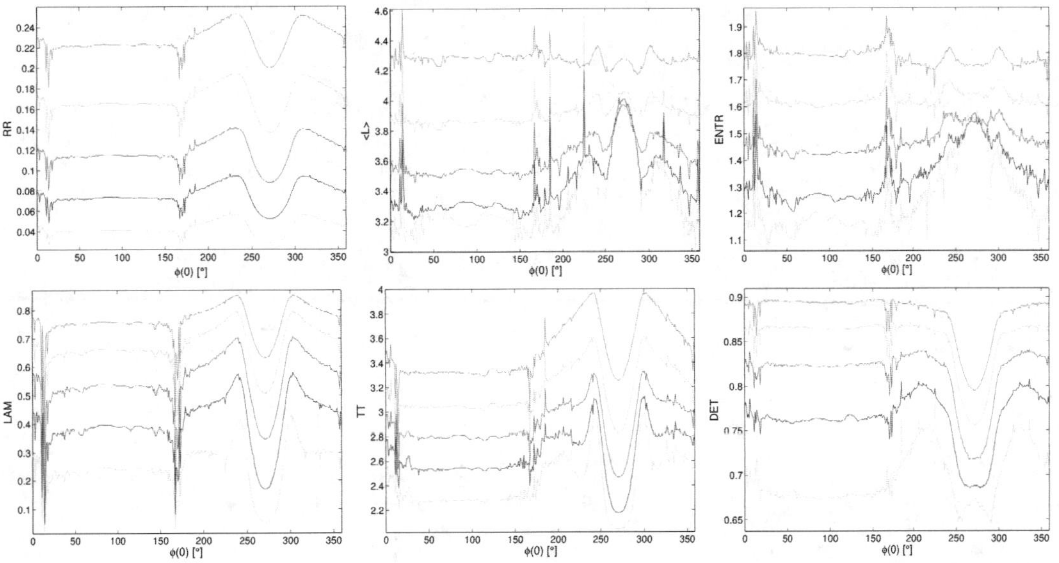

Figure 3. Several R A measures as a function of $\varphi(0)$ (with step $\Delta\varphi = 1^\circ$) of trajectories analyzed in Fig. 2 are shown. Each of them is computed for five distinct values of the recurrence threshold ε (red: $\varepsilon = 2.2$, green: $\varepsilon = 1.95$, blue: $\varepsilon = 1.7$, black: $\varepsilon = 1.45$, cyan: $\varepsilon = 1.2$) with time series of each coordinate normalized separately to zero mean and standard mean deviation equal to unity (see [13] for details).

while only the case of $\varphi(0) = 3\pi/2$ leads to a qualitatively different behaviour and χ converges to the significantly smaller value. With $\varphi(0) = 3\pi/2$ the trajectory visibly differs from others also in direct comparison of their three-dimensional structures (top left panel of Fig. 2). To further explore and compare its dynamics we employ the technique of recurrence quantification analysis (R A) [13]. First we construct recurrence plots (RPs) of trajectories with different values of $\varphi(0)$. In bottom panels of Fig. 2 we compare RPs for $\varphi(0) = 0$, $\varphi(0) = \pi/3$ and $\varphi(0) = 3\pi/2$. The former two RPs are almost identical while the latter has slightly different and more diagonally oriented pattern which is a general hallmark of less divergent dynamics. Thus the visual survey of RPs is in agreement with the observation that $\varphi(0) = 3\pi/2$ leads to significantly lower value of χ, i.e. is less chaotic in this sense. However, recurrences of the trajectory in the phase space may also be quantified in terms of various R A measures (see [13] for their definitions). In Fig. 3 we present plots of several of these indicators as a function of $\varphi(0)$. We observe that profound change of dynamics for $\varphi(0) = 3\pi/2$ is detected by all these recurrence measures. Although the interpretation of R A measures is not straightforward, we note that R A represents a suitable tool to systematically search the parameter space for any anomalies in resulting dynamics. In particular, here it proves that the case $\varphi(0) = 3\pi/2$ is dynamically distinct from all other values of $\varphi(0)$.

Finally, in Fig. 4 we further investigate the case $\varphi(0) = 3\pi/2$ and calculate the evolution of χ for several values of inclination. We observe that with decreasing B_x/B_z the extraordinary behaviour represented by damped oscillations of χ quickly disappears and the dynamics gradually approaches regular behaviour which is reached for zero inclination. Ordered sequence of asymptotical values of χ with respect to the inclination angle confirms the results of our previous analysis [4] which was perfomed for the trajectory launched at $\varphi(0) = 0$.

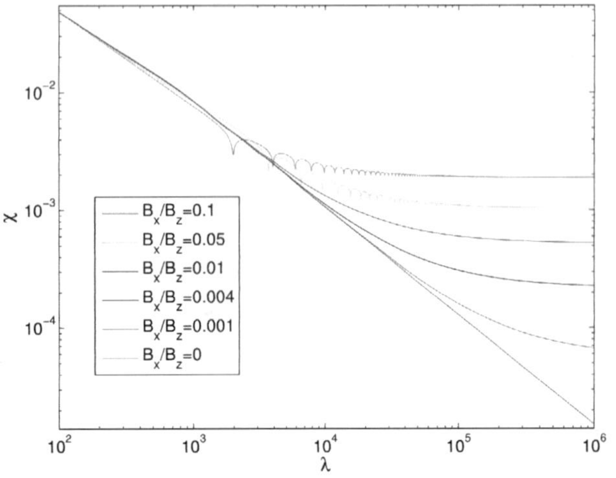

Figure 4. Comparison of Lyapunov exponents of trajectories with $\varphi(0) = 3\pi/2$ which only differ in inclination of the field. For the aligned field the trajectory remains regular, however, if the field even only slightly inclines the chaos sets on immediately. Then the gradual growth of the largest Lyapunov exponent χ and also the emergence of damped oscillations is observed as the inclination of the field increases. Logarithmic scale is used for the both axes.

4. Conclusions

We have numerically analyzed the role of initial azimuthal angle in determining the dynamic properties of charged particles in slightly (but not negligibly) non-axisymmetric setup of rotating black hole immersed in the oblique magnetic field. Demonstrating its effect on the suitably chosen example we have seen that even for small inclination of the field, the choice of $\varphi(0)$ may lead to substantially different behaviours of particles. Employing the largest Lyapunov exponent χ and the recurrence quantification analysis, we have shown that although most cases may lead to the similar results, we can also find some prominent mutual orientation of the field and the initial velocity of the particle which produces markedly different trajectory with distinct evolution and significantly different asymptotic value of χ. This change of dynamics was also clearly detected by various R A measures. We conclude that even in a weakly non-axisymmetric system the role of initial azimuthal angle becomes highly important and its effect cannot be neglected.

Acknowledgments

OK acknowledges the postdoctoral program of the Czech Academy of Sciences. VK thanks the Czech Science Foundation for support via the project GAČR 14-37086G.

References

[1] Kovář J, Kopáček O, Karas V and Stuchlík Z 2010 *Class. Quantum Grav.* **27** 135006 (arXiv:1005.3270)
[2] Kopáček O, Karas V, Kovář J and Stuchlík Z 2010 *Astrophys. J.* **722** 1240–1259 (arXiv:1008.4650)
[3] Kopáček O, Kovář J, Karas V and Stuchlík Z 2010 (*American Institute of Physics Conference Series* vol 1283) ed de León M, de Diego D M and Ros R M pp 278–287 (arXiv:1011.2707)
[4] Kopáček O and Karas V 2014 *Astrophys. J.* **787** 117 (arXiv:1404.5495)
[5] Misner C W, Thorne K S and Wheeler J A 1973 *Gravitation* (San Francisco: W. H. Freeman & Co.)
[6] Wald R M 1974 *Phys. Rev. D* **10** 1680–1685
[7] Bičák J and Janiš V 1985 *Mon. Not. R. Astron. Soc.* **212** 899–915
[8] Karas V, Kopáček O and Kunneriath D 2012 *Class. Quantum Grav.* **29** 035010 (arXiv:1201.0009)
[9] Karas V and Kopáček O 2009 *Class. Quantum Grav.* **26** 025004 (arXiv:0811.1772)
[10] Lichtenberg A J and Lieberman M A 1992 *Regular and chaotic dynamics* (New York: Springer)
[11] Skokos C 2010 (*Lecture Notes in Physics, Berlin Springer Verlag* vol 790) ed Souchay J and Dvorak R pp 63–135 (arXiv:0811.0882)
[12] Kaltchev D and Dragt A J 2013 *Phys. D* **242** 1–15 (arXiv:1102.3394)
[13] Marwan N, Carmen Romano M, Thiel M and Kurths J 2007 *Phys. Rep.* **438** 237–329

AUTHOR INDEX

AUTHOR INDEX